PURE AND APPLIED MATHEMATICS

A Series of Texts and Monographs

Vol. I:	**Supersonic Flow and Shock Waves** By R. Courant and K. O. Friedrichs
Vol. II:	**Nonlinear Vibrations in Mechanical and Electrical Systems** By J. J. Stoker
Vol. III:	**Dirichlet's Principle, Conformal Mapping, and Minimal Surfaces** By R. Courant
Vol. IV:	**Water Waves** By J. J. Stoker
Vol. V:	**Integral Equations** By F. G. Tricomi
Vol. VI:	**Differential Equations: Geometric Theory** By Solomon Lefschetz
Vol. VII:	**Linear Operators—Parts I and II** By Nelson Dunford and Jacob T. Schwartz
Vol. VIII:	**Modern Geometrical Optics** By Max Herzberger
Vol. IX:	**Orthogonal Functions** By G. Sansome
Vol. X:	**Lectures on Differential and Integral Equations** By K. Yosida
Vol. XI:	**Representation Theory of Finite Groups and Associative Algebras** By C. W. Curtis and I. Reiner
Vol. XII:	**Electromagnetic Theory and Geometrical Optics** By Morris Kline and Irvin W. Kay
Vol. XIII:	**Combinatorial Group Theory** By W. Magnus, A. Karrass, and D. Solitar
Vol. XIV:	**Asymptotic Expansions for Ordinary Differential Equations** By Wolfgang Wasow
Vol. XV:	**Tchebycheff Systems: With Applications in Analysis and Statistics** By Samuel Karlin and William J. Studden
Vol. XVI:	**Convex Polytopes** By Branko Grunbaum
Vol. XVII:	**Fourier Analysis in Several Complex Variables** By Leon Ehrenpreis
Vol. XVIII:	**Generalized Integral Transformations with Applications** By Armen H. Zemanian
Vol. XIX:	**Introduction to the Theory of Categories and Functors** By Ion Bucur and Aristide Deleanu
Vol. XX:	**Differential Geometry** By J. J. Stoker
Vol. XXI:	**Ordinary Differential Equations** By Jack K. Hale
Vol. XXII:	**Introduction to Potential Theory** By L. L. Helms

PURE AND APPLIED MATHEMATICS

A Series of Texts and Monographs

VOLUME XVII

FOURIER ANALYSIS IN SEVERAL COMPLEX VARIABLES

LEON EHRENPREIS
BELFER SCHOOL OF SCIENCE
YESHIVA UNIVERSITY
NEW YORK CITY

WILEY-INTERSCIENCE PUBLISHERS
A Division of John Wiley & Sons
New York · London · Sydney · Toronto

10 9 8 7 6 5 4 3 2 1

Library of Congress Catalog Card Number: 68–8755
SBN 471 23400 1
Printed in the United States of America

To

My Parents

Rejoice! Rejoice!
The Course is set
The time for word is past
Let Strength and Speed be bared.

Preface

This book represents our way of looking at the central problem of Fourier analysis. Let f be a function and F its Fourier transform. We are given some properties of f and some properties of F, and it is up to us to combine these in such a way as to derive more information about each one. In this book we are concerned with linear properties. Thus we want to determine which topological vector spaces contain f and F, respectively. Results of this kind are called *comparison theorems*. This approach leads us to be rather abstract in our reasoning, although we tend to emphasize the classical content of our work.

The plan of the book is as follows: Part A has for its objective the establishment of the quotient structure theorem or *Fundamental Principle* (Theorems 4.1 and 4.2) which shows, essentially, that Fourier analysis on the subspace of the linear space $\mathscr{W}$ defined as the simultaneous solutions of $D_1 f = 0$, $D_2 f = 0, \ldots, D_r f = 0$ is very similar to Fourier analysis on $\mathscr{W}$ except that the frequencies must be restricted to the set V of common solutions of $P_1(z) = 0, \ldots, P_n(z) = 0$. In Chapter I we study Fourier analysis on $\mathscr{W}$ itself. We introduce the concept of *multiplicity variety* which plays a large role in our work. Chapter II deals with the question of the geometric structure of ideals and modules in the ring of local holomorphic functions. Chapter III adds some quantitative estimates, and Chapter IV pieces the local bits together. In Chapter V we give some examples of spaces $\mathscr{W}$ to which the theory can be applied.

Part B is concerned with applications to partial differential equations. In Chapter VI we show how to solve general inhomogeneous systems. Chapter VII deals with the above mentioned result on Fourier representation of solutions of homogeneous systems, namely that the frequencies may be restricted to lie in V. In Chapter VIII we examine comparison theorems of the type mentioned above. Chapter IX is a study of existence and uniqueness questions related to Cauchy's problem, and Chapter X is a similar study of boundary value problems for solutions in a cube. Chapter XI deals with assorted questions, for example, a partial extension of the quotient structure theorem, questions of removable singularities such as

(a) When can a solution of a homogeneous system defined on a domain be extended to a larger domain?

(b) When is a solution of a homogeneous system on a domain G which is regular on a subdomain G' of G regular on G?

We study also the question of characterizing domains on which we can solve inhomogeneous systems.

In Part C we study functions f for which the frequencies arising in the Fourier representation are restricted to lie in a set λ which is much larger than the set of zeros of polynomials, or, more generally, that in the Fourier representation of f those frequencies outside λ should occur with small coefficients. Chapter XII deals with questions of type (b) above for such functions, while in Chapter XIII uniqueness properties are considered.

Each chapter begins with a detailed summary and ends, when the occasion permits, with general remarks, bibliographical remarks, and problems for further study.

I should like to express my sincere appreciation to Drs. Pierre van Goethem, Milos Dostal, Christopher Henrich, and Mr. Carlos Berenstein for the valuable aid they have given me in the preparation of this work. I should also like to thank Miss Anne Paniccia for the extreme care she has taken in typing and proof-reading the manuscript.

L. Ehrenpreis

June 1969
New York, New York

Contents

Chapter I. Introduction, Analytically Uniform Spaces, and Multiplicity Varieties

Summary 1
I.1 Introduction 2
I.2 Prerequisites 6
I.3 Analytically Uniform Spaces 8
I.4 Algebroid Functions and Multiplicity Varieties . . 14
Remarks 25
Problems 27

PART A. QUOTIENT STRUCTURE THEOREMS

Chapter II. The Geometric Structure of Local Ideals and Modules

Summary 29
II.1 L'Hospital Representation 29
II.2 Formulation of the Main Results 36
II.3 The Main Results for Principal Ideals 49
II.4 Conclusion of the Proofs of Theorems 2.3, 2.5, and 2.6 59
Remarks 70
Problems 71

Chapter III. Semilocal Theory

Summary 72
III.1 Formulation of the Main Results 72
III.2 L'Hospital Representation 77
III.3 Connectability 78

III.4 Principal Ideals 81
III.5 Completion of the Proofs 88
Remarks 94
Problems 94

Chapter IV. Passage from Local to Global

Summary 95
IV.1 Localizeable Spaces 96
IV.2 Formulation of the Quotient Structure Theorem (Fundamental Principle) 98
IV.3 Cohomology with Growth Conditions 98
IV.4 Proof of the Main Result 101
IV.5 The Oka Embedding 116
Remarks 119
Problems 120

Chapter V. Examples

Summary 121
V.1 Example 1. The Space $\mathscr{H}$ of Entire Functions . . 122
V.2 Example 2. The space $\mathscr{H}'$ 138
V.3 Example 3. The Space $\mathscr{D}_F'$ of Distributions of Finite Order 139
V.4 Example 4. The Space $\mathscr{D}'$ of all Distributions . . 148
V.5 Example 5. The Space $\mathscr{E}$ of Indefinitely Differentiable Functions 152
V.6 Example 6. Non-Quasianalytic Classes 163
V.7 Example 7. Rapidly Increasing Functions and Distributions 169
V.8 Example 8. Formal Power Series and Polynomials 172
Remarks 172
Problems 173

PART B. SYSTEMS OF PARTIAL DIFFERENTIAL EQUATIONS WITH CONSTANT COEFFICIENTS

Chapter VI. Inhomogeneous Equations

Summary 175
VI.1 General Inhomogeneous Systems 175
VI.2 Fundamental Solutions 179
Remarks 187
Problems 187

Chapter VII. Integral Representation of Solutions of Homogeneous Equations

Summary 188
Remarks 196
Problems 197

Chapter VIII. Extension and Comparison Theorems. Elliptic and Hyperbolic Systems

Summary 198
VIII.1 Comparison Theorems 199
VIII.2 Elliptic Systems 204
VIII.3 Hyperbolic Systems 206
VIII.4 Quasihyperbolicity 210
Remarks 212
Problems 213

Chapter IX. General Theory of Cauchy's Problem

Summary 214
IX.1 Formulations of the Problems 217
IX.2 Generalization of the Cauchy-Kowalewski Theorem 224
IX.3 Hyperbolic Systems 231
IX.4 Parabolic Systems 231
IX.5 Groups and Semigroups Associated with Cauchy and Initial Value Problems 232
IX.6 Structure of Cauchy Data 235
IX.7 The Fundamental Solution and Nonlinear Initial Surfaces 240
IX.8 Domain of Dependence and Lacunas 243
IX.9 Uniqueness of the Cauchy Problem 244
IX.10 More Refined Uniqueness Results 265
IX.11 The Zeros of Solutions of Elliptic Systems . . . 271
IX.12 General Systems 276
Remarks 276
Problems 279

Chapter X. Balayage and General Boundary Value Problems

Summary 282
X.1 Balayage, General Boundary Value Problems, and Interpolation 283
X.2 An Example 287
X.3 The General Case 293

X.4 The Dirichlet Problem 295
X.5 Convex Polyhedra 303
X.6 The Fredholm Alternative. 305
X.7 The Wiener-Hopf and Goursat Problems 309
X.8 The Newman-Shapiro Parametrization Problem . 311
X.9 General Systems 314
Remarks 314
Problems 315

Chapter XI. Miscellanae

Summary 316
XI.1 Extension to Convolution Systems 317
XI.2 Singularities 325
XI.3 General Domains 367
XI.4 Special Functions and Group Representations. . 379
XI.5 Variable Coefficient Equations 385
Remarks 388
Problems. 391

PART C. SEQUENCES OF OPERATORS

Chapter XII. Lacunary Series. Refined Comparison Theorems

Summary 394
XII.1 Formulation of the Problem 395
XII.2 $(\mathscr{W}, \mathscr{W}_1)$ Density 406
XII.3 Analytic $(\mathscr{W}, \mathscr{W}_1)$ Density 410
XII.4 Geometric Density 413
XII.5 Example 1. The Classical Fabry Gap Theorem 416
XII.6 Example 2. Analog of the Fabry Gap Theorem for $\mathscr{E}$ 419
XII.7 Example 3. Relation to the Riemann ζ Function 422
XII.8 Example 4. An Analog of the Fabry Gap Theorem for Several Variables and Complex a^j 432
XII.9 Grouping of Terms 436
XII.10 Natural Boundaries 438
XII.11 The Converse Problem 439
Remarks 442
Problems. 445

Chapter XIII. General Theory of Quasianalytic Functions

Summary 446
XIII.1 General Quasianalyticity and Approximation . 447
XIII.2 A Nonsymmetric Generalization of the Denjoy-Carleman Classes 459

XIII.3 The Punctual and Local Images of Quasianalytic and Non-Quasianalytic Classes . . . 473
XIII.4 Quasianalytic Functions on Lines 484
Remarks 486
Problems 487

Bibliography 490

Index of Special Notation 497

Index 501

CHAPTER I

Introduction, Analytically Uniform Spaces, and Multiplicity Varieties

Summary

In Section I.1 we describe the type of Fourier analysis developed in this book. For many of the applications which we treat, a Fourier representation of functions of large growth is necessary. The large growth of the functions forces us to use unbounded exponentials, whereas bounded exponentials are sufficient for the classical representation of L^p functions. If we think of the dual of an Abelian group as the group of those characters necessary for Fourier analysis on the group, the "dual" of a real vector space is a complex vector space $C = C^n$ whose complex dimension is equal to the real dimension of the original space. In general, we do not need the whole complex space in order to obtain a Fourier representation; parts of the complex space which are large enough for this purpose are called *sufficient*.

In Section I.2 we explain what the reader should know in order to understand this book. We prove an inequality on polynomials which plays an important role in our work.

Section I.3 is a precise description of the type of generalized Fourier analysis which we use. This is essentially related to the notion of analytically uniform (AU) space which is a space $\mathscr{W}$ of functions or distributions [of $x = (x_1, \ldots, x_n)$] that contains all exponentials $\exp(ix \cdot z)$ for $z = (z_1, \ldots, z_n)$, z_j complex, such that the map $z \to \exp(ix \cdot z)$ is analytic, and such that the space $\mathbf{W}'$ of Fourier transforms

$$\hat{S}(z) = S \cdot \exp(ix \cdot z)$$

for $S \in \mathscr{W}'$ ($\mathscr{W}$ is the dual of $\mathscr{W}'$) is a space of entire functions which can be described by growth conditions. These growth conditions are of the following form: There exists a family $K = \{k\}$ of continuous positive functions on C (called *an AU structure for* $\mathscr{W}$) such that for any $S \in \mathscr{W}'$

and $k \in K$ we have $\hat{S}(z) = o(k(z))$. Moreover, the topology of $\mathscr{W}'$ can be described by the seminorms

$$\|S\|_k = \max |\hat{S}(z)|/k(z).$$

Given an AU space $\mathscr{W}$ we show that any $w \in \mathscr{W}$ has a Fourier representation of the form

$$w(x) = \int \exp(ix \cdot z)\, d\mu(z)/k(z),$$

where $\mu(z)$ is a complex measure on $C = C^n$ which has bounded total variation and $k(z)$ belongs to an AU structure for $\mathscr{W}$. The support of μ need not be all of C, but can be any sufficient set for $\mathscr{W}$.

In Section I.4 we introduce the concept of multiplicity variety, which is a refinement of the classical notion of algebraic variety. Multiplicity varieties are necessary if we want to give a geometric structure to an arbitrary polynomial ideal. For example, if $n = 1$, the ideal generated by z^2 can be described by the multiplicity variety (origin identity; origin, d/dz) which means that any polynomial P is determined modulo the ideal generated by z^2 by $(P(0), P'(0))$, which we can think of as the restriction of P to this multiplicity variety. In general ($n \geqslant 1$), an *algebraic multiplicity variety* $\mathfrak{V}$ is a collection

$$\mathfrak{V} = (V_1, \partial_1; \ldots; V_q, \partial_q),$$

where V_j is a complex algebraic variety (the set of common zeros of polynomials) and ∂_j are differential operators of a suitable type. There is a restriction map $\rho_{\mathfrak{V}}$ which maps a polynomial P into

$$\rho_{\mathfrak{V}}(P) = (\partial_1 P|_{V_1}, \ldots, \partial_q P|_{V_q})$$

which is an element of $\mathscr{P}(\mathfrak{V})$, the ring of polynomials on $\mathfrak{V}$. (Here $\partial_j P|_{V_j}$ is the restriction of $\partial_j P$ to V_j.)

An analogous concept for the ring of local analytic functions is introduced.

I.1. Introduction

The theory of Fourier integrals, as originally envisaged, was a method of representing a more or less arbitrary function $f(x)$ of one or several real variables as a linear combination of exponentials. The classical theory dealt with the situation in which f is assumed to be small at infinity; in particular L^p, $1 \leqslant p \leqslant 2$, were the most common spaces studied. The next stage, which was pioneered by the work of N. Wiener, culminated in the theory of distributions of L. Schwartz [1] in which f is (essentially) allowed to be of polynomial growth. The methods of Schwartz extend to somewhat larger functions, but they cannot penetrate the class of functions which are allowed to grow exponentially. The reason for this is the following: If f is of polynomial growth we can represent it as a Fourier integral in which the frequencies that enter are real, that is, $f(x)$ is represented as a suitably convergent integral involving $e^{ix \cdot z}$ with z real. This possibility no longer exists (at least if we are to interpret the Fourier integral in essentially its usual manner) if f is of exponential growth.

A reasonable way to overcome this is to allow exponentials with complex frequencies in the Fourier representation. If we do this we are quickly led to the realization that the Fourier integral representation will not be unique. The reason is that the complex exponentials satisfy many linear relations; for example, if x (or z) is a single real (or complex) parameter, then Cauchy's formula shows that for any z_0 we may write

$$\exp(ixz_0) = \frac{1}{2\pi i} \int_\gamma \frac{\exp(ixz)}{z - z_0} \, dz, \tag{1.1}$$

where the path of integration γ is a closed curve surrounding z_0. Thus, $\exp(ixz_0)$ is a limit of linear combinations of $\exp(ixz)$ for $z \in \gamma$.

For this reason we might ask if there exist subsets of complex space such that we could write the functions we are interested in as Fourier integrals with frequencies in σ. Such a set σ is called a *sufficient set*. Examples of sufficient sets are given below. It is interesting to note that there are many simple examples of sufficient sets whose dimension is equal to the dimension of the real space. For example, for indefinitely differentiable functions of one variable x, the union of the real and imaginary axes is a sufficient set. There are also examples of sufficient sets of dimension zero. Unfortunately, it does not seem possible in the usual examples to find sets σ which are sufficient and also have the property that the Fourier integral representation with frequencies in σ is unique.

What we have said above refers to "general" functions f which might be restricted only by growth and regularity conditions. If we put further restrictions on f then we might expect to find smaller sufficient sets. A well-known example of this type is the representation of the solution f of a linear ordinary differential equation with constant coefficients, $Df = 0$, as a linear combination of the exponential polynomial solutions. Neglecting for the moment the polynomial terms, we may say that a sufficient set for solutions of $Df = 0$ is the finite set of solutions of the polynomial equation $P(z) = 0$ where D is obtained from P by replacing z by $-id/dx$, z^2 by $-d^2/dx^2$, etc. One of the principal results of this book is that a similar result holds for linear partial differential operators with constant coefficients, more generally, if $D_1, \ldots, D_r$ are linear partial differential operators with constant coefficients then a sufficient set for those f which satisfy $D_1 f = 0, \ldots, D_r f = 0$, is the set of common zeros of the polynomials $P_1(z), \ldots, P_r(z)$ where P_j is obtained from D_j by replacing $-i\partial/\partial x_l$ by z_l. It turns out that if this set has positive dimension then there are proper subsets which are still sufficient (see Chapter VII).

The above statement is only approximate because, as we mentioned before, we have ignored the exponential polynomial solutions. For ordinary differential equations, these arise because of multiple zeros of P. In the general case, we must consider, in addition to the algebraic variety of common zeros of the P_j, certain multiplicities. This leads us to the concept of multiplicity varieties, one of our chief tools.

In addition to the intrinsic interest in the question of finding sufficient sets, they are important in many problems. For example, the regularity of solutions of linear elliptic equations with constant coefficients and the properties of linear hyperbolic equations with constant coefficients are immediate consequences of the aforementioned results on sufficient sets for such functions (see Chapter VIII).

It is of interest to look at Fourier representations from the following point of view: In general we may think of "arbitrary" functions as being "parametrized" by the values they assign to each point. In giving Fourier representations we "parametrize" functions by their Fourier transforms. If f satisfies $D_1 f = 0, \ldots, D_r f = 0$ as before, we could try to compare the Fourier parametrization of f with the parametrization by means of boundary data, Cauchy data, etc. This point of view is pursued in Chapters IX and X and leads to conditions under which boundary value problems, the Cauchy problem, etc., make sense for the system of equations.

One of the most important types of problems with which we shall deal in this book is the following: We place on f conditions of the following form:

1. f has a Fourier representation in which the frequencies are restricted to some fixed set τ which is not sufficient for general functions.
2. f has certain regularity and growth conditions.

What do these imply about f?

The kind of regularity and growth conditions we shall consider in 2 are linear, that is, the set of f satisfying them will form a linear space $\mathscr{W}$ over the complex numbers. In addition we shall have to assume that the conditions are of such a form that $\mathscr{W}$ can be topologized in a way such that $\mathscr{W}$ is an analytically uniform (AU) space (see Section I.2). We feel that AU spaces form a natural class of spaces to consider when studying the Fourier analysis of large functions.

Of course, there are many interesting problems in Fourier analysis which involve conditions of growth and regularity which are *not* in the framework of analytically uniform spaces. For example, the space of analytic functions in the open unit disk with radial limits at the point one is of interest in the study of Tauberian theorems, but is not analytically uniform. It would be of interest to extend the methods of this book so as to apply to such spaces.

The deductions we should like to make from Conditions 1 and 2 are of the following form:

(a) f is determined by its values on a certain subset λ of the x space. It is natural to call this a uniqueness property.

(b) f satisfies an additional growth or regularity condition. This will be expressed by saying that f belongs to $\mathscr{W}_1$ where $\mathscr{W}_1$ is another AU space.

(a) is closely related to the parametrization question mentioned above. It says that functions satisfying 1 and 2 are determined by their restrictions to λ. More specifically, as mentioned above, Condition 1 is satisfied if f satisfies a system of linear differential equations with constant coefficients. In Chapter IX we show how to obtain uniqueness properties for Cauchy's problem for such f. This is extended somewhat in Chapter XI. In Chapter XIII we consider an amelioration of Condition 1: We assume that the frequencies which do not belong to τ occur with very small coefficients in the Fourier representation of f. Then it is still possible to derive uniqueness properties. This type of uniqueness is called *quasianalyticity*, since it generalizes the classical uniqueness properties of analytic functions.

Examples of conclusions of type (b) which we shall obtain for solutions of linear partial differential equations with constant coefficients are regularity properties of elliptic equations, and the fact that solutions of some equations which are defined on subsets of x space (real or complex) can be extended to be solutions in larger subsets (see Chapter VIII). In Chapter XI we show that solutions in the whole space which are regular in some subsets must be regular in the whole space. In Chapter XII we derive similar results when we generalize the class of f considered from solutions of a system of linear constant coefficient equations to a class of functions for which certain *a priori* conditions are placed on τ. An interesting by-product of these investigations is that, among all functions $g(s)$ represented by the series

$$\sum_{n=0}^{\infty} a_n n^{-s}$$

which have analytic (meromorphic) continuations to the whole complex s plane, the Riemann ζ function

$$\zeta(s) = \sum n^{-s}$$

has, except for trivial exceptions, essentially the smallest growth.

This type of problem has the following structure: We are given Condition 1 and, in addition, we know that f belongs to AU spaces $\mathscr{T}, \mathscr{T}_1, \ldots, \mathscr{T}_l$. We want to conclude that f belongs to some other AU space $\mathscr{W}_1$. That is, we are comparing functions having Fourier representations with frequencies in τ which belong to $\mathscr{T}, \mathscr{T}_1, \ldots, \mathscr{T}_l$ with similar functions in $\mathscr{W}_1$. For this reason we refer to this as a *comparison problem.*

One of the original reasons for the introduction of Fourier representations was that it would enable one to solve equations

$$Df = g, \tag{1.2}$$

where D is a linear partial differential operator with constant coefficients. The reason for this is that the exponentials are eigenfunctions of $\partial/\partial x_j$ so that (1.2) is transformed formally into

$$P\hat{f} = \hat{g}, \tag{1.3}$$

where $\hat{f}$ and $\hat{g}$ are the Fourier transforms of f and g, respectively, and P is a polynomial. It is easy to solve (1.3) by dividing by P, provided that the zeros V of P do not cause any problems. This will be the case if we can find a sufficient set σ on which, say, $|P| \geq 1$, for then $\hat{g}$ can be chosen to be zero except on σ (see Section I.3).

If we want to go further and study, for example, solutions f of the system

$$D_1 f = g_1, \quad D_2 f = g_2, \ldots, \quad D_r f = g_r,$$

where the g_j must satisfy some compatibility conditions, then we meet algebraic and analytic difficulties. These are handled in Chapters II, III, and IV, and their solution enables us to deduce in Chapter VI a general existence theorem for inhomogeneous systems. This is extended somewhat in Chapter XI.

I.2. Prerequisites

The reader is expected to have some knowledge of topological vector spaces. Except for the details of some proofs, familiarity with the Hahn-Banach theorem, the concept of dual space, and reflexivity would be sufficient.

It would be helpful if he had some acquaintance with classical Fourier analysis, e.g., L^2 theory, although this is not essential. Some knowledge of the theory of distributions (see Schwartz [1]) would be very useful.

The reader is also expected to have some familiarity with polynomial ideals, especially the Hilbert basis theorem. Also some knowledge of the theory of analytic functions of several complex variables including the Weierstrass preparation theorem is important, and is indispensable for Chapter II. This can be found in Bochner and Martin [1] or in Gunning and Rossi [1].

In addition we shall often use the following important results:

THEOREM 1.1. *Let $P(s, w)$ be a polynomial in two variables of the form*

$$P(s, w) = a_m(w)s^m + a_{m-1}(w)s^{m-1} + \cdots + a_0(w),$$

where $m \geqslant 1$ and $a_m \not\equiv 0$. Then we can write

$$P(s, w) = a_m(w) \prod_{j=1}^{m} (s - s_j(w)).$$

Here each $s_j(w)$ can be represented in $0 < |w| < \delta$ by a series of the form

$$s_j(w) = \sum_{k=l}^{\infty} b_k w^{k/p}$$

for some integer l and some positive integer p. If a_m is a constant then $l \geqslant 0$.

The expansion for $s_j(w)$ described above is called the *Puiseux expansion.* A proof can be found in Van der Waerden [1] pp. 50–55, except for the fact that $l \geqslant 0$ if a_m is a constant. However, if $l < 0$ and $b_l \neq 0$, then s_j^m contains a power $w^{lm/p}$. This does not appear in any $a_k(w)s_j{}^k$ for $k < m$. This implies that $P(s_j, w)$ cannot vanish identically if a_m is a nonzero constant.

More detail on the Puiseux expansion is found in Theorem 1.13.

The next results are essentially classical. Their application to problems in Fourier analysis seems to have been pointed out for the first time by Ehrenpreis [1] and Malgrange [1].

LEMMA 1.2. *Let $P(z) = a_m z^m + \cdots + a_0$ be a polynomial in one variable. Given any $u > 0$ and any z^0 we can find a circle γ surrounding z^0 of radius $\leqslant u$ on which*

$$|P(z)| \geqslant |a_m|[u/2(m + 1)]^m.$$

PROOF. On replacing z by $z - z^0$, we may assume $z^0 = 0$. Since the result is trivial if $a_m = 0$, we assume $a_m \neq 0$. We write

$$P(z) = a_m \prod_{j=1}^{m} (z - z^j),$$

where z^j are the zeros of P. Since there are, at most, m zeros z^j, there exists a subinterval of $[0, u]$ of length at least $u/(m+1)$ which does not contain any $|z^j|$. Let γ be the circle with center 0 which bisects this interval. Then for any $z \in \gamma$ and any j we have $|z - z^j| \geqslant \big| |z| - |z^j| \big| \geqslant u/2(m+1)$ which gives the result.

See Remark 1.1.

From Lemma 1.2 and the maximum modulus theorem, we derive immediately

COROLLARY 1.3. *Let $F(z)$ be a holomorphic function in $|z - z^0| \leqslant u$ and P as in Lemma 1.2. If $|P(z)F(z)| \leqslant c$ in $|z - z^0| \leqslant u$ then*

$$|a_m F(z^0)| \leqslant c[u/2(m+1)]^{-m}.$$

THEOREM 1.4. *Let $z = (z_1, \ldots, z_n)$, and let*

$$P(z) = a_m(z_1, \ldots, z_{n-1})z_n^m + \cdots + a_0(z_1, \ldots, z_{n-1})$$

be a polynomial in n variables, and let $F(z)$ be holomorphic in $|z_j - z_j^0| \leqslant u$ for all j, where z^0 is some fixed point. If $|P(z)F(z)| \leqslant c$ in

$$\{z \,|\, |z_j - z_j^0| \leqslant u \text{ for all } j\},$$

then there are constants a, b depending only on P such that

$$|F(z^0)| \leqslant acu^{-b}.$$

PROOF. We fix $z_1, \ldots, z_{n-1}$ with $|z_j - z_j^0| \leqslant u$. Then we can think of $P(z_1, \ldots, z_{n-1}, z_n)$ and $F(z_1, \ldots, z_{n-1}, z_n)$ as functions of the single variable z_n. Applying Corollary 1.3, we find

$$|a_m(z_1, \ldots, z_{n-1})F(z_1, \ldots, z_{n-1}, z_n^0)| \leqslant c[u/2(m+1)]^{-m}.$$

We may clearly assume that $a_m \not\equiv 0$. Our result now follows by iteration. □

I.3. Analytically Uniform Spaces

By R (C) we denote the real (or complex) euclidean space of dimension n with coordinates $x = (x_1, \ldots, x_n)$ (or $z = (z_1, \ldots, z_n)$). We shall write $x \cdot z$ or $z \cdot x$ for $\sum x_j z_j$. For any complex number a, we denote by Re a (or Im a) the real (or imaginary) part of a. Let $\mathscr{W}$ be a space of functions or more general objects such as the distributions of L. Schwartz [1] on R. For any $f \in \mathscr{W}$, we want to establish a representation

$$f(x) = \int \exp(ix \cdot z)\, d\mu(z)/k(z), \tag{1.4}$$

where k is a suitable measurable function, and where μ is a bounded measure (i.e., total variation finite) on C. To give (1.4) meaning in general, we require that $\mathscr{W}$ be a topological vector space and that the integral on the right side should converge in the topology of $\mathscr{W}$.

In order to make a "good" theory, we have found it necessary to restrict our considerations to those $\mathscr{W}$ which satisfy the following conditions:

(a) There exists a locally convex topological vector space $\mathscr{W}'$ so that $\mathscr{W}$ is the dual of $\mathscr{W}'$, that is, $\mathscr{W}$ is the space of continuous linear maps of $\mathscr{W}'$ into the complex numbers and the topology of $\mathscr{W}$ is that of uniform convergence on the bounded sets of $\mathscr{W}'$.

For $f \in \mathscr{W}$, $S \in \mathscr{W}'$ we shall write $S \cdot f$ or $f \cdot S$ for the value $f(S)$. In most of the examples we treat, $\mathscr{W}$ is reflexive, that is, $\mathscr{W}'$ is the dual of $\mathscr{W}$.

(b) For each z we have $\exp(ix \cdot z) \in \mathscr{W}$ and the map $z \to \exp(ix \cdot z)$ is complex analytic from $C \to \mathscr{W}$. Moreover, the linear combinations of the exponentials are dense in $\mathscr{W}$.

An analytic map is a continuous map whose integral around any closed curve in any z_i plane is zero. Condition (b) means that for any $S \in \mathscr{W}'$ the *Fourier transform*

$$\hat{S}(z) = S \cdot \exp(ix \cdot z) \tag{1.5}$$

is defined for all z, is an entire function, and determines S. We denote by $\mathbf{W}'$ the space of Fourier transforms of elements of $\mathscr{W}'$. $\mathbf{W}'$ is topologized so as to make the Fourier transform a topological isomorphism.

We require in addition:

(c) There exists a family K of continuous positive functions k on C (which are allowed to take the value $+\infty$) so that, for each $F \in \mathbf{W}'$ and each $k \in K$,

$$|F(z)|/k(z) \to 0 \qquad \text{as } |z| \to \infty, \tag{1.6}$$

and such that the sets

$$N_k = \{F \in \mathbf{W}' \mid |F(z)| \leqslant k(z) \qquad \text{for all } z\} \tag{1.7}$$

form a fundamental system of neighborhoods of zero in $\mathbf{W}'$, that is, each N_k is a neighborhood of zero, and every neighborhood of zero contains some N_k.

A space $\mathscr{W}$ satisfying Conditions (a), (b), and (c) will be called *analytically uniform*, or briefly, *AU*. K will be called an *AU structure* for $\mathscr{W}$.

We require K to have the property that if $k \in K$ and α is a positive real constant, then $\alpha k \in K$. It should be observed that there are many AU structures for a given AU space. These AU structures may even be disjoint. Examples of AU structures and spaces are given in Chapter V. In particular, the spaces $\mathscr{H}$ of entire functions (when x is thought of as complex variables), $\mathscr{E}$ of indefinitely differentiable functions, $\mathscr{D}'$ of distributions, and $\mathscr{D}'_F$ of distributions of finite order are AU. On the other hand, the space of real analytic functions is *not* AU (see Ehrenpreis [4], pp. 556–559).

See Problem 1.1.

Now, for $\mathscr{W}$ analytically uniform, let us consider any $\hat{T} \in \mathbf{W}$, the dual of $\mathbf{W}'$. Then $\hat{T}$ is bounded on some neighborhood of zero in $\mathbf{W}'$. Thus, there exists a $k \in K$ so that

$$F \in \mathbf{W}'$$

$$\max_{z \in C} |F(z)|/k(z) \leqslant 1$$

imply $|\hat{T} \cdot F| \leqslant 1$. By the Hahn-Banach theorem, $\hat{T}$ can be extended to a continuous linear function $\hat{T}$ on the space L_k of continuous functions G on C such that $|G(z)|/k(z)$ is (for this k) bounded on C and zero at infinity, with the norm $\max |G(z)/k(z)|$. Hence, there exists a measure μ on C with total variation finite such that, for any such G, we have

$$\hat{T} \cdot G = \int G(z)\, d\mu(z)/k(z). \tag{1.8}$$

A measure whose total variation is finite will be referred to as a *bounded measure*.

We want to use this representation to define the Fourier transform on $\mathbf{W}$. First we notice that for any $S \in \mathscr{W}'$, the integral

$$\int S \cdot \exp\,(ix \cdot z)\, d\mu(z)/k(z) = \int \hat{S}(z)\, d\mu(z)/k(z) \tag{1.9}$$

certainly exists in the sense of Lebesgue because of our assumption (1.6) on $\mathbf{W}'$. Moreover, these integrals are uniformly bounded for S in the neighborhood of zero defined by $|\hat{S}(z)|/k(z) \leqslant 1$ for $z \in C$. Thus, by the definition of $\mathscr{W}$, the map

$$S \to \int \hat{S}(z)\, d\mu(z)/k(z)$$

defines an element of $\mathscr{W}$ which we call the Fourier transform of $\hat{T}$ and which we denote by T. Symbolically, we write

$$T(x) = \int \exp\,(ix \cdot z)\, d\mu(z)/k(z). \tag{1.10}$$

This expression is only symbolic because T may be a more general type of object than a function so that the value $T(x)$ may not be defined for fixed x. (1.10) means that, for any $S \in \mathscr{W}'$, we have

$$S \cdot T = \int \hat{S}(z)\, d\mu(z)/k(z). \tag{1.11}$$

Moreover, the integral converges in the topology of $\mathscr{W}$ [this makes sense because $\exp(ix \cdot z) \in \mathscr{W}$]. Note that when k is large at infinity, T will be a function.

Finally, it is clear that the correspondence $T \leftrightarrow \hat{T}$ is one-to-one between $\mathscr{W}$ and $\mathbf{W}$. (Of course, there are many measures μ which represent T.) Thus we have proved the following theorem:

THEOREM 1.5. *For any $T \in \mathscr{W}$, we can find a bounded measure μ on C and a $k \in K$ such that*

$$T \cdot S = \int \hat{S}(z)\, d\mu(z)/k(z)$$

for any $S \in \mathscr{W}'$, where the integral exists in the sense of Lebesgue-Stieltjes.

Thus every element of $\mathscr{W}$ can be represented by a symbolic integral

$$\int \exp(ix \cdot z)\, d\mu(z)/k(z) \tag{1.12}$$

for some $k \in K$ and some bounded measure μ. This symbolic integral is to be understood in the sense of (1.11) for any $S \in \mathscr{W}'$. The integral converges in the topology of $\mathscr{W}$, that is, integrals (1.11) converge uniformly for S in any bounded set of $\mathscr{W}'$ (even uniformly for S in some neighborhood of zero).

For many important function spaces, the symbolic integral will actually exist as a Lebesgue-Stieltjes integral. This is in particular true of the spaces $\mathscr{E}$ and $\mathscr{H}$.

Theorem 1.5 follows from the fact that $\mathscr{W}$ is AU. One might wonder whether the converse holds, that is, if every $T \in \mathscr{W}$ has a Fourier representation in the form (1.10), is $\mathscr{W}$ an AU space? Under certain conditions this converse is true. For details, see the proof of theorem 13.8.

The measure μ of Theorem 1.5 is not unique. In many cases we can say much more. For example, if T satisfies a partial differential equation, then the support of μ can be chosen to be an algebraic variety. This result is very deep and is proven in Chapter VII. We content ourselves here with showing that in certain cases the support of μ can be chosen to be on certain subsets of C.

We explain how these results are derived: First, let us note that the measure μ was found because of the representation of the elements of the dual of L_k as bounded measures on C divided by k. Now let σ be a closed subset of C and denote by $L_k(\sigma)$ the space of continuous functions G on σ, for which $|G(z)|/k(z)$ is bounded on σ, and tends to zero as $|z| \to \infty$; the norm in $L_k(\sigma)$ is defined as $\max_{z \in \sigma} |G(z)|/k(z)$.

It is easy for many k to produce sets σ so that the conditions $G \in \mathbf{W}'$ and

$$\max_{z \in \sigma} |G(z)|/k(z) \leqslant 1$$

imply

$$\max_{z \in C} |G(z)|/k(z) \leqslant 1.$$

Even if this cannot be done, then we can usually produce a σ and a continuous positive function k_1 on σ so that the conditions $G \in \mathbf{W}'$,

$$\max_{z \in \sigma} |G(z)|/k_1(z) \leqslant 1$$

imply

$$\max_{z \in C} |G(z)|/k(z) \leqslant 1.$$

Moreover $|G(z)|/k_1(z) \to 0$ as $|z| \to \infty$, $z \in \sigma$. If this can be done, then we say that (σ, k_1) is k *sufficient.*

Suppose that (σ, k_1) is k sufficient. Then, if $\hat{T} \in \mathbf{W}$ is continuous on $\mathbf{W}'$ in the topology of L_k, then $\hat{T}$ is also continuous on $\mathbf{W}$ in the topology induced by $L_{k_1}(\sigma)$. The method of proof of Theorem 1.5 then leads to the representation

$$T(x) = \int \exp\,(ix \cdot z)\, d\mu(z)/k_1(z), \tag{1.13}$$

where μ is a bounded measure with support $\mu \subset \sigma$. In particular, we have

THEOREM 1.6. *Suppose there exists a closed set $\sigma \subset C$ so that, for each k in an AU structure for $\mathscr{W}$, we can find a k_1 as above so that (σ, k_1) is k sufficient, and the restriction to σ is one-one on $\mathbf{W}'$; then every $T \in \mathscr{W}$ has the Fourier representation*

$$T(x) = \int \exp\,(ix \cdot z)\, d\mu(z)/k_1(z) \tag{1.14}$$

for some $k_1 \in K$, where μ is a bounded measure with support on σ.

A set σ having the properties required by Theorem 1.6 will be called *sufficient for $\mathscr{W}$*. The corresponding set of k_1 will be said to be σ *sufficient for $\mathscr{W}$*.For example, the maximum modulus theorem shows that for any $\mathscr{W}$ a sufficient set is C minus any compact set.

In Chapter V, we shall give examples of sufficient sets. In particular, if

$n = 1$, then the union of the real and imaginary axes is sufficient for $\mathscr{E}$, $\mathscr{H}$, $\mathscr{D}'_F$, but not for $\mathscr{D}'$.

See Problem 1.2.

The representation (1.10) for T has its origin in the representation of $\hat{T}$ in the form $d\mu/k$. There are often other interesting types of representations for $\hat{T}$. Suppose, for example, that $n = 1$. Since $\mathbf{W}'$ is a space of entire functions, for certain infinite sequences $\{a_j\}$ the linear function

$$S \to \sum a_j \hat{S}^{(j)}(0) = \hat{T} \cdot \hat{S}$$

is continuous on $\mathbf{W}$. If this is the case then we may think of $\hat{T}$ as being an *infinite differential operator in* $\mathbf{W}$. The theory of infinite derivatives is discussed in detail in Ehrenpreis [7]. We shall meet this concept only slightly in the present work.

One important use of sufficient sets is in division problems. Let F be an entire function. We say that F is a *multiplier* for $\mathbf{W}'$ if $F\mathbf{W}' \subset \mathbf{W}'$ and if the map $G \to FG$ is continuous on $\mathbf{W}'$.

Let us denote by $S \to f * S$ the map of $\mathscr{W}'$ into $\mathscr{W}'$ which is the Fourier transform of $G \to FG$. By duality we get a continuous map $w \to f *' w$ of $\mathscr{W}$ into itself. More precisely, $S \cdot (f *' w) = (f * S) \cdot w$. We call $f * S$ the *convolution* of f and S. It should be noted that f is defined only in a formal sense. We have not stated to which space f belongs.

DEFINITION. We say that F is *slowly decreasing* for $\mathscr{W}$ if there is a $\mathscr{W}$ sufficient set σ such that F has no zeros on σ, and, if $K = \{k\}$ is an AU structure for $\mathscr{W}$, then the set of restrictions of the functions $k/|F|$ to σ is σ sufficient for $\mathscr{W}$.

THEOREM 1.7. *Suppose F is slowly decreasing. Then $w \to f *' w$ maps $\mathscr{W}$ onto itself.*

PROOF. Suppose we want to solve the equation

$$f *' w = w_1. \tag{1.15}$$

This means that for any $S \in \mathscr{W}'$

$$w \cdot (f * S) = S \cdot w_1. \tag{1.16}$$

Thus w is determined on the subspace $f * \mathscr{W}'$ of $\mathscr{W}'$. By the Hahn-Banach theorem, it suffices to show that, if $f * S \to 0$, then also $S \to 0$. But this is an immediate consequence of the definition of "slowly decreasing." ∎

Instead of considering a single multiplier F, we could consider a family of multipliers F_λ depending continuously on a parameter λ in a suitable sense. Then, if we assume that the F_λ are uniformly slowly decreasing, that is, the set σ can be chosen independently of λ, and the set of restrictions of

$$k(z)/\min_{\lambda} |F_\lambda(z)|$$

to σ coincides with the set of restrictions of an AU structure, then we could show that if w_1^λ depend continuously on λ, and we can find w^λ depending continuously on λ such that

$$f_\lambda *' w^\lambda = w_1^\lambda, \tag{1.17}$$

provided that $\mathscr{W}$ satisfies certain topological conditions. We shall omit the details as we shall make no use of this result in this book. The argument is given in Ehrenpreis [10] for the space $\mathscr{D}$ and the same method applies in general.

If F is a polynomial, then for all of the spaces $\mathscr{W}$ which we shall consider, the set where $|F| \geqslant 1$ is sufficient. From this it follows that we can solve inhomogeneous linear constant coefficient equations in these spaces.

For the spaces $\mathscr{D}$, $\mathscr{E}$, and several others, the set of all slowly decreasing multipliers is determined in Ehrenpreis [4]. Some other examples are given in Hörmander [6]. It turns out that this is the same as the set of f for which $w \to f *' w$ is onto.

See Problem 1.3.

I.4. Algebroid Functions and Multiplicity Varieties

For $z^0 \in C$ and $\lambda > 0$, we denote by $\mathcal{O}(n; z^0; \lambda)$ the ring of functions which are analytic (holomorphic) on $|z - z^0| < \lambda$. Here $|z|$ is any convenient norm of z, for example, $\max |z_j|$. We shall sometimes suppress the z^0 and n. We shall denote by $\mathcal{O}^l(n; z^0; \lambda)$ the direct sum of l copies of $\mathcal{O}(n; z^0; \lambda)$ with itself. By $\mathcal{O}(n; z^0)$ we denote the ring of germs of functions analytic at z^0, that is, the space of functions analytic in some neighborhood of z^0, two such functions being identified if they agree on some neighborhood of z^0. We shall often suppress the word "germ." $\mathcal{O}^l(n; z^0)$ is the direct sum of l copies of $\mathcal{O}(n; z^0)$ with itself. Again we shall often repress z^0 and n.

Let $F_1, \ldots, F_r$ be functions in $\mathcal{O}(\lambda)$ or in $\mathcal{O}$. Then we call the set of common zeros of the F_i an *analytic variety* (resp. *germ. of local analytic variety*); we shall often write simply "variety." If the F_i are polynomials, we call the variety *algebraic*. We allow the possibility of the variety being empty or of dimension n. In $\mathcal{O}$ we consider two varieties to be equal if they agree on a neighborhood of z^0.

A (local) variety V is called *irreducible* if whenever the product $FG = 0$ on V where $F, G \in \mathcal{O}$, then either $F = 0$ on V or $G = 0$ on V. It is known (see, e.g., Bochner and Martin [1]) that every local variety W can be written as a finite union of local irreducible varieties, none of which is contained in the union of the others. These are called the *components* of W.

A *Z variety* V' is a variety V minus a subvariety. If we write $V = V_1 \cup \cdots \cup V_b$ where the V_j are the irreducible components of V, then we can write (by changing the numbering if necessary) $V' = V_1' \cup \cdots \cup V_{b'}'$, where $b' \leqslant b$, and each V_j' is a Z subvariety of V_j, and no V_j' is empty. We call $\tilde{V} = V_1 \cup \cdots \cup V_{b'}$ the *Z closure* of V'. (This is the same as its ordinary closure, but the concept of Z closure makes sense over an arbitrary field.) The components of $\tilde{V}$ will be called the *components* of V'.

Note that the union of varieties is a variety but the union of Z varieties need not be a Z variety, as the example

$$[C^2 - \{z_1 = 0\}] \cup \{(0, 0)\}$$

shows. However, when no confusion can arise, we shall use the term "Z variety" for a union of Z varieties (which is the same as a Z variety minus a Z subvariety). (In conformity with the notation of algebraic geometry (see Weil [1]), we could use the term "bunch of Z varieties," but we shall not do this.)

Let S be any set in C, and let $z \in S$. The function f on S is said to be *analytic* (*regular, holomorphic*) on S at p if, on some neighborhood of p in S it coincides with the restriction to S of a function which is analytic in a neighborhood in C of p. f is called *regular on* S if it is regular at each point of S.

For any set S we denote by $G \to G \mid S$ the restriction to S of a function G defined on a neighborhood of S.

If V is a variety, the space of functions analytic at each point of V will be denoted by $\mathcal{O}(V)$. If V' is a Z variety then $\mathcal{O}(V')$ will denote $\mathcal{O}(\tilde{V})$, where $\tilde{V}$ is the Z closure of V'. Note that $\mathcal{O}(V')$ is *not* the same as the set of functions which are regular on V'.

Let V' be a variety and $\tilde{V}$ its Z closure. A *meromorphic function* on $\tilde{V}$ (or on V') is a quotient g/h where $g, h \in \mathcal{O}(\tilde{V})$ and h does not vanish identically on any component of $\tilde{V}$. The space of meromorphic functions on $\tilde{V}$ will be denoted by $\mathcal{M}(\tilde{V})$ or $\mathcal{M}(V')$.

In Chapters I and II we shall have to deal with multivalued functions. The exact way in which we think of these is not too important, since, eventually, we shall deal with expressions formed from multivalued functions which are single-valued. We shall follow an approach which closely resembles the classical theory of Riemann (see e.g. Weyl [1]). It should be pointed out that arguments involving finite multivaluedness can usually be reduced to single-valuedness arguments by using symmetric functions.

We could have made our presentation more "elegant" by eliminating the use of multivalued functions and dealing exclusively with covering spaces. However, we feel that an analyst likes to have a "concrete" formula even though it involves multivalued functions. For example, if z is a complex number $\sqrt{z}$ seems to have more concrete meaning—even though there is an ambiguity in sign—than a definition of square root involving covering surfaces. In any case, the interested reader will find no difficulty in "translating" from our language to more modern language.

DEFINITION. Let $s = (s_1, \ldots, s_b)$. Let W be a local analytic variety in C^{n+b} (coordinates (s, z)). Suppose that for each i there is a function $J_i(s_i, z)$ of the form

$$J_i(s_i, z) = s_i^{m_i} + J_{i1}(z)s_i^{m_i-1} + \cdots + J_{im_i}(z), \tag{1.18}$$

where the J_{ij} are in $\mathcal{O}$, such that $J_i(s_i, z) = 0$ for any $(s, z) \in W$. Let V be a local Z variety in z space such that each $z^0 \in V$ has at least one point in W *above* it, that is, there exists an s^0 such that $(s^0, z^0) \in W$, and, moreover, whenever $(s, z) \in W$ we have $z \in V$. Then, W is called a *covering variety* of V.

By an *abstract algebroid function on* V is meant a meromorphic function $f = g/h$ with $g, h \in \mathcal{O}(W)$ and $h \not\equiv 0$ on W. In case h can be chosen $\equiv 1$ we call f *integral*.

Let V be a variety or Z variety. Suppose for each point z^0 of V we are given a finite number of continuous functions $f_{z^0}^1, \ldots, f_{z^0}^l$ defined on some neighborhood $N(z_0)$. (In general, we allow a continuous function to take the value ∞.) These $f_{z^0}^j$ are called *function elements*. The function elements define a *multivalued locally continuous function* f on V if, for each parametrized curve in V (that is, a continuous function $\lambda(t)$ from $0 \leq t \leq 1$ into V, say with $\lambda(0) = z^0$, $\lambda(1) = z^1$) there exists an ε small enough so that for all t, $N(\lambda(t))$ and $N(\lambda(t+\varepsilon))$ intersect if $t + \varepsilon \leq 1$, and for each $\varepsilon' \leq \varepsilon$ and each j there is a j' so that

$$f_{\lambda(t)}^j = f_{\lambda(t+\varepsilon')}^{j'} \text{ on } N(\lambda(t)) \cap N(\lambda(t+\varepsilon')).$$

See Fig. 1.

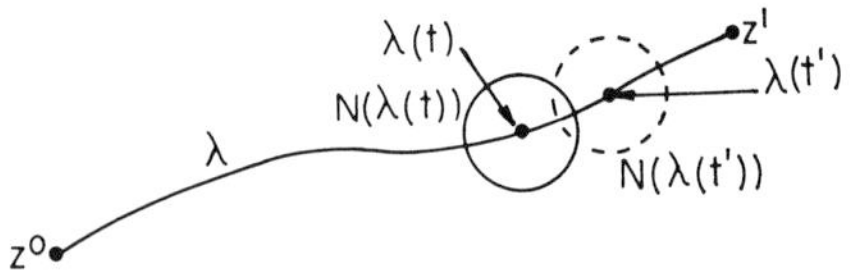

FIGURE 1.

Note that given λ and a function element $f^j_{z^0}$, going along the curve λ yields a unique function element at z^1 called the function element deduced by *direct continuation* along λ. f is called *irreducible* if, for each z_0 and each j, every $f^{j'}_{z^0}$ can be deduced from $f^j_{z^0}$ by direct continuation along a closed curve starting and ending at z^0. f is called *analytic* if each $f^j_{z^0}$ is analytic.

By an *algebroid function on V* is meant a multivalued locally continuous function f on V for which there exists an abstract algebroid function g/h on W and a choice mechanism such that for each $z \in V$ and each local determination of f, that is, each function element defining f, we can write

$$f(z) = g(s, z)/h(s, z).$$

Here (s, z) is the chosen point in W; we assume it depends continuously on z and that $h(s, z) \neq 0$. Of course, (s, z) depends on which function element of f we use. See Fig. 2.

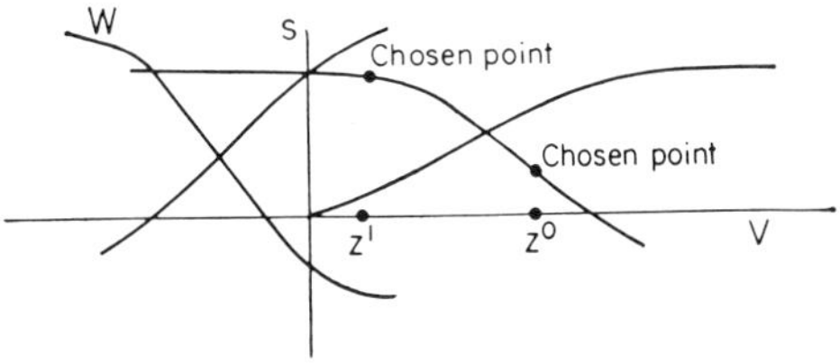

FIGURE 2.

f is called *regular* at $z \in V$ if all function elements of f can be represented as holomorphic functions in a neighborhood of z.

In case h can be chosen $\equiv 1$, we call f an *integral algebroid* function.

If V and W are algebraic varieties and g and h polynomials in all variables, we call f an *algebraic function.*

In addition to algebroid functions we shall deal with algebroid maps. By this we mean that we form the various local determinations of an algebroid function f, say f^j, into a vector $\vec{f}$ which we regard as a mapping of V into some C^l. The mapping $z \to \vec{f}(z)$ is, of course, multi-valued, but we shall be concerned, eventually, only with those functions of $\vec{f}(z)$ which are single valued functions of z.

The meaning of *algebraic map* or *integral map* is clear.

For example, the function elements defined by $\pm\sqrt{z}$ form an algebraic function in the complex plane minus the origin (the definition of which square root is $+\sqrt{z}$ being made in some convenient manner, for example by starting on the positive real axis). The map $z \to (+\sqrt{z}, -\sqrt{z})$ is an integral algebroid map.

See Remarks 1.2 and 1.3.

We may assume that the roots of the polynomials $J_i(s_i, z)$ in s_i are distinct at some point of V. If $J_i(s_i, z)$ had multiple roots on all of V we could replace J_i by $\partial J_i/\partial s_i$ since the zeros of $\partial J_i/\partial s_i$ above V would then be the same as those of J_i above V. Thus in case V is irreducible we may assume J_i has distinct roots except on a subvariety of V of lower dimension than V.

We may add or multiply two algebroid functions f and f' and obtain algebroid functions $f + f'$ and ff'. To see this, we suppose that f is defined using a variety W in the space with coordinates $(s_1, \ldots, s_b, z)$ and W' in the space with coordinates $(s_1', \ldots, s_{b'}', z)$. Let $s'' = (s, s')$. Let W'' be the variety in (s'', z) space defined by using all the equations of W and W'. If g/h and g'/h' are the respective abstract algebroid functions on W and W' defining f, f', then $g/h + g'/h'$ and gg'/hh', with the choice (s, s', z) where (s, z) and (s', z) are all the choices for f and f' respectively, define $f + f'$ and ff', respectively. In forming the sum or product of multivalued functions f and f', we thus add or multiply arbitrary pairs of function elements at a given point.

In the same manner we can show that given any finite number of algebroid functions $f_1, \ldots, f_c$ on V we can find a variety W over V such that each f_j can be represented by an abstract algebroid function g_j/h_j on W.

Suppose in the above notation that f is an algebroid function which is regular on V. Let $f = g/h$ be a representation as an abstract algebroid function. Let $z^0 \in V$, and let N be a neighborhood of z^0 in V on which the function elements of f are continuous and single valued. The choice function gives us a partition of N into sets N_i where N_i consists of all $z \in N$ for which the chosen pair (s, z) belongs to some fixed local irreducible subvariety W_i of W. This partition will, of course, depend on the function element of f. We call W_i an *essential component of W above z^0* if N_i has the property that a function $k(z)$ which is analytic near z^0 and vanishes on N_i must vanish identically on V.

Though it is not essential for us, it is not difficult to show that the N_i are Z varieties. In any case, since we are assuming that the $J_i(s_i, z)$ have distinct roots for some z, hence for all z outside a subvariety, there is at least one N_{i_0} which contains an open set of V; it is the W_{i_0} corresponding to these N_{i_0} which are essential. The reason for the existence of N_{i_0} is as follows. Let z be a point in N above which all the $J_i(s_i, z)$ have distinct roots. Then above a small neighborhood M of z, W has distinct sheets, that is, the roots s_i of the J_i can be made into single-valued functions of z (compare Proposition 1.10). Since the number of sheets is finite, there is some sheet S above M for which the set of choice points (s, z) lying in S is a set of uniqueness for analytic functions. Since f agrees with g/h on these chosen points and since, by the analyticity of the $s_i(z)$, f "lifts" to an analytic function on S, $f = g/h$ everywhere on M.

A slight modification of the above shows that the essential W_i lie above N_i which are neighborhoods minus Z subvarieties.

PROPOSITION 1.8. *Let V be an irreducible local Z variety. Let f be an algebroid function which is regular on V which can be represented by an abstract algebroid function g/h. Suppose $f \equiv 0$. Then $g = 0$ on all irreducible components $\tilde{W}_j$ of W for which there is a $z^0 \in V$ such that $\tilde{W}_j$ contains an essential component of W above z^0.*

PROOF. Let W_j be an essential irreducible component of W above z^0 and let $\tilde{W}_j$ contain W_j. Then, in the above notation, for each $z \in N_j$ there is an $(s, z) \in W_j$, hence in $\tilde{W}_j$, such that $g(s, z) = 0$. Since, given z, there are only a finite number of possibilities for (s, z), and since N_j is a set of uniqueness for analytic functions, for at least one "sheet" in W_j the chosen points form a set of uniqueness. Since W_j is irreducible this implies that the chosen points in W_j are a set of uniqueness. Since $g = 0$ on the chosen points, $g \equiv 0$ on W_j hence on $\tilde{W}_j$. ▯

PROPOSITION 1.9. *Let V be a variety, V' a Z dense subvariety, and f an algebroid function which is regular on V'. Then there exists a (local) holomorphic function k which is not identically zero on any component of V' such that $kf \to 0$ on $V - V'$.*

PROOF. By definition, V is V' minus some subvarieties V_j'. Since V' is Z dense, each V_j' lies in a component V_j of V, which is of higher dimension than V_j'. Since V_j' is a proper algebraic subvariety of V_j, there is a holomorphic function which is zero on V_j' and not identically zero on any V_i. It follows that there is a $\tilde{k}$ which is not identically zero on any component of V' such that $\tilde{k}$ vanishes on each V_j'.

The symmetric functions of the function elements of f are single-valued functions on V' which are regular on V'. We claim they are in $\mathscr{M}(V')$. This follows from the fact that they are algebroid functions by Galois theory (see Van de Waerden [2]); we shall omit the details. Being in $\mathscr{M}(V')$ it follows from the definitions that they are the restrictions to V' of functions which are meromorphic on a full neighborhood of zero.

Now f satisfies an equation (on V)

$$f^m - S_1 f^{m-1} \pm \cdots \pm S_m = 0,$$

where S_i are the elementary symmetric functions. Clearing denominators yields an equation of the form

$$S_0' f^m + S_1' f^{m-1} + \cdots + S_m' = 0,$$

where the $S_i' \in O$. Thus $S_0'f = f'$ satisfies

$$(f')^m + S_1'(f')^{m-1} + S_2'S_0'(f')^{m-2} + \cdots + S_m'(S_0')^{m-1} = 0.$$

This is an equation with coefficients in $\mathcal{O}$, leading coefficient 1. By a simple argument (see, e.g., Marden [1]), $S_0'f$ must be bounded on V.

It follows from the algebraic method of construction that S_0' does not vanish identically on any component of V'. Thus $k = \tilde{k}S_0'$ satisfies the desired property. ▯

See Remark 1.4.

LEMMA 1.10. *Let $J(s, z) = s^m + J_1(z)s^{m-1} + \cdots + J_m(z)$ be holomorphic on $|z - z^0| < \lambda$ for all s. Then there is a Z variety V' of dimension n in $|z - z^0| < \lambda$ (that is, V' is the whole of $|z - z^0| < \lambda$ minus an analytic subvariety) on which the roots $\{s_j(z)\}$ of J define a regular function.*

PROOF. Suppose first that the discriminant of J is not identically zero. Let V' be the whole of $|z - z^0| < \lambda$ minus the set on which the discriminant is zero. Since the roots of J are distinct on V', they certainly define locally single-valued continuous functions. We claim that they are actually regular, that is, the $s_j(z)$ are locally holomorphic on V'. For this we note that, by the formula for implicit differentiation, the derivatives

$$\frac{\partial s_j(z)}{\partial z_k} = -\frac{\partial J(s_j(z), z)/\partial z_k}{\partial J(s_j(z), z)/\partial s}.$$

At the points we are considering. J has no multiple roots, that is, $\partial J(s_j(z), z)/\partial s \neq 0$. Thus, $\partial s_j(z)/\partial z$ is a continuous function of z, and hence, by a usual argument on implicit functions, s_j is locally holomorphic. Thus our assertion is proved in case the discriminant of J is not identically zero.

In case the discriminant of J is identically zero, we cannot use the above, since the denominator on the right side is identically zero. However (see Bochner and Martin [1]), the ring of local holomorphic functions is a unique factorization domain. Thus, in the neighborhood of any point of $|z - z^0| < \lambda$ we can write J as a product of locally irreducible factors. These can again be chosen to be polynomials in s with leading coefficient 1 (see Lemma 2.10). Now for an irreducible function G, the discriminant does not vanish identically because otherwise G and G_s would have a common factor. Thus, we can treat G as in our previous case of J and we have the result. ▯

PROPOSITION 1.11. *Let V be all z space minus a subvariety; let f be an algebroid function regular on V. Suppose f is defined by W and g/h. For any* *point z in V for which all the roots of the $J_i(s_i, z)$ are simple we have*

$$(1.19)\quad \frac{\partial f}{\partial z_j}(z) = [h(s, z)]^{-2}$$

$$\times \left\{ h(s, z)\left[\frac{\partial g}{\partial z_j}(s, z) - \sum_i \frac{\partial g}{\partial s_i}(s, z)\,\frac{(\partial J_i(s_i, z)/\partial z_j)}{(\partial J_i(s_i, z/\partial s_i)}\right]\right.$$

$$\left. - g(s, z)\left[\frac{\partial h}{\partial z_j}(s, z) - \sum_i \frac{\partial h}{\partial s_i}(s, z)\,\frac{(\partial J_i(s_i, z)/\partial z_j)}{(\partial J_i(s_i, z)/\partial s_i)}\right]\right\}.$$

Here (s, z) is the chosen point above z. For those $z \in V$ above which some J_i have multiple roots, the value $\partial f(z)/\partial z_j$ is a limit of expressions (1.19).

PROOF. The first part of Proposition 1.11 is just the formula for implicit differentiation. The second part follows from the first and the fact, noted above, that we may assume that the J_i have distinct roots off a proper subvariety of V so the denominators in (1.19) are zero on a proper subvariety. □

By the *zeros* of an algebroid function we mean the set of points at which some determination vanishes.

PROPOSITION 1.12. *Let f be an algebroid function on the Z variety V. The zeros of f form a subvariety of V. The zeros of an algebraic function form an algebraic variety.*

See Remark 1.5.

PROOF. We shall use some properties of local analytic varieties. These can be found in Bochner and Martin [1] and Gunning and Rossi [1].

It is clear that we may assume f and V are irreducible. Let g/h be an abstract representation for f on a Z variety W over V. Let V' be the set of zeros of f. Since f is analytic on V, the zeros of f form a "general subvariety" of V, that is they are defined locally on V by the zeros of a local analytic function on V rather than by a function in $\mathcal{O}$.

Let W' be the variety of zero of g in W. Consider all components W'_i of W' such that W'_i contains an open set of chosen points above V'. Here "chosen" refers to the particular branch of f which vanishes at the given point of V'. By Samuel [1] the projection V'_i of W'_i on V is a Z subvariety. Since W is a finite covering of V, the projection is an open map. Thus f vanishes on an open subset of V'_i, hence on the union of those components of V'_i which contain an open set upon which f vanishes. The union of these components is a Z variety. It is easily seen, using the remarks preceding Proposition 1.8, that f vanishes exactly on the closure in V of this set. Thus our result for algebroid functions is proven.

The result for algebraic functions is similar. □

THEOREM 1.13. *Let* $J(s, z) \in \mathcal{O}$ *be of the form*

$$J(s, z) = s^m + J_1(z)s^{m-1} + \cdots + J_m(z),$$

where $J_i \in \mathcal{O}$. *Write* $z = (t, w)$, *where* $t = z_1$, $w = (z_2, \ldots, z_n)$. *Consider the Puiseux expansion (see Theorem* 1.1) *of the roots* $s_j(t, w_0)$ *of* $J(s, t, w_0) = 0$ *thought of as functions of* t *for each fixed* $w = w_0$. *For* t_0 *near zero, we expand these roots in terms of a suitable power* $(t - t_0)^{1/p}$.

Assertion 1. *The set* V_p *of* (t_0, w_0) *where* p *has a fixed value is a union of* Z *varieties.*

Assertion 2. *If we write*

$$s_j(t, w_0) = \sum b_k(t - t_0)^{k/p},$$

then $k \geqslant 0$, *and the* b_k *are algebroid functions on* V_p.

PROOF. We examine carefully the proof of the Puiseux theorem as given in Van der Waerden [1], pp. 52–54. By replacing s by $s - J_1/m$ we may assume $J_1 = 0$. Call $\lambda = t - t_0$. For each j such that $J_j \not\equiv 0$, let the power series for J_j in λ begin with $a_{m_j}\lambda^{m_j}$ with $a_{m_j} \neq 0$. The set where all $m_j = \infty$ is a variety, and there $b_k = 0$ for all k. For those j with $m_j \neq \infty$, call σ the minimum of m_j/j. Thus

$$m_j - \sigma j \geqslant 0$$

and equality holds for at least one j.

We now introduce a new variable $\zeta = s\lambda^{-\sigma}$. Thus

$$J(s, \lambda) = \tilde{J}(\zeta, \lambda) = \lambda^{m\sigma}(\zeta^m + J_2\lambda^{-2\sigma}\zeta^{m-2} + \cdots + J_m\lambda^{-m\sigma}).$$

Write $\sigma = p/q$ with $q > 0$ and set

$$\xi = \lambda^{1/q}$$

Note that σ and hence q are constant on unions of Z varieties, since the set for which the power series of J_j begins with const $\cdot\, \lambda^{m_j}$ is clearly a Z variety, and the intersection of Z varieties is a Z variety.

We use variables ξ and ζ and write

$$\Phi(\xi, \zeta) = \zeta^m + B_2(\xi)\zeta^{m-2} + \cdots + B_m(\xi)$$

with

$$B_j(\xi) = J_j(\xi)\xi^{-jp}.$$

For those $B_j \not\equiv 0$ the power series begins with

$$a_{m_j}\xi^{qm_j - jp} = a_{m_j}\xi^{q(m_j - j\sigma)}.$$

For at least one j we have $B_j(0) \neq 0$. The polynomial

$$\varphi(\zeta) = \Phi(0, \zeta) = \zeta^m + \cdots + a_{m_j} \zeta^{m-j} + \cdots$$

is therefore not identically ζ^m. Since the coefficient of ζ^{m-1} is zero, $\varphi(\zeta)$ cannot be $(\zeta - \alpha)^m$, that is, $\varphi(\zeta)$ has at least two distinct roots. We can therefore write

$$\varphi(\zeta) = g_0(\zeta) h_0(\zeta),$$

where g_0 and h_0 are relatively prime polynomials in ζ of degrees β and γ, respectively, with leading coefficient one. The coefficients of g_0 and h_0 can be expressed simply in terms of the roots of φ, and hence are algebroid functions of the a_j, hence of z.

Having factored $\varphi(\zeta) = \Phi(0, \zeta)$ we apply the argument of Hensel's lemma to factor Φ itself. We try to write

$$\Phi(\xi, \zeta) = G(\xi, \zeta) H(\xi, \zeta).$$

Write

$$\begin{aligned} \Phi(\xi, \zeta) &= \varphi(\zeta) + \xi \varphi_1(\zeta) + \cdots \\ G(\xi, \zeta) &= g_0(\zeta) + \xi g_1(\zeta) + \cdots \\ H(\xi, \zeta) &= h_0(\zeta) + \xi h_1(\zeta) + \cdots. \end{aligned}$$

The equation $GH = \Phi$ yields, for each k (calling $\varphi_0 = \varphi$),

$$g_0(\zeta) h_k(\zeta) + g_1(\zeta) h_{k-1}(\zeta) + \cdots + g_k(\zeta) h_0(\zeta) = \varphi_k(\zeta).$$

Now g_0 and h_0 are given. Suppose $g_0, \ldots, g_{k-1}$ and $h_0, \ldots, h_{k-1}$ have been determined with the g_j polynomials of degree $<\beta$ and h_j of degree $<\gamma$. Then we must determine g_k and h_k by

$$g_0(\zeta) h_k(\zeta) + h_0(\zeta) g_k(\zeta) = c_k(\zeta),$$

where c_k is a given polynomial in ζ of degree $<\beta, \gamma$. Since g_0 and h_0 are relatively prime, we can solve this equation. Using the explicit euclidean algorithm we see easily that g_k and h_k can be chosen polynomials of degrees $<\beta, \gamma$. Moreover, g_k and h_k can be so chosen that their coefficients are polynomials in the coefficients of c_k, and g_0, and h_0, hence by induction are polynomials in the coefficients of g_0, h_0. By the above this means that the coefficients of g_k, h_k are algebroid functions of z.

We have thus factored Φ as

$$\Phi = GH.$$

This yields a corresponding factorization of J. However, the Puiseux expansion is a factorization of $J(s, \lambda)$ into linear factors. Thus we must factor G and H again if they are not linear. We see easily that, because of Proposition 1.12, this leads to no new difficulties. Theorem 1.13 is thus completely proved. $\square$.

By a *differentiation* (of order m) on C^n we mean an expression of the form $\partial^m/\partial z_1^{m_1} \cdots \partial z_n^{m_n}$. Let V be a (local) variety or (local) Z variety in C^n. By a *differential operator* on V, we mean a linear combination of differentiations whose coefficients (written to the left) are regular on V. These coefficients will be, depending on the context, local analytic functions, polynomials, algebroid functions, algebraic functions, etc. When we wish to specify the coefficients of the differential operators, we shall write "algebroid differential operator," "algebraic differential operator," etc. When the coefficients of the ∂_j are multivalued, we must explain how to add the various multivalued functions that appear in the sum. This is done in accordance with the above definition of addition of multivalued functions. Let $V_1, \ldots, V_a$ be varieties, and for each j let ∂_j be a differential operator on V_j. Then we call $\mathfrak{V} = (V_1, \partial_1; V_2, \partial_2; \ldots; V_a, \partial_a)$ a *multiplicity variety.* If the V_j are Z varieties and the ∂_j are differential operators on V_j then we call $\mathfrak{V}$ a *multiplicity Z variety.* We shall often identify a variety or Z variety V with the multiplicity (Z) variety (V, identity).

In case the V_j are algebraic varieties and the ∂_j are algebraic differential operators, we call $\mathfrak{V}$ an *algebraic multiplicity variety.* Similarly, we define an *algebraic multiplicity Z variety, polynomial multiplicity variety,* etc.

For our purposes we shall deal with various types of multiplicity varieties and Z varieties. Thus "multiplicity variety" will mean one of them, depending on the context. A similar convention will hold for differential operators. Of course, if there is any possibility of confusion, we shall be more explicit.

Let $\mathfrak{V} = (V_1, \partial_1; V_2, \partial_2; \ldots; V_a, \partial_a)$. For each i, let V_i' be an irreducible component of V_i. Then $\mathfrak{V}' = (V_1', \partial_1; V_2', \partial_2; \ldots; V_a', \partial_a)$ is called an *irreducible component of* $\mathfrak{V}$. This holds for $\mathfrak{V}$ a multiplicity variety or a multiplicity Z variety.

For $j = 1, \ldots, l$ let $\mathfrak{V}^j$ be multiplicity varieties or Z varieties. Then $(\mathfrak{V}^1, \ldots, \mathfrak{V}^l)$ is called a *vector multiplicity (Z) variety.*

Let $\mathfrak{V}$ be a multiplicity variety. By an *analytic function on* $\mathfrak{V}$ we mean an a-tuple $(G_1, \ldots, G_a)$, where G_j is an analytic (multivalued if the coefficient of the operators are multivalued) function on V_j, and where the G_j satisfy the following compatibility condition: There exists a function G analytic (always single-valued) in a neighborhood in C^n of

$$V_1 \cup V_2 \cup \cdots \cup V_a$$

such that the restriction $\partial_j G|_{V_j} = G_j$ for all j. We write

$$G|_{\mathfrak{V}} = (\partial_1 G|_{V_1}, \ldots, \partial_a G|_{V_a}). \tag{1.20}$$

Let $\mathfrak{V}' = (V_1', \partial_1'; \ldots; V_a', \partial_a')$ be a multiplicity Z variety. We denote by $\mathcal{O}(\mathfrak{V}'; n; z^0; \lambda)$ or $\mathcal{O}(\mathfrak{V}'; n; z^0)$ the space of a-tuples $(G_1, \ldots, G_a)$, where each G_j is algebroid on V_j' and is regular on V_j' and the G_j satisfy the same type of compatibility condition as above. Namely, we require that there exist a single-valued G, regular in a neighborhood in C^n of the union of the closures of the V_j', such that $\partial_j G|_{V_j'} = G_j$ for all j. Thus the only singularities we allow in $\mathcal{O}(\mathfrak{V}'; n; z^0)$ are those of the coefficients of the differential operators of $\mathfrak{V}'$. The only multivaluedness is that of the coefficients of the differential operators of $\mathfrak{V}'$. Note that this notation agrees with our previous notation if we identify a Z variety V' with the multiplicity Z variety $(V', \text{identity})$.

We shall denote by V the underlying variety of $\mathfrak{V}$; that is,

$$V = V_1 \cup \cdots \cup V_a.$$

We shall sometimes use the expression "point of $\mathfrak{V}$" to mean a point of V. If we use variables $z = (s, w)$, then by a *point of* $\mathfrak{V}$ *above* w_0, we mean a point of the form (s_0, w_0) which is in $\mathfrak{V}$. Let $\mathfrak{V}$ and $\mathfrak{V}^1$ be two multiplicity varieties. Suppose

$$\mathfrak{V} = (V_1, \partial_1; \ldots; V_a, \partial_a) \quad \text{and} \quad \mathfrak{V}^1 = (V_1^1, \partial_1^1; \ldots; V_{a_1}^1, \partial_{a_1}^1).$$

We write $\mathfrak{V} \supset \mathfrak{V}^1$ if each V_i^1 is of the form

$$V_i^1 \subset \cup V_{j_i} \tag{1.21}$$

in such a way that for each j_i that occurs in (1.21) we can find a differential operator $\delta_{j_i}^i$ whose coefficients are regular on V_i^1 so that $\delta_{j_i}^i \partial_{j_i} = \partial_i^1$.

Remarks

Remark 1.1. See page 8. Using Pólya and Szegö [1], p. 86, problem 66, we can improve the inequality to

$$|P(z)| \geqslant 2a_m(u/4)^m.$$

However, the explicit form of the constant is of no importance to us.

Remark 1.2. See page 17. We have not stated the definition of algebroid or algebraic function in the most economical way. For example it is possible to drop the assumption that the choice function be continuous. It is even possible to modify the assumption that f be locally analytic; we need only assume that f be sufficiently differentiable ("sufficient" depending on the singularities of V) and that the chosen points (s, z) stay in one locally irreducible subvariety of W. Other definitions are also possible. We shall not go into this matter since it is not germane to our subject. We leave it as a problem for the reader to study the question in more detail.

Remark 1.3. See page 17. We have used functions of the form (1.18) to construct roots of equations. It can be proved that any local analytic function $P(s_i, z)$ can be multiplied by a local analytic function Q with no zeros such that QP is of the form

$$(QP)(s_i, z) = a_0(z)s_i^m + a_1(z)s_i^{m-1} + \cdots + a_m(z),$$

where the $a_k(z)$ are local analytic functions. (If P is a polynomial, we can clearly choose $Q \equiv 1$. The general case follows from Theorem 1, p. 308 of Ehrenpreis [6] by setting, in the notation of that theorem, $B = w^s$.) We can rewrite the above as

$$a_0^{m-1}(QP)(s_i, z) = (a_0 s_i)^m + a_1(a_0 s_i)^{m-1} + a_0 a_1(a_0 s_i)^{m-2} + \cdots + a_0^{m-1}a_m.$$

Since we allow rational expressions in the definitions of algebroid and algebraic functions, it makes no difference whether we use s_i or $a_0 s_i$.

Thus the use of roots of equations of type (1.18) involves no restriction.

Remark 1.4. See page 20. It is possible to avoid Galois theory and give a purely analytic proof of Proposition 1.9. This is based on the fact that an algebroid function regular on V' cannot grow faster than a reciprocal power of the distance to $V - V'$ as we approach $V - V'$. Thus we could choose $k = \tilde{k}^l$ for sufficiently large l (notation as in proof). We have preferred to give some proofs from the algebraic viewpoint and some from the analytic viewpoint so as to acquaint the reader with both.

The rate at which an analytic function can approach zero is the subject of a very profound study made by Łojasiewicz (see Łojasiewicz [1]).

Remark 1.5. See page 21. It is possible to give an algebraic proof of this result based on the fact that the zeros of f are essentially the same as the zeros of a suitable *norm* of f. We give an alternative analytic proof.

Remark 1.6. For the case $r = 1$ most of the main results of Chapters II–X were found in 1957. At that time a resume was submitted to the Bulletin of the American Mathematical Society; unfortunately the manuscript was lost. There is a brief mention of the results in Gårding's Congress note [2]. The complete results were obtained in 1960 and appeared in Ehrenpreis [16].

The proof of the main results falls into two parts: local and semilocal (see Chapters II and III), and passage from local to global (Chapter IV). If we are interested in proving the results of Chapter VI and a weakened version of the results of Chapter VII in which we represent solutions of $\vec{D}f = 0$ as (a) Fourier transforms of distributions on V without giving the local structure of these distributions or as (b) Fourier transforms of suitable measures defined in a small neighborhood of V, then we can dispense with the study of the multiplicity variety and most of the details of Chapters II and III can be avoided. Thus Malgrange in [4] proved (a) using his formulation of the Łojasiewicz division results. He also found an ingenious device which gives another method for passing from local to global. His method for this cohomology problem seems applicable only to special $\mathscr{W}$.

If one is willing to settle for the weaker (b) then we need only that part of Theorem 4.2 which states that the map λ_F is a topological isomorphism. No mention of $\mathfrak{V}$ is required and those local and semilocal properties which are necessary can be derived by power series methods. (A similar method is used in Ehrenpreis, Guillemin, and Sternberg [1].)

Our method of passage from local to global has its origin in Oka's proof of the vanishing of cohomology groups, except that we need to be careful of bounds. Recently J. J. Kohn [1], [2], following a program of Spencer and Morrey, has given another proof of the vanishing of cohomology groups. Using Kohn's method Hörmander [5] gave a very general result on the vanishing of cohomology groups with bounds; Hörmander's result covers many of the examples given in Chapter V below, as well as many examples not treated in this book. The exact relation of Hörmander's results to ours is not clear. Kohn's results lead to differentiability on the boundary for domains with smooth boundaries, whereas Oka's method seems better for analytic polyhedra. This seems to indicate that more complete results could be obtained by a combination of the approaches.

While either (a) or (b) above can be used to derive the "general Poincaré lemma" of Chapter VI (Theorem 6.1) and most of the applications in Chapter VIII, neither seem sufficient for most of the results of Chapters IX, X, or Theorem 11.4. Also, the concept *sufficient in V* cannot be formulated in (a) or (b).

Most of the results of Chapter XII were found in 1961 and were presented in an address to the American Mathematical Society at Stillwater.

In conclusion we should like to stress that this book represents an introduction to the subject of Fourier analysis in several complex variables and its application to differential operators. There are many subtle questions which require investigation. Perhaps the most important of these involves an extension of our results to spaces which are not analytically uniform, for example the space of indefinitely differentiable functions on a nonconvex set. It is the hope of the author that this book will stimulate further study. A possible approach to this problem is to cut the nonconvex set into convex pieces. One then meets problems of compatibility which are of somewhat similar nature to those discussed in Section X.5. Instead of looking at compatibility conditions there is a method employing tensor products discussed briefly in Section XII.1. Another possible approach is to replace AU spaces in which we have Fourier representation by absolutely convergent integrals by spaces in which we have a more subtle kind of convergence, for example, by means of grouping of terms (see Section XII.9).

Problems

PROBLEM 1.1. See page 10.

Let K be a family of continuous positive functions k on C. Define the space $\mathbf{W}'$ to consist of all entire functions satisfying (1.6) for all $k \in K$ with topology defined using the sets (1.7) as a fundamental system of neighborhoods of zero.

1. Under what conditions on K is $\mathbf{W}'$ the Fourier transform of a space whose dual is a reflexive AU space?
2. When do two families K and K' define the same space $\mathbf{W}'$?

When the functions k depend only on the distance from the origin, these problems have been studied by Taylor [1].

PROBLEM 1.2. See page 13.

Give a general procedure for constructing sufficient sets in terms of the AU structure. (By considering examples like those of Chapter V, we see that for $n = 1$ the only sets τ such that $C - \tau$ is sufficient for *every* $\mathscr{W}$ are unions of compact sets with sets having no interior. For $n > 1$ the situation is more complicated because of the phenomenon of envelopes of holomorphy.)

PROBLEM 1.3. See page 14.

What conditions on $\mathscr{W}$ imply that the converse of Theorem 1.7 holds?

CHAPTER II

The Geometric Structure of Local Ideals and Modules

Summary

The concept of L'Hospital representation is introduced in Section II.1. This is a way of expressing the value of a regular function $f = g/h$ where $g(z)$ and $h(z)$ are analytic functions of $z = (z_1, \ldots, z_n)$ at a point where $h = 0$. This generalizes the classical L'Hospital rule for $n = 1$.

In Section II.2 we formulate the main results of this chapter. These consist in giving a geometric structure to the quotient of the ring $\mathcal{O}$ of (germs of) local holomorphic functions by an ideal: *Let I be an ideal in $\mathcal{O}$. Then there exists a (local) multiplicity variety $\mathfrak{B}$ such that the restriction $f \to \rho_{\mathfrak{B}} f$ is one-to-one from $\mathcal{O}/I$ onto the space of (local) analytic functions on* $\mathfrak{B}$. We call this result the *(local) quotient structure theorem*. We can regard this result as a sharpening of the Hilbert's nullstellensatz (see Van der Waerden [2]). We also formulate a structure for $\mathcal{O}/I$ as a direct sum of several copies of rings $\mathcal{O}$ depending on fewer than n variables. Similar results are formulated for the quotient of $\mathcal{O}^l$ (the l-fold direct sum of $\mathcal{O}$ with itself) by a module.

In Section II.3 we prove our main results for principal ideals. The algebraic part of the proof rests essentially on the division algorithm. The geometric part depends on the unique factorization theorem in the ring $\mathcal{O}$ and the Lagrange interpolation formula.

The form of our results is such that we can use induction to reduce them to the case of principal ideals. This is carried out in Section II.4.

II.1. L'Hospital Representation

In this chapter we shall give a geometric description for the quotient of $\mathcal{O}$, the ring of local holomorphic functions, by an arbitrary ideal. More generally we shall describe the quotient of $\mathcal{O}^l$ (the l-fold direct sum of $\mathcal{O}$ with itself) by an arbitrary submodule.

Let $I \subset \mathcal{O}^l$ be a module over $\mathcal{O}$, that is, I is a subset of $\mathcal{O}^l$ which is closed under addition and under multiplication by $\mathcal{O}$. It is well known (see Bochner and Martin [1]) that I is finitely generated, that is, there exist $\vec{\mathrm{F}}^1, \vec{\mathrm{F}}^2, \ldots, \vec{\mathrm{F}}^r \in I$ such that I is the set of sums $G_1\vec{\mathrm{F}}^1 + \cdots + G_r\vec{\mathrm{F}}^r$ where the G_j are in $\mathcal{O}$. We think of elements in $\mathcal{O}^l$ as column vectors and we form the $l \times r$ matrix $\boxed{\mathrm{F}} = (\vec{\mathrm{F}}^1, \vec{\mathrm{F}}^2, \ldots, \vec{\mathrm{F}}^r)$. Then we can think of I as the module $I = \boxed{\mathrm{F}}\,\mathcal{O}^r$. This will be the point of view adopted throughout this book.

Unless stated otherwise we shall assume in this chapter that the functions we deal with are in $\mathcal{O}$, the vectors are vectors of functions in $\mathcal{O}$, etc.

Now, let $\boxed{\mathrm{F}}$ be an $l \times r$ matrix of functions in $\mathcal{O}$. In this chapter we shall give a structure for the quotient module $\mathcal{O}^l/\boxed{\mathrm{F}}\,\mathcal{O}^r$. For $\vec{\mathrm{H}} \in \mathcal{O}^l$ we shall always write $\vec{\mathrm{H}} = (H_1, \ldots, H_l)$. By an $\mathcal{O}$ isomorphism of two $\mathcal{O}$ modules we shall mean an isomorphism which commutes with multiplication by $\mathcal{O}$.

One of the main tools which we shall use in deriving our results is an extension of L'Hospital's rule. In its original form, L'Hospital's rule gives a way of expressing the quotient f/g of two smooth functions of a single variable at points where both f and g vanish. We formulate now an extension to functions of several variables in $\mathcal{O}$.

THEOREM 2.1. *Let $F \in \mathcal{O}$. Then we can find $U_0, \ldots, U_p$ which are local analytic Z varieties with the following property:*

For each $j = 0, \ldots, p$ we can find a differential operator d_j with coefficients which are analytic on U_j such that (1) *the U_j are disjoint and their union contains a neighborhood of the origin;* (2) *let $H \in \mathcal{O}$ be divisible by F. For each z close to* 0, *if $z \in U_j$*, we have

$$H(z)/F(z) = (d_j H)(z). \tag{2.1}$$

Here the U_j and d_j depend only on F.

In case $G = H/F$ can be described by the method of Theorem 2.1, we say that G is F *L'Hospital describable in terms of H.*

PROOF. By a linear change of variables and Weierstrass' preparation theorem (see Bochner and Martin [1]) we may assume that F is a polynomial in z_n. We write for simplicity $w = (z_1, \ldots, z_{n-1})$. Then we may write

$$F(z) = F_0(w) + F_1(w)z \ + \cdots + z_n^m. \tag{2.2}$$

We define U_0 as the set where $F \neq 0$. We define U_j as the set of (w^0, z_n^0) such that $F(w^0, z_n)$, considered as a polynomial in z_n with coefficients fixed at w^0, has z_n^0 as a zero of order j. (A zero of order zero means not a zero at all.)

It is clear that the U_j are disjoint and cover a neighborhood of the origin. Since U_j can also be described as the set of $z = (w, z_n)$ for which $F(z) = 0$, $\partial F(z)/\partial z_n = 0, \ldots, \partial^{j-1} F(z)/\partial z_n^{j-1} = 0$, but $\partial^j F(z)/\partial z_n^j \neq 0$, we see that U_j is a local analytic variety minus a local analytic subvariety, that is, a local Z variety.

Now suppose that $H \in \mathcal{O}$ is divisible by F so write $H = FG$, where $G \in \mathcal{O}$. Then for $z \in U_j$ we have by L'Hospital's rule

$$G(z) = \frac{\partial^j H(z)/\partial z_n^j}{\partial^j F(z)/\partial z_n^j}$$

which is the desired result. ▯

Actually, Theorem 2.1 is too weak for our purposes. We need the analog for modules rather than for principal ideals. That is, given $\boxed{\mathrm{F}}$ which is an $l \times r$ matrix (F_{ij}) of functions in $\mathcal{O}$, we want to be able in some manner using varieties to describe $\vec{\mathrm{G}} \in \mathcal{O}^r$ in terms of $\vec{\mathrm{H}} \in \mathcal{O}^l$ and $\boxed{\mathrm{F}}$ in case $\vec{\mathrm{H}} = \boxed{\mathrm{F}}\vec{\mathrm{G}}$. Rather, since $\vec{\mathrm{G}}$ is not unique, we want to be able to describe some $\vec{\mathrm{G}}$. Now the L'Hospital rule of Theorem 2.1 is a "punctual" result in the sense that the value G at a point z^0 depends only on the values of some derivatives of H at z^0. In the general case of $l, r > 1$ we no longer know how to obtain such a result. Rather the value of G at z^0 will depend on the values of derivatives of H at a finite number of points which will depend in an algebroid manner on z^0. More precisely, we have (suppose first that $l = 1$)

THEOREM 2.2. *Let $\vec{\mathrm{F}} = (F_j) \in \mathcal{O}^r$. Then there exist $U^1, \ldots, U^p$ which are local analytic Z varieties, local integral algebroid functions f_{ak}^t for $a = 1, \ldots, n$, $k = 1, \ldots, p$, $t = 1, \ldots, q$, which are regular on U^k, and (algebroid) differential operators d_{jk}^t whose coefficients are regular on $\vec{\mathrm{f}}_k^t(U^k)$ such that*

1. *The U^k are disjoint and their union is a neighborhood of zero in C^n.*
2. *Let $H \in \mathcal{O}$ be of the form $\vec{\mathrm{F}}\,\vec{\mathrm{G}}'$ where $\vec{\mathrm{G}}' \in \mathcal{O}^r$. Then we can find a $\vec{\mathrm{G}} \in \mathcal{O}^r$ such that*

$$H = \vec{\mathrm{F}}\,\vec{\mathrm{G}} \tag{2.3}$$

and such that for $z \in U^k$ we have

$$G_j(z) = \sum_t (d_{jk}^t H)[f_{1k}^t(z), \ldots, f_{nk}^t(z)]. \tag{2.4}$$

The U^k, f_{ik}^t, and d_{jk}^t depend on $\vec{\mathrm{F}}$.

Because of the multivalued nature of the coefficients of d_{jk}^t, the meaning of the right side of (2.4) is not completely clear. It will become clear in the course of the proof how the right side of (2.4) and similar expressions which appear below are precisely defined.

When such an expression is given we say that $\vec{\mathrm{G}}$ is $\vec{\mathrm{F}}$ *L'Hospital describable* (or *expressible*) *in terms of H*. Note that this means that in U^k we can write

$$\vec{\mathrm{G}} = \left(\sum_t \vec{\mathrm{f}}_k^t \circ \vec{\mathrm{d}}_k^t\right) \cdot H,$$

where $\vec{\mathrm{d}}_k^t$ is the vector of operators d_{jk}^t. We shall see that $z \to \vec{\mathrm{f}}_k^t(z)$ is an *algebroid map* in the sense of I.4.

Theorem 2.2 will be proved below. Let us use it now to derive the generalization to $l > 1$.

In order to understand the method, let us consider the simple "scalar" analog: Let $\boxed{\mathrm{a}} = (a_{ij})$ be an $l \times r$ matrix of complex numbers; suppose we want to solve the equation

$$\boxed{\mathrm{a}}\,\vec{\mathrm{u}} = \vec{\mathrm{v}}$$

for the complex vector $\vec{\mathrm{u}}$. Our method is to solve the equations for the components v_i of $\vec{\mathrm{v}}$, one at a time.

We start by trying to solve

$$\sum a_{1j} u_j = v_1 .$$

If this equation has a solution (u_j^0), then all solutions are of the form $(u_j^0 + g_{1j})$, where

$$\sum a_{1j} g_{1j} = 0.$$

We express this last equation by saying that $\vec{\mathrm{g}}_1 = (g_{1j})$ belongs to the *module of relations* of the first row of $\boxed{\mathrm{a}}$.

To solve simultaneously

$$\sum a_{1j} u_j = v_1, \qquad \sum a_{2j} u_j = v_2$$

is equivalent to solving

$$\sum a_{2j}(u_j^0 + g_{1j}) = v_2$$

or

$$\sum a_{2j} g_{1j} = v_2 - \sum a_{2j} u_j^0$$

for a vector $\vec{\mathrm{g}}_1$ which is no longer arbitrary, but belongs to the module of relations of the first row of $\boxed{\mathrm{a}}$.

The continuation of the method to solve for $v_3, \ldots, v_l$ is clear.

In case we cannot solve $\boxed{\mathrm{a}}\,\vec{\mathrm{u}} = \vec{\mathrm{v}}$, we can use the method to find the "obstruction" to solving this equation. In the present case of trying to solve $\boxed{F}\vec{G} = \vec{H}$, the argument takes the following form: Let $\boxed{F} = (F_{ij})$ and let $\vec{H} \in \mathcal{O}^l$ be of the form $\boxed{F}\tilde{\vec{G}}$ for $\tilde{\vec{G}} \in \mathcal{O}^r$. First we can replace $\tilde{\vec{G}}$ by $\vec{G}_1$ say where $(\boxed{F}\vec{G}_1)_1 = H_1$ and $\vec{G}_1$ satisfies the conclusion of Theorem 2.2.

Next we consider the module M_1 of relations of the vector (F_{1j}), that is, the set of $\vec{N}_1 = (N_{1j})$ with $\sum F_{1j}N_{1j} = 0$. We know by assumption that there exists a $\tilde{\vec{G}}_2$ satisfying

$$(\boxed{F}\tilde{\vec{G}}_2)_1 = H_1, \quad (\boxed{F}\tilde{\vec{G}}_2)_2 = H_2. \tag{2.5}$$

Equation (2.5) shows that $(\tilde{\vec{G}}_2 - \vec{G}_1) \in M_1$, so that

$$H_2 \in (\boxed{F}\vec{G}_1)_2 + (\boxed{F}M_1)_2. \tag{2.6}$$

We should like now to conclude that we can find a $\vec{G}_2 \in M_1$ which can be described in terms of varieties, so that

$$H_2 = (\boxed{F}\vec{G}_1)_2 + (\boxed{F}\vec{G}_2)_2. \tag{2.7}$$

(2.7) is almost of the form dealt with in Theorem 2.2 except that we want $\vec{G}_2 \in M_1$ rather than $\vec{G}_2$ arbitrary. This difficulty can be overcome by choosing a basis $\vec{N}_1^1, \ldots, \vec{N}_1^q$ for the module M_1. Then the desired $\vec{G}_2 \in M_1$ is of the form

$$\vec{G}_2 = \sum G_2^i \vec{N}_1^i \tag{2.8}$$

with $G_2^i \in \mathcal{O}$. Thus, (2.7) becomes

$$H_2 = (\boxed{F}\vec{G}_1)_2 + \sum G_2^i(\boxed{F}\vec{N}_1^i)_2 \tag{2.9}$$

which we regard as an equation for the vector $\vec{G}_2' = (G_2^i) \in \mathcal{O}^q$. We can now use Theorem 2.2 to assert that $\vec{G}_2$ exists and can be expressed in a manner similar to the expression of $\vec{G}$ in Theorem 2.2.

We continue this process. Suppose $\vec{G}_k$ has been defined with the property that $(\boxed{F}\vec{G}_k)_i = H_i$ for $i = 1, 2, \ldots, k$ and $\vec{G}_k$ is $\boxed{F}$ L'Hospital describable in terms of $\vec{H}$. We define M_k as the module of simultaneous relations among the first k rows of $\boxed{F}$, that is, M_k consists of those $\vec{N}_k = (N_{kj})$ with $\sum F_{ij}N_{kj} = 0$ for $i = 1, 2, \ldots, k$. Our hypothesis asserts the existence of $\tilde{\vec{G}}_{k+1}$ satisfying

$$(\boxed{F}\tilde{\vec{G}}_{k+1})_j = H_j \quad \text{for } j = 1, 2, \ldots, k+1. \tag{2.10}$$

Thus, $\tilde{\vec{G}}_{k+1} - \vec{G}_k \in M_k$ so that

$$H_{k+1} \in (\boxed{F}\vec{G}_k)_{k+1} + (\boxed{F}M_k)_{k+1}. \tag{2.11}$$

We can now conclude as above that there exists a $\vec{G}_{k+1} \in M_k$ which can be described by varieties so that

$$H_{k+1} = (\boxed{F}\vec{G}_k)_{k+1} + (\boxed{F}\vec{G}_{k+1})_{k+1}. \tag{2.12}$$

Hence we have

THEOREM 2.3. *Suppose there exists a $\vec{\mathrm{G}}$ satisfying $\boxed{F}\vec{\mathrm{G}} = \vec{\mathrm{H}}$. Then there exists a $\vec{\mathrm{G}} \in \mathcal{O}^r$, a disjoint covering $\{U^k\}$ of a neighborhood of zero in C^n by Z varieties, integral algebroid functions f^t_{ak} which are regular on U^k, and differential operators d^{it}_{jk} regular on $\vec{\mathrm{f}}^t_k(U^k)$ such that, on U^k*

$$\vec{\mathrm{G}} = \left(\sum_t \vec{\mathrm{f}}^t_k \circ \boxed{\mathrm{d}}^t_k\right) \cdot \vec{\mathrm{H}}.$$

Here $\boxed{\mathrm{d}}^t_k$ is the matrix of the d^{it}_{jk}, and each $\vec{\mathrm{f}}^t_k$ defines an algebroid map.

We can generalize the L'Hospital representation

$$\vec{\mathrm{G}} = \left(\sum_t \vec{\mathrm{f}}^t_k \circ \boxed{\mathrm{d}}^t_k\right) \cdot \vec{\mathrm{H}}$$

somewhat as follows:

DEFINITION. Suppose V is a local analytic Z variety. We say that $\vec{\mathrm{G}} \in \mathcal{O}^r(V)$ is $\boxed{F}$ *L'Hospital expressible in terms of* $\vec{\mathrm{H}} \in \mathcal{O}^l$ *(in the general sense)* if there exists a covering $\{U^k\}$ of a neighborhood of the origin in V by disjoint Z varieties, integral algebroid maps $w \to \vec{\mathrm{f}}^t_k(w)$ of V into C^n which are regular on U^k, and (algebroid) differential operators d^{it}_{jk} on C^n which are regular on $\vec{\mathrm{f}}^t_k(U^k)$ such that, on U^k

$$\vec{\mathrm{G}} = \left(\sum_t \vec{\mathrm{f}}^t_k \circ \boxed{\mathrm{d}}^t_k\right) \cdot \vec{\mathrm{H}}.$$

Here the U^k, d^{it}_{jk}, and $\vec{\mathrm{f}}^t_k$ depend only on $\boxed{F}$.

We can write the expression

$$\vec{\mathrm{G}} = \left(\sum_t \vec{\mathrm{f}}^t_k \circ \boxed{\mathrm{d}}^t_k\right) \cdot \vec{\mathrm{H}}$$

in another form. Consider first the nongeneralized L'Hospital expression. As in Section I.4, let all the coefficients of $\boxed{\mathrm{d}}^t_k$ and all the integral algebroid functions in $\vec{\mathrm{f}}^t_k$ be defined on some variety W. Then we can write

$$\left[\left(\sum_t \vec{\mathrm{f}}^t_k \circ \boxed{\mathrm{d}}^t_k\right) \cdot \vec{\mathrm{H}}\right](z) = \vec{\mathrm{k}}[\vec{\mathrm{f}}(z)] = (\vec{\mathrm{k}} \circ \vec{\mathrm{f}})(z),$$

where $\vec{k}$ is an algebroid function defined by W such that the components of $\vec{k}$ are linear combinations of some $D_i H_j$ with coefficients rational on W. Here D_i is a linear differential operator with constant coefficients.

We want to show how to write $\vec{k} \circ \vec{f}$ as an algebroid function. Since the components f_i are integral algebroid functions, they are represented by functions g_i on W where g_i are regular in a whole neighborhood of zero. We now add new coordinates t_i to the (s, z) coordinates of W. We consider W embedded in (t, s, z) space, and we add to the ideal defining W the equations $t_i - g_i(s, z) = 0$. We obtain a new local variety W' such that each z has at least one point in W' above it. Moreover, $(\vec{k} \circ \vec{f})(z)$ is just the value of $\vec{k}$ at the point $(t, s, z) \in W'$ above z where (s, z) is the chosen point in W above z, and t is determined by $t_i = g_i(s, z)$ for this chosen point (s, z).

The only problem remaining is that the equations $t_i = g_i(s, z)$ are not of the form (1.18) because of the appearance of s in the g_i. Using (1.18) for the s_j, as in the proof of Theorem 2.8 (see (2.22) and following), we can reduce all powers of s_j to finitely many. Thus we may assume g_i is a polynomial in s with coefficients in $\mathcal{O}(z)$. Since the s_j are integral over $\mathcal{O}(z)$ we may appeal to a standard theorem in algebra (see, e.g., Van der Waerden [2], Vol. 2, p. 78) to conclude that the t_i are integral over $\mathcal{O}(z)$, that is, they satisfy equations with coefficients in $\mathcal{O}(z)$, with leading coefficient one (just like (1.19) for the s_j).

An analogous analysis for generalized L'Hospital representation leads to the conclusion that L'Hospital representation can be described as follows: $\vec{G} \in \mathcal{O}^r(V)$ *is* $\boxed{F}$ *L'Hospital expressible in terms of* $\vec{H} \in \mathcal{O}^l(z)$ *if there exists a covering* $\{U^k\}$ *of a neighborhood of the origin in* V *by disjoint Z varieties, such that* $\vec{H}$ *can be represented on* U^k *by an abstract algebroid function* $\vec{h}^k$ *on a variety* W^k, *where* $\vec{h}^k$ *is a linear combination of derivatives of the components of* $\vec{H}$ *with coefficients rational on* W^k. *Here* $\{U^k\}$, $\{W^k\}$ *and the coefficients in* $\vec{h}^k$ *of the derivatives of* $\vec{H}$ *depend only on* $\boxed{F}$.

In our definition of differential operator we have required that the coefficients precede the differentiation. Also in our definition of the L'Hospital representation we have required that

$$\vec{G} = \left(\sum_t \vec{f}^t_k \circ \boxed{d}^t_k\right) \cdot \vec{H},$$

that is, that we first apply $\boxed{d}^t_k$, then the map $\vec{f}^t_k$. The reason for this is that it is clear how to form $\boxed{d}^t_k \vec{H}$ and $(\vec{f}^t_k \circ \boxed{d}^t_k) \cdot \vec{H}$, whereas there might be some confusion if we used the reverse order. However, in the course of our proofs we shall often meet a situation in which the operations are reversed (but the definition will be clear). We need a result which will allow us to rearrange such expressions and put them in the desired order. This can be thought of as an "implicit differentiation" result. For this purpose we shall prove (see Theorem 2.15) that if $\vec{A}$ is $\boxed{F}_1$ L'Hospital expressible (in the general sense) in terms of $\vec{B}$ and $\vec{B}$ is $\boxed{F}_2$ L'Hospital expressible (in the general sense) in terms of $\vec{C}$, then $\vec{A}$ is $(\boxed{F}_1, \boxed{F}_2)$ expressible in terms of $\vec{C}$ (in the general sense). (For this to be meaningful, $\vec{B}$ must be defined on a neighborhood of zero in some C^b.)

II.2. Formulation of the Main Results

We wish now to give a geometric description of $\mathcal{O}^l/\boxed{F}\mathcal{O}^r$. We shall consider first the case $l = 1$, so $\boxed{F}$ is a vector $\vec{F}$ and $\vec{F}\mathcal{O}^r$ is an ideal. Before discussing the general formulation we wish to give some examples. It is recommended that the reader study these examples before going into the general formulation.

EXAMPLE 1. $n = 1$. Then $\mathrm{F}\mathcal{O}^r$ is a principal ideal, say $\mathrm{F}\mathcal{O}^r = F\mathcal{O}$. Let m be the order of the zero of F at the origin. Then the quotient $\mathcal{O}/F\mathcal{O}$ is described by the point $\{0\}$ with the multiplicity m, or, what is the same thing, by the multiplicity variety $\mathfrak{V} = (0,\ \text{identity};\ 0, d/dz; \ldots ; 0, d^{m-1}/dz^{m-1})$ in the sense that the restriction map

$$H \to H \mid \mathfrak{V} = (H(0), dH(0)/dz, \ldots, d^{m-1} H(0)/dz^{m-1})$$

is an isomorphism of $\mathcal{O}/F\mathcal{O}$ onto $\mathcal{O}(\mathfrak{V})$ which is the space of m-tuples of complex numbers (which are thought of as functions whose domain is the origin).

EXAMPLE 2. n arbitrary, $r = 1$, $\vec{F} = F = z_1^m$. Then the multiplicity variety

$$\mathfrak{V} = (z_1 = 0,\ \text{identity};\ z_1 = 0,\ \partial/\partial z_1; \ldots ; z_1 = 0,\ \partial^{m-1}/\partial z_1^{m-1})$$

describes the quotient $\mathcal{O}/F\mathcal{O}$ in the sense that the restriction map

$$H \to H \mid \mathfrak{V} = (H(0, z_2, \ldots, z_n), \partial H(0, z_2, \ldots, z_n)/\partial z_1, \ldots, \partial^{m-1} H(0, z_2, \ldots, z_n)/\partial z_1^{m-1})$$

is an isomorphism of $\mathcal{O}/F\mathcal{O}$ onto $\mathcal{O}(\mathfrak{V})$ which is the space of m-tuples of local holomorphic functions on $z_1 = 0$.

EXAMPLE 3. $n > 1$, $r = 1$, $\vec{\mathrm{F}} = F = z_1^m(z_1 - z_2)^{m'}$. Then the multiplicity variety

$$\mathfrak{B}' = (z_1 = 0, \text{identity}; \ldots; z_1 = 0, \partial^{m-1}/\partial z_1^{m-1}; \quad z_1 = z_2, \text{identity}; \ldots; z_1 = z_2, \partial^{m'-1}/\partial z_1^{m'-1})$$

describes the quotient $\mathcal{O}/F\mathcal{O}$ in the sense that the restriction map

$$H \to H \mid \mathfrak{B}' = (H|_{z_1=0}, \ldots, \partial^{m-1}H/\partial z_1^{m-1}|_{z_1=0}, \quad H|_{z_1=z_2}, \ldots, \partial^{m'-1}H/\partial z_1^{m'-1}|_{z_1=z_2})$$

is an isomorphism of $\mathcal{O}/F\mathcal{O}$ onto $\mathcal{O}(\mathfrak{B}')$ which is the subset of $\mathcal{O}^m(z_1 = 0) + \mathcal{O}^{m'}(z_1 = z_2)$ defined by certain compatibility conditions at $z_1 = z_2 = 0$ (see Chapter I). (We have used the notation $\mathfrak{B}'$ rather than $\mathfrak{B}$ so as to be in conformity with the notation used below.)

In Examples 1, 2, and 3 we had $r = 1$. We wish now to give an example in which $r > 1$:

EXAMPLE 4. $n = 2$ (write $z = (w, s)$), $r = 2$. $F_1 = s^2 - w$, $F_2 = w^2$. The appropriate multiplicity variety is

$$\mathfrak{B} = (z = 0, \text{identity}; z = 0, \partial/\partial s; \quad z = 0, \partial/\partial w + \tfrac{1}{2}\partial^2/\partial s^2; z = 0, \partial^2/\partial s\,\partial w + \tfrac{1}{6}\partial^3/\partial s^3).$$

(In order to understand how we arrived at this expression of $\mathfrak{B}$, see the further remarks on Example 4 preceding the statement of Theorem 2.5.) This means that the restriction map $H \to H \mid \mathfrak{B}$ has kernel equal to the ideal generated by F_1 and F_2. To see this we check first that $F_1\mathcal{O} \mid \mathfrak{B} = 0$ and $F_2\mathcal{O} \mid \mathfrak{B} = 0$. Conversely, suppose $H \mid \mathfrak{B} = 0$. We expand H in a power series, say

$$H(w, s) = \sum H_{ij} w^i s^j.$$

Since $H(0, 0) = 0$, we have $H_{00} = 0$. Since

$$\frac{\partial H}{\partial s}(0, 0) = 0$$

we have $H_{01} = 0$.

$$\frac{\partial H}{\partial w}(0, 0) + \frac{1}{2}\frac{\partial^2 H}{\partial s^2}(0, 0) = 0$$

gives $H_{10} + H_{02} = 0$. Finally the fact that

$$\frac{\partial^2 H}{\partial s\,\partial w}(0, 0) + \frac{1}{6}\frac{\partial^3 H}{\partial s^3}(0, 0) = 0$$

implies $H_{11} + H_{03} = 0$.

Now, we can write

$$H(w, s) = w^2 \sum_{i \geqslant 2} H_{ij} w^{i-2} s^j + w \sum H_{1j} s^j + \sum H_{0j} s^j.$$

We have

$$ws^2 = w(s^2 - w) + w^2 \in \vec{\mathrm{F}}\mathcal{O}^2$$
$$s^4 = (s^2 + w)(s^2 - w) + w^2 \in \vec{\mathrm{F}}\mathcal{O}^2$$
$$s^3 = s(s^2 - w) + sw \equiv sw \bmod \vec{\mathrm{F}}\mathcal{O}^2$$
$$s^2 = (s^2 - w) + w \equiv w \bmod \vec{\mathrm{F}}\mathcal{O}^2.$$

Thus,

$$H(w, s) \equiv H_{10}\, w + H_{11} ws + H_{00} + H_{01}\, s + H_{02}\, w + H_{03}\, sw \bmod \vec{\mathrm{F}}\mathcal{O}^2$$
$$= 0$$

by the above.

Our main result for ideals is

THEOREM 2.4. *For any* $\vec{\mathrm{F}}$ *there exists a local analytic multiplicity variety* $\mathfrak{V}$ *such that the restriction map* $H \to H \mid \mathfrak{V}$ *is an* $\mathcal{O}$ *isomorphism of* $\mathcal{O}/\vec{\mathrm{F}}\mathcal{O}^r$ *onto* $\mathcal{O}(\mathfrak{V})$.

Theorem 2.4 follows from the apparently weaker result:

THEOREM 2.4′. *For any* $\vec{\mathrm{F}}$ *there exists a local algebroid multiplicity Z variety* $\mathfrak{V}'$ *such that* $H \to H \mid \mathfrak{V}'$ *is an* $\mathcal{O}$ *isomorphism of* $\mathcal{O}/F\mathcal{O}^r$ *onto* $\mathcal{O}(\mathfrak{V}')$.

Note that the apparent weakness of Theorem 2.4′ stems from the fact that the coefficients of the differential operators that appear in $\mathfrak{V}'$ may be multivalued and may not be regular on the closure of the associated varieties.

PROOF THAT THEOREM 2.4′ IMPLIES THEOREM 2.4. Suppose that $\mathfrak{V}' = \{(V_j', \partial_j')\}$. If V_j' is reducible, say $V_j' = V_{j1}' \cup \ldots \cup V_{jt}'$, then we can replace $\mathfrak{V}'$ by the multiplicity variety $\mathfrak{V}''$ in which (V_j', ∂_j') is replaced by $(V_{j1}', \partial_j'; V_{j2}', \partial_j'; \ldots; V_{jt}', \partial_j')$. It is clear that the kernels of the restriction maps to $\mathfrak{V}'$ and $\mathfrak{V}''$ are the same. Thus we may assume that each V_j' is irreducible.

Let j be fixed and write (V', ∂') for this (V_j', ∂_j'). By the remarks in Section I.4, we may assume that all the (algebroid) coefficients of the differentiations in ∂' are defined by a fixed variety W' and a fixed choice mechanism. Let $H \in \mathcal{O}$ with $\partial' H = 0$ on V'. We can write

$$\partial' H = \sum f_j \dot{H}_j,$$

where f_j are algebroid functions and $\dot{H}_j$ are of the form $D_j H$ where D_j is a linear partial differential operator with constant coefficients.

Now each f_j can be represented by g_j/h_j where g_j and h_j are in $\mathcal{O}(W')$ and h_j does not vanish at the chosen points above V'. Thus we may clear denominators, that is, the vanishing of $\partial' H$ on V' is equivalent to the vanishing of $h\partial' H$ on V', where h is the algebroid function defined by the abstract algebroid function Πh_j on W'. (It may be assumed by replacing V' by a dense Z open subset that all the roots of the $J_j(s_i, z)$ of (1.18) are distinct all over V' so that h, together with the choice mechanism, defines an analytic function on V'.) Thus we may assume that each $h_i = 1$.

We use the equations (1.18) to reduce all the powers of the s_i to finitely many. (This type of argument is carried out in detail in (2.22) and following.) Thus we may assume that the f_j are polynomials in s.

By Proposition 1.8 the vanishing of $\partial' H$ on V' implies (and hence is equivalent to) the vanishing of $\sum g_j H_j$ on W'' which is a well-defined subvariety of W'.

Let $v_1, \ldots, v_p$ be a maximal linearly independent set of polynomials in s on W'' over the field $\mathcal{M}(V')$ of meromorphic functions on V'. (Since V' is irreducible, $\mathcal{M}(V')$ is a field.) By this we mean that $v_1, \ldots, v_p$ is a maximal set of polynomials in s so that no nontrivial relation

$$\sum u_j v_j \equiv 0 \quad \text{on} \quad W''$$

can hold with $u_j \in \mathcal{M}(V')$.

We may write $g_j = \sum u_{jk} v_k$ on W'' with $u_{jk} \in \mathcal{M}(V')$ by the above remarks. Thus $\partial' H = 0$ on V' means $\sum g_j H_j \equiv 0$ on W'' which means

$$\sum_{jk} u_{jk} v_k H_j = \sum_k \left(\sum_j u_{jk} H_j \right) v_k \equiv 0 \quad \text{on} \quad W''.$$

By the independence of the v_k this is equivalent to

$$\sum u_{jk} H_j = 0 \quad \text{on} \quad V'$$

for every k.

Call V the closure of V'. By multiplying the u_{jk} by the least common multiple of their denominators we may assume they are in $\mathcal{O}$. Thus

$$\partial' H = 0 \quad \text{on } V' \Leftrightarrow \sum_j u_{jk} H_j = 0 \quad \text{on } V \quad \text{for all } k.$$

The last condition involves a (closed) variety V and a differential operator

$$\sum u_{jk} D_j$$

which has coefficients in $\mathcal{O}(V)$. This completes the proof. □

Remark 2.1. We shall not distinguish Theorem 2.4 from Theorem 2.4′. See Problem 2.1.

Theorem 2.4 will be called the *local quotient structure theorem* (for ideals).

In addition to the geometric description of $\mathcal{O}/\vec{\mathbf{F}}\mathcal{O}^r$ we shall need an algebraic description. This will be an isomorphism of $\mathcal{O}/\vec{\mathbf{F}}\mathcal{O}^r$ onto a direct sum of spaces $\mathcal{O}(j)$, which is the space of local holomorphic functions of j variables. We shall need an explicit expression for this isomorphism, which will be similar to the L'Hospital expression of Section 1 above except that the L'Hospital expression will be complicated by the fact that $\mathfrak{V}$ enters. In addition we shall want to know how multiplication by $G \in \mathcal{O}$ on $\mathcal{O}/\vec{\mathbf{F}}\mathcal{O}^r$ behaves under this isomorphism. We illustrate what this means for the four examples described above.

EXAMPLE 1. $\mathcal{O}/F\mathcal{O}$ is isomorphic to $\mathcal{O}(0)^m$. The isomorphism we have in mind here is *not* the restriction to $\mathfrak{V}$ but rather the map which assigns to each $H \in \mathcal{O}$ the unique polynomial K of degree $m-1$ whose restriction to $\mathfrak{V}$ is the same as the restriction of H to $\mathfrak{V}$. Explicitly,

$$K(z) = H(0) + H'(0)z + \frac{H''(0)}{2!} z^2 + \cdots + \frac{H^{(m-1)}(0)}{(m-1)!} z^{m-1},$$

so that the Taylor coefficients of K are expressible in terms of the values of H on $\mathfrak{V}$. Multiplication by z on H transforms the coefficients of K (thought of as a column vector) by left multiplication by the matrix

$$\begin{pmatrix} 0 & & & & & & \\ 1 & 0 & & & & & 0 \\ 0 & 1 & 0 & & & & \\ \cdot & & \cdot & \cdot & & & \\ \cdot & & & \cdot & \cdot & & \\ \cdot & & & & \cdot & \cdot & \\ 0 & \cdot & \cdot & \cdot & 0 & 1 & 0 \end{pmatrix}.$$

EXAMPLE 2. $\mathcal{O}/F\mathcal{O}$ is isomorphic to $\mathcal{O}(n-1)^m$. The isomorphism with which we are here concerned is the map of $H \in \mathcal{O}$ into $K(z)$ which is the unique polynomial of degree $m-1$ in z_1 whose restriction to $\mathfrak{V}$ is the same as $H \mid \mathfrak{V}$. Explicitly,

$$K(z) = H(0, z_2, \ldots, z_n) + \cdots + \frac{z_1^{m-1}}{(m-1)!} \frac{\partial^{m-1} H}{\partial z_1^{m-1}} (0, z_2, \ldots, z_n).$$

Again, the Taylor coefficients of K are expressible in terms of $H \mid \mathfrak{V}$. Multiplication by z_j for $j \neq 1$ on H corresponds to scalar multiplication by z_j on the vector of Taylor coefficients (in z_1) of K. Multiplication by z_1

corresponds to left multiplication by the matrix

$$\begin{pmatrix} 0 & & & & & \\ 1 & 0 & & & 0 & \\ 0 & 1 & 0 & & & \\ \cdot & & \cdot & \cdot & & \\ \cdot & & & \cdot & \cdot & \\ \cdot & & & & \cdot & \cdot \\ 0 & \cdot & \cdot & \cdot & 0 \;\; 1 & 0 \end{pmatrix}.$$

EXAMPLE 3. $\mathcal{O}/F\mathcal{O}$ is isomorphic to $\mathcal{O}(n-1)^{m+m'}$. The isomorphism in question is the map of $H \in \mathcal{O}$ into $K(z)$ which is the unique polynomial of degree $m + m' - 1$ in z_1 whose restriction to $\mathfrak{V}'$ is the same as $H \mid \mathfrak{V}'$. We write

$$K(z) = K_0(z_2, \ldots, z_n) + K_1(z_2, \ldots, z_n)z_1 + \cdots + K_{m+m'-1}(z_2, \ldots, z_n)z_1^{m+m'-1}.$$

Now for any fixed $z_2 \neq 0$, the polynomial $F(z_1, z_2)$ (thought of as a polynomial in z_1) has an m-fold at $z_1 = 0$ and an m'-fold zero at $z_1 = z_2$. Since $K(z) = H(z)$ on $\mathfrak{V}'$, this means that for $z_2 \neq 0$ we can write the system of $m + m'$ equations for the $m + m'$ unknown numbers $K_j(z_2, \ldots, z_n)$:

$$K_0(z_2, \ldots, z_n) = H(0, z_2, \ldots, z_n)$$

$$K_1(z_2, \ldots, z_n) = \frac{\partial H}{\partial z_1}(0, z_2, \ldots, z_n)$$

$$\vdots \qquad\qquad \vdots$$

$$(m-1)!K_{m-1}(z_2, \ldots, z_n) = \frac{\partial^{m-1} H}{\partial z_1^{m-1}}(0, z_2, \ldots, z_n)$$

$$K_0(z_2, \ldots, z_n) + K_1(z_2, \ldots, z_n)z_2 + \cdots + K_{m+m'-1}(z_2, \ldots, z_n)z_2^{m+m'-1} = H(z_2, z_2, \ldots, z_n)$$

$$K_1(z_2, \ldots, z_n) + \cdots + (m+m'-1)K_{m+m'-1}(z_2, \ldots, z_n)z_2^{m+m'-2} = \frac{\partial H}{\partial z_1}(z_2, z_2, \ldots, z_n)$$

$$\vdots \qquad\qquad \vdots$$

$$(m'-1)!K_{m'-1}(z_2, \ldots, z_m) + \cdots + \frac{(m+m'-1)!}{m!} K_{m+m'-1}(z_2, \ldots, z_n)z_2^m = \frac{\partial^{m'-1} H}{\partial z_1^{m'-1}}(z_2, z_2, \ldots, z_n).$$

We regard this as a sytem of $m + m'$ equations for the $m + m'$ unknowns $K_j(z_2, \ldots, z_n)$. The determinant of the coefficients can be thought of as a generalization of the Vandermonde determinant. It does not vanish for $z_2 \neq 0$. This can be seen directly; it also follows from Lemma 2.12 below.

When $F(z_1, z_2)$ is thought of as a polynomial in z_1, it has a zero of order $m + m'$ at the origin for $z_2 = 0$. The $K_j(0, z_3, \ldots, z_n)$ can be calculated as in Example 2 above. However, there is a difference between the situations here and in Example 2. For, according to Example 2, we have

$$K_j(0, z_3, \ldots, z_n) = \frac{1}{j!} \frac{\partial^j H}{\partial z_1^j} (0, 0, z_3, \ldots, z_n).$$

In the present situation, $\partial^j H(0, 0, z_3, \ldots, z_n)/\partial z_1^j$ is *not* obviously expressible in terms of the restriction $H \mid \mathfrak{V}'$ at points of $\mathfrak{V}'$ above $(0, z_3, \ldots, z_n)$ for $j \geq \max(m, m')$. For the only point of $\mathfrak{V}'$ above $(0, z_3, \ldots, z_n)$ is $(0, 0, z_3, \ldots, z_n)$ and at this point the only z_1 derivatives of H which enter into the definition of $\mathfrak{V}'$ are of order $\max(m, m') - 1$.

In order to see how to overcome this difficulty, let us suppose for simplcity that $m = m' = 1$, as the general case is similar. Then we can write

$$\frac{\partial H}{\partial z_1}(0, 0, z_3, \ldots, z_n)$$

$$= \left[\frac{\partial H}{\partial z_1}(0, 0, z_3, \ldots, z_n) + \frac{\partial H}{\partial z_2}(0, 0, z_3, \ldots, z_n)\right] - \left[\frac{\partial H}{\partial z_2}(0, 0, z_3, \ldots, z_n)\right].$$

The term in the first brackets is a derivative of the restriction of H to $z_1 = z_2$; the second term is a derivative of the restriction of H to $z_1 = 0$. Thus, $\partial H(0, 0, z_3, \ldots, z_n)/\partial z_1$ can be expressed in terms of $H \mid \mathfrak{V}'$ and derivatives "on $\mathfrak{V}'$" of $H \mid \mathfrak{V}'$.

Let us denote by $\mathfrak{V}$ the multiplicity variety

$$\begin{aligned}\mathfrak{V} = (z_1 = 0, z_2 = 0, \text{identity}; z_1 = 0, z_2 = 0, \partial/\partial z_1; \\ \ldots; z_1 = 0, z_2 = 0, \partial^{m+m'-1}/\partial z_1^{m+m'-1}; \\ z_1 = 0, \text{identity}; \ldots; z_1 = 0, \partial^{m-1}/\partial z_1^{m-1}; \\ z_1 = z_2, \text{identity}; \ldots; z_1 = z_2, \partial^{m'-1}/\partial z_1^{m'-1}).\end{aligned}$$

Then the above shows that $\mathfrak{V}'$ is "equivalent" to $\mathfrak{V}$ in the sense that $\mathcal{O}(\mathfrak{V}') = \mathcal{O}(\mathfrak{V})$. Thus, for the description of the multiplicity variety associated to the principal ideal $F\mathcal{O}$, $F = z_1^m(z_1 - z_2)^{m'}$, we could use either $\mathfrak{V}$ or $\mathfrak{V}'$. This point is essential for the understanding of the statement of Theorem 2.5.

Thus we see that in all three examples the values of $K_j(z_2, \ldots, z_n)$ can be calculated in terms of the values of H on the points of $\mathfrak{V}'$, with mutliplicity which lie above $(z_2, \ldots, z_n)$ (with the proviso that we may have to change $\mathfrak{V}'$ to an "equivalent" multiplicity variety). However, we may have to use several different types of formulas for different values of $(z_2, \ldots, z_n)$. In Example 3 the formulas for $z_2 = 0$ and for $z_2 \neq 0$ are different. The reason for this is that the part of $\mathfrak{V}'$ above $(z_2, \ldots, z_n)$ is different for $z_2 = 0$ and for $z_2 \neq 0$. In the former case it consists of the single point $(0, 0, z_3, \ldots, z_n)$ while in the latter case it consists of the two points $(0, z_2, z_3, \ldots, z_n)$ and $(z_2, z_2, z_3, \ldots, z_n)$, everything with appropriate multiplicity.

Another way of looking at this is to say that the analog of the Vandermonde determinant which arises in Example 3 vanishes at $z_2 = 0$. For this reason we might expect, in accordance with the L'Hospital representation discussed in Section II.1 above that we should have a different formula at $z_2 = 0$, one which involves suitable derivatives of H "along $\mathfrak{V}'$."

We can also regard the function K as constructed by a generalized Lagrange interpolation formula. For, if $z_2 \neq 0$ then for fixed $(z_3, \ldots, z_n)$, when K is regarded as a polynomial in z_1, it is the unique polynomial of degree $m + m' - 1$ whose values as well as the values of its first $m - 1$ derivatives are prescribed at $z_1 = 0$ and whose value as well as the values of its first $m' - 1$ derivatives are prescribed at $z_1 = z_2$. (In the usual Lagrange interpolation formula we prescribe the values of a polynomial of degree $l - 1$ at l *distinct* points. Thus the above corresponds to the situation in which some of the points coincide.) Of course, it is important to know that K is holomorphic in the parameters $(z_2, \ldots, z_n)$.

In the examples given above, we can think of the isomorphism in question as being determined by the map $H \to K$. K is a particular representative in the class of H modulo the ideal $F\mathcal{O}$. Moreover, K has a particularly simple form, namely, it is a polynomial in z_1 whose coefficients are arbitrary functions in $\mathcal{O}(n-1)$. Thus, the monomials $1, z_1, \ldots, z_1^t$ for suitable t form a basis for $\mathcal{O}/F\mathcal{O}$ thought of as an $\mathcal{O}(n-1)$ module.

In the general case of $\mathcal{O}/\vec{\mathrm{F}}\mathcal{O}^r$, we shall no longer be able to give such a simple basis, so the form of our isomorphism will be more complicated.

The effect of multiplication of H by z_j on the Taylor coefficients (in z_1) of K can be described as in Examples 1 and 2.

EXAMPLE 4. We note first, as in Examples 1, 2, and 3, that $\mathcal{O}/F_1\mathcal{O}$ is isomorphic to $\mathcal{O}(1)^2$. The isomorphism is the map of $H \in \mathcal{O}$ into $K(z)$ which is the unique polynomial of degree 1 in s which agrees with H on $s^2 - w = 0$. We write

$$K(z) = K_0(w) + K_1(w)s.$$

As in Example 3 we can express $K_0(w)$ and $K_1(w)$ in terms of the values of H at the points on $s^2 - w = 0$ which lie above w if $w \neq 0$, while for $w = 0$ we need to use $K(0, 0)$ and $\partial K(0, 0)/\partial s$.

To compute $\mathcal{O}/\vec{\mathrm{F}}\mathcal{O}^2$, we note that under the isomorphism $\mathcal{O}/F_1\mathcal{O} \to \mathcal{O}(1)^2$ multiplication by F_2 gets mapped into scalar multiplication by F_2, because F_2 depends only on w. We note that $\mathcal{O}/\vec{\mathrm{F}}\mathcal{O}^2$ is in a natural way isomorphic to $(\mathcal{O}/F_1\mathcal{O})/(F_2[\mathcal{O}/F_1\mathcal{O}])$ where $F_2[\mathcal{O}/F_1\mathcal{O}]$ is the space obtained by the natural multiplication action of F_2 on $\mathcal{O}/F_1\mathcal{O}$. Thus we are led to the problem of studying the quotient of $\mathcal{O}(1)^2$ by $F_2\,\mathcal{O}(1)^2$ where F_2 acts by scalar multiplication.

The problem we have is essentially the one treated in Example 1 except that we must now deal with $r = 1, l = 2$ rather than $r = 1, l = 1$. However, this presents no difficulty because F_2 acts by scalar multiplication so we can treat $K_0(w)$ and $K_1(w)$ independently. It follows that $(K_0(w), K_1(w))$ is congruent mod $F_2\mathcal{O}(1)^2$ uniquely to

$$(K_{00} + K_{01}w,\ K_{10} + K_{11}w)$$

where K_{ij} are constants. Since

$$H(w, s) \equiv K_0(w) + K_1(w)s \bmod F_1\mathcal{O}(2)$$

this means that

$$H(w, s) \equiv K_{00} + K_{01}w + K_{10}\,s + K_{11}ws \bmod \vec{\mathrm{F}}\mathcal{O}(2)^2.$$

Thus,

$$\mathcal{O}/\vec{\mathrm{F}}\mathcal{O}^2 \approx \mathcal{O}(0)^4,$$

the isomorphism being

$$H \to K_{00} + K_{01}w + K_{10}\,s + K_{11}ws.$$

We wish to explain how the multiplicity variety $\mathfrak{V}$ for Example 4 is constructed (see the remarks on Example 4 preceding the statement of Theorem 2.4). Let $\mathfrak{V}_1$ be the multiplicity variety associated with F_1:

$$\mathfrak{V}_1 = \{s^2 - w = 0,\ \text{identity};\ z = 0,\ \partial/\partial s\}.$$

Thus, as mentioned above, $K_0(w)$ and $K_1(w)$ can be constructed from the values of H at the points of $\mathfrak{V}_1$ above w. Precisely, for $w \neq 0$,

$$K_0(w) = \frac{H(w, \sqrt{w}) + H(w, -\sqrt{w})}{2}$$

$$K_1(w) = \frac{H(w, \sqrt{w}) - H(w, -\sqrt{w})}{2\sqrt{w}}$$

as is easily derived from the fact that $K_0(w) + K_1(w)s = H(w, s)$ when $s = \pm\sqrt{w}$. (This gives two equations in two unknowns.) Moreover,

$$K_0(0) = H(0, 0)$$
$$K_1(0) = \partial H(0, 0)/\partial s.$$

Now clearly,

$$K_{00} = K_0(0)$$
$$K_{01} = dK_0(0)/dw$$
$$K_{10} = K_1(0)$$
$$K_{11} = dK_1(0)/dw.$$

The only common zero of F_1, F_2 is $z = 0$, so, as in the case $r = 1$, we try to express the K_{ij} in terms of the values of certain derivatives of H at the origin.

We have

$$K_{00} = K_0(0) = H(0, 0)$$
$$K_{10} = K_1(0) = \partial H(0, 0)/\partial s.$$

The expressions for K_{01} and K_{11} are much more difficult to obtain. Namely

$$\begin{aligned} K_{01} &= \left.\frac{dK_0}{dw}\right|_{w=0} \\ &= \left.\frac{d}{dw}\left[\frac{H(w, \sqrt{w}) + H(w, -\sqrt{w})}{2}\right]\right|_{w=0} \\ &= \left.\frac{\partial H}{\partial w}(0, 0) + \frac{1}{4}\, w^{-1/2}\left[\frac{\partial H}{\partial s}(w, \sqrt{w}) - \frac{\partial H}{\partial s}(w, -\sqrt{w})\right]\right|_{w=0}. \end{aligned}$$

To evaluate the last term we expand H in a power series, then divide by $w^{1/2}$ and compute the constant term. This yields

$$K_{01} = \left.\left(\frac{\partial}{\partial w} + \frac{1}{2}\frac{\partial^2}{\partial s^2}\right)H\right|_{z=(0,0)}.$$

A similar method shows that

$$K_{11} = \left.\left(\frac{\partial^2}{\partial s\, \partial w} + \frac{1}{6}\frac{\partial^3}{\partial s^3}\right)H\right|_{z=(0,0)}.$$

All this indicates that if we define $\mathfrak{B}$ to be

$$\mathfrak{B} = (z = 0, \text{identity};\ z = 0,\ \partial/\partial s;\ z = 0,\ \partial/\partial w + \tfrac{1}{2}\partial^2/\partial s^2;$$
$$z = 0,\ 6\partial^2/\partial s \partial w + \partial^3/\partial s^3)$$

then we could express the K_{ij} in terms of the restriction of H to $\mathfrak{B}$. However, we must know that $\vec{\mathrm{F}}\mathcal{O}^2 = 0$ on $\mathfrak{B}$, that is, $\mathfrak{B}$ represents $\vec{\mathrm{F}}\mathcal{O}^2$

in the sense of Theorem 2.4. Of course, this can be readily verified directly, but it follows from our construction that $\vec{\mathrm{F}}\mathcal{O}^2 \mid \mathfrak{B} = 0$. For, if there were an $H \in \vec{\mathrm{F}}\mathcal{O}^2$ with $H \mid \mathfrak{B} \neq 0$ then some K_{ij} would be different from zero, contradicting the fact that H is determined uniquely mod $\vec{\mathrm{F}}\mathcal{O}^2$ by the K_{ij}.

We wish to compute the effect on the vector $(K_{00}, K_{01}, K_{10}, K_{11})$ of multiplication on H. Let us consider first multiplication by w. We note that

$$wH(z) \equiv wK_0(w) + wK_1(w)s \bmod F_1\mathcal{O}(2).$$

Moreover,

$$K_0(w) \equiv K_{00} + K_{01}w \bmod F_2\,\mathcal{O}(1)$$

so

$$wK_0(w) \equiv K_{00}\,w \bmod F_2\,\mathcal{O}(1).$$

Similarly,

$$wK_1(w) \equiv K_{10}\,w \bmod F_2\,\mathcal{O}(1).$$

This means that multiplication by w on H corresponds to the action of left multiplication by the matrix

$$\begin{pmatrix} 0 & 0 & 0 & 0 \\ 1 & 0 & 0 & 0 \\ 0 & 0 & 0 & 0 \\ 0 & 0 & 1 & 0 \end{pmatrix}$$

on the (column) vector $(K_{00}, K_{01}, K_{10}, K_{11})$.

Let us now compute the effect of multiplication by s. We have

$$sH \equiv wK_1(w) + K_0(w)s \bmod F_1\mathcal{O}(2).$$

Now,

$$K_1(w) \equiv K_{10} + K_{11}w \bmod F_2\,\mathcal{O}(1)$$
$$K_0(w) \equiv K_{00} + K_{01}w \bmod F_2\,\mathcal{O}(1).$$

Thus, by the above,

$$sH \equiv wK_{10} + K_{00}\,s + K_{01}ws \bmod \vec{\mathrm{F}}(\mathcal{O}(2))^2.$$

This means that $H \to sH$ corresponds to left multiplication by

$$\begin{pmatrix} 0 & 0 & 0 & 0 \\ 0 & 0 & 1 & 0 \\ 1 & 0 & 0 & 0 \\ 0 & 1 & 0 & 0 \end{pmatrix}$$

on the column vector $(K_{00}, K_{01}, K_{10}, K_{11})$.

See Problem 2.2.

THEOREM 2.5. *For any* $\vec{F}$, *there is an isomorphism of linear spaces*

$$\mathcal{O}(n)/\vec{F}\mathcal{O}^r(n) \approx \mathcal{O}^{k_0}(0) \oplus \mathcal{O}^{k_1}(1) \oplus \cdots \oplus \mathcal{O}^{k_n}(n). \tag{2.13}$$

More precisely, we can make a linear transformation in z *(depending on* $\vec{F}$*) with the following properties: For each* j *there exist* k_j *polynomials* Q_{ij} $(z_{j+1}, \ldots, z_n)$ *so that every* $H \in \mathcal{O}(n)$ *is congruent modulo* $\vec{F}\mathcal{O}^r(n)$ *to a unique function* $K \in \mathcal{O}$ *of the form*

$$K(z) = \sum_{i,j} Q_{ij}(z_{j+1}, \ldots, z_n) L_{ij}(z_1, \ldots, z_j), \tag{2.14}$$

where $L_{ij} \in \mathcal{O}(j)$. *We write* $w^j = (z_1, \ldots, z_j)$ *and* $s^{n-j} = (z_{j+1}, \ldots, z_n)$.

For each j *we can find local* Z *varieties* $T_{tj} \subset C^j$ *which are disjoint and which cover a neighborhood of zero, so that if* T^q_{tj} *denote* Z *varieties of the form* $w^j \in T_{tj}$, $(w^j, s^{n-j}) \in V_q$ *(where* $\mathfrak{V} = (V_1, \partial_1; \ldots; V_s, \partial_s)$ *is a suitable multiplicity variety satisfying the conclusion of Theorem* 2.4*) then we can find functions* f_{iqtj} *which are algebroid on* T^q_{tj} *so that for* $w^j \in T_{tj}$,

$$L_{ij}(w^j) = \sum_q \sum_{s^{n-j}} f_{iqtj}(w^j, s^{n-j})(\partial_q H)(w^j, s^{n-j}) \tag{2.15}$$

the sum being over those s^{n-j} *with* $(w^j, s^{n-j}) \in T^q_{tj}$ (the sum being finite).

We can obtain another expression for L_{ij} *in terms of* H *by a finite sequence (depending only on* $\vec{F}$ *and* (i, j)*) of applications of the division algorithm (see the proof of Theorem* 2.8 *for more details).*

The effect of multiplication by $B \in \mathcal{O}$ *on* $\mathcal{O}/\vec{F}\mathcal{O}$ *via the isomorphism* (2.13) *or* (2.14) *can be described by a finite sequence of processes (determined by* $\vec{F}$*), each one of which consists in studying the corresponding problem for* $r = 1$ *which is explicitly described in Theorem* 2.8.

Note by (2.14) that to find BH mod $\vec{F}\mathcal{O}^r$ we need only find the coefficient of Q_{ij} corresponding to BH. However this *cannot* be equal to a matrix applied to the vector of all L_{ij} (corresponding to H) because they depend on different numbers of variables.

Let us note that (2.13) or (2.14) gives us an explicit method of selecting a unique element from each equivalence class in $\mathcal{O}/\vec{F}\mathcal{O}^r$. We shall refer to this as a *parametrization* of $\mathcal{O}/\vec{F}\mathcal{O}^r$. In Chapters IX and X we shall show how to relate similar parametrization problems to the problem of parametrizing the solutions of systems of linear partial differential equations with constant coefficients.

DEFINITION. If the parametrization (2.14) has the property that the $L_{ij}(w^j)$ can be expressed as in (2.15) then we say that the parametrization is *L'Hospital*.

The reason for this terminology is clear from Section II.1 above. It implies that $\boxed{L}$ is $\vec{F}$ L'Hospital expressible in terms of K.

We shall now explain how to extend Theorem 2.4 to the case $l > 1$, that is, we shall define a map ρ which is a generalization of the restriction map and will allow us to extend Theorem 2.4 to the case of $l > 1$ (see the remarks following the statement of Theorem 2.2 for the formalism of the argument). Let $\boxed{\mathrm{F}}$ be the matrix (F_{ij}). $\mathfrak{V}_1$ will have the property that a function will belong to the ideal generated in $\mathcal{O}$ by F_{1j} if and only if its restriction to $\mathfrak{V}_1$ is zero. That is, $\mathfrak{V}_1$ is a multiplicity variety associated to the ideal generated by the F_{1j} according to Theorem 2.4.

Let $\vec{\mathrm{H}} \in \mathcal{O}^l$ and write $\vec{\mathrm{H}} = (H_1, \ldots, H_l)$. We define $\rho\vec{\mathrm{H}} = ((\rho\vec{\mathrm{H}})_1, \ldots, (\rho\vec{\mathrm{H}})_l)$ as follows: Let $(\rho\vec{\mathrm{H}})_1$ be the restriction of H_1 to $\mathfrak{V}_1$. Using Theorem 2.5 above, we can choose a unique extension of $(\rho\vec{\mathrm{H}})_1$ of the form

$$\sum Q^1_{ik}(s^{n-i}) L^1_{ik}(w^i).$$

(More precisely, the L^1_{ik} are uniquely determined once the Q^1_{ik} are given). Then we choose $\vec{\mathrm{G}}_1 = (G_{1j})$ satisfying

$$\sum F_{1j} G_{1j} = H_1(z) - \sum Q^1_{ik}(s^{n-i}) L^1_{ik}(w^i). \tag{2.16}$$

This can be done, since $\sum Q^1_{ik}(s^{n-i}) L^1_{ik}(w^i) - H_1(z)$ has zero restriction to $\mathfrak{V}_1$. (The proof of Theorem 2.5 given below will show that we can actually choose the same way of writing the coordinates $z = (w_j, s^{n-j})$ for all the ideals considered in the present argument.)

Next, let M_1 be the module of relations of the vector (F_{1j}) in $\mathcal{O}$. Let $\mathfrak{V}_2$ be a multiplicity variety associated by Theorem 2.4 to the ideal of sums $\sum F_{2j} N_{1j}$ for $(N_{1j}) \in M_1$. Then set

$$(\rho\vec{\mathrm{H}})_2 = \text{restriction of } H_2 - \sum F_{2j} G_{1j} \text{ to } \mathfrak{V}_2. \tag{2.17}$$

Note that, by the characteristic property of $\mathfrak{V}_2$, the choice of $\vec{\mathrm{G}}_1$ satisfying (2.16) is immaterial.

We now define $\vec{\mathrm{G}}_2 = (G_{2j})$ as follows: According to (2.14) we may write $(\rho\vec{\mathrm{H}})_2$ uniquely as the restriction to $\mathfrak{V}_2$ of

$$\sum Q^2_{ik}(s^{n-i}) L^2_{ik}(w^i).$$

Hence, by the property of $\mathfrak{V}_2$, we can find $\vec{\mathrm{G}}_2 \in M_1$ such that

$$H_2 - \sum F_{2j} G_{1j} - \sum Q^2_{ik}(s^{n-i}) L^2_{ik}(w^i) = \sum G_{2j} F_{2j}. \tag{2.18}$$

Now suppose $\mathfrak{V}_1, \ldots, \mathfrak{V}_t$, and $(\rho\vec{\mathrm{H}})_1, \ldots, (\rho\vec{\mathrm{H}})_t$, and $\vec{\mathrm{G}}_1, \ldots, \vec{\mathrm{G}}_t$ have been defined. Here $\vec{\mathrm{G}}_t \in M_{t-1}$, the module of simultaneous relations among the first $t-1$ rows of $\boxed{\mathrm{F}}$, that is, M_{t-1} consists of those $\vec{\mathrm{N}}_{t-1}$ for which

$$\sum F_{ij} N_{t-1,j} = 0 \quad \text{for } i = 1, \ldots, t-1.$$

$\mathfrak{V}_{t+1}$ will be a multiplicity variety associated by Theorem 2.4 to the ideal of sums $\sum F_{t+1,j}N_{tj}$ with $\vec{\mathrm{N}}_t \in M_t$. We define

$$(2.19)\quad (\rho\vec{\mathrm{H}})_{t+1} = \text{restriction to } \mathfrak{V}_{t+1} \quad \text{of } H_{t+1} - \sum_{i \leqslant t} \sum_j F_{i+1,j}G_{ij}.$$

By (2.14) we may write $(\rho\vec{\mathrm{H}})_{t+1}$ uniquely as the restriction to $\mathfrak{V}_{t+1}$ of

$$\sum Q_{ik}^{t+1}(s^{n-i})L_{ik}^{t+1}(w^i).$$

Hence, we can find $\vec{\mathrm{G}}_{t+1} \in M_t$ such that

$$(2.20)\quad H_{t+1} - \sum_{i \leqslant t} \sum_j F_{i+1,j}G_{ij} - \sum_{i,k} Q_{ik}^{t+1}(s^{n-i})L_{ik}^{t+1}(w^i) = \sum G_{t+1,j}F_{t+1,j}.$$

This completes the description of the map ρ. Note that at each stage there is a choice in the definition of $\vec{\mathrm{G}}_t$, and an arbitrary element of M_t can be added to $\vec{\mathrm{G}}_t$. By (2.19) this choice does not alter $(\rho\vec{\mathrm{H}})_{t+1}$. However, by (2.20) this choice *does* alter $\vec{\mathrm{G}}_{t+1}$ significantly. In fact, we could, by an appropriate choice of $\vec{\mathrm{G}}_t$, make $\vec{\mathrm{G}}_{t+1}$ an arbitrary element of M_t. This means that $(\rho\vec{\mathrm{H}})_{t+2}$ is altered by a change of $\vec{\mathrm{G}}_t$, so that ρ is not uniquely determined by $\boxed{\mathrm{F}}$. We shall discuss this question later (see Section II.4).

We may now state the general

THEOREM 2.6. *The map ρ defines an $\mathcal{O}$ isomorphism between $\mathcal{O}^l/\boxed{\mathrm{F}}\mathcal{O}^r$* and the direct sum of the $\mathcal{O}(\mathfrak{V}_j)$.

Theorem 2.6 will be proved below.

II.3. The Main Results for Principal Ideals

Let $H \in \mathcal{O}$. We shall say that H is *distinguished in z_n* if, on writing H as a power series

$$H(z) = \sum H_j(z_1, \ldots, z_{n-1})z_n^j,$$

where the H_j are analytic in the neighborhood of the origin, $H_j(0) \neq 0$ for some j. The Weierstrass preparation theorem (see Bockner and Martin [1]) says that if j_0 is the smallest j for which $H_j(0) \neq 0$, then we can multiply H by a unit $Q \in \mathcal{O}$ (that is, $Q(0) \neq 0$) in such a way that

$$Q(z)H(z) = \sum_{<j_0} H_j'(z_1, \ldots, z_{n-1})z_n^j + z_n^{j_0}$$

where each $H_j'(0) = 0$.

We call QH a *distinguished polynomial* in z_n. We say also that z_n is *distinguished* for H.

The well-known methods (see e.g., Bochner and Martin [1]) show

LEMMA 2.7. *Given any $H \in \mathcal{O}$, $H \not\equiv 0$, we can make a nonsingular linear change of variables to make z_n distinguished for H. Given any finite number of functions $H_1, \ldots, H_q$ we can make a nonsingular linear change of variables to make z_n distinguished for $H_1, \ldots, H_q$.*

Remark 2.2. The general method of proof of our main results is by induction on n. In the proof a finite number of successive applications of Lemma 2.7 must be appealed to. One might be concerned by the fact that the various nonsingular linear changes of variable are not independent since the linear changes in $n-1$ variables which occur in the induction argument depend, to some extent, on those in n variables. However, this dependence can cause no difficulties, because once we make z_n distinguished for $H_1, \ldots, H_q$, *any* linear transformation in $z_1, \ldots, z_{n-1}$ preserves the fact that z_n is distinguished. Thus there is no interference in the linear changes of variables.

The first step in the proof of our main results is the case of principal ideals.

Let $F \in \mathcal{O}$ and suppose z_n is distinguished. As far as the ideal $F\mathcal{O}$ is concerned we may assume that F is a distinguished polynomial in z_n. We write for simplicity w for $(z_1, \ldots, z_{n-1})$ and s for z_n.

THEOREM 2.8. *Suppose F is a distinguished polynomial in s of degree m. Let $F = P_1^{a_1} P_2^{a_2} \cdots P_b^{a_b}$ be the decomposition of F in $\mathcal{O}$ into prime factors and call V_j' the local analytic variety of zeros of P_j.*

Call $\mathfrak{V}'$ the multiplicity variety:

$$\mathfrak{V}' = \{V_1', \text{identity}; V_1', \partial/\partial s; \ldots; V_1', \partial^{a_1-1}/\partial s^{a_1-1}; \ldots; V_b', \partial^{a_b-1}/\partial s^{a_b-1}\}.$$

Assertion 1. The restriction Map $H \to H|\mathfrak{V}'$ is an isomorphism of $\mathcal{O}/F\mathcal{O}$ onto $\mathcal{O}(\mathfrak{V}')$. Every $H \in \mathcal{O}(\mathfrak{V}')$ can be extended to $\mathcal{O}$ uniquely in the form

$$K(w, s) = L_0(w) + L_1(w)s + \cdots + L_{m-1}(w)s^{m-1},$$

where $L_i \in \mathcal{O}$.

There exists a multiplicity Z variety $\mathfrak{V} = \{(V_q, \partial_q)\}$ such that

Assertion 2. The restriction map $H \to H \mid \mathfrak{V}$ is an isomorphism of $\mathcal{O}/F\mathcal{O}$ onto $\mathcal{O}(\mathfrak{V})$ (that is, $\mathfrak{V}$ is equivalent to $\mathfrak{V}'$).

Assertion 3. We can find disjoint local Z varieties $T_t \subset C^{n-1}$ which cover a neighborhood of zero so that if T_t^q denote the Z varieties of the form $w \in T_t$, $(w, s) \in V_q$, then we can find functions f_{iqt} which are algebroid on T_t^q such that

$$L_i(w) = \sum_q \sum_{(w,s) \in T_t^q} f_{iqt}(w, s)(\partial_q H)(w, s),$$

that is, the parametrization of $\mathcal{O}(\mathfrak{V}) \cong \mathcal{O}(\mathfrak{V}')$ given by extending H to $K(w, s) = L_0(w) + \cdots + L_{m-1}(w)s^{m-1}$ is a L'Hospital parametrization.

The map $\psi: H \to \{L_i\}$ is one-to-one from $\mathcal{O}/F\mathcal{O}$ onto $\mathcal{O}^m(n-1)$. Under *this map, multiplication by* $B \in \mathcal{O}$ *gets transformed into multiplication by a matrix* $\boxed{B}$ *whose coefficients lie on* $\mathcal{O}(n-1)$.

We see that Theorem 2.8 gives Theorems 2.4 and 2.5 for principal ideals. For the proof we need several lemmas:

LEMMA 2.9. *Let* P *be an irreducible function in* $\mathcal{O}$ *which is a distinguished polynomial in* s *of degree* c. *Then, for fixed* w *outside a local analytic variety in* C^{n-1}, $P(w, s)$ *has* c *distinct roots. If* F *is any distinguished polynomial in* s *of degree* m *then, for* w *small,* F *has* m *zeros (with multiplicity) which are small.*

PROOF. Let $D(w)$ be the discriminant of P considered as a polynomial in s. Since P is irreducible, $D \not\equiv 0$. Hence, except for the local analytic variety of zeros of D, $P(w, s) = 0$ has c distinct zeros. The statement about F is a consequence of the fact that $F(0, s) \equiv s^m$, and the roots of a polynomial depend continuously on its coefficients (see Marden [1]). ▯

LEMMA 2.10. *Let* $F \in \mathcal{O}$ *and let* $F = P_1^{a_1} \cdots P_b^{a_b}$ *be the prime factorization of* F *in* $\mathcal{O}$. *By Lemma 2.7 we choose a linear transformation so that* $z_n = s$ *is distinguished for* F *and all the* P_j. *We multiply* F *by a suitable unit* Q *and we multiply the* P_j *by suitable units* Q_j *so that* QF *and* $Q_j P_j$ *are distinguished polynomials in* s *of degrees* m *and* p_j, *respectively. Then* $m = \sum a_j p_j$, *and* $Q = Q_1^{a_1} \cdots Q_b^{a_b}$.

PROOF. From $F = P_1^{a_1} \cdots P_b^{a_b}$ we derive

$$QF = QQ_1^{-a_1} \cdots Q_b^{-a_b}(Q_1 P_1)^{a_1} \cdots (Q_b P_b)^{a_b}. \tag{2.21}$$

Now, the left side of (2.21) is a polynomial in s of degree m. By Lemma 2.9, we see that for fixed w close to zero the left side of (2.21) has m zeros, with multiplicity, close to zero. Hence, the right side does also. Also, $Q_j P_j$ has p_j zeros (with multiplicity) by Lemma 2.9. Since Q and the Q_j are units, this implies that $m = \sum a_j p_j$. Since QF and the $Q_j P_j$ have leading coefficient 1, we have also $Q = Q_1^{a_1} \cdots Q_b^{a_b}$. ▯

PROOF OF THEOREM 2.8. By Lemma 2.10 we may assume also that each P_j is a distinguished polynomial in s of degree p_j. Write

$$F(w, s) = F_0(w) + F_1(w)s + \cdots + s^m. \tag{2.22}$$

Then

$$s^m = -F_0(w) - F_1(w)s - \cdots - F_{m-1}(w)s^{m-1} + F(w, s), \tag{2.23}$$

that is,

$$s^m \equiv -F_0(w) - F_1(w)s - \cdots - F_{m-1}(w)s^{m-1} \bmod F\mathcal{O}. \tag{2.24}$$

Multiplying (2.24) by s and using (2.24), we have

$$(2.25)\quad s^{m+1} \equiv F_{m-1}(w)F_0(w) + [F_{m-1}(w)F_1(w) - F_0(w)]s + \cdots + [F_{m-1}(w)F_{m-1}(w) - F_{m-2}(w)]s^{m-1} \bmod F\mathcal{O}.$$

Proceeding in this fashion, we can write, for any j,

$$(2.26)\quad s^j \equiv A_0^j(w) + A_1^j(w)s + \cdots + A_{m-1}^j(w)s^{m-1} \bmod F\mathcal{O}.$$

(Of course $A_i^j = \delta_i^j$ for $j \leqslant m-1$.) This means that, given any $H \in \mathcal{O}$, if $H(w, s) = \sum H_i(w)s^i$ then we can by use of (2.26) (formally) reduce all powers of s higher than the $(m-1)$st and get H congruent mod $F\mathcal{O}$ to a polynomial in s of degree $m-1$. In this way we shall derive the isomorphism of $\mathcal{O}/F\mathcal{O}$ with $\mathcal{O}^m(n-1)$. However, we must show convergence of our process.

Since the series $\sum H_i(w)s^i$ converges for w, s sufficiently small, we can find constants C, C' so that for w sufficiently small we have $|H_i(w)| < C'C^i$. By the definition of a distinguished polynomial, all $F_j(0) = 0$, so we can choose a sufficiently small neighborhood of zero in which $|F_j(w)| < \delta$ where $0 < \delta < 1$; δ will be prescribed later.

Suppose that $j \geqslant m$. We write (2.26) in the form

$$(2.27)\quad s^j = \sum_{i=0}^{m-1} A_i^j s^i + A^j F.$$

Then we derive

$$(2.28)\quad s^{j+1} = \sum_{i=0}^{m-1} A_i^j s^{i+1} + A^j s F.$$

Thus

$$(2.29)\quad A_i^{j+1} = \begin{cases} A_{i-1}^j + A_{m-1}^j A_i^m & \text{for } i > 0 \\ A_{m-1}^j A_0^m & \text{for } i = 0 \end{cases}$$

$$(2.30)\quad A^{j+1} = A^j s + A_{m-1}^j.$$

Now (2.29) is not good enough for our purposes, because the first term for $i > 0$ does not decrease as $j \to \infty$. (Compare (2.33) where we are helped by the factor δ.) However, we may iterate the procedure and obtain

$$(2.31)\quad A_i^{j+2} = \begin{cases} A_{i-2}^j + A_{m-1}^j A_{i-1}^m + A_i^m(A_{m-2}^j + A_{m-1}^j A_{m-1}^m) & \text{for } i \geqslant 2 \\ A_{m-1}^j A_0^m + A_1^m(A_{m-2}^j + A_{m-1}^j A_{m-1}^m) & \text{for } i = 1 \\ A_0^m(A_{m-2}^j + A_{m-1}^j A_{m-1}^m) & \text{for } i = 0 \end{cases}$$

if $m > 1$. We iterate this expression and find that

$$A_i^{j+m} = \sum_{k=0}^{m-1} B_i^k A_k^j, \tag{2.32}$$

where each B_i^k is a sum of the A_l^m and products of them. Let q be the maximum number of such products that occur. Then if γ_j is the maximum of the $|A_i^j|$ for w in some neighborhood of zero, since $|A_l^m| = |F_l| < \delta < 1$, we have by (2.32), $\gamma_{j+m} = \max |A_i^{j+m}| < q\delta\gamma_j$. Also, for $t < m$ we have $|A_i^{m+t}| \leqslant q$. This proves that for w sufficiently small

$$\gamma_j = \max |A_i^j| < q(q\delta)^{[j/m]-1}, \tag{2.33}$$

where $[j/m]$ is the greatest integer in j/m.

Now, using (2.30) and (2.33), we see that for $|s| < \delta$, for w small and for j large,

$$|A^{j+1}| < |A^j|\,\delta + q(q\delta)^{[j/m]-1}. \tag{2.34}$$

This gives easily: Given any δ' we can find δ sufficiently small so that

$$|A^j| < \text{const } (\delta')^j. \tag{2.35}$$

If we combine (2.33) and (2.35) with the inequality $|H_j(w)| < C'C^j$, it follows that for δ small enough, the series $\sum H_j A_i^j$ and $\sum H_j A^j$ converge in the topology of $\mathcal{O}$. This means that

$$H(w, s) = \sum_{i=0}^{m-1} s^i(\sum H_j A_i^j) + F \sum H_j A^j. \tag{2.36}$$

Equation (2.36) shows that there is a linear map ψ of $\mathcal{O}/F\mathcal{O}$ into $\mathcal{O}^m(n-1)$. This map is one-to-one and *onto*, for, by the minimality condition on m, no polynomial in s of degree smaller than m can be divisible by F.

Next let $B \in \mathcal{O}$. We want to find the image under ψ of multiplication by B. We note first from (2.36) that if B is independent of s, then B commutes with ψ. Next, we compute the case $B = s$:

Multiplying (2.36) by s and using (2.27) when $i + 1 = m$, we derive

$$\begin{aligned} sH &= \sum_{i=0}^{m-1} s^{i+1}(\sum H_j A_i^j) + Fs \sum H_j A^j \\ &= \sum_{i=1}^{m-1} s^i(\sum H_j(A_{i-1}^j + A_{m-1}^j A_i^m)) \\ &\quad + \sum H_j A_{m-1}^j A_0^m + Fs \sum H_j A^j + F \sum H_j A_{m-1}^j \\ &= \sum_{i=1}^{m-1} (\psi H)_{i-1} s^i + \sum_{i=0}^{m-1} A_i^m (\psi H)_{m-1} s^i + Fs \sum H_j A^j + F \sum H_j A_{m-1}^j . \end{aligned} \tag{2.37}$$

Thus, the matrix for multiplication by s is

$$\begin{pmatrix} 0 & 0 & \cdot & \cdots & \cdot & \cdot & A_0^m \\ 1 & 0 & \cdot & \cdots & \cdot & \cdot & A_1^m \\ 0 & 1 & 0 & \cdots & \cdot & \cdot & \cdot \\ \cdot & \cdot & & \cdots & \cdot & \cdot & \cdot \\ \cdot & 0 & \cdot & \cdots & \cdot & \cdot & \cdot \\ \cdot & \cdot & & \cdots & \cdot & \cdot & \cdot \\ \cdot & \cdot & \cdot & \cdots & \cdot & \cdot & \cdot \\ \cdot & \cdot & \cdot & \cdots & \cdot & \cdot & \cdot \\ 0 & \cdot & \cdot & \cdots & 0 & 1 & A_{m-1}^m \end{pmatrix}. \tag{2.38}$$

Note the similarity of (2.38) [which is the "companion matrix" of the polynomial $-F(u) = \sum A_j^m u^j$ (see Birkhoff and MacLane [1])] and the matrices obtained by changing a linear differential equation of order m into a first order system. The reason for this similarity will appear in Chapter IX.

We may obtain similar matrices corresponding to multiplication by s^t for any t. Writing $B = \sum B_t s^t$, the method used above to show convergence for (2.36) shows that we obtain a matrix $\boxed{\mathrm{B}} = (B_{ij})$ corresponding to multiplication by B, where $B_{ij} \in \mathcal{O}$, that is,

$$\psi(BH) = \boxed{\mathrm{B}}\,\psi H. \tag{2.39}$$

Thus, $\mathcal{O}^m(n-1)$ has this structure as an $\mathcal{O}$ module which makes ψ an $\mathcal{O}$ isomorphism.

In order to complete the proof of Theorem 2.8 we shall need to express ψH in terms of the restriction of H to $\mathfrak{B}$. (Of course, we must first define $\mathfrak{B}$.) For this purpose we need the Lagrange interpolation formula. We write $F = P_1^{a_1} \cdots P_b^{a_b}$, where F and the P_j are polynomials in s with leading coefficients 1 of degrees m and p_j, respectively. For each small w we denote by $s_{jk}(w)$ for $k = 1, 2, \ldots, p_j$ the zeros of $P_j(w, s)$. By Lemma 2.9 for fixed j these zeros are distinct except when w belongs to a local analytic variety of complex dimension $< n-1$. By the same type of reasoning, all the $s_{jk}(w)$ (for varying j and k) are distinct except for a local analytic variety β of dimension $< n-1$.

LEMMA 2.11. *Suppose we are given distinct points $s_1, \ldots, s_a$ in the complex s plane and integers $m_1, \ldots, m_a > 0$. Suppose for each $i = 1, 2, \ldots, a$ we are given m_i complex numbers α_i^l, $l = 0, 1, \ldots, m_i - 1$. Then there exists a unique polynomial $R(s)$ of degree $(\sum m_i) - 1$ such that*

$$\frac{\partial^l R}{\partial s^l}(s_i) = \alpha_i^l \tag{2.40}$$

for all i, l. If we write

$$R(s) = \sum \gamma_j s^j, \tag{2.41}$$

then for each j

$$\gamma_j = D_j/D, \tag{2.42}$$

where D is a polynomial in the s_i which does not vanish. Moreover, D^2 is invariant under any permutation of those s_i for which the corresponding m_i are equal. D_j is of the form

$$D_j = \sum \eta_{ij}^l \alpha_i^l \tag{2.43}$$

where the η_{ij}^l are polynomials in the s_i.

We defer the proof of Lemma 2.11 until later. We conclude the proof of Theorem 2.8.

Consider the restriction map $\rho_{\mathfrak{B}'} : H \to H \mid \mathfrak{B}'$ of $\mathcal{O}$ into $\mathcal{O}(\mathfrak{B}')$. The map is onto by the definition of $\mathcal{O}(\mathfrak{B}')$. It is clear from the definitions that $\rho_{\mathfrak{B}'}(F\mathcal{O}) = 0$. Conversely, suppose that $\rho_{\mathfrak{B}'}(H) = 0$. We want to show that H is divisible by F. First of all, H vanishes on each V_j', so H is divisible by $P_1 P_2 \cdots P_b$. Also, the method of proof of Theorem 2.1 combined with Lemma 2.9 shows that, if $a_j > 1$, H/P_j vanishes on V_j' except on a subvariety of codimension at least 1, that is, H/P_j^2 is regular in all of C except on a variety of complex codimension >1. Hence, by well-known results (see Bochner and Martin [1]) H/P_j is divisible by P_j. Proceeding in this way, we find that H is divisible by $P_j^{a_j}$ for all j so that H is divisible by F.

We can prove that $H \in F\mathcal{O}$ by another method: We want to show that $\psi H = 0$. We know from the above that

$$H = (\psi H)_0 + (\psi H)_1 s + \cdots + (\psi H)_{m-1} s^{m-1} + FT. \tag{2.44}$$

For fixed $w \notin \beta$ we have therefore for every i, by Lemma 2.11

$$(\psi H)_i(w) = D_i(w)/D(w), \tag{2.45}$$

where $D(w) \neq 0$. But by (2.43) (for $D_i(w)$ in place of D_j) it follows from $H \mid \mathfrak{B}' = 0$ that $(\psi H)_i(w) = 0$ for $w \notin \beta$ (since the α_{jk}^l are, by (2.40), expressed in terms of $H \mid \mathfrak{B}'$). Since the $(\psi H)_i$ are regular they are identically zero.

For any H we have (2.44) where instead of (2.45) we may write

$$(\psi H)_i(w) = D_i(w) D(w)/D^2(w). \tag{2.46}$$

We claim that $D^2(w) \in \mathcal{O}(n-1)$. To see this we note by Lemma 2.11 that $D^2(w)$ is invariant under $s_{jk}(w) \to s_{jk'}(w)$ for any w, k, k' with $w \notin \beta$. Since the $s_{jk}(w)$ are all the zeros of P_{jk}, the theory of symmetric functions shows that $D^2(w)$ can be expressed as a quotient of two functions which are analytic in a neighborhood of the origin minus β. But the expression for $D^2(w)$ given by Lemma 2.12 (see below) shows that $D^2(w)$ is bounded. Thus, by the theorem of removable singularities (see Bochner and Martin [1]), $D^2(w) \in \mathcal{O}(n-1)$. It follows also that

$$D_i(w)D(w) \in \mathcal{O}(n-1).$$

If we apply Theorem 2.1 and (2.46) we obtain a covering $\{U_j'\}$ of a neighborhood of zero in $\mathcal{O}(n-1)$ and differential operators d_j depending only on $D^2(w)$ (hence, on F) such that for $w \in U_j'$, $w \notin \beta$, we have

$$(\psi H)_i(w) = d_j[D_i(w)D(w)].$$

Applying (2.43) to the right side of the above yields

$$(\psi H)_i(w) = d_j\left[\sum_{jkl} \eta_{ijk}^l \frac{\partial^l K}{\partial s^l}(w, s_{jk}(w))\right].$$

Here η_{ijk}^l is a polynomial in $s_{jk}(w)$.

However, the expression for $(\psi H)_i(w)$ is *not* sufficient for Assertion 3 of Theorem 2.8. What we must do is "commute" d_j and η_{ijk}^l. Rather than do this, we shall follow a different procedure, which is in conformity with our treatment of Example 3 of Section II.2. We could actually perform the commutation using Theorem 2.15 below. However, the present approach is somewhat simpler and somewhat more explicit.

For each w_0 near zero, let us denote as above by $s_{jk}(w_0)$ the zeros of P_j which lie above w_0. We denote the multiplicity of these zeros by $m_{jk}(w_0)$.

We define $T_{jk}(w_0)$ as that irreducible component of the variety of common zeros of P_j, $\partial P_j/\partial s, \ldots, \partial^{m_{jk}(w_0)-1}P_j/\partial s^{m_{jk}(w_0)-1}$ which contains w_0 *minus* the subvariety on which $\partial^{m_{jk}(w_0)}P_j/\partial s^{m_{jk}(w_0)}$ vanishes. We define $T(w_0)$ as the set of points w near zero such that the multiplicities and the coincidences of roots of all the P_j above w "look the same" as they do above w_0, that is, the multiplicities as roots of P_j are the same and the coincidence of roots of P_j, $P_{j'}$ is the same. Precisely,

$$T(w_0) = \bigcap_{jk} \text{proj}\,[T_{jk}(w_0)] - \bigcap_{(j,k),(j',k')}' \text{proj}\,[T_{jk}(w_0) \cap T_{j'k'}(w_0)],$$

where we have written $\cap'$ to exclude those (j, k), (j', k') for which $T_{jk}(w_0)$ and $T_{j'k'}(w_0)$ meet above w_0, that is, $s_{jk}(w_0) = s_{j'k'}(w_0)$. (By "projection" we mean "projection on $s = 0$."). See Fig. 3.

CLAIM. $T_{jk}(w_0)$ and $T(w_0)$ are (unions of) Z varieties.

PROOF. The $T_{jk}(w_0)$ are Z varieties by definition. The fact that $T(w_0)$ is a union of Z varieties is a consequence of this and the fact that (see Samuel [1]) the projection of a variety is a Z variety.

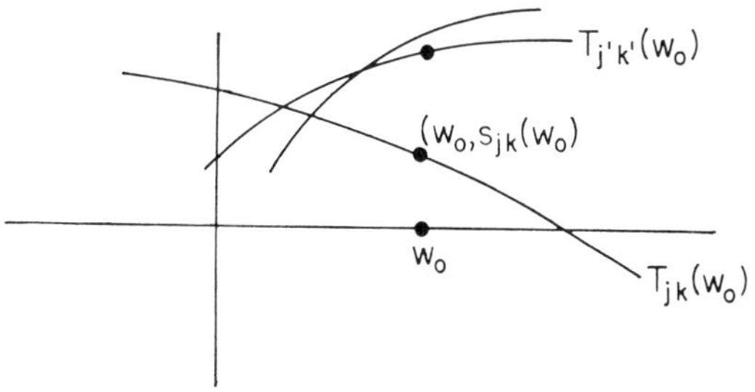

FIGURE 3.

As mentioned in Chapter I, we shall often use the term " Z variety " for "union of Z varieties."

We now define the multiplicity Z variety $\mathfrak{B}$ as follows: The $T_{jk}(w_0)$ were defined as irreducible components of the varieties of common zeros of the P_j and various s derivatives *minus* subvarieties of a similar nature. Thus, the $T_{jk}(w_0)$ are finite in number for varying j, k, w_0 because the P_j are distinguished polynomials in s and so only a finite number of s derivatives are nontrivial. To define the V_q we start with the $T_{jk}(w_0)$ minus any nontrivial intersections of two $T_{jk}(w_0)$. We then take the intersections of two $T_{jk}(w_0)$ and subtract nontrivial varieties formed by intersections of more than two. Continuing in this way we obtain all V_q. Since the number of $T_{jk}(w_0)$ is finite, there are only finitely many V_q. To each V_q we associate the operators *identity*, $\partial/\partial s. \ldots, \partial^{c_q-1}/\partial s^{c_q-1}$, where

$$c_q = \sum a_j m_{jk},$$

the sum being extended over all j for which V_q is contained in T_{jk}. Note that c_q is just the order of vanishing of F at any point of V_q. (Note the analogy between this and Example 3 of Section II.2.)

We define the T_t required by Theorem 2.8 as the distinct $T(w_0)$. The Claim says that the T_t are local Z varieties. Since each w_0 is contained in $T(w_0)$, the union of the T_j is a neighborhood of zero in C^{n-1}. It is clear that the T_j are disjoint.

PROOF OF ASSERTION 2 OF THEOREM 2.8. The vanishing of H on $\mathfrak{B}$ implies its vanishing on $\mathfrak{B}'$ because, by definition, the varieties of $\mathfrak{B}$ cover the varieties of $\mathfrak{B}'$ and we have at least as many differential conditions at any point given by $\mathfrak{B}$ as by $\mathfrak{B}'$. Thus, by the first part of Theorem 2.8, the kernel of $H \to H \mid \mathfrak{B}$ is contained in $F\mathcal{O}$.

On the other hand, F vanishes on $\mathfrak{B}$. For, given any V_q of $\mathfrak{B}$ and any point $(w_0, s_0) \in V_q$, by definition the differential operators associated to V_q are *identity*, $\ldots, \partial^{c_q-1}/\partial s^{c_q-1}$ where c_q is the sum of the orders of the zeros of $P_j^{a_j}(w_0, s)$ at $s = s_0$. Thus, $F = P_1^{a_1} \cdots P_b^{a_b}$ has a zero in s of order c_q at

(w_0, s_0), so F vanishes on $\mathfrak{B}$. Hence, $F\mathcal{O}$ vanishes on $\mathfrak{B}$. Finally, the definition of $\mathcal{O}(\mathfrak{B})$ is such that $H \to H \mid \mathfrak{B}$ is *onto* $\mathcal{O}(\mathfrak{B})$. ▯

PROOF OF ASSERTION 3 OF THEOREM 2.8. We use Lemma 2.11. Let $w_0 \in T_t = T(w_0)$. Then we can write

$$L_i(w_0) = D_i^t(w_0)/D^t(w_0)$$
$$= D^t(w_0)D_i^t(w_0)/D^{t2}(w_0).$$

Here $D^t(w_0)$ is a polynomial in the roots of $F(w_0, s) = 0$ and $D_i^t(w_0)$ is a linear expression in $H(w_0, s_{jk}(w_0))$ and suitable derivatives with respect to s, the coefficients being polynomials in the $s_{jk}(w_0)$. The proof of Lemma 2.11 shows that the form of (2.43) (and the analogous one for D) depends only on the m_j. Thus, in our application the polynomials that appear in the $D_i^t(w_0)$ and $D^t(w_0)$ do not vary as w_0 varies in $T(w_0)$. Moreover, since the distinct $s_{jk}(w_0)$ stay distinct as w_0 varies in $T(w_0)$, they define algebroid functions on $T(w_0)$.

Assertion 3 is therefore proved. ▯

PROOF OF LEMMA 2.11. We write

$$R(s) = \sum \gamma_j s^j.$$

Call $m = \sum m_j$. Then we can solve our problem uniquely if the determinant

$$D = \begin{vmatrix} 1 & s_1 & s_1^2 & \cdot & \cdot & & & \cdot & \cdots & s_1^{m-1} \\ 0 & 1 & 2s_1 & \cdot & \cdot & & & \cdot & \cdots & (m-1)\, s_1^{m-2} \\ 0 & & & & & 0 & 1 & (m_1-1)!s_1 & \cdots & (m-1)!s_1^{m-m_1}/(m-m_1)! \\ 1 & s_2 & \cdot & \cdot & \cdot & & & \cdot & \cdots & s_2^{m-1} \\ \cdot & \cdot & \cdot & \cdot & \cdot & & & \cdot & \cdots & \cdot \end{vmatrix}$$

is not zero. D is clearly a generalization of the Vandermonde determinant (which is the case when all $m_i = 1$). We claim:

LEMMA 2.12

$$D = \pm c \prod_{i<j} (s_i - s_j)^{m_i m_j}.$$

(If there is only one s_i, then $D = 1$.) Here

$$c = \prod_{i=1}^{m-m_1} (m_1 + i - 1)!/(m_1!)^{m_2}(m_1 + m_2)!^{m_3} \cdots (m_1 + \cdots + m_{n-1})!^{m_n}.$$

PROOF. Consider D as a polynomial in the s_j. Then D is easily seen to be homogeneous of degree

$$\sum_{i<j} m_i m_j .$$

Moreover, if we set $s_i = s_j$ then two rows of D become equal, so D must vanish. Thus, D is divisible by each $s_i - s_j$. To show that D is actually divisible by $(s_i - s_j)^{m_i m_j}$, we differentiate D regarding the s_i as functions of some parameter. We use the classical formula for the derivative of a determinant, namely, it is the sum of determinants obtained from the original one by replacing successively each row by its derivative. We differentiate D several times, so we obtain a sum of determinants from D by replacing certain rows by their derivatives. It is easily seen that if we differentiate D less than $m_i m_j$ times, then in the resulting sum each determinant will have two rows equal if $s_i = s_j$. Thus, D is divisible by

$$\prod_{i<j} (s_i - s_j)^{m_i m_j} .$$

Since D and the product are homogeneous of degree

$$\sum_{i<j} m_i m_j ,$$

we have

$$D = \text{const.} \prod_{i<j} (s_i - s_j)^{m_i m_j} .$$

On examining the diagonal of D, we can find the value of the constant. Thus, Lemma 2.12 is proven. ▯

The invariance of D^2 under the requisite permuations is clear so Lemma 2.11 is proved. ▯

See Remark 2.3.

II.4. Conclusion of the Proofs of Theorems 2.3, 2.5, and 2.6

Before going into the details of the proofs, we wish to give the idea. Let $\boxed{a}$ be an $l \times r$ matrix of elements a_{ij} belonging to some ring S; suppose we want to solve the equation

$$\boxed{a}\,\vec{u} = \vec{v} .$$

As mentioned in the remarks preceding the statement of Theorem 2.6, we try to solve this equation step by step for the components of $\vec{v}$. We start by trying to solve

$$\sum a_{1j} u_j = v_1 .$$

To solve this equation we try first to solve

$$a_{11}u_1 = v_1.$$

If this is possible we go on to the problem of solving the equation involving v_2. If not, we solve the equation $a_{11}u_1 = v_1$ "to the best of our ability" which means we pass to the quotient of S by the principal ideal (a_{11}) generated by a_{11}. If we denote by $\bar{a}_{1j}, \bar{u}_j, \bar{v}_1$ the respective images of a_{1j}, u_j, v_1 in this quotient space, then we are left with the problem of solving

$$\sum_{j \geqslant 2} \bar{a}_{1j}\bar{u}_j = \bar{v}_1.$$

In the present situation, when S is the ring $\mathcal{O}(n)$, Theorem 2.8 shows that $S/(a_{11})$ is isomorphic to $\mathcal{O}^m(n-1)$. Thus, when we express $\bar{a}_{1j}, \bar{u}_j, \bar{v}_1$ in terms of their components in $\mathcal{O}^m(n-1)$, we see that

$$\sum_{j \geqslant 2} \bar{a}_{1j}\bar{u}_j = \bar{v}_1$$

is an equation of the same type as we started with, namely $\boxed{\mathrm{a}}\,\vec{\mathrm{u}} = \vec{\mathrm{v}}$, except that we have lowered n to $n-1$. Thus we can continue the step-by-step procedure and it will eventually stop.

The essential point to notice is that the procedure solves matrix equations $\boxed{\mathrm{a}}\,\vec{\mathrm{u}} = \vec{\mathrm{v}}$, and finds the obstructions in $\vec{\mathrm{v}}$ to solving such equations, by studying the solutions of scalar equations such as $a_{11}u_1 = v_1$ and the quotient of S by *principal* ideals such as (a_{11}).

It is strongly recommended that the reader familiarize himself with our remarks on Example 4 preceding the statements of Theorems 2.4 and 2.5 (pages 37, 43) before reading this section.

We wish now to complete the proofs of Theorems 2.3, 2.5, and 2.6. These will be proved by induction. We shall denote the theorems more precisely by

$$\mathrm{T2.5}(n), \quad \mathrm{T2.6}(n, l), \quad \text{and } \mathrm{T2.3}(n, l).$$

PROOF OF THEOREM 2.3. We have shown on pages 32ff that Theorem 2.2 implies Theorem 2.3; i.e., T2.3(n, 1) implies T2.3(n, l). Since T2.3(0, 1) is trivial, it remains to show that for $n > 0$, T2.3($n-1$, l) for all l implies T2.3(n, 1).

Let $\vec{\mathrm{F}} = (F_j)$, where $F_j \in \mathcal{O}$. Let $H \in \mathcal{O}$ be of the form $\vec{\mathrm{F}}\,\vec{\mathrm{G}}'$, where $\vec{\mathrm{G}}' = (G'_j)$, $G'_j \in \mathcal{O}$, that is, $H = \sum F_j G'_j$. We may assume for simplicity that $\mathrm{F}_1 \not\equiv 0$. If F_1 divides H, then we can write $H = F_1 G_1$, and we are finished by Theorem 2.1. *But*, then the varieties, operators, and algebroid maps might depend on this fact (that F_1 divides H), hence not wholly on

$\vec{\mathrm{F}}$. For this reason, even in this case we must apply the following process—which will lead to the same result. We apply Theorem 2.8, that is, we assume as we may that $z_n = s$ is distinguished for F_1. Then we apply the map $\psi : \mathcal{O}/F_1\mathcal{O} \to \mathcal{O}^m(n-1)$ described in Theorem 2.8. Each F_j operates on $\mathcal{O}^m(n-1)$ by a well-defined $m \times m$ matrix $\boxed{\mathrm{F}}_j$.

We denote by $\boxed{\mathrm{F}}^1$ the $m \times (r-1)m$ matrix

$$\boxed{\mathrm{F}}^1 = (\boxed{\mathrm{F}}_2, \ldots, \boxed{\mathrm{F}}_r). \tag{2.47}$$

We then have

LEMMA 2.13. *There exists a* $\vec{\mathrm{G}}'' \in \mathcal{O}^{(r-1)m}(n-1)$ *with* $\boxed{\mathrm{F}}^1\vec{\mathrm{G}}'' = \psi H$.

PROOF. We apply ψ to the equation $H = \sum F_j G_j'$ and the lemma follows if we set

$$\vec{\mathrm{G}}'' = \begin{pmatrix} \psi G_2' \\ \cdot \\ \cdot \\ \cdot \\ \psi G_r' \end{pmatrix}. \quad \square$$

We continue the proof of Theorem 2.3. By T2.3$(n-1, m)$ we can find local Z varieties and differential operators depending only on $\boxed{\mathrm{F}}^1$, hence on $\vec{\mathrm{F}}$, such that we can find $\vec{\mathrm{G}}''' \in \mathcal{O}^{(r-1)m}(n-1)$ satisfying

$$\boxed{\mathrm{F}}^1\vec{\mathrm{G}}''' = \psi H, \tag{2.48}$$

and $\vec{\mathrm{G}}'''$ can be expressed in terms of the differential operators applied to ψH on these Z varieties. That is, $\vec{\mathrm{G}}'''$ is $\vec{\mathrm{F}}$ L'Hospital describable in terms of ψH.

We want to apply ψ^{-1} to (2.48). Of course, ψ^{-1} does not exist, since ψ is not one-to-one on $\mathcal{O}$, but there exists a "good" inverse χ of ψ, namely,

$$\chi(K_0, \ldots, K_{m-1}) = \sum K_j s^j. \tag{2.49}$$

LEMMA 2.14. *χ is a right inverse for ψ. That is, $\psi\chi =$ identity. For any $L \in \mathcal{O}$ we have $\chi\psi L \equiv L$ mod $F_1\mathcal{O}$. Let $B \in \mathcal{O}$ and let $\boxed{\mathrm{B}}$ be the $m \times m$ matrix associated with B by Theorem 2.8. Then for any $\vec{\mathrm{K}} \in \mathcal{O}^m(n-1)$*

$$\chi(\boxed{\mathrm{B}}\vec{\mathrm{K}}) \equiv B\chi(\vec{\mathrm{K}}) \bmod F_1\mathcal{O}. \tag{2.50}$$

PROOF. The statement that $\psi\chi =$ identity is an immediate consequence of the definitions. To prove that $\chi\psi L \equiv L$ mod $F_1\mathcal{O}$ and (2.50) we must merely show that the image of both sides under ψ are equal. But this follows from the fact that $\psi\chi =$ identity and that ψ changes multiplication by B into multiplication by $\boxed{\mathrm{B}}$. $\square$

We now write $\vec{G}'''$ in the form

$$\vec{G}''' = \begin{pmatrix} \vec{G}_2''' \\ \cdot \\ \cdot \\ \cdot \\ \vec{G}_r''' \end{pmatrix} \tag{2.51}$$

where $\vec{G}_k'''$ is a vector with components

$$G'''_{(k-2)m+1},\ G'''_{(k-2)m+2},\ \ldots,\ G'''_{(k-1)m}.$$

Then (2.48) becomes

$$\sum_{k=2}^{r} \boxed{F}_k \vec{G}_k''' = \psi H. \tag{2.52}$$

We apply χ to (2.52) and use Lemma 2.14. This gives

$$\sum F_k \chi(\vec{G}_k''') \equiv H \bmod F_1 \mathcal{O} \tag{2.53}$$

or

$$\sum F_k \chi(\vec{G}_k''') - H = F_1 G_1''' \tag{2.54}$$

where $G_1''' \in \mathcal{O}$. Applying Theorem 2.1, we see that G_1''' can be described by derivatives of values of $\sum F_k \chi(\vec{G}_k''') - H$ on Z varieties determined by F_1. In order to complete the proof of Theorem 2.3 we need

THEOREM 2.15. *Let V be a local irreducible Z variety and V' a Z subvariety of V of maximum dimension, that is,* dim $V' =$ dim V. *Suppose $\boxed{d}$ is an algebroid differential operator with coefficients regular on V'. Let $\vec{B}$, $\vec{C}$, and $\boxed{E}$ be regular on full neighborhoods of zero and suppose $\vec{B}$ is $\boxed{E}$ L'Hospital representable in terms of $\vec{C}$ on a Z variety V'' of maximum dimension, that is, a full neighborhood of zero minus a proper subvariety. Suppose $\vec{A} = \boxed{d} \cdot \vec{B}$ is regular on V. Then $\vec{A}$ is $\boxed{E}'$ L'Hospital representable in terms of $\vec{C}$ on all of V. We write $\boxed{E}'$ for $\boxed{E}$ and $\boxed{d}$. In all cases we mean L'Hospital representable in the general sense.*

The conclusion that $\vec{A}$ is $\boxed{E}'$ L'Hospital representable in terms of $\vec{C}$ on V holds also if instead of $\vec{A}$ being of the form $\boxed{d} \cdot \vec{B}$, we require that $\vec{A}$ be regular on V and that $\vec{A}$ be $\boxed{E}''$ L'Hospital representable by $\vec{B}$ on V' for some $\boxed{E}''$ which is regular on a neighborhood of zero. In this case $\boxed{E}'$ depends on $\boxed{E}$ and $\boxed{E}''$.

It is suggested that the reader study Example 4 preceding the statement of Theorem 2.5. above before reading the proof.

PROOF. We use coordinates z and (s', z) for the respective ambient spaces of $\vec{B}$ and $\vec{C}$. We may, by making V'' smaller if necessary, assume that there is a single expression for $\vec{B}(z)$ in terms of $\vec{C}$ for all $z \in V''$. From Section II.1, this expression is of the following form: There exists a local variety W in $(s'', s', z) = (s, z)$ space such that $\vec{B}(z)$ is an algebroid function defined by W and an abstract algebroid (vector) function $\vec{g}'/\vec{h}'$ on W. Here $\vec{g}'$ is a linear expression in certain derivatives of $\vec{C}$ with coefficients depending only on $\boxed{E}$ and $\vec{h}'$ depends only on $\boxed{E}$. $\vec{g}'/\vec{h}'$ is the vector with components g_i'/h_i'.

We should like to apply L'Hospital's rule, as in the proof of Theorem 2.1, to obtain the desired conclusion for $\vec{A}$. This is, unfortunately, impossible because we would have to take derivatives of the s_i which could involve denominators that are zero (compare (1.19)). We shall therefore follow a different approach based on the use of a suitable local parameter.

By use of Proposition 1.11 we can obtain an expression for $\boxed{d}\vec{B}$ as an abstract algebroid function. By making V' smaller, if necessary, we may assume that this expression is valid to express $\vec{A}(z)$ for $z \in V'$. We can thus write

$$\vec{A}(z) = \vec{g}(s, z)/\vec{h}(s, z) \tag{2.55}$$

for a suitable choice of $(s'', s', z) \in W'$ where $\vec{g}$ is a linear expression in derivatives of $\vec{C}$ with coefficients in $\mathcal{O}$ depending only on $\boxed{E}$ and $\boxed{d}$ and $\vec{h} \in \mathcal{O}$ depends only on $\boxed{E}$ and $\boxed{d}$.

If z were a single variable and V' were one dimensional (hence, a complex line minus a finite set of points), then we could use an argument based on expanding all the s_i in terms of a suitable fractional power of z using the Puiseux expansion of Theorem 1.1. In the general case we proceed as follows: Suppose V' is of dimension d. By a modification of Lemma 2.7 we can show that it is possible, after making a linear transformation in the z variables, to write $z = (z', z'')$, where $z'' = z_1'', \ldots, z_d''$ and each z_i' satisfies on V' an equation of the form

$$^iJ(z) \equiv z_i'^{m_i} + {}^iJ_1(z'')z_i'^{m_i-1} + \cdots + {}^iJ_{m_i}(z'') = 0 \tag{2.56}$$

(compare (1.18)). The reason for this is that the fact that V' is of dimension d means that we can choose d coordinates $z_1'', \ldots, z_d''$ so that every other coordinate z_i' satisfies on V' an algebraic equation over the field generated by the z''. By using Lemma 2.7 we can make these equations of the form (2.56). Using (2.56) it is possible to show that we can assume that in (1.18) the functions $J_{ik}(z)$ can be chosen to depend only on z''.

Now, using Lemma 2.7 we can write each ${}^iJ_k(z'')$ in the form

$${}^iJ_k(z'') = {}^iJ'_k(z''){}^iJ''_k(z''),$$

where ${}^iJ'_k \neq 0$ and ${}^iJ''_k(z'')$ may be assumed to be a distinguished polynomial in z''_1, say of degree m_k. This means that for $z''_2, \ldots, z''_d$ fixed close to zero, ${}^iJ''_k(z''_1, z''_{21} \ldots, z''_d)$ has at most m_k zeros. Since ${}^iJ'_k(0) \neq 0$, ${}^iJ''_k$ and ${}^iJ_k$ have the same zeros. A similar remark applies to the J_{ik}.

Now let Λ be a straight line (one complex dimension) in the z'' space of the form: $(z''_2, \ldots, z''_d)$ fixed close to zero. Call λ the coordinate on Λ and set $\lambda_a \to \lambda - a$ for any small complex number a. On Λ each z''_j for $j > 1$ is constant and $z''_1 = \lambda$. By Theorem 1.1, (2.56), and (1.18), each z'_i and s_i is, on Λ, expressible as a series in positive powers of $\lambda_a^{1/p}$ for some p which can be chosen independent of Λ. We may make a linear transformation so that, for each Λ, all the J_i of (1.18) have distinct roots above any Λ at all but a finite number of points. For, if δ is the products of the discriminants of all the J_i, we need only perform a linear transformation to make δ the product of a unit by a distinguished polynomial in z''_1 which is possible by Lemma 2.7. Thus we can use (1.19) above Λ at all but a finite number of points.

Let a be a point in Λ for which some J_i has a multiple root above a. We use (2.55), expanding everything in a power series in $\lambda_a^{1/p}$. Since $\vec{C}$ is regular on a whole neighborhood of zero, only positive powers of $\lambda_a^{1/p}$ will come from terms involving $\vec{C}$. Also, when we expand C in a Taylor series around the chosen point on W above a, the high order monomials will give high order powers of $\lambda_a^{1/p}$ in $\vec{g}$. Now suppose that no component of $\vec{h}$ vanishes identically above any Λ, unless that component vanishes identically. This can be accomplished by a proper choice of coordinates as for δ above. Then $\vec{h}$ can contain a factor of $\lambda_a^{1/p}$ to a *bounded* power. All this means that the coefficient of $(\lambda_a^{1/p})^0$ on the right side of (2.55), which is the value of the left side at the z point which corresponds to a, is a linear expression in a bounded number of derivatives of $\vec{C}$ at a.

We must now know two things:

1. For each fixed t the set of a where $\vec{h}$ has a factor $\lambda^{t/p}$ but not $\lambda^{(t+1)/p}$ is a Z variety.
2. Each coefficient in the expansion of s_i or z'_i in powers of $\lambda_a^{1/p}$ is an algebroid function of $z \in V$.

Both these are consequences of Lemma 1.13. The result now follows from this since the type of expression for $\vec{A}(z)$ for $z \in V$ corresponding to a depends on the power of $\lambda_a^{1/p}$ which divides $\vec{h}$ and the coefficients in the expression for $\vec{A}$ are expressible in terms of the expansion coefficients of the z_i, s_i.

The second part of Theorem 2.15 is contained in the first by the remarks of Section II.1. ▯

LEMMA 2.16. *Suppose* $\vec{\mathrm{L}} \in \mathcal{O}^m(n-1)$ *is* $\vec{\mathrm{F}}$ *L'Hospital expressible in terms of* ψH. *Then* $\chi(\vec{\mathrm{L}})$ *is* $\vec{\mathrm{F}}$ *L'Hospital expressible in terms of* H.

PROOF. Let U^k be local Z varieties which are disjoint and whose union is a neighborhood of zero in C^{n-1}; let g^t_{ik} be integral algebroid functions which are regular on U^k, and let d^{it}_{jk} be differential operators whose coefficients are regular on $g^t_k(U^k)$ such that for $w \in U^k$

$$L_j(w) = \sum_{i,t} (d^{it}_{jk} H_i)[g^t_{1k}(w), \ldots, g^t_{nk}(w)].$$

Define $\tilde{U}^k = U^k \times C^1$ so the $\tilde{U}^k$ are disjoint and their union is a neighborhood of zero in C^n. For $z \in \tilde{U}^k$,

$$(\chi\vec{\mathrm{L}})(z) = \sum L_j(w)s^j = \sum_i \left[\left(\sum_j d^{it}_{jk} s^j\right) H_i\right][g^t_k(w)]. \tag{2.57}$$

We now apply Assertion 3 of Theorem 2.8. This allows us to replace H_i on the right side of (2.57) by use of

$$H_i(w) = \sum_q \sum_{(w,s')\in T^q_t} f_{iqt}(w, s')(\partial_q H)(w, s').$$

The result now follows from Theorem 2.15. ▯

We can now conclude the proof of Theorem 2.3. We examine (2.54). We already know that $\vec{\mathrm{G}}'''_k$ is $\vec{\mathrm{F}}$ L'Hospital describable in terms of ψH. By Lemma 2.16 $\chi(\vec{\mathrm{G}}'''_k)$ is $\vec{\mathrm{F}}$ L'Hospital describable in terms of H for $k > 1$. By Theorem 2.1, G'''_1 is F_1 L'Hospital describable in terms of $\sum F_k\chi(\vec{\mathrm{G}}'''_k) - H$. Hence, by Theorem 2.15, G'''_1 is $\vec{\mathrm{F}}$ L'Hospital describable in terms of H which is the desired result. ▯

PROOF OF THEOREMS 2.5 AND 2.6. We shall use induction. First we note that in the definition of the map ρ (preceding the statement of Theorem 2.6) there was an arbitrariness in the choice of the $\vec{\mathrm{G}}_j$. We now require that the $\vec{\mathrm{G}}_j$ be chosen so as to be $\boxed{\mathrm{F}}$ L'Hospital expressible in terms of $\vec{\mathrm{H}}$.

More precisely, we examine the step-by-step definition of $\vec{\mathrm{G}}$. For example, $\vec{\mathrm{G}}_1$ must satisfy (2.16). By Theorem 2.3 we can choose such a $\vec{\mathrm{G}}_1$ which is $\boxed{\mathrm{F}}$ L'Hospital expressible in terms of

$$H_1(z) - \sum Q^1_{ik}(s^{n-1})L^1_{ik}(w^i).$$

We now use the fact that the expression $\sum Q^1_{ik}(s^{n-i})L^1_{ik}(w^i)$ is, by Theorem 2.5, a L'Hospital expression in terms of the restriction of H_1 to $\mathfrak{B}_1$. Thus,

$\vec{G}_1$ is $\boxed{F}$ L'Hospital expressible in terms of $\vec{H}$. Similarly, for $\vec{G}_2, \ldots$. Thus, we may choose G to be $\boxed{F}$ L'Hospital expressible in terms of $\vec{H}$ [if we assume Theorem 2.5(n)].

T2.5(0) and T2.6(0, 1) are clear. We claim that T2.5(n) and T2.6(n, 1) imply T2.6 (n, l) for all $l > 0$. To prove this we return to the notation preceding the statement of Theorem 2.6. It is clear from the definitions that if $\vec{H} \in \boxed{F}\mathcal{O}^r$ then $\rho\vec{H} = 0$, so ρ is defined on $\mathcal{O}^l/\boxed{F}\mathcal{O}^r$. Conversely, if $\rho\vec{H} = 0$ then first H_1 is of the form $\sum F_{1j} G_{1j}$ and, in fact, by (2.20)

$$H_{k+1} - \sum F_{i+1,j} G_{ij} = \sum G_{k+1,j} F_{k+1,j}, \tag{2.58}$$

where $\vec{G}_{k+1} \in M_k$. (Note that the other term in (2.20) vanishes because $\rho\vec{H} = 0$.) This gives

$$\vec{H} = \boxed{F}\vec{G},$$

where $\vec{G} = \vec{G}_1 + \vec{G}_2 + \cdots + \vec{G}_l$, since $\vec{G}_{k+1} \in M_k$. Hence ρ is one-to-one. It is clearly onto. Thus, T2.5(n) and T2.6(n, 1) imply T2.6(n, l) for all r.

Thus, it remains to prove that T2.5($n-1$) and T2.6($n-1$, l) for all l imply T2.5(n) and T2.6(n, 1). We begin with T2.6(n, 1). We may suppose that $F_1 \not\equiv 0$. We make a linear transformation so that $s = z_n$ is distinguished for F_1. Then, as in the above proof of Theorem 2.3, we pass to the quotient $\mathcal{O}/F_1\mathcal{O}$.

LEMMA 2.17. *Using the notation of Lemma 2.13, ψ defines an isomorphism ψ' of*

$$\mathcal{O}/\vec{F}\mathcal{O}^r \quad \text{onto} \quad \mathcal{O}^m(n-1)/\boxed{F}^1\mathcal{O}^{(r-1)m}(n-1).$$

PROOF. By Lemma 2.13 the map is defined on the quotient $\mathcal{O}/\vec{F}\mathcal{O}^r$. Suppose $\psi' L = 0$. This means that $\psi L = \boxed{F}^1\vec{G}'''$ for some $\vec{G}''' \in \mathcal{O}^{(r-1)m}(n-1)$. Proceeding as in the part of the proof of Theorem 2.3 following Lemma 2.14, we deduce that $L \in \vec{F}\mathcal{O}^r$, so that ψ' is one-to-one. Finally, since ψ is onto $\mathcal{O}^m(n-1)$, ψ' is onto. ∎

Thus, to describe $\mathcal{O}/\vec{F}\mathcal{O}^r$, we begin with a description of $\mathcal{O}^m(n-1)/\boxed{F}^1\mathcal{O}^{(r-1)m}(n-1)$. By T2.6($n-1$, m) we can describe this quotient fully: Let $\mathfrak{W}_j$ be the multiplicity variety associated by T2.6($n-1$, 1) to the ideal of all sums $\sum_k F^1_{jk} N^1_{j-1,k}$ where F^1_{jk} is the jth row of $\boxed{F}^1$ and $\vec{N}^1_{j-1} \in M^1_{j-1}$, the module of simultaneous relations of the first $j-1$ rows of $\boxed{F}^1$. Then, for any $\vec{H} \in \mathcal{O}^m(n-1)$ and any j, we define ρ^1 by

$$(\rho^1\vec{H})_k = \text{restriction of } H_k - \sum_{i<k} \sum_j F_{i+1,j} G_{ij} \quad \text{to} \quad \mathfrak{W}_k, \tag{2.59}$$

where G_{ij} is described as in the remarks preceding the statement of Theorem 2.6, and satisfies the L'Hospital expressibility required above.

We need now

THEOREM 2.18. *Let* $\boxed{F}$ *be an* $l \times r$ *matrix of functions in* $\mathcal{O}$, *and let* $\vec{H} \in \mathcal{O}^l$. *Then* $\vec{H}$ *lies in the image* $\boxed{F}\mathcal{O}^r$ *if and only if the* H_j *satisfy the same differential relations as the rows of* $\boxed{F}$. *More precisely, there exist local Z varieties* X_i *and differential operators* d_{ij} *regular on* X_i *so that* $\vec{H} \in \boxed{F}\mathcal{O}^r$ *if and only if for any* i

$$\sum d_{ij} H_j(z) = 0 \quad \text{for } z \in X_i. \tag{2.60}$$

Here,

$$\sum d_{ij}\,(F_{jk}P) = 0 \quad \text{for } z \in X_i, \quad \text{all } i, k, \text{ and all } P \in \mathcal{O}.$$

The X_i *and* d_{ij} *depend only on* $\boxed{F}$.

PROOF. We shall prove Theorem 2.18 by induction on n simultaneously with Theorems 2.5 and 2.6. Thus, we denote the statement of the theorem more precisely by T2.18(n, l). T2.18(0, 0) is trivial. We shall complete our proofs of Theorems 2.5, 2.6, and 2.18 by showing

(2.61) T2.6(n, l) for all l implies T2.18(n, l) for all l.

(2.62) T2.18($n-1$, l) for all l implies T2.6(n, 1).

(2.63) T2.5($n-1$) and T2.6($n-1$, l) for all l imply T2.5(n).

Since we already know that T2.6(n, 1) and T2.5(n) imply T2.6(n, l) for all l, this gives the proof of all theorems T2.5(n), T2.6(n, l) and T2.18(n, l).

Proof of (2.61). For $l = 1$ the result is obvious; we use for the X_i and the d_i (note there is only one j) the Z varieties and operators used in the definition of the multiplicity variety corresponding to $\vec{F}$ by T2.6(n, 1).

For $l > 1$ we examine the step by step definition of ρ in Theorem 2.6. We use the same notation as there. If there is a $\vec{G}$ satisfying $\boxed{F}\vec{G} = \vec{H}$ then $\rho(\vec{H}) = 0$. Thus, (2.16) becomes

$$\sum F_{1j} G_{1j} = H_1. \tag{2.64}$$

By Theorem 2.3 we know that $\vec{G}_1$ is L'Hospital expressible by $\vec{F}_1$ (the first row of $\boxed{F}$) hence, $\boxed{F}$, in terms of H_1. (2.18) becomes

$$H_2 - \sum F_{2j} G_{1j} = \sum G_{2j} F_{2j}. \tag{2.65}$$

Again by Theorem 2.3, $\vec{G}_2$ is $\boxed{F}$ L'Hospital expressible in terms of H_2 and H_1.

Now, (2.64) is satisfied if and only if $H_1 \mid \mathfrak{V}_1 = 0$, which means that certain derivatives of H_1 are zero on certain varieties. Similarly, (2.65) says that certain derivatives of $H_2 - \sum F_{2j} G_{1j}$ are zero on appropriate varieties. Taking account of the L'Hospital expression of $\vec{G}_1$ by $\vec{F}_1$ in terms of H_1, this means that (2.65) holds if and only if certain linear differential relations hold between H_1 and H_2, where these differential relations depend only on $\boxed{F}$. Proceeding in this way and making use of Theorem 2.15 we derive varieties X_i and operators d_{ij} depending only on $\boxed{F}$ so that (2.60) is a necessary and sufficient condition for $\vec{H} \in \mathcal{O}^l$ to be in $\boxed{F}\mathcal{O}^r$.

We note that for fixed k the vector (F_{jk}) is in $\boxed{F}\mathcal{O}^r$, since it is $\boxed{F}\varepsilon_k$, where ε_k is the vector which is one in the kth place and zero elsewhere. Thus, for fixed k, F_{jk} satisfies (2.60). For the same reason, $F_{jk}P$ satisfies (2.60). This completes the proof of (2.61). ▯

See Remark 2.4.

Proof of (2.62). Let X_i, d_{kj} be the Z varieties and differential operators which correspond to $\boxed{F}^1$ by T2.18$(n-1, m)$. Let $K \in \mathcal{O}$ and consider ψK. By T2.18$(n-1, m)$ $K \in \vec{F}\mathcal{O}^r$ if and only if for every i

$$\sum d_{ij}(\psi K)_j(w) = 0 \quad \text{for } w \in X_i. \tag{2.66}$$

We now apply Assertion 3 of Theorem 2.8 to (2.66). This allows us, on a Z variety of maximal dimension, to express $(\psi K)_j$ in terms of values of K on a suitable variety. We now apply Theorem 2.15 which enables us to rewrite (2.66) in the form

$$\left[\sum_t (\vec{f}_k^t \circ d_k^t) \cdot K\right](w) = 0, \tag{2.67}$$

where $\vec{f}_k^t$ are algebroid maps regular on U^k and d_k^t are differential operators with coefficients regular on $\vec{f}_k^t(U^k)$. Here U^k are suitable Z varieties.

Now the sum in (2.67) is finite. We claim that for each fixed w and each fixed t the sum

$$\left[\sum_{t'}{}' (\vec{f}_k^{t'} \circ d_k^{t'}) \cdot K\right](w) = 0. \tag{2.68}$$

Here we sum only over those t' for which $\vec{f}_k^{t'}(w)$ has a fixed value, say $\vec{f}_k^t(w)$. For, since (2.67) holds for all K in an ideal, we need merely multiply K by a polynomial whose Taylor series at $\vec{f}_k^t(w)$ is $1 +$ high powers of $(z - \vec{f}_k^t(w))$ and which has a high order of vanishing at the other images $\vec{f}_k^{t''}(w)$, to see that (2.67) implies (2.68). Now (2.68) can clearly be rewritten as

$$\left\{\left[\vec{f}_k^t \circ \left(\sum{}' d_k^{t'}\right)\right] K\right\}(w) = 0. \tag{2.69}$$

(2.69) is thus a necessary and sufficient condition for K to belong to the ideal. (2.69) means that the restriction of certain differential operators (namely, $\sum' d_k^{t'}$) applied to K vanish on certain multiplicity varieties (namely, $\vec{\mathfrak{f}}_k^t(U^k)$), that is, the restriction of K to a multiplicity Z variety is zero. However, the differential operators $\sum' d_k^{t'}$ are not the same on all of U^k. Rather, they are the same on sets $W^t(w_0)$ where $W^t(w_0)$ is the set of all $w \in U^k$ at which the same set of t' satisfy $\vec{\mathfrak{f}}_k^{t'}(w) = \vec{\mathfrak{f}}_k^t(w)$ as satisfy $\vec{\mathfrak{f}}_k^{t'}(w_0) = \vec{\mathfrak{f}}_k^t(w_0)$. Now, given t, t' the set $W^{tt'}$ of w satisfying $\vec{\mathfrak{f}}_k^{t'}(w) = \vec{\mathfrak{f}}_k^t(w)$ is a Z variety by Proposition 1.12. Moreover, there are only finitely many $W^{tt'}$. Since $W^t(w_0)$ is the intersection of those $W^{tt'}$ (t' varying) which contain w_0, it follows that $W^t(w_0)$ is a Z variety and there are finitely many of them. This completes the proof of (2.62). ▯

Proof of (2.63). We return to the notation preceding the statement of Theorem 2.6, except that the $\boxed{\mathrm{F}}$ there is to be replaced by $\boxed{\mathrm{F}}^1$ here. (We shall use Lemma 2.17 and the result for principal ideals, namely, Theorem 2.8.) By T2.5($n-1$) we can write each element of $\mathcal{O}(\mathfrak{V}_t)$ uniquely as a restriction to $\mathfrak{V}_t$ in the form

$$\sum Q_{ik}^t(s_t^{n-k-1}) L_{ik}^t(w_t^k).$$

As usual, set $s = z_n$. We have three things to show:

(2.63a) Every $K \in \mathcal{O}/\vec{\mathrm{F}}\mathcal{O}^r$ can be written uniquely in the form

$$\sum Q_{ik}^t(s_t^{n-k-1}) L_{ik}^t(w_t^k) s^t.$$

(Note that s^t is the tth power of s while t is a superscript in Q_{ik}^t and L_{ik}^t and $\mathrm{n-k-1}$ is a superscript in s_t^{n-k-1}.)

(2.63b) We can choose the way of writing $z = (w_t^k, s_t^{n-k-1})$ independent of t. [So we may write merely (w^k, s^{n-k-1}).]

(2.63c) $L_{ik}^t(w^k)$ can be expressed in terms of K as in (2.15).

Proof of (2.63a). The method of definition of ρ shows how to define the L_{ik}^t in terms of $\vec{\mathrm{H}} \in \mathcal{O}^m$. It is clear that they are defined in terms of $H \in \mathcal{O}^m/\boxed{\mathrm{F}}^1\mathcal{O}^{m(r-1)}$ and that the map $H \to \{L_{ik}^t\}$ is one-to-one from $\mathcal{O}^m/\boxed{\mathrm{F}}^1\mathcal{O}^{m(r-1)}$ onto the direct sum of the $\mathcal{O}^{k_j}(j)$. Now, using Lemma 2.17 we have our result. ▯

Proof of (2.63b). To prove this we actually prove a result which is stronger than Theorem 2.5, namely, that given m ideals we can choose the corresponding splittings of z into (w_t^k, s_t^{n-k}) independently of $t = 1, 2, \ldots, m$. This result can be proved by the above procedure of reduction to principal ideals if we note the result of Lemma 2.7 that any finite number of $K_\alpha \in \mathcal{O}$ have a common distinguished direction. ▯

Proof of (2.63c). We first express $L^t_{ij}(w^j)$ by T2.5($n-1$) and the definitions, in terms of

$$K_t - \sum_{i<t} \sum_j F_{i+1,j} G_{ij}$$

and the multiplicity variety $\mathfrak{V}_t$, where $(K_t) = \psi K$. Using, as in the proof of (2.61), the L'Hospital expression of $\vec{G}_i$ by $\vec{F}$ in terms of K, we are led to expressions of the form

$$L^t_{ij}(w^j) = \sum_{a,q} \sum_{s^{n-j+1}} \tilde{f}^t_{iqpja}(w^j, s^{n-j-1})(\tilde{\partial}^t_{qa} K_a)(w^j, s^{n-j-1}) \tag{2.70}$$

where the $\tilde{f}^t_{iqpja}$ are algebroid on suitable Z varieties T^{tq}_{pja} and the sum is over (w^j, s^{n-j-1}) belonging to T^{tq}_{pja}.

Note that (2.70) is very similar to (2.66): (2.66) gives the condition, in terms of ψK, for K to belong to $\boxed{\mathrm{F}}^1 \mathcal{O}^r$ whereas (2.70) gives an expression for the L^t_{ij} in terms of ψK. We can now follow step by step the argument following (2.72) to prove (2.63c).

Finally, to complete the proof of (2.63) we must examine the action of $\mathcal{O}$ on the vectors $L^t_{ij}(w^i)$. Let $H \in \mathcal{O}$. Theorem 2.8 shows that there is a matrix $\boxed{\mathrm{H}} = (H_{tl})$ so that $\psi(HK) = \boxed{\mathrm{H}}\psi(K)$ for $K \in \mathcal{O}$. Thus, we need examine only how each H_{tl} acts on the set of $L^t_{ij}(w^i)$. But this action is given by T2.5($n-1$). The inductive step is now clear.

This completes the proof of (2.63) and, hence, of all the results of this chapter. □

See Remark 2.5.

Remarks

Remark 2.1. See page 39.

Remark 2.2. See page 50.

Remark 2.3. See page 59. Presumably it should not be difficult to find an explicit formula for D^2 in terms of the $F_j(w)$ when the $s_i = s_i(w)$ are the roots of $F(s, w) = 0$ [see (2.22)]. This would generalize the usual formula for the resolvant.

Remark 2.4. See page 68. A modification of the above method could be used to show that $\vec{\mathrm{H}} \in \boxed{\mathrm{F}}\mathcal{O}^r$ if $\vec{\mathrm{H}} \in \boxed{\mathrm{F}}\mathcal{O}^r$ where $\mathcal{O}$ is the ring of formal power series. This result is known by the theory of local rings (see Samuel [1]).

Remark 2.5. See page 70. The above method shows that, except for (2.15), Theorem 2.5 can be proved by induction without recourse to Theorem 2.6 or to the use of multiplicity varieties. As we shall see in Chapter IX, this part of Theorem 2.5 is related to a Cauchy-Kowalewski problem for overdetermined systems of partial differential equations. The fact that the proof is independent of multiplicity varieties will allow us to extend our results on the Cauchy-Kowaleski problem to some systems with variable coefficients.

Problems

Problem 2.1. See page 39.

Determine the conditions on a multiplicity variety $\mathfrak{V}$ in order that the kernel of the restriction map $H \to H \mid \mathfrak{V}$ be an ideal.

Determine conditions on multiplicity varieties $\mathfrak{V}$, $\mathfrak{V}'$ in order that the kernels of $H \to H \mid \mathfrak{V}$ and $H \to H \mid \mathfrak{V}'$ be equal.

Problem 2.2. (*Exercise*). See page 46.

Compute the multiplicity varieties corresponding to the ideal in C^3 generated by z_2^2, $(z_3 - 1)z_1^2 - z_2$, and also to the ideal generated by z_2^2, $z_2 - z_1 z_2$. Show that the corresponding differential operators *cannot* be chosen to have constant coefficients.

These examples were communicated to the author by Louis Boutet de Monvel. A similar type of example was constructed by Palamodov.

CHAPTER III

Semilocal Theory

Summary

We shall quantify the results in Chapter II. In order to do this we shall restrict our considerations to ideals and modules generated by polynomials. The quantifications we shall consider are of two kinds: (1) We shall study the geometric structure of ideals and modules generated by polynomials in the ring of holomorphic functions on a cube of *fixed size* (not, as in Chapter II, of arbitrarily small size.) (2) We shall find bounds on the representatives which we choose from the equivalence classes modulo such ideals and modules.

The results and methods are very similar to those of Chapter II except that for quantification (1) above we must make a careful study of the geometry of the roots of a polynomial in several variables, while for quantification (2) we must make extensive use of Theorem 1.4 which asserts that a polynomial is "usually" large.

III.1. Formulation of the Main Results

In this chapter we shall prove the analogs of the main results of Chapter II for cubes of a fixed size, rather than for arbitrarily small neighborhoods as in the local theory of Chapter II. One such possible extension can be obtained by the "piecing together" process of Oka and Cartan-Serre (see e.g., Cartan and Serre [1]). However, this is not of interest to us here because we need quantitative results. For this purpose we must assume that all the ideals and modules we deal with are generated by polynomials. "Variety" ("Z variety") will therefore mean "algebraic variety," ("algebraic Z variety"). "Differential operator" will mean "differential operator with algebraic or polynomial coefficients." Similarly for multiplicity variety. Also, in this chapter L'Hospital representation will be defined by algebraic Z varieties and algebraic or rational factors, etc.

The main results in this chapter will be derived in essentially the same manner as the corresponding results in Chapter II except that we must estimate certain bounds. This estimation will usually be based on a modification of Theorem 1.4. In order to understand our calculations the reader should bear in mind the fact that we shall be interested in making estimates modulo polynomial factors, provided that these polynomial factors depend only on the module considered (that is, only on $\boxed{\mathrm{F}}$).

For each $\alpha \in C^t$, $\beta > 0$ we denote by $\gamma(t; \alpha; \beta)$ the open cube in C^t center α, side 2β. When there is no possible confusion we write $\gamma(\alpha; \beta)$, or $\gamma(\alpha)$, or simply γ. More generally, if β is a t-tuple, $\beta = (\beta_1, \ldots, \beta_t)$, then we define $\gamma(\alpha; \beta)$ as the set of z such that $|z_j - \alpha_j| < \beta_j$ for all j. In particular, we might let some of the $\beta_j = \infty$. We call $\mathcal{O}(t; \alpha; \beta)$ the space of functions which are holomorphic on $\gamma(t; \alpha; \beta)$. The subspace of bounded functions in $\mathcal{O}(t; \alpha; \beta)$ is a Banach space and for any $A > 0$ we denote by $\mathcal{O}(t; \alpha; \beta; A)$ the set of those functions whose maximum modulus is less than or equal to A.

If V is an algebraic variety, then we define $\mathcal{O}(V; t; \alpha; \beta)$ as the space of functions which are holomorphic (single valued) on $V \cap \gamma(t; \alpha; \beta)$. If $\mathfrak{V}$ is a multiplicity variety $\mathfrak{V} = (V_1, \partial_1; \ldots; V_a, \partial_a)$, we define $\mathcal{O}(\mathfrak{V}; t; \alpha; \beta)$ as the space of $\vec{\mathrm{G}} = (G_1, \ldots, G_a)$, where each G_i is an algebroid function regular on $V_i \cap \gamma(t; \alpha; \beta)$ and the G_i satisfy the compatibility conditions described in Section I.4. Of course, if $\mathfrak{V}$ is a polynomial multiplicity variety, we do not have to consider multivalued functions. $\mathcal{O}(V; t; \alpha; \beta; A)$ consists of all holomorphic functions on $\gamma(t; \alpha; \beta) \cap V$ which are $\leqslant A$ there. $\mathcal{O}(\mathfrak{V}; t; \alpha; \beta; A)$ consists of all $\vec{\mathrm{G}} = (G_1, \ldots, G_a) \in \mathcal{O}(\mathfrak{V}; t; \beta; \alpha)$ for which $|\vec{\mathrm{G}}| = \sum |G_i| \leqslant A$ on $V_i \cap \gamma(t; \alpha; \beta)$. Here when we write $\sum |G_i| \leqslant A$ we mean an inequality on each of the possible choices of the multivalued function $\vec{\mathrm{G}}$.

For V' a Z variety we define $\mathcal{O}(V'; t; \alpha; \beta)$ as equal to $\mathcal{O}(\tilde{V}; t; \alpha; \beta)$ where $\tilde{V}$ is the Z closure of V' (as in Section I.4). As in I.4, we can also define $\mathcal{O}(\mathfrak{V}; t; \alpha; \beta)$ if $\mathfrak{V}$ is a multiplicity Z variety.

It is important to define $\mathcal{O}(\mathfrak{V}; t; \alpha; \beta; A)$ when $\mathfrak{V}$ is a multiplicity Z variety, say $\mathfrak{V} = (V_1, \partial; \ldots; V_a, \partial_a)$. Since the functions in $\mathcal{O}(\mathfrak{V})$ can have singularities at the holes in V_i, we cannot define this directly, since the functions can be unbounded on $V \cap \gamma(t; \alpha; \beta)$. For example, for $n = 1$, $\mathfrak{V} = (C - \{0\}, 1/z)$, we want the function $1/z$ which belongs to $\mathcal{O}(\mathfrak{V})$ to belong to $\mathcal{O}(\mathfrak{V}; 1; 0; 1; A)$ for some A. The nature of the bounds we need is such that we can allow division by a fixed set of polynomials even though they may have zeros. The reason for this will become clear later, but its nature is similar to Theorem 1.4.

Let R_i be polynomials such that $R_i\, \partial_i$ has locally bounded algebraic coefficients in all of C for each i. These exist by the algebraic analog of Proposition 1.9. Let $H \in \mathcal{O}$ so $(H_i) = (\partial_i H) \in \mathcal{O}(\mathfrak{B})$. Note, however, that $R_i \, \partial_i H$ is locally bounded on V_i. Thus, $(R_i H_i)$ is in $\mathcal{O}(\tilde{\mathfrak{B}})$ where $\tilde{\mathfrak{B}}$ is the multiplicity Z variety

$$\tilde{\mathfrak{B}} = (V_1, R_1 \, \partial_1; \ldots; V_a, R_a \, \partial_a).$$

If, in addition, R_i does not vanish identically on any component of V_i for any i, then the map $(H_i) \to (R_i H_i)$ is one-to-one from $\mathcal{O}(\mathfrak{B})$ onto $\mathcal{O}(\tilde{\mathfrak{B}})$. We shall denote $\tilde{\mathfrak{B}}$ by $R\mathfrak{B}$, where $R = (R_1, \ldots, R_a)$. We shall call R *admissible* (for $\mathfrak{B}$) if the R_i are as above, namely, $R_i \, \partial_i$ have locally bounded algebraic coefficients and R_i does not vanish identically on any component of V_i.

For our purpose it will be sufficient for bounds on a multiplicity Z variety $\mathfrak{B}$ to use $\mathcal{O}(R\mathfrak{B}; A)$ where R is admissible for $\mathfrak{B}$; we shall, in fact, denote this by $\mathcal{O}(\mathfrak{B}; A)$ when no confusion is possible.

As in Chapter II, we formulate our results first for ideals, that is, $l = 1$:

THEOREM 3.1. *Let $\vec{F}$ be a vector with r components which are polynomials. Then, given any α and any $\beta > 0$, there exists a polynomial multiplicity variety $\mathfrak{B}$ (sometimes denoted by $\mathfrak{B}(\alpha; \beta)$) with the following properties: Given any $A > 0$ there exist $\beta' > 0$, $A' > 0$ so that, if $\rho_{\mathfrak{B}}$ denotes the restriction map to $\mathfrak{B}$, then*

$$\rho_{\mathfrak{B}} \mathcal{O}(\alpha; \beta; A) \supset \mathcal{O}(\mathfrak{B}; \alpha; \beta'; A'). \tag{3.1}$$

There exist c, $c' > 0$ so that we may choose

$$\beta \leqslant \beta' < c(1 + \beta + |\alpha|)^c \beta^c \tag{3.2}$$

$$A \geqslant A' > \frac{A}{c(1 + |\alpha|)^c (1 + \beta + \beta^{-1})^c}. \tag{3.3}$$

Here c, c' depend only on $\vec{F}$. Moreover, if $0 < \beta'' < \beta$, then

$$\rho_{\mathfrak{B}} \mathcal{O}(\alpha; \beta; A) \subset \mathcal{O}(\mathfrak{B}; \alpha; \beta''; A''), \tag{3.4}$$

where

$$A'' \leqslant cA(1 + |\alpha|)^c (1 + |\beta - \beta''|^{-1})^c. \tag{3.5}$$

The $\mathfrak{B}(\alpha; \beta)$ for varying α and β can be chosen from a finite set of polynomial multiplicity varieties.

The kernel of $\rho_{\mathfrak{B}}$ on $\mathcal{O}(\alpha; \beta)$ contains $\vec{F}\mathcal{O}^r(\alpha; \beta)$ while the kernel of $\rho_{\mathfrak{B}}$, considered as a map into $\mathcal{O}(\mathfrak{B}; \alpha; \beta)$, on $\mathcal{O}(\alpha; \beta')$ is exactly $\vec{F}\mathcal{O}^r(\alpha; \beta) \cap \mathcal{O}(\alpha; \beta')$.

Note that $\mathcal{O}(\mathfrak{B}; \alpha; \beta'; A')$ increases as β' decreases and A' increases. Thus (3.2) and (3.3) say that the right side of (3.1) is large.

The fact that (3.4) and (3.5) hold is a simple consequence of the definitions and Cauchy's formnla. The proof of (3.1) is very difficult and will take the remainder of this chapter. It is patterned after the proof of Theorem 2.4 except that we have the complication of bounds.

Theorem 3.1 will be called the *semilocal quotient structure theorem* (for ideals).

Remark 3.1. Just as in the case of Theorem 2.4, in proving Theorem 3.1 we shall actually prove the apparently weaker result in which we obtain an algebraic multiplicity Z variety $\mathfrak{B}'$ instead of a polynomial multiplicity variety $\mathfrak{B}$. The proof that this implies Theorem 3.1 goes as follows: The fact that the kernel of $\rho_{\mathfrak{B}'}$ can be replaced by the kernel of $\rho_{\mathfrak{B}}$ for a suitable $\mathfrak{B}$ is proven in exactly the same fashion as the fact that Theorem 2.4′ implies Theorem 2.4. For that $\mathfrak{B}$ there is a way of mapping $\mathcal{O}(\mathfrak{B}'; \alpha; \beta)$ one-to-one onto $\mathcal{O}(\mathfrak{B}; \alpha; \beta)$ obtained by expanding the coefficients of the operators for $\mathfrak{B}'$ in a suitable basis and then multiplying by suitable polynomials.

As far as bounds (that is, A') go, the polynomials which we multiply by play no role because of the definition of $\mathcal{O}(\mathfrak{B}'; \alpha; \beta; A')$. On the other hand, *a priori*, the expansion of the coefficients of the operators for $\mathfrak{B}'$ might play a role. If we go step by step through the proof that Theorem 2.4′ implies Theorem 2.4 (following the statement of Theorem 2.4), making the modifications needed for the semilocal situation, we see that what we have to know is:

$$\left|\sum_k \left(\sum_j u_{jk} H_j\right) v_k\right| \leqslant A' \quad \text{on} \quad W'' \cap \mathcal{O}(\alpha; \beta)$$

is implied by

$$\left|\sum_j u_{jk} H_j\right| \leqslant A''' \quad \text{on} \quad V \cap \mathcal{O}(\alpha: \beta)$$

for every k. Here A''' satisfies (3.3) with A''', A' replacing A', A. For then we would have

$$\mathcal{O}(\mathfrak{B}'; \alpha; \beta'; A') \supset \mathcal{O}(\mathfrak{B}; \alpha; \beta'; A''')$$

which means we can replace $\mathfrak{B}'$ on the right side of (3.1) by $\mathfrak{B}$.

In the present case we can assume that the v_k are polynomials, depending only on $\vec{\mathrm{F}}$, so the desired implication is clear. The rest of Theorem 3.1 presents no difficulties. ∎

We shall not distinguish between Theorem 3.1 and the corresponding result for algebraic multiplicity Z varieties.

See Remark 3.2.
See Problem 3.1.

THEOREM 3.2. *With the notations as above, given any α and any $\beta > 0$ there exists $\beta' > 0$ with the following properties:*

There exist integers $k_0, k_1, \ldots, k_n$ so that $\mathcal{O}(\mathfrak{B}; n; \alpha; \beta')$ is contained in a certain well-defined subspace $\mathcal{W}(k_0, \ldots, k_n; \alpha; \beta)$ of the direct sum

$\mathcal{O}^{k_0}(0;\alpha;\beta)\oplus\mathcal{O}^{k_1}(1;\alpha;\beta)\oplus\cdots\oplus\mathcal{O}^{k_n}(n;\alpha;\beta)$. *For any A there is an A' so that (with the usual notation)*

$$\mathcal{O}(\mathfrak{V};n;\alpha;\beta';A')\subset \mathscr{W}(k_0,\ldots,k_n;\alpha;\beta;A). \tag{3.6}$$

We may choose β' and A' to satisfy (3.2) *and* (3.3).

Let us use the notation of Chapter II. For each j there exist k_j polynomials $Q_{ij}(s^{n-j})$ so that every $H\in\mathcal{O}(\mathfrak{V};n;\alpha;\beta')$ can be written uniquely on $\gamma(n;\alpha;\beta)$ as the restriction to $\mathfrak{V}$ of a function

$$K\in\mathcal{O}(n;\alpha;\beta)$$

of the form

$$K(z)=\sum_{i,j}Q_{ij}(s^{n-j})L_{ij}(w^j), \tag{3.7}$$

where $L_{ij}\in\mathcal{O}(j;\alpha;\beta)$ and $\{L_{ij}\}\in\mathscr{W}(k_0,\ldots,k_n;\alpha;\beta)$. Moreover if $H\in\mathcal{O}(\mathfrak{V};n;\alpha;\beta';A')$, then we have $\{L_{ij}\}\in\mathscr{W}(k_0,\ldots,k_n;\alpha;\beta;A)$.

The L_{ij} can be expressed in terms of K in the manner of (2.15). *The action of multiplication of $H\in\mathcal{O}(n;\alpha;\beta)$ on $\mathcal{O}(\mathfrak{V};\alpha;\beta)$ can be described as in the statement of Theorem* 2.5. *The subspace $\mathscr{W}$ can be described as follows: There exists a $J\in\mathcal{O}(n;\alpha;\beta)$ so that, with the above described action of multiplication of J on $\mathcal{O}^{k_0}(0;\alpha;\beta)\oplus\cdots\oplus\mathcal{O}^{k_n}(n;\alpha;\beta)$, $\mathscr{W}$ is the image space of J.*

$\mathfrak{V}$, or more precisely $\mathfrak{V}(\alpha,\beta)$, depends on α, β but we can choose a finite number from which all $\mathfrak{V}(\alpha,\beta)$ can be chosen. The numbers $k_0,\ldots,k_n$ depend on α and β but can be bounded independently of α and β. Even more: There exists a finite sequence of polynomials $\{Q\}$ from which all $\{Q_{ij}\}$ for all α, β, i, j can be chosen. Similarly, the number of f_{iqtj} of our analog of (2.15) *is finite.*

We can proceed as in the remarks following Theorem 2.5 to describe a map $\rho_{\vec{\mathfrak{V}}}(\alpha;\beta)$. We use the fact that (see Serre [1]) for any α, β the module of relations of a module generated by polynomials in any $\mathcal{O}(\alpha;\beta)$ is generated by the (global) module of relations in the ring of polynomials.

THEOREM 3.3. *Let the notation be as in Theorem* 3.1.

$$\rho_{\vec{\mathfrak{V}}}(\alpha;\beta)\mathcal{O}^l(\alpha;\beta;A)\supset\sum_{\oplus}\mathcal{O}(\mathfrak{V}_j;\alpha;\beta';A'). \tag{3.8}$$

(Here $\mathfrak{V}_j$ are constructed as in Theorem 2.6). *Moreover*

$$\rho_{\vec{\mathfrak{V}}}(\alpha;\beta)\mathcal{O}^l(\alpha;\beta;A)\subset \text{the direct sum of } \mathcal{O}(\mathfrak{V}_j;\alpha;\beta'';A''). \tag{3.9}$$

The $\vec{\mathfrak{V}}(\alpha;\beta)$ can be chosen from a finite set of vector polynomial multiplicity varieties.

The kernel of $\rho_{\vec{\mathfrak{V}}}(\alpha;\beta)$, considered as a map into $\mathcal{O}_l(\alpha;\beta)$, on $\mathcal{O}^l(\alpha;\beta')$, is exactly $\boxed{F}\,\mathcal{O}^r(\alpha;\beta)\cap\mathcal{O}^l(\alpha;\beta')$. The kernel of $\rho_{\vec{\mathfrak{V}}}(\alpha;\beta)$ on $\mathcal{O}^l(\alpha;\beta)$ contains $\boxed{F}\,\mathcal{O}^r(\alpha;\beta)$.

The proof of (3.9) is simple (the same as the proof of (3.4)) and will be omitted. To prove the remaining parts of our results we shall use simultaneous induction as in Chapter II.

III.2. L'Hospital Representation

THEOREM 3.4. *Let F be a polynomial. Then we can find $U_0, \ldots, U_p$; $d_0, \ldots, d_p$ where U_j is a Zariski open set of an algebraic variety, and where d_j is a linear partial differential operator with rational coefficients, these coefficients being regular on U_j such that*

1. *The U_j are distinct and their union is C.*
2. *Let α, β be given and let $H \in \mathcal{O}(\alpha; \beta)$ be divisible by F in the ring $\mathcal{O}(\alpha; \beta)$. Then for each $z \in \gamma(\alpha; \beta)$, if $z \in U_j$,*

$$H(z)/F(z) = (d_j H)(z). \tag{3.10}$$

3. *There exists a $c > 0$ depending only on F so that if $0 < \beta'' < \beta$ and $H \in \mathcal{O}(\alpha; \beta; A)$, then $H/F \in \mathcal{O}[\alpha; \beta''; cA(1 + |\beta - \beta''|^{-1})^c]$.*

PROOF. By a linear change of variables we may assume that F has the form (2.2) where the F_j are polynomials. Then the construction of U_j and d_j is as in the proof of Theorem 2.1. Statement (3) is a consequence of Theorem 1.4. ▯

Again, as in Chapter II (see Theorem 2.27) we need extensions of Theorem 3.4 to modules:

THEOREM 3.5. *Let $\vec{F} = (F_j)$ where the F_j are polynomials. Then there exists U^k, d^t_{jk}, f^t_{ak} where the U^k are Zariski open subsets of algebraic varieties, the f^t_{ak} are integral algebraic functions which are regular on U^k, and the d^t_{jk} are differential operators with algebraic coefficients which are regular on $\vec{f}^t_k U^k$ such that*

1. *The U^k are disjoint and their union is C.*
2. *Let α, β, A be given. Then there exist β', A' satisfying (3.2) and (3.3) so that if $H \in \mathcal{O}(\alpha; \beta'; A')$ is of the form $\vec{F}\vec{G}'$ where $\vec{G}' \in \mathcal{O}^r(\alpha; \beta')$, we can choose $\vec{G} \in \mathcal{O}^r(\alpha; \beta; A)$ such that*

$$H = \vec{F}\vec{G}.$$

Suppose $z \in \gamma(\alpha; \beta)$. Let $z \in U^k$. Then

$$\vec{G} = (\sum_t \vec{f}^t_k \circ d^t_k) \cdot H \tag{3.11}$$

When $\vec{G}$ is so representable we say that $\vec{G}$ is $\vec{F}$ *L'Hospital describable in terms of H.*

Now, using the procedure of Chapter II following Theorem 2.2 and using the above-mentioned result that the module of relations in $\mathcal{O}(\alpha; \beta)$ of a polynomial module is generated by polynomials, we deduce

THEOREM 3.6. *Let* α, β, A *be given. Then there exist* β', A' *satisfying* (3.2) *and* (3.3) *so that if* $\vec{H} \in \mathcal{O}^l(\alpha; \beta'; A')$ *is of the form* $\boxed{F}\vec{G} = \vec{H}$, *where* $\vec{G} \in \mathcal{O}^r(\alpha; \beta')$, *then we can find a* $\vec{G} \in \mathcal{O}^r(\alpha; \beta; A)$ *with* $\boxed{F}\vec{G} = \vec{H}$ *such that* $\vec{G}$ *is* $\boxed{F}$ *L'Hospital describable in terms of* $\vec{H}$.

As in Chapter II the main step in the induction argument will be the proof of the main results for principal ideals.

III.3. Connectability

DEFINITION. We say that z_n is *noncharacteristic* for the polynomial K if $K(z)$ is of the form

$$az_n^q + K_1(z_1, \ldots, z_{n-1})z_n^{q-1} + \cdots + K_q(z_1, \ldots, z_{n-1}),$$

where $q =$ degree K, that is, degree $K_j \leqslant j$. Here a is a nonzero constant. (We have defined noncharacteristic in terms of a fixed coordinate system. If we were interested in an intrinsic definition, we should define noncharacteristic as a direction rather than as a hyperplane.)

DEFINITION. Let α, β be given and let $\alpha' = (\alpha_1, \ldots, \alpha_{n-1})$. Let K be a polynomial and let V be the algebraic variety of zeros of K. Let z', z'' be two points in V such that $(z_1', \ldots, z_{n-1}')$ and $(z_1'', \ldots, z_{n-1}'') \in \gamma(n-1; \alpha'; \beta)$. We say that z' and z'' are *connectable over* $\gamma(n-1; \alpha'; \beta)$ if there exists a curve lying in V joining z' and z'', all of whose points z satisfy $(z_1, \ldots, z_{n-1}) \in \gamma(n-1, \alpha'; \beta)$.

See Fig. 4.

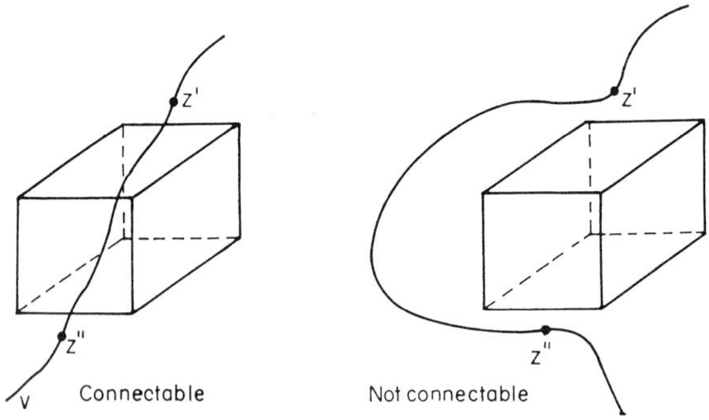

FIGURE 4.

LEMMA 3.7. *Given any polynomial $K \not\equiv 0$ we can find nonsingular linear transformations Λ of C such that*

1. *z_n is noncharacteristic for $K\Lambda$.*

Let us apply Λ so we assume z_n is noncharacteristic for K. Then

2. *For any $\beta > 0$ and any $c'' \geqslant 0$ and any $\alpha \in C$, we can find a $\beta' > \beta$ satisfying (3.2) such that for any β'' with $0 \leqslant \beta' - c'' \leqslant \beta'' \leqslant \beta'$, if $z' \in \gamma(\alpha; \beta'') \cap V$ and $(z'_1, \ldots, z'_{n-1}) \in \gamma(n-1; \alpha'; \beta)$ is connectable over $\gamma(n-1; \alpha'; \beta)$ to a z'', where $(z''_1, \ldots, z''_{n-1}) \in \gamma(n-1; \alpha'; \beta)$, then $|z''_n - \alpha_n| \leqslant \beta''$. Here $\alpha' = (\alpha_1, \ldots, \alpha_{n-1})$.*

Moreover, if we are given countably many polynomials $K_1, K_2, \ldots$ then we can choose Λ so that statement 1 holds simultaneously for all K_j. If the number of K_j is finite, we can then choose β' satisfying statement 2 for all K_j.

See Fig. 5.

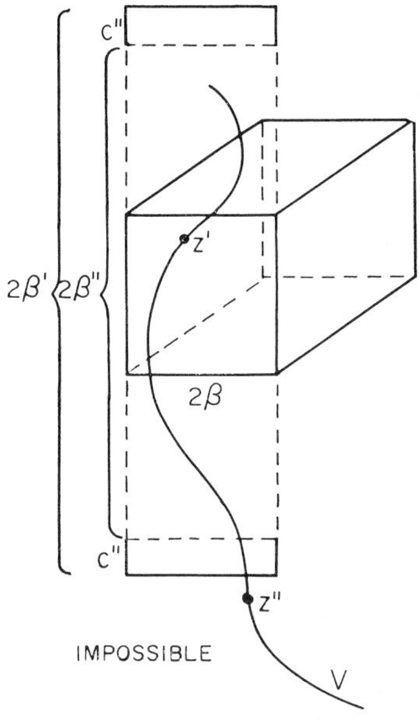

FIGURE 5.

Remark 3.3. We shall have to apply statement (1) of Lemma 3.7 many times. These applications will not interfere with each other for the same reason as that given in Remark 2.2.

Note, however, that a linear transformation does not take a rectangular parallelepiped into a reactangular parallelepiped. On the other hand, Theorems 3.1, 3.2, etc. are stated in terms of cubes (or rectangular parallelepipeds) and hence are *not* invariant under linear transformation. This difficulty can be overcome as follows: In the proofs we shall apply statement (1) of Lemma 3.7 simultaneously to translates by α of polynomials for all lattice points α. Since the number of polynomials that enter is finite and the number of α is countable, this is possible by the last paragraph of Lemma 3.7. (Actually, it is clear that if z_n is noncharacteristic for K then it is noncharacteristic for any translate of K.) On the other hand, by this means we shall actually prove a slightly different version of Theorems 3.1, 3.2, etc., in which the sets $\gamma(\alpha; \beta)$ are replaced by the images under the translate by α of a *fixed* nonsingular linear transformation Λ (independent of α and β) of $\gamma(0; \beta)$. Since Λ is nonsingular, there exists a constant $a \geqslant 1$ such that

$$a^{-1}\Lambda[\gamma(0; \beta)] \subset \gamma(0; \beta) \subset a\Lambda[\gamma(0; \beta)].$$

Since the sizes of the rectangular parallelepipeds in Theorems 3.1, 3.2, etc. are insensitive to multiplication by a or a^{-1}, the transformation Λ introduces no difficulties.

In the formulation of statement (2) of Lemma 3.7 we could have been more precise: Instead of considering z_n as noncharacteristic for K, we could have applied the translate to α of $\Lambda(K)$. The same results and, in fact, the same proof hold in this case. The situation is similar as regards all results proved in this chapter. For this reason we shall be somewhat cavalier in our usage of statement (1) of Lemma 3.7, that is, we have written the proof as though the theorems were invariant under linear transformation. It is always to be understood that what we have in mind precisely is the simultaneous application of the translate to α of $\Lambda(K)$ for all lattice points α.

PROOF. Statement 1 is well known (see Bochner and Martin [1]). In fact, all nonsingular linear transformations except those belonging to a proper subvariety will transform K into a polynomial for which z_n is noncharacteristic. The statement about countably many K_j follows.

To prove statement 2 we fix $w^0 = (w_1^0, \ldots, w_{n-1}^0) \in C^{n-1}$ and study the roots of $K(w^0, z_n) = 0$. There are q of them, say $z_n^i(w^0)$. Then the roots may be, heuristically, divided into classes consisting of those that are close together. We now move w away from w^0, but we keep $|w - w^0| < \beta$. Then the roots $z_n^i(w)$ move continuously, and, in fact, in a Lipschitz fashion. More precisely we must have

$$|z_n^i(w) - z_n^i(w^0)| \leqslant 4q\lambda\delta^{1/n} \tag{3.12}$$

as follows from Ostrowski [1]. Here

$$\lambda = \max_{|w - w^0| \leqslant \beta} (1, |K_0(w)|, \ldots, |K_q(w)|^{1/q})$$

and

$$\delta = [\sum |K_i(w^0) - K_i(w)|^2]^{1/2}.$$

We now define the rectangular parallelepipeds

$$S_i(\beta) = \{z | \, |w - w^0| \leqslant \beta \quad \text{and} \quad |z_n - z_n^i(w^0)| \leqslant 4q\lambda\delta^{1/n}\}.$$

We call $T_i(\beta)$ the connected component of $\cup_i S_i(\beta)$ which contains $(w^0, z^i(w^0))$.

The above shows that for each i there is an a_i so that for any w with $|w - w^0| \leqslant \beta$ there are exactly a_i zeros (with multiplicity) of $K(w, z_n) = 0$ in $T_i(\beta)$. Thus, there are no zeros outside $\cup\, T_i(\beta)$. Since the $T_i(\beta)$ are closed disjoint sets and since the roots $z_n(w)$ of $K(w, z_n)$ depend continuously on w, this means that, for $i \neq j$, the roots in $T_i(\beta)$ are not connectable to those in $T_j(\beta)$ over $|w - w^0| \leqslant \beta$. (See Fig. 6.)

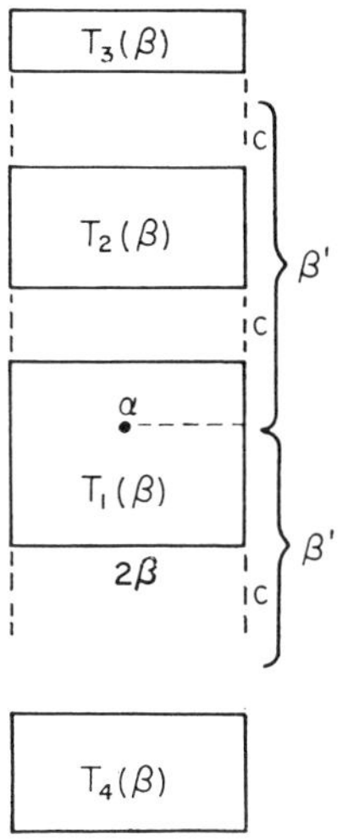

FIGURE 6.

We now define β' so that the situation is as in the figure, namely, no $T_j(\beta)$ meets $|z_n - \alpha_n| < \beta'$ unless that $T_j(\beta)$ is contained in the rectangular parallelepiped L: $|w - \alpha'| < \beta$, $|z_n - \alpha_n| < \beta'$. Moreover, every $T_j(\beta)$ which is contained in L is also contained in $|w - \alpha'| < \beta$, $|z_n - \alpha_n| < \beta' - c''$. It is clear from the above inequalities that β' exists satisfying (3.2). □

We shall often (when no confusion can arise) denote by β' the vector $(\beta, \beta, \ldots, \beta, \beta')$. Thus $\gamma(\alpha, \beta')$ is the rectangular parallelepiped $|w - \alpha'| < \beta$, $|z_n - \alpha_n| < \beta'$.

III.4 Principal Ideals

THEOREM 3.8. *Let F be a polynomial of degree m and let $\beta > 0$. Let $F = P_1^{a_1} \cdots P_b^{a_b}$ be the prime power decomposition of F in the ring of polynomials. For each j call V_j' the algebraic variety of zeros of P_j. Let β' be associated to β and to $c'' = 2$ and simultaneously to all the P_j by Lemma 3.7. We assume, by Lemma 3.7, that each P_j has $z_n = s$ as a noncharacteristic.*

For each α, $\beta > 0$, $\mathcal{O}(n; \alpha; \beta')/(F\mathcal{O}(n; \alpha; \beta'))$ *is isomorphic as a vector space to a well-defined subspace* $\mathcal{W}^m(n-1; \alpha; \beta)$ *of* $\mathcal{O}^m(n-1; \alpha'; \beta)$ *where* $\alpha' = (\alpha_1, \ldots, \alpha_{n-1})$. ($\mathcal{W}^m$ *depends on* α *and not just* α'.) *Under this isomorphism* $\psi(\alpha, \beta')$, *multiplication by* $B \in \mathcal{O}$ *gets mapped into multiplication by a matrix* $\boxed{B} = (B_{ij})$ *which is not uniquely determined by* B, F, α, β. *If* B *is a polynomial, then* $\boxed{B}$ *may be chosen to be a matrix of polynomials which is independent of* α, β.

For each α, β *there exists a function* $J(\alpha; \beta') \in \mathcal{O}(n; \alpha; \beta')$ *such that* $\mathcal{W}^m$ *is the subspace of* $\mathcal{O}^m$ *which is the image under multiplication by* $\boxed{J}(\alpha; \beta')$ *which is a projection.* $J(\alpha, \beta')$ *is an algebraic function and the number of* $J(\alpha, \beta')$ *for all* α, β *is finite.*

For any $A > 0$ *there is an* A' *satisfying* (3.3) *such that*

$$\psi\mathcal{O}(n; \alpha; \beta'; A') \subset \mathcal{W}^m(n-1; \alpha; \beta - 1; A). \tag{3.13}$$

Let $\mathfrak{V}'$ *be the multiplicity variety*

$$\mathfrak{V}' = \{(V_1', \text{identity}), (V_1', \partial/\partial s), \ldots, (V_1', \partial^{a_1-1}/\partial s^{a_1-1}), \ldots, (V_b', \partial^{a_b-1}/\partial s^{a_b-1})\}. \tag{3.14}$$

For any A *there is an* A' *satisfying* (3.3) *such that*

$$\rho_{\mathfrak{V}'}\mathcal{O}(\alpha; \beta' - 1; A) \supset \mathcal{O}(\mathfrak{V}'; \alpha; \beta'; A'). \tag{3.15}$$

The kernel of $\rho_{\mathfrak{V}'}$ *on* $\mathcal{O}(\alpha; \beta')$ *is just* $F\mathcal{O}(\alpha; \beta')$.

Let $K \in \mathcal{O}(\mathfrak{V}'; n; \alpha; \beta'; A')$. *Then there exists an extension of* K *to* $L \in \mathcal{O}(n; \alpha; \beta'; A)$, *where*

$$L(w, s) = L_0(w) + L_1(w)s + \cdots + L_{m-1}(w)s^{m-1} \tag{3.16}$$

with $(L_0, \ldots, L_{m-1}) \in \mathcal{W}^m(n; \alpha; \beta' - 1; A)$.

There exist $T_1, \ldots, T_p$ *which are disjoint algebraic* Z *varieties in* C^{n-1} *such that*

1. *The union of* T_t *is all of* C^{n-1}.

There exists a multiplicity Z *variety* $\mathfrak{V} = \{(V_q, \partial_q)\}$ *such that*

2. (3.15) *holds with* $\mathfrak{V}'$ *replaced by* $\mathfrak{V}$. *Again the kernel of* $\rho_{\mathfrak{V}}$ *on* $\mathcal{O}(\alpha; \beta')$ *consists of multiples of* F.
3. *Let* T_t^q *denote the* Z *varieties of the form* $w \in T_t$, $(w, s) \in V_q$. *Then we can find algebraic functions* f_{iqt} *such that*

$$L_i(w) = \sum_q \sum_{(w,s)\in T_t^q \cap \gamma(\alpha;\beta')} f_{iqt}(w, s)(\partial_q K)(w, s). \tag{3.17}$$

Here again, the number of $\mathfrak{V}$, T_t, *and* f_{iqt} *for varying* α, β *is finite.*

PROOF. We shall follow the lines of proof of Theorem 2.8. The first point is the definition of $J(\alpha, \beta')$. For this we need

LEMMA 3.9. *The number of zeros with multiplicity of $F(w, s)$ for $w \in \gamma(n-1; \alpha'; \beta)$ and $(w, s) \in \gamma(n; \alpha; \beta')$ is constant.*

PROOF. This is an immediate consequence of Lemma 3.7 and the fact that the roots of a polynomial in one variable depend continuously on the coefficients.

Let us call $\delta = (\beta, \ldots, \beta, \infty)$. Then we define $\tilde{J}(\alpha; \beta')$ on $\mathfrak{V}' \cap \gamma(\alpha; \delta)$ by

$$(3.18) \qquad \tilde{J}(\alpha, \beta'; w, s) = \begin{cases} \text{restriction of 1 to } \mathfrak{V}' \text{ for } (w, s) \in \gamma(\alpha, \beta') \cap \mathfrak{V}' \\ 0 \text{ otherwise.} \end{cases}$$

LEMMA 3.10. $\tilde{J} \in \mathcal{O}(\mathfrak{V}'; \alpha; \delta)$.

PROOF. By definition, we must show that $\tilde{J}$ is the restriction to $\mathfrak{V}'$ of a function in $\mathcal{O}(\alpha; \delta)$. It is well known (see Cartan and Serre [1], Oka's theorem) that it is sufficient to show that $\tilde{J}$ is locally (that is, in the ring of holomorphic functions at each point of $\gamma(\alpha; \delta)$) the restriction to $\mathfrak{V}'$ of a local holomorphic function. For points in $\gamma(\alpha; \beta')$ it is the restriction of 1, and for points not in $\gamma(\alpha; \beta')$ it is the restriction of 0. There is no compatibility condition because

$$(3.19) \qquad [\bar{\gamma}(\alpha; \beta') - \gamma(\alpha; \beta')] \cap \mathfrak{V} \cap \gamma(\alpha; \delta) = \varnothing,$$

where $\bar{\gamma}$ is the closure of γ and $\varnothing$ is the empty set, which is a consequence of the definition of connectable. ▯

We now define $J(\alpha, \beta')$ by interpolating $\tilde{J}(\alpha, \beta')$ using Lemma 2.11. That is, we consider $\tilde{J}(\alpha, \beta')$ for fixed w as a function on the zeros (with multiplicity) of $F(w, s)$ and we interpolate by means of Lemma 2.11. We thus obtain (formally)

$$(3.20) \qquad J(\alpha, \beta'; w, s) = \sum_{l=0}^{m-1} J_l(\alpha, \beta'; w) s^l.$$

Of course, this result is only formal because we need

LEMMA 3.11. *The $J_l(\alpha, \beta'; w)$ are algebraic functions. More precisely, we can divide $\gamma(n-1; \alpha'; \beta')$ into a finite number of Z varieties on each of which the J_l are regular algebraic functions. The number of algebraic functions so obtained for varying α, β is finite. [Here $\alpha' = (\alpha_1, \ldots, \alpha_{n-1})$.]*

PROOF. Let H be a function in $\mathcal{O}(n; \alpha; \delta)$ for which $H \mid \mathfrak{V}' = \tilde{J}$. ($H$ exists by Lemma 3.10.) We write $H(w, s)$ as a power series in s with coefficients which are in $\mathcal{O}(n-1; \alpha'; \beta)$. Next we apply (2.27) to our situation. Since F is now a polynomial, this relation becomes

$$(3.21) \qquad s^j \equiv A_0^j(w) + A_1^j(w)s + \cdots + A_{m-1}^j(w)s^{m-1} \bmod F\mathscr{P},$$

where $\mathscr{P}$ denotes the ring of polynomials and the A_i^j are polynomials. This means that we can reduce all powers of s in the expansion of H to the first m. We must show that this process converges suitably.

We show first that there is a $C > 0$ so that

$$|A_i^j(w)| < C^j \tag{3.22}$$

for all i, j, $w \in \gamma(\alpha'; \beta)$. To this end, let $C_0 > 1$ be a bound for all $|A_i^j(w)|$ for all $i, j \leqslant m$, and $w \in \gamma(\alpha'; \beta)$. We choose $C = 2C_0$. Using (2.29) we have for $j > m$ by induction

$$\begin{aligned} |A_i^{j+1}(w)| &\leqslant \max\,(C^j + C_0 C^j, C^j C_0) \\ &= C^j(1 + C_0) \\ &< 2C_0 C^j \\ &= C^{j+1} \end{aligned} \tag{3.23}$$

which is the desired result (3.22).

Now, we write $H(s, w) = \sum H_j(w)s^j$. Using (3.21), this gives (formally)

$$H(s, w) = J_0(w) + J_1(w)s + \cdots + J_{m-1}(w)s^{m-1} + F(w, s)R(w, s), \tag{3.24}$$

where

$$J_i(w) = \sum A_i^j(w)H_j(w). \tag{3.25}$$

Now we know that H is holomorphic in $\gamma(\alpha; \delta)$ so the series $\sum H_j(w)s^j$ converges on the compact sets of $\gamma(\alpha; \delta)$. This means that for any $C' > 0$ the series $\sum |H_j(w)|\,(C')^j$ converges uniformly on the compact sets of $\gamma(\alpha'; \beta)$. Using (3.22), this means that (3.24) holds analytically, that is, the J_i are holomorphic on $\gamma(\alpha'; \beta)$, and R is holomorphic on $\gamma(\alpha; \delta)$, and (3.24) holds.

It follows from the definition of J_i and (2.42) and (2.43), since in our case $\alpha_{jk}^l = 1$ for $l = 0$, $\alpha_{jk}^l = 0$ for $l \geqslant 1$, that the $J_i = J_i(\alpha; \beta')$ are the restrictions to $\gamma(\alpha'; \beta)$ of algebraic functions in C^{n-1}, which we denote again by $J_i(\alpha; \beta')$. Since the roots of F define only a finite number of algebraic functions, the number of $J_i(\alpha; \beta')$ is finite for all α, β'. Thus, $J(\alpha; \beta')$ is algebraic and the number of $J(\alpha; \beta')$ is finite. Hence, Lemma 3.11 is proved. ▯

Next let K be any function in $\mathcal{O}(\alpha; \beta')$, and denote by $\tilde{L}$ the restriction of K to $\mathfrak{V}' \cap \gamma(\alpha; \beta')$. We can extend $\tilde{L}$ to $\mathfrak{V}' \cap \gamma(\alpha; \delta)$ by defining the extension to be zero outside $\gamma(\alpha; \beta')$. If we denote this extension again by $\tilde{L}$, then just as above (for $\tilde{J}$) we see that $\tilde{L} \in \mathcal{O}(\mathfrak{V}'; \alpha; \delta)$. Again, as above this means we can extend $\tilde{L}$ to a function $L \in \mathcal{O}(\alpha; \delta)$, where

$$L(w, s) = L_0(w) + L_1(w)s + \cdots + L_{m-1}(w)s^{m-1}. \tag{3.26}$$

We define $\psi = \psi(\alpha; \beta')$ by setting $\psi K = (L_0, \ldots, L_{m-1})$.

We must show that ψ is a one-to-one map of $\mathcal{O}(\alpha; \beta')/F\mathcal{O}(\alpha; \beta')$. First it is clear from the definitions that ψ is defined on $\mathcal{O}/F\mathcal{O}$. Suppose now that $\psi K = 0$. Then $K = 0$ on $\mathfrak{V}' \cap \gamma(\alpha; \beta')$, in particular, for any point

$z^0 \in \gamma(\alpha; \beta')$, $K = 0$ on the intersection of $\mathfrak{B}'$ with a neighborhood of z^0. We now need

LEMMA 3.12. *Let $z \in C$, and denote by $\mathfrak{B}'(z)$ the local analytic multiplicity variety defined by Theorem 2.8 for F at z (in that theorem $z = 0$ and it is called $\mathfrak{B}'$). Then $\mathfrak{B}'(z)$ coincides with $\mathfrak{B}'$ on sufficiently small neighborhoods of z. More precisely, it is possible to make a choice of $\mathfrak{B}'(z)$ so that this is the case.*

PROOF. First we note that F is distinguished in s at every point z because the coefficient of the highest power of $(s - z_n)^m$ is always one in the Taylor expansion of F around z

Next it is known (see Bochner and Martin [1] p. 193, Theorem 3; the result is stated there for formal power series but is the same for convergent power series) that each P_i factors in the ring $\mathcal{O}(z)$ of local holomorphic functions at z into distinct irreducible factors, say

$$P_i = P_{i1}(z) \cdots P_{il_i}(z). \tag{3.27}$$

(Note that P_i is globally irreducible though perhaps not locally irreducible.)

The intersection of V' (the underlying variety of $\mathfrak{B}'$) with a sufficiently small neighborhood of z is just the union of the $V'_{ij}(z)$ which are the zeros of the $P_{ij}(z)$; hence, this intersection is just $V'(z)$. The prime decomposition of F in $\mathcal{O}(z)$ becomes

$$F = [P_{11}(z) \cdots P_{1l_1}(z)]^{a_1} \cdots [P_{b1_b}(z) \cdots P_{bl_b}(z)]^{a_b} \tag{3.28}$$

as follows by applying (3.27) to the representation $F = \prod P_i^{a_i}$. The result now follows immediately from the definitions and Lemma 2.10. ▯

Proof of Theorem 3.8 continued. We return to the notations preceding Lemma 3.12. By Theorem 2.8, K is divisible in the neighborhood of each point of $\gamma(\alpha; \beta')$ by F. Thus (see Cartan and Serre [1]), K is divisible by F in $\mathcal{O}(\alpha; \beta')$. This proves that ψ is one-to-one.

Next we calculate the image of ψ. We denote by $\chi = \chi(\alpha; \beta)$ an "inverse" of ψ, namely,

$$\chi(K_0, \ldots, K_{m-1}) = \sum K_j s^j. \tag{3.29}$$

We show first that the image of ψ consists exactly of those $(K_0, \ldots, K_{m-1})$ for which $\chi(K_0, \ldots, K_{m-1})$ is zero on $\mathfrak{B}' \cap [\gamma(\alpha; \delta) - \gamma(\alpha; \beta')]$. First the definition of ψ shows that each element of the image of ψ has this property. Conversely, let $K = \chi(K_0, \ldots, K_{m-1})$ be zero on $\mathfrak{B}' \cap [\gamma(\alpha; \delta) - \gamma(\alpha; \beta')]$. Call K' the restriction of K to $\gamma(\alpha; \beta')$, call K'' the restriction of K to $\mathfrak{B}' \cap \gamma(\alpha; \beta')$, and call K''' the extension of K'' to $\mathfrak{B}' \cap \gamma(\alpha; \delta)$ defined as being zero outside $\gamma(\alpha; \beta')$. Then

$$K''' = K \mid \mathfrak{B}' \cap \gamma(\alpha; \delta).$$

The definition of ψ now shows that necessarily $(K_0, \ldots, K_{m-1}) = \psi(K')$.

Call $\mathscr{W}^m(\alpha; \beta)$ the subspace of $\mathcal{O}^m(\alpha'; \beta)$ consisting of those $(K_0, \ldots, K_{m-1})$ as above, so ψ is a linear map of $\mathcal{O}(\alpha; \beta')/F\mathcal{O}(\alpha; \beta')$ onto $\mathscr{W}^m(\alpha; \beta)$.

We next examine the behavior of ψ under multiplication. Let $B \in \mathcal{O}(\alpha; \beta')$. To B we associate the matrix $\boxed{B} = (B_{ij})$ where the jth column of $\boxed{B}$ is $\psi(Bs^j)$.

ASSERTION.   For any $K \in \mathcal{O}(\alpha; \beta)$ we have

(3.30) $\psi(BK) = \boxed{B}\,\psi(K)$.

PROOF OF ASSERTION.   We note first that $\boxed{B}\psi(K) = \psi[B\chi(\psi(K))]$; this is an easy consequence of the definitions and the linearity of ψ over functions of w (which follows from the definition of ψ).

To see this in detail, write

$$\psi K = (K_0, \ldots, K_{m-1}).$$

Then

$$\chi\psi K = \sum K_j(w)s^j$$
$$B\chi\psi K = \sum Bs^j K_j(w)$$
$$\psi B\chi\psi K = \sum \psi(Bs^j)K_j(w)$$
$$= \boxed{B}\,\psi K.$$

Thus, we must show that $\psi[B\chi\psi K] = \psi(BK)$, which by the above means that $B\chi\psi K = BK$ on $\mathfrak{V}' \cap \gamma(\alpha; \beta')$. But, by definition, $\chi\psi K = K$ on $\mathfrak{V}' \cap \gamma(\alpha; \beta')$ so that the result follows. $\square$

We sum up our results concerning the relation of ψ and χ as

LEMMA 3.13. χ is a right inverse of ψ, that is, $\psi\chi = $ identity on $\mathscr{W}^m$. For any $L \in \mathcal{O}(\alpha; \beta')$ we have $\chi\psi L \equiv L \mod F\mathcal{O}(\alpha; \beta')$. Let $B \in \mathcal{O}(\alpha; \beta')$ and let $\boxed{B}$ be the matrix associated with B by the above; then for any $\vec{K} \in \mathscr{W}^m(n-1; \alpha'; \beta)$ we have

(3.31) $\chi(\boxed{B}\vec{K}) \equiv B\chi(\vec{K}) \mod F\mathcal{O}(\alpha; \delta)$.

Suppose B is a polynomial; we want to show that $\boxed{B}$ can be chosen to be a matrix of algebraic functions. For this purpose we shall not use the previous description of $\boxed{B}$ in terms of B, but rather a new one: We can certainly find polynomials $B'_{ij}(w)$, such that

(3.32) $Bs^j \equiv \sum_{i=0}^{m-1} B'_{ij}(w)s^i \mod F\mathscr{P}$, where, as above, $\mathscr{P}$ is the ring of polynomials in z. (3.32) is an immediate consequence of the fact that the A^j_k of (3.21) are polynomials. Thus, by the linearity of ψ over function of w

$$\boxed{\text{B}} = \boxed{\text{I}}\,\boxed{\text{B}}', \tag{3.33}$$

where $\boxed{\text{I}}$ is the matrix whose columns are $\psi(s^i)$. As in Lemma 3.11, the coefficients of $\boxed{\text{I}}$ are algebraic functions. This proves our assertion. ∎

Let us note further that if $B \in \mathscr{O}(\alpha; \delta)$ vanishes on the part of $\mathfrak{B}'$ outside $\gamma(\alpha; \beta')$, then $\boxed{\text{B}}$ is equal to $\boxed{\text{B}}'$, defined by (3.32). (B is no longer assumed to be a polynomial.) For, by definition, $\psi(Bs^j)$ is equal to (B'_{ij}). In particular, the matrix denoted by $\boxed{\text{I}}$ above is just $\boxed{\text{J}}$ [compare (3.20)]. Moreover, $\boxed{\text{J}} = \boxed{\text{J}}'$.

For $\vec{\text{H}} = \psi(H) \in \mathscr{W}^m$ we have by definition

$$\boxed{\text{J}}\,\vec{\text{H}} = \psi(JH) = \psi(H) = \vec{\text{H}}.$$

Thus multiplication by $\boxed{\text{J}}$ is the identity on $\mathscr{W}^m$. For any $\vec{\text{L}} \in \mathscr{O}^m$ we have, by the above

$$\begin{aligned}\boxed{\text{J}}\,\vec{\text{L}} &= \boxed{\text{J}}'\vec{\text{L}} \\ &= \psi'[J\chi(\vec{\text{L}})],\end{aligned}$$

where ψ' is the map sending a function M in $\mathscr{O}(\alpha; \delta)$ into (M_i) where the expression

$$M \equiv \sum_{i=0}^{m-1} M_i s^i \bmod F\mathscr{O}(\alpha; \delta)$$

is obtained using (3.21) to reduce powers of s higher than s^{m-1}. Since $J\chi(\vec{\text{L}})$ vanishes on the part of $\mathfrak{B}'$ not in $\gamma(\alpha; \beta')$ this means that

$$\boxed{\text{J}}\,\vec{\text{L}} = \psi'[J\chi(\vec{\text{L}})] = \psi[J\chi(\vec{\text{L}})]$$

is in $\mathscr{W}^m$. This proves that $\mathscr{W}^m$ is the image under multiplication on $\mathscr{O}^m$ by the idempotent $\boxed{\text{J}}$.

We wish now to prove the quantitative statements involved in Theorem 3.8. To prove (3.13) we suppose that $A'' > 0$ and $K \in \mathscr{O}(n; \alpha; \beta'; A'')$. Then the above procedure shows that for each i we can write, for w in a Z variety of maximum dimension,

$$(\psi K)_i(w) = D_i(w)D(w)/[D(w)]^2, \tag{3.34}$$

where $[D(w)]^2$ is a polynomial in w which is not identically zero and where $D_i(w)$ is a linear expression in the values of K on $\mathfrak{B}'$ above w with coefficients which are algebraic functions. Since, as is easily seen from the definitions and simple estimates on the zeros of polynomials, an algebraic function is majorized by a power of $(1+|w|)$, it follows that

$$\max_{w \in \gamma(\alpha'; \beta)} |D_i(w)D(w)| < \varkappa(1+|\alpha'|+|\beta'|)^e A'' \tag{3.35}$$

for suitable e, $\varkappa$. Using Theorem 1.4 this gives

$$\max_{w \in \gamma(\alpha'; \beta-1)} |(\psi K)_i(w)| \leqslant \varkappa'(1+|\alpha'|+|\beta'|)^{e'} A'' \tag{3.36}$$

for suitable $\varkappa'$, e' (depending only on F). The last inequality implies (3.13) immediately.

The same reasoning and the definitions give the inequality following (3.16), and hence also (3.15).

Finally, we construct the multiplicity variety $\mathfrak{V}$ just as the corresponding multiplicity variety in Theorem 2.8 was constructed. Then (3.17) is proved the same way as (3) of Theorem 2.8 was proved.

The second statement of (2) follows from (2) of Theorem 2.8 and Lemma 3.12. The first statement of (2) is proved by examining the proof of (2) of Theorem 2.8 and using the same type of quantitative estimates as in the above proof of (3.13).

With this the whole of Theorem 3.8 is proved. □

See Remark 3.4.

III.5. Completion of the Proofs

We wish now to complete the proofs of Theorems 3.2, 3.3, and 3.6. As in the case of their analogs (Theorems 2.5, 2.6, and 2.3, respectively) in Chapter II, these theorems will be proved simultaneously by induction, so we denote them more precisely by T3.2(n), T3.3(n, l) and T3.6(n, l).

PROOF OF THEOREM 3.6. Theorems 3.6 differs from Theorem 2.3 in that we require bounds on G. We shall now repeat the argument of Theorem 2.3 with the appropriate modifications. Again, T3.6(0, 1) is trivial and there is no difficulty in modifying the parallel argument in Chapter II to show that T3.6(n, 1) implies T3.6(n, l) for all l. Thus, we assume $n > 0$ and T3.6($n-1$, l) for all l. We want to prove T3.6(n, 1).

Let $\vec{\mathrm{F}} = (F_j)$ where F_j are polynomials and $F_1 \not\equiv 0$. Let β'', A'' be unspecified now; let β', A' be related to F_1, β'', A'' as in Theorem 3.8 (with F_1, β'', A'' replacing F, β, A of Theorem 3.8), and let $H \in \mathcal{O}(n; \alpha; \beta'; A')$ be of the form $\vec{\mathrm{F}}\vec{\mathrm{G}}'$ where $\vec{\mathrm{G}}' = (G_j')$ with $\mathrm{G}_j' \in \mathcal{O}(n; \alpha; \beta'')$.

As in the proof of Theorem 2.3, we apply Theorem 3.8. Then we obtain a map

$$\psi : \mathcal{O}(n; \alpha; \beta')/F_1\mathcal{O}(n; \alpha; \beta') \to \mathscr{W}^m(n-1; \alpha; \beta'').$$

Each F_j operated on $\mathscr{W}^m$ by a matrix $\boxed{\mathrm{F}}_j$ of algebraic functions which depends only on $\vec{\mathrm{F}}$.

Denote by $\boxed{\mathrm{F}}^1$ the $m \times (r-1)m$ matrix

$$\boxed{\mathrm{F}}^1 = (\boxed{\mathrm{F}}_2, \ldots, \boxed{\mathrm{F}}_r). \tag{3.37}$$

In analogy to Lemma 2.13 we have

LEMMA 3.14. *There exists a* $\vec{\mathrm{G}}'' \in \mathcal{O}^{(r-1)m}(n-1; \alpha; \beta')$ *with* $\boxed{\mathrm{F}}^1\vec{\mathrm{G}}'' = \psi H$.

PROOF. We apply ψ to the equation $H = \sum F_j G'_j$, and we set

$$\vec{\mathrm{G}}'' = \begin{pmatrix} \psi G'_2 \\ \cdot \\ \cdot \\ \cdot \\ \psi G'_r \end{pmatrix}.$$

This gives the result. ▯

We continue the proof of Theorem 3.6. By Theorem 3.8 we have $\psi H \in \mathscr{W}^m(n-1; \alpha'; \beta''-1; A'')$. We may also assume that $\beta''-1 > \beta$. Let β'', A'' be related to $\beta+2$, A''' by T3.6$(n-1, m)$ for $\boxed{\mathrm{F}}^1$. Then by T3.6$(n-1, m)$ we can find Z varieties and differential operators which depend only on $\boxed{\mathrm{F}}^1$, hence, on $\vec{\mathrm{F}}$, such that we can find $\vec{\mathrm{G}}''' \in \mathcal{O}^{(r-1)m}(n-1; \alpha; \beta+1; A''')$ satisfying

$$\boxed{\mathrm{F}}^1\vec{\mathrm{G}}''' = \psi H \tag{3.38}$$

and G''' can be expressed in terms of the differential operators applied to ψH on these Z varieties.

We now want to apply χ to (3.38). We write $\vec{\mathrm{G}}'''$ in the form

$$\vec{\mathrm{G}}''' = \begin{pmatrix} \vec{\mathrm{G}}'''_2 \\ \cdot \\ \cdot \\ \cdot \\ \vec{\mathrm{G}}'''_r \end{pmatrix}, \tag{3.39}$$

where $\vec{\mathrm{G}}'''_k$ is a vector with components $\mathrm{G}'''_{(k-2)m+1}, \mathrm{G}'''_{(k-2)m+2}, \ldots, \mathrm{G}'''_{(k-1)m}$. Then (3.38) becomes

$$\sum_{k=2}^{r} \boxed{\mathrm{F}}_k \vec{\mathrm{G}}'''_k = \psi H. \tag{3.40}$$

We cannot apply χ yet because the $\vec{\mathrm{G}}'''_k$ may not be in $\mathscr{W}^m$. Thus, we multiply (3.40) by $\boxed{\mathrm{J}}$ of Theorem 3.8 (for F_1 in place of F). Since $\boxed{\mathrm{J}}$ is a projection on $\mathscr{W}^m$, we may now assume the $\vec{G}_k''' \in \mathscr{W}^m$. It follows from the definitions that equation (3.40) holds after this multiplication. Since the coefficients of $\boxed{J}(\alpha; \beta'')$ are algebraic functions and finite in number, the maximum of the $\vec{G}_k'''(w)$ can be increased by at most a factor $c(1 + |w|)^c$ where c depends only on $\vec{F}$. Also, it is clear that multiplication by $\boxed{J}$ does not destroy the F L'Hospital description of $\vec{G}_k'''$ in terms of ψH.

We now apply χ to (3.40) and use Lemma 3.13. This gives

$$\sum_{k=2}^{r} F_k \chi(\vec{G}_k''') \equiv H \bmod F_1 \mathscr{O}(\alpha; \beta'') \tag{3.41}$$

or

$$\sum_{k=2}^{r} F_k \chi(\vec{G}_k''') - H = F_1 G_1''', \tag{3.42}$$

where $G_1''' \in \mathscr{O}(\alpha; \beta'')$.

We want to apply Theorem 3.4. The left side of (3.42) is in $\mathscr{O}(\alpha; \beta + 1; c'(1 + |\alpha| + |\beta|)^{c'} A''' + A')$ where c' depends only on $\vec{F}$. Thus, by Theorem 3.4, G_1''' can be chosen in $\mathscr{O}(\alpha; \beta; c[c'(1 + |\alpha| + |\beta|)^{c'} A''' + A'])$ for some c which depends only on F_1, hence on $\vec{F}$. This gives the desired inequality on G_1''', namely, that it is in $\mathscr{O}(n; \alpha; \beta; A)$, if A''' and A' are suitably chosen.

Finally, we must pass from $\vec{G}'''$ to $\vec{G}$. This is done by

LEMMA 3.15. *Suppose* $\vec{L} \in \mathscr{O}^m(n - 1; \alpha; \beta + 1)$ *is* $\vec{F}$ *L'Hospital describable in terms of* ψH. *Then* $\chi(\vec{L})$ *is* $\vec{F}$ *L'Hospital describable in terms of* H.

This is just a restatement of Lemma 2.16 in "semilocal" form. It is proved by the same method as the proof of Lemma 2.16. With this, and the observation made as at the end of the proof of Theorem 2.3 that G_1''' is $\vec{F}$ L'Hospital describable in terms of H, the proof of Theorem 3.6 is concluded by setting

$$\vec{G} = \begin{pmatrix} G_1''' \\ \chi(\vec{G}_2''') \\ \cdot \\ \cdot \\ \cdot \\ \chi(\vec{G}_r''') \end{pmatrix} \square.$$

PROOF OF THEOREMS 3.2 AND 3.3. As in Chapter II we use induction. T3.2(0) and T3.3(0, 1) are clear. Again it is readily verified that T3.2(n) and T3.3(n, 1) imply T3.3(n, l) for all $l > 0$. Thus, it remains to prove that T3.2($n - 1$) and T3.3($n - 1$, l) for all l imply T3.2(n) and T3.3(n, 1). We begin with T3.3(n, 1). We may suppose that $F_1 \not\equiv 0$ and that $s = z_n$ is noncharacteristic for F_1.

LEMMA 3.16. *Using the notation of Lemma 3.14, ψ defines an isomorphism ψ' of $\mathcal{O}(\alpha; \beta')/\vec{F}\mathcal{O}^r(\alpha; \beta')$ onto $\mathcal{W}^m(n-1; \alpha; \beta)/\boxed{F}^1\mathcal{W}^{(r-1)m}(n-1; \alpha; \beta)$, where $\mathcal{W}^{(r-1)m}$ is the direct sum of $r-1$ copies of $\mathcal{W}^m$. Namely, for $K \in \mathcal{O}(\alpha; \beta')$ we define $\psi'(K)$ as the class of $\psi(K)$ mod $\boxed{F}^1\mathcal{W}^{(r-1)m}(\alpha; \beta)$. $K \in \mathcal{O}(\alpha; \beta')$ is in $\vec{F}\mathcal{O}^r(\alpha; \beta')$ if and only if $\psi K \in \boxed{F}^1\mathcal{O}^{(r-1)m}(\alpha; \beta)$. Moreover, there is an A''' satisfying (3.3) such that*

$$\psi'\mathcal{O}(\alpha; \beta'; A''') \subset \mathcal{W}^m(\alpha; \beta-1; A). \tag{3.43}$$

PROOF. By Lemma 3.14 ψ' is defined on the quotient $\mathcal{O}/\vec{F}\mathcal{O}^r$. Suppose $\psi'L = 0$. Then $\psi L = \boxed{F}^1\vec{G}'''$ for some $\vec{G}''' \in \mathcal{O}^{(r-1)m}(\alpha; \beta'')$. Applying the projection J, we may assume $\vec{G}''' \in \mathcal{W}^{(r-1)m}(\alpha; \beta'')$. Applying χ and using Lemma 3.13 we see that $L \in \vec{F}\mathcal{O}^r$, so ψ' is one-to-one and remains one-to-one when we replace the image space by $\mathcal{O}^m(\alpha; \beta'')/\boxed{F}^1\mathcal{O}^{(r-1)m}(\alpha; \beta'')$. Since ψ is onto so is ψ', and (3.43) is just (3.13). ☐

We now use T3.3($n-1$, m) to describe $\mathcal{W}^m/\boxed{F}^1\mathcal{W}^{(r-1)m}$. Let $\mathfrak{U}_j$ be the multiplicity Z variety associated by T.3.3($n-1$, 1) to the ideal of all sums $\sum_k F^1_{jk} N^1_{j-1,k}$, where (F^1_{jk}) is the jth row of $\boxed{F}^1$ and $(N^1_{j-1,k}) \in M^1_{j-1}(\alpha; \beta'')$ which is the module of simultaneous relations of the first $j-1$ rows of $\boxed{F}^1$. Then for any $\vec{H} \in \mathcal{W}^m(\alpha; \beta')$ and any j let

$$(\rho^1\vec{H})_k = \text{restriction of } H_k - \sum_{i<k}\sum_j F_{i+1,j}G_{ij} \text{ to } \mathfrak{A}_k, \tag{3.44}$$

where the G_{ij} are described in the remarks preceding the statement of Theorem 2.6, suitably modified to apply to the semilocal situation of Theorem 3.3. Moreover, the G_{ij} are $\vec{F}$ L'Hospital describable in terms of H and satisfy certain growth conditions which will be discussed later.

Before continuing we need

THEOREM 3.17. *Let $\beta''' > 0$; let $\boxed{F}$ be an $l \times r$ matrix of polynomials. Then we can find $\beta'''' > 0$, with $\beta'''' < c(\beta''')^{c'}(1+|\alpha|+|\beta'''|)^c$ such that $\vec{H} \in \mathcal{O}^r(\alpha; \beta'''')$ is in the image $\vec{F}\mathcal{O}^r(\alpha; \beta''')$ if and only if the H_j satisfy the same differential relations as the rows of $\boxed{F}$. That is, there exist algebraic Z varieties X_i and differential operators d_{ij} so that $\vec{H} \in \boxed{F}\mathcal{O}^r(\alpha; \beta''')$ if and only if*

$$\sum d_{ij}H_j(z) = 0 \qquad \text{for } z \in X_i \cap \gamma(\alpha; \beta''''). \tag{3.45}$$

Here,

$$\sum d_{ij}(F_{jk}P) = 0 \qquad \text{for } z \in X_i, \text{ all } i, k, \text{ and all polynomials P.} \tag{3.46}$$

The $X_i = X_i(\alpha; \beta''')$ and $d_{ij} = d_{ij}(\alpha; \beta''')$ depend only on $\boxed{F}$, α, β''', and there are only a finite number of them for $\boxed{F}$ fixed.

We prove Theorem 3.17 simultaneously with Theorems 3.2 and 3.3. Thus, we denote the statement by T3.17(n, l). Our results will be complete if we show

(3.47) T3.3(n, l) for all l implies T3.17(n, l) for all l.

(3.48) T3.17$(n-1, l)$ for all l implies T3.3$(n, 1)$.

(3.49) T3.2$(n-1)$ and T3.3$(n-1, l)$ for all l imply T3.2(n).

Proof of (3.47). This, except for obvious modifications, is the same as the proof of (2.61) and so will be omitted. ▯

Proof of (3.48). Let X_i, d_{ij} be the Z varieties and differential operators which correspond to $\boxed{\mathrm{F}}^1$ by T3.17$(n-1, m)$ with $\boxed{\mathrm{F}}$ replaced by $\boxed{\mathrm{F}}^1$, β''' replaced by β, and β'''' replaced by β''. Here β' is related to β'' as β' is related to β in the statement of Theorem 3.8. Let $K \in \mathcal{O}(\alpha; \beta')$ and consider ψK. By Lemma 3.16, $K \in \vec{\mathrm{F}}\mathcal{O}^r(\alpha; \beta')$ if and only if

$$\psi K \in \boxed{\mathrm{F}}^1\mathcal{O}^{(r-1)m}(n-1; \alpha; \beta''),$$

which by T3.17$(n-1, m)$ is equivalent to: For each i

$$\sum d_{ij}(\psi K)_j(w) = 0 \qquad \text{for } w \in X_i \cap \mathcal{O}(\alpha; \beta''). \tag{3.50}$$

We wish now to apply Theorem 3.8 to (3.50), that is, to express the $(\psi K)_j$ in terms of K so that (3.50) becomes the statement that certain derivatives of K must vanish on certain Z varieties.

To this end we use (3.17). Combining (3.17) with (3.50) and applying Theorem 2.15 shows that (3.50) implies

$$\left[\sum (\vec{\mathrm{f}}_k^t \circ d_k^t) \cdot K\right](w) = 0, \tag{3.51}$$

where $\vec{\mathrm{f}}_k^t$ are algebraic maps (compare (2.67)).

We can reverse our steps, that is, (3.50) is equivalent to (3.51). Thus, (3.51) is a set of necessary and sufficient conditions for $K \in \mathcal{O}(\alpha; \beta')$ to be in $\vec{\mathrm{F}}\mathcal{O}^r(\alpha; \beta')$. Since the sum (3.51) is finite, we reason as in the proof of (2.62) to deduce that each term corresponding to a distinct point (w, s) must be zero. We can now define $\mathfrak{V}$ as in the proof of (2.62) and again $K \to K \mid \mathfrak{V}$ is one-to-one for $K \in \mathcal{O}(\alpha; \beta')/\vec{\mathrm{F}}\mathcal{O}^r(\alpha; \beta) \cap \mathcal{O}(\alpha; \beta')$.

To complete the proof of (3.48) we must derive the desired bounds given by (3.8), which for $l = 1$ is the same as (3.1). Before proving this we give the

Proof of (3.49) (except for (3.6)). This is the same as the proof of the corresponding result (2.63) as given in Chapter II with obvious modifications; note that this proof does not use the quantitative statements (3.8). ▯

Proof of (3.48) concluded. It suffices for the qualitative result (3.8) (or (3.1)) to show that each of the L_{ij} of (3.7) can be bounded in a suitable manner in terms of the restriction of K to $\mathfrak{B}$. To show this, we use the espression (2.15), the analog of which holds in our case because of Theorem 3.2, where the f_{iqtj} are algebraic functions. Thus we can bound $f_{ij}L_{ij}$, where f_{ij} is the greatest common denominator of the norms (products of conjugates) of the f_{iqtj}. Using Theorem 1.4 we obtain the desired bounds for L_{ij}, noting that there are only a finite number of possible f_{ij}. This completes the proof of (3.8) (and also of (3.6)), and hence of all Theorems 3.2, 3.3, and 3.6. ▯

For later applications we shall need also

THEOREM 3.18. *Let* $\boxed{\mathrm{F}}$ *be a matrix of polynomials. Then there exists a* $c > 0$ *so that if* b *is sufficiently large and if*

$$\left|\boxed{\mathrm{F}}\vec{\mathrm{H}}\right| \leqslant a \quad \text{on} \quad |z - z_0| \leqslant b,$$

then for a suitable $\vec{\mathrm{H}}_1$ *we have*

$$|\vec{\mathrm{H}} - \boxed{\mathrm{M}}\vec{\mathrm{H}}_1| \leqslant ca(1 + |z_0| + b)^c$$

for

$$|z - z_0| \leqslant \frac{1}{c(1 + |z_0| + |b|)^c}.$$

Here $\boxed{\mathrm{M}}$ *is a matrix representing the module of relations of the columns of* $\boxed{\mathrm{F}}$. $\vec{\mathrm{H}}$ *and* $\vec{\mathrm{H}}_1$ *are assumed to be holomorphic on the sets in question.*

PROOF. Set $\vec{\mathrm{G}} = \boxed{\mathrm{F}}\vec{\mathrm{H}}$. Thus $\vec{\mathrm{G}}$ is in the range of $\boxed{\mathrm{F}}$ and $\vec{\mathrm{G}}$ is small on $|z - z_0| \leqslant b$. We claim there exists an $\vec{\mathrm{H}}'$ which is small on

$$|z - z_0| \leqslant \frac{1}{c(1 + |z_0| + |b|)^c}$$

satisfying $\boxed{\mathrm{F}}\vec{\mathrm{H}}' = \vec{\mathrm{G}}$. If we could find such an $\vec{\mathrm{H}}'$, then

$$\boxed{\mathrm{F}}(\vec{\mathrm{H}} - \vec{\mathrm{H}}') = 0$$

so that we can write

$$\vec{\mathrm{H}} = \vec{\mathrm{H}}' + \boxed{\mathrm{M}}\vec{\mathrm{H}}_1$$

for a suitable H_1, which is the desired result.

The existence of $\vec{\mathrm{H}}'$ can be proved by our step by step construction, as explained in the remarks preceding the statement of Theorem 2.6, and the remarks at the beginning of Section II.4. As this type of argument has been used several times in this book, we shall omit the relevant details. ▯

Remarks

Remark 3.1. See page 75.

Remark 3.2. See page 75.

In Theorem 3.1 we stated that the kernel of $\rho_{\mathfrak{W}}$, considered as a map into $\mathcal{O}(\mathfrak{V}; \alpha; \beta)$ on $\mathcal{O}(\alpha; \beta')$, is $\overrightarrow{F}\mathcal{O}^r(\alpha; \beta) \cap \mathcal{O}(\alpha; \beta')$. Actually, the kernel of $\rho_{\mathfrak{W}}$, considered as a map into $\mathcal{O}(\mathfrak{V}; \alpha; \beta')$, is $\overrightarrow{F}\mathcal{O}^r(\alpha; \beta')$. Similar improvements can be made in Theorems 3.3, 3.6, and 3.17. These improved results can be made along the lines of the present proof except that, in several places, we must be more careful in the selection of constants marked β', β'', ... than we have been. We leave the details to the interested reader.

Remark 3.3. See page 79.

Remark 3.4. See page 88.

We set $c'' = 2$ in the statement of Theorem 3.8. Actually any $c'' > 0$ would do.

Problems

PROBLEM 3.1. See page 75.

In Remark 3.1 we showed that $\mathcal{O}(\mathfrak{V}'; \alpha; \beta'; A') \supset \mathcal{O}(\mathfrak{V}; \alpha; \beta'; A''')$. On the other hand (3.4) shows that a reverse inclusion is possible, if the constants are changed somewhat. The only proof we know of this reverse inclusion is by means of Theorem 3.1. Find a direct proof.

CHAPTER IV

Passage from Local to Global

Summary

In Section IV.1 we introduce the concept of product localizeable analytically uniform (PLAU) space. Roughly speaking, an AU space $\mathscr{W}$ is PLAU if

1. The AU structure can be defined by products of functions of one variable.
2. For any $z^0 \in C$ we can find a function in $\mathbf{W}'$ which is one at z^0 and suitably small in certain other parts of C.

In Section IV.2 we formulate the quotient structure theorem or fundamental principle (Theorem 4.1) which gives, for PLAU spaces, a global geometric structure of ideals generated by polynomials:

Let $\vec{\mathrm{F}}$ *be a vector of polynomials with* r *components. There exists a polynomial multiplicity variety* $\mathfrak{V}$ *depending only on* $\vec{\mathrm{F}}$ *such that the restriction map* $G \to G|\mathfrak{V}$ *is a topological isomorphism of* $\mathbf{W}'$ *(the Fourier transform of* $\mathscr{W}'$*) modulo* $\vec{\mathrm{F}}\mathbf{W}'^r$ *onto* $\mathbf{W}'(\mathfrak{V})$ *which is the space of analytic functions on* $\mathfrak{V}$ *with growth conditions "induced" by* $\mathbf{W}'$.

In case $\mathbf{W}'$ is the ring of polynomials, the fact that $\mathrm{G} \to G|\mathfrak{V}$ is one-one can be regarded as a refinement of the Hilbert Nullstellensatz.

Section IV.3 introduces (Čech) cohomology with bounds. The bounds are of two kinds: Bounds (from below) on the size of the sets on which the cochains are defined, and bounds (from above) on the functions on these sets. It is for this reason that we needed two types of bounds in Chapter III. The main result of our book (Theorem 4.2) is formulated in Section IV.3. This contains Theorem 4.1 and the statement that allows the quotient space $\mathbf{W}'^l/\boxed{\mathrm{F}}\mathbf{W}'^r$ to be computed semilocally (hence, the name "localizeable"). Here $\boxed{\mathrm{F}}$ is an $l \times r$ matrix of polynomials and $\mathbf{W}'^r$ is the r-fold direct sum of $\mathbf{W}'$ with itself.

In Section IV.4 we prove Theorem 4.2. The essential point is the proof of the vanishing of the groups of cohomology with bounds. For this we use the results of Chapter III and a method which for cohomology without bounds is due to Oka.

In Section IV.5 we show how to extend the main theorem to some spaces which are AU but not PLAU.

IV.1. Localizeable Spaces

We shall now use a "piecing together" process to derive global results from the semilocal results of Chapter III. Our procedure is somewhat analogous to that of Oka-Cartan (see Cartan and Serre [1], Gunning and Rossi [1]) except that we have the complication of bounds. In this chapter, as in Chapter III, "variety" will be algebraic etc.

Let $\mathscr{W}$ be an analytically uniform space and let K be an analytic uniform structure for $\mathscr{W}$. Our passage from local to global can be accomplished only for certain $\mathscr{W}$, namely, those which are *localizeable.*

DEFINITION. $\mathscr{W}$ is called *product localizeable* if

(a) Any entire function which is $0(k(z))$ for all $k \in K$ is in $\mathbf{W}'$;

(b) For any $N > 0$ if we replace the analytic uniform structure $K = \{k\}$ by $K_N = \{k_N(z)\}$ where

$$k_N(z) = \max_{|z'-z| \leqslant N} |k(z')|(1 + |z'|)^N \tag{4.1}$$

then K_N is again an analytic uniform structure for $\mathscr{W}$;

(c) We assume that in addition to the analytic uniform structure $K = \{k\}$ there is given a bounded analytic uniform structure (BAU structure) $M = \{m\}$ where $m(z)$ are all the positive continuous functions on C such that for each $k \in K$ and $m \in M$ we have

$$m(z) = 0(k(z)). \tag{4.2}$$

Moreover, for every bounded set B in $\mathbf{W}'$ there is an $m \in M$ so that all $F \in B$ satisfy $|F(z)| \leqslant m(z)$ for all $z \in C$.

We assume in addition that the AU structure K and the BAU structure M can be chosen (see the definition of sufficient BAU structure below) to consist of products of functions of single variables, that is, each k is of the form $k_1(z_1)k_2(z_2) \cdots k_m(z_m)$ where the k_j are continuous and positive. (A similar property holds for the functions $m(z)$.) We shall usually write K for K_j, M for M_j. For each of these sets K, M of functions of single variables we require:

(d) K and M can be chosen so that: For each $k \in K$ and $\eta > 0$ there is a $k' \in K$ and a $c > 0$ so that for any $m \in M$ with $m(z) \leqslant k'(z)$ for all $z \in C^1$ and for any $z^0 \in C^1$, we can find an entire function $\varphi(z; z^0, \eta)$ such that

$$\frac{m(z^0)|\varphi(z; z^0, \eta)|}{\min\limits_{|\zeta - z^0| \leqslant \eta} |\varphi(\zeta; z^0, \eta)|} \leqslant c\, k(z). \tag{4.3}$$

Moreover, there is a $c'' > 0$ and an $m' \in M$ so that the left side of (4.3) is $\leqslant c''m'(z)$. Here c and k' depend only on k and η but φ and c'' and m' can depend also on m.

(e) K and M can be chosen so that for each $k \in K$ and $\eta > 0$ there is a $k' \in K$ and a $c > 0$, so that for any $m \in M$ with $m(z) \leqslant k'(z)$ for all $z \in C^1$ and for any $z^0 \in C^1$, we can find an entire function $\psi(z; z^0, \eta)$ such that

$$\sup_{-\infty < \xi < \infty} \left\{ \frac{(1 + |\xi|^2) m(\xi + i \operatorname{Im} z^0)}{\min\limits_{|t - \operatorname{Im} z^0| \leqslant \eta} |\psi(\xi + it; z^0, \eta)|} \right\} |\psi(z; z^0, \eta)| \leqslant c\, k(z). \tag{4.4}$$

Moreover, there is a $c'' > 0$ and an $m' \in M$ so that the left side of (4.4) is $\leqslant c''m'(z)$. Again c and k' depend only on k and η, while ψ and c'' and m' may depend also on m.

See Remark 4.1.

We shall write "PLAU" for "product localizeable analytically uniform." A generalization in which the product assumptions are removed will be discussed in Section IV.5. Often different analytic uniform structures can be used to define the same AU space. Suppose one such $K = \{k\}$ is given and a bounded analytic uniform structure $M = \{m\}$ is given. Often there are too many functions in M, and we want to get a subset which can be handled. We say that a subset $M_1 = \{m_1\}$ of M is a *sufficient bounded analytically uniform structure* if for any $\tilde{k}' \in K$ there is a $k' \in K$ such that if $m \in M$ satisfies $m(z) \leqslant k'(z)$ for all z then there is an $m_1 \in M_1$ with

$$m(z) \leqslant m_1(z) \leqslant \tilde{k}'(z) \qquad \text{for all } z.$$

Remark 4.2. The reasons for these definitions of BAU structure and sufficient BAU structure are as follows: We need to know that a BAU structure M has many functions in order to know that the function denoted by m immediately following (4.38) is in M. Thus, in order to be able to replace M by a suitable subset, say M_2, which will do just as well, we must know that for each $m \in M$ there is an $m_2 \in M_2$ with $m \leqslant cm_2$ for some $c > 0$.

However, this condition on M_2 is not enough. For, in order to replace M by M_2 and have Lemma 4.7 hold for M_2 in place of M, we need to know that for any $\tilde{k}' \in K$ there is a $k' \in K$ so that if $m(z) \leqslant k'$, then there is an $m_2 \in M$ with $m \leqslant m_2 \leqslant \tilde{k}'$. For in this case (in the notation of Lemma 4.7) we start with $\tilde{k}'$ and choose k' by this requirement. Then choose k'' in accordance with Lemma 4.7. It follows that the function denoted by b preceding the statement of Lemma 4.7 has a continuous majorant $m \in M$ with $m \leqslant ck'$. Hence b has a continuous majorant $m_2 \in M_2$ with $m_2 \leqslant c\tilde{k}'$.

Hence if M_2 is sufficient we can replace M by M_2 for all the purposes of this book.

Roughly speaking, we can say that a sufficient BAU structure M_2 has the property that there is an AU structure K for which each $k \in K$ is essentially the upper envelope of all $m_2 \in M_2$ which are $\leqslant k$.

In conditions (d) and (e) of product localizeability we may assume that the functions m which occur belong to some sufficient bounded AU structure.

IV.2. Formulation of the Quotient Structure Theorem (Fundamental Principle)

Let $\mathscr{W}$ be an AU space and $\mathfrak{V} = (V_1, \partial_1; \ldots; V_a, \partial_a)$ a polynomial multiplicity variety. We define $\mathbf{W}'(\mathfrak{V})$ as the space of $\vec{\mathrm{G}} = (G_1, \ldots, G_a)$ which are regular on all of $\mathfrak{V}$ and satisfy the compatibility condition: There exists an entire function H on all of C^n with $\partial_j H = G_j$ on V_j for all j. Moreover, for any k in *any* AU structure for $\mathscr{W}$ we have $G_j(z)/k(z) \to 0$ as $|z| \to \infty$, $z \in V_j$. The topology of $\mathbf{W}'(\mathfrak{V})$ is defined by means of the seminorms

$$\|\vec{\mathrm{G}}\|_k = \sup_{\text{all } j,\, z \in V_j} |G_j(z)|/k(z)$$

for all k as above.

THEOREM 4.1. *Quotient Structure Theorem* (*Fundamental Principle*). *Let* $\vec{\mathrm{F}}$ *be a vector of polynomials and let* $\mathscr{W}$ *be* PLAU. *Then there exists a polynomial multiplicity variety* $\mathfrak{V}$ *such that the restriction map*

$$\rho_{\mathfrak{V}}: H \to H \mid \mathfrak{V}$$

is a topological isomorphism of $\mathbf{W}'/\vec{\mathbf{F}}\mathbf{W}'^r$ *onto* $\mathbf{W}'(\mathfrak{V})$. $\mathfrak{V}$ depends on $\vec{\mathrm{F}}$ but not on $\mathscr{W}$.

This theorem and a slight generalization (Theorem 4.2) are the main results of this chapter and play a central role in all our later work.

IV.3. Cohomology with Growth Conditions

In this chapter, α, α', ... will denote lattice points in C^r, that is, points whose real and imaginary coordinates are integers.

DEFINITION. Let $\beta > 1$ and suppose for each α we are given a function $f(\alpha; z)$ which is holomorphic on $\gamma(\alpha; \beta)$. Then we call f a *cochain* (for β). The space of such cochains is denoted by $\mathcal{O}(\beta)$. We say that f is a *nice cochain* if for every $k \in K$

$$\sup_{z \in \gamma(\alpha;\beta)} |f(\alpha; z)|/k(z) \to 0 \qquad \text{as } |\alpha| \to \infty. \tag{4.5}$$

The nice cochains form a space $\mathbf{W}'(\beta)$ with a natural topology, namely, we use for seminorms the supremum over α (for β fixed) of the left side of (4.5). We say that the nice cochain f is in $\mathbf{W}'(\beta; \vec{\mathrm{F}})$ if, whenever α and α' are lattice points all of whose components except one are equal and such that $\gamma(\alpha; \beta) \cap \gamma(\alpha'; \beta)$ is nonempty,

$$f(\alpha; z) - f(\alpha'; z) = \sum G_j(\alpha, \alpha'; z) F_j(z) \tag{4.6}$$

for $z \in \gamma(\alpha; \beta) \cap \gamma(\alpha'; \beta)$, where the G_j are holomorphic on $\gamma(\alpha; \beta) \cap \gamma(\alpha'; \beta)$.

This implies that (4.6) holds whenever α, α' have the property that $\gamma(\alpha; \beta) \cap \gamma(\alpha'; \beta)$ is nonempty. To see this, observe that if α and ${}_1\alpha$ are lattice points such that $\gamma(\alpha; \beta) \cap \gamma({}_1\alpha; \beta) \neq \varnothing$, then we can find a finite sequence of lattice points $\alpha^0 = \alpha, \alpha^1, \alpha^2, \ldots, \alpha^p = {}_1\alpha$ such that, for each $i < p$, $\alpha_j^i = \alpha_j^{l+1}$ except for a single j, and for that j, $|\alpha_j^i - \alpha_j^{i+1}| = 1$; moreover

$$\gamma(\alpha; \beta) \cap \gamma({}_1\alpha; \beta) \subset \gamma(\alpha^i; \beta) \cap \gamma(\alpha^{i+1}; \beta)$$

for all i. Now apply (4.6) to write

$$f(\alpha^i) - f(\alpha^{i+1}) = \sum G_j(\alpha^i, \alpha^{i+1}; z) F_j(z).$$

Our assertion follows since

$$f(\alpha; z) - f({}_1\alpha; z) = \sum [f(\alpha^i; z) - f(\alpha^{i+1}; z)]. \;\square$$

We define $\mathcal{O}(\beta; \vec{\mathrm{F}})$ in a similar manner for cochains which are not necessarily nice. Finally, for any c with $0 \leqslant c < \beta - 1$ we define the quotient space $\mathbf{W}'(\beta; \vec{\mathrm{F}})/\mathbf{W}'(\beta; \vec{\mathrm{F}}) \cap \vec{\mathrm{F}}\mathcal{O}^r(\beta - c)$. All these spaces have the natural topologies defined by the functions $k \in K$.

In analogy to the results of Chapters II and III we should expect to be able to pick a "natural" extension to invert $\rho_{\mathfrak{B}}$. However, this seems to be quite difficult in general; this is related to the question of general boundary value problems and will be discussed to some extent in Chapters IX and X.

Let $\mathfrak{B}$ be a multiplicity variety satisfying the conclusions of Theorem 4.1. The restriction map $\rho_{\mathfrak{B}}$ defines naturally a local restriction map $\rho_{\mathfrak{B}}^L$: $\mathbf{W}'(\beta; \vec{\mathrm{F}}) \to \mathbf{W}'(\mathfrak{B})$: For any nice cochain f in $\mathbf{W}'(\beta; \vec{\mathrm{F}})$ we define $\rho_{\mathfrak{B}}^L f$

on $\mathfrak{V} \cap \gamma(\alpha; \beta)$ to be the restriction of f to $\mathfrak{V} \cap \gamma(\alpha; \beta)$. Since $f \in \mathbf{W}'(\beta; \vec{\mathrm{F}})$, there is no difficulty on the overlap. (We must know that the ideal generated by the F_j in the rings $\mathcal{O}(\alpha; \beta)$ vanish on the $\mathfrak{V}$ of theorem 4.1. This will become clear below.) By construction $\rho^L_{\mathfrak{V}}$ is seen to be a continuous map of $\mathbf{W}'(\beta; \vec{\mathrm{F}})/\vec{\mathrm{F}}\mathbf{W}'^r(\beta) \to \mathbf{W}'(\mathfrak{V})$. There is also a natural continuous map

$$\lambda_{\vec{\mathrm{F}}}\colon \mathbf{W}'/\vec{\mathrm{F}}\mathbf{W}'^r \to \mathbf{W}'(\beta; \vec{\mathrm{F}})/\vec{\mathrm{F}}\mathbf{W}'^r(\beta). \tag{4.7}$$

Actually, even to an $l \times r$ matrix $\boxed{\mathrm{F}}$ of polynomials, Theorem 3.3 associates vector polynomial multiplicity varieties $\vec{\mathfrak{V}}(\alpha; \beta)$ for any α, β. However, if $l > 1$ the restriction to $\vec{\mathfrak{V}}(\alpha; \beta)$ depends on choosing bases for quotient spaces in a suitable manner. (Compare the remarks preceding the statement of Theorem 2.6). Now, $\vec{\mathfrak{V}}(\alpha; \beta)$ is a vector polynomial multiplicity variety, hence is defined in all of C^n. However, the definition of the restriction map $\rho(\alpha; \beta)$ depends on certain properties of $\vec{\mathfrak{V}}(\alpha; \beta)$ and the bases in question; we know only that these properties hold near α. Thus, for example, in (2.16) the choice of $\vec{\mathrm{G}}_1$ depends on the Q^1_{ik}, and the fact that $\vec{\mathrm{G}}_1$ exists depends on the fact (see Theorem 2.5) that the Q^1_{ik} form a basis for the quotient of $\mathcal{O}$ modulo the ideal generated by the first row of $\boxed{\mathrm{F}}$. Theorem 3.2 demonstrates the semilocal analog of Theorem 2.5 *near* α. However the semilocal $Q^1_{ik}(\alpha; \beta)$ may not be a basis far from α. Since $\vec{\mathrm{G}}_1$ is used in the definition of the restriction map ρ [see (2.17)], this shows the difficulty in defining the local restriction map $\rho^L_{\vec{\mathfrak{V}}(\alpha;\beta)}$. We also need to know the bounds in Theorem 3.2.

It is possible however, to overcome this difficulty by choosing the semilocal bases with more care. We shall not enter into the details as the result is needed in only one place on our work, namely, Theorem 7.3. We shall assume the result. In fact, it is possible to modify the construction in Chapter III to show that all the $\vec{\mathfrak{V}}(\alpha; \beta)$ may be assumed to be equal, and, in fact, the meaning of semi-local restriction is independent of α and β. Without going into the construction let us remark that the idea is already set forth in Theorem 3.8. In that theorem we employed the polynomials (of s) $1, s, s^2, \ldots, s^{m+1}$ as a "basis" for the quotient of $\mathcal{O}$ even though they are not independent in the semi-local theory. In the case of general r, l, we replace the $Q^j_{ik}(\alpha; \beta)$ by sets of polynomials which do not have the requisite linear independence but which do not depend on α and β.

The reader who is not interested in going through the details can content himself with thinking of the $\rho_{\vec{\mathfrak{V}}(\alpha;\beta)}$ as having only a semilocal meaning and, in fact, functions on $\vec{\mathfrak{V}}$ should be thought of as being semi-locally defined.

The reason for this is that, if $\vec{G}$ is globally defined, then $\rho_{\vec{\mathfrak{B}}(\alpha;\,\beta)}\,\vec{G}$ is a collection of semi-local functions on $\vec{\mathfrak{B}}$ (even though $\vec{\mathfrak{B}}$ itself is globally defined).

See Problem 4.1.

The definition of the global $\rho_{\vec{\mathfrak{B}}}$ is much more difficult. The reason is that, in general, there is no analog of Theorem 2.5 in the global case. This will be made clear in Chapter IX where the global analog of Theorem 2.5 will be shown to be the Cauchy problem. On the other hand, it may be possible to replace the Cauchy problem by a suitable parametrization problem (see Chapter X) and so to obtain a global analog of $\rho_{\vec{\mathfrak{B}}}$.

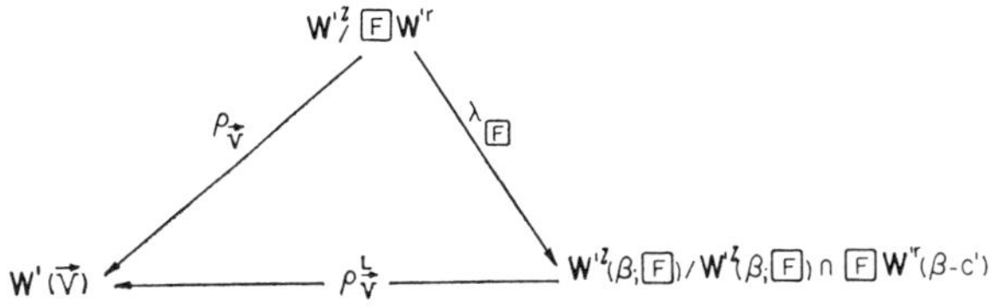

FIGURE 7.

THEOREM 4.2. *Given any c′ and any* β *sufficiently large; all maps in the diagram are topological isomorphisms onto and the diagram is commutative* ($\rho_{\mathfrak{B}}$ *is defined only for* $l = 1$).

Clearly Theorem 4.2 contains Theorem 4.1.

DEFINITION. We say that the AU space $\mathscr{W}$ is *localizeable* if Theorem 4.2 holds and if conditions (a) and (b) for PLAU hold.

In this case many problems involving $\mathbf{W}'$ can be reduced to semilocal problems. We write "$\mathscr{W}$ is LAU" if this is the case. Thus, Theorem 4.2 states that PLAU implies LAU.

IV.4. Proof of the Main Result

The commutativity of the diagram is obvious. The proof of Theorem 4.2 is divided into three main parts:

(4.8) Define $\vec{\mathfrak{B}}$ and $\rho^{L}_{\vec{\mathfrak{B}}}$.

(4.9) $\rho^{L}_{\vec{\mathfrak{B}}}$ is one-to-one and has a continuous inverse.

(4.10) $\lambda_{\boxed{F}}$ is one-to-one and has a continuous inverse.

DEFINITION OF $\vec{\mathfrak{V}}$ AND $\rho^{L}_{\vec{\mathfrak{V}}}$. For any α, β there is a vector multiplicity variety $\vec{\mathfrak{V}}(\alpha; \beta)$ satisfying the conclusions of Theorem 3.3.

Theorem 3.3 asserts that we may assume there are only finitely many. We let $\vec{\mathfrak{V}}$ be the collection of these $\vec{\mathfrak{V}}(\alpha^i; \beta^i)$, $i = 1, 2, \ldots, p$. The remarks preceding the statement of Theorem 4.2 explained how to define $\rho^{L}_{\vec{\mathfrak{V}}(\alpha;\beta)}$ for each fixed $(\alpha; \beta)$. We define

$$\rho^{L}_{\vec{\mathfrak{V}}}(\vec{\mathrm{L}}) = [\rho^{L}_{\vec{\mathfrak{V}}(\alpha^1;\beta^1)}(\vec{\mathrm{L}}), \ldots, \rho^{L}_{\vec{\mathfrak{V}}(\alpha^p;\beta^p)}(\vec{\mathrm{L}})].$$

The definitions of Section I.4 show what compatibility conditions must be satisfied on $\vec{\mathfrak{V}}$.

Proof of (4.9). We shall actually prove that $\rho^{L}_{\vec{\mathfrak{V}}}$ is one-to-one and has a continuous inverse whenever β is sufficiently large

We wish to construct the inverse to $\rho^{L}_{\vec{\mathfrak{V}}}$. Let $\vec{\vec{\mathrm{H}}} \in \mathbf{W}'(\vec{\mathfrak{V}})$. By the definition of $\vec{\mathfrak{V}}$, for each α, β'', $\vec{\vec{\mathrm{H}}}$ defines a function $\vec{\mathrm{H}}$ [or, more precisely, $\vec{\mathrm{H}}(\alpha; \beta'')$] in $\mathbf{W}'(\vec{\mathfrak{V}}(\alpha; \beta''))$. We need the following result:

Let A', $\beta' > 0$ be given. Then for each α there are A'', $\beta'' = \beta''(\alpha, \beta') \geqslant \beta'$ such that

(4.11) the β'' are bounded by $c(1 + \beta' + |\alpha|)^c(\beta')^{c'}$.

(4.12) For any $\vec{\mathrm{H}} \in \mathcal{O}(\vec{\mathfrak{V}}(\alpha; \beta'); \alpha; \beta''; A'')$ there is an $\vec{\mathrm{L}} \in \mathcal{O}^l(\alpha; \beta'; A')$ with

$$\rho^{L}_{\vec{\mathfrak{V}}(\alpha;\beta')}\vec{\mathrm{L}} = \vec{\mathrm{H}} \text{ on } \vec{\mathfrak{V}}(\alpha; \beta') \cap \gamma(\alpha; \beta').$$

(4.13) $$A'' \geqslant A'(c + |\alpha| + \beta'')^{-c}.$$

Here c, c' depend only on $\boxed{\mathrm{F}}$.

These statements are merely a transcription of Theorem 3.3.

$\vec{\mathrm{L}}$ is clearly a cochain for β'. We wish to conclude

(a) $\rho^{L}_{\vec{\mathfrak{V}}}(\vec{\mathrm{L}}) = \vec{\vec{\mathrm{H}}}$.
(b) $\vec{\mathrm{L}} \in \mathcal{O}^l(\beta; \boxed{\mathrm{F}})$.
(c) $\vec{\mathrm{L}}$ is a nice cochain.

Here β is assumed to be fixed and β' depends on β.

For the proofs we shall need

LEMMA 4.3. *For any $z^0 \in C$, the intersection of $\vec{\mathfrak{V}}$ with a sufficiently small neighborhood of z^0 is a local multiplicity variety $\vec{\mathfrak{V}}'(z^0)$ satisfying the conclusions of Theorem 2.6. (In Theorem 2.6 $z^0 = 0$ and $\vec{\mathfrak{V}}'(z^0)$ is denoted by $\vec{\mathfrak{V}}$.)*

PROOF OF LEMMA 4.3. If $l = 1$ and $\vec{\mathfrak{B}}$ (which is written $\mathfrak{B}$) is a hypersurface, then for $z^0 \in \gamma(\alpha; \beta)$ Lemma 3.12 shows that $\vec{\mathfrak{B}}(\alpha; \beta)$ coincides with $\vec{\mathfrak{B}}'(z^0)$ in a neighborhood of z^0. The inductive procedure of Chapter III shows that, for general $\vec{\mathrm{F}}$, $\vec{\mathfrak{B}}(\alpha; \beta)$ coincides with $\vec{\mathfrak{B}}'(z^0)$ in a neighborhood of zero. As noted in the remarks preceding Theorem 4.2, we shall not discuss further the case $l > 1$. Thus the only question that arises is for points z^0 which lie in some variety $\vec{\mathfrak{B}}(\tilde{\alpha}; \tilde{\beta})$, with $\vec{\mathfrak{B}}(\tilde{\alpha}; \tilde{\beta}) \neq \vec{\mathfrak{B}}(\alpha; \beta)$. These occur, for example, if $\tilde{\alpha}$ is far from z^0.

Denote by $\vec{\mathfrak{B}}(z^0)$ the restriction of $\vec{\mathfrak{B}}$ to a small neighborhood of z^0. The above shows that $\vec{\mathfrak{B}}(z^0)$ contains some $\vec{\mathfrak{B}}'(z^0)$. On the other hand, the construction of Chapter III shows that anything in the range of $\boxed{\mathrm{F}}$ certainly vanishes on any $\vec{\mathfrak{B}}(\alpha'; \beta')$; in fact

$$\boxed{\mathrm{F}}\,\mathcal{O}^r(z^0) = 0 \text{ on } \vec{\mathfrak{B}}(\alpha'; \beta') \cap N,$$

where N refers to a "germ" of a neighborhood of z^0. Here $\mathcal{O}(z^0)$ is the space of "germs" of functions regular at z^0. Thus the kernel of restriction to $\vec{\mathfrak{B}}(z^0)$ contains $\boxed{\mathrm{F}}\,\mathcal{O}^r(z^0)$. This means that the kernel of restriction to $\vec{\mathfrak{B}}(z^0)$ is exactly $\boxed{\mathrm{F}}\,\mathcal{O}^r(z^0)$, which is the desired result.

PROOF OF (a). The difficulty involved here is that certain $\vec{\mathfrak{B}}(\tilde{\alpha}; \tilde{\beta})$ which are used in the makeup of $\vec{\mathfrak{B}}$ and are different from $\vec{\mathfrak{B}}(\alpha; \beta')$ may meet $\gamma(\alpha; \beta')$. This difficulty is overcome as follows: We claim first that there is an entire function $\vec{\mathrm{G}}$ (that is, $\vec{\mathrm{G}}$ is regular in all of C^n) such that $\rho_{\vec{\mathfrak{B}}}\vec{\mathrm{G}} = \vec{\tilde{\mathrm{H}}}$. Because, by definition, for each point $z^0 \in C$ there is a function $\vec{\mathrm{G}}_{z^0}$ regular near z^0 whose restriction to $\vec{\mathfrak{B}}$ is equal to $\vec{\tilde{\mathrm{H}}}$ in a neighborhood of z^0. By Lemma 4.3, $\vec{\mathrm{G}}_{z^0}$ is determined up to the addition of an element of $\boxed{\mathrm{F}}\,\mathcal{O}^r(z^0)$. By Oka's theorem (see Gunning and Rossi [1]; also the proof of (4.10) below gives a proof of the result) such a global $\vec{\mathrm{G}}$ exists.

We know that

$$\rho^L_{\vec{\mathfrak{B}}(\alpha;\, \beta')}(\vec{\mathrm{L}}) = \vec{\tilde{\mathrm{H}}} \text{ on } \gamma(\alpha; \beta') \cap \vec{\mathfrak{B}}(\alpha; \beta')$$

because that is the way $\vec{\mathrm{L}}$ was constructed. Thus

$$\rho^L_{\vec{\mathfrak{B}}(\alpha;\, \beta')}(\vec{\mathrm{L}} - \vec{\mathrm{G}}) = 0 \qquad \text{on} \qquad \gamma(\alpha; \beta'). \tag{4.14}$$

We wish to deduce from this that, for any $\vec{\mathfrak{B}}(\tilde{\alpha}; \tilde{\beta})$ we have

$$\rho^L_{\vec{\mathfrak{B}}(\tilde{\alpha};\, \tilde{\beta})}(\vec{\mathrm{L}}) = \vec{\tilde{\mathrm{H}}} \qquad \text{on} \qquad \gamma(\alpha; \beta') \cap \vec{\mathfrak{B}}(\tilde{\alpha}; \tilde{\beta}). \tag{4.15}$$

For then, by definition of $\rho_{\vec{\mathfrak{B}}}^{L}$,

$$\rho_{\vec{\mathfrak{B}}}^{L}(\vec{\mathrm{L}}) = \vec{\tilde{\mathrm{H}}}.$$

Now, (4.14) would imply (4.15) if we knew that

$$\vec{\mathrm{L}} - \vec{\mathrm{G}} = \boxed{\mathrm{F}}\vec{\mathrm{K}} \qquad \text{on} \qquad \gamma(\alpha; \beta') \tag{4.16}$$

for a $\vec{\mathrm{K}}$ which is regular on $\gamma(\alpha; \beta')$, since the range of $\boxed{\mathrm{F}}$ is zero on $\vec{\mathfrak{B}}$ (see the proof of Lemma 4.3). Now, as in the proof of Lemma 4.3, $\vec{\mathfrak{B}}(\alpha, \beta')$ coincides near each point z^0 of $\gamma(\alpha, \beta')$ with some $\vec{\mathfrak{B}}'(z^0)$. Thus, by (4.14) and Theorem 2.6, $\vec{\mathrm{L}} - \vec{\mathrm{G}} \in \boxed{\mathrm{F}}\mathcal{O}^r(z^0)$ for each such z^0. By the above-mentioned theorem of Oka, (4.16) holds. This completes the proof of (a).

PROOF OF (b). Let α and α' be lattice points for which

$$\gamma(\alpha; \beta) \cap \gamma(\alpha'; \beta) \neq \varnothing.$$

Then, by (a)

$$\vec{\mathrm{L}}(\alpha; \beta) = \vec{\mathrm{L}}(\alpha'; \beta) \qquad \text{on} \qquad \vec{\mathfrak{B}} \cap \gamma(\alpha; \beta) \cap \gamma(\alpha'; \beta).$$

By Lemma 4.3 and Theorem 2.6 this means that, in the neighborhood of any $z^0 \in \gamma(\alpha; \beta) \cap \gamma(\alpha'; \beta)$, we can write

$$\vec{\mathrm{L}}(\alpha; \beta) - \vec{\mathrm{L}}(\alpha'; \beta) = \boxed{\mathrm{F}}\vec{\mathrm{M}}(z^0),$$

where $\vec{\mathrm{M}}$ is regular in a neighborhood of z^0. Oka's theorem, cited above, allows us to write

$$\vec{\mathrm{L}}(\alpha; \beta) - \vec{\mathrm{L}}(\alpha'; \beta) = \boxed{\mathrm{F}}\vec{\mathrm{M}}(\alpha, \alpha'; \beta)$$

on $\gamma(\alpha; \beta) \cap \gamma(\alpha'; \beta)$, where $\vec{\mathrm{M}}(\alpha,\alpha'; \beta)$ is regular on the whole of $\gamma(\alpha; \beta) \cap \gamma(\alpha'; \beta)$. This completes the proof of (b). □

PROOF OF (c). The difficulty we encounter is that $\vec{\mathrm{L}}$ may be too large. The reason for this is, by (4.13), the bounds we obtain for $\vec{\mathrm{L}}$ depend on the bounds for $\vec{\mathrm{H}}$ on $\gamma(\alpha; \beta'')$. By (4.11), β'' may be unbounded. On the other hand, property (b) of localizeability shows that if β'' were bounded, even if $\beta'' > \beta'$, the bounds on $\vec{\tilde{\mathrm{H}}}$ on $\gamma(\alpha; \beta'')$ would be allowable bounds for a good cochain for β'. The polynomial factor $(c + |\alpha| + \beta'')^c$ of (4.13) is not important because of (4.1).

Thus the only difficulty that remains to prove that $\vec{\mathrm{L}}$ is a good cochain is the size of β''. We overcome this difficulty as follows: Instead of supposing that β' is of fixed size, suppose that β' is very small, so small that the corresponding β'' satisfying (4.11) is $\leqslant 1$. Thus β' will be of size $\leqslant (1 + |\alpha|)^{-c''}/c''$ for a suitable c''. Hence β must be $\leqslant (1 + |\alpha|)^{-c'''}/c'''$ for a suitable c''' (larger than c''). We divide the cube $\gamma(\alpha; 1)$ into a "grid," that is, small cubes, sides parallel to the axes, the length of the sides being $(1+|\alpha|)^{-c'''}/20nc'''$. We think of the centers of these cubes as "lattice points" and apply the same sort of construction of $\vec{L}$ from $\vec{\tilde{H}}$ as above. (There is no difficulty in applying Theorem 3.3 because we never used the fact that the coordinates of the α are actually integers.) Note that the number of cubes involved is $\leqslant c''''(1+|\alpha|)c''''$. We must now "piece together" these $\vec{L}$ to obtain a single function on the whole cube $\gamma(\alpha; \beta')$ whose restriction to $\vec{\mathfrak{B}}(\alpha; \beta')$ is $\vec{\tilde{H}}$. As this piecing together process will be used in a much more complicated fashion in the proof of (4.10) below, we shall omit the details. We remark only that the fact that the numbers of small cubes involved is less than a power of $|\alpha|$ and the size of the cubes is greater than an inverse power of $|\alpha|$ is essential.

It remains to prove that $\rho^L_{\mathfrak{B}}$ is one-to-one. We should like to apply Theorem 3.3. However, this does not give a strong enough conclusion, for if $\vec{G} \in \mathbf{W}'^l(\beta; \boxed{\mathrm{F}})$ and if $\rho^L_{\mathfrak{B}}(\vec{G}) = 0$, then we can deduce from Theorem 3.3 that $\vec{G} \in \boxed{\mathrm{F}}\mathbf{W}'^r(\beta''')$ where β''' might be much smaller than β and, in addition, β might have to be chosen as growing with $|\alpha|$. This difficulty can be overcome just as the largeness of β'' was overcome on the previous page. Thus, (4.9) is proved. ▯

OUTLINE OF PROOF OF (4.10). Since the proof of (4.10) is somewhat involved, especially to those unfamiliar with the Oka-Cartan-Serre theory, we shall give an outline of the proof. We shall consider only the case $n=1$, $l=1$ as this is sufficient to give the general idea.

We are given a nice cochain $f \in \mathbf{W}'(\beta, \vec{F})$, and we want to modify f by addition of a cochain of the form $\vec{F}\vec{g}$ where $\vec{g} \in \mathbf{W}'^r(\beta)$ and obtain a globally defined function $h \in \mathbf{W}'$. Now for each lattice point α, $f(\alpha)$ is a function on $\gamma(\alpha; \beta)$. The difficulty is that $f(\alpha)$ may not be equal to $f(\alpha')$ on $\gamma(\alpha; \beta) \cap \gamma(\alpha'; \beta)$; for if $f(\alpha) = f(\alpha')$ on $\gamma(\alpha; \beta) \cap \gamma(\alpha'; \beta)$ for all α, α', then f would be a global function. We want to add $\vec{F}\vec{g}$ to f so that it is global. We do this a little at a time, that is, we gradually increase the set on which f is a function, not a general cochain.

This is done as follows: We start with $f(0)$ and $f(1)$. They may not be equal on $\gamma(0; \beta) \cap \gamma(1; \beta)$. Consider $f(0) - f(1)$ on $\gamma(0; \beta) \cap \gamma(1; \beta)$. By assumption, we can write

$$f(0) - f(1) = \vec{F}\vec{g}(0, 1),$$

where $\vec{g}$ is regular on $\gamma(0; \beta) \cap \gamma(1; \beta)$. Suppose we can write

$$\vec{g}(0, 1) = \vec{g}(0) - \vec{g}(1),$$

where $\vec{g}(0)$ is regular on $\gamma(0;\beta)$ and $\vec{g}(1)$ is regular on $\gamma(1;\beta)$. Then if we replace f by $(\vec{f}-\vec{F}\vec{g})=f'$, we have on $\gamma(0;\beta)\cap\gamma(1;\beta)$

$$\begin{aligned} f'(0)-f'(1) &= f(0)-\vec{F}\vec{g}(0)-f(1)+\vec{F}\vec{g}(1) \\ &= f(0)-f(1)-\vec{F}[\vec{g}(0)-\vec{g}(1)] \\ &= 0 \end{aligned}$$

by the construction.

Thus f' defines a regular function on $\gamma(0;\beta)\cap\gamma(1;\beta)$. We can continue the process and replace f' by f'' which defines a regular function on $\gamma(-1;\beta)\cup\gamma(0;\beta)\cup\gamma(1;\beta)$. Continuing in this way we can replace f by f^* which is a cochain for which $f^*(p)$ is defined and regular on the strip $|\text{Im } z|<\beta$, and $f^*(p)$ is independent of the integer p. We can perform a similar process starting with $f^*(ji)$ and $f^*(ji+1)$ for any integer j. We obtain finally a cochain f^{**} whose functions are defined on the strips $|\text{Im } z-j|<\beta$.

We now start with $f^{**}(0)$ and $f^{**}(i)$ and we want to piece them together. We want to write

$$f^{**}(0)-f^{**}(i)=\vec{F}\vec{g}^{**}(0,i)$$

on

$$\{|\text{Im } z|<\beta\}\cap\{|\text{Im } z-1|<\beta\}.$$

Suppose we can write $\vec{g}^{**}(0,i)=\vec{g}^{**}(0)-\vec{g}^{**}(i)$, where $\vec{g}^{**}(0)$ is regular on $|\text{Im } z|<\beta$ and $\vec{g}^{**}(i)$ is regular on $|\text{Im } z-1|<\beta$. Replacing f^{**} by $f^{**}-\vec{F}\vec{g}^{**}$, we obtain a cochain which is a function on

$$\{|\text{Im } z|<\beta\}\cup\{|\text{Im } z-1]|<\beta\}.$$

Continuing in this way we obtain a global function f^{***}.

There are two things that must be shown:

(a) How to split $\vec{g}(0,1)$ into $\vec{g}(0)-\vec{g}(1)$, and $\vec{g}^{**}(0,i)$ into $\vec{g}^{**}(0)-\vec{g}^{**}(i)$.

(b) How to obtain bounds for $f', f'', \ldots, f^*, \ldots, f^{**}, \ldots, f^{***}$.

We obtain (a) by use of a suitable version of the Cauchy integral formula, that is, we express $\vec{g}(0,1)$ (or $\vec{g}^{**}(0,i)$) as the Cauchy integral (with a somewhat modified kernel) over the boundary of $\gamma(0;\beta)\cap\gamma(1;\beta)$ (respectively $\{|\text{Im } z|<\beta\}\cap\{|\text{Im } z-1|<\beta\}$) and then "split" the integral. To achieve (b) we must first obtain bounds for $\vec{g}(0,1)$, ... Now $\vec{g}(0,1)$ is not uniquely determined. Theorem 3.18 says that we can obtain bounds for some choice (perhaps changing β). The modified Cauchy kernel used to achieve (a) is also used to obtain bounds for $\vec{g}(0)$ and $\vec{g}(1)$ from bounds for $\vec{g}(0,1)$.

We need also bounds for $\vec{g}^{**}(0, i)$. These *cannot* be obtained from Theorem 3.18 because $\vec{g}^{**}(0, i)$ is defined in a whole strip. In order to overcome this difficulty, we consider the restriction of $f^{**}(0) - f^{**}(i)$ to $\gamma(j; \beta'')$ for any real integer j for a suitable β''. By Theorem 3.18 we can write on $\gamma(j; \beta'')$

$$f^{**}(0) - f^{**}(i) = \vec{F}\vec{g}^{**}(0, i; j)$$

where we can estimate the size of $\vec{g}^{**}(0, i; j)$. Of course $\vec{g}^{**}(0, i; j)$ is not uniquely determined; may we add to it any $\boxed{F}^1\vec{h}^{**}(0, i; j)$ where $\boxed{F}^1$ generates the module of relations of the columns of $\vec{F}$. We can now piece together the $\vec{g}^{**}(0, i; j)$ just as we pieced together the $f(j)$ to obtain $\vec{g}^{**}(0, i)$ with bounds.

From this outline the following points should be emphasized (they apply for $n \geqslant 1$):

1. The piecing together of f proceeds one real variable at a time.
2. Simultaneously with piecing together f, we must piece together cochains for which $\vec{F}$ is replaced by $\boxed{F}^1$, the module of relations of the columns of $\vec{F}$.
3. The cochains that appear for $\boxed{F}^1$ are defined on sets $\gamma(\alpha; \beta'')$. Thus β must be sufficiently large in order that the process make sense.
4. We shall apply a modified Cauchy integral formula to write a function holomorphic on a set S as the difference of functions holomorphic on S^+ and S^- where $S^+ \cap S^- = S$. The modification of the Cauchy kernel is used to obtain bounds for the functions constructed.

PROOF OF (4.10). We define the modules $\boxed{F}^j$ for $j \geqslant 0$ as follows: Each $\boxed{F}^j$ is generated by a matrix which we denote again by $\boxed{F}^j$, which is an $l_j \times r_j$ matrix (compare Section II.1). $\boxed{F}^0$ is the module in $\mathscr{P}^l$ generated by $\boxed{F}$; $\boxed{F}^j$, for $j > 0$, is the module in $\mathscr{P}^{l_j}$ of relations of the columns of $\boxed{F}^{j-1}$. (Here $\mathscr{P}$ is the ring of polynomials.) The Hilbert syzygy theorem (see Nagata [1], p. 102) asserts that $\boxed{F}^j = \{0\}$ if $j > n$. (This could be proved by the methods of Chapter II but we shall not discuss the details). It is clear that

$$l_{j+1} = r_j, \quad \boxed{F}^j\boxed{F}^{j+1} = 0. \tag{4.17}$$

LEMMA 4.4. *Given any* $\beta > 1$ *and any* j, *if*

$$\vec{H} \in \mathbf{W}'^{l_j}(\beta) \cap \boxed{F}^j\mathscr{O}^{r_j}(\beta), \tag{4.18}$$

then also $\vec{H} \in \boxed{F}^j \mathbf{W}'^{r_j}(\beta - c'')$ *whenever* $0 < c'' < \beta - 1$. *Moreover, if* $\vec{H}$ *is as above and if we write* $\vec{H} = \boxed{F}^j \vec{G}$, *then* $\vec{G}$ *can be chosen to vary continuously with* $\vec{H}$.

By "$\vec{G}$ varying continuously with $\vec{H}$" is meant that for any $k \in K$ we can find a $k' \in K$ so that if $\|\vec{H}\|_{k'} \leqslant 1$, we can choose $\vec{G} \in \mathbf{W}'^{r_j}(\beta - c'')$ with $\|\vec{G}\|_k \leqslant 1$.

The proof of Lemma 4.4 is essentially the same as the proof of the statements (c) in the proof of (4.9) and so will be omitted.

We shall now write $\alpha = (\alpha_1, \alpha_2, \ldots, \alpha_{2n})$, where the α_j are integers which are the coordinates of α in C^n considered as real $2n$-dimensional space R^{2n}. Moreover, the coordinates x_j of R^{2n} are chosen so that $z_j = x_{2j-1} + ix_{2j}$. For any a with $0 \leqslant a \leqslant 2n$ and any β we denote by $\mathcal{O}_a(\beta)$ the space of functions (cochains) $f(\alpha; z)$ which, for each α with $\alpha_1 = 0, \ldots, \alpha_a = 0$, are holomorphic (in z) on the strip $\gamma_a(\alpha; \beta) = \{x \mid |x_j - \alpha_j| < \beta \text{ for } j > a\}$. Thus, $\mathcal{O}_0(\beta) = \mathcal{O}(\beta)$ and $\mathcal{O}_{2n}(\beta)$ is the space of entire functions. We define $\mathbf{W}'_a(\beta)$ as the set of cochains in $\mathcal{O}_a(\beta)$ which satisfy the required growth conditions.

We wish now to show that for each j and each β the spaces $\mathbf{W}'^{l_j}_a(\beta; \boxed{F}^j)/\boxed{F}^j \mathbf{W}'^{r_j}_a(\beta)$ for varying a are all topologically isomorphic, the isomorphisms being very similar to to $\lambda_{\boxed{F}}$. This result for $j = 0$, $a = 0$, and $a = 2n$ will give the desired (4.10). Unfortunately, we cannot show that all these spaces are topologically isomorphic directly; we must let the β change slightly. This is, of course, unimportant for the final result, since $\mathbf{W}'^{l_j}_{2n}(\beta)$ is independent of β. For this reason we shall not keep too careful account of the size of β.

THEOREM 4.5. *There are* $c_1, c_1' > 0$ *which may be chosen arbitrarily small so that, for any* β *sufficiently large, for any* a, j, *the natural map*

$$\lambda_a^j\colon \mathbf{W}'^{l_j}_{a+1}(\beta; \boxed{F}^j)/\boxed{F}^j \mathbf{W}'^{r_j}_{a+1}(\beta - c_1) \cap \mathbf{W}'^{l_j}_{a+1}(\beta) \to \mathbf{W}'^{l_j}_a(\beta; \boxed{F}^j)/\boxed{F}^j \mathbf{W}'^{r_j}_a(\beta - c_1) \cap \mathbf{W}'^{l_j}_a(\beta) \tag{4.19}$$

is continuous and λ_a^j *is "one-to-one", that is, if* $\lambda_a^j \vec{Q} = 0$ *then* $\vec{Q} = \boxed{F}^j \vec{R}$ where $\vec{R} \in \mathbf{W}'^{r_j}_{a+1}(\beta - c_1')$. *Moreover, there exists a continuous inverse to* λ_a^j *on the smaller space*

$$\mathbf{W}'^{l_j}_a(\beta + c_1'; \boxed{F}^j)/\boxed{F}^j \mathbf{W}'^{r_j}_a(\beta) \cap \mathbf{W}'^{l_j}_a(\beta + c_1'). \tag{4.20}$$

In particular, λ_a^j *is "essentially" onto.*

Remark 4.3. See also (4.20*) below for a slight modification of the above theorem. This modification will also be referred to as "Theorem 4.5."

It is clear that (4.10) is a consequence of Theorem 4.5 and the fact that $\lambda_{\boxed{F}}$ is the composition of the λ_a^j.

THEOREM 4.6. *For β sufficiently large, and for any a, c', and for any j, if*

$$(4.21)\qquad \vec{H} \in \mathbf{W}'^{l_j}_a(\beta - c') \cap \boxed{F}^j \mathcal{O}^{r_j}_a(\beta - c'),$$

then also $\vec{H} \in \boxed{F}^j \mathbf{W}'^{r_j}_a(\beta - c'')$ *for a suitable c'' depending only on c'. If c' is arbitrarily small, c'' may also be chosen arbitrarily small. Moreover, if $\vec{H}$ is as above and if we write $\vec{H} = \boxed{F}^j\vec{G}$, then $\vec{G}$ can be chosen to vary continuously with $\vec{H} \in \mathbf{W}'^{l_j}_a(\beta - c')$.*

We shall prove Theorems 4.5 and 4.6 by simultaneous induction on a, so we denote the theorems by T4.5(a) and T4.6(a). Now T4.6(0) is just Lemma 4.4. Thus we must show that

(4.22) T4.5(a) and T4.6(a) imply T4.6($a + 1$);

(4.23) T4.6(a) implies T4.5(a).

PROOF OF (4.22). Let $\vec{H} \in \mathbf{W}'^{l_j}_{a+1}(\beta - c') \cap \boxed{F}^j \mathcal{O}^{r_j}_{a+1}(\beta - c')$. Then $\vec{H}$ defines naturally an element $\lambda^j_a \vec{H} \in \mathbf{W}'^{l_j}_a(\beta - c') \cap \boxed{F}^j \mathcal{O}^{r_j}_a(\beta - c')$. By T4.6($a$) we can write $\lambda^j_a \vec{H} = \boxed{F}^j \vec{G}$ where $\vec{G} \in \mathbf{W}'^{r_j}_a(\beta - c'')$ can be chosen to vary continuously with $\vec{H}$. We want to replace this $\vec{G}$ by a

$$\vec{G}'' \in \mathbf{W}'^{r_j}_{a+1}(\beta - c'' - c_1').$$

DEFINITION. The lattice points α, α' are called *a semiadjacent* if the first a components of α and α' are zero, and the last $2n - a - 1$ components are equal. If in addition $|\alpha_{a+1} - \alpha'_{a+1}| = 1$, we call α and α' *a adjacent*. If α and α' are a semiadjacent, we write $\alpha > \alpha'$ or $\alpha < \alpha'$ depending on the relative size of α_{a+1} and α'_{a+1}. When we write $\alpha < \alpha'$ it is to be understood that α and α' are semiadjacent.

For each α with $\alpha_1, \ldots, \alpha_a = 0$, let α' be a semiadjacent lattice point $> \alpha$. We consider such pairs α, α' for which $\alpha'_{a+1} - \alpha_{a+1}$ is fixed. Consider

$$\vec{G}(\alpha') - \vec{G}(\alpha) = \vec{L}(\alpha).$$

Then $\vec{L}(\alpha)$ is a cochain in $\mathbf{W}'^{r_j}_a(\beta - c'' - |\alpha_{a+1} - \alpha'_{a+1}|)$. By the definition of $\vec{G}$, we have

$$(4.24)\qquad \boxed{F}^j \vec{L}(\alpha) = (\lambda^j_a \vec{H})(\alpha') - (\lambda^j_a \vec{H})(\alpha) = 0$$

because $\vec{H} \in \mathbf{W}'^{l_j}_{a+1}(\beta - c')$. Thus, by definition,

$$(4.25\qquad \vec{L}(\alpha) \in \boxed{F}^{j+1} \mathcal{O}^{r_{j+1}}_a (\beta - c'' - |\alpha'_{a+1} - \alpha_{a+1}|).$$

We should like to conclude from this that

$$(4.26)\qquad \vec{G} \in \mathbf{W}'^{r_j}_a(\beta - c''; \boxed{F}^{j+1}).$$

For then we could apply the "onto" part of T4.5(a) (for $j+1$ in place of j) and use the fact that $\boxed{\mathrm{F}}^{j}\boxed{\mathrm{F}}^{j+1}=0$ to obtain (4.22). However, we cannot prove (4.26). For in order to verify (4.26) we need to study analogs of $\vec{\mathrm{L}}(\alpha)$ for points α' such that $\alpha_b=\alpha'_b$ for all but one $b>a$ without this b being necessarily equal to $a+1$. This is because in the definition of $\mathbf{W}_a^{\prime r_j}(\beta;\ \boxed{\mathrm{F}}^{j+1})$ we require that for any $\vec{\mathrm{f}}\in\mathbf{W}_a^{\prime r_j}(\beta;\ \boxed{\mathrm{F}}^{j+1})$ and any lattice points α'', α''' all of whose coordinates are equal except one (which may *not* be the $a+1$ coordinate), we have $\vec{\mathrm{f}}(\alpha'')-\vec{\mathrm{f}}(\alpha''')\in\boxed{\mathrm{F}}^{j+1}\mathcal{O}^{r_{j+1}}(\gamma(\alpha'';\beta)\cap\gamma(\alpha''';\beta))$. As is clear from (4.24) we could not verify this for $\vec{\mathrm{f}}=\vec{\mathrm{G}}$.

We therefore define the space ${}_W\mathbf{W}_a^{\prime r_j}(\beta;\ \boxed{\mathrm{F}}^{j+1})$ as consisting of all $\vec{\mathrm{K}}\in\mathbf{W}_a^{\prime r_j}(\beta)$ such that if α, α' are a semiadjacent, then

$$\vec{\mathrm{K}}(\alpha')-\vec{\mathrm{K}}(\alpha)\in\boxed{\mathrm{F}}^{j+1}\mathcal{O}_a^{r_{j+1}}(\beta-|\alpha_{a+1}-\alpha'_{a+1}|).$$

Then the same proof as we shall give below for Theorem 4.5 [namely, (4.23)] will show that there exists a continuous inverse to the map λ_a^j going from

$$\text{(4.20*)}\qquad {}_W\mathbf{W}_a^{\prime l_j}(\beta+c'_1;\ \boxed{\mathrm{F}}^{j})/\boxed{\mathrm{F}}^{j}\mathbf{W}_a^{\prime r_j}(\beta)\cap\mathbf{W}_a^{\prime l_j}(\beta+c'_1)\to \mathbf{W}_{a+1}^{\prime l_j}(\beta;\ \boxed{\mathrm{F}}^{j})/\boxed{\mathrm{F}}^{j}\mathbf{W}_{a+1}^{\prime r_j}(\beta-c_1)\cap\mathbf{W}_{a+1}^{\prime l_j}(\beta).$$

When referring to Theorem 4.5 we shall refer to either the original statement or to this modified one.

We can now apply T4.5(a), with (4.20) replaced by (4.20*), and j replaced by $j+1$, to $\vec{\mathrm{G}}$. (4.25) shows that $\vec{\mathrm{G}}\in{}_W\mathbf{W}_a^{\prime l_{j+1}}(\beta-c'';\ \boxed{\mathrm{F}}^{j+1})$. This shows that we can write, for some $\vec{\mathrm{G}}'$ (depending on $\vec{\mathrm{G}}$) in

$$\boxed{\mathrm{F}}^{j+1}\mathbf{W}_a^{\prime r_{j+1}}(\beta-c''-c'_1)\cap\mathbf{W}_a^{\prime r_{j+1}}(\beta-c'')$$

$$\text{(4.27)}\qquad \vec{\mathrm{G}}+\vec{\mathrm{G}}'=\lambda_a^{j+1}\vec{\mathrm{G}}'',$$

where $\vec{\mathrm{G}}''\in\mathbf{W}_{a+1}^{\prime l_{j+1}}(\beta-c''-c'_1)$ can be chosen to vary continuously with $\vec{\mathrm{G}}$. We have

$$\text{(4.28)}\qquad \boxed{\mathrm{F}}^{j}\vec{\mathrm{G}}=\boxed{\mathrm{F}}^{j}\lambda_a^{j+1}\vec{\mathrm{G}}''-\boxed{\mathrm{F}}^{j}\vec{\mathrm{G}}'.$$

Since $\vec{\mathrm{G}}'\in\boxed{\mathrm{F}}^{j+1}\mathbf{W}_a^{\prime r_{j+1}}$, we must have $\boxed{\mathrm{F}}^{j}\vec{\mathrm{G}}'=0$. Thus,

$$\boxed{\mathrm{F}}^{j}\vec{\mathrm{G}}=\boxed{\mathrm{F}}^{j}\lambda_a^{j+1}\vec{\mathrm{G}}''.$$

But clearly $\boxed{\mathrm{F}}^{j}$ commutes with λ_a^j, that is, $\boxed{\mathrm{F}}^{j}\lambda^{j+1}=\lambda^j\boxed{\mathrm{F}}^{j}$, so that $\boxed{\mathrm{F}}^{j}\vec{\mathrm{G}}=\lambda_a^j\boxed{\mathrm{F}}^{j}\vec{\mathrm{G}}''$. This means that

$$\text{(4.29)}\qquad \lambda_a^j\boxed{\mathrm{F}}^{j}\vec{\mathrm{G}}''-\lambda_a^j\vec{\mathrm{H}}=0$$

from which it follows from the definition that $\boxed{\mathrm{F}}^j\vec{\mathrm{G}}'' = \vec{\mathrm{H}}$ which proves (4.22). □

PROOF OF (4.23). We shall prove first that λ_a^j is onto and has a "continuous inverse."

Let $\vec{\mathrm{H}}$ be given in $\mathbf{W}_a'^{l_j}(\beta; \boxed{\mathrm{F}}^j)$. Then we consider, whenever α, α' are a adjacent with $\alpha < \alpha'$,

$$\vec{\mathrm{L}}(\alpha) = \vec{\mathrm{H}}(\alpha') - \vec{\mathrm{H}}(\alpha) \tag{4.30}$$

so $\vec{\mathrm{L}} \in \mathbf{W}_a'^{l_j}(\beta - 1) \cap \boxed{\mathrm{F}}^j \mathcal{O}^{r_j}(\beta - 1)$ by the definition of $\mathbf{W}_a'^{l_j}(\beta; \boxed{\mathrm{F}}^j)$. By T4.6($a$) we can write $\vec{\mathrm{L}} = \boxed{\mathrm{F}}^j\vec{\mathrm{G}}$ where $\vec{\mathrm{G}} \in \mathbf{W}_a'^{r_j}(\beta - c'')$. Moreover, $\vec{\mathrm{L}}$ varies continuously with $\vec{\mathrm{H}}$, so $\vec{\mathrm{G}}$ may also be so chosen.

We wish now to apply the Cauchy integral formula to $\vec{\mathrm{G}}$. There are two cases to distinguish:

Case 1. *a is even.* We consider the complex $z_{a'}$ plane where $a' = 1 + a/2$. To each lattice point α whose first a coordinates are zero, we shall associate a function $\varphi(\alpha) = \varphi(\alpha; z_{a'}) = \varphi(\alpha; z)$, which depends only on a single complex variable $z_{a'}$, whose properties will be listed below. (The existence of functions $\varphi(\alpha)$ will depend on the fact that $\mathscr{W}$ is product localizeable.) We write the Cauchy integral

$$\vec{\mathrm{G}}(\alpha; z) = \frac{\varphi(\alpha; z)}{2\pi i}\int_{\Gamma_a(\alpha)} \frac{\vec{\mathrm{G}}(\alpha; z_1, \ldots, z_{a'-1}, w, z_{a'+1}, \ldots, z_n)}{(w - z_{a'})\varphi(\alpha; w)}\, dw, \tag{4.31}$$

where $\Gamma_a(\alpha)$ is the boundary of the rectangle $\gamma(1; \alpha_{a+1} + i\alpha_{a+2}; \beta - c'')$ in the complex $z_{a'}$ plane. Here we assume that $\varphi(\alpha; z)$ has no zeros in or on this contour. See Fig. 8.

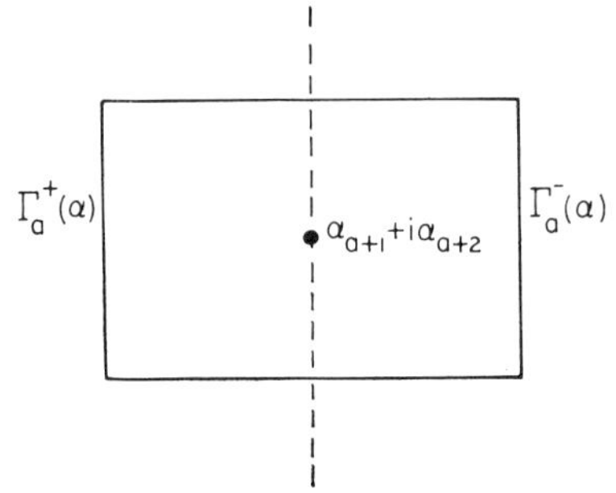

FIGURE 8

We now split $\Gamma_a(\alpha)$ into $\Gamma_a^+(\alpha) \cup \Gamma_a^-(\alpha)$, where $\Gamma_a^+(\alpha)$ is the part of $\Gamma_a(\alpha)$ where $Rw \leqslant \alpha_{a+1}$ and $\Gamma_a^-(\alpha)$ is the other part. From this splitting we derive a splitting $\vec{\mathrm{G}}(\alpha) = \vec{\mathrm{G}}^+(\alpha) + \vec{\mathrm{G}}^-(\alpha)$.

Now for any α whose first a components are zero, and for those z with $z_{a'} \in \gamma(1; \alpha_{a+1} + i\alpha_{a+2}; \beta - c'' - 2)$ we define

$$(4.32) \qquad \vec{\tilde{H}}(\alpha; z) = \vec{H}(\alpha; z) + \sum_{\alpha' \geq \alpha} \boxed{F}^j \vec{G}^-(\alpha'; z) - \sum_{\alpha'' < \alpha} \boxed{F}^j \vec{G}^+(\alpha''; z)$$

(α' and α'' are semiadjacent to α). Of course, it must be shown that the sums on the right side converge; we leave this question aside now and operate formally.

For $\alpha' \geqslant \alpha$ it is clear that $\vec{G}^-(\alpha') \in \mathcal{O}_a^{r_j}(\alpha; \beta - c'' - 2)$ and similarly $G^+(\alpha'') \in \mathcal{O}_a^{r_j}(\alpha; \beta - c'' - 2)$ if $\alpha'' < \alpha$. Thus, $\vec{\tilde{H}} \in \mathcal{O}_a^{l_j}(\beta - c'' - 2)$.

Properties of $\vec{\tilde{H}}$ *(formal).*

1. $\vec{\tilde{H}} \in \mathcal{O}_a^{l_j}(\beta - c'' - 2)$ (just shown).
2. Let α' be a adjacent to α with $\alpha' > \alpha$ and let $z \in \gamma_a(\alpha; \beta - c'' - 2) \cap \gamma_a(\alpha'; \beta - c'' - 2)$. Then we can write

$$\begin{aligned}(4.33) \quad \vec{\tilde{H}}(\alpha; z) - \vec{\tilde{H}}(\alpha'; z) &= \vec{H}(\alpha; z) - \vec{H}(\alpha'; z) \\ &\quad + \sum_{\alpha''' \geqslant \alpha} \boxed{F}^j \vec{G}^-(\alpha'''; z) - \sum_{\alpha'''' < \alpha} \boxed{F}^j \vec{G}^+(\alpha''''; z) \\ &\quad - \sum_{\alpha''''' \geqslant \alpha'} \boxed{F}^j \vec{G}^-(\alpha'''''; z) + \sum_{\alpha'''''' < \alpha'} \boxed{F}^j \vec{G}^+(\alpha''''''; z) \\ &= \vec{H}(\alpha; z) - \vec{H}(\alpha'; z) \\ &\quad + \boxed{F}^j \vec{G}^-(\alpha; z) + \boxed{F}^j \vec{G}^+(\alpha; z) \\ &= -\boxed{F}^j \vec{G}(\alpha; z) + \boxed{F}^j [\vec{G}^+(\alpha; z) + \vec{G}^-(\alpha; z)] \\ &= 0\end{aligned}$$

because $\vec{G} = \vec{G}^+ + \vec{G}^-$. Thus, $\vec{\tilde{H}} \in \mathcal{O}_{a+1}^{l_j}(\beta - c'' - 2)$, i.e., the cochain $\vec{H}(\alpha; z)$ does not depend on the a coordinate of α.

3. $\vec{\tilde{H}} - \vec{H} \in \boxed{F}^j \mathcal{O}_a^{r_j}(\beta - c'' - 2)$ (clear from the definitions).

Property 3 shows that

$$(4.34) \qquad \lambda_a^j \vec{\tilde{H}} = \vec{H}.$$

It remains to improve Properties 2 and 3 to show that

$$\vec{\tilde{H}} \in \mathbf{W}'^{l_j}_{a+1}(\beta - c'' - 2)$$

and that

$$\vec{\tilde{H}} - \vec{H} \in \boxed{F}^j \mathbf{W}'^{r_j}_a(\beta - c'' - 2).$$

We must also show that $\vec{\tilde{H}}$ can be chosen to vary continuously with $\vec{H}$. This, of course, depends on the choice of $\varphi(\alpha; z)$. The ability to choose a proper $\varphi(\alpha; z)$ will depend on the fact that $\mathscr{W}$ is product localizeable.

We now use (4.31) and the definitions of $\Gamma^{\pm}$ and $\vec{G}^{\pm}$ to obtain estimates for $\vec{G}^{-}(\alpha')$ on $\gamma_a(\alpha; \beta - c'' - 2)$ for $\alpha' \geqslant \alpha$ and for $\vec{G}^{+}(\alpha'')$ on $\gamma_a(\alpha; \beta - c'' - 2)$ for $\alpha'' < \alpha$. We note that for such z we have $|w - z_{a'}| \geqslant \frac{1}{2}$ for $w \in \Gamma_a(\alpha)$. Thus, up to a simple constant factor which we shall ignore,

$$(4.35)\quad |G^{-}(\alpha'; z)| \leqslant \frac{(\beta - c'')|\varphi(\alpha'; z)| \max_{w \in \Gamma_a(\alpha')} |\vec{G}(\alpha'; z_1, \ldots, z_{a'-1}, w, z_{a'+1}, \ldots, z_n)|}{\min_{w \in \Gamma_a(\alpha')} |\varphi(\alpha'; w)|}$$

Let us denote the ratio of the max to the min on the right side of (4.35) by $\sigma(\alpha; \vec{G}; \varphi; z)$, so σ is, of course, independent of $z_{a'}$. Thus, (4.35) becomes

$$(4.36)\qquad |\vec{G}^{-}(\alpha; z)| \leqslant \beta |\varphi(\alpha'; z)\sigma(\alpha'; G; \varphi; z)|.$$

To investigate the convergence in (4.32), we must check the convergence of

$$\sum_{\alpha' \geqslant \alpha} \vec{G}^{-}(\alpha')$$

and of

$$\sum_{\alpha'' < \alpha} \vec{G}^{+}(\alpha'').$$

Using (4.36) we must estimate

$$(4.37)\qquad \sum_{\alpha' \geqslant \alpha} |\varphi(\alpha'; z)\sigma(\alpha'; \vec{G}; \varphi; z)|.$$

Thus we need:

We can choose the cochain φ in such a way that the series (4.37) *converges uniformly for z in compact subsets of $\gamma_a(\alpha; \beta - c'' - 2)$. Moreover, the value of the sum is* (*uniformly in* α) $0(k(z))$ *for any $k \in K$. Given any $k \in K$ we can find a $k' \in K$ so that if $|\vec{G}(\alpha; z)|/k'(z) \leqslant 1$ for all α, z, then the value of the series* (4.37) *is* $\leqslant k(z)$.

The possibility of choosing such a cochain φ comes from property (d) of product localizeability. The right side of (4.35) is exactly of the form for which we can get estimates from (4.3). Namely, call

$$(4.38)\qquad b(z^0) = \max_{\alpha', w \in \Gamma_a(\alpha')} |\vec{G}(\alpha'; z_1^0, \ldots, z_{a'-1}^0, w, z_{a'+1}^0, \ldots z_n^0)|.$$

Here the max over α' means over all α' for which $z_{a'}^0$ is in the interior of $\Gamma_a(\alpha')$. (These α' are finite in number.) Since $\vec{G}$ is a nice cochain it follows from property (b) of product localizeability that $b(z^0) = 0(k(z^0))$ for all $k \in K$. Moreover, it is clear from property (b) of product localizeability that b has a continuous majorant $m(z^0)$ which is still $0(k(z^0))$ for all $k \in K$. Thus, $m \in M$. Again using property (b) of localizeability we can establish easily the following

LEMMA 4.7. *There exists an analytic uniform structure $K = \{k\}$ so that for any $k' \in K$ there is a $k'' \in K$ and a $c' > 0$ such that if $\|\vec{G}\|_{k''} < 1$ then $m(z^0)$ can be chosen $\leqslant c'k'(z^0)$.*

Let $k \in K$. By our assumptions we may replace k by products of functions k_i of single variables z_i. Let $k'_{a'}$ be related to $k_{a'}$ as in property (d) of localizeability (where $k'_{a'}$ and $k_{a'}$ are denoted simply by k', k). Let $k'_j = k_j$ for $j \neq a'$ and call k' the product of the k'_p for all p. Thus k' is in an AU structure for $\mathscr{W}$. Let k'' and m be related to k' according to Lemma 4.7. We now define $\varphi(\alpha; w)$ to be a function of the type $\varphi(w; \alpha'_{a+1} + i\alpha'_{a+2}; \beta)$ associated to $m_{a'}$ and $k'_{a'}$ by property (d) of product localizeability.

It follows from Lemma 4.7 and (4.3) that $\|\vec{G}\|_{k''} \leqslant 1$ implies that the right side of (4.35) is $\leqslant c'''k(z)$ and is also $\leqslant c''m'(z)$ for some $m' \in M$. (m' may depend on m in addition to k.)

We now examine the right side of (4.32). We know that each term has suitable growth and we should like to conclude that the same holds for the sum. It is clear that the factors $\boxed{F}^j$ play no important role in the estimates, since they could introduce polynomial factors in the estimates which are unimportant by property (b) of localizeability. For the same reason, the sum of the terms can be estimated, essentially, by the maximum of the terms which we just estimated. Thus, our estimates allow us to conclude that $\vec{\tilde{H}} \in \mathbf{W}'^{l_j}_{a+1}(\beta - c'' - 2)$ and that $\vec{\tilde{H}}$ varies continuously with $\vec{H}$. It is clear from (4.32) and the above that

$$\vec{H} - \vec{\tilde{H}} \in \boxed{F}^j \mathbf{W}'^{r_j}_a(\beta - c'' - 2).$$

Thus, in case a is even, we have produced a continuous map μ^j_a from

$$\mathbf{W}'^{l_j}_a(\beta + c'_1; \boxed{F}^j)/\boxed{F}^j \mathbf{W}'^{r_j}_a(\beta) \cap \mathbf{W}'^{l_j}_a(\beta + c'_1)$$

into

$$\mathbf{W}'^{l_j}_{a+1}(\beta; \vec{F}^j)/\boxed{F}^j \mathbf{W}'^{r_j}_{a+1}(\beta - c_1) \cap \mathbf{W}'^{l_j}_{a+1}(\beta)$$

which is an inverse to λ^j_a in the sense that

$$\lambda^j_a \mu^j_a = \text{identity}.$$

It remains to show that λ^j_a is one-to-one. This means that, given any $\vec{K} \in \mathbf{W}'^{l_j}_{a+1}(\beta; \boxed{F}^j)$, if $\lambda^j_a \vec{K} \in \boxed{F}^j \mathbf{W}'^{r_j}_a(\beta - c_1)$, then we can find an

$\vec{L} \in \mathbf{W}_{a+1}^{\prime r_j}(\beta - c_1)$ with $\vec{K} = \boxed{F}^j \vec{L}$. Let us write

$$\lambda_a^j \vec{K} = \boxed{F}^j \vec{Q}.$$

Thus $\vec{Q} \in \mathbf{W}_a^{\prime r_j}(\beta - c_1)$ and we want to replace $\vec{Q}$ by $\tilde{\vec{Q}} \in \mathbf{W}_{a+1}^{\prime r_j}(\beta - c_1')$ and still have

$$\lambda_a^j \vec{K} = \boxed{F}^j \tilde{\vec{Q}}.$$

Since $\vec{K} \in \mathbf{W}_{a+1}^{\prime l_j}(\beta; \boxed{F}^j)$, $\vec{Q}$ satisfies the same kind of conditions that are satisfied by the cochain denoted by $\vec{G}$ in the proof of (4.22). Thus, as in the proof of (4.22) we can produce $\tilde{\vec{Q}}$. (The sharp reader will note that we have not used the induction argument correctly, since the proof of (4.22) assumed T4.5(a). However, the proof of (4.22) does not use the fact that λ_a^j is one-to-one. Thus there is no difficulty in our logic.) This completes the proof of (4.23) in case a is even.

Case 2. *a is odd.* We want again to apply the Cauchy integral formula to G in the $z_{a'}$ plane where now $a' = (a+1)/2$. Instead of using the contours $\Gamma_a(\alpha)$ that we used in Case 1, we use the boundaries of suitable infinite strips parallel to the real $z_{a'}$ axis. This is carried out exactly as in Case 1, and we conclude that (4.23) holds in this case also. ▯

Thus Theorems 4.5 and 4.6, hence also Theorems 4.1 and 4.2, are completely proved. ▯

For some applications the conditions (d) and (e) for product localizeability are too strong. We shall replace them by (d′) and (e′) below; a space satisfying (a), (b), (c), (d′), and (e′) will be called *weakly product localizeable.*

We require again that k and m can be chosen to consist of products of functions of a single complex variable. For the corresponding spaces of functions of one variable (for which we use the same notation) we require

(d′) For each $k \in K$, $a > 0$, there is a $k' \in K$ so that for any $A > 2a$ and any $m \in M$ with $m(z) \leqslant k'(z)$ for all z and any z^0 with $|z^0| < A$, we can find a function $\varphi(z; z^0, a, A)$ which is holomorphic on $|z| < 2A$ such that

$$\frac{m(z^0)|\varphi(z; z^0, a, A)|}{\min\limits_{|\zeta - z^0| \leqslant a} |\varphi(\zeta; z^0, a, A)|} \leqslant c\, k(z) \tag{4.3′}$$

whenever $|z| < A$. Moreover, there is a $c'' > 0$ and an $m' \in M$ so that the left side of (4.3′) is $\leqslant c'' m'(z)$. Here c and k' depend only on k and a but c'' and m' may depend also on m but *not* on A.

(e′) The same modification of (e).

We claim that all the results of this chapter hold if we use (d′) and (e′) in place of (d) and (e). The only place where (d) and (e) were used above is in the proof of (4.23). If we examine our proof of (4.23) we see that (d′) and (e′) would enable us to prove the analog of (4.23) with the modification that the functions and cochains are considered as defined and holomorphic on large cubes in C instead of on all of C itself. Using a simple argument involving normal families, we conclude again that (4.23) holds. Since we shall not use (d′) and (e′) in any essential way, we shall omit the details.

See Problem 4.2.

For later applications we need a strengthening of Theorem 3.18.

THEOREM 4.8. *Let* $\boxed{\mathrm{F}}$ *be a matrix of polynomials. Then for any* $\varepsilon > 0$, *there exists a* $c > 0$, *so that if* b *is sufficiently large and if*

$$|\boxed{\mathrm{F}}\vec{\mathrm{H}}| \leqslant a \qquad \text{on} \qquad |z - z_0| \leqslant b$$

then for a suitable $\vec{\mathrm{H}}_1$ *we have*

$$|\vec{\mathrm{H}} - \boxed{\mathrm{F}}^1\vec{\mathrm{H}}_1| \leqslant ca(1 + |z_0| + b)^c \tag{4.39}$$

for

$$|z - z_0| \leqslant b - \varepsilon \tag{4.40}$$

Here $\boxed{\mathrm{F}}^1$ *is the matrix representing the modules of relations of the columns of* $\boxed{\mathrm{F}}$. $\vec{\mathrm{H}}$ *and* $\vec{\mathrm{H}}_1$ *are holomorphic on the sets in question.*

PROOF. Theorem 3.18 gives this result for a much smaller $|z - z_0|$ than that given by (4.40). We fill the cube of (4.40) by a finite number of small cubes of side $\varepsilon b/c^1(1 + |z_0| + b)^c$, where say $c^1 > 10nc(1 + b)$. These cubes are put together in a manner similar to the way the cubes $|z - \alpha| < \beta$, where α is a lattice point and $\beta > 1$, are put together in C. Then a simple modification of the method used to prove Theorem 4.2 (actually in a much simpler form, since we don't need any convergence factors φ or ψ) gives the result. (Compare the proof of (c) in the proof of (4.9) above.) ▯

IV.5. The Oka Embedding

Up to now we have used weighting functions k and m which are products of functions of a single complex variable. Thus (see Chapter V, Example 5), our methods will apply to the space $\mathscr{W} = \mathscr{E}(\Omega)$ of indefinitely differentiable functions on Ω when Ω is an open or closed cube. We want to extend the results to the case when Ω is an arbitrary open or closed convex set, namely, we want to show that $\mathscr{E}(\Omega)$ is LAU in the general case. We show (see Theorem 4.9 below) that the fact that $\mathscr{E}(\Omega)$ is PLAU when Ω is a cube implies the general case.

For this purpose we use an idea introduced by Oka in a somewhat different context. Let Ω be an open, convex, bounded, symmetric polyhedron in R^n.

We may represent Ω as follows: There exist real linear functions $L_i(x)$ for $i = 1, 2, \ldots, T$ so that Ω is the set of x which satisfy

$$|L_i(x)| < 1 \qquad \text{for} \quad i = 1, 2, \ldots, T. \tag{4.41}$$

We map $\Omega \to R^T$ by sending $x \to (L_i(x)) = \lambda(x)$. Since Ω is bounded, there are n linearly independent linear functions among the L_i, so λ is one-to-one. Moreover, Ω is mapped linearly onto a convex polyhedron on an n-dimensional linear variety $\Lambda \subset R^T$. Here Λ is defined as $\lambda(R^n)$ or

$$\sum b_j^i \omega_j = 0, \tag{4.42}$$

where $\sum b_j^i L_j(x) = 0$ for all x; $\omega = (\omega_1, \ldots, \omega_T)$ are the coordinates in R^T. Call Ω_1 the unit cube: $|\omega_i| < 1$ for all i. It is clear that

$$\lambda(\Omega) = \Omega_1 \cap \Lambda. \tag{4.43}$$

We use now the notation of Example 5 of Chapter V. Under the map λ the space $\mathscr{E}'(\Omega)$ of distributions whose support is contained in a compact subset of Ω gets mapped isomorphically on the space $\mathscr{E}'(\lambda(\Omega))$. We can identify $\mathscr{E}'(\lambda(\Omega))$ with the space of distributions in $\mathscr{E}'(\Omega_1)$ which are "the δ function in directions orthogonal to Λ". More precisely, each $S \in \mathscr{E}'(\lambda(\Omega))$ can be identified with $S \times \delta(\Lambda_\perp)$, where $\delta(\Lambda_\perp)$ is the unit mass at the origin in $\Lambda_\perp$, which is some complementary space to Λ in R^T and $\times$ denotes the direct product. We have thus "embedded" $\mathscr{E}'(\Omega)$ as a well-defined subspace of $\mathscr{E}'(\Omega_1)$.

Denote by $\zeta = (\zeta_1, \ldots, \zeta_T)$ the complex dual coordinates to ω. We shall write $\zeta = (\xi, \eta)$, where ξ is a complex variable in Λ^C, the complexification of Λ, and η is a complex variable in $\Lambda_\perp^C$. For any $w \in \mathscr{E}'(\Omega_1)$ we form the Fourier transform in the usual manner (compare Chapter I)

$$\hat{w}(\zeta) = w \cdot \exp(i\omega \cdot \zeta).$$

The space $\mathbf{E}'(\Omega_1)$ of Fourier transforms of $\mathscr{E}'(\Omega_1)$ consists of all entire functions $H(\zeta)$ satisfying

$$H(\zeta) \leqslant A(1 + |\zeta|)^A \exp(B \sum |\mathrm{Im}\zeta_i|) \tag{4.44}$$

for some $A > 0$ and some $B < 1$. An AU structure for $\mathscr{E}'(\Omega_1)$ consists of all continuous positive functions which dominate the right side of (4.44) for all $A > 0$, $B < 1$.

From the above embedding of $\mathscr{E}'(\Omega)$ in $\mathscr{E}'(\Omega_1)$ we conclude that $\mathscr{E}'(\lambda(\Omega))$ consists of all $H = H(\xi, \eta)$ satisfying (4.44) which are constant in η. In view of the isomorphism between $\mathscr{E}'(\Omega)$ and $\mathscr{E}'(\lambda(\Omega))$ this can be shown to yield the following description of $\mathscr{E}'(\Omega)$:

$\mathbf{E}'(\Omega)$ *consists of all entire functions* $F(z)$ *for which there is an* $A > 0$ *and an* $\varepsilon > 0$ so that

$$|F(z)| \leq A(1 + |z|)^A \exp[(1 - \varepsilon)\theta(\text{Im}(z))]. \tag{4.45}$$

Here $\theta(\text{Im } z)$ is defined as

$$\theta(\text{Im } z) = \sup_{x \in \Omega} |x \cdot \text{Im } z|.$$

This description is due to Lions [1].

PROBLEM 4.3. (Exercise).

Derive (4.45) from (4.44).

The same argument can be used to show that $\mathscr{E}(\Omega)$ is AU. An AU structure consists of all continuous positive functions which dominate the right sides of (4.45) for all $A > 0$, $\varepsilon > 0$. Of course, if Ω is not a cube, we would not expect $\mathscr{E}(\Omega)$ to be PLAU, since properties (d) and (e) of product localizeability certainly do not hold. Nevertheless, $\mathscr{E}(\Omega)$ is LAU.

The fact that $\mathscr{E}(\Omega)$ is localizeable follows from :

THEOREM 4.9. *Let* $\mathscr{W}$ *be an AU space. Suppose there exists an LAU space* $\mathscr{W}_1$ *and a plane* Λ^C *in* C^T *such that* $\mathbf{W}'$ *is the space of restrictions of* $\mathbf{W}_1'$ *from* C^T *to* Λ^C. *(The functions in* $\mathbf{W}_1'$ *depend on* T *complex variables while the functions in* $\mathbf{W}'$ *depend on fewer, namely, the complex dimension of* Λ^C.) *Then* $\mathscr{W}$ *is localizeable.*

PROOF. The idea is that we want to extend functions, cochains, etc., from Λ^C to C^T, then use the result assumed for $\mathbf{W}_1'$ in C^T, and then restrict back to Λ^C. We extend polynomials on Λ^C to be constant in the directions complementary to Λ. In this way we extend polynomial ideals and modules from Λ^C to C^T. We then add to the generators of all ideals and modules on C^T the linear functions $\sum b_j^i \omega_j$. We can now extend all cochains from Λ^C to a finite distance from Λ^C by making them constant in the orthogonal directions. We make the cochains zero outside this neighborhood of Λ^C. It is clear that all coherence properties (that is, properties on the intersections of the $\gamma(\alpha; \beta)$) are preserved. Now, using the fact that $\mathscr{W}_1$ is localizeable and restricting to Λ^C, we have our result. ▯

To apply Theorem 4.9 to the space $\mathscr{E}(\Omega)$, we set $\mathscr{W}_1 = \mathscr{E}(\Omega_1)$ and $\mathscr{W} = \mathscr{E}(\Omega)$ and apply our above remarks.

It should be noted that, although in our above remarks we assumed that Ω was symmetric and had compact closure, we could eliminate these assumptions by a slight modification. The only nontrivial point to note is that if Ω is not symmetric, the representation (4.41) for Ω must be altered. This does not seriously affect the rest of the proof.

See Remark 4.4.

Of course, the above does *not* apply to $\mathscr{E}(\Omega)$ if Ω is an arbitrary convex set which is not a convex polygon. For this we need a suitable limiting argument. Now there is no difficulty in proving the part of Theorem 4.2 referring to the vanishing of cohomology groups. In fact, we see easily that this follows from the result for a convex polygon. However, the continuity requires a more delicate argument which we have not carried out.

See Remark 4.5.

Remarks

Remark 4.1. See page 97.

In general it will not be difficult to verify that a given space $\mathscr{W}$ is product localizeable. The functions φ and ψ can be chosen to be exponential polynomials in some of the simpler examples. The examples are discussed in Chapter V. We shall give below a generalization of these conditions which will enable us to treat many more examples.

Remark 4.2. See page 97.

Remark 4.3. See page 108.

Remark 4.4. See page 119.

While we have used the space $\mathscr{E}(\Omega)$ for illustration it is clear that our method applies to many other spaces which are not PLAU.

Remark 4.5. See page 119.

It seems that this type of argument is perhaps better carried out using the Spencer-Morrey-Kohn (see Kohn [1]) method to prove the vanishing of the cohomology groups rather than the Oka-Cartan-Serre method which we have used here. The reason for this is that the latter method, for domains of holomorphy, says nothing about boundary behavior whereas Kohn has shown that one can get boundary results. Hörmander in [5] has shown how to use the Spencer-Morrey-Kohn method to obtain results on the vanishing of cohomology groups with bounds.

Another possible approach would be to replace Ω_1 by a "unit cube" in a suitable Hilbert or Banach space. This is because any convex body is the intersection of countably many half-planes. However, we have not attempted to carry out the details.

Problems

PROBLEM 4.1. (Exercise). See page 101.
Carry out the details.

PROBLEM 4.2. See page 116.

Can the continuity statements in (4.9) and (4.10) be proved by a suitable version of the open mapping theorem?

PROBLEM 4.3. (Exercise). See page 118.

CHAPTER V

Examples

Summary

We give examples of spaces which are AU, LAU, and PLAU. For some of these we give examples of sufficient sets.

Example 1 (p. 122) is the space $\mathscr{H}$ of entire functions. A sufficient set σ can be obtained by taking

$$\sigma = (A_1 \cup B_1) \times \cdots \times (A_n \cup B_n)$$

where A_j, B_j are two distinct straight lines through the origin in the complex z_j plane. An example is given of a discrete set which is sufficient.

Example 2 (p. 138) is the space $\mathbf{H}'$ of entire functions of exponential type.

Example 3 (p. 139) is the space $\mathscr{D}'_F$ of distributions of finite order. A sufficient set σ can be taken as

$$\sigma = (R_1 \cup A_1) \times \cdots \times (R_n \cup A_n)$$

where R_j is the real axis through the complex z_j plane and A_j is any other line through the origin.

Example 4 (p. 148) is the space $\mathscr{D}'$ of distributions.

In Example 5 (p. 152) we deal with the spaces $\mathscr{E}$ of indefinitely differentiable functions and $\mathscr{E}(\Omega)$ of indefinitely differentiable functions on the convex polyhedron Ω (open, closed, or partially open and closed). Using the fact that $\mathscr{E}(\Omega)$ is LAU and Theorem 4.1, we show that $\mathscr{H}(\Omega)$ is LAU.

Let $j = (j_1, \ldots, j_n)$, where the j_k are non-negative integers. Let $A = \{a_j\}$ be a multisequence of positive numbers. Then we define (Example 6, p. 163.) $\mathscr{E}_A$ to be space of those $f \in \mathscr{E}$ such that, essentially, uniformly on compact sets,

$$|f^{(j)}(x)| \leqslant ac^{|j|}a_j$$

where $|j| = j_1 + \cdots + j_n$. For suitable A we show that $\mathscr{E}_A$ is LAU.

Let Φ be a continuous convex function on $R = R^n$. The space (Example 7, p. 169) $\mathscr{E}(\Phi)$ consists of all $f \in \mathscr{E}$ which, together with each of its derivatives, is $0[\exp(\Phi(bx))]$ for all $b > 0$. For suitable Φ we show that $\mathscr{E}(\Phi)$ is PLAU. An analogous space of distributions $\mathscr{D}'(\Phi)$ is also considered.

Example 8 (p. 172) is the space of formal power series. This example may be of some algebraic interest, since the Fourier transform of its dual is the ring $\mathscr{P}$ of polynomials.

V.1. EXAMPLE 1. THE SPACE $\mathscr{H}$ OF ENTIRE FUNCTIONS. We think of $x = (x_1, \ldots, x_n)$ as complex variables. We define $\mathscr{H}$ as the space of entire functions of x with the topology of uniform convergence on compact sets. Since the closed bounded sets of $\mathscr{H}$ are compact, it follows from general topological results that $\mathscr{H}$ is reflexive. It is readily verified that $z \to \exp(ix \cdot z)$ is a complex analytic map of C into $\mathscr{H}$. That the linear combinations of the exponentials are dense in $\mathscr{H}$ can be seen as follows: Since the power series of an entire function converges in the topology of $\mathscr{H}$, it suffices to show how to approximate polynomials. For example, if $n = 1$ we write $x = \lim_{\varepsilon \to 0} [1 - \exp(\varepsilon x)]/\varepsilon$. This limit exists in the topology of $\mathscr{H}$. The general case is similar.

From the Hahn-Banach theorem, we deduce that for each $S \in \mathscr{H}'$ there is a measure μ of compact support on C such that for any $f \in \mathscr{H}$ we may write

$$S \cdot f = \int f(x)\, d\mu(x). \tag{5.1}$$

By $\|x\|$ we shall denote $\max |x_j|$ and by $\|z\|$ we shall denote $\sum |z_j|$. Although the notation may seem confusing at first, no problems will arise because we could have reversed the definitions. What is needed is a pair of convenient norms on x and z which are dual in the sense that $\|z\| \leqslant 1$ if and only if $|x \cdot z| \leqslant 1$ whenever $\|x\| \leqslant 1$. If no confusion is possible, we shall sometimes write $|z|$ or $|x|$ for $\|z\|$ and $\|x\|$.

Since the measure μ of (5.1) has compact support, its support will be contained in a set $\|x\| \leqslant a$. Since by definition (see Chapter I) $\hat{S}(z) = S \cdot \exp(iz \cdot \quad)$, we will have

$$|\hat{S}(z)| \leqslant \int |\exp(ix \cdot z)|\, d|\mu(x)| \leqslant A \exp(a\,\|z\|) \tag{5.2}$$

where A is the total variation of μ. An entire function $\hat{S}$ satisfying (5.2) is said to be of *exponential type* $\leqslant a$. The infimum of all a for which there exists an A satisfying (5.2) is called the *exponential type* of $\hat{S}$.

We see thus that $\mathbf{H}'$ consists of entire functions of exponential type. A classical result of E. Borel (see Titchmarsh [2]) asserts that, conversely, every entire function of exponential type belongs to $\mathbf{H}'$. The simplest way of seeing this is as follows: Let $f \in \mathscr{H}$, $S \in \mathscr{H}'$. Then if

$$f(x) = \sum f_j x^j$$

is the Taylor expansion of f at the origin, where $j = (j_1, \ldots, j_n)$ and $x^j = x_1^{j_1} x_2^{j_2} \ldots x_n^{j_n}$, then the series $\sum f_j X^j$ converges to f in the topology of $\mathscr{H}$. Here X^j is the function $x \to x^j$. Thus,

$$S \cdot f = \sum S_j f_j, \tag{5.3}$$

where

$$S_j = S \cdot X^j. \tag{5.4}$$

(5.4) combined with the definition of $\hat{S}$ shows that

$$(i^{j_1+\cdots+j_n}/j_1!\,j_2!\ldots j_n!)S_j$$

are the Taylor coefficients in the expansion of $\hat{S}$ at zero.

We can now reverse the argument and show that if $F(z)$ is an entire function of exponential type whose Taylor coefficients at zero are F_j, then the map $f \to \sum f_j F_j\, j!\, i^{-|j|}$ defines an element $T \in \mathscr{H}'$ with $\hat{T} = F$. [In general we shall write $|j| = j_1 + \cdots + j_n$ and $j! = j_1!\,j_2!\ldots j_n!$. We denote by $j + 1$ the multiindex $(j_1 + 1, \ldots, j_n + 1)$.] ▯

In order to prove that $\mathscr{H}$ is AU we need

LEMMA 5.1. *Let $\{F^l\}$ be a sequence of entire functions of exponential type. A necessary and sufficient condition that $F^l \to 0$ in the topology of $\mathbf{H}'$ is that there exist an $a > 0$ so that*

$$\max_{z \in C} |\exp(-a\,\|z\|)F^l(z)| \to 0. \tag{5.5}$$

Denote by S^l the inverse Fourier transform of F^l. (5.5) implies that, for any $a' > a$, $S^l \cdot h \to 0$ uniformly on the set of h satisfying

$$\max_{\|x\| \leqslant a'} |h(x)| \leqslant 1. \tag{5.6}$$

PROOF. Suppose first that $S^l \to 0$ in the topology of $\mathscr{H}'$. Then (see Schwartz [1], Vol. 1, p. 91 for an analogous result which is proved in the same manner; see also Dieudonné and Schwartz [1]) there exists a neighborhood N of zero in $\mathscr{H}$ such that $S^l \cdot h \to 0$ uniformly for $h \in N$. By the definition of the topology of $\mathscr{H}$, we may assume that N is defined by (5.6). Now for any $z \in C$ we see that $\exp(-a'\,\|z\|)\exp(iz \cdot \quad) \in N$ from which (5.5) follows immediately (with $a' = a$).

To prove the converse, denote by Γ the product of the circles $|z_t| = a'$. Then for any h satisfying (5.6) we write Cauchy's formula in the form

$$h_j = 1/(2\pi i)^n \int_\Gamma (h(z)/z^{j+1})\, dz$$

so that

$$(5.7) \qquad |h_j| \leqslant (a')^{-|j|}.$$

We now use a similar method to estimate the Taylor coefficients F_j^l of F^l, except that the contour must be chosen to vary with j. By a simple argument we find that for any $\varepsilon > 0$

$$(5.8) \qquad |F_j^l| \leqslant \varepsilon a^{|j|} (j_1 \cdots j_n)^{1/2}/j\,!$$

for l sufficiently large. On the other hand, by (5.3) and the remark following (5.4) we have

$$|S^l \cdot h| = \left|\sum S_j^l h_j\right| \leqslant \sum |j\,! F_j^l h_j| \leqslant \varepsilon \sum (a/a')^{-|j|} (j_1 \cdots j_n)^{1/2}.$$

Since $a/a' < 1$ we have our result. ▯

A similar argument shows

LEMMA 5.2. *A set $B \subset \mathbf{H}'$ is bounded if and only if there exist $a, b > 0$ so that*

$$\max_{\substack{z \in C \\ F \in B}} |F(z)| \exp(-a\,\|z\|) \leqslant b.$$

THEOREM 5.3. *$\mathscr{H}$ is AU. An AU structure for $\mathscr{H}$ can be chosen to consist of all functions $k(z)$ which are continuous and positive and satisfy*

$$(5.9) \qquad \exp(t\,\|z\|)/k(z) \to 0 \qquad \text{as } \|z\| \to \infty$$

for any t

A function k satisfying (5.9) will be said to dominate all linear exponentials. A similar expression will be used when the class of linear exponentials is replaced by another class of functions.

PROOF. Properties (a) and (b) of AU are clear (see Section I.3). We wish to verify (c). Let N_k be the set of $F \in \mathbf{H}'$ with $|F(z)| \leqslant k(z)$ for all $z \in C$. By Lemma 5.2, given any bounded set $B \subset \mathbf{H}'$, there is a $b > 0$ so that $bB \subset N_k$, that is, N_k *swallows* every bounded set. Now, (see Grothendieck [1]) $\mathscr{H}'$ is *bornologic*, that is, any convex circled set which swallows every bounded set is a neighborhood of zero. Thus, N_k is a neighborhood of zero.

Conversley, let N be a neighborhood of zero in $\mathbf{H}'$; we want to produce a k satisfying (5.9) such that $N_k \subset N$. Call $Q \subset \mathscr{H}'$ the set of inverse Fourier transforms of functions in N. By the definition of the topology of a dual space we can find a bounded set $B \subset \mathscr{H}$ so the conditions $S \in \mathscr{H}'$, $|S \cdot f| \leqslant 1$ for all $f \in B$ imply $S \in Q$.

Since B is bounded in $\mathscr{H}$, for each $l = (l_1, \ldots, l_n)$, where the l_i are positive integers, we can find a positive number c_l such that every $f \in B$

satisfies

(5.10) $$|f(x)| \leqslant c_l \quad \text{for } |x_1| \leqslant l_1, \ldots, |x_n| \leqslant l_n.$$

From this and Cauchy's formula we deduce as in (5.7) that, for all l, j,

(5.11) $$|f_j| \leqslant c_l l^{-j},$$

where $l^{-j} = l_1^{-j_1} l_2^{-j_2} \ldots l_n^{-j_n}$. We may clearly assume that $c_{1,1,\ldots,1} = 1$ and that $c_l < c_{l'}$ if $l_1 \leqslant l'_1, \ldots, l_n \leqslant l'_n$.

The idea of the proof is to construct k in such a way that we can use the fact that $G \in H$, $|G(z)| \leqslant k(z)$ to deduce, by means of Cauchy's formula, that $\sum |G_j f_j| \leqslant 1$, if $\{f_j\}$ satisfies (5.11). For this purpose, we must choose the contour on which we apply Cauchy's formula to vary with j.

Let $\{\gamma_s\}$ be a strictly increasing sequence of numbers with $\gamma_0 = 0$. Let $\{d_s\}$ be a sequence of numbers with $d_0 = 0$, $d_s > 0$ for $s > 0$. We require further that

(a) $$d_s + \gamma_s \leqslant \gamma_{s+1}$$

for any s.

For given $s_1, \ldots, s_n$ we denote by $\alpha_{s_1 s_2 \cdots s_n}$ the region in C consisting of those y for which $d_{s_i} + \gamma_{s_i} \leqslant |y_i| \leqslant \gamma_{s_i+1}$ for all i. We require that

(b) $$\exp\left[\tfrac{1}{2}(s_1|y_1| + s_2|y_2| + \cdots + s_n|y_n|)\right]/c_{s_1+1, s_2+1, \cdots s_n+1} \geqslant \exp\left[\tfrac{1}{2}((s_1-1)|y_1| + \cdots + (s_n-1)|y_n|)\right]$$

whenever $s_1 \geqslant 1, \ldots, s_n \geqslant 1$ and whenever $|y_i| \geqslant d_{s_i} + \gamma_{s_i}$ for all i.

(c) There is no point r/m in the interval $\gamma_s \leqslant x \leqslant \gamma_s + d_s$ for any s, where r, m are integers and $m \leqslant 2(s+1)$.

The existence of the sequences $\{\gamma_s\}$, $\{d_s\}$ is obvious. Moreover, it is clear that the regions $\alpha_{s_1 s_2 \cdots s_n}$ do not overlap.

We define $k(z)$ as follows: For $z \in \alpha_{s_1 s_2 \cdots s_n}$ we set

(5.12) $$k(z) = \exp\left[\tfrac{1}{2}(s_1|z_1| + \cdots + s_n|z_n|)\right]/c_{s_1+1, \cdots s_n+1}.$$

The definition of k is completed by requiring that it be continuous and larger than the right side of (5.12) whenever $|z_i| \geqslant d_{s_i} + \gamma_{s_i}$ for all i. It follows from (5.12) and (*b*) that k satisfies (5.9).

It is not difficult to see, using (*c*), that given $j = (j_1, \ldots, j_n)$ we can find an n-tuple of positive numbers b such that, for suitable l,

(5.13) $$j_1/l_1 \geqslant b_1 \geqslant j_1/(l_1+1), \ldots, j_n/l_n \geqslant b_n \geqslant j_n/(l_n+1)$$

and such that the chain Γ in C defined by $|z_i| = b_i$ for all i lies in $\alpha_{l_1 \cdots l_n}$. Indeed, we can deal with each coordinate separately. Using (c) we check that, for any integer j_i, there is an integer l_i such that the intervals $[j_i/(l_i+1), j_i/l_i]$ and $[d_{l_i}+\gamma_{l_i}, \gamma_{l_{i+1}}]$ intersect; b_i is chosen in the intersection.

Let $G \in N_k$ and let G_j be the coefficients of the Taylor expansion of G at zero. Then Cauchy's formula may be written as

$$G_j = (2\pi i)^{-n} \int_\Gamma G(z)/z^{j+1}\, dz.$$

Now, on Γ we have

$$|G(z)| \leqslant k(z) = \exp\left(\tfrac{1}{2}b \cdot l\right)/c_{l_1+1,\ldots,l_n+1}$$

by (5.12). Thus, on making use of (5.13) and Cauchy's formula,

$$c_{l_1+1,\cdots l_n+1} j^j \, |G_j| \leqslant \exp\left(\tfrac{1}{2}\,|j|\right)(l_1+1)^{j_1}\cdots(l_n+1)^{j_n}.$$

By Stirling's formula we can find a $\theta > 0$ so that

$$(5.14)\qquad j\,!\, c_{l_1+1,\ldots,l_n+1}\,|G_j| \leqslant \theta j_1 \cdots j_n \exp\left(-\tfrac{1}{2}\,|j|\right)(l_1+1)^{j_1}\cdots(l_n+1)^{j_n}.$$

Using (5.3) and following with $G = \hat{S}$, (5.11) for l replaced by $l+1$, and (5.14) it is obvious that we can find a $\theta' > 0$ such that every $G \in N_k$ satisfies

$$|S \cdot f| \leqslant \theta' \qquad \text{for all } f \in B.$$

This proves that $(1/\theta')N_k \subset N$ which completes the proof of Theorem 5.3. ▯

We wish to show

THEOREM 5.4. *$\mathscr{H}$ is PLAU.*

PROOF. To verify condition (a) for PLAU (see Section IV.1), let F be an entire function which is not of exponential type. Thus, for each positive integer l there is a point z^l at which $|F(z^l)| > \exp(l\,\|z^l\|)$. By passing to a subsequence we may assume that $\|z^{l+1}\| \geqslant \|z^l\| + 1$ for all l. We define k by setting $k(z^l) = |F(z^l)|^{1/2}$. The definition of k is completed by requiring that

1. k is a function of $\|z\|$.
2. $k(0) \neq 0$.
3. $\log k$ is linear between $\|z^l\|$ and $\|z^{l+1}\|$.

It is clear that k is in the AU structure described in Theorem 5.3. Moreover our construction shows that $|F(z^l)|/k(z^l) = |F(z^l)|^{1/2}$ so F/k is unbounded. Thus, condition (a) is verified.

Condition (b) is an immediate consequence of Theorem 5.3. Condition (c) is an immediate consequence of Lemma 5.2.

To verify (d) and (e) we need only consider the case $n = 1$. Using Theorem 5.3, we can easily show that an AU structure can be given as follows: The functions k are continuous positive functions of $|z|$ which on each segment $i \leqslant |z| \leqslant i+1$ are of the form $k(z) = \exp(A_i |z| - B_i)$ for suitable constants $A_i \to \infty$ and B_i. (That is, any k in the AU structure described in Theorem 5.3 has a minorant in that AU structure whose log is piecewise linear.) Moreover by Lemma 5.2 if $m \in M$ (a BAU structure for $\mathscr{H}$) then $m(z) < B' \exp(A'|z|)$ for some A', B'.

We shall construct k' presently. The function $\varphi(z; z^0, \eta)$ will be $\exp[A'(\theta^0 z - |z^0|)]$, where $\theta^0 = \exp(-i \arg z^0)$.

$$(5.15)\qquad m(z^0)\,|\exp A'(\theta^0 z - |z^0|)|/\min_{|\zeta - z^0| \leqslant \eta} |\exp A'(\theta^0 \zeta - |z^0|)|$$
$$\leqslant m(z^0)\exp(A'\eta + A'|z| - A'|z^0|)$$

This will be $\leqslant ck(z)$ if

$$(5.16)\qquad m(z^0) \leqslant ck(z)\exp(-A'\eta - A'|z| + A'|z^0|)$$

By our construction

$$(5.17)\qquad m(z^0) \leqslant B'\exp(A'|z^0|).$$

Thus we need

$$(5.18)\qquad B'\exp(A'|z^0|) \leqslant ck(z)\exp(-A'\eta - A'|z| + A'|z^0|).$$

that is,

$$(5.19)\qquad B'\exp A'(|z| + \eta) \leqslant ck(z).$$

Now choose

$$(5.20)\qquad k'(z) = k[\max(0, |z| - \eta)]$$

and the result is clear.

It remains to verify condition (*e*). For this purpose it is simplest to use functions of the form $\tilde{k}(\text{Re } z)\tilde{k}(\text{Im } z)$ for the analytic uniform structure of $\mathscr{H}$. We may assume that each function $\tilde{k}$ has the following properties:

1. $\tilde{k}(t)$ is a positive, even, monotonically increasing continuous function.
2. For each A we have $\exp(At)/\tilde{k}(t) \to 0$ as $t \to \infty$.
3. There exists an increasing sequence of positive integers a_l with $a_{l+1} \geqslant l a_l^2$ such that for $a_l \leqslant t \leqslant 2a_l$ we have

$$(5.21)\qquad \tfrac{1}{2} \leqslant \tilde{k}(t)\exp(-lt) \leqslant 2,$$

while for $2a_l \leqslant t \leqslant a_{l+1}$ we have

$$(5.22)\qquad \tfrac{1}{2}\exp(lt) \leqslant \tilde{k}(t) \leqslant 2\exp(l+1)t.$$

The fact that we may assume k to be of this form follows from the (easily verified) fact that every function satisfying (1) and (2) has a minorant which satisfies (1), (2), and (3).

Next we see that we can choose a bounded uniform structure $\{m\}$ such that each $m(z) = \tilde{m}(\text{Re } z)\tilde{m}'(\text{Im } z)$ where the $\tilde{m}(t)$, $\tilde{m}'(t)$ satisfy

4. $\tilde{m}(t)$ and $\tilde{m}'(t)$ are positive, even, monotonically increasing, continuous piecewise differentiable functions which are equal to one for $|t| \leqslant 1$.
5. There exists an increasing finite sequence of positive integers a_l, say, for $1 \leqslant l \leqslant l_0$, such that $a_{l+1} \geqslant (l+4)a_l^2$. We can impose any further conditions on $\{a_l\}$ we want which make $a_l \to \infty$ rapidly. Moreover, for $a_l \leqslant t \leqslant 2a_l$ we have

$$\tfrac{1}{2} \leqslant \tilde{m}(t) \exp(-lt) \leqslant 2. \tag{5.23}$$

For $2a_l \leqslant t \leqslant a_{l+1}$ we have

$$\tfrac{1}{2} \exp(lt) \leqslant \tilde{m}(t) \leqslant 2 \exp(l+1)t \tag{5.24}$$

while for $t \geqslant a_{l_0}$ we have

$$\tfrac{1}{2} \leqslant \tilde{m}(t) \exp(-l_0 t) \leqslant 2. \tag{5.25}$$

The same inequalities hold for $\tilde{m}'$.

To establish property (e) we must (roughly) do the following: Given m as above and given a complex number z^0, we must try to approximate $m(z)$ in a strip $|\text{Im } z - \text{Im } z^0| \leqslant \eta$ by the modulus of a function $\psi \in \mathbf{H}'$ such that $\psi(z)$ falls off as rapidly as possible as $|\text{Im } z|$ decreases. For the fact that $|\psi|$ approximates m in the strip $|\text{Im } z - \text{Im } z^0| \leqslant \eta$ means that the term in brackets on the left side of (4.4) is close to one so the left side of (4.4) will be close to $|\psi(z; z^0)|$. The second condition on ψ will guarantee that (4.4) is satisfied.

We explain first the approximation procedure. For this we assume for simplicity that $\text{Im } z^0 = 0$, as the general case is similar. Again, to obviate some notational difficulties we shall assume in (4.4) that $\eta = \frac{1}{2}$. We shall define $\psi(z)$ as a product, namely

$$\psi(z) = \prod_{j=1}^{l_0} [\cosh z/b_j]^{b_j}, \tag{5.26}$$

where the b_j will be chosen suitably in terms of the a_l.

Let us note the following: For z/b small, $[\cosh z/b]^b$ behaves like $(1 + z^2/b^2)^b$, which is close to one. For $|\text{Im } z| < \frac{1}{2}$ and $\text{Re } z$ large it is clear that $[\cosh z/b]^b$ behaves like $\exp(|\text{Re } z|)$. More precisely, given any $\varepsilon > 0$ and any $d > 0$ we can find a positive integer $b = b(\varepsilon, d)$ and an $a^1 = a^1(\varepsilon, b)$

so that (always $|\text{Im } z| \leqslant \frac{1}{2}$)

$$(5.27)\qquad 1 - \varepsilon \leqslant |\cosh z/b|^b \leqslant 1 + \varepsilon \qquad \text{for } |\text{Re } z| \leqslant d$$

$$(5.28)\quad 1 - \varepsilon \leqslant \exp(-(1-\varepsilon)|\text{Re } z|)\,|\cosh z/b|^b \leqslant 1 + \varepsilon \qquad \text{for } |\text{Re } z| \geqslant a^1.$$

In terms of these functions b, a^1 we define the sequence $\{b_l\}$: We set $a_1 = 1$, and, for any l, $b_l = b(1/(l+1)^2, 2a_l)$. We require, as we may by the above remark that we can make $a_l \to \infty$ as fast as we want, that $a_{l+1} \geqslant \max\{2a_l, a^1(1/(l+1)^2, 2b_l)\}$. With this choice we define $\psi(z)$ by (5.26).

It is clear that $\psi \in \mathbf{H}'$. We examine the size of ψ in the strip $|\text{Im } z| \leqslant \frac{1}{2}$. For such z and $a_l \leqslant |\text{Re } z| \leqslant a_{l+1}$ we have, by (5.27) and (5.28)

$$(5.29)\qquad 1 - \frac{1}{(j+1)^2} \leqslant |\cosh z/b_j|^{b_j} \leqslant 1 + \frac{1}{(j+1)^2}$$

for $j \geqslant l+1$ since (5.27) holds for $|\text{Re } z| \leqslant 2a_j$, hence for $|\text{Re } z| \leqslant a_j$. For $j \leqslant l-1$ we have, again for $a_l \leqslant |\text{Re } z| \leqslant a_{l+1}$,

$$(5.30)\quad 1 - \frac{1}{(j+1)^2} \leqslant \exp\left(-\left(1 - \frac{1}{(j+1)^2}\right)|\text{Re } z|\right)|\cosh z/b_j|^{b_j} \leqslant 1 + \frac{1}{(j+1)^2}$$

since (5.28) holds for $|\text{Re } z| \geqslant a_{j+1}$ and $a_l \geqslant a_{j+1}$. Finally, for $a_l \leqslant |\text{Re } z| < 2a_l$ we have by (5.27).

$$(5.31)\qquad 1 - \frac{1}{(l+1)^2} \leqslant |\cosh z/b_l|^{b_l} \leqslant 1 + \frac{1}{(l+1)^2},$$

while for $2a_l \leqslant |\text{Re } z| \leqslant a_{l+1}$ we have, for $c_l' = |\cos 1/2b_l|^{b_l}$,

$$(5.32)\qquad c_l' \leqslant |\cosh z/b_l|^{b_l} \leqslant \exp(|\text{Re } z|).$$

The second inequality in (5.32) is obvious. The first is a consequence of

$$|\cosh z/b_l| \geqslant |\text{Re}\cosh z/b_l| = |\cosh \text{Re } z/b_l|\,|\cos \text{Im } z/b_l|$$

and the fact that $|\text{Im } z| \leqslant \frac{1}{2}$.

Putting these results together gives, for $a_l \leqslant |\text{Re } z| \leqslant 2a_l$, for any l,

$$(5.33)\qquad \prod_{j=1}^{l_0}\left(1 - \frac{1}{(j+1)^2}\right)\exp((l-2)|\text{Re } z|) \leqslant |\psi(z)| \leqslant \prod_{j=1}^{l_0}\left(1 + \frac{1}{(j+1)^2}\right)\exp((l-1)|\text{Re } z|),$$

while for $2a_l \leqslant |\text{Re } z| \leqslant a_{l+1}$

$$(5.34) \qquad c_l' \prod_{j=1}^{l_0} \left(1 - \frac{1}{(j+1)^2}\right) \exp\,((l-2)|\text{Re } z|) \leqslant |\psi(z)|$$
$$\leqslant \prod_{j=1}^{l_0} \left(1 + \frac{1}{(j+1)^2}\right) \exp\,(l|\text{Re } z|).$$

These results together with (5.23), (5.24), and (5.25) imply that there exists a universal constant c so that for $|\text{Im } z| \leqslant \frac{1}{2}$ we have

$$(5.35) \qquad c^{-1} \exp\,(-3\,|z|) \leqslant |\psi(z)|/m(z) \leqslant c \exp\,(3\,|z|).$$

All this was relevant to the strip $|\text{Im } z| \leqslant \frac{1}{2}$. If z^0 is an arbitrary complex number, then we can construct $\psi_{z0}(z)$ so that

$$(5.36) \quad c^{-1} \exp\,[-3(|z| + |\text{Im } z^0|] \leqslant |\psi_{z^0}(z)|/m(z) \leqslant c \exp\,[3(|z| + |\text{Im } z^0|]$$

for $|\text{Im } z - \text{Im } z^0| \leqslant \frac{1}{2}$. Namely, we just set $\psi_{z^0}(z) = \tilde{m}'(\text{Im } z^0)\psi(z - i \text{ Im } z^0)$.

Let us note the following: On any line $\text{Re } z = \text{const}$ the function $\cosh z$ takes its maximum at $\text{Im } z = 0$. For, $|2 \cosh z|^2 = (e^z + e^{-z})(e^{\bar{z}} + e^{-\bar{z}}) = 2 \cosh 2 \text{ Re } z + 2 \cos 2 \text{ Im } z$, from which our contention follows. This means that $|\psi(z)| \leqslant |\psi(\text{Re } z)|$.

Suppose for simplicity that $\text{Im } z^0 > 0$. Suppose that $a_l < \text{Im } z^0 \leqslant a_{l+1}$. Then consider

$$\tilde{\psi}_{z^0}(z) = \exp\,(-i(l+4)(z - z^0))\psi_{z^0}(z).$$

Since $l \leqslant \text{Im } z^0$, using (5.36) we have

$$(5.37) \qquad e^{-2}c^{-1} \exp\,(-4(|z| + \text{Im } z^0)) \leqslant |\tilde{\psi}_{z^0}(z)|/m(z)$$
$$\leqslant ce^2 \exp\,(4(|z| + \text{Im } z^0))$$

for $|\text{Im } z - \text{Im } z^0| \leqslant \frac{1}{2}$. Moreover, for any z with $0 \leqslant \text{Im } z \leqslant \text{Im } z^0$

$$|\tilde{\psi}_{z^0}(z)|/m(z) = \frac{|\psi(z - i \text{ Im } z^0)|\, \tilde{m}'(\text{Im } z^0)|\, \exp\,[-i(l+4)(z - z^0)]|}{m(z)}$$
$$\leqslant \frac{|\psi(\text{Re } z)|\, \tilde{m}'(\text{Im } z^0) \exp\,[(l+4)(\text{Im } z - \text{Im } z^0)]}{\tilde{m}(\text{Re } z)\tilde{m}'(\text{Im } z)}.$$

Now by (5.35) $|\psi(\text{Re } z)|/\tilde{m}(\text{Re } z) \leqslant c \exp\,(3\,|z|)$ and by (5.23) and (5.24) we have

$$\tilde{m}'(\text{Im } z^0) \leqslant 2 \exp\,((l+1) \text{ Im } z^0).$$

Thus,

$$(5.38) \quad |\tilde{\psi}_{z^0}(z)|/m(z) \leqslant 2c \exp\,(3\,|z|) \exp\,((l+4) \text{ Im } z - 3 \text{ Im } z^0)/\tilde{m}'(\text{Im } z)$$

Suppose that $a_j \leqslant \text{Im } z \leqslant a_{j+1}$. Then by (5.23) and (5.24) $\tilde{m}'(\text{Im } z) \geqslant \frac{1}{2} \exp(j \text{ Im } z)$. Thus,

$$\exp((l+4) \text{ Im } z - 3 \text{ Im } z^0)/\tilde{m}'(\text{Im } z) \leqslant 2 \exp((l+4-j) \text{ Im } z - 3 \text{ Im } z^0).$$

Suppose that $\text{Im } z \leqslant (3/(l+4)) \text{ Im } z^0$. Then the above is $\leqslant 2$. In case $\text{Im } z > (3/(l+4)) \text{ Im } z^0$ (which by our conditions on the sequence $\{a_i\}$ can happen only if $j = l$ or $j = l-1$) then the above is $\leqslant 2 \exp(5 \text{ Im } z) \leqslant 2 \exp(5|z|)$. Thus, we have shown that

$$|\tilde{\psi}_{z^0}(z)|/m(z) \leqslant 4c \exp(9|z|) \tag{5.39}$$

whenever $0 \leqslant \text{Im } z \leqslant \text{Im } z^0$.

For $\text{Im } z < 0$ the above argument (in an even simpler form) shows again that (5.39) holds.

Finally, we consider the case $\text{Im } z > \text{Im } z^0$. (5.38) is proved as above. Then we note that $\tilde{m}'(\text{Im } z) \geqslant \frac{1}{2} \exp(l \text{ Im } z)$, so we again derive (5.39) in this case. Thus, (5.39) holds for all z.

We are now in a position to verify condition (e). In our case inequality (4.4) becomes

$$\sup_{-\infty < \xi < \infty} \left\{ \frac{(1+|\xi|^2) m(\xi + i \text{ Im } z^0)}{\min\limits_{|t - \text{Im } z^0| \leqslant \frac{1}{2}} |\psi(\xi + it; z^0)|} \right\} |\psi(z; z^0)| \tag{5.40}$$
$$\leqslant ck(z).$$

Here k is given and $m(z) \leqslant k'(z)$, where k' is chosen depending on k. Now (5.40) closely resembles (5.37) and (5.39) with $\psi(z; z^0) = \tilde{\psi}_{z^0}(z)$, and, in fact, would be implied by them except for certain exponential and polynomial factors. Since these factors play a trivial role in the topology of $\mathbf{H}'$, we can easily construct k' so they do not appear.

The final statement of property (e) of localizeability follows from the same argument.

With this we have completed the proof that the space $\mathscr{H}$ is product localizeable. ▯

See Remark 5.1.

We wish to give some examples of sufficient sets for $\mathscr{H}$.

THEOREM 5.5. *For each j let A_j and B_j be any two distinct straight lines through the origin in the complex z_j plane. A sufficient set for $\mathscr{H}$ is the cartesian product $(A_1 \cup B_1) \times \cdots \times (A_n \cup B_n)$.*

PROOF. For simplicity, we consider first the case $n = 1$; the passage to $n > 1$ will be discussed below. We write A for A_1 and B for B_1. We may clearly assume that A is the real axis; let α be the counterclockwise angle between the positive real axis and B.

We map the sector of angle α onto the upper halfplane by means of the mapping $z \to z^{\pi/\alpha} = \lambda$, say. For any $k \in K$, we have $k(|z|) = k(|\lambda|^{\alpha/\pi})$, so k is larger at infinity than $\exp(t|\lambda|^{\alpha/\pi})$ for any $t > 0$. We need

LEMMA 5.6. *Let M be a continuous function defined on R such that $M(x) \geqslant 1$ and $M(x) \geqslant |x|^{\alpha/\pi}$ for all x and such that, at infinity, $M(\lambda)$ is larger than $t|\lambda|^{\alpha/\pi}$ for any $t > 0$. Then there is a positive monotonic even function $\tilde{M}$ on R which is real analytic except for a discrete sequence of jumps, with $\tilde{M}$ larger than $t|\lambda|^{\alpha/\pi}$ at infinity for each $t > 0$, and satisfying*

$$\tilde{M}(x) \leqslant M(x), \tag{5.41}$$

$$\tilde{M}(x+y) \leqslant 4\tilde{M}(x) + 4\tilde{M}(y), \tag{5.42}$$

$$\tilde{M}(xy) \leqslant 2\tilde{M}(x)\tilde{M}(y), \tag{5.43}$$

$$\int_{-\infty}^{\infty} \frac{\tilde{M}(x)\,dx}{1+x^2} < \infty. \tag{5.44}$$

Before proving Lemma 5.6, we shall use it to complete the proof of Theorem 5.5. The idea of the proof is as follows: From Theorem 5.3 it follows that we can find an AU structure for $\mathscr{H}$ consisting of functions $k(|z|)$ such that $\log k$ dominates all linear functions. We map some sector γ formed by the lines A, B onto the upper halfplane and $\log k$ gets mapped into a function M satisfying the hypothesis of Lemma 5.6. Let $\tilde{M}$ be the minorant of M described by Lemma 5.6. Let M^* be the Poisson integral of $\tilde{M}$ in the upper halfplane and let M^{**} be a conjugate of M^*. Then we shall show that (we write $\lambda = x + iy$) $M^*(\lambda)$ is essentially of the same order of magnitude as $\tilde{M}(|\lambda|)$. Using the maximum modulus theorem it will then follow (roughly) that if $F \in \mathbf{H}'$ and $F/k \to 0$ uniformly on $A \cup B$, then also

$$F(z)/\exp[M^*(z^{\pi/\alpha}) + iM^{**}(z^{\pi/\alpha})] \to 0$$

uniformly in γ. Since $M^*(\lambda)$ is essentially of the same order of magnitude as $\tilde{M}(\lambda)$, this will give the result.

Now for the details: We construct the Poisson integral M^* of $\tilde{M}$ for the upper halfplane:

$$M^*(x, y) = \frac{1}{\pi}\int_{-\infty}^{\infty} \frac{y\tilde{M}(x')}{(x-x')^2 + y^2}\,dx'. \tag{5.45}$$

We shall show first that $M^*(x, y) \leqslant \text{const}\,[\tilde{M}(x) + \tilde{M}(y)]$. For this purpose we write

$$M^* = \int_{-\infty}^{x} + \int_{x}^{\infty}$$

Now

$$\int_{-\infty}^{x} \frac{y\tilde{M}(x')}{(x-x')^2+y^2}\,dx' = \int_{-\infty}^{0} \frac{y\tilde{M}(x'+x)}{x'^2+y^2}\,dx'$$

$$\leqslant 4\int_{-\infty}^{0} \frac{y[\tilde{M}(x')+\tilde{M}(x)]}{x'^2+y^2}\,dx'$$

$$= 4\tilde{M}(x)\int_{-\infty}^{0} \frac{y}{x'^2+y^2}\,dx' + 4\int_{-\infty}^{0} \frac{y\tilde{M}(x')}{x'^2+y^2}\,dx'.$$

The first term is just $2\pi\tilde{M}(x)$. The second is

$$4\int_{-\infty}^{0} \frac{\tilde{M}(x')\,dx'/y}{1+(x'/y)^2} \leqslant \text{const}\int_{-\infty}^{0} \frac{\tilde{M}(x'/y)\tilde{M}(y)\,dx'/y}{(1+x'/y)^2}$$

$$= \text{const}\,\tilde{M}(y)\int_{-\infty}^{0} \frac{\tilde{M}(x'/y)\,dx'/y}{1+(x'/y)^2}$$

$$= \text{const}\,\tilde{M}(y).$$

On the other hand, we have

$$\int_{x}^{\infty} \frac{y\tilde{M}(x')}{(x'-x)^2+y^2}\,dx' = \int_{0}^{\infty} \frac{y\tilde{M}(x'+x)}{x'^2+y^2}\,dx',$$

which is handled as above. Combining the above calculations, we deduce

$$M^*(x, y) \leqslant \text{const}\,[\tilde{M}(x) + \tilde{M}(y)], \tag{5.46}$$

which is the desired result.

In particular we derive the fact that

$$M^*(x, y) \leqslant \text{const}\,\tilde{M}(|\lambda|), \tag{5.47}$$

where $\lambda = x + iy$.

Next, we want to show that M^* is large at infinity; in fact, we want to show that, at infinity, $M^*(x, y) > t|\lambda|^{\alpha/\pi}$ for any $t > 0$. For this we use the fact that $\tilde{M}$ is positive. By the above, for $x \geqslant 0$,

$$\pi M^*(x, y) \geqslant \int_{0}^{\infty} \frac{y\tilde{M}(x+x')}{x'^2+y^2}\,dx'. \tag{5.48}$$

For any t there is a t' (possibly $t' < 0$) so that

$$\tilde{M}(x + x') \geqslant t' + t(x + x')^{\alpha/\pi}.$$

Thus,

$$\begin{aligned}\pi M^*(x, y) &\geqslant \frac{\pi}{2} t' + t \int_0^\infty \frac{y(x + x')^{\alpha/\pi}}{x'^2 + y^2} dx' \\ &\geqslant \frac{\pi}{2} t' + c''t \int_0^\infty \frac{y(x^{\alpha/\pi} + x'^{\alpha/\pi})}{x'^2 + y^2} dx' \\ &\geqslant \frac{\pi}{2} t' + c''' \frac{\pi}{2} t x^{\alpha/\pi} + c''' t y^{\alpha/\pi}\end{aligned}$$

for suitable c'', c''' depending on α. Since $x, y \geqslant 0$ this means that, at infinity,

$$\begin{aligned}M^*(x, y) &\geqslant c't(x^{\alpha/\pi} + y^{\alpha/\pi}) \\ &\geqslant ct\,|\lambda|^{\alpha/\pi},\end{aligned} \tag{5.49}$$

where c is some constant independent of t. This gives our desired result for $x \geqslant 0$ and the case $x < 0$ is similar.

We wish now to complete the proof of Theorem 5.5 for the case $n = 1$. It is easily seen that we may assume the functions $k \in K$ are functions of $|z|$. Let $k \in K$ be given; it is sufficient to show that we can find, for each of the four sectors Λ_j into which the lines A and B divide C, a $k_j \in K$, or rather, a k_j which is positive and tends to infinity faster than any exponential so that the conditions $F \in \mathbf{H}'$,

$$|F(z)|/k_j(z) \leqslant 1 \quad \text{for } z \text{ on boundary } \Lambda_j$$

imply

$$|F(z)|/k(z) \leqslant \text{const (independent of } F) \text{ for } z \in \Lambda_j.$$

(Actually the k_j we construct is not continuous, but the construction can be easily modified to produce a continuous k_j.)

For simplicity of notation we restrict ourselves to Λ_1 which is the sector formed by the figure below.

See Fig. 9.

We use the notation of Lemma 5.6: (Recall that $\lambda = z^{\pi/\alpha}$.) Set $M(\lambda) = (1/c) \log k(z)$, where c is the constant occurring in (5.47) and then define $k_1(z) = \exp(\tilde{M}(\lambda))$. Let $M^{**}(\lambda)$ be a conjugate of $M^*(\lambda)$ and set

$$\tilde{k}_1(z) = \exp[M^*(\lambda) + iM^{**}(\lambda)]$$

for $z \in \Lambda_1$.

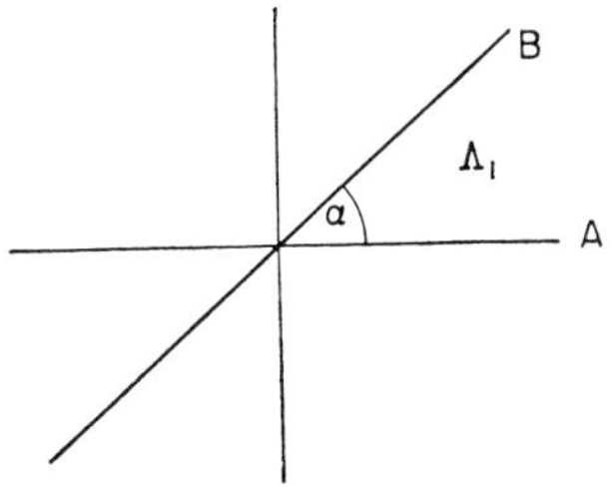

FIGURE 9

From (5.49) we deduce that on the boundary of Λ_1 we have $|k_1(z)| > \exp(t|z|)$ at infinity for any $t > 0$.

Suppose $F \in \mathbf{H}'$ and $|F(z)|/k_1(z) \leqslant 1$ for $z \in$ boundary Λ_1. Now, $|\tilde{k}_1(z)| = k_1(z)$ for $z \in$ boundary Λ_1 except at the discontinuities of $\tilde{M}$, but these points do not matter because F and k are continuous everywhere, so we can neglect them. [It is also possible to modify $\tilde{M}$ on the sets $x_j - \frac{1}{2} \leqslant x \leqslant x_j + \frac{1}{2}$ (see the construction below) to make $\tilde{M}$ indefinitely differentiable, by changing slightly the constants that occur in the conditions on $\tilde{M}$. We have thus $|F(z)/\tilde{k}_1(z)| \leqslant 1$ for $z \in$ boundary Λ_1.] The function $F(z)/\tilde{k}_1(z)$ is analytic in Λ_1; by (5.49) and the fact that F is exponential type, it follows that $|F(z)/\tilde{k}_1(z)| \to 0$ as $|z| \to \infty$, $z \in \Lambda_1$. Thus by the maximum modulus principle, $|F(z)/\tilde{k}_1(z)| \leqslant 1$ for $z \in \Lambda_1$. That is, for $z \in \Lambda_1$ we have

$$\begin{aligned} |F(z)| &\leqslant |\tilde{k}_1(z)| \\ &= \exp(M^*(\lambda)) \\ &\leqslant \exp(c\tilde{M}(|\lambda|)) \quad \text{by (5.47)} \\ &\leqslant \exp(cM(\lambda)) \\ &= k(|z|). \end{aligned}$$

This is the desired result.

Passage to $n > 1$. There is no essential difficulty involved. We notice that the functions $k \in K$ could be chosen to be products $k(z) = k^1(z_1) \ldots k^n(z_n)$. In each of the complex z_j planes we have four sectors Λ_l^j, $l = 1, 2, 3, 4$. Then given any cartesian product of sectors

$$\prod \Lambda_{l_j}^j$$

we must produce a $k_1 \in K$ such that the conditions $F \in \mathbf{H}'$,

$$|F(z)|/k_1(z) \leqslant 1, \quad z \in \text{distinguished boundary of } \prod \Lambda_{l_j}^j, \tag{5.50}$$

imply

$$|F(z)|/k(z) \leqslant \text{const} \qquad z \in \prod \Lambda_{l_j}^j. \tag{5.51}$$

We construct k_1 as a product $k_1 = k_1^1(z_1) \dots k_1^n(z_n)$ where k_1^j is constructed for k^j and $\Lambda_{l_j}^j$ by our above one-dimensional argument. The corresponding functions $\tilde{k}_l^j$ are constructed as above. The one-dimensional maximum modulus principle is replaced by the fact that a function which is holomorphic in

$$\prod \Lambda_{l_j}^j$$

and tends to zero at infinity will take its maximum on the distinguished boundary. This concludes the passage to $n > 1$.

PROOF OF LEMMA 5.6. The proof of Theorem 5.5 will be complete when we complete the proof of Lemma 5.6, that is the construction of $\tilde{M}$ with the desired properties. Since $\tilde{M}$ can be chosen to be even, we construct $\tilde{M}(x)$ for $x > 0$. For simplicity we assume $\tilde{M}(x)$ is equal to const $x^{\alpha/\pi}$ for long intervals. More precisely, we shall construct a sequence $\{x_j\}$ with $x_{j+1} > x_j > 0$ and we define

$$\tilde{M}(x) = \begin{cases} 1 & \text{for } 0 \leqslant x < x_1 \\ jx^{\alpha/\pi} & \text{for } x_j \leqslant x < x_{j+1}. \end{cases} \tag{5.52}$$

We must construct $\{x_j\}$ so that $\tilde{M}$ will have all the desired properties.

It is clear that $\tilde{M}$ is positive, monotonic, even, and larger at infinity than any $tx^{\alpha/\pi}$. Next we note that for any α

$$(x + y)^{\alpha/\pi} \leqslant x^{\alpha/\pi} + y^{\alpha/\pi}$$

We choose $x_1 = 1$ and we choose the x_j so lacunary that $\tilde{M}(x) \leqslant M(x)$, and

$$x_{j+1} > 2 \sum_{l \leqslant j} x_l$$

and $x_{j+1} > (x_j)^2$. Moreover, we assume that $jx_j^{(\alpha/\pi)-1} < j^{-2}$. (Such a choice is certainly possible, since $\alpha < \pi$.) Then we have for $\tilde{M}(x + y)$ the following inequality: Suppose $x \geqslant y$; then, if j_0 is the smallest j such that $x < x_{j_0}$, then $x + y < x_{j_0+1}$. Thus, if $j_0 \geqslant 2$,

$$\tilde{M}(x) = (j_0 - 1)x^{\alpha/\pi} \tag{5.53}$$

and

$$\begin{aligned} \tilde{M}(x + y) &\leqslant j_0(x + y)^{\alpha/\pi} \\ &\leqslant 2^{\alpha/\pi} j_0 \tilde{M}(x)/(j_0 - 1) \\ &\leqslant 4\tilde{M}(x) \\ &\leqslant 4\tilde{M}(x) + 4\tilde{M}(y). \end{aligned} \tag{5.54}$$

If $j_0 = 1$, then $\tilde{M}(x+y) \leqslant 2^{\alpha/\pi} < 2 < 4\tilde{M}(x) + 4\tilde{M}(y)$. Thus, condition (5.42) is satisfied.

To verify condition (5.43), suppose $x, y > 1$. Suppose again that $x \geqslant y$ and that j_0 is the smallest j so that $x < x_{j_0}$. Then $xy \leqslant x^2 < x_{j_0}^2 < x_{j_0+1}$. Thus, $\tilde{M}(xy) \leqslant j_0 x^{\alpha/\pi} y^{\alpha/\pi}$, while $\tilde{M}(x) = (j_0 - 1)x^{\alpha/\pi}$. This means that

$$\tilde{M}(xy) \leqslant 2\tilde{M}(x)\tilde{M}(y).$$

In case either x or y is $\leqslant 1$, say $y \leqslant 1$, then $xy \leqslant x$, so $\tilde{M}(xy) \leqslant \tilde{M}(x) = \tilde{M}(x)\tilde{M}(y)$. Thus, condition (5.43) is verified.

Finally, for condition (5.44) we note that for $j \geqslant 1$,

$$\begin{aligned} \int_{x_j}^{x_{j+1}} \frac{\tilde{M}(x)}{1+x^2}\,dx &= j \int_{x_j}^{x_{j+1}} \frac{x^{\alpha/\pi}}{1+x^2}\,dx \\ &\leqslant j \int_{x_j}^{x_{j+1}} x^{(\alpha/\pi)-2}\,dx \\ &\leqslant \frac{j}{(\alpha/\pi)-1}\left[x_{j+1}^{(\alpha/\pi)-1} - x_j^{(\alpha/\pi)-1}\right] \\ &< \text{const}/j^2 \end{aligned} \tag{5.55}$$

by construction of the x_j. Thus,

$$\sum_{j=1}^{\infty} \int_{x_j}^{x_{j+1}} \frac{\tilde{M}(x)}{1+x^2}\,dx < \infty,$$

which is the desired result. □

See Remarks 5.2 and 5.3.

We should like to reformulate Theorem 1.2 for the sufficient set σ of Theorem 5.5 in classical language. For this purpose we restrict ourselves to the case $n = 1$, and we choose $A = A_1$ as the real axis and $B = B_1$ as the imaginary axis. Then any entire function f can be represented in the form

$$\begin{aligned} f(x) &= \int \exp(ixz)\,d\mu_1(z)/k_1(z) + \int \exp(ixz)\,d\mu_2(z)/k_2(z) \\ &= f_1 + f_2, \quad \text{say,} \end{aligned} \tag{5.56}$$

where k_1 and k_2 are continuous, positive, monotonic functions which grow faster than any exponential and μ_1 and μ_2 are bounded measures with support $\mu_1 \subset A$ and support $\mu_2 \subset B$.

We want to investigate f_1 and f_2. It is easily seen that (writing $x = \xi + i\eta$)

$$\begin{aligned} |f_1(\xi + i\eta)| &\leqslant \int \exp(|\eta z|)\,|d\mu_1(z)|/k_1(z) \\ &= g(\eta), \quad \text{say.} \end{aligned} \tag{5.57}$$

This means that f_1 is uniformly bounded in strips parallel to the real axis. Similarly, f_2 is uniformly bounded in strips parallel to the imaginary axis. We do not know if one can give a direct proof of such a representation of an arbitrary entire function.

It should be noted that the above is not quite a characterization of the functions f_j. However, it is "almost" a characterization, for if f_1 were L_2 on lines parallel to the real axis and f_2 were L_2 on lines parallel to the imaginary axis, then we would have a representation

$$f_j(x) = \int \exp(ixz)\, d\mu_j(z)/k_j(z). \tag{5.58}$$

We wish to give next an example of a discrete set which is sufficient for $\mathscr{H}$:

THEOREM 5.7. *Let σ be the set of points in C each of whose real and imaginary coordinates is of the form $l + j \exp(-2l^2)$ where l is a positive integer and j is a non-negative integer $< \exp(2l^2)$. Then σ is sufficient for $\mathscr{H}$.*

PROOF. We shall give the proof for $n = 1$, as the general case is similar. Let $k(z)$ be a continuous function dominating all linear exponentials which we may assume to be a monotonic function of $|z|$. We may also assume that

$$k(z) \leqslant \exp(|z|^2 - 2|z|) \tag{5.59}$$

Let $F \in \mathbf{H}'$, $|F(z)| \leqslant k(z)$ for all $z \in \sigma$.

Since $F \in \mathbf{H}'$, F is entire. Using Cauchy's formula for $F'(z)$. We see that $F'(z)$ is smaller than $e \exp(|z|^2)$. Now any z is at a distance not exceeding $2 \exp(-|z|^2)$ from a point $z_0 \in \sigma$ with $|z_0| \leqslant |z|$. Hence, for any z we have by Taylor's formula with remainder

$$|F(z)| \leqslant k(z_0) + 2\,|F'(z')| \exp(-|z|^2) \tag{5.60}$$

for some z' between z and z_0 which gives

$$|F(z)| \leqslant k(z_0) + 2e \tag{5.61}$$

(5.61) implies

$$|F(z)| \leqslant k(|z|) + 2e. \tag{5.62}$$

Now, given any k' in the usual AU structure for $\mathscr{H}$, it is clear that we can find a k in this AU structure with $k(|z|) + 2e \leqslant ak'(z)$ for a suitable $a > 0$. Thus the proof of Theorem 5.7 is complete. ∎

See Problem 5.1.

V.2. EXAMPLE 2. THE SPACE $\mathbf{H}'$. If we set $\mathscr{W} = \mathbf{H}'$, then $\mathbf{W}'$ becomes just $\mathscr{H}$ with its usual topology. $\mathbf{H}'$ is AU. An AU structure consists of all continuous positive functions which are $+\infty$ outside of a compact set. To verify that $\mathbf{H}'$ is PLAU, we verify (*a*), (*b*), and (*c*) easily. (*d*) can be satisfied by choosing $\varphi(z, z^0, \eta)$ to be the constant 1. Condition (*e*) could be verified with some difficulty, but (*e'*) (as stated in the definition of weakly PLAU on page 115) is fairly easy. We shall omit the details as this is essentially the classical case of Oka [2]. (See also Bers [1].)

V.3. Example 3. The Space $\mathscr{D}'_F$ of Distributions of Finite Order. Now let $x = (x_1, \ldots, x_n)$ be real variables. For any integer $s \geqslant 0$ and any $l > 0$ we denote by $\mathscr{D}^s_l$ the space of (complex valued) functions on $R = R^n$ which are s times continuously differentiable and which vanish for $\|x\| \geqslant l$, that is, they have their *supports* (the closure of the set where they are different from zero) contained in $\|x\| \leqslant l$. The topology is defined by means of the seminorms $v_\partial(f) = \max |(\partial f)(x)|$ for ∂ any partial differential operator with constant coefficients of order $\leqslant s$. By $\mathscr{D}^s$ we denote the *inductive limit* of the $\mathscr{D}^s_l$, that is, set theoretically $\mathscr{D}^s = \cup \mathscr{D}^s_l$ and a fundamental system of neighborhoods of zero in $\mathscr{D}^s$ consists of those convex sets N for which $N \cap \mathscr{D}^s_l$ is a neighborhood of zero in $\mathscr{D}^s_l$ for each l.

We introduce a topology on $\mathscr{D}_F = \cap \mathscr{D}^s$ by saying that $N \subset \mathscr{D}_F$ is a neighborhood of zero if it is the intersection with $\mathscr{D}_F$ of a neighborhood of zero in some $\mathscr{D}^s$. $\mathscr{D}_F$ consists of all indefinitely differentiable functions of compact support. Its dual $\mathscr{D}'_F$ is the space of *distributions of finite order*. The Hahn-Banach theorem shows that $\mathscr{D}'_F = \cup (\mathscr{D}^s)'$. Moreover, each $T \in \mathscr{D}'_F$ can be written as a finite sum

$$T = \sum \partial_j \mu_j, \tag{5.63}$$

where ∂_j are linear partial differential operators with constant coefficients and μ_j are measures. The reason for calling elements of $\mathscr{D}'_F$ distributions of finite order will be clear on contrasting (5.63) with the analogous representation for elements in the space $\mathscr{D}'$ of all distributions (see Example 4, expression (5.85) below) in which certain infinite sums are allowed.

We denote by $\mathscr{D}_l$ the intersection $\cap \mathscr{D}^s_l$. A set $N \subset \mathscr{D}_l$ is a neighborhood of zero if it is the intersection with $\mathscr{D}_l$ of a neighborhood of zero in some $\mathscr{D}^s_l$.

At this point we must make some notational conventions. We have chosen the present form of notation because we want to treat AU spaces in a uniform manner and their duals in a uniform manner.

We identify a locally integrable function h on R with the element of $\mathscr{D}'_F$ (or other spaces of distributions) defined by $f \to \int fh\, dx$ for $f \in \mathscr{D}_F$. Here dx refers to the usual Lebesgue measure. Thus, choosing for h the exponential function, we see that the Fourier transform of f is

$$F(z) = \int f(x) \exp(ix \cdot z)\, dx \tag{5.64}$$

and the inverse is

$$f(x) = \int_{z\,\text{real}} F(z) \exp(-ix \cdot z)\, dz. \tag{5.65}$$

Here dz is the Lebesgue measure divided by $(2\pi)^n$. It must be admitted that there is a slight inconsistency in the notation at this point, namely, by Theorem 1.5 when representing elements of $\mathscr{D}'_F$ we use an integral involving $\exp(ix \cdot z)$ while, by the above, we use $\exp(-ix \cdot z)$ for functions in $\mathscr{D}_F$. The minus sign is inherent in the theory because, for $f, g \in \mathscr{D}_F$, if F, G denote their Fourier transforms, Parseval's relation reads

$$\int f(x)g(x)\, dx = \int F(z)G(-z)\, dz.$$

Thus there is an ambiguity in terminology: The Fourier transform of f considered as an element of $\mathscr{D}_F$ is $F(z)$ while the Fourier transform of f when it is thought of as an element of $\mathscr{D}'_F$ is $F(-z)\, dz$. We shall sometimes call $F(-z)$ the generalized Fourier transform of f.

Our normalization of dz to be the Lebesgue measure divided by $(2\pi)^n$ has the consequence that the factor 2π in Cauchy's residue formula is dropped.

If D is a linear constant coefficient partial differential operator, we shall denote by $D'f$ the result of applying to f the differential operator obtained from D by replacing each $\partial/\partial x_j$ by $-\partial/\partial x_j$ (the formal adjoint of D). For any $T \in \mathscr{D}'_F$ we define $DT \in \mathscr{D}'_F$ by

$$f \cdot DT = D'f \cdot T.$$

(The reason for using this notation is that we regard the AU space as the primary object of study.) If T is identified with a sufficiently smooth function, then clearly DT is the function which results by applying D to T in the usual manner.

For $f \in \mathscr{D}_F$, the Fourier transform of $D'f$ is

$$D'f \cdot \exp(ix \cdot z) = P(z)F(z)$$

where F is the Fourier transform of f and

$$P(z) = \exp(-ix \cdot z) D \exp(ix \cdot z)$$

is the polynomial obtained from D by replacing $\partial/\partial x_j$ by iz_j. We call $P(z)$ the *Fourier transform* of D'. The Fourier transform of Df (thought of as an element of $\mathscr{D}_F$) is just $P(-z)F(z)$; $P(-z)$ is the Fourier transform of D.

Let $\vec{f} = (f_1, \ldots, f_r)$ be a (column) vector in $\mathscr{D}^r_F$. The Fourier transform of $\vec{f}$ is defined to be $\vec{F} = (F_1, \ldots, F_r)$, where F_j is the Fourier transform of f_j. Let $\boxed{D}$ be an $r \times l$ matrix of linear constant coefficient operators.

Then we have

$$\boxed{\mathrm{D}}'\vec{\mathrm{f}}=[\sum (D_{j1})'f_j, \ldots, \sum (D_{jl})'f_j].$$

Thus the Fourier transform of $\boxed{\mathrm{D}}'\vec{\mathrm{f}}$ is

$$(\sum P_{j1}F_j, \ldots, \sum P_{jl}F_j).$$

This means that the Fourier transform of $\boxed{\mathrm{D}}'$ is $\boxed{\mathrm{P}}$ which is the $l \times r$ matrix obtained from $\boxed{\mathrm{D}}'$ be replacing $(D_{ji})'$ by P_{ij}.

For any $f, g \in \mathscr{D}_F$, by the definition of convolution given in Section I.3, $f * g$ is the function whose Fourier transform is FG. This gives

$$(f * g)(x) = \int f(x-y)g(y)\, dy$$

because

$$\begin{aligned}\iint f(x-y)g(y) \exp(ix \cdot z)\, dx\, dy &= \int g(y) \exp(iy \cdot z)\, dy \int f(x-y) \exp(i(x-y)\cdot z)\, dx \\ &= F(z)G(z).\end{aligned}$$

For any $T \in \mathscr{D}'_F$, if T is a function, then we define $f * T$ by

$$(f * T)(x) = \int f(x-y)T(y)\, dy.$$

This gives

$$\begin{aligned}g \cdot (f * T) &= \iint g(x)f(x-y)T(y)\, dx\, dy \\ &= \int T(y)\, dy \int f(x-y)g(x)\, dx \\ &= \int T(y)\, dy \int \check{f}(y-x)g(x)\, dx \\ &= (\check{f} * g) \cdot T.\end{aligned}$$

Here $\check{f}(x) = f(-x)$. We can use the formula

$$g \cdot (f * T) = (\check{f} * g) \cdot T$$

to define $f * T$ for an arbitrary $T \in \mathscr{D}'_F$.

If T is a function, we denote by $\tau_y T$ the function $(\tau_y T)(x) = T(x-y)$, so τ_y denotes translation by y. For any f we have clearly

$$f \cdot (\tau_y T) = (\tau_{-y} f) \cdot T.$$

For a general distribution T we use the above expression to define $\tau_y T$. We can write, in general,

$$(f * T)(x) = (\tau_x \check{f}) \cdot T = f \cdot \tau_x \check{T}$$

Here $\check{T}$ is defined by $f \cdot \check{T} = \check{f} \cdot T$. It is easily seen (see Schwartz [1]) that $f * T$ is always an indefinitely differentiable function. Moreover, if D is any linear partial differential operator with constant coefficients,

$$Df * T = f * DT = D(f * T).$$

In Section I.3 we defined the adjoint of convolution $f *' T$ for $f \in \mathscr{D}_F$, $T \in \mathscr{D}'_F$ by

$$g \cdot (f *' T) = (f * g) \cdot T$$

for any $g \in \mathscr{D}_F$. Thus, if T is a function,

$$\begin{aligned} g \cdot (f *' T) &= \iint f(x-y) g(y) T(x) \, dx \, dy \\ &= \int g(y) \, dy \int f(x-y) T(x) \, dx \\ &= \int g(y) \, dy \int f(x) T(x+y) \, dx. \end{aligned}$$

Thus

$$f *' T = \check{f} * T = (f * \check{T})\check{}\,.$$

Either of these formulas expresses $f *' T$ for an arbitrary $T \in \mathscr{D}'_F$.

Unlike the convolution, which is commutative, the adjoint of convolution is not commutative. By the above, if $f, g \in \mathscr{D}_F$,

$$\begin{aligned} f *' g &= \check{f} * g \\ &= (f * \check{g})\check{}\, \\ &= (\check{g} * f)\check{}\, \\ &= (g *' f)\check{}\,. \end{aligned}$$

If D is any linear partial differential operator with constant coefficients, we find

$$D(f *' g) = \check{f} * Dg = f *' Dg$$

and

$$D(f *' g) = D\check{f} * g = (D'f)\check{} * g = D'f *' g.$$

We denote by fT the element of $\mathscr{D}'_F$ defined by

$$fT \cdot g = T \cdot fg.$$

Thus if T is a function, then fT is the function which is the product of f and T. Note that $fT \cdot g$ makes sense even if g is not of compact support, that is, it is meaningful for any $g \in \mathscr{E}$, the space of all indefinitely differentiable functions (see Example 5 below). Thus we can identify fT with an element of $\mathscr{E}'$.

In general, we can identify $\mathscr{E}'$ with a subspace of $\mathscr{D}'_F$. This may lead to a slight confusion in notation; we wish to explain how to avoid it. Thus if $S \in \mathscr{E}'$ and D is a linear constant coefficient partial differential operator, then $D'S$ is defined if S is thought of as belonging to $\mathscr{E}'$ and DS is defined if S is thought of as belonging to $\mathscr{D}'_F$, since D is defined on an AU space and D' is defined on its dual. This confusion can be avoided if we interpret D' as the differential operator obtained from D by replacing $\partial/\partial x_j$ by $-\partial/\partial x_j$ (as in the situation described above for $D'f$, $f \in \mathscr{D}_F$). Thus, if $S \in \mathscr{E}'$ and $f \in \mathscr{D}_F$, then

$$D'S \cdot f = S \cdot Df$$

whether we think of $S \in \mathscr{E}'$ or $\in \mathscr{D}'_F$, and $f \in \mathscr{E}$, or $\in \mathscr{D}_F$.

The celebrated Paley-Wiener-Schwartz theorem (see Schwartz [1]) asserts that $\mathbf{D}_l$ consists of all entire functions of exponential type $\leqslant l$ which approach zero for real z as $|z| \to \infty$ faster than the reciprocal of any power of the distance from the origin. Thus $\mathbf{D}_F$ consists of all entire functions of exponential type which are $0(1 + \|z\|)^{-p}$ for any p for real z.

We shall now study the topology of $\mathbf{D}_F$.

LEMMA 5.8. *Let $\{G^j\}$ be a sequence of functions in $\mathbf{D}_F$. A necessary and sufficient condition that $G^j \to 0$ in the topology of $\mathbf{D}_F$ is that the G^j should all belong to some $\mathbf{D}_l$ and for any p*

$$\max_{z \text{ real}} (1 + \|z\|)^p |G^j(z)| \to 0. \tag{5.66}$$

PROOF. Suppose first that $G^j \to 0$ in $\mathbf{D}_F$. If we call g^j the inverse Fourier transform of G^j, then we see easily from the definition of the topology of $\mathscr{D}_F$ that all g^j must belong to some $\mathscr{D}_l$. Moreover, for any differential operator ∂ with constant coefficients, $(\partial g^j)(x) \to 0$ uniformly. (5.66) results immediately.

Conversely, suppose all $G^j \in \mathbf{D}_l$ and (5.66). Then each $g^j \in \mathscr{D}_l$, and writing g^j in terms of G^j by means of (5.65) shows that $g^j(x) \to 0$ uniformly. Similarly, $(\partial g^j)(x) \to 0$ uniformly because the Fourier transform of ∂g^j is a polynomial times G^j. Thus, $g^j \to 0$ in the topology of $\mathscr{D}_F$ which gives the result. ▯

A similar method shows

LEMMA 5.9. *A subset B of $\mathbf{D}_F$ is bounded if and only if $B \subset \mathbf{D}_l$ for some l and for any p there is an A so that for all $G \in B$*

$$\max_{z \text{ real}} (1 + \|z\|)^p |G(z)| \leqslant A. \tag{5.67}$$

THEOREM 5.10. *$\mathscr{D}'_F$ is AU. An AU structure for $\mathscr{D}'_F$ can be chosen to consist of all functions of the form $k(z) = k_1(\operatorname{Im} z)(1 + \|z\|)^{-q}$ where q is any integer and k_1 is any continuous positive function which dominates all linear exponentials.*

PROOF. Properties (a) and (b) of AU are easy to establish so we concentrate on (c) (see Section I.3). Let $k(z) = k_1(\operatorname{Im} z)(1 + \|z\|)^{-q}$ be as in the statement of the theorem; call N_k the set of $G \in \mathbf{D}_F$ with $|G(z)| \leqslant k(z)$ for all $z \in C$. We show first that N_k is a neighborhood of zero in $\mathbf{D}_F$. This means that N_k is the intersection with $\mathbf{D}_F$ of a neighborhood of zero in some $\mathbf{D}^s$, which, in turn, means that $N_k \cap \mathbf{D}_l^s$ is a neighborhood of zero in $\mathbf{D}_l^s$ for all l.

Let s be an integer $>q$. Let M_l^s be the set of $g \in \mathscr{D}_l^s$ all of whose derivatives of order $\leqslant s$ are in modulus $\leqslant 1$. By definition, M_l^s is a neighborhood of zero in $\mathscr{D}_l^s$. Then a trivial computation shows that any $G \in \hat{M}_l^s$ (the set of Fourier transforms of M_l^s) satisfies

$$|G(z)| \leqslant a(1 + \|z\|)^{-s} \exp (l \|z\|)$$

where a is a constant independent of G. Thus, for some $a' > 0$ we have $a'\hat{M}_l^s \subset N_k$ because k_1 dominates all linear exponentials. Thus, N_k is a neighborhood of zero in $\mathbf{D}_F$.

Conversely, let M be a neighborhood of zero in $\mathscr{D}_F$; we want to construct k satisfying the above requirements so that, if N'_k denotes the set of inverse Fourier transforms of N_k, then for a suitable $a > 0$ we have $aN'_k \subset M$. By the definition of the topology of $\mathscr{D}_F$, we may assume M is given as follows: There exists an integer s and a decreasing sequence of positive numbers $b_l < 1$, $b_l \to 0$ so that if ∂_j denote the differentiations of order $\leqslant s$ then M consists of all $g \in \mathscr{D}_F$ which satisfy

$$\max_{\|x\| \geqslant l} |(\partial_j g)(x)| \leqslant b_l \tag{5.68}$$

for all j, l and also

$$\max_{\|x\| \leqslant 1} |(\partial_j g)(x)| \leqslant 1. \tag{5.69}$$

We define k_1 so that k_1 is a function of $\|\operatorname{Im} z\|$. For $l > 0$ let $a_l = -\log b_l$. Let ($n = \dim C$)

$$k_1(na_l) = \exp ((l - 1)a_l) \tag{5.70}$$

and let k_1 be defined to be continuous and log k_1 linear between the points na_l. It is clear that k_1 dominates all linear exponentials. Call $q = s + n + 1$.

Let $g \in N'_k$, $\|x\| \geqslant l$. Then some component of x is $\geqslant l$. By Cauchy's formula we may write

$$g(x) = \int_{\alpha_l} G(z) \exp(-ix \cdot z)\, dz \tag{5.71}$$

where α_l is the plane $Iz_i = -a_l \operatorname{sgn} x_i$ for all i, where sgn x is the signum of x. On α_l we have

$$|\exp(-ix \cdot z)| \leqslant \exp(-la_l). \tag{5.72}$$

Thus, by the definition of k, we have on α_l

$$|G(z) \exp(-ix \cdot z)| \leqslant k(z) \exp(-la_l) \leqslant (1 + \|z\|)^{-s-n-1} b_l \tag{5.73}$$

by (5.70) because for $z \in \alpha_l$ we have $\|\text{Im } z\| = na_l$. Now, using (5.71) we have our result for $\|x\| \geqslant 1$. The result for $\|x\| < 1$ is clear. ▯

THEOREM 5.11. *$\mathscr{D}'_F$ is PLAU.*

PROOF. Conditions (a), (b), and (c) for product localizeability are clearly satisfied (see Section IV.1). Thus, we may restrict our considerations to $n = 1$. As is clear from Example 1, Condition (d) is easier to verify than Condition (e) so we consider only the latter. Actually it is easy to modify our construction to obtain (d).

We must again give a "good" description of suitable AU structures and bounded AU structures for $\mathscr{D}'_F$. The former is easy to describe, namely, functions $k(z) = k_1(Rz)k_2(Iz)$ where k_2 satisfies Conditions 1, 2, and 3 of Example 1 (for $\tilde{k}$) and $k_1(t) = (1 + |t|)^{-l}$ for some l. It is not difficult to verify that a bounded AU structure can be chosen to consist of functions $m(z) = m_1(Rz)m_2(Iz)$ where m_2 satisfies Conditions 4 and 5 of Example 1 (for $\tilde{m}$), and m_1 is the absolute value of a function of the form

$$\psi(z) = \prod_{j=1}^{\infty} \frac{\sin[(z + ie_j)/d_j]}{(z + ie_j)/d_j}. \tag{5.74}$$

Here $\{d_j\}$, $\{e_j\}$ are infinite sequences of positive integers with $\sum 1/d_j < \infty$, $\sum 1/e_j < \infty$, and $\sum e_j/d_j < \infty$.

In order to complete the proof that $\mathscr{D}'_F$ is localizeable, we need to derive certain inequalities [see (5.79) and (5.80) below] which will serve the same purpose as (5.33) and (5.34) do for the space $\mathscr{H}$.

We have

$$|\sin(a + ib)| \leqslant e^{|b|}(|\sin a| + |b|\,|\cos a|).$$

This gives

(5.75)

$$\left|\frac{\psi(x+iy)}{\psi(x)}\right| \leqslant \frac{\prod\left|\dfrac{|\sin[(x+ie_j)/d_j]| + |[iy/d_j]\cos[(x+ie_j)/d_j]|}{(x+ie_j+iy)/d_j}\right| \exp(|y|/d_j)}{\prod\left|\dfrac{\sin(x+ie_j)/d_j}{(x+ie_j)/d_j}\right|}$$

$$= \exp c|y| \prod |(x+ie_j)/(x+ie_j+iy)| \prod \left[1 + \left|\frac{y}{d_j}\cot\left(\frac{x+ie_j}{d_j}\right)\right|\right].$$

Now for b small $|\cos(a+ib)| \leqslant e^{|b|} < 1 + 2|b|$ while $|\sin(a+ib)| \geqslant |b|$. The first of these inequalities is obvious; the second follows easily from

$$\begin{aligned} |\sin(a+ib)|^2 &= \sin^2 a \cosh^2 b + \cos^2 a \sinh^2 b \\ &\geqslant \sin^2 a + b^2 \cos^2 a \\ &\geqslant b^2. \end{aligned}$$

This gives $|\cot(a+ib)| \leqslant (1/|b|)(1+2|b|)$. Thus, except for constant factors,

$$\text{(5.76)} \quad \left|\frac{\psi(x+iy)}{\psi(x)}\right| \leqslant \exp c|y| \prod \left[1 + \frac{|y|}{d_j}\cdot\frac{d_j}{e_j}\left(1 + 2\frac{e_j}{d_j}\right)\right] \prod \left|1 + \frac{iy}{x+ie_j}\right|^{-1}$$

$$= \exp c|y| \prod \left[1 + \frac{|y|}{e_j} + 2\frac{|y|}{d_j}\right] \prod \left|1 + \frac{iy}{x+ie_j}\right|^{-1}.$$

Now, our assumptions are that

$$\sum \frac{1}{e_j} \quad \text{and} \quad \sum \frac{1}{d_j}$$

are convergent. Hence also

$$\prod \left|1 + \frac{iy}{x+ie_j}\right|^{-1} \qquad \text{and} \qquad \prod \left[1 + \frac{|y|}{e_j} + \frac{2|y|}{d_j}\right]$$

are convergent. Thus,

$$\text{(5.77)} \qquad \left|\frac{\psi(x+iy)}{\psi(x)}\right| \leqslant c(y),$$

where $c(y)$ is some function of y. We wish to estimate $c(y)$. Clearly,

$$\text{(5.78)} \qquad \prod \left(1 + \frac{y}{e_j} + 2\frac{y}{d_j}\right) \qquad \text{and} \qquad \prod \left|1 + \frac{iy}{x+ie_j}\right|^{-1}$$

are majorized by const $\exp(\alpha|y|)$.

Thus, we have proved the important inequality

$$|\psi(x+iy)/\psi(x)| \leqslant C \exp(C|y|). \tag{5.79}$$

If $|y| \leqslant \frac{1}{2}$, it is easy to reverse the estimates used in (5.75) and (5.77). We obtain

$$C^{-1} \leqslant |\psi(x+iy)/\psi(x)| \leqslant C \quad \text{for} \quad |y| \leqslant \tfrac{1}{2}. \tag{5.80}$$

We can now complete the proof that $\mathscr{D}'_F$ is product localizeable. Let k be given in the AU structure described above. k' will be chosen in the AU structure. Then, if m is in the BAU structure defined above and if $m(z) \leqslant k'(z)$ for all z, we set

$$\tilde{\psi}_{z^0}(z) = m_2(\operatorname{Im} z^0) \exp[-i(l+C+4)(z-z^0)]\psi(z - i \operatorname{Im} z^0)$$

if

$$a_l < \operatorname{Im} z^0 \leqslant a_{l+1}.$$

As in (5.37), we have by (5.80)

$$C_1^{-1} \exp(-3|\operatorname{Im} z^0|) \leqslant \frac{\min\limits_{|t-\operatorname{Im} z^0| \leqslant \frac{1}{2}} |\tilde{\psi}_{z^0}(\xi+it)|}{\max\limits_{|t-\operatorname{Im} z^0| \leqslant \frac{1}{2}} m(\xi+it)} \tag{5.81}$$

$$\leqslant \frac{\max\limits_{|t-\operatorname{Im} z^0| \leqslant \frac{1}{2}} |\tilde{\psi}_{z^0}(\xi+it)|}{\min\limits_{|t-\operatorname{Im} z^0| \leqslant \frac{1}{2}} m(\xi+it)}$$

$$\leqslant C_1 \exp(3|\operatorname{Im} z^0|).$$

Moreover, for any z we find by (5.79), for $a_l < \operatorname{Im} z^0 \leqslant a_{l+1}$

$$|\tilde{\psi}_{z^0}(z)|/m(z) \leqslant \tag{5.82}$$

$$\frac{2C \exp[C|\operatorname{Im}(z-z^0)| + (l+1)\operatorname{Im} z^0 + \operatorname{Im}(z-z^0)(l+C+4)]}{m_2(\operatorname{Im} z)},$$

which is the counterpart of (5.38). We may now repeat the argument following (5.38) to obtain

$$|\tilde{\psi}_{z^0}(z)|/m(z) \leqslant c' \exp(c'|\operatorname{Im} z|). \tag{5.83}$$

We can now conclude the proof of (e) as in Example 1, since the factor $\exp(c'|\operatorname{Im} z|)$ is harmless because we can easily absorb it into the k and m. ▯

By a method similar to the proof of Theorem 5.5 we can show

THEOREM 5.12. *For each j let R_j be the real axis in the z_j plane and let*

A_j be any other line through the origin. Then, $(R_1 \cup A_1) \times \cdots \times (R_n \cup A_n)$ *is sufficient for* $\mathscr{D}'_F$.

See Problem 5.2.

It is easy to find analogs of Theorem 5.7 for $\mathscr{D}'_F$.

It is easily seen that R is sufficient for $\mathscr{D}'_l$ for any l.

If we apply Theorem 5.12 to Theorem 1.2, then we derive the fact that, for $n = 1$, any $T \in \mathscr{D}'_F$ can be written in the form $h + T_1$, where h is represented by an integral along A_1 and is in $\mathscr{H}$ and T_1 is represented by an integral along R and belongs to the space $\mathscr{S}'$ of tempered distributions (see Schwartz [1]).

V.4. EXAMPLE 4. THE SPACE $\mathscr{D}'$ OF ALL DISTRIBUTIONS. We define $\mathscr{D}$ as the inductive limit of the spaces $\mathscr{D}_l$, that is, set theoretically $\mathscr{D} = \cup \mathscr{D}_l = \mathscr{D}_F$; a convex set $N \subset \mathscr{D}$ is a neighborhood of zero if its intersection with each $\mathscr{D}_l$ is a neighborhood of zero in $\mathscr{D}_l$ (see Schwartz [1], Dieudonné and Schwartz [1]). In regard to bounded sets and convergence of sequences the topologies of $\mathscr{D}$ and $\mathscr{D}_F$ are the same, since bounded sets and convergent sequences in $\mathscr{D}$ and $\mathscr{D}_F$ are contained in some $\mathscr{D}_l$, and both $\mathscr{D}$ and $\mathscr{D}_F$ induce the topology of $\mathscr{D}_l$ on the elements of $\mathscr{D}_l$. But as topological vector spaces $\mathscr{D}$ and $\mathscr{D}_F$ are *not* the same. For example, for $n = 1$, $S = \sum d^j\delta_j/dx^j$ belongs to $\mathscr{D}'$ but not to $\mathscr{D}'_F$. Here δ_j is the unit mass at the point j. The fact that $S \in \mathscr{D}'$ is an easy consequence of the definition of the topology of $\mathscr{D}$. The fact that $S \notin \mathscr{D}'_F$ is left as an exercise for the reader [compare (5.85) and (5.63)].

It can be shown without much difficulty that a fundamental system of neighborhoods N of zero in $\mathscr{D}$ can be given as follows: For each integer $j \geqslant 0$ we choose an integer e_j and a constant $c_j > 0$. Then N consists of all $f \in \mathscr{D}$ which satisfy

$$\max_{\|x\| \geqslant j} |(\partial f)(x)| \leqslant c_j \tag{5.84}$$

for all differentiations ∂ of order $\leqslant e_j$ (see Schwartz [1]). By use of the Hahn-Banach theorem we see that any $T \in \mathscr{D}'$ can be represented in the form

$$T = \sum \partial_l \mu_l . \tag{5.85}$$

Here ∂_l are linear partial differential operators with constant coefficients and $\{\mu_l\}$ is an infinite sequence of measures which is *locally finite*, that is, for any compact set L in R, the support of μ_l meets L for only a finite number of l. (5.85) should be contrasted to (5.63).

Since $\mathscr{D}$ and $\mathscr{D}_F$ have the same functions, so do $\mathbf{D}$ and $\mathbf{D}_F$. By our above remarks, Lemmas 5.8 and 5.9 describe also the bounded sets and sequential convergence in $\mathbf{D}$.

THEOREM 5.13. *$\mathscr{D}'$ is AU. An AU structure for $\mathscr{D}'$ consists of all continuous positive functions k such that any continuous function $h(z)$ which for some A satisfies*

$$h(z) = 0[\exp(A\,|\mathrm{Im}\, z|)(1 + |z|)^{-l}] \tag{5.86}$$

for all l is also $0(k(z))$.

PROOF. Given any k as above, call N_k the set of $F \in \mathbf{D}$ which satisfy $|F(z)| \leqslant k(z)$ for all z. We claim that N_k swallows every bounded set $B \subset \mathbf{D}$. For, by Lemma 5.9, if we set

$$h(z) = \max_{F \in B} |F(z)|,$$

then h satisfies (5.86). Thus, there is a $b > 0$ so that $bB \subset N_k$. Now, $\mathscr{D}_l$ is a complete metrizable space, since its topology is defined by a countable number of seminorms. Thus, $\mathscr{D}_l$ is bornologic. It is easily seen that the inductive limit of bornologic spaces is bornologic. Thus, $\mathscr{D}$ is bornologic, so N_k is a neighborhood of zero.

To prove that the sets N_k are fundamental, we shall prove the following apparently stronger result which implies it.

THEOREM 5.13*. *An AU structure for $\mathscr{D}'$ can be chosen to consist of functions k of the following form: Let $\{a_j\}$ be a strictly increasing sequence of integers with $a_0 = a_1 = a_2 = 0$, $a_{j+1} > 2a_j$, and let l be a positive integer. Set*

$$k(z) = (1 + |\mathrm{Re}\, z|)^{-l}(1 + |\mathrm{Im}\, z|)^{-l} \exp((j-2)|\mathrm{Im}\, z|) \tag{5.87}$$

for

$$a_j + a_j \log(1 + |\mathrm{Re}\, z|) \leqslant |\mathrm{Im}\, z| \leqslant \tfrac{1}{2}(a_{j+1} + a_{j+1} \log(1 + |\mathrm{Re}\, z|)).$$

The definition of k is completed by requiring that k be a function of $|\mathrm{Re}\, z|$, $|\mathrm{Im}\, z|$ which is continuous and such that, for fixed $|\mathrm{Re}\, z|$,

$$\log k(|\mathrm{Re}\, z|, |\mathrm{Im}\, z|) + l[\log(1 + |\mathrm{Re}\, z|) + \log(1 + |\mathrm{Im}\, z|)]$$

is linear in $|\mathrm{Im}\, z|$ in the regions in which it is not already defined above.

PROOF. It is clear that such k satisfy the conditions of Theorem 5.13 so the N_k are neighborhoods of zero. To prove they are fundamental we shall restrict our considerations to the case $n = 1$ to avoid unnecessary complications in notation. There is no essential difficulty in passing to $n > 1$.

Let N be a neighborhood of zero in $\mathscr{D}$, which we may assume to be as in (5.84), where the sequence $\{c_j\}$ is strictly decreasing with $c_0 = 1$ and $\{e_j\}$ is strictly increasing. We want to find the conditions on the sequence $\{a_j\}$ in order that $N_k \subset$ Fourier transform of N. For any $F \in c_2 N_k$ we write the inverse Fourier transform

$$f(x) = \int_{-\infty}^{\infty} \exp(-ixz)F(z)\,dz, \tag{5.88}$$

where we have normalized the Haar measure dz so the unit interval has measure $1/2\pi$. If we choose $l = 1 + 4e_2$, $a_0 = a_1 = a_2 = 0$, then (5.87) and (5.88) show that (5.84) holds for $j = 0, j = 1, j = 2$.

To see what the proper choice of a_j is for $j > 2$, we proceed as follows: For $x > 0$ we shift the contour in (5.88) to the curve γ_j defined by

$$\gamma_j\colon \operatorname{Im} z = -a_j - a_j \log(1 + |\operatorname{Re} z|). \tag{5.89}$$

On γ_j we have

$$|\exp(-ixz)| = (1 + |\operatorname{Re} z|)^{-a_j x} \exp(-a_j x).$$

Hence, every derivative with respect to x of $\exp(-ixz)$ of order $\leqslant e_j$ is majorized by

$$(1 + |z|)^{e_j}(1 + |\operatorname{Re} z|)^{-a_j x} \exp(-a_j x).$$

Now,

$$(1 + |z|) \leqslant (1 + |\operatorname{Re} z|)(1 + |\operatorname{Im} z|).$$

For any e_j we can make a_j large enough so that on γ_j we have

$$(1 + |\operatorname{Im} z|)^{e_j} \leqslant \exp(|\operatorname{Im} z|).$$

This means that on γ_j all derivatives with respect to x of $\exp(-ixz)F(z)$ of order $\leqslant e_j$ are majorized by

$$(1 + |\operatorname{Re} z|)^{e_j - a_j x} \exp[(j-1)|\operatorname{Im} z| - a_j x]$$
$$= (1 + |\operatorname{Re} z|)^{e_j + a_j(j-x-1)} \exp[a_j(j - x - 1)]$$

by (5.89). It is clear from this that for any $j \geqslant 3$ we can choose a_j sufficiently large so that (5.84) is satisfied. ▯

THEOREM 5.14. *$\mathscr{D}'$ is LAU.*

PROOF. As usual we restrict our considerations to $n = 1$ and Condition (e).

Before describing the bounded AU structure, it is useful to give some heuristic remarks: The descrption of $\mathbf{D}_F$ given in Example 3 is an illustration of the principle that we can make the inverse Fourier transform f of an $F \in \mathbf{D}_F$ small for x large, say for $|x| \geqslant a$, by making F small for sufficiently many Im z, say for $|\operatorname{Im} z| \leqslant b$. This is indeed the guiding principle of the proof of Theorem 5.10 that $\mathscr{D}'_F$ is AU with the stated AU structure.

A similar principle is relevant to $\mathscr{D}'$, except that now because of (5.84) we want to make many derivatives of f small for $|x| \geqslant a$; to accomplish this we must make F small on a set of the form

$$|\operatorname{Im} z| \leqslant b + b \log (1+|x|).$$

The description of suitable bounded sets of $\mathscr{D}$ which can be used to construct a sufficient BAU structure is given by the same principle. For $\mathscr{D}_F$ a sufficient number of bounded sets is given as follows: Let $\{b_j\}$ be any sequence of positive numbers, let u be a positive function on R, and let l be any positive number. Then we form the bounded set B_F consisting of all f with $|f^{(j)}(x)| \leqslant b_j u(x)$ for all x, and support f contained in $|x| \leqslant l$. More precisely, it can be shown without difficulty that a fundamental system of neighborhoods of zero in $\mathscr{D}_F$ consists of those set N_F for which there exists an l_0 and a continuous positive function u so that N_F consists of all $f \in \mathscr{D}_F$ satisfying $|f^{(j)}(x)| \leqslant u(x)$ for $j \leqslant l_0$ and all x. Thus, the suitable bounded sets which "define" N_F in the sense that N_F is the union of these bounded sets, are those B_F described above for varying l and with $\{b_j\}$ arbitrarily large for $j > l_0$ and $b_j = 1$ for $j \leqslant l_0$. This leads, by Fourier transform, to the description of a sufficient bounded AU structure given in Example 3.

On the other hand, by (5.84) a fundamental system of neighborhoods of zero in $\mathscr{D}$ can be described as follows: Let u be a continuous positive function on R and let $\{d_j\}$ be an increasing sequence of positive integers Then we consider the set N of all functions $f \in \mathscr{D}$ with $|f^{(k)}(x)| \leqslant u(x)$ for $|x| \geqslant j$, for $k \leqslant d_j$. From this, a "sufficient" collection of bounded sets in $\mathscr{D}$ can be described as follows: Let $\{b_j\}$ be any sequence of positive numbers, and let l be a positive integer. Then we form the set B of all f with $|f^{(k)}(x)| \leqslant b_{k-d_j} u(x)$ for $|x| \geqslant j$ and support f contained in $|x| \leqslant l$. Here b_a is defined to be 1 if $d \leqslant 0$. This description will lead to the description of a sufficient bounded AU structure given below.

It is important to note that while the bounded sets of $\mathscr{D}$ and $\mathscr{D}_F$ are the same, by our above remarks, we *cannot* choose sufficient bounded AU structures to be the same.

See Remark 4.2.

The above strongly influences the construction of the analog for $\mathscr{D}'$ of the function ψ_{z^0} of Example 3. For $z^0 = 0$ we see from (5.74) and the fact that $|(\sin z)/z| \leqslant \exp(|z|)$ that $\psi = \psi_{z^0}$ is of exponential type 1. For other z^0 we obtain ψ_{z^0} by multiplying ψ by a suitable exponential and translating through complex values. The Fourier transform of multiplication by an exponential is translation. By our above principle, the effect of the translation will be to make the resulting function smaller. Note that this fits exactly into our description of the sufficient bounded sets B_F of $\mathscr{D}_F$, namely, the regularity conditions that a function f must possess in order to belong to B_F, that is, the type of inequalities placed on the derivatives of f, are similar at all $x \in R$, the only difference between various x is multiplication by suitable constants.

For $\mathscr{D}$, however, there is a difference in the regularity of $f \in B$ at different $x \in R$, that is, we place more stringent conditions on more derivatives as $x \to \infty$. This will be reflected in the fact that the integration cannot be taken over lines $\text{Im } z = \text{const}$ but over more complicated curves.

We can now define a bounded AU structure for $\mathscr{D}'$. Let ψ be as in Example 3, (5.74). Let $s_0 = 0$ and let a_j for $j \geqslant 0$ be an increasing sequence of integers with $a_0 0 =$. Then there is an l_0 so that for $j \leqslant l_0 - 1$

$$m(z) = \psi(z) \exp (j \,|\text{Im } z|) \tag{5.90}$$

for

$$a_j + a_j \log (1 + |\text{Re } z|) \leqslant |\text{Im } z| < a_{j+1} + a_{j+1} \log (1 + |\text{Re } z|)$$

and (5.90) holds for $j = l_0$ for all z with $|\text{Im } z| \geqslant a_{l_0} + a_{l_0} \log (1 + |\text{Re } z|)$. The proof that these m are sufficient relies on inequality (5.79) and a somewhat weaker reverse inequality.

In order to prove that $\mathscr{D}'$ is localizeable, we first make a slight modification in the proof in Chapter IV of the passage from local to global. Namely, we want to replace strips parallel to the real axis by strips of the form

$$b + c \log (1 + |\text{Re } z|) \leqslant \text{Im } z \leqslant b^1 + c^1 \log (1 + |\text{Re } z|) \tag{5.91}$$

for suitable b, c, b^1, c^1. Thus, for example (in the notation of that proof) for $n = 1$, for $a = 0$ we would have to take squares of increasing size as $\text{Re } z \to \infty$, while for $a = 1$ we would integrate over the boundaries of strips of the form (5.91). The definition of product localizeability would then have to be slightly altered to suit this situation. All this can be done in a very simple manner. Then, by a slight modification of the method of Example 3, we can show that $\mathscr{D}'$ fulfills this type of product localizeability. We shall omit the details as they are reasonably straightforward. □

Similar results hold for the space of distributions on a convex set.

Problem 5.3. (Exercise). Carry out the details.

We shall discuss some examples of sufficient sets for $\mathscr{D}'$ in conjunction with similar examples for $\mathscr{E}$ in Example 5.

V.5. EXAMPLE 5. THE SPACE $\mathscr{E}$ OF INDEFINITELY DIFFERENTIABLE FUNCTIONS. We denote by $\mathscr{E}$ the space of indefinitely differentiable complex-valued functions on $R = R^n$ with the topology of uniform convergence of functions and their derivatives on compact sets. The dual space $\mathscr{E}'$ is the space of distributions of compact support. Any $T \in \mathscr{E}'$ can be represented by a finite sum $T = \sum \partial_j \mu_j$, where ∂_j are linear differential operators with constant coefficients and μ_j are measures of compact support.

In order to prove that $\mathscr{E}$ is AU we shall need the following description of $\mathscr{E}'$:

THEOREM 5.15. *$\mathscr{E}'$ consists of all distributions $S \in \mathscr{D}'$ whose convolution $S * f$ with any element of $\mathscr{D}$ again belongs to $\mathscr{D}$. The topology of $\mathscr{E}'$ is that given by considering each $S \in \mathscr{E}'$ as defining the transformation $f \to S * f$ of $\mathscr{D}$ into $\mathscr{D}$ and giving this set of transformations the compact open topology.*

PROOF. We show first that $S * \mathscr{D} \subset \mathscr{D}$ implies $S \in \mathscr{E}'$.

Suppose that $S * \mathscr{D} \subset \mathscr{D}$ and $S \notin \mathscr{E}'$. We shall construct a sequence $\{g_i\}$ in $\mathscr{D}$ such that

1. $\sum g_i$ converges to g in the topology of $\mathscr{D}$.
2. There is a strictly increasing sequence of positive integers m_i such that

$$\text{support } S * g_i \quad \subset \{x \mid \|x\| \leqslant m_i\}$$

$$\text{support } S * g_{i+1} \not\subset \{x \mid \|x\| \leqslant m_i\}.$$

3. There is a sequence $\{a_i\}$ in R^n with $m_{i-1} \leqslant \|a_i\| \leqslant m_i$ with

$$(S * g_i)(a_i) \neq 0$$

$$|(S * g_{i+j})(a_i)| \leqslant 3^{-j} |(S * g_i)(a_i)|$$

for any $j \geqslant 1$.

If $\{g_i\}$ can be found, then for any $i > 1$ we will have $(S * g_k)(a_i) = 0$ for $k < i$ so that

$$\begin{aligned} |(S * g)(a_i)| &\geqslant |(S * g_i)(a_i)| - \sum |(S * g_{i+j})(a_i)| \\ &\geqslant |(S * g_i)(a_i)| [1 - \sum 3^{-j}] \\ &> 0. \end{aligned}$$

Since $\{a_i\}$ is not bounded, this contradicts our assumptions that $S * g \in \mathscr{D}$.

To construct $\{g_i\}$, we note (see Schwartz [1]) that S is the limit in the topology of $\mathscr{D}'$ of $S * f_l$, where $f_l \in \mathscr{D}_1$, since such $S * f_l$ are regularizations of S. Let g_1 be the first f_l for which $S * f_l \neq 0$. Let a_i be any point in R for which $(S * g_1)(a_1) \neq 0$, and choose m_1 so that

$$\text{support } (S * g_1) \subset \{x \mid \|x\| \leqslant m_1\}.$$

Suppose $g_1, \ldots, g_i$, $a_1, \ldots, a_i$, $m_1, \ldots, m_i$ have been defined. Since $S * f_l$ converges to S which by assumption is not of compact support, the $S * f_l$ cannot have their supports in a fixed compact set. Thus, there is an l such that

$$\text{support } S * f_l \not\subset \{x| \; \|x\| \leqslant m_i + 1\}.$$

Let m_{i+1} be chosen so that

$$\text{support } S * f_l \subset \{x| \; \|x\| \leqslant m_{i+1}\}$$

and choose a_{i+1} with $m_i < \|a_{i+1}\| < m_{i+1}$ such that $(S * f_l)(a_{i+1}) \neq 0$. Let $\{\partial_j\}$ be an enumeration of the differentiations with $\partial_1 = \text{identity}$. Define

$$g_{i+1} = \frac{f_l \min_{p \leqslant i} [1, |(S * g_p)(a_p)|]}{3^{i+1} \max_{j \leqslant i+1} [1, \max_{x \in R} |(\partial_j f_l)(x)|] \max_{j \leqslant i} [1, |(S * f_l)(a_j)|]}.$$

It is readily verified that 1, 2, and 3 are satisfied.

We wish now to prove that the topology of $\mathscr{E}'$ is the compact open topology. It is easily seen that if $f \in \mathscr{D}$, then the maps $S \to S * f$ are of $\mathscr{E}' \to \mathscr{D}$ which are equicontinuous for f in any bounded set, that is, the topology of $\mathscr{E}'$ is at least as strong as the compact open topology.

Let N be a neighborhood of zero in $\mathscr{E}'$; there is a bounded set $B \in \mathscr{E}$ so that N contains the set of $S \in \mathscr{E}'$ with $|S \cdot b| \leqslant 1$ for all $b \in B$. For each lattice point a in R let $h_a \in \mathscr{D}$ vanish for $\|x - a\| \geqslant 3$ and be such that $\sum h_a = 1$. Then the set $B' = \{h_a f\}$ for $f \in B$, all a is again bounded in $\mathscr{E}$. For each a there is a sequence $s_a = \{M_{ai}\}$ of positive numbers so that B' is contained in the bounded (in $\mathscr{E}$) set of all $g \in \mathscr{E}$ with

$$\max_{\|x-a\| \leqslant 3} |(\partial_i g)(x)| \leqslant M_{a_i}$$

for all i, a. From the denumerable collection of sequences s_a we construct a single sequence $s = \{M_i\}$ of positive numbers so that, for each a, $M_{ai} \leqslant M_i$ for all but a finite number of i. Hence, $e_a M_{ai} \leqslant M_i$ for all i for a suitable $e_a > 0$. We may clearly assume $e_a = e_{-a}$.

Call A the set of $f \in \mathscr{D}$ with

1. $f(x) = 0$ for $\|x\| \geqslant 3$.
2. $\max_{x \in R} |(\partial_i f)(x)| \leqslant M_i$ for all i.

Thus A is bounded in $\mathscr{D}$. Let M be the neighborhood of zero in $\mathscr{D}$ consisting of those $f \in \mathscr{D}$ with

$$\max_{\|x-a\| \leqslant 3} |f(x)| \leqslant e_a d_a.$$

Here d_a are positive numbers with $\sum d_a = 1$.

Call N' the set of $S \in \mathscr{E}'$ with $S * A \subset M$. We claim that $N' \subset N$. Assume this is not the case. Then there is an $S \in N'$ which is not in N, that is,

$$|S \cdot f| > 1$$

for some $f \in B$. But the support of S meets only a finite number of sets of the form $\|x - a\| \leqslant 3$. Thus,

$$|S \cdot f| \leqslant |S \cdot h_{a_1} f| + \cdots + |S \cdot h_{a_r} f|$$

for suitable $a_1, \ldots, a_r$.

It is clear from the definitions that each $h_{a_i} f$ is of the form

$$h_{a_i} f = \frac{1}{e_{a_i}} \tau_{a_i} g$$

for a suitable $g \in A$, where τ denotes translation. Thus,

$$\begin{aligned} |S \cdot h_{a_i} f| &= \frac{1}{e_{a_i}} |S \cdot \tau_{a_i} g| \\ &= \frac{1}{e_{a_i}} |(S * \check{g})(a_i)| \\ &\leqslant d_{a_i} \end{aligned}$$

by the definition of A. Here $\check{g}(x) = g(-x)$ so $\check{g} \in A$. Hence,

$$|S \cdot f| \leqslant \sum d_{a_i} < 1$$

which is a contradiction. The proof of Theorem 5.15 is thus complete. ▯

The proof of Theorem 5.15 yields also

THEOREM 5.16. *The conclusions of Theorem* 5.15 *hold if the space* $\mathscr{D}$ *is replaced by* $\mathscr{D}_F$.

We denote by $\mathbf{E}'$ the Fourier transform of $\mathscr{E}'$. Just as in the case of $\mathscr{D}_F$, the Paley-Wiener-Schwartz theorem describes $\mathbf{E}'$: It is the set of entire functions of exponential type which are bounded by a polynomial on R.

LEMMA 5.17. *Let* $\{F^j\}$ *be a sequence in* $\mathbf{E}'$. *A necessary and sufficient condition that* $F^j \to 0$ *in* $\mathbf{E}'$ *is that there exist an* $a > 0$ *so that*

$$\sup_{z \in C} |F^j(z)| \exp(-a |\text{Im } z|)(1 + \|z\|)^{-a} \to 0.$$

PROOF. If $\{F^j\}$ satisfies the condition, then Theorems 5.16, 5.10, and Lemma 5.8 show that $F^j \to 0$.

Conversely, suppose $F^j \to 0$ and let F^j be the Fourier transform of S^j. By a result of Schwartz [1], Vol. I, p. 91, there exists a neighborhood N

of zero in $\mathscr{E}$ so that $S^j \cdot f \to 0$ uniformly for $f \in N$. By the definition of the topology of $\mathscr{E}$ we may assume there is an $a > 1$ so that N consists of all $f \in \mathscr{E}$ with $|\partial f(x)| \leqslant a$ for $\|x\| \leqslant a$ and ∂ a differentiation of order $\leqslant a$. Thus, for any z we have

$$\exp(-a|\text{Im } z|)(1 + \|z\|)^{-a} \exp(ix \cdot z) \in N$$

which gives the desired result. □

In a similar manner we can prove

LEMMA 5.18. *A set $B \subset \mathbf{E}'$ is bounded if and only if there exists an $a > 0$ so that for all $F \in B$*

$$\sup_{z \in C} |F(z)| \exp(-a|\text{Im } z|)(1 + \|z\|)^{-a} \leqslant a.$$

THEOREM 5.19. *$\mathscr{E}$ is AU. An AU structure consists of all continuous positive functions $k(z) = k_1(\text{Re } z)k_2(\text{Im } z)$, where k_1 dominates all polynomials and k_2 dominates all linear exponentials.*

PROOF. Let k satisfy the above conditions and denote by N_k the set of $F \in \mathbf{E}'$ with $|F(z)| \leqslant k(z)$ for all z. Lemma 5.18 shows that N_k swallows every bounded set. Since $\mathscr{E}'$ is bornologic, N_k is a neighborhood of zero.

Conversely, let N' be a neighborhood of zero in $\mathbf{E}'$. We want to find a k so that $N_k \subset N'$. By Theorem 5.16 we may assume N' is given as follows: There is a bounded (compact) set B in $\mathbf{D}_F$ and a neighborhood M of zero in $\mathscr{D}_F$ so that N' consists of all $F \in \mathbf{E}'$ satisfying $FG \in M$, whenever $G \in B$. Using Lemma 5.9 and the Phragmén-Lindelöf theorem (or using the proof of Lemma 5.9) we can describe B as follows: There is a continuous positive function $k_3(Rz)$ which dominates all polynomials and an $l > 0$ so that B consists of all $G \in \mathbf{D}$ with

$$|G(z)| \leqslant \exp(l|\text{Im } z|)/k_3(\text{Re } z)$$

for all z. By Theorem 5.10 we may assume that M consists of all $H \in \mathbf{D}$ with $|H(z)| \leqslant k_4(\text{Im } z)(1 + |\text{Re } z|)^{-q}$ for all z; here q is some positive number and k_4 dominates all linear exponentials.

We set $k(z) = k_1(\text{Re } z)k_2(\text{Im } z)$ where

$$k_1(\text{Re } z) = k_3(\text{Re } z)(1 + |\text{Re } z|)^{-q}$$
$$k_2(\text{Im } z) = k_4(\text{Im } z) \exp(-l|\text{Im } z|).$$

We claim that $N_k \subset N'$. For, given any $F \in N_k$, $G \in B$, we have

$$|F(z)G(z)| \leqslant k(z) \exp(l|\text{Im } z|)/k_3(\text{Re } z) = k_4(\text{Im } z)(1 + |\text{Re } z|)^{-q}$$

which means that $FG \in M$; this is the desired result. □

THEOREM 5.20. *$\mathscr{E}$ is* PLAU.

PROOF. As usual, we will consider only Condition (e) for the case $n = 1$.

Again, as in Example 1, we want to give a suitable analytic uniform structure for $\mathscr{E}$. It is readily verified that we can choose one consisting of all functions $k(z) = k_1(\operatorname{Re} z)k_2(\operatorname{Im} z)$, where k_2 satisfies the Conditions 1, 2, 3 of Example 1 (for $\tilde{k}$), and k_1 can be extended to an analytic function of a complex variable of the form

$$k_1(t) = \prod_{j=1}^{\infty} (1 + t^2/d_j^2), \tag{5.92}$$

where $\{d_j\}$ is an increasing sequence of positive integers with $\sum 1/d_j < \infty$.

We note the following inequalities:

$$\begin{aligned} |1 + z^2|^2 &= |1 + (\operatorname{Re} z)^2 - (\operatorname{Im} z)^2 + 2i(\operatorname{Re} z)(\operatorname{Im} z)|^2 \\ &\leqslant 1 + (\operatorname{Re} z)^4 + (\operatorname{Im} z)^4 + 2(\operatorname{Re} z)^2 + 2(\operatorname{Im} z)^2 \\ &\qquad + 2(\operatorname{Re} z)^2 (\operatorname{Im} z)^2 \\ &\leqslant (1 + (\operatorname{Re} z)^2)^2 (1 + (\operatorname{Im} z)^2)^2. \end{aligned} \tag{5.93}$$

We have

$$\left|\frac{1 + z^2/a^2}{1 + (\operatorname{Re} z)^2/a^2}\right| = \frac{|z + ia|\,|z - ia|}{|\operatorname{Re} z + ia|\,|\operatorname{Re} z - ia|}.$$

Now,

$$\frac{|z + ia|}{|\operatorname{Re} z + ia|} = \left|1 + \frac{i \operatorname{Im} z}{\operatorname{Re} z + ia}\right|$$

which gives immediately, for $a \geqslant 1$,

$$\begin{aligned} \frac{|z + ia|}{|\operatorname{Re} z + ia|} &\geqslant 1 - \frac{a|\operatorname{Im} z|}{(\operatorname{Re} z)^2 + a^2} \\ &\geqslant 1 - \frac{1}{2a} \end{aligned} \tag{5.94}$$

for $|\operatorname{Im} z| \leqslant \frac{1}{2}$.

From (5.93) we deduce, for all z,

$$\left|\frac{1 + z^2/a^2}{1 + (\operatorname{Re} z)^2/a^2}\right| \leqslant 1 + (\operatorname{Im} z)^2/a^2. \tag{5.95}$$

These inequalities imply (since $\cos \pi z = \Pi\,(1 - z^2/j^2)$)

$$|k_1(z)| \leqslant |k_1(\operatorname{Re} z)k_1(\operatorname{Im} z)| \leqslant |k_1(\operatorname{Re} z)| \exp(\pi\,|\operatorname{Im} z|) \tag{5.96}$$

for all z and

$$C^{-1} \leqslant |k_1(z)|/k_1(\operatorname{Re} z) \leqslant C \tag{5.97}$$

for $|\operatorname{Im} z| \leqslant \frac{1}{2}$ for some absolute constant C.

Next we need a "good" bounded analytic uniform structure for $\mathscr{E}$. It is a simple consequence of Lemma 5.18 and Theorem 5.19 that the set of $m(z)$ of the form

$$m(z) = \prod_{j=1}^{l_0} (1 + (\operatorname{Re} z)^2/d_j^2) m_2(\operatorname{Im} z) \tag{5.98}$$

for a suitable m_2 satisfying the Conditions 4 and 5 of Example 1 (for $\tilde{m}$) is a sufficient BAU structure for $\mathscr{E}$.

We can now conclude the proof that $\mathscr{E}$ is product localizeable. Let k be given as above and let k' be chosen in the AU structure described above. Then, given $m \leqslant k'$ we may assume that m is of the form (5.98). Then, for any z^0 with (see Condition 5 of Example 1)

$$a_l < \operatorname{Im} z^0 \leqslant a_{l+1}$$

we define

$$\tilde{\psi}_{z^0}(z) = m_2(\operatorname{Im} z^0) \exp[-i(l + \pi + 4)(z - z^0)] \times \prod_{j=1}^{l_0} [1 + (z - i \operatorname{Im} z^0)^2/d_j^2].$$

Then, by (5.97) or rather its analog for finite products, we have as in (5.37)

$$C_1^{-1} e^{-1} \exp(-3|\operatorname{Im} z^0|) \leqslant \frac{\min\limits_{|t - \operatorname{Im} z^0| \leq \frac{1}{2}} |\tilde{\psi}_{z^0}(\xi + it)|}{\max\limits_{|t - \operatorname{Im} z^0| \leq \frac{1}{2}} m(\xi + it)} \tag{5.99}$$

$$\leqslant \frac{\max\limits_{|t - \operatorname{Im} z^0| \leq \frac{1}{2}} |\tilde{\psi}_{z^0}(\xi + it)|}{\min\limits_{|t - \operatorname{Im} z^0| \leq \frac{1}{2}} m(\xi + it)}$$

$$\leqslant C_1 e \exp(3|\operatorname{Im} z^0|).$$

Moreover, for any z, by (5.96), or rather its analog for a finite product,

$$|\tilde{\psi}_{z^0}(z)|/m(z) \leqslant \frac{\exp[\pi|\operatorname{Im}(z - z^0)| + (l+1) \operatorname{Im} z^0] \exp[\operatorname{Im}(z - z^0)(l + \pi + 4)]}{m_2(\operatorname{Im} z)} \tag{5.100}$$

which is the counterpart of (5.38). We may now repeat the argument following (5.38) step by step to obtain

$$|\tilde{\psi}_{z^0}(z)|/m(z) \leqslant 4c' \exp(c'|\operatorname{Im} z|). \tag{5.101}$$

The verification of Condition (e) is now concluded exactly as in Example 1, since the factor $\exp(c'|\text{Im } z|)$ is easily absorbed in the functions k and m. ▯

We wish to show how to modify the construction so as to apply to the space $\mathscr{E}(\Omega)$ of indefinitely differentiable functions on Ω, where Ω is a *closed* parallelepiped. As usual, it suffices to treat the case $n = 1$, so we assume Ω is the closed unit interval. Then by the Paley-Wiener-Schwartz theorem (see Schwartz [1]) $\mathscr{E}'(\Omega)$ consists of all entire functions of exponential type which are $0(1 + |z|)^l \exp(|\text{Im } z|)$ for some l. An AU structure consists of all $k(z) = k_1(z) \exp(|\text{Im } z|)$ where k_1 dominates all polynomials. We may assume that k_1 is of the form (5.92). For a suitable bounded analytic uniform structure we use functions of the form [compare (5.98)]

$$m(z) = \prod_{j=1}^{l_0} [1 + (\text{Re } z)^2/d_j^2] \exp(|\text{Im } z|). \tag{5.98*}$$

For any k we choose k' as above; if $m \leqslant k'$ then m can be assumed to be of the form (5.98*). Given any z^0 (say for $\text{Im } z^0 \geqslant 0$) set

$$\tilde{\Psi}_{z^0}(z) = \exp(-iz) \cdot \prod_{j=1}^{l_0} [1 + (z - i \text{ Im } z^0)^2/d_j^2].$$

Then for $|\text{Im } z - \text{Im } z^0| \leqslant \frac{1}{2}$ we have by (5.97) or rather its analog for finite products.

$$C^{-1} \leqslant \frac{\min\limits_{|t - \text{Im } z^0| \leq \frac{1}{2}} |\tilde{\Psi}_{z^0}(\xi + it)|}{\max\limits_{|t - \text{Im } z^0| \leq \frac{1}{2}} m(\xi + it)} \leqslant \frac{\max\limits_{|t - \text{Im } z^0| \leq \frac{1}{2}} |\tilde{\Psi}_{z^0}(\xi + it)|}{\min\limits_{|t - \text{Im } z^0| \leq \frac{1}{2}} m(\xi + it)} \leqslant C. \tag{5.99*}$$

Moreover, for any z by the analog of (5.96) for finite products we have

$$|\tilde{\Psi}_{z^0}(z)|/m(z) \leqslant \prod_{j=1}^{l_0} [1 - (\text{Im } z - \text{Im } z^0)^2/d_j^2]. \tag{5.100*}$$

We have thus derived suitable analogs of (5.99) and (5.100), and we can now proceed as before to obtain the result. ▯

The methods of Section IV.5 (Oka embedding) now give the extension of these results to the spaces $\mathscr{E}(\Omega)$ of indefinitely differentiable functions on Ω, where Ω is either the interior of a polyhedron or the closure of the interior. Now, suppose n is even, say $n=2m$; we then think of $R^n = C^m$, where we use complex coordinates $y_j = x_{2j-1} + ix_{2j}$ for $j = 1, 2, \ldots, m$. Then $\mathscr{H}(\Omega)$ is exactly the subspace of $\mathscr{E}(\Omega)$ consisting of those f which satisfy the Cauchy-Riemann equations

$$\frac{\partial f}{\partial \bar{y}_j} = \frac{\partial f}{\partial x_{2j-1}} + i\,\frac{\partial f}{\partial x_{2j}} = 0$$

for $j = 1, 2, \ldots, m$. It is trivial to verify that the multiplicity variety $\mathfrak{V}$ for these equations is just the ordinary variety defined by $z_{2j-1} + iz_{2j} = 0$ for $j = 1, 2, \ldots, m$ with the identity operator. Theorem 4.1 now shows that $\mathscr{H}(\Omega)$ is AU with the usual type of AU structure. (We are using the identification of $\mathbf{H}'(\Omega)$ defined in the usual manner, with the Fourier transform of the dual of $\mathscr{H}(\Omega)$, when this is considered as the subspace of $\mathscr{E}(\Omega)$ satisfying the Cauchy-Riemann equations. It is after this identification is made that Theorem 4.1 implies that $\mathscr{H}(\Omega)$ is AU.) This seems to be difficult to establish directly when Ω is not a product of domains Ω_i in the complex y_i plane.

THEOREM 5.21. *Let Ω be an open bounded convex set in C^m (not necessarily a polyhedron). Then $\mathbf{H}'(\Omega)$ consists of all entire functions $F(w)$ of exponential type which satisfy , for some $\varepsilon > 0$*

$$|F(w)| \leqslant C \exp\left(\Psi(w) - \varepsilon\,|w|\right).$$

Here,

$$\Psi(w) = \sup_{x \in \Omega} I(x \cdot w)$$

PROOF. By the Hahn-Banach theorem each $S \in \mathscr{H}'(\Omega)$ can be extended to a continuous linear function on $\mathscr{H}(\Omega_0)$ for some open convex polyhedron $\Omega_0 \subset \Omega$. It follows that it is sufficient to prove the result for Ω an open convex polyhedron, in which case it is contained in the above remarks on the structure of $\mathbf{H}'(\Omega)$. ▯

In Section IV.5 we showed how to deduce properties of $\mathscr{E}(\Omega)$, for Ω an arbitrary polyhedron, from those of $\mathscr{E}(\Omega_1)$, where Ω_1 is the unit cube. The case of Theorem 5.21 for the unit cube is easily reduced to the case $n=1$. One might therefore expect a simple reduction of Theorem 5.21 to the case $n=1$. However the method of Section IV.5 does not seem to work directly for the space $\mathscr{H}(\Omega)$ in place of $\mathscr{E}(\Omega)$ because the map λ is *real* linear but not complex linear.

See Problem 5.4.

By a method similar to the proof of Theorem 5.5 we have

THEOREM 5.22. *For each j let R_j be the real axis in the z_j plane and let A_j be any other line through the origin. Then $(R_1 \cup A_1) \times \cdots \times (R_n \cup A_n)$ is sufficient for $\mathscr{E}$.*

For $n = 1$ the content of combining Theorems 5.22 and 1.5 is that any $f \in \mathscr{E}$ can be written in the form $h_1 + h_2$, where h_1 is entire, and h_2 and all its derivatives are uniformly bounded on R.

As in Theorem 5.7 we could find discrete sets which are sufficient for $\mathscr{E}$.

We wish now to give an example of a sufficient set for $\mathscr{D}'$. Let us notice the following:

LEMMA 5.23. *Let $w(y)$ be any positive, continuous monotonic function of the real variable $y \geqslant 0$ such that, at infinity, $\log(y)/w(y) \to 0$. Get R_w be the set in C defined by*

$$|\text{Im}\, z| \leqslant w(|\text{Re}\, z|). \tag{5.102}$$

*Then, outside of R_w "the topologies of **D** and **E**$'$ are the same," that is, given any k of the set K defining the analytically uniform structure for $\mathscr{D}'$ there is a k' of the set defining the analytically uniform structure for $\mathscr{E}$ such that $k = k'$ outside of R_w.*

PROOF. We give the proof for $n = 1$; there are no difficulties involved in the extension to $n > 1$. We define k' as in the diagram below. On the line joining z with z' we define k' by linear interpolation. We wish to verify that $k'(z)$ satisfies

$$\exp(t|\text{Im}\, z|)(1 + |z|)^p = 0(k'(z)), \tag{5.103}$$

for all t, p so k' satisfies the properties required by Theorem 5.19. (Actually k' does not belong to the AU structure described in Theorem 5.19, but k' dominates some k'' in that AU structure.)

See Fig. 10.

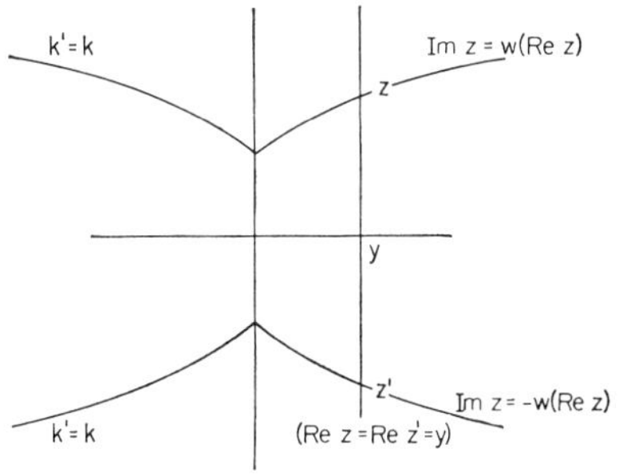

FIGURE 10

Before proving this, let us explain why it should be so. On $|\text{Im } z| = w(|\text{Re } z|)$ we have

$$\exp(t|\text{Im } z|) = \exp[tw(|\text{Re } z|)].$$

Since $w(y)/\log(y) \to \infty$, the function $\exp[tw(|\text{Re } z|)]$ is larger at infinity than any polynomial. Since, in general, the only difference between the functions defining the analytically uniform structures for $\mathbf{E}'$ or $\mathbf{D}$ is a question of polynomial growth, we should expect our result.

We shall prove (5.103) outside R_w as the inequality for points inside R_w is easier. For simplicity we assume that $w(y) > 2$ for all y. Then for any p, t let

$$f_{p,t}(z) = \begin{cases} 0 & \text{for } |\text{Im } z| < w(|\text{Re } z|) - 1 \\ \exp(t|\text{Im } z|)(1+|z|)^p & \text{for } z \in R_w. \end{cases} \tag{5.104}$$

The definition of $f_{p,t}$ is completed by requiring that $f_{p,t}$ be continuous and linear on the part of the line $\text{Re } z = y$ at which it is not yet defined (see Fig. 10).

We claim that for any q, we have

$$f_{p,t}(z) = 0(\exp[(t+1)|\text{Im } z|]/(1+|z|)^q). \tag{5.105}$$

Note that (5.105) implies Lemma 5.23 since, by the description of the AU structure for $\mathscr{D}'$ given in Theorem 5.13, (5.105) implies that $f_{p,t}(z) = 0(k(z))$ which clearly gives (5.103) outside R_w.

Now (5.105) is certainly satisfied if $|\text{Im } z| < w(|\text{Re } z|) - 1$, for then the left side is zero. For $|\text{Im } z| \geqslant w(|\text{Re } z|) - 1$ we need only prove that for any q

$$(1+|z|)^{p+q} = 0[\exp(|\text{Im } z|)],$$

that is,

$$\log(1+|z|) \leqslant \text{const} + \frac{|\text{Im } z|}{p+q}$$

or, what is the same thing,

$$\log(1+|z|)/|\text{Im } z| \to 0 \tag{5.106}$$

as $|z| \to \infty$, $|\text{Im } z| \geqslant w(|\text{Re } z|) - 1$. If (5.106) were violated, then there would exist a sequence of points z_j such that

$$|\text{Im } z_j| \geqslant w(|\text{Re } z_j|) - 1 \tag{5.107}$$

and

$$\log(1+|z_j|)/|\text{Im } z_j| \geqslant 1, \text{ say.}$$

But for $|z_j|$ large, $\log(1+|z_j|) > |\text{Im } z_j|$ implies $\log|\text{Re } z_j| > \frac{1}{2}|\text{Im } z_j|$ which is impossible if $|\text{Im } z_j| \geqslant w(|\text{Re } z|) - 1$. This completes the proof of Lemma 5.23. ▯

As an immediate consequence of Theorem 5.22 and Lemma 5.23 we have

THEOREM 5.24. *Given any w as in Lemma 5.23, the set of points each of whose coordinates is either pure imaginary or belongs to R_w is sufficient for $\mathscr{D}'$.*

See Remark 5.5.

V.6. EXAMPLE 6. NON-QUASIANALYTIC CLASSES. Let $j = (j_1, \ldots, j_n)$ where j_k are non-negative integers. Let $A = \{a_j\}$ be a multisequence of positive numbers. Then we define $\mathscr{E}_A$ to be the space of all functions f on R such that f and all its derivaties ∂f satisfy, on every compact set, for every $\varepsilon > 0$

$$|(\partial f)^{(j)}(x)| \leqslant a\varepsilon^{|j|}a_j \tag{5.108}$$

for all j. (Here a depends on ε, and on the compact set, and which derivative ∂f of f we take. We have written $|j| = j_1 + \cdots + j_n$.) $\mathscr{D}_A$ is the subspace of $\mathscr{E}_A$ consisting of functions of compact support.

We shall consider the case when $\{a_j\}$ is a "product" multisequence, that is, there are sequences $\{a_l^p\}$ for $p = 1, 2, \ldots, n$ such that $a_j = a_{j_1}^1 \cdots a_{j_n}^n$. In this case the considerations for $n > 1$ are essentially the same as those for $n = 1$, so we shall restrict ourselves to the latter.

The topology of $\mathscr{D}_A$ is defined as follows: A fundamental system of neighborhoods of zero consists of those sets N for which we can find a continuous positive function $u(x)$, an integer $l > 0$, and a positive number b so that N consists of all $f \in \mathscr{D}_A$ which satisfy

$$|f^{(j')}(x)| \leqslant b^j a_j u(x) \quad \text{for} \quad j' = j, j+1, \ldots, j+l. \tag{5.109}$$

See Remark 5.6.

Set $\lambda(x) = \sum |x|^j/a_j$ (where $\lambda(x) = \infty$ if the series diverges). A classical theorem of Denjoy and Carleman (see e.g., Paley and Wiener [1]) shows that the space $\mathscr{D}_A$ is not reduced to $\{0\}$ if and only if

$$\int_{-\infty}^{\infty} \log \lambda(x)(1+x^2)^{-1}\, dx < \infty.$$

We say in this case that $\mathscr{E}_A$ is non-quasianalytic.

Actually the condition in Paley and Wiener [1], p. 14, is stated in a slightly different form, namely, the function $\lambda(x)$ is replaced by

$\sum |x|^{2j}/(a_j)^2 = \tilde{\lambda}(x)$, say. However, it is clear that $\tilde{\lambda}(x) \leqslant [\lambda(x)]^2$. On the other hand,

$$\begin{aligned}[\lambda(x/2)]^2 &= [\sum |x|^j/2^j a_j]^2 \\ &\leqslant \{[\max |x^j|/a_j] \sum 2^{-j}\}^2 \\ &= 4 \max x^{2j}/(a_j)^2 \\ &\leqslant 4\tilde{\lambda}(x).\end{aligned}$$

Thus, our condition is equivalent to (7.04) of Paley and Wiener [1]. The above argument shows also that we could have expressed the condition for quasianalyticity in terms of the function $\lambda_1(x) = \max |x^j|/a_j$. It is often easier to use $\lambda_1(x)$ than $\lambda(x)$ in computations. Also the property of quasianalyticity remains invariant under "small perturbations" of the sequence A.

We shall assume in what follows that A is *convex*, that is,

$$a_j^2 \leqslant a_{j-1}\, a_{j+1}$$

for all $j \geqslant 1$ (though this is probably not necessary for most of our results). This has the simple consequence that $\mathscr{E}_A$ and $\mathscr{D}_A$ are algebras under multiplication. From this it follows, in case $\mathscr{E}_A$ is non-quasianalytic, that $\mathscr{D}_A$ is dense in $\mathscr{E}_A$.

Suppose that $\mathscr{E}_A$ is non-quasianalytic. Then we can extend the theorem of Paley-Wiener-Schwartz to show that $\mathbf{D}_A$ consists of all entire functions F of exponential type for which there exists an $a > 0$ so that for all l and all $b > 0$

$$F(z) = 0\{\exp (a\,|\text{Im } z|)[\lambda(b\,|z|)]^{-1}(1 + |z|)^{-l}\}. \tag{5.110}$$

Using this result and the fact that $\mathscr{D}_A$ is dense in $\mathscr{E}_A$, we deduce from the representation of $f \in \mathscr{D}_A$ as an inverse Fourier transform that the linear combinations of the exponentials are dense in $\mathscr{E}_A$.

The methods of Example 3 can be used to show

THEOREM 5.25. *$\mathscr{D}'_A$ is PLAU. An AU structure for $\mathscr{D}'_A$ consists of all continuous positive functions $k(z) = k_1(\text{Re } z)k_2(\text{Im } z)$ where k_2 dominates all linear exponentials and k_1 is of the form $(1 + |\text{Re } z|)^{-l}/\lambda(b\,|\text{Re } z|)$ for some $b, l > 0$.*

The only comment we should like to make on the proof that $\mathscr{D}'_A$ is product localizeable is that Mandelbrojt [1] has already noticed that functions like ψ of Example 3 [see (5.74)] can be used to give good approximations to $1/\lambda$ if the sequences $\{d_j\}$, $\{e_j\}$ are suitably chosen. We should also like to warn the reader that the proof of Theorem 5.25 is quite involved. In case that $a_j \leqslant C\,(mj)!$ for some C, m, then Theorem 5.25 can be proved in the manner of Theorem 5.26. The construction of suitable functions like ψ can best be handled by methods of potential theory as, for example, in Ehrenpreis and Malliavin [1].

We could now deduce that $\mathscr{E}_A$ is AU in a manner similar to the proof of Theorem 5.19 in case

$$\int \log \lambda(x)(1+x^2)^{-1}\, dx < \infty.$$

For them we could derive an analog of Theorem 5.16. However, if we drop this condition, then no such proof is possible since $\mathscr{D}_A$ would then be reduced to zero so no analog of Theorem 5.16 is possible. Nevertheless we have

THEOREM 5.26. *Suppose that either $\mathscr{E}_A$ is non-quasianalytic or that for some $C>0$ and some integer $m>0$ we have $a_j \leqslant C(mj)!$ for all j. Suppose moreover in the quasianalytic case, that $\mathscr{E}_A$ contains the space of entire functions and A satisfies the conditions imposed below. Then $\mathscr{E}_A$ is LAU. An AU structure for $\mathscr{E}_A$ consists of all continuous positive functions $k(z)$ of the form*

$$k(z) = k_1(\text{Re } z)k_2(\text{Im } z)\tilde{\lambda}(z),$$

where k_1 dominates all polynomials, k_2 dominates all linear exponentials and $\tilde{\lambda}$ dominates $\lambda(az)$ for all $a>0$.

PROOF. We have already explained how to prove that $\mathscr{E}_A$ is AU with the stated AU structure in case $\mathscr{E}_A$ is non-quasianalytic.

In the quasianalytic case we proceed as follows: Since $a_j \leqslant C(mj)!$, for any $f \in \mathscr{E}_A$ the series

$$g(x_1, x_2) = \sum_j f^{(j)}(x_1)(ix_2)^{(m+1)j}/[(m+1)j]! \tag{5.111}$$

converges in the topology of the space $\mathscr{E}$. Moreover, g satisfies a "heat equation"

$$\frac{\partial g}{\partial x_1} = (-i)^{m+1}\frac{\partial^{m+1} g}{\partial x_2^{m+1}} \tag{5.112}$$

as is clear from the definition. In addition we have an estimate for the size of g. Namely, for x_1 in a compact set, we have

$$\begin{aligned} |g(x_1, x_2)| &\leqslant \sum a\varepsilon^j a_j |x_2|^{(m+1)j}/[(m+1)j]! \\ &= a \sum a_j |\varepsilon' x_2|^{(m+1)j}/[(m+1)]j!, \end{aligned}$$

where $\varepsilon' = \varepsilon^{1/(m+1)}$. Let us denote by $\alpha(t)$ the function

$$\alpha(t) = \sum a_j |t|^{(m+1)j}/[(m+1)j]!.$$

Then the above shows

$$|g(x_1, x_2)| \leqslant a\alpha(\varepsilon' x_2) \tag{5.113}$$

uniformly for x_1 in a compact set. Here a may depend on the compact set and on ε.

Call $\Phi(t) = \log \alpha(t)$. Let $\tilde{\mathscr{E}}(\Phi)$ be the space of functions $h(x_1, x_2)$ which are in $\mathscr{E}$ such that h and all its derivatives satisfy, for every $\varepsilon > 0$,

$$h(x_1, x_2) = O[\exp(\Phi(\varepsilon x_2))] \tag{5.114}$$

uniformly for x_1 in compact sets. Here the constant in O may depend on ε, on the compact set, and which derivative of h we take.

We claim that g defined by (5.111) is in $\tilde{\mathscr{E}}(\Phi)$. (5.113) verifies (5.114) for g itself. We need to verify a similar inequality for the derivatives of g. By the definition of the space $\mathscr{E}_A$ there is no difficulty with x_1 derivatives. Thus we need consider only x_2 derivatives. By (5.112) we can obtain suitable inequalities for $\partial^{m+1}g/\partial x_2^{m+1}$. We need

Conditions on A:

1. Φ satisfies the conditions imposed in Example 7 below so that $\tilde{\mathscr{E}}(\Phi)$ is PLAU.
2. Let Ψ denote the conjugate of Φ as in Example 7 below. There exists $c' > 0$ so that

$$\exp[\Psi(|t|^{1/(m+1)})] \geqslant \lambda(c't) \tag{5.115}$$

for t sufficiently large.

By Condition 1, Φ is convex and larger at infinity than any linear function. This implies that, for any $a > 0$, there are $c, a' > 0$ so that $a' \to 0$ as $a \to 0$ and for $x_2 > 0$

$$\int_0^{x_2} \exp[\Phi(at)]\, dt \leqslant c \exp[\Phi(a'x_2)].$$

From this and the above, $g \in \tilde{\mathscr{E}}(\Phi)$. We claim that, conversely, for every $g \in \tilde{\mathscr{E}}(\Phi)$ the function $f = g(x_1, 0)$ belongs to $\mathscr{E}_A$. To see this we apply Theorem 7.1 to the solutions of the equation

$$\frac{\partial g}{\partial x_1} = (-i)^{m+1} \frac{\partial^{m+1} g}{\partial x_2^{m+1}}$$

By Example 5 (which is needed for the x_1 direction) and Condition 3, the space $\tilde{\mathscr{E}}(\Phi)$ is PLAU, so Theorem 7.1 can be applied. We can write such a g in the form

$$g(x_1, x_2) = \int_V e^{ix_1z_1 + ix_2z_2}\, d\mu(z_1, z_2)/k(z_1, z_2). \tag{5.116}$$

Here V is the variety $iz_1 = z_2^{m+1}$ (which is irreducible), μ is a measure on V whose total variation is bounded, and k is in an AU structure for $\tilde{\mathscr{E}}(\Phi)$. From Examples 5 and 7 this means that k is of the form

$$k(z_1, z_2) = k_1(\text{Re } z_1, \text{Re } z_2)k_2(\text{Im } z_1)k_3(\text{Im } z_2)$$

where k_1 dominates all polynomials, k_2 dominates all linear exponentials, and k_3 dominates all $\exp [\Psi(b \text{ Im } z_2)]$.

Now, from (5.116) we deduce

$$g(x_1, 0) = \int_V e^{ix_1z_1} d\mu(z_1, z_2)/k(z_1, z_2). \tag{5.117}$$

We write the integral in terms of z_1, that is, we parametrize V by the z_1 coordinate. Of course, this cannot be done exactly since to each z_1 there are $m+1$ points on V. However, we obtain $m+1$ integrals each of which is essentially of the same form, namely

$$g(x_1, 0) = \int e^{ix_1z_1} d\mu^1(z_1)/k_1^1(z_1)k_2^1(\text{Im } z_1)k_3^1[\text{Im } (iz_1)^{1/m+1}]. \tag{5.118}$$

Here k_1^1 dominates all polynomials, k_2^1 dominates all linear exponentials, and k_3^1 dominates all $\exp [\Psi(b \text{ Im } z)]$. For $|\arg z_1| \leqslant \pi/4(m+1)$ or $|\arg (z_1 - \pi)| \leqslant \pi/4(m+1)$ we must have $|\arg [(iz_1)^{1/m+1}]| \geqslant \pi/4(m+1)^2$ and $|\arg [(iz_1)^{1/m+1}] - \pi)| \geqslant \pi/4(m+1)^2$, as can be checked easily by direct computation. (These inequalities hold for all determinations of $(iz_1)^{1/m+1}$.)

See Fig. 11.

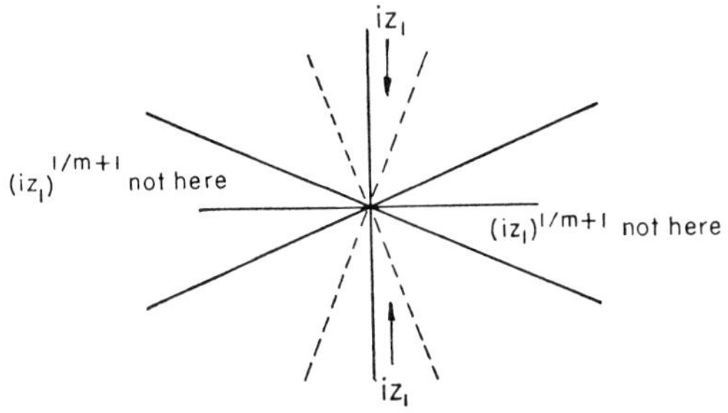

FIGURE 11

This means that, for any z_1, either

$$|\text{Im } z_1| \geqslant \pi |z_1|/8(m+1) \tag{5.119}$$

or

$$|\text{Im } (iz_1)^{1/m+1}| \geqslant \pi |z_1|^{1/m+1}/8(m+1)^2. \tag{5.120}$$

In either of the regions (5.119) or (5.120) we claim that

$$(5.121)\qquad k_1^1(z_1)k_2^1(\operatorname{Im} z_1)\exp[\Psi(\operatorname{Im} b[iz_1]^{1/m+1})] \geqslant k_1^2(z_1)k_2^2(\operatorname{Im} z_1)\lambda(a|z_1|)$$

for suitable $a > 0$. Here, as usual, k_1^2 dominates all polynomials and k_2^2 dominates all linear exponentials. In the region (5.119), (5.121) holds because our assumption that $\mathscr{E}_A$ contains the space of entire functions implies easily that $\lambda(|z_1|)$ is dominated by $\exp(c|z_1|)$ for some c. In the region (5.120) we have (5.121) because of Condition 4 imposed above.

It follows easily from (5.121) and (5.118) that $g(x_1, 0) \in \mathscr{E}_A$. Thus $f \in \mathscr{E}_A$ is equivalent, for solutions of (5.112), to $g \in \tilde{\mathscr{E}}(\Phi)$.

We would be able to complete the proof of Theorem 5.26 if we knew that the correspondence $f \leftrightarrow g$ were one-to-one. This is not the case because equation (5.112) is of degree $m + 1$ in x_2. However, if we start with $m + 1$ functions $f_0, \ldots, f_m \in \mathscr{E}_A$ and form g defined by

$$g(x_1, x_2) = \sum_{j,l} f_l^{(j)}(x_1)(ix_2)^{(m+1)j+l}/[(m+1)j+l]!$$

then g satisfies the "heat equation" (5.112) and, moreover,

$$(5.122)\qquad \frac{\partial^l g(x_1, 0)}{\partial x_2^l} = f_l(x_1) \qquad \text{for} \quad l = 0, 1, \ldots, m.$$

By the Holmgren uniqueness theorem (see Section IX.9), a solution of (5.112) is uniquely determined by $f_0, \ldots, f_m$ satisfying (5.122). The same argument as above shows that $g \in \tilde{\mathscr{E}}(\Phi)$ and conversely, for any $g \in \tilde{\mathscr{E}}(\Phi)$ satisfying (5.112), the functions $\partial^l g(x_1, 0)/\partial x_2^l$ belong to $\mathscr{E}_A$.

The correspondence

$$(f_0, \ldots, f_m) \leftrightarrow g$$

gives an isomorphism between $\mathscr{E}_A^{m+1}$ [the $(m+1)$-fold direct sum of $\mathscr{E}_A$ with itself] and the subspace $\tilde{\mathscr{E}}_D(\Phi)$ of $\tilde{\mathscr{E}}(\Phi)$ of solutions of (5.112). (The above argument can easily be modified to show that this correspondence is bicontinuous.) Now Theorem 4.1 says that $\tilde{\mathscr{E}}_D(\Phi)$ is very much like an AU space except that euclidean space C^n is replaced by the variety V: $ix_1 = x_2^{m+1}$. Thus $\mathscr{E}_A^{m+1}$ is very much like an AU space. If we use Theorem 4.1 and the explicit formulas for the solution of the Cauchy Problem, that is, the isomorphism of $\mathscr{E}_A^{m+1}$ with $\tilde{\mathscr{E}}_D(\Phi)$ as given in Section IX.1, then we find that $\mathscr{E}_A^{m+1}$ is AU with the stated AU structure. Actually (5.118) with the subsequent estimate (5.121) and similar results for the other f_l show that we can obtain the required Fourier representation (compare (1.12)) for elements of $\mathscr{E}_A^{m+1}$. We shall omit the additional details required to show that $\mathscr{E}_A^{m+1}$ is actually AU.

If we now use the same argument as in the Oka embedding (see Theorem

4.9) except that the plane Λ^C of that theorem is to be replaced by V, we deduce that $\mathscr{E}_A^{m+1}$ is localizable. This implies immediately that $\mathscr{E}_A$ is localizable. □

Illustration. Let $A = \{j!\}$. Then we can choose any $m \geqslant 1$. We have

$$\lambda(t) = \sum |t|^j/j! = e^{|t|}$$
$$\alpha(t) = \sum j!\,|t|^{(m+1)j}/[(m+1)j]!.$$

By Stirling's formula we have (aside from a factor c^j which plays no role)

$$j!/[(m+1)j]! \sim 1/(mj)!.$$

Thus

$$\alpha(t) \sim \sum |t|^{(m+1)j}/(mj)!$$
$$\sim \exp\left(|t|^{(m+1)/m}\right).$$

This means that

$$\Phi(t) \sim |t|^{(m+1)/m}$$

so

$$\Psi(t) \sim |t|^{m+1}.$$

Hence

$$\exp\left[\Psi(t^{1/m+1})\right] \sim e^{|t|} = \lambda(t).$$

This verifies Condition 4. All the other conditions are easy to verify.

See Remark 5.7.

In more refined theories of quasianalyticity (see Beurling [2] and Chapter XIII) we have a situation which is quite analogous to the above except that the function λ which occurs need not be even. It is then more difficult to generalize the method of Example 3, for the function $(\sin z)/z$ which plays an essential role in that example must be replaced by suitable functions of the form $(\sin z)/\Pi\,(1 - z/j\pi)$. We shall not enter into the details here.

V. Example 7. Rapidly Increasing Functions and Distributions. Let Φ be a continuous, positive, increasing, non constant convex function on $R = R^n$. We define the space $\mathscr{E}(\Phi)$ to consist of all indefinitely differentiable complex-valued functions f on R such that, for every j and every $\varepsilon > 0$.

$$f^{(j)}(x) = 0[\exp(\Phi(\varepsilon x))]. \tag{5.123}$$

(Here j is a multi-index.) We give $\mathscr{E}(\Phi)$ the natural topology. We can show that $\mathscr{E}(\Phi)$ is a reflexive Frechet space. To prove the reflexivity we use the fact that $\Phi(2x) - \Phi(x) \to \infty$ as $|x| \to \infty$ which follows from the fact that Φ is convex and increasing.

Note that for

$$\Phi(x) = \begin{cases} (1 - |x|)^{-1} & \text{for } |x| < 1 \\ \infty & \text{for } |x| \geqslant 1 \end{cases}$$

the space $\mathscr{E}(\Phi)$ is just $\mathscr{E}$.

We assume that Φ dominates all linear functions. Then the elements of $\mathbf{E}'(\Phi)$ are entire functions. More precisely, let Ψ denote the conjugate of Φ in the sense of Young (sometimes called the Legendre transform)

$$\Psi(y) = \max\,[(x \cdot y) - \Phi(x)]. \tag{5.124}$$

It is known (see Hormander [2]) that Ψ is again continuous and convex. It can be shown (see e.g., Hörmander [2] or Ehrenpreis [7]) that $\mathbf{E}'(\Phi)$ consists of all entire functions F for which there exist, a, b, such that

$$F(z) = 0[(1 + |z|)^a \exp \Psi(b \operatorname{Im} z)]. \tag{5.125}$$

Again, our methods show that $\mathscr{E}(\Phi)$ is AU. An AU structure can be chosen to consist of functions $k(z) = k_1(\operatorname{Re} z)k_2(\operatorname{Im} z)$, where $k_1(t)$ dominates all $(1 + |t|)^a$ and $k_2(t)$ dominates all $\exp(\Psi(bt))$ at infinity (see Berenstein and Dostal [1]).

We shall now find some conditions on Φ in order to guarantee that $\mathscr{E}(\Phi)$ be LAU. We restrict ourselves to the case in which Φ, hence also Ψ, is a sum of functions of one variable. As usual we may assume that Φ and Ψ are functions of a single variable. We assume that for any a there are b, c so that

$$c^{-1}\Psi(b^{-1}\operatorname{Im} z) \leqslant \Psi(\operatorname{Im} z + a) \leqslant c\Psi(b \operatorname{Im} z). \tag{5.126}$$

We claim that $\mathscr{E}(\Phi)$ is product localizeable. To prove this we use an analog of the method of Example 5. The important point to observe is that a sufficient bounded AU structure can be chosen to consist of functions of the form

$$m(z) = \prod_{j=1}^{l_0} [1 + (\operatorname{Re} z)^2/d_j^2]\, m_2(\operatorname{Im} z)$$

where $\{d_j\}$ is a suitable sequence and $\log m_2$ is *convex*. For any $z^0 \in C$, denote by $L(m_2; z^0)$ the slope of the linear function of $\operatorname{Im} z$ defining a plane of support of $\log m_2$ at $\operatorname{Im} z^0$, that is,

$$\log m_2(\operatorname{Im} z^0) + L(m_2; z^0)(\operatorname{Im} z - \operatorname{Im} z^0) \leqslant \log m_2(\operatorname{Im} z) \tag{5.127}$$

for all Im z. Then we define

$$(5.128)\quad \tilde{\psi}_{z^0}(z) = m_2(\text{Im } z^0)\prod_{j=1}^{l_0}[1 + (z - i\text{ Im } z^0)^2/d_j^2] \times \exp[L(m_2; z^0)(\text{Im }(z - z^0))].$$

Using (5.127) and (5.93) we have, for all z

$$(5.129)\quad \begin{aligned} |\tilde{\psi}_{z_0}(z)| &\leqslant m_2(\text{Im } z)\left|\prod_{j=1}^{l_0}[1 + (z - i\text{ Im } z^0)^2/d_j^2]\right| \\ &\leqslant m(z)\prod_{j=1}^{l_0}[1 + (\text{Im }(z - z^0))^2/d_j^2] \\ &\leqslant m(z)p[\text{Im }(z - z^0)], \end{aligned}$$

where

$$p(t) = \prod_{j=1}^{\infty}(1 + t^2/d_j^2).$$

In order to verify (4.4) for the present situation we need to know that $\tilde{\psi}_{z_0}(\xi + it)$ does not vary much for $|t - \text{Im } z^0| \leqslant 1/2$. The second inequality in (5.129) shows that it does not get larger than a constant times m. To derive a lower bound, we observe that from (5.94) and (5.95) we derive easily [compare (5.97)] that the polynomial factor in (5.128) does not vary by more than a constant factor. On the other hand, the exponential factor satisfies

$$(5.130)\quad \exp[-\tfrac{1}{2}L(m_2; z^0)] \leqslant |\exp[L(m_2; z^0)\text{ Im }(z - z^0)]| \leqslant \exp[\tfrac{1}{2}L(m_2; z^0)].$$

This leads easily to

$$(5.131)\quad \frac{m(\xi + i\text{ Im } z^0)}{\min\limits_{|t - \text{Im } z^0| \leq \frac{1}{2}} |\tilde{\psi}_{z^0}(\xi + it)|} \leqslant c\exp[\tfrac{1}{2}L(m_2; z^0)].$$

Thus the left side of (4.4) is majorized by

$$(5.132)\quad c'\exp[\tfrac{1}{2}L(m_2; z^0)]m(z)$$

in the strip $|\text{Im }(z - z^0)| \leqslant \frac{1}{2}$ because of (5.131) and the second inequality in (5.129). Now, by construction [see (5.127)]

$$(5.133)\quad \exp[\tfrac{1}{2}L(m_2; z^0)]m_2(\text{Im } z^0) \leqslant m_2(\tfrac{1}{2} + \text{Im } z^0).$$

By (5.126) we see that translation by a constant amount in the imaginary direction does not affect the topology of $\mathbf{E}'(\Phi)$. Thus (5.133) shows that the factor $\exp[\frac{1}{2}L(m_2; z^0)]$ in (5.132) is unimportant.

We can now conclude as in the previous examples that $\mathscr{E}(\Phi)$ is PLAU.

We could also give an analogous treatment of rapidly increasing distributions. Namely, we define $\mathscr{D}(\Phi)$ to consist of all indefinitely differentiable functions f such that there is an $a > 0$ so that for all j

$$f^{(j)}(x) = 0[\exp(-\Phi(ax))]. \tag{5.134}$$

Then $\mathscr{D}'(\Phi)$ consists of "rapidly increasing" distributions. The above methods apply to $\mathscr{D}'(\Phi)$ with slight modification.

Finally, we can combine Examples 6 and 7 to obtain spaces $\mathscr{D}'_A(\Phi)$ and $\mathscr{E}_A(\Phi)$ which can be proven to be PLAU. For example, $\mathscr{D}_A(\Phi)$ consists of all indefinitely differentiable functions f for which there exist constants a, b so that f and all its derivatives satisfy, for every $\varepsilon > 0$

$$|f^{(j)}(x)| \leqslant b\varepsilon^{|j|}a_j \exp(-\Phi(ax)). \tag{5.135}$$

These spaces will be used in Chapter XIII.

See Remark 5.8.

V.8. EXAMPLE 8. FORMAL POWER SERIES AND POLYNOMIALS. In this case W′ is the ring $\mathscr{P}$ of polynomials. Our methods show that $\mathscr{W}$, which is the ring of formal power series, is LAU so that our Theorems 4.1 and 4.2 may be thought of as giving a geometric description of an arbitrary ideal or module.

See Remark 5.9.

Remarks

Remark 5.1. See page 131.
By use of the methods of Section IV.5 we could prove an analogous result for the space $\mathscr{H}(\Omega)$ of holomorphic functions on the convex polygon Ω if we knew that $\mathscr{H}(\Omega)$ were AU. This is discussed at the end of Example 3 below.

Remark 5.2. See page 137.
L. Rubel has pointed out that another proof of Theorem 5.5 can be obtained by replacing the function $\tilde{M}(z)$ by a series of the form $\sum a_j \exp(jz)$, where the a_j are suitably chosen.

Remark 5.3. See page 137.
We could get an even more precise result by replacing the sets $A_j \cup B_j$ by any three half-lines through the origin such that each of the sectors formed is of opening $< \pi$. The proof of this is the same as the above proof.

Remark 5.4. See page 151.
Theorem 5.21 for $m = 1$ is due to Pólya. The extension to $n > 1$ was found independently by Martineau [1] and the author.

Remark 5.5. See page 163.
By considering spaces between $\mathscr{D}'_F$ and $\mathscr{D}'$ we could obtain sufficient sets with regions smaller than R_w.

Remark 5.6. See page 163.

We can define other spaces of functions of a similar nature. For example, we can replace the condition "for all $\varepsilon > 0$" in (5.108) by "for some $\varepsilon > 0$." In this case there are two possibilities:

(α) ε depends only on f.

(β) ε can depend on the compact set.

Now (α) does not seem to affect matters very much, but (β) does. A simple modification of the proof of Theorem 5.26 shows that (α) defines an LAU space. However this is not generally the case for (β). For example, if $n = 1$, $a_j = j!$, then (α) defines the space of functions analytic in a strip around R in the complex plane, while (β) defines the space of real analytic functions, which is not AU (see Ehrenpreis [4]). (See Problem 5.5.)

Remark 5.7. See page 169.

The idea of relating the spaces $\mathscr{E}_A$ and $\tilde{\mathscr{E}}_D(\Phi)$ is due to Täcklind [1]. His methods show that Condition 4 holds for "nice" A. A somewhat different approach to the verification of Condition 4 is found in the beginning of the proof of Theorem 9.30.

Remark 5.8. See page 172.

By modifying the construction of Chapter IV, we could weaken assumption (5.126) somewhat. We do not need to require that the norms in $\mathbf{E}'(\Phi)$ are insensitive to translations by constant amounts a in the imaginary direction, but rather that they are insensitive to translations by amounts $a(\operatorname{Im} z)$, where $a(\operatorname{Im} z)$ can fall off like the reciprocal of a polynomial. In this case, the lattice points used in Chapter IV have to be suitably modified so that the imaginary coordinates, instead of differing by integers, differ say by $1/20\, a(\operatorname{Im} z)$. We leave to the reader the details of the construction.

In particular, the above applies to the space of functions f which satisfy

$$(1 + |x|)^l f^{(j)}(x) = 0(\exp \Sigma\, x_i^2)$$

for any j, l.

Remark 5.9. See page 172.

It would be of interest to prove an analogous result for the ring of polynomials over an abstract field. We hope to return to this question at a future date, as well as to algebraic consequences of our results.

Problems

Problem 5.1. See page 138.

For $n = 1$, are the lattice points (that is, $m + ij$, m, j integers) sufficient for $\mathscr{H}$? We suspect that this set is sufficient.

Problem 5.2. See page 148.

Is $R \cup (A_1 \times \cdots \times A_n)$ suffiicent for $\mathscr{D}'_F$?

We conjecture that the answer to this problem is in the affirmative. The difficulty in the problem stems from our inability to find suitable analogs of the principle of harmonic majorant for functions of several complex variables. Using Phragmén-Lindelöf theorems we can show easily that the union of R with any set sufficient for $\mathscr{H}$ is sufficient for $\mathscr{D}'_F$.

PROBLEM 5.3. See page 152.

PROBLEM 5.4. See page 160.
Is there an analog of the Oka embedding which applies directly to the space $\mathscr{H}(\Omega)$?

PROBLEM 5.5. See Remark 5.6.
Study solutions of systems of constant coefficient linear partial differential equations in these spaces. In particular, does the analog of Theorem 6.1 hold for the space of real analytic functions? (We don't know the result even for $l = r = 1$.)

CHAPTER VI

Inhomogeneous Equations

Summary

The main result proved here is the following: Let $\boxed{\mathrm{D}} = (D_{ij})$ be an $r \times l$ matrix of linear constant coefficient partial differential operators. Thus, $\boxed{\mathrm{D}}$ defines a continuous linear map of $\mathscr{W}^l$ into $\mathscr{W}^r$. Let $\partial = (\partial_i)$ be any vector with r components which are linear differential operators with constant coefficients. We say that (∂_i) is in the *module of relations of the rows of* $\boxed{\mathrm{D}}$ *if* $\sum \partial_i D_{ij} = 0$ *for all* j. Then we prove: *Suppose* $\mathscr{W}$ *is LAU. The image of* $\boxed{\mathrm{D}}$ *on* $\mathscr{W}^l$ *consists exactly of those* $\vec{\mathrm{w}} = (w_i) \in \mathscr{W}^r$ *which satisfy* $\sum \partial_i w_i = 0$ *whenever* $\partial = (\partial_i)$ *is in the module of relations of the rows of* $\boxed{\mathrm{D}}$.

This result may be regarded as the general Poincaré lemma for $\mathscr{W}$.

In Section VI.2 we introduce (for $l = 1$) the concept of a fundamental solution for $\vec{\mathrm{D}}$. This is an $\vec{\mathrm{e}} = (e_1, \ldots, e_r) \in \mathscr{W}^r$ such that $\sum D_i e_i = \delta$ (the Dirac δ function). We explain a method of constructing "good" fundamental solutions.

VI.1. General Inhomogeneous Systems

Let $\mathscr{W}$ be a LAU space. Let $\boxed{\mathrm{D}} = (D_{ij})$ be an $r \times l$ matrix of linear partial differential operators with constant coefficients. Denote by $\mathscr{W}^r$ the r-fold direct sum of $\mathscr{W}$ with itself. We consider the map $\vec{\mathrm{w}} \to \boxed{\mathrm{D}}\,\vec{\mathrm{w}}$ of $\mathscr{W}^l \to \mathscr{W}^r$. In much of this book we study properties of this map. In this chapter we determine the image of the map, that is, those $\vec{\mathrm{g}} \in \mathscr{W}^r$ for which we can solve

$$\boxed{\mathrm{D}}\,\vec{\mathrm{f}} = \vec{\mathrm{g}}$$

for $\vec{\mathrm{f}} \in \mathscr{W}^l$. In Chapter VII we determine the kernel of the map.

Let $\boxed{\mathrm{P}}$ denote the matrix of polynomials which is the Fourier transform of $\boxed{\mathrm{D}}'$, that is (see Example 3 of Chapter V) $\boxed{\mathrm{P}}$ is obtained from

$\boxed{\mathrm{D}}'$ by replacing each $(\partial/\partial x_j)'$ by iz_j. Let $\vec{\mathfrak{B}}$ be a multiplicity variety for $\boxed{\mathrm{P}}$ (see Chapter IV; in case $l > 1$, $\vec{\mathfrak{B}}$ is defined only semilocally).

DEFINITION. The system $\boxed{\mathrm{D}}\vec{\mathrm{f}} = \vec{\mathrm{g}}$ is called *determined* if the determinant of the matrix $\boxed{\mathrm{D}}$ is $\not\equiv 0$. It is called *over-determined* if each variety in $\vec{\mathfrak{B}}$ is of dimension $< n - 1$. It is called *underdetermined* if some variety in $\vec{\mathfrak{B}}$ is of dimension n.

It should be noted that the system might not be determined, or over-determined, or underdetermined.

Let (∂_i) be any element of the module of relations of the rows of $\boxed{\mathrm{D}}$, that is, ∂_i are constant coefficient linear partial differential operators with

$$\sum_i \partial_i D_{ij} = 0$$

for all j. If $\vec{\mathrm{w}} \in \mathscr{W}^r$ is in the range of $\boxed{\mathrm{D}}$, then clearly $\sum \partial_i w_i = 0$. We can prove that the converse also holds, namely,

THEOREM 6.1. *The image of* $\boxed{\mathrm{D}}$ *consists exactly of those* $\vec{\mathrm{w}} \in \mathscr{W}^r$ *which satisfy* $\sum \partial_i w_i = 0$ *whenever* (∂_i) *is in the module of relations of the rows of* $\boxed{\mathrm{D}}$.

This subspace of $\mathscr{W}^r$ will be denoted by ${}_{\boxed{\mathrm{D}}}\mathscr{W}^r$.

Closely allied to Theorem 6.1. is

THEOREM 6.2. *Let* $\boxed{\mathrm{D}}'$ *denote the adjoint of* $\boxed{\mathrm{D}}$ *so* $\boxed{\mathrm{D}}' : \mathscr{W}'^r \to \mathscr{W}'^l$. *Then the image* $\boxed{\mathrm{D}}'\mathscr{W}'^r$ *is closed in* $\mathscr{W}'^l$.

PROOF OF THEOREM 6.2. We must prove that if $\vec{\mathrm{K}} = (K_i) \in \mathbf{W}'^r$ and $\boxed{\mathrm{P}}\vec{\mathrm{K}}$ converges in the topology of $\mathbf{W}'^l$, say to $\vec{\mathrm{G}}$, then $\vec{\mathrm{G}}$ is of the form $\boxed{\mathrm{P}}\vec{\mathrm{H}}$ with $\vec{\mathrm{H}} \in \mathbf{W}'^r$.

We apply Theorem 4.2 (with $\boxed{\mathrm{F}}$ replaced by $\boxed{\mathrm{P}}$). This tells us that an element $\vec{\mathrm{G}} \in \mathbf{W}'^l$ is in the image of $\boxed{\mathrm{P}}$ if and only if $\lambda\,\vec{\mathrm{G}}$ belongs to the image of $\boxed{\mathrm{P}}$ on $\mathbf{W}'^r(\beta - c')$. To show that this defines a closed set of $\vec{\mathrm{G}}$ we apply Theorem 4.8. Thus, if $\vec{\mathrm{G}} = \boxed{\mathrm{P}}\vec{\mathrm{H}}$ converges, then also $\vec{\mathrm{H}} - \boxed{\mathrm{P}}^1\vec{\mathrm{H}}_1$ converges so that

$$\begin{aligned}\lim \vec{\mathrm{G}} &= \lim \boxed{\mathrm{P}}\vec{\mathrm{H}}\\ &= \lim \boxed{\mathrm{P}}(\vec{\mathrm{H}} - \boxed{\mathrm{P}}^1\vec{\mathrm{H}}_1)\\ &= \boxed{\mathrm{P}} \lim (\vec{\mathrm{H}} - \boxed{\mathrm{P}}^1\vec{\mathrm{H}}_1)\end{aligned}$$

is in the range of $\boxed{\mathrm{P}}$. This completes the proof of Theorem 6.2. □

THEOREM 6.3. *The kernel of* $\boxed{\mathrm{D}}$ *is the dual of* $\mathscr{W}'^l/\boxed{\mathrm{D}}'\mathscr{W}'^r$.

PROOF. By a general result in the theory of locally convex topological vector spaces the kernel of $\boxed{D}$ is the dual of the quotient of $\mathscr{W}'^l$ by the closure of $\boxed{D}'\mathscr{W}'^r$. (For, any $\vec{w} \in \mathscr{W}^l$ which is zero on all $\boxed{D}'\vec{S}$ with $\vec{S} \in \mathscr{W}^r$ satisfies $\vec{S} \cdot \boxed{D}\vec{w} = 0$ for all $\vec{S} \in \mathscr{W}^r$, so $\boxed{D}\vec{w} = 0$.) The result follows from Theorem 6.2. ▯

PROOF OF THEOREM 6.1. It is clear that we can find an $m \times r$ matrix $\boxed{D}^1$ so that the subspace of $\mathscr{W}^r$ satisfying the same relations as the rows of $\boxed{D}$ is just the kernel of $\boxed{D}^1$. (We merely note that the module of relations has a finite basis.) Thus, by Theorem 6.3, this kernel is the dual of $\mathscr{W}'^r / \boxed{D}^{1'} \mathscr{W}'^m$.

We want to solve

$$\boxed{D}\vec{f} = \vec{g}$$

for $\vec{f} \in \mathscr{W}^l$ with $\vec{g}$ given in $\boxed{D}\mathscr{W}^r$. This means that, for any $\vec{S} \in \mathscr{W}'^r$,

$$\vec{f} \cdot \boxed{D}' \vec{S} = \vec{g} \cdot \vec{S}.$$

Thus, $\vec{f}$ is determined on the subspace $\boxed{D}'\mathscr{W}'^r$ of $\mathscr{W}'^l$. Moreover, changing $\vec{S}$ by addition of an element of $\boxed{D}^{1'}\mathscr{W}'^m$ does not effect the value $\vec{g} \cdot \vec{S}$. We want to extend this determination to all of $\mathscr{W}'^l$.

By the Hahn-Banach theorem we must show the following: Let $\vec{S} \in \mathscr{W}'^r$ and let $\boxed{D}'\vec{S} \to 0$ in $\mathscr{W}'^l$; then $\vec{S} \to 0$ in $\mathscr{W}'^r / \boxed{D}^{1'}\mathscr{W}'^m$. Let $\boxed{P}^1$ denote the Fourier transform of $\boxed{D}^{1'}$. Then this means: Let $\vec{G} \in \mathbf{W}'^r$ with $\boxed{P}\vec{G} \to 0$ in the topology of $\mathbf{W}'^l$, then $\vec{G} \to 0$ in $\mathbf{W}'^r / \boxed{P}^1 \mathbf{W}'^m$.

To prove our assertion we apply Theorem 4.2. Then we apply Theorem 4.8. as in the proof of Theorem 6.2. The result follows as in that proof. ▯

The above argument shows also

THEOREM 6.4. *Suppose $\mathscr{W}'$ is reflexive. Let $\{\vec{w}_p\}$ be a sequence in $\boxed{D}\mathscr{W}^r$ which converges to zero. Then we can find $\vec{f}_p \in \mathscr{W}^l$ such that $\boxed{D}\vec{f}_p = \vec{w}_p$ and $\vec{f}_p \to 0$.*

See Remark 6.1.

We wish to show how to give a somewhat constructive method for the proof of Theorem 6.1. We shall illustrate in case $r = 2$, $l = 1$, though the general case can be handled by similar methods.

We want, therefore, to solve

$$D_1 w = u_1 \tag{6.1}$$

$$D_2 w = u_2 . \tag{6.2}$$

We solve (6.1) first as follows: Since $\mathscr{W}$ is LAU, we see easily from the proof of Theorem 1.4 that the set of points where $|P_1(z)| \geqslant 1$ is sufficient (see Section I.3) for $\mathscr{W}$. Thus, we can write

$$u_1(x) = \int_{|P_1(z)| \geq 1} \exp(ix \cdot z)\, dv_1(z)/k_1(z), \tag{6.3}$$

where v_1 is a bounded measure and k_1 belongs to an AU structure for $\mathscr{W}$. Then we define w_1 by

$$w_1(x) = \int_{|P_1(z)| \geq 1} \exp\,(ix \cdot z)\, dv_1(z)/P_1(z)k_1(z). \tag{6.4}$$

It is clear that $w_1 \in \mathscr{W}$.

Subtracting w_1 from w, (6.1) and (6.2) become

$$D_1 w_2 = 0, \quad D_2 w_2 = u_2 - D_2 w_1, \tag{6.5}$$

where $w_2 = w - w_1$. Here

$$D_3(u_2 - D_2 w_1) = 0, \tag{6.6}$$

where D_3 is D_1 divided by the greatest common divisor of D_1, D_2.

To solve (6.5) subject to (6.6) we use a representation for $u_2 - D_2 w_1$ which is analogous to (6.3) except that we integrate over the part of the variety V_3' where $P_3(z) = 0$ on which $|P_2(z)| \geqslant 1$. The possibility of such a representation follows easily from the methods of Chapter VII (see Theorem 7.2 and Problem 7.1) since P_3 and P_2 have no common factor. We may therefore write

$$u_2(x) - (D_2 w_1)(x) = \sum_{j=1}^{p} \int_{V_j \cap |P_2(z)| \geq 1} [\partial_j \exp(ix \cdot z)]\, dv_j(z)/k_2(z) \tag{6.7}$$

in the notation of Theorem 7.1. Here P_3 vanishes on all the V_j and k_2 is in an AU structure for $\mathscr{W}$. Now ∂_j is a differential operator in z so the operators ∂_j, D_3 commute. Therefore if we define w_2 by

$$w_2(x) = \sum_{j=1}^{p} \int_{V_j \cap |P_2(z)| \geq 1} [\partial_j \exp(ix \cdot z)] dv_j/P_2(z)k_2(z) \tag{6.8}$$

we see that w_2 satisfies (6.5), even with the apparently stronger (but actually equivalent) condition $D_3 w_2 = 0$.

We can also describe the kernel of $\boxed{\mathrm{D}}'$ on $\mathscr{W}'^r$, namely,

THEOREM 6.5. *The kernel of* $\boxed{\mathrm{D}}'$ *on* $\mathscr{W}'^r$ *is the image of* $\boxed{\mathrm{D}}^{1\prime}$ *on* $\mathscr{W}'^m$. *Here* $\boxed{\mathrm{D}}^1$ *is an* $m \times r$ *matrix which generates the relations of the rows of* $\boxed{\mathrm{D}}$. *Moreover, if* $\vec{\mathrm{T}} \in \mathscr{W}'^r$ *and* $\boxed{\mathrm{D}}'\vec{\mathrm{T}} \to 0$ *in the topology of* $\mathscr{W}'^l$, *then we can write* $\vec{\mathrm{T}} = \vec{\mathrm{T}}_1 + \vec{\mathrm{T}}_2$ *where* $\vec{\mathrm{T}}_1, \vec{\mathrm{T}}_2 \in \mathscr{W}'^r$, $\vec{\mathrm{T}}_1 \to 0$ *in the topology of* $\mathscr{W}'^r$, *and* T_2 *is of the form* $\boxed{\mathrm{D}}^{1\prime}\vec{\mathrm{S}}$.

PROOF. Since $\boxed{\mathrm{D}}^1 \boxed{\mathrm{D}} = 0$, the image of $\boxed{\mathrm{D}}^{1\prime}$ is contained in the kernel of $\boxed{\mathrm{D}}'$. Conversely, if $\boxed{\mathrm{D}}'\vec{\mathrm{w}} = 0$, then by Fourier transform $\boxed{\mathrm{P}}\vec{\hat{\mathrm{w}}} = 0$. Now, $\boxed{\mathrm{P}}^1$ generates the module of right relations of $\boxed{\mathrm{P}}$ so we can write $\vec{\hat{\mathrm{w}}} = \boxed{\mathrm{P}}^1 \vec{\hat{\mathrm{w}}}_1$, where $\vec{\hat{\mathrm{w}}}_1$ is some vector of entire functions. This is, of course, not good enough for our purposes. However, by Theorem 3.18 we can write $\vec{\hat{\mathrm{w}}} = \boxed{\mathrm{P}}^1 \vec{\hat{\mathrm{w}}}_2$, where $\vec{\hat{\mathrm{w}}}_2$ is a nice cochain. The proof of Theorem 4.2 shows that we may assume that actually $\vec{\hat{\mathrm{w}}}_2 \in \mathbf{W}'^m$ which is the desired result.

The second part of Theorem 6.5 is proved by a similar type of reduction to the semilocal theory. We leave the details to the reader. ▯

VI.2. Fundamental Solutions

In case $r = l = 1$, an important role in the theory of partial differential equations is played by a fundamental solution. This is a distribution e satisfying $De = \delta$, where δ is the unit mass at the origin, often called the "Dirac δ function" or the "Dirac measure." We wish to define an analogous concept for systems. We shall consider first the case $l = 1$. We shall suppose that $\delta \in \mathscr{W}$.

DEFINITION. *A fundamental solution for $\vec{\mathrm{D}}$ is an r-tuple $(e_1, \ldots, e_r) \in \mathscr{W}$ satisfying*

$$\sum D_j e_j = \delta. \tag{6.9}$$

Fundamental solutions exist. We could take for example, $e_1 =$ fundamental solution for D_1, all other $e_j = 0$. The existence of such an e_1 follows from Theorem 6.1. However, we might expect to be able to construct fundamental solutions with nicer properties (e.g., smoother, vanishing on certain sets) than those we could obtain by considering merely fundamental solutions for the D_j individually.

For example, if $n = 2$, $r = 2$, $D_j = \partial/\partial x_j$, we could choose

$$e_j = x_j/(x_1^2 + x_2^2).$$

This is a fundamental solution because $\sum D_j e_j = \frac{1}{2}\Delta \log(x_1^2 + x_2^2)$, where Δ is the Laplacian $\partial^2/\partial x_1^2 + \partial^2/\partial x_2^2$.

The general procedure for construction of suitable fundamental solutions is the following: Let Θ be a set in C with the following properties:

1. δ can be represented as a Fourier transform

$$\delta = \int \exp(iz \cdot \quad)\, d\mu(z)/k(z), \tag{6.10}$$

where k is in an AU structure for $\mathscr{W}$ and μ is a bounded measure whose support lies in Θ.

2. We can find functions $g_j(z)$ which are defined and locally integrable on Θ (with respect to $d|\mu(z)|$) and satisfy

$$\sum P_j(z)g_j(z) = 1 \tag{6.11}$$

for $z \in \Theta$, except for a set of measure zero with respect to $d|\mu(z)|$ and also

$$|g_j(z)|/k'(z) \to 0 \quad \text{as} \quad z \in \Theta, \qquad |z| \to \infty, \tag{6.12}$$

whenever k' is in an AU structure for $\mathscr{W}$.

Then we can define e_j by

$$e_j = \int g_j(z)\exp(iz \cdot \quad)\, d\mu(z)/k(z). \tag{6.13}$$

In case the AU structure for $\mathscr{W}$ contains the square root of any of its elements (which is true in all the examples of Chapter V), it is clear that $e_j \in \mathscr{W}$ and that (e_j) is a fundamental solution for $\vec{\mathrm{D}}$.

Now by (6.11) the existence of the g_j implies that the P_j do not vanish simultaneously on a set of positive measure for $d|\mu(z)|$. We may assume, by removing parts of Θ of $d|\mu(z)|$ measure 0, that the P_j do not vanish at any point of Θ. The method we shall use to actually construct the g_j below with suitable properties (see for example the proof of Theorem 11.10) is as follows: Suppose we know that the P_j have no common zeros on a large neighborhood Θ' of Θ. Then the technique of Chapter IV would show that we can find functions $G_j(z)$ which are holomorphic on a somewhat smaller neighborhood Θ'' of Θ, satisfy (6.11) for $z \in \Theta''$, and (6.12) for $z \in \Theta''$, $|z| \to \infty$.

If, in addition, we construct Θ, μ, and k in such a way that certain analytic properties hold so that we can shift contours, then we can verify (6.10). Our construction of fundamental solutions with suitable properties is thus somewhat similar to the usual case $r = 1$, $G_1 = 1/P_1$.

In general, we shall consider fundamental solutions which belong to the space $\mathscr{D}'$ or $\mathscr{D}'_F$ (see Examples 3 and 4 of Chapter V). When nothing further is said, this will be assumed to be the case.

Fundamental solutions are often used to study the solutions of homogeneous and inhomogeneous systems. For example, suppose we want to solve the system.

$$D_j f = g_j \tag{6.14}$$

for $f \in \mathscr{D}$, given $g_j \in \mathscr{D}$. Suppose that the compatibility conditions for the g_j are satisfied (see Section VI.1). Let $\vec{\mathrm{e}}$ be a fundamental solution for $\vec{\mathrm{D}}$.

Then we write (see Chapter V, Example 3)

$$
\begin{aligned}
(6.15) \qquad f &= f * \delta \\
&= \sum f * D_j e_j \\
&= \sum D_j f * e_j \\
&= \sum g_j * e_j .
\end{aligned}
$$

Thus, if a solution $f \in \mathscr{D}$ of (6.14) exists, it is given by

$$f = \sum g_j * e_j .$$

(Of course, there may not exist a solution $f \in \mathscr{D}$ of (6.14).)

In case the f or g_j of (6.14) are in $\mathscr{D}'$ or $\mathscr{E}$ we cannot apply the above argument since we cannot commute D_j in (6.15). Nevertheless we can often use the fundamental solution to advantage by "cutting" off g_j and f, that is, by multiplying them by functions in $\mathscr{D}$ which are $\equiv 1$ for $|x| \leqslant R$. This process will be used several times in this book.

It is interesting to observe that there is a certain "duality" associated to fundamental solutions. Note that the construction of fundamental solutions described in (6.13) uses an integral defined in the complement of the variety V of common zeros of the P_j. On the other hand, Theorems 7.1 and 7.2 show that solutions of the homogeneous system $D_j f = 0$, $j = 1, \ldots, r$ are represented as integrals on suitable subsets of V. Now, in ordinary topology, Alexander duality defines a duality between the homology of a set and that of its complement. Homology is a study of cycles, and the above remarks indicate that solutions of homogeneous systems and fundamental solutions are expressible in terms of suitable integrals. Some of the results described in the chapters below indicate that there should be a type of "Alexander duality" between solutions of homogeneous systems and fundamental solutions.

See Problem 6.1.

We can use the fundamental solution in some cases where results like Theorem 4.1 cannot be applied. The reason for this is that we can often use a *single* fundamental solution to advantage. Such a fundamental solution can be constructed as an integral of a suitable type, and it is possible that "cancellation" properties can make this fundamental solution "better" than we might have hoped from purely absolute value estimates. An example of this is the Huygens principle which depends (see Petrowski [2] and Section IX.8) on being able to suitably deform a contour. On the other hand, Theorem 4.1 describes a whole class of solutions and depends on estimates of the modulus of the functions involved.

Thus far we have considered systems for which $l = 1$. Let us try to define the concept of fundamental solution for the case $l > 1$. The reason that this case is more difficult is that no argument like (6.15) can hold, since there might not be uniqueness for $\vec{f} \in \mathscr{D}^l$ satisfying $\boxed{D}\vec{f} = \vec{g}$. For example, let $n = 2$, $r = 1$, $l = 2$, $\boxed{D} = (\partial/\partial x_1, \partial/\partial x_2)$. Then the system is

$$\partial f_1/\partial x_1 + \partial f_2/\partial x_2 = g.$$

There is no uniqueness for $\vec{f} \in \mathscr{D}^2$ since we could add to $\vec{f}$ anything of the form $(\partial h/\partial x_2, -\partial h/\partial x_1)$ for any $h \in \mathscr{D}$.

We shall try to define a fundamental solution in such a way that it gives a solution $\vec{f} \in \mathscr{D}^l$ to the system $\boxed{D}\vec{f} = \vec{g}$ for $\vec{g} \in \mathscr{D}^r$ when such an $\vec{f}$ exists. Let us write the system in longhand as

$$\sum D_{ij} f_j = g_i. \tag{6.16}$$

According to the method suggested in Section II.2, we try to solve this system step by step. Thus we try to solve

$$D_{i1} f_1 = g_i - \sum_{j>1} D_{ij} f_j \tag{6.17}$$

for f_1 in terms of the g_i and f_j for all i and for all $j > 1$. (It may be that some $D_{i1} = 0$). (6.17) is a system for one unknown function, so we can express f_1 in terms of the g_i and f_j for $j > 1$ by means of a fundamental solution for $(D_{11}, D_{21}, \ldots, D_{r1})$.

In order to solve (6.17) we must know that the right side of (6.17) satisfies suitable compatibility conditions (see Theorem 6.1). These involve differential relations among the f_j and g. Thus we are in essentially the same situation as we started with except that l has been reduced to $(l - 1)$. Proceeding in this way we can finally solve for $\vec{f}$ in terms of $\vec{g}$. It should be pointed out that if, at a certain point, no compatibility conditions for the f_j on the right side of (6.17) are needed, then we can make them arbitrary.

DEFINITION. By a *fundamental solution* for the system $\boxed{D}\vec{f} = \vec{g}$ is meant a sequence $\vec{e}_1, \vec{e}_2, \ldots, \vec{e}_l$, where $\vec{e}_1$ is a fundamental solution for $(D_{11}, \ldots, D_{r1})$, $\vec{e}_2$ is a fundamental solution for the equations expressing f_2 in terms of $g_1, \ldots, g_r$ and $f_3, \ldots, f_l$ obtained from the compatibility conditions on the right side of (6.17), and $\vec{e}_3, \ldots, \vec{e}_l$ are defined successively in a similar manner. In case, in this process, there is no equation for some f_j, we set the corresponding $\vec{e}_j = 0$.

EXAMPLE. Let $n = 3$; consider the system

$$\operatorname{curl} \vec{f} = \vec{g} \tag{6.18}$$

where div $\vec{g} = 0$. Here

$$\text{curl} = \begin{pmatrix} 0 & \partial/\partial x_3 & -\partial/\partial x_2 \\ -\partial/\partial x_3 & 0 & \partial/\partial x_1 \\ \partial/\partial x_2 & -\partial/\partial x_1 & 0 \end{pmatrix}.$$

(6.17) becomes in this case

$$(6.19) \qquad \begin{aligned} 0 &= \quad g_1 - \partial f_2/\partial x_3 + \partial f_3/\partial x_2 . \\ \partial f_1/\partial x_3 &= -g_2 + \partial f_3/\partial x_1 \\ \partial f_1/\partial x_2 &= \quad g_3 + \partial f_2/\partial x_1 \end{aligned}$$

The compatibility conditions for (6.19) give, for f_2, the equations

$$(6.20) \qquad \begin{aligned} \partial f_2/\partial x_3 &= g_1 + \partial f_3/\partial x_2 , \\ \partial^2 f_2/\partial x_1 \partial x_3 &= -\partial g_3/\partial x_3 - \partial g_2/\partial x_2 + \partial^2 f_3/\partial x_1 \partial x_2 . \end{aligned}$$

Note that the second equation of (6.20) is a consequence of the first since div $\vec{g} = 0$. Thus we need only solve the first equation of (6.20), so there are no compatibility conditions for f_3. A fundamental solution for (6.18) is therefore

$$\vec{e}_1 = (0, E_3\, 0,); \quad \vec{e}_2 = E_3; \quad \vec{e}_3 = 0,$$

where E_3 is a fundamental solution for $\partial/\partial x_3$.

It should be pointed out that if $\vec{g} \in \mathscr{D}^r$, then the fundamental solution produces an $\vec{f} \in \mathscr{E}^l$ satisfying $\boxed{D}\vec{f} = \vec{g}$. However, even if there is an $\vec{f} \in \mathscr{D}^l$ satisfying this equation it might not be given by the above described construction using the fundamental solution.

If we denote by $\boxed{e}$ the matrix

$$\boxed{e} = \begin{pmatrix} 0 & -E_3 & 0 \\ E_3 & 0 & 0 \\ 0 & 0 & 0 \end{pmatrix},$$

then the solution of curl $\vec{f} = \vec{g}$ (formally) is

$$\vec{f} = \boxed{e} * \vec{g}.$$

Note that, if we denote by curl $\boxed{e}$ the matrix product, then

$$\text{curl}\ \boxed{e} = \begin{pmatrix} \delta & 0 & 0 \\ 0 & \delta & 0 \\ -\partial E_3/\partial x_1 & -\partial E_3/\partial x_2 & 0 \end{pmatrix}.$$

Thus, if $\boxed{\delta}$ denotes the diagonal matrix whose diagonal entries are δ, then curl $\boxed{e} = \boxed{\delta} + \Lambda$ where

$$\Lambda = \begin{pmatrix} 0 & 0 & 0 \\ 0 & 0 & 0 \\ -\partial E_3/\partial x_1 & -\partial E_3/\partial x_2 & -\delta \end{pmatrix},$$

so Λ is formally zero on vectors $\vec{g}$ whose divergence is zero. (Since, for such a $\vec{g}$, formally,

$$-E_3 * \partial g_1/\partial x_1 - E_3 * \partial g_2/\partial x_2 - g_3 = E_3 * \partial g_3/\partial x_3 - g_3 = 0.)$$

Of course, we could not hope to make curl $\boxed{e} = \boxed{\delta}$ since the determinant of curl is $\equiv 0$.

We wish now to discuss another type of fundamental solution for general systems. This method will be used in Chapter XI for the solution of systems on nonconvex domains. The idea is the following: Fundamental solutions are used to solve inhomogeneous equations when there is some hope of obtaining a unique solution. We shall define *semifundamental solutions* which will extract much information about solutions of the inhomogeneous equation even when there is no uniqueness.

For example, if curl $\vec{f} = \vec{g}$, then the integral of $\vec{f}$ around any closed curve is determined by $\vec{g}$ (if there are no topological difficulties). We can interpret this result as follows: Let U_3 be the distribution which is the product of the constant 1 on the x_1, x_2 plane by the δ function in the x_3 direction. Then

$$\text{curl}(0, 0, U_3) = 0.$$

Let T_3 be the product of U_3 by the characteristic function of the cylinder over a smooth compact set Ω in the x_1, x_2 plane. Then

$$\text{curl}(0, 0, T_3) = (\partial\Omega),$$

where by $(\partial\Omega)$ we denote the element of $\mathscr{E}'^3$ whose value on any $\vec{g} \in \mathscr{E}^3$ is

$$(\partial\Omega) \cdot \vec{g} = \int_{\partial\Omega} g.$$

Now, if curl $\vec{f} = \vec{g}$, then

$$(6.21) \qquad (\partial\Omega) \cdot \vec{f} = [\text{curl}(0, 0, T_3)] \cdot \vec{f} = (0, 0, T_3) \cdot \vec{g}$$

(by the antisymmetry of curl). Thus,

$$(6.22) \qquad \int_{\partial\Omega} \vec{f} = T_3 \cdot g_3$$

is determined, and also the integral of $\vec{f}$ over the boundary of any translate of Ω is determined. However, there is more information contained in curl $\vec{f} = \vec{g}$, namely the integrals of $\vec{f}$ around other closed curves.

In the present situation we can apply linear transformations to $(0, 0, T_3)$ and obtain a dense set of distributions in $\mathscr{E}'^3$. More precisely, consider the group $G = G_1 \times G_2$, where G_1 is the general linear group operating on x and G_2 is the general linear group operating on 3-vectors. For $\lambda \in G$ we write $\lambda = (\lambda_1, \lambda_2)$. G operates on functions $\vec{h} \in \mathscr{E}^3$ by

$$(\lambda\vec{h})(x) = \lambda_2[\vec{h}(\lambda_1 x)].$$

G operates on distributions by adjoint action, denoted by λ'. Now, certain $\lambda \in G$ have the property that they map solutions of curl $\vec{h} = 0$ into solutions of the same equation. It is these λ which we consider now; we term them "suitable."

We leave to the reader the task of showing that the set of $\lambda'(0, 0, T_3)$ we obtain from suitable λ are dense in $\mathscr{E}'^3$ (see below). Thus, all the information to be obtained from curl $\vec{f} = \vec{g}$ can be obtained by applying to $\vec{f}$ the curl of suitable linear transformations of $(0, 0, T_3)$. This information is the integral of $\vec{f}$ along any linear transform of $\partial\Omega$.

Now, suppose that we are given the integrals of $\vec{f}$ over the linear transforms of $\partial\Omega$. These are equal to the suitable linear transforms of $(0, 0, T_3)$ applied to $\vec{g}$. Call M the set of elements of $\mathscr{E}'^3$ which are linear combinations of curl$[\lambda'(0, 0, T_3)]$. Let $\vec{f}$ be the linear function on M defined by

$$\vec{f} \cdot \text{curl}\,[\lambda'(0, 0, T_3)] = \lambda'(0, 0, T_3) \cdot \vec{g}. \tag{6.23}$$

We claim that $\vec{f}$ is continuous on M in the topology induced by $\mathscr{E}'^3$. Let $\vec{S} \in M$ and $\vec{S} \to 0$ in the topology of $\mathscr{E}'^3$. We write $\vec{S} = \text{curl}\,\vec{T}$ with $\vec{T} \in \mathscr{E}'^3$. Thus, curl $\vec{T} \to 0$ so, by Theorem 6.5 we can write $\vec{T} = \vec{T}_1 + \vec{T}_2$ where $\vec{T}_1 \to 0$ and $\vec{T}_2 = \text{grad}\,U$. By (6.23) this gives

$$\begin{aligned} \vec{f} \cdot \text{curl}\,\vec{T} &= \vec{T} \cdot \vec{g} \\ &= \vec{T}_1 \cdot \vec{g} + \text{grad}\,U \cdot \vec{g} \\ &= \vec{T}_1 \cdot \vec{g} - U \cdot \text{div}\,\vec{g} \\ &= \vec{T}_1 \cdot \vec{g} \ (\text{since div}\,\vec{g} = 0) \\ &\to 0 \ (\text{since } \vec{T}_1 \to 0). \end{aligned} \tag{6.24}$$

This is our continuity assertion.

Hence, by the Hahn-Banach theorem we can extend $\vec{f}$ to $\vec{\mathrm{f}} \in \mathscr{E}^3$. We have, since the linear combinations of $\lambda'(0, 0, T_3)$ are dense in $\mathscr{E}'^3$,

$$\text{(6.25)} \qquad \operatorname{curl} \vec{\mathrm{f}} \cdot \mathrm{V} = \vec{f} \cdot \operatorname{curl} \vec{\mathrm{V}} = \vec{\mathrm{V}} \cdot \vec{\mathrm{g}}$$

for all $\vec{\mathrm{V}} \in \mathscr{E}'^3$. Thus $\operatorname{curl} \vec{\mathrm{f}} = \vec{\mathrm{g}}$ which is the desired conclusion.

It is interesting to note that while the classical solution to $\operatorname{grad} \mathrm{f} = \vec{\mathrm{g}}$ is obtained by taking the line integral of $\vec{\mathrm{g}}$, there does not seem to exist any classical solution of $\operatorname{curl} \vec{\mathrm{f}} = \vec{\mathrm{g}}$ using surface integrals. The above procedure does this to some extent.

This leads us to make the following general

Definition. Let $\boxed{\mathrm{D}}$ be an $r \times l$ matrix of linear partial differential operators with constant coefficients. Denote by $\boxed{\mathrm{D}}'$ the adjoint of $\boxed{\mathrm{D}}$. A collection $\{\vec{\mathrm{U}}\} \subset \mathscr{E}^r$ is called a *semifundamental solution* for $\boxed{\mathrm{D}}$ if $\boxed{\mathrm{D}}'\vec{\mathrm{U}} = 0$ for each $\vec{\mathrm{U}}$, and if the set M obtained by taking linear combinations of translates of the products $\alpha\vec{\mathrm{U}}$ for a suitable $\alpha \in \mathscr{E}'$ is dense in $\mathscr{E}'^r$.

Let us explain how to obtain semifundamental solutions. We shall illustrate the case $r = 1$ and leave the general case to the reader. Suppose we are given a collection $\{U\} \subset \mathscr{E}$ and we ask whether the linear combinations of the translates of $\{\alpha U\}$ are dense in $\mathscr{E}'$. Since convolution is a limit of linear combinations of translates (see Schwartz [1]), this is the same as the question of whether the ideal in the convolution algebra $\mathscr{E}'$ generated by $\{\alpha U\}$ is dense in $\mathscr{E}'$. By Fourier transform this is the same as the question of whether the ideal generated by $\{V\}$, where V is the Fourier transform of αU, is dense in the multiplication algebra $\mathbf{E}'$. We try to find $\{U\}$ so the corresponding V have no common zero. Thus we have an ideal with no common zeros. Theorem 4.1 and Section XI.1 suggest that this ideal must be dense in $\mathbf{E}'$. (In case $n = 1$, this is proved in Schwartz [4].)

If we are given any semifundamental solution, we can proceed as in the case of curl to solve the equation

$$\boxed{\mathrm{D}}\vec{\mathrm{f}} = \vec{\mathrm{g}}.$$

See Problem 6.2. (Exercise)

One might think that the fundamental solution has an advantage over the semifundamental solution because when solving inhomogeneous equations by use of the fundamental solution we obtain the pointwise value of the solution. The semifundamental solution only gives the value on the set $M \subset \mathscr{E}'^r$. However, from the general point of view of topological vector spaces, there is no essential difference, since we regard pointwise values as the values on $\{\delta_x\}$ which is considered as a subset of $\mathscr{E}'$.

Remarks

Remark 6.1. See page 177.

It is of interest to note that Theorems 6.1, 6.2, 6.3, and 6.4 do not rely on the concept of multiplicity variety but only on Theorem 4.8 and the part of Theorem 4.2 which refers to the map $\lambda_{\boxed{F}}$. These can be proved by using the method of Chapter IV combined with the methods of Chapter III which do not involve the multiplicity variety but only the ideas involved in proving Theorem 3.2 and its analog for modules. Malgrange [4] uses a somewhat similar method expect that he replaces the spaces $\mathbf{W}'(\beta)$ by a space of differential forms with growth conditions. Moreover

$$\mathbf{W}'^{l}(\beta;\ \boxed{F})/\mathbf{W}'^{l}(\beta;\ \boxed{F}) \cap \boxed{F}\mathbf{W}'^{r}(\beta - c')$$

is replaced by the space of distributions on V.

Problems

PROBLEM 6.1. See page 181.

Find a precise formulation of this duality.

PROBLEM 6.2. (Exercise) See page 186.

Show how to use the semilocal solution in case $r = l = 1$ to solve the inhomogeneous equation. Note the differences between solutions obtained by use of the fundamental and semifundamental solutions.

CHAPTER VII

Integral Representation of Solutions of Homogeneous Equations

Summary

Let $\mathscr{W}$ be an LAU space. Let $\vec{D} = (D_1, \ldots, D_r)$ where the D_j are linear constant coefficient partial differential operators and $\vec{D}$ is thought of as a column vector, and call $\vec{P} = (P_1, \ldots, P_r)$ the Fourier transform of the adjoint of $\vec{D}$. Let $\mathfrak{V} = (V_1, \partial_1; V_2, \partial_2; \ldots; V_p, \partial_p)$ be the multiplicity variety associated to $\vec{P}$ by the results of Chapter IV. Then we prove the following Fourier representation theorem: Let $w \in \mathscr{W}$ *satisfy* $\vec{D}w = 0$. *Then there exists a* k *in an* AU *structure for* $\mathscr{W}$ *and bounded measures* μ_j *with support on* V_j *such that*

$$w(x) = \sum \int [\partial_j \exp(iz \cdot x)] \, d\mu_j(z)/k(z).$$

The integrals converge in the topology of $\mathscr{W}$. *Conversely, any* w *of the above form satisfies* $\vec{D}w = 0$.

We can often restrict the support of μ_j to be a "sufficient" subset of V_j. This is related to the balayage problem discussed in Chapter X below. An example of a discrete sufficient set is given.

A partial extension is given to the case $l > 1$.

VII. Integral Representation of Solutions of Homogeneous Equations

In the beginning of this chapter we shall restrict ourselves for simplicity to systems in which $l = 1$, that is, one unknown function. The simplicity occurs because Theorem 4.2 is incomplete for general systems.

We shall assume in the rest of this chapter and in the rest of this book that differentiation and multiplication by polynomials are continuous on the LAU spaces which we consider.

Let $\mathscr{W}$ be LAU. Let $\vec{\mathrm{D}} = (D_j)$ be an $r \times 1$ matrix of constant coefficient linear partial differential operators. We are interested in the kernel of $\vec{\mathrm{D}}$ on $\mathscr{W}$, that is, the set of $w \in \mathscr{W}$ with $D_j w = 0$ for all j. By Theorem 6.3 this kernel is dual to $\mathscr{W}'/\vec{\mathrm{D}}'\mathscr{W}'^r$. By Fourier transform w is mapped into an element of the dual of $\mathbf{W}'/\vec{\mathrm{P}}\mathbf{W}'^r$ where $\vec{\mathrm{P}}$ is the Fourier transform of $\vec{\mathrm{D}}'$. Theorem 4.2 tells us that $\mathbf{W}'/\vec{\mathrm{P}}\mathbf{W}'^r$ is isomorphic to $\mathbf{W}'(\mathfrak{V})$, the isomorphism being given by restriction. Let $\mathfrak{V} = (V_1, \partial_1; \ldots; V_p, \partial_p)$, where the V_j are complex algebraic varieties and ∂_j are linear partial differential operators with polynomial coefficients.

By a "bounded measure" on a locally compact space we mean an element of the dual of the space of continuous bounded functions which are zero at infinity, that is, a measure with finite total variation.

THEOREM 7.1. *For any $w \in \mathscr{W}$ satisfying $\vec{\mathrm{D}}w = 0$ and any* AU *structure K for $\mathscr{W}$, we can find a $k \in K$ and bounded measures μ_j on V_j such that*

$$w = \sum_{j=1}^{p} \int [\partial_j \exp(iz \cdot \quad)]\, d\mu_j(z)/k(z). \tag{7.1}$$

The integrals converge in the topology of $\mathscr{W}$. Conversely, any w of the form (7.1) *satisfies $\vec{\mathrm{D}}w = 0$.*

PROOF. Suppose w is of the form (7.1). First we prove the convergence of the integrals in (7.1). Let $S \in \mathscr{W}'$, and let G be the Fourier transform of S. Then, by definition, the convergence of the integral means that $\int (\partial_j G)(z)\, d\mu_j(z)/k(z)$ converges absolutely and, moreover, if S varies in some neighborhood N of zero in $\mathscr{W}'$, then these integrals converge absolutely and uniformly for $S \in N$. By the definition of an AU space, we can find a neighborhood N_1 of zero so that $S \in N_1$ implies that $|G(z)| \leqslant k(z)$ for all z. From our assumption that differentiation and multiplication by polynomials are continuous on $\mathscr{W}$, we can find a neighborhood N of zero so that for all j and all $S \in N$ we have $|(\partial_j G)(z)| \leqslant k(z)$ for all z.

Next we note that for any q,

$$D_q w = \sum_{j=1}^{p} \int \partial_j[P_q(z) \exp(iz \cdot \quad)]\, d\mu_j(z)/k(z) \tag{7.2}$$

where P_q is the Fourier transform of D_q' because $T \to D_q T$ is continuous on $\mathscr{W}$ and because differentiations in x and z commute. (7.2) means that

$$S \cdot D_q w = \sum \int \partial_j[G(z) P_q(z)]\, d\mu_j(z)/k(z). \tag{7.3}$$

But by Theorem 4.1, the restriction of any $G(z)P_q$ to $\mathfrak{V}$ is zero, that is, $\partial_j[G(z)P_q(z)] = 0$ for all j, q for $z \in V_j$. Thus, $D_q w = 0$, that is, $\vec{\mathrm{D}}w = 0$.

Conversely, let $\vec{D}w = 0$. We consider the function on $\mathbf{W}'$:

$$G \to S \cdot w. \tag{7.4}$$

This is clearly a continuous function on $\mathbf{W}'$; hence, it defines an element $\hat{w} \in \mathbf{W}$, the dual of $\mathbf{W}'$. Since $\vec{D}w = 0$, we have $(S + \sum D_j' T_j) \cdot w = S \cdot w$ for any $T_j \in \mathscr{W}'$, whence, $\hat{w} \cdot (G + \sum P_j H_j) = \hat{w} \cdot G$ for any $H_j \in \mathbf{W}'$. Thus, $\hat{w} \cdot G$ depends only on the class of G modulo the module generated by $\{P_j\}$. By Theorem 4.2, this means that w defines an element $\hat{w}^1 \in \mathbf{W}(\mathfrak{B})$, the dual of $\mathbf{W}'(\mathfrak{B})$; more precisely

$$\hat{w} \cdot G = \hat{w}^1 \cdot \rho_{\mathfrak{B}} G. \tag{7.5}$$

Next we examine the elements of $\mathbf{W}(\mathfrak{B})$. It is an immediate consequence of the Hahn-Banach theorem that we can find a $k \in K$ and, for each j, a bounded measure μ_j on V_j, such that

$$\hat{w}^1 \cdot \rho_{\mathfrak{B}} G = \sum \int \partial_j G \, d\mu_j / k. \tag{7.6}$$

Now, let us consider the expression on the right side of (7.1); we claim it is equal to w. We have

$$\begin{aligned} S \cdot \sum \int \partial_j [\exp(iz \cdot \quad)] \, d\mu_j(z)/k(z) &= \sum \int (\partial_j G)(z) \, d\mu_j(z)/k(z) \\ &= \hat{w}^1 \cdot \rho_{\mathfrak{B}} G \\ &= \hat{w} \cdot G \quad \text{(by (7.5))} \\ &= S \cdot w. \end{aligned}$$

Thus, (7.1) is established, and the proof of Theorem 7.1 is complete. □

In case $\vec{D} \equiv 0$, that is, the condition $\vec{D}w = 0$ is empty, then $\mathfrak{B}$ is just $(C, \text{identity})$ and Theorem 7.1 reduces to the Fourier representation of elements of $\mathscr{W}$, which is Theorem 1.5. In Section I.3 it was shown how that result can be improved, namely, the support of the measure μ_j can be reduced. We now show that a similar result holds in our situation.

Let us note first that the existence of the measures μ_j comes from the Hahn-Banach theorem and the fact that the topology of $\mathbf{W}'(\mathfrak{B})$ is in terms of sup norms on the V_j. Thus, to obtain an improvement on the support of μ_j, we should have to show that the topology of $\mathbf{W}'(V_j)$ can be defined on a subset of V_j. More precisely, we make the

DEFINITION. Let σ, τ be closed subsets of C with $\sigma \subset \tau$. Denote by $\mathbf{W}'(\tau)$ the space of holomorphic functions H on τ such that $|H(z)|/k(z) \to 0$ as $|z| \to \infty, z \in \tau$, for any $k \in K$ which is an AU structure for $\mathscr{W}$. We say that σ is *$\mathscr{W}$-sufficient for* τ if for any $k \in K$ there is a $k_1 \in K$ and a $c > 0$ so that the conditions $G \in \mathbf{W}'(\tau)$ and

$$\max_{z \in \sigma} |G(z)|/k_1(z) \leqslant 1$$

imply

$$\max_{z \in \tau} |G(z)|/k(z) \leqslant c.$$

Let $\mathfrak{V}$ be a multiplicity variety, $\mathfrak{V} = (V_1, \partial_1; \ldots; V_p\, \partial_p)$. For each j let σ_j be a $\mathcal{W}$-sufficient set for V_j. Then we say that the collection $\{\sigma_j\}$ is *$\mathcal{W}$-sufficient for $\mathfrak{V}$.*

The proof of Theorem 7.1, with simple modifications, gives

THEOREM 7.2. *Let $\{\sigma_j\}$ be a $\mathcal{W}$-sufficient collection for $\mathfrak{V}$. Then every $w \in \mathcal{W}$ which satisfies $\vec{D}w = 0$ can be represented in the form* (7.1), *where support $\mu_j \subset \sigma_j$.*

For any V_j which is a union of noncompact (that is, of dimension > 0) irreducible varieties we can choose $\sigma_j = V_j - B$, where B is a compact subset of V_j. This follows easily from the maximum modulus theorem. To allow suitable non-compact B we must apply the Phragmén-Lindelöf theorem (which is, of course, a modification of the maximum modulus theorem). For $\tau = C$ we have given examples of $\mathcal{W}$ sufficient sets in Chapter V. Those methods can be modified so as to apply to the case in which τ is a complex algebraic variety of complex dimension 1. Several other cases can be treated satisfactorily, but a suitable treatment for arbitrary algebraic varieties seems quite difficult.

See Problem 7.1.

We wish now to extend the results somewhat to general systems ($l \geqslant 1$). As mentioned above, the difficulty is that we know of no analog of $\rho_{\mathfrak{V}}$ for general systems. Thus, to derive the analog of Theorem 7.1 for $l > 1$, we use a slightly different approach based on the fact that $\rho_{\vec{\mathfrak{V}}}^{L}$ of Theorem 4.2 is defined:

As before, let $\boxed{\mathrm{D}}\vec{\mathrm{w}} = 0$. Then, as in (7.4), $\vec{\mathrm{w}}$ defines a $\vec{\hat{\mathrm{w}}} \in \mathbf{W}^l$, namely,

$$\vec{\mathrm{G}} \to \vec{\mathrm{S}} \cdot \vec{\mathrm{w}}. \tag{7.7}$$

Again we see that $\vec{\hat{\mathrm{w}}} \cdot \vec{\mathrm{G}}$ depends only on the class of $\vec{\mathrm{G}}$ modulo $\boxed{\mathrm{P}}\mathbf{W}'^r$. By Theorem 4.2 this means that $\vec{\hat{\mathrm{w}}}$ defines an element $\hat{\mathrm{w}}^1 \in \mathbf{W}(\vec{\mathfrak{V}})$, more precisely,

$$\vec{\hat{\mathrm{w}}} \cdot \vec{\mathrm{G}} = \hat{\mathrm{w}}^1 \cdot \rho_{\vec{\mathfrak{V}}}^{L} \lambda_{\boxed{\mathrm{P}}} \vec{\mathrm{G}}. \tag{7.8}$$

Now we recall that $\vec{\mathfrak{B}}$ is a "vector" of algebraic multiplicity varieties $\mathfrak{B}^i = \{(V_j^i, \partial_j^i)\}$. Using the Hahn-Banach theorem, we see that we can find bounded measures μ_j^i with supports on V_j^i and a $k \in K$ so that

$$\vec{\mathrm{w}}^1 \cdot \rho_{\vec{\mathfrak{B}}}^L \lambda_{\boxed{\mathrm{P}}} \vec{\mathrm{G}} = \sum_{i,j} \int (\rho_{\vec{\mathfrak{B}}}^L \lambda_{\boxed{\mathrm{P}}} \vec{\mathrm{G}})_j^i \, d\mu_j^i / k, \tag{7.9}$$

where $(\rho_{\vec{\mathfrak{B}}}^L \lambda_{\boxed{\mathrm{P}}} \vec{\mathrm{G}})_j^i$ is the element of $\mathbf{W}'(V_j^i)$ defined by $\rho_{\vec{\mathfrak{B}}}^L \lambda_{\boxed{\mathrm{P}}} \vec{\mathrm{G}}$.

The reasoning of the proof of Theorem 7.1 gives

THEOREM 7.3. *For any $\vec{\mathrm{w}} = (w^i) \in \mathscr{W}^l$ satisfying $\boxed{\mathrm{D}}\vec{\mathrm{w}} = 0$ we can find a $k \in K$ and bounded measures μ_j^i on V_j^i such that*

$$w^i = \sum_j \int [\rho_{\vec{\mathfrak{B}}}^L \lambda_{\boxed{\mathrm{P}}} \{\exp(iz \cdot \quad)\}]^i \, d\mu_j^i(z)/k(z) \tag{7.10}$$

where $\{\exp(iz \cdot \quad)\}$ is the vector whose l components are all $\exp(iz \cdot \quad)$. *The integrals converge in the topology of $\mathscr{W}$. Conversely any $\vec{\mathrm{w}}$ of the form (7.10) satisfies $\boxed{\mathrm{D}}\vec{\mathrm{w}} = 0$.*

It is clear that Theorem 7.3 reduces to Theorem 7.1 for the case of one unknown function ($l = 1$).

Naturally, it is possible to derive an extension of Theorem 7.2 to $l > 1$.

See Remark 7.1.

As an illustration of Theorem 7.2, consider the heat equation $r = l = 1$, $n = 2$, $D = \partial/\partial x_1 - \partial^2/\partial x_2^2$. The multiplicity variety is (V, identity), where V is the variety $iz_1 + z_2^2 = 0$. We identify a point $(z_1, z_2) \in V$ with its z_2 coordinate. This gives a map ν from functions on V into functions of z_2, namely $(\nu G)(z_2) = G(iz_2^2, z_2)$. The inverse ξ of ν is given by $(\xi H)(iz_2^2, z_2) = H(z_2)$ for any function $H(z_2)$.

For any $a > 0$ denote by $\mathscr{E}(|x_1| \leqslant a)$ the space of indefinitely differentiable functions on $|x_1| \leqslant a$. Using Example 5 of Chapter V we find easily that $\nu\mathscr{E}'(|x_1| \leqslant a)(V)$ is the space of entire functions $H(z_2)$ satisfying

$$H(z_2) \leqslant \text{const } (1 + |z_2|)^N \exp[a \,|\, \mathrm{Re}\,(z_2^2)| + N\,|\mathrm{Im}\, z_2|]$$

for some N. The functions of an AU structure dominate the right sides of the above inequality. Using the method of proof of Theorem 5.5 we see that the lines $\mathrm{Re}\, z_2 = \pm\, \mathrm{Im}\, z_2$ are sufficient for $\nu\mathscr{E}'(|x_1| \leqslant a)(V)$. Hence

$$\mathrm{Re}\, z_2 = \pm\, \mathrm{Im}\, z_2\,,\ z_1 = iz_2^2$$

is sufficient for $\mathscr{E}'(|x_1| \leqslant a)(V)$.

Note that this set is *not* sufficient for $\mathscr{E}'(|x_1| < a)(V)$.

We wish to given an example of Theorem 7.2 for which the sufficient set is discrete. This discrete set will be much smaller than the one obtained in Theorem 5.7 because we shall consider functions on a compact set.

Let Ω (or sometimes $\Omega^{\vec{a}}$) be a closed symmetric rectangular parallelepiped in R, center origin, sides $2\vec{a} = (2a_1, 2a_2, \ldots, 2a_n)$; denote by $\mathscr{E}(\Omega)$ the space of indefinitely differentiable functions on Ω. $\mathscr{E}'(\Omega)$ is the space of distributions with support on Ω. The technique of Chapter V shows that $\mathscr{E}(\Omega)$ is LAU. $\mathbf{E}'(\Omega)$ consists of all entire functions $F(z)$ with

$$|F(z)| \leqslant c(1 + |z|)^c \exp(\sum a_j |\operatorname{Im} z_j|). \tag{7.11}$$

An AU structure for $\mathscr{E}(\Omega)$ consists of all continuous positive functions k which dominate

$$(1 + |z|)^c \exp(\sum a_j |\operatorname{Im} z_j|) \text{ for all } c. \tag{7.12}$$

We claim that for any $b > 1$ the set of points

$$\sigma = \{(m_1\pi/a_1b, m_2\pi/a_2b, \ldots, m_n\pi/a_nb)\}, m_1, \ldots, m_n \text{ integers} \tag{7.13}$$

is sufficient for $\mathscr{E}(\Omega)$. To see this we note that there is a natural topological map of $\mathscr{E}'(\Omega)$ into the space ${}^0\mathscr{D}'(b\Omega)$ of periodic distributions with periods $2b\vec{a}$. The theory of periodic distributions (see Schwartz [1]) shows how to describe the topology of ${}^0\mathscr{D}'(b\Omega)$ in terms of Fourier series coefficients. The Poisson summation formula (see Schwartz [1]) shows that the Fourier series coefficients are just the values of an $F \in \mathscr{E}'(\Omega)$ on σ. Thus, σ is sufficient for $\mathscr{E}(\Omega)$.

We wish now to describe an analogous result for $\mathscr{E}_{\vec{D}}(\Omega)$, the kernel of $\vec{D}$ on $\mathscr{E}(\Omega)$. Some restriction on $\vec{a}$ seems necessary in this case, and we shall prove the result only for a dense set of $\vec{a}$. For simplicity we shall suppose that D is a single operator which we denote by D. Let P be the Fourier transform of D', and let $\mathfrak{V}$ be the multiplicity variety of P. We shall assume for simplicity of writing that $\mathfrak{V}$ is an ordinary (nonmultiplicity) variety V though there is no difficulty in extending our argument to the general case. We assume that each z_i is non-characteristic for P.

For each i and for each point of the form ${}_iz \in C^{n-1}$, where ${}_iz_j = \pi m_j/a_j$ for $j \neq i$, where the m_j are half integers, let $z_{i1}, \ldots, z_{id}$ be the z_i coordinates of the points in V above ${}_iz$. See Fig. 12. By a slight variation of the a_j we may assume that the z_{il} are all distinct (compare Lemma 1.8). Let us denote by $H_{ikl}({}_iz)$ the numbers defined by

$$\sum_{k=0}^{d-1} H_{ikl}({}_iz)t^k = \begin{cases} 1 & \text{if } t = z_{il} \\ 0 & \text{if } t = z_{il'},\ l' \neq l. \end{cases} \tag{7.14}$$

The H_{ikl} are obtained by Lagrange interpolation (see Chapter II). It can be seen, from the explicit formula for the interpolation coefficients (see Lemma 2.11) and the proof of Theorem 1.4 which shows how to handle the denominator in Lemma 2.11, that after a slight variation of the a_j we may

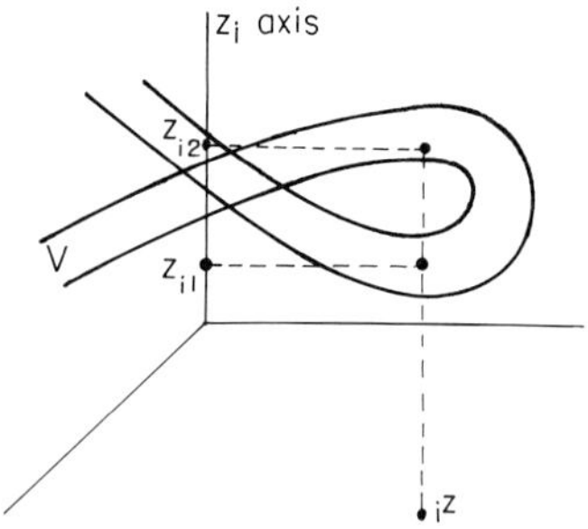

FIGURE 12

assume the existence of a constant so that

(7.15a) $$|H_{ikl}({}_iz)| \leqslant \text{const}(1 + |{}_iz|)^{\text{const}}$$

(7.15b) $$|H_{ikl}({}_iz)| \geqslant \text{const}(1 + |{}_iz|)^{-\text{const}}$$

Let us define the functions $K_{ikl}({}_iz, z)$ by

(7.16)

$$K_{ikl}({}_iz, z) = H_{ikl}({}_iz) \frac{\cos a_1 z_1 \cdots \cos a_n z_n}{(z_1 - {}_iz_1) \cdots (z_{i-1} - {}_iz_{i-1})(z_{i+1} - {}_iz_{i+1}) \cdots (z_n - {}_iz_n)}$$

and call

(7.17) $$H_{il}({}_iz, z) = \sum K_{ikl}({}_iz, z) z_i^k.$$

We see from the definitions and (7.14) above that $H_{il}({}_iz, z) = 0$ whenever z is of the form ${}_{jm}z = ({}_jz_1, \ldots, {}_jz_{j-1}, z_{jm}, {}_jz_{j+1}, \ldots, {}_jz_n)$ with $j \neq i$ or if $j = i$ if $m \neq l$. On the other hand, we may, by a slight variation of the a_j, assume that

(7.18) $$|H_{il}({}_iz, {}_{il}z)| \geqslant c(1 + |z|)^{-c'} \exp(a_i |\text{Im } z_{il}|).$$

For, if (7.18) did not hold, then, as in the proof of (7.15a), the $K_{ikl}({}_iz, {}_{il}z)$ would be small. But (7.15b) shows that $H_{ikl}({}_iz)$ is large. The other factor on the right side of (7.16) is just $\pm a_1 \cdots a_{i-1} a_{i+1} \cdots a_n \cos a_i z_{il}$. Thus (7.18) holds. By a slight variation we can make z_{il} not too close to half integers divided by a_i. In addition the set $\{H_{il}({}_iz, z)\}$ for vaying i, l, ${}_iz$ is bounded in $\mathbf{E}'(\Omega)$.

Next let $G \in \mathbf{E}'(\Omega)(V)$. Suppose first that

(7.19) $$|G({}_{il}z)| \leqslant c''(1 + |z|)^{-c'-n-2} \exp(a_i |\text{Im } z_{il}|).$$

Then the series

$$\sum_{i,l,\iota z} \frac{G({}_{il}z)}{H_{il}({}_{i}z, {}_{il}z)} H_{il}({}_{i}z, z) \tag{7.20}$$

converges in the topology of $\mathbf{E}'(\Omega)$ to a function $\tilde{G}(z)$ which agrees with $G(z)$ at the points ${}_{il}z$ for all i, l.

If G does not satisfy (7.19), then for a suitable polynomial Q, G/Q satisfies (7.19). We then form $\widetilde{G/Q}$ as in (7.20) and set $\tilde{G} = Q(\widetilde{G/Q})$ to obtain a function with the same properties.

In general, G may not be equal to $\tilde{G}$. For example, if G is the product of $\cos a_1 z_1 \cdots \cos a_n z_n$ by a polynomial the resulting interpolating function $\tilde{G}$ is identically zero. In general, $G = \tilde{G}$ at the points ${}_{il}z$. Thus, $G - \tilde{G} = 0$ at all points of V above all ${}_{i}z$. Again by varying the a_j slightly we may assume that these points are all simple points of V, that is, V is a manifold in the neighborhood of any such point. Thus, $G - \tilde{G}$ is divisible by $\cos a_1 z_1 \cdots \cos a_n z_n$. By comparing growth conditions we see easily that

$$|(G - \tilde{G})/\cos a_1 z_1 \cdots \cos a_n z_n| \leq c'''(1 + |z|)^{c'''}. \tag{7.21}$$

By Theorem 4.1 and Chapter V, Example 8, this means that $(G - \tilde{G})/\cos a_1 z \cdots \cos a_n z$ is the restriction of a polynomial $Q(z)$ to V, that is,

$$G - \tilde{G} = \text{restriction to } V \text{ of } Q(z) \cos a_1 z_1 \cdots \cos a_n z_n . \tag{7.22}$$

By going through the above argument carefully we can see that $\tilde{G}$ can be chosen so as to depend continuously (though not linearly) on G. Thus, except for polynomial factors the set of points ${}_{il}z$ is sufficient for $\mathbf{E}'(\Omega)(V)$ (the space of restrictions of functions in $\mathbf{E}'(\Omega)$ to V).

It is easy to handle the polynomial factors by using, for example, the points ${}^{il}z$ which are those points in V which lie over points ${}^{i}z$ with ${}^{i}z_k = \pi n_k/a_k$ with n_k integers, $k \neq i$. Thus, we have proved

THEOREM 7.4. *For a dense set of* $\vec{a}$ *the points* ${}_{il}z$ *and* ${}^{il}z$ *are sufficient for* $\mathbf{E}'(\Omega)(V)$.

It should be noted that the method of proof of Theorem 7.4 could be used to give another proof of Theorem 4.1 for the space $\mathscr{E}(\Omega)$.

It is of interest to remark that we can obtain a representation for solutions of $\vec{\mathbf{D}}w = 0$ even in certain spaces which are *not* AU. For example, let Ω be an open convex set and denote by $\mathscr{E}(R - \Omega)$ the space of indefinitely differentiable functions on a neighborhood of $R - \Omega$. Suppose for simplicity that $l = 1$. Let $f \in \mathscr{E}(R - \Omega)$, $\vec{\mathbf{D}}f = 0$. Let $h \in \mathscr{E}(R)$ be 1 on $R - \Omega$ and 0 outside the set where $\vec{\mathbf{D}}f = 0$. Then $\vec{\mathbf{D}}(hf) = g$ has support contained in Ω. More precisely, we write

$$D_j(hf) = g_j . \tag{7.23}$$

The g_j satisfy the obvious compatibility conditions

$$D_i g_j = D_j g_i \tag{7.24}$$

for all i, j. We wish to construct an explicit $m \in \mathscr{E}(R)$ satisfying

$$\vec{\mathrm{D}}m = g. \tag{7.25}$$

There are three cases to distinguish.

Case 1. *All D_j are equal.* We are then reduced to the case of only one equation ($r = 1$). Then we can determine by Fourier transform an explicit $m \in \mathscr{E}(R)$ satisfying (7.25) in many ways.

Case 2. *The D_j have no common factor.* By Fourier transform (7.24) becomes, for G_j the inverse Fourier transform of g_j,

$$P_i G_j = P_j G_i \tag{7.26}$$

or

$$M = G_j/P_j \quad \text{is independent of } j. \tag{7.27}$$

Since the G_j are entire functions, M is an entire function whose singularities are contained in the common zeros of the P_j. Since the P_j have no common divisor, the theorem of removable singularities shows that M is entire. A simple argument shows that $M \in \mathbf{D}(\Omega)$, so we have found an m satisfying (7.25) which is in $\mathscr{D}(\Omega)$ (cf. Section XI.2.)

Case 3. *$D_j = D_j^1\, \partial$ where the D_j^1 have no common factor.* We determine $\partial m \in \mathscr{D}(\Omega)$ satisfying

$$D_j^1\, \partial m = g_j \tag{7.28}$$

by Case 2. Then we use Case 1 to construct m.

Thus, in all cases we have an explicit representation of an m satisfying (7.25). Now clearly $hf - m \in \mathscr{E}(R)$ and

$$\vec{\mathrm{D}}(hf - m) = 0. \tag{7.29}$$

We may use Theorem 7.1 to obtain a representation for $hf - m$, and hence, we obtain a representation for f on a neighborhood of $R - \Omega$.

Remarks

Remark 7.1. See page 192.

If we wanted a weaker statement than Theorem 7.1 (which would suffice for most applications), namely, that any $w \in \mathscr{W}$ satisfying $\vec{\mathrm{D}}w = 0$ can be represented in the form

$$w = \int \exp(iz \cdot \quad)\, d\mu(z)/k(z) \tag{7.30}$$

where $k \in K$ and μ has its support on the set of z whose distance from V is $\leqslant 1$, then we would not need the full strength of Theorem 4.2 but only the part dealing with the map $\lambda_{\vec{F}}$, which can be proved without the use of multiplicity varieties and is, in fact, much easier. (Compare Remark 6.1.)

Problems

PROBLEM 7.1. (Exercise)

For all the spaces $\mathscr{W}$ dealt with in Chapter V, if P is a polynomial, $P \not\equiv 0$, a $\mathscr{W}$ sufficient set on any $\mathfrak{V}$ which contains no varieties of dimension 0 or varieties on which $P \equiv 0$ consists of those points where $|P| \geqslant 1$.

CHAPTER VIII

Extension and Comparison Theorems. Elliptic and Hyperbolic Systems

Summary

Let $\mathscr{W}$ and $\mathscr{W}_1$ be LAU spaces. Let $\vec{\mathrm{D}} = (D_1, \ldots, D_r)$ and call $\vec{\mathrm{P}} = (P_1, \ldots, P_r)$ the Fourier transform of $\vec{\mathrm{D}}'$. Then we want to know when the solutions of $\vec{\mathrm{D}}f = 0$ in $\mathscr{W}$ are the same as those in $\mathscr{W}_1$. In Section VIII.1 several general methods are given to decide this question.

In Section VIII.2 we consider the example $\mathscr{W} = \mathscr{D}'$, $\mathscr{W}_1 = \mathscr{E}$. If the kernels of $\vec{\mathrm{D}}$ in $\mathscr{W}$ and in $\mathscr{W}_1$ are the same, then we call $\vec{\mathrm{D}}$ *hypoelliptic*. We prove: *A necessary and sufficient condition that D be hypoelliptic is that there exist an $a > 0$ so that for any $z \in V$, the variety of common zeros of the P_j, we have*

$$|\operatorname{Im} z| \geqslant a(1 + |\operatorname{Re} z|^a).$$

Another example of the above concerns hyperbolic systems. We say that $\vec{\mathrm{D}}$ is *hyperbolic in* $x_1, \ldots, x_q$ if there exists a $b > 0$ so that every f which is defined and indefinitely differentiable for $|x_1| < b, \ldots, |x_q| < b$ and for all values of $x_{q+1}, \ldots, x_n$ and satisfies $\vec{\mathrm{D}}f = 0$ for these values can be uniquely extended to an $\tilde{f} \in \mathscr{E}$ which satisfies $\vec{\mathrm{D}}\tilde{f} = 0$ on all of R. We prove: *A necessary and sufficient condition that $\vec{\mathrm{D}}$ be hyperbolic in $x_1, \ldots, x_q$ is the existence of an $a > 0$ so that for all $z \in V$ we have for $t = 1, \ldots, q$*

$$|\operatorname{Im} z_t| \leqslant a(1 + |\operatorname{Im} z_{q+1}| + \cdots + |\operatorname{Im} z_n|).$$

The property of hyperbolicity is independent of b.

In Section VIII.4 we consider a generalization of hyperbolicity in which we drop the condition that the extension be unique.

VIII.1. Comparison Theorems

Let $\mathscr{W}$ and $\mathscr{W}_1$ be two LAU spaces and let $\boxed{\mathrm{D}}$ be an $r \times l$ matrix of linear constant coefficient partial differential operators. We want to compare the kernels of $\boxed{\mathrm{D}}$ in $\mathscr{W}^l$ and in $\mathscr{W}_1^l$. We ask when the kernel of $\boxed{\mathrm{D}}$ in $\mathscr{W}^l$ is contained in the kernel of $\boxed{\mathrm{D}}$ in $\mathscr{W}_1^l$, that is, when every solution $\vec{\mathrm{w}} \in \mathscr{W}^l$ of $\boxed{\mathrm{D}}\vec{\mathrm{w}} = 0$ can be extended to a solution in $\mathscr{W}_1^l$. For example, $\mathscr{W}$ and $\mathscr{W}_1$ could be the spaces of indefinitely differentiable functions on the convex domains Ω and Ω_1 respectively, with $\Omega \cap \Omega_1 \neq \phi$ (see Chapter V, Example 5). We call this the (*Special*) *Comparison Problem*. A refinement of this will be discussed in Chapter XII. We assume that $\mathscr{W}$ and $\mathscr{W}_1$ depend on the same number of variables.

In order to give a precise meaning to the Comparison Problem, we must explain what is meant by $\mathscr{W} \cap \mathscr{W}_1$. Let $f \in \mathscr{W}$; we want to say when f is also in $\mathscr{W}_1$. By $\mathscr{W}' \cap \mathscr{W}_1'$ we denote the subspace of $\mathscr{W}'$ formed by those elements whose Fourier transform belongs to $\mathbf{W}_1'$. Then we say that $f \in \mathscr{W} \cap \mathscr{W}_1$ if the restriction of f to $\mathscr{W}' \cap \mathscr{W}_1'$ is continuous in the topology induced by $\mathscr{W}_1'$. We say also that f *extends to* $\mathscr{W}_1$ if this is the case. This extension is unique if and only if $\mathscr{W}' \cap \mathscr{W}_1'$ is dense in $\mathscr{W}_1'$, in which case the topology of $\mathscr{W} \cap \mathscr{W}_1$ is defined as follows: $\mathscr{W} \cap \mathscr{W}_1$ is the dual of the space of sums $T = S + S_1$ where $S \in \mathscr{W}'$, $S_1 \in \mathscr{W}_1'$. Here $T \to 0$ if it is possible to write $T = S + S_1$ in such a way that $S \to 0$ and $S_1 \to 0$ in the respective topologies of $\mathscr{W}'$ and $\mathscr{W}_1'$. Thus $(\mathscr{W} \cap \mathscr{W}_1)'$ is the quotient of $\mathscr{W}' \oplus \mathscr{W}_1'$ by $\mathscr{W}' \cap \mathscr{W}_1'$. Note that $\mathscr{W}' \cap \mathscr{W}_1'$ is defined since $\mathbf{W}'$ and $\mathbf{W}_1'$ are spaces of entire functions of the same number of variables.

More generally, let $\mathscr{W}_2$ be a closed subspace of $\mathscr{W}_1$ (in applications the kernel of $\boxed{\mathrm{D}}$ on $\mathscr{W}_1$). We want to say when a given $f \in \mathscr{W}$ is in $\mathscr{W} \cap \mathscr{W}_2$, which we shall sometines write more precisely as $f \in \mathscr{W} \cap \mathscr{W}_2(\mathscr{W}_1)$. Denote by $\mathscr{W}_2^\perp$ the subset of $\mathscr{W}_1'$ consisting of those S which vanish on $\mathscr{W}_2$. Then we say that $f \in \mathscr{W} \cap \mathscr{W}_2(\mathscr{W}_1)$ if, whenever $S \in \mathscr{W}' \cap \mathscr{W}_1'$ converge to an $S_0 \in \mathscr{W}_2^\perp$ in the topology of $\mathscr{W}_1'$, we have $S \cdot f \to 0$. (The Hahn-Banach theorem then shows that the restriction of f to $\mathscr{W}' \cap \mathscr{W}_1'$ extends to an $\tilde{f} \in \mathscr{W}_1$ which vanishes on $\mathscr{W}_2^\perp$ and so is in $\mathscr{W}_2$.) We say in this case that f is *extendible to* $\mathscr{W}_2$ (*over* $\mathscr{W}_1$). In particular, if $\mathscr{W}_2 = 0$, then such an $f = 0$ on $\mathscr{W}' \cap \mathscr{W}_1'$. If $\mathscr{W}_3$ is a closed subspace of $\mathscr{W}$, we denote by $\mathscr{W}_3 \cap \mathscr{W}_2$ the set of $f \in \mathscr{W}_3$ which are in $\mathscr{W} \cap \mathscr{W}_2$. The intersection of more than two spaces is defined similarly.

Note that in order for these definitions to have meaning, the spaces $\mathbf{W}'$ and $\mathbf{W}_1'$ must be entire functions of the same number of variables.

These definitions have meaning even if the spaces $\mathscr{W}$, $\mathscr{W}_1$, $\mathscr{W}_2$ are not AU. We require that there be a one-to-one map i of a subspace of $\mathscr{W}_1'$

into $\mathscr{W}'$. We identify the domain and range of i and denote them by $\mathscr{W}' \cap \mathscr{W}_1'$. Then the definition $f \in \mathscr{W} \cap \mathscr{W}_2(\mathscr{W}_1)$ is made as before.

We can now explain why we termed the above Comparison Problem "Special". There are two reasons:

(a) The spaces in question are AU (or direct sums of AU spaces).
(b) The $f \in \mathscr{W}$ which we want to show are in $\mathscr{W}_2$ form the kernel of $\boxed{\mathrm{D}}$ on $\mathscr{W}$, and $\mathscr{W}_2$ is the kernel of $\boxed{\mathrm{D}}$ on $\mathscr{W}_1$.

In Chapter XII we shall study the General Comparison Problem for which $\mathscr{W}$, instead of being AU, is the intersection of a finite number of AU spaces. (b) is generalized to the study of functions which can be represented as sums of solutions of a countable set of equations $\boxed{\mathrm{D}}^i \vec{\mathrm{f}}^i = 0$.

In fact, one can go still further: If $\mathscr{W}$ is a topological vector space, we can define $\mathscr{W} \cap \mathscr{W}_1$ or $\mathscr{W} \cap \mathscr{W}_2$ even if $\mathscr{W}_1$ is *not* a topological vector space. We define the *generalized extension* $\mathscr{W} \cap \mathscr{W}_1$ as follows: We say $\mathscr{W}_1$ is given if we are given a collection of subsets $\{\sigma\}$ of $\mathscr{W}'$. If $f \in \mathscr{W}$, we say that $f \in \mathscr{W} \cap \mathscr{W}_1$ if f is bounded on each σ. (In case $\mathscr{W}_1$ is a topological vector space as above, then $\{\sigma\}$ is the collection of subsets of $\mathscr{W}' \cap \mathscr{W}_1'$ which are bounded in the topology of $\mathscr{W}_1'$. To define $\mathscr{W} \cap \mathscr{W}_2$ we choose for $\{\sigma\}$ the collection of subsets $\mathscr{W}' \cap \mathscr{W}_1'$ whose image in $\mathscr{W}_1'/\mathscr{W}_2^\perp$ is bounded.)

In general, there are three methods to verify that a given $f \in \mathscr{W}$ is in $\mathscr{W} \cap \mathscr{W}_2$.

(a) We show that if $S \in \mathscr{W}' \cap \mathscr{W}_1'$ converges to an $S_0 \in \mathscr{W}_2^\perp$ in the topology of $\mathscr{W}_1'$, then we can find $T \in \mathscr{W}'$ such that $T \cdot f = 0$ and $S - T$ converge to zero in the topology of $\mathscr{W}'$.
(b) Using the fact that $f \in \mathscr{W}$ we find a Fourier series or Fourier integral representation. We then verify that this series or integral converges in the topology of $\mathscr{W}_1$ to an element of $\mathscr{W}_2$. Instead of using a Fourier representation, we might use a different type of representation such as one obtained by use of a fundamental solution (see Section VI.2).
(c) We do not try directly to extend f to $\mathscr{W}_2$ over $\mathscr{W}_1$, but we extend it first to a larger space, say $\mathscr{W}_3$, where $\mathscr{W}_2 \subset \mathscr{W}_3 \subset \mathscr{W}_1$. We then try to correct the "error" we have made. This usually involves solving a type of inhomogeneous or cohomology problem.

See Remark 8.1.

The extension problem is a linear problem, that is, we try to find conditions on an $f \in \mathscr{W}$ to be in $\mathscr{W} \cap \mathscr{W}_2$. We could also consider a corresponding "affine" problem in which we ask when an $f \in \mathscr{W}$ belongs to a given set of cosets of $\mathscr{W} \cap \mathscr{W}_2$. We shall refer to this as the problem of *specification of singularities*. This terminology comes from the simplest example: $\mathscr{W} = \mathscr{W}_1 =$ space of functions analytic in $|z| < 1$ except at 0, $\mathscr{W}_2 =$ space of functions analytic in whole disk. A type of condition we might impose on an $f \in \mathscr{W}$ is $f(z) \sim 1/z$ as $z \to 0$. Then we conclude that $f(z) = z^{-1} + g(z)$ where $g \in \mathscr{W}_2$.

Another way of looking at the problem of specification of singularities is that we think of $\mathscr{W} \cap \mathscr{W}_1$, as the "trivial" part of $\mathscr{W}$ and we want to specify $f \in \mathscr{W}$ modulo a trivial element.

We can also consider the problem of *limitation of singularities* in which we try to find conditions on an $f \in \mathscr{W}$ to be in $\mathscr{W}_3$ which contains $\mathscr{W} \cap \mathscr{W}_2$. In the above illustration, if $\mathscr{W}_3$ is the space of functions in $\mathscr{W}$ which have poles of order at most one at the origin, then if $f \in \mathscr{W}, f(z) = 0(z^{-2})$ as $z \to 0$ then $f \in \mathscr{W}_3$. We may think of the image of f on the quotient $\mathscr{W}_3/\mathscr{W} \cap \mathscr{W}_2$ as the singular part of f.

Very often, in the applications, the type of information given about f is that certain maps applied to f have special properties. Often, this information can be used to prove that $f \in \mathscr{W} \cap \mathscr{W}_2$ by application of a *parametrix*. In order to explain this concept, let us recall the classical notion of parametrix for elliptic operators. Let ∂ be a linear elliptic partial differential operator with constant coefficients. A distribution p of compact support is called a parametrix for ∂ if $\partial p = \delta + \zeta$, where $\zeta \in \mathscr{D}$ (see Chapter V, Example 4). Let us see how the existence of a parametrix guarantees that all distribution solutions g of $\partial g = f, f \in \mathscr{E}$ are themselves in $\mathscr{E}$. We write

$$\begin{aligned} g &= g * \delta \\ &= g * \partial p - g * \zeta \\ &= \partial g * p - g * \zeta. \end{aligned}$$

Thus, $g \in \mathscr{E}$, because $\partial g * p$ and $g * \zeta$ are in $\mathscr{E}$.

The important thing to notice is that there are two assumptions on g, namely, $g \in \mathscr{D}'$ and $\partial g \in \mathscr{E}$. These are used, respectively, to show that $g * \zeta$ and $\partial g * p$ belong to $\mathscr{E}$. This in turn is a consequence of the fact that $g \to g * \zeta$ and $f \to f * p$ are maps of $\mathscr{D}' \to \mathscr{E}$ and $\mathscr{E} \to \mathscr{E}$, respectively. We are thus led to make the following

DEFINITION. Let $\mathscr{U}, \mathscr{U}_1, \mathscr{U}_2, \mathscr{V}_1, \mathscr{V}_2$ be topological vector spaces, where $\mathscr{V}_1, \mathscr{V}_2$ are subspaces of $\mathscr{U}$. Let $L_1 : \mathscr{V}_1 \to \mathscr{U}_1$ and $L_2 : \mathscr{V}_1 \to \mathscr{U}_2$ be linear maps. Then we say that (M_1, M_2) is *an* $(L_1, L_2; \mathscr{V}_2)$ *parametrix* if $M_1 : \mathscr{U}_1 \to \mathscr{V}_2$ and $M_2 : \mathscr{U}_2 \to \mathscr{V}_2$ are linear maps such that

$$M_1 L_1 + M_2 L_2 = \text{identity of } \mathscr{V}_1. \tag{8.1}$$

See Fig. 13.

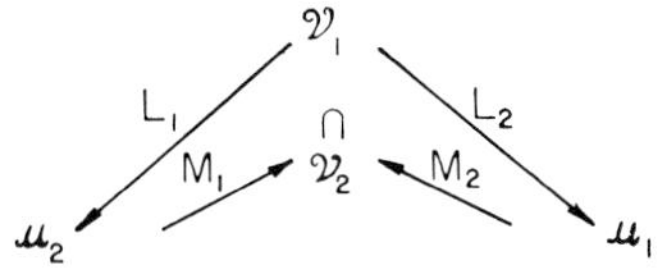

FIGURE 13

Thus, the existence of an $(L_1, L_2; \mathscr{V}_2)$ parametrix implies that $\mathscr{V}_1 \subset \mathscr{V}_2$. Note that this conclusion is obtained by using *both* pieces of information that we have about $\mathscr{V}_1$, namely, that $L_1\mathscr{V}_1 \subset \mathscr{U}_1$ and $L_2\mathscr{V}_1 \subset \mathscr{U}_2$. Moreover, (M_1, M_2) is, essentially, an inverse for the map (L_1, L_2) of $\mathscr{V}_1$ into $\mathscr{U}_1 \oplus \mathscr{U}_2$, though the range is different. Thus the existence of a parametrix implies that (L_1, L_2) is one-to-one.

In our example, $\mathscr{U} = \mathscr{U}_1 = \mathscr{D}'$, $\mathscr{U}_2 = \mathscr{V}_2 = \mathscr{E}$, and $\mathscr{V}_1$ is the space of $g \in \mathscr{D}'$ with $\partial g \in \mathscr{E}$. L_1 is the identity and $L_2 = \partial$. Then M_1 is convolution by $-\zeta$ and M_2 is convolution by p.

As in our example, we shall usually consider the case when $\mathscr{V}_2$ is a subspace of $\mathscr{V}_1$.

If (L_1, L_2) is one-to-one then the inverse will map the image $\mathscr{U}_3$ of (L_1, L_2) onto $\mathscr{V}_1$. We can then extend $(L_1, L_2)^{-1}$ linearly to all of $\mathscr{U}_1 \oplus \mathscr{U}_2$ and, if $\mathscr{V}_1 = \mathscr{V}_2$, obtain a parametrix (M_1, M_2). (Note that M_1 and M_2 are *not* required to be continuous.)

Often we can use the closed graph theorem to show that $(L_1, L_2)^{-1}$ is continuous. We can then often use topological methods (properties of nuclear spaces) to show that (M_1, M_2) can be chosen to be continuous. In this case we can form the adjoint of the parametrix. This allows us, after identifying $\mathscr{V}_1$ with $\mathscr{V}_2$, to express any $v_1' \in \mathscr{V}_1'$ in the form

$$v_1' = L_1' M_1' v_1' + L_2' M_2' v_1'$$

which we can write in the form

$$v_1' = L_1'u_1' + L_2' u_2', \tag{8.2}$$

where $u_1' \in \mathscr{U}_1'$ and $u_2' \in \mathscr{U}_2'$. Moreover, u_1' and u_2' depend continuously on v_1'.

Conversely, if $\mathscr{V}_2 \subset \mathscr{V}_1$ and if we can write (8.2) where u_1' and u_2' depend continuously in the respective topologies of $\mathscr{U}_1'$, $\mathscr{U}_2'$ on v_1' in the topology of $\mathscr{V}_2'$, then it follows easily from the Hahn-Banach theorem that $\mathscr{V}_1 = \mathscr{V}_2$; we assume that $\mathscr{V}_j$ is the dual of $\mathscr{V}_j'$. For, given $v_2 \in \mathscr{V}_2$, we need to know that v_2 is continuous on $\mathscr{V}_1'$. From (8.2) we have

$$v_2 \cdot v_1' = (L_1 v_2) \cdot u_1' + (L_2 v_2) \cdot u_2'$$

from which our assertion follows. Thus, in most interesting circumstances,

(8.2) can be used in place of the parametrix to show that $\mathscr{V}_1 = \mathscr{V}_2$. Conversely, we can sometimes show the impossibility of (8.2) which implies that $\mathscr{V}_1 \neq \mathscr{V}_2$.

In the example of elliptic equations (8.2) is

$$\begin{aligned} S &= S * \delta \\ &= -S * \zeta + \partial(S * p) \\ &= S_1 + \partial S_2 . \end{aligned}$$

Here $S \in \mathscr{E}'$, and $S_1 \in \mathscr{D}$, $S_2 \in \mathscr{E}'$ depend continuously on S.

Actually, we can often use (8.2) to prove that the map (L_1, L_2) is one-to-one. We must show that the set of v_1' represented by the right side of (8.2) is dense in the topology of $\mathscr{V}_1'$ to conclude that (L_1, L_2) is one-to-one on $\mathscr{V}_1$.

In general, we shall speak of a method based on the parametrix or the use of (8.2) as a *parametrix method.*

We shall now give some examples of extension theorems. The examples in this chapter use Methods (a) and (b). They are examples of the Special Comparison Problem. Many other Comparison Problems appear in later chapters.

THEOREM 8.1. *Let $\mathscr{W}$ and $\mathscr{W}_1$ be* LAU. *The kernel of $\boxed{\mathrm{D}}$ in $\mathscr{W}^l$ is contained in the kernel of $\boxed{\mathrm{D}}$ in $\mathscr{W}_1^l$ if the following condition is satisfied: Let $K = \{k\}$ and $K_1 = \{k_1\}$ be* AU *structures for $\mathscr{W}$ and $\mathscr{W}_1$, respectively; then for each $k \in K$ there is a $k_1 \in K_1$ and an $a > 0$ so that*

$$k_1(z) \leqslant ak(z) \tag{8.3}$$

for $z \in \vec{\mathfrak{V}}$.

Here $\vec{\mathfrak{V}}$ is the (semi-local) vector multiplicity variety associated to $\boxed{\mathrm{P}}$ by Theorem 4.2 *and* (8.3) *is to hold for z in any of its components.*

PROOF. For $l = 1$ this is a consequence of Theorem 7.1, if we observe that for any bounded measure μ on V_j we have $\mu/k = \mu_1/k_1$, where $\mu_1 = \mu k_1/k$ is again a bounded measure on V_j by (8.3).

For $l > 1$ the result follows from Theorem 7.3 in the same manner. For later purposes we wish to show also how to derive the result by use of the isomorphism $\lambda_{\boxed{\mathrm{P}}}$ of Theorem 4.2. From the fact that $\rho_{\vec{\mathfrak{V}}}^{L}$ in Theorem 4.2 is a topological isomorphism, we see that the cochains in

$$\mathbf{W}'^{l}(\beta; \boxed{\mathrm{P}})/\mathbf{W}'^{l}(\beta; \boxed{\mathrm{P}}) \cap \boxed{\mathrm{P}}\mathscr{W}'^{r}(\beta - c')$$

can be assumed to be zero except for those lattice points α which are within a distance of 2β from $\vec{\mathfrak{V}}$, since no other lattice points can contribute to the restriction to $\vec{\mathfrak{B}}$. The same holds for $\mathbf{W}_1'$ in place of $\mathbf{W}'$. By property (b) of localizability and our hypothesis, the natural map

$$\mathbf{W}_1'^{l}(\beta; \boxed{\mathrm{P}})/\mathbf{W}_1'^{l}(\beta; \boxed{\mathrm{P}}) \cap \boxed{\mathrm{P}}\mathbf{W}_1'^{r}(\beta - c') \to \mathbf{W}'^{l}(\beta; \boxed{\mathrm{P}})/\mathbf{W}'^{l}(\beta; \boxed{\mathrm{P}}) \cap \boxed{\mathrm{P}}\mathbf{W}'^{r}(\beta - c')$$

is continuous. We now apply the type of Hahn-Banach argument used above, together with Theorems 4.2 and 6.3 and we obtain the desired result. □

THEOREM 8.2. *The kernel of* $\boxed{\mathrm{D}}$ *in* $\mathscr{W}^l$ *is contained in its kernel in* $\mathscr{W}_1^l$ *if there exists a* $\mathfrak{s} = \{\mathfrak{s}_j\}$ *which is* $\mathscr{W}$*-sufficient for* $\vec{\mathfrak{B}}$ *such that for any* $k \in K$ *there is a* $k_1 \in K_1$ *for which* (8.3) *holds for* $z \in \mathfrak{s}$.

This follows by a simple modification of the proof of Theorem 8.1.

We shall now give several applications of the above theorems:

VIII.2. Elliptic Systems

The general elliptic (or hypoelliptic) problem is concerned with finding what additional regularity properties a solution of an equation $\vec{D}w = 0$ with $w \in \mathscr{W}$ must have. Thus, if $\mathscr{W}$ is given, we let $\mathscr{W}_1$ be some subspace of $\mathscr{W}$ which has stronger regularity conditions than $\mathscr{W}$ and we ask whether $\vec{D}w = 0$, $w \in \mathscr{W}$ implies $w \in \mathscr{W}_1$. This is, of course, a Special Comparison Problem.

See Remark 8.2.

We shall concern ourselves here only with some simple examples, since our method applies with almost no change to more general situations.

The classical problem of ellipticity is for $\mathscr{W} = \mathscr{E}$, the space of indefinitely differentiable functions on R, and $\mathscr{W}_1 = \mathscr{H}$, the space of entire functions on the complexification of R. The results of Chapter V show that these spaces are LAU. We note that the topologies of $\mathbf{H}'$ and of $\mathbf{E}'$ are different only near the real part of C. Thus if $\mathfrak{B}$ is far enough away from this real part then the norms are the same. More precisely, we have

THEOREM 8.3. *A necessary and sufficient condition that* $\vec{\mathrm{f}} \in \mathscr{E}^l$, $\boxed{\mathrm{D}}\vec{\mathrm{f}} = 0$ *imply* $\vec{\mathrm{f}} \in \mathscr{H}^l$ *is the existence of an* $a > 0$ *so that, if* $\mathfrak{B}^i = (V_1^i, \partial_1^i; \ldots; V_{a_i}^i, \partial_{a_i}^i)$, *then for any* z *in any* V_j^i

$$|\mathrm{Im}\, z| \geqslant a(1 + |\mathrm{Re}\, z|).$$

PROOF. The necessity of the condition is well known and standard (see, e.g., Petrowski [1]). We can use the method of lacunary series (see Chapter XII) to show the necessity of the condition. This is carried out in detail in Ehrenpreis [4]. The sufficiency follows immediately from Theorem 8.1 and our description (see Chapter V) of AU structures for $\mathscr{E}$, $\mathscr{H}$. ▯

See Problem 8.1.

THEOREM 8.4. *A necessary and sufficient condition that* $\vec{\mathrm{f}} \in \mathscr{D}'^{l}$ *(or* $\vec{\mathrm{f}} \in \mathscr{D}'^{l}_{\mathrm{F}}$*),* $|\mathrm{D}|\vec{\mathrm{f}} = 0$ *imply that* $\vec{\mathrm{f}} \in \mathscr{E}^{l}$ *is the existence of an* $a > 0$ *so that, if* $\mathfrak{V}^i = (V_1^i, \partial_1^i, \ldots; V_{a_i}^i, \partial_{a_i}^i)$ *then for any* z *in any* V_j^i

$$|\mathrm{Im}\, z| \geqslant a(1 + |\mathrm{Re}\, z|^a). \tag{8.4}$$

PROOF. The sufficiency of the condition follows immediately from Theorem 8.1 and our descriptions of the AU structures for $\mathscr{D}'$, $\mathscr{E}$. The necessity of a condition which seems weaker than (8.4), namely

$$|\mathrm{Im}\, z| \geqslant b \log(1 + |z|) \tag{8.4$'$}$$

for all b, for $|z|$ large, can be proved by the method of lacunary series. The equivalence of (8.4) and (8.4′) is proven by Hörmander [1] using Seidenberg's decision process. We shall omit the details.

See Remarks 8.3 and 8.4.

We wish to give a different type of example. Namely, we give an example in which we can obtain additional regularity by imposing growth conditions. Let $r = l = 1$; let D be the heat operator $D = \partial/\partial x_1 - \partial^2/\partial x_2^2$. (Although the heat operator satisfies the hypothesis of Theorem 8.4, we shall make no use of this fact. Indeed, the type of results we obtain for the heat operator apply to some D which do not satisfy that hypothesis.) We choose for $\mathscr{W}$ the space of indefinitely differentiable functions which grow in x_2 like $\exp(x_2^2)$, and we set $\mathscr{W}_1 = \mathscr{H}$. More precisely, in the notation of Chapter V, Example 7 we let $\Phi_2(x_2) = x_2^2$, let

$$\Phi_1(x_1) = \begin{cases} 1 & \text{for } |x_1| \leqslant 1 \\ \infty & \text{for } |x_1| > 1, \end{cases}$$

and we set $\Phi(x_1, x_2) = \Phi_1(x_1) + \Phi_2(x_2)$. We then define $\mathscr{E}(\Phi)$ as in Example 7 of Chapter V. (The fact that Φ_1 is discontinuous gives no difficulty in this case.) The space $\mathscr{E}(\Phi)$ is again LAU. $\Psi_1(y_1) = |y_1|$, $\Psi_2(y_2) = y_2^2/4$, and $\Psi(y_1, y_2) = \Psi_1(y_1) + \Psi_2(y_2)$.

We have $\mathfrak{V} = (V, \text{identity})$ where V is the set of zeros of $iz_1 + z_2^2$. If we represent a point on V in the form (iz_2^2, z_2), then the functions k of an AU structure for $\mathscr{E}(\Phi)$ are characterized on V by the fact that they dominate all

$$(1 + |z_2|)^a \exp\{a[|\mathrm{Re}\, z_2^2| + |\mathrm{Im}\, z_2|^2]\},$$

while those of an AU structure for $\mathscr{H}$ are characterized on V by the fact that they dominate all

$$(1 + |z_2|)^a \exp [a(|z_2|^2 + |z_2|)]. \tag{8.5}$$

Since these conditions are clearly equivalent, we derive the (well-known) result that a solution of the heat equation is entire in x_1 and x_2 if for fixed x_1 it is $0(\exp(\varepsilon\, x_2^2))$ for every $\varepsilon > 0$. (Compare the proof of Theorem 5.26 which shows that the converse is also true.)

The above example, as mentioned above, can be extended to obtain a general result applying to an arbitrary D. This general result is actually a corollary of the proof of Theorem 5.2.6.

VIII.3. Hyperbolic Systems

The classical theory of hyperbolic partial differential equations is concerned, to a large extent, with Cauchy's problem. It is noted in Ehrenpreis [5] that a suitable generalization of hyperbolicity to convolution equations can be made in terms of an extension problem ($r = l = 1$): The equation $Dw = 0$ is hyperbolic in x_1 if, for some $a > 0$, every w which is indefinitely differentiable in the strip $|x_1| < a$ and satisfies $Dw = 0$ there can be uniquely extended to a $\tilde{w}$ which is indefinitely differentiable for all x and satisfies $D\tilde{w} = 0$ in all of R. It is shown in Ehrenpreis [5] that this condition is independent of a and is equivalent to the classical definition of hyperbolicity.

We are, therefore, led to the following

DEFINITION. The system $(r, l \geqslant 1)\, \boxed{\mathrm{D}}\vec{\mathrm{w}} = 0$ is called *hyperbolic in* $x_1, \ldots, x_q (q < n)$ if there exists an $a > 0$ so that every $\vec{\mathrm{w}}$ which is defined and indefinitely differentiable for all $x_{q+1}, \ldots, x_n$ and for $|x_1| < a, \ldots, |x_q| < a$ and satisfies $\boxed{\mathrm{D}}\vec{\mathrm{w}} = 0$ for these values can be uniquely extended to a $\vec{\tilde{w}} \in \mathscr{E}^r$, which satisfies $\boxed{\mathrm{D}}\vec{\tilde{w}} = 0$ on all of R. The $x_1, \ldots, x_q$ are called *time variables* for $\boxed{\mathrm{D}}$ and $x_{q+1}, \ldots, x_n$ are called *space variables*.

Another concept of hyperbolicity for systems based on a Cauchy problem is considered in Chapter IX below.

We regard hyperbolicity as a comparison property. Namely, denote by $\mathscr{E}(q; a)$ the space of indefinitely differentiable functions on

$$|x_1| < a, \ldots, |x_q| < a \quad \text{all} \quad x_{q+1}, \ldots, x_n .$$

Then $\boxed{\mathrm{D}}$ is hyperbolic in $x_1, \ldots, x_q$ if and only if the kernel of $\boxed{\mathrm{D}}$ in $\mathscr{E}^l(q; a)$ is equal to its kernel in $\mathscr{E}^l$ (for some a). Thus, we want equality of the kernels rather than having one contained in the other. (For the latter concept see Section VIII.4)

THEOREM 8.5. *A necessary and sufficient condition that* $\boxed{\mathrm{D}}$ *be hyperbolic in* $x_1, \ldots, x_q$ *is the existence of a* $b > 0$ *so that, if* $\mathfrak{V}^i = (V_1^i, \partial_1^i, \ldots; V_p^i, \partial_p^i)$, *for any* $z \in V_j^i$ *and any* j *we have for* $t = 1, \ldots, q$

$$(8.6) \qquad |\mathrm{Im}\, z_t| \leqslant b(1 + |\mathrm{Im}\, z_{q+1}| + \cdots + |\mathrm{Im}\, z_n|).$$

The property of hyperbolicity is independent of a.

PROOF. We use Example 5 of Chapter V. As mentioned at the end of that example, the method applies to show also that $\mathscr{E}(q; a)$ is LAU. An AU structure for $\mathscr{E}(q; a)$ consists of all k which dominate all functions of the form

$$(1 + |z|)^c \exp [c'(|\mathrm{Im}\, z_1| + \cdots + |\mathrm{Im}\, z_q|) + c(|\mathrm{Im}\, z_{q+1}| + \cdots + |\mathrm{Im}\, z_n|)]$$

for all c and for $c' < a$. Theorem 8.1 then shows that every solution in $\mathscr{E}^l(q; a)$ can be extended to a solution in $\mathscr{E}^l$—but perhaps not uniquely.

In order to prove the uniqueness we use, Theorem 4.2 directly: It follows from (8.6) and the proof of Theorem 8.1 that $\lambda_{\boxed{\mathrm{P}}}\mathbf{E}'^l(q; a) = \lambda_{\boxed{\mathrm{P}}}\mathbf{E}'^l$. Thus, by the method of proof of Theorem 8.1 we see that the kernels of $\boxed{\mathrm{D}}$ in $\mathscr{E}^l(q; a)$ and in $\mathscr{E}^l$ are the same.

For the converse, we can apply the method of lacunary series (see Chapter XII) to deduce the necessity of the apparently weaker condition

$$(8.6') \qquad |\mathrm{Im}\, z_t| \leqslant b[1 + |\mathrm{Im}\, z_{q+1}| + \cdots + |\mathrm{Im}\, z_n| + \log (1 + |z|)].$$

To prove the equivalence of (8.6) with (8.6′) the method of Gårding applies easily (see Gårding [1]). Since we have nothing to add, we shall omit the details.

Since (8.6) is independent of a, the property of hyperbolicity must be also. ▯

See Remark 8.5.

Actually, the case of several functions ($l > 1$) can be reduced to the case $l = 1$ by linear algebra. Moreover, the case $l = 1$, $r > 1$ can be reduced to the case $l = 1$, $r = 1$ (that is, one equation, one operator) by

PROPOSITION 8.6. *Let* $\vec{\mathrm{D}}$ $(l = 1)$ *satisfy* (8.6). *For any linear variety* Λ *in* R *of dimension* $n - q + 1$ *which contains the set* $x_1 = x_2 = \cdots = x_q = 0$, *there is a* $\delta = \delta_\Lambda$ *which belongs to the ideal generated by the* D_i *such that* δ *depends only on the variables along* Λ *and, as such, is hyperbolic with space variables* $x_{q+1}, \ldots, x_n$. *In particular,* $\dim V \leqslant n - q$ *so* $r \geqslant q$.

Thus, a solution w of the system $\vec{\mathrm{D}}w = 0$ satisfies the hyperbolic equation $\delta_\Lambda w = 0$ on Λ.

PROOF. We normalize the coordinates so that Λ is the linear variety determined by $x_1 = x_2 = \cdots = x_q = 0$ and the x_1 axis, that is, Λ is defined by $x_2 = \cdots = x_q = 0$.

It is an immediate consequence of (8.6) that the projection proj(V) of V on Λ omits an open set, namely it omits points with $|\text{Im } z_1|$ large and $|\text{Im } z_{q+1}|, \ldots, |\text{Im } z_n|$ small. However, it is known that the closure of proj(V) is an algebraic variety (see Samuel [1]). Thus, there is a polynomial $\hat{\delta}'(z_1, z_{q+1}, z_{q+2}, \ldots, z_n)$ whose zeros contain proj(V), and hence, when $\hat{\delta}'$ is thought of as a polynomial in $(z_1, \ldots, z_n)$, its zeros contain V. By Hilbert's Nullstellensatz some power of $\hat{\delta}'$, say $\hat{\delta}$, belongs to the ideal generated by the P_i. Finally, by using the results of Lech [1] we can make sure that δ', and hence, δ is hyperbolic in x_1. Lech's method shows how to construct a polynomial whose variety $\tilde{V}$ of zeros contains a given algebraic variety V_1 and such that on $\tilde{V}$ the coordinates satisfy similar types of inequalities as do the coordinates on V_1. In the present case we set $V_1 = \text{proj}(V)$. Lech's method applies because (8.6) still holds on proj(V). We shall omit the details as they are somewhat complicated, and we have nothing to add to Lech's construction.

The statement about the dimension is a consequence of the fact that the condition given by Theorem 8.5 allows Λ to be varied somewhat, Thus (see Samuel [1]) if dim $V > n - q$, then the projection of V on some Λ would be all of Λ except a proper subvariety which contradicts the above argument. ▯

It is a classical result (see Gårding [1]) that, if δ is a differential operator in $x_1, x_{q+1}, \ldots, x_n$ which is hyperbolic in x_1, there exist fundamental solutions $E^{\pm}$ with support in forward (backward) light cones respectively:

$$x_1 \geqslant 0 \qquad (x_1 \leqslant 0), \qquad (|x_{q+1}| + \cdots + |x_n|) \leqslant c|x_1|$$

for some $c > 0$. (This can be shown by the method of construction of fundamental solutions used in the proof of Theorem 11.10.)

Conversely, the existence of such fundamental solutions implies that δ is hyperbolic with time variable x_1. The proof of the converse goes as follows: Let λ be a point in Λ with $\lambda_1 \neq 0$, say $\lambda_1 > 0$. Let χ^- be the characteristic function (on Λ) of $x_1 \leqslant 0$. We form $\chi^-(\tau_{-\lambda} E^+)$ where $\tau_{-\lambda} E^+$ is obtained from E^+ by translating the origin to $-\lambda$. (It is not difficult to show that the product $\chi^- \tau_{-\lambda} E^+$ is defined). Now, it is easily seen that

$$\delta(\chi^- \tau_{-\lambda} E^+) = \delta_{-\lambda} - \eta$$

where $\delta_{-\lambda}$ is the unit mass at $-\lambda$ and η has its support on $x_1 = 0$.

If $f \in \mathscr{E}(\Lambda)$ and $\delta f = 0$, then

$$f(\lambda) = (f * \delta_{-\lambda})(0) = (f * \eta)(0).$$

Since η has its support on $x_1 = 0$, $(f * \eta)(0)$ depends only on the values of f on $x_1 = 0$. Thus, the above formula shows how to express $f(\lambda)$ in terms of f (and some derivatives) on $x_1 = 0$. Conversely, the formula shows how to extend f to all of Λ if it is a solution near $x_1 = 0$. □

This method of extension is a form of the parametrix method described in Section 1 above.

PROBLEM 8.2. (EXERCISE)

Carry out the details of this argument. Use it to show that Proposition 8.6 implies Theorem 8.5.

Remark 8.6. Suppose $r = l = 1$. One might think that the existence of a fundamental solution with support in a forward (or backward) cone contradicts Theorem 8.5. Let D be hyperbolic in x_1 and let E^- be a fundamental solution for D with support in a backward light cone. Let $x_1^0 > 0$ and consider $\tau_{x^0} E^-$. The restriction of $\tau_{x^0} E^-$ to the strip $|x_1| < x_1^0/2$ satisfies $D\tau_{x^0} E^- = 0$. (The fact that $\tau_{x^0} E^-$ is a distribution rather than a function in $\mathscr{E}$ presents no difficulties; we can regularize E^- by convolving it with a function in $\mathscr{D}$ of small support. We could also prove an analog of Theorem 8.5 for $\mathscr{D}'$ replacing $\mathscr{E}$.) What is the extension of this restriction to all of R? One might think it would have to be $\tau_{x^0} E^-$ which is not in the kernel of D. The answer is that the extension is $\tau_{x^0}E^- - \tau_{x^0}E^+$.

It is interesting to observe that if we give up our insistance that the fundamental solution belong to the space $\mathscr{D}'$, then we can give nonhyperbolic operators which have fundamental solutions having supports in cones. For example, let $n = 2$, $r = l = 1$, $D = \partial^2/\partial x_1^2 - \partial/\partial x_2$. A "fundamental solution" for D is

$$e^+ = \sum_{m=0}^{\infty} \frac{x_1^{2m}}{(2m)!} \frac{\partial^m}{\partial x_2^m} S,$$

where S is the product of the δ function in x_2 by the function which is 0 for $x_1 < 0$ and x_1 for $x_1 \geqslant 0$. The fact that e^+ is a fundamental solution for D is verified by a simple direct computation. The existence of such an e^+ whose support is on the halfline $x_2 = 0$, $x_1 \geqslant 0$ is related to the fact that there is a well-posed Cauchy problem for D in a suitable framework (compare the proof of Theorem 5.26).

There is an analog of Theorem 8.5 for analytic functions. Denote by $\mathscr{H}(q; a)$ the space of functions which are analytic for all complex $x_{q+1}, \ldots, x_n$, and for $|x_j| < a$ for $j = 1, \ldots, q$. Reasoning as in Theorem 8.5, we have

THEOREM 8.7. *A necessary and sufficient condition that every solution $\vec{w}$ of $\boxed{D}\,\vec{w} = 0$ in $\mathscr{H}^l(q; a)$ can be extended to $\mathscr{H}^l$ is that there exists a $b > 0$ so that*

$$|z_t| \leqslant b(1 + |z_{q+1}| + \cdots + |z_n|) \tag{8.7}$$

for $t = 1, q, \ldots, q$, whenever $z \in$ some V_j^i.

The same type of reasoning can be used to determine when solutions in $\mathscr{E}(\Phi)$ must belong to $\mathscr{E}(\Phi')$ for given Φ, Φ'. (We are using the notation of Example 7 of Chapter V.) Since $\mathscr{E}$ is like $\mathscr{E}(\Phi)$ except that now Φ is infinite outside $|x| < 1$, we may regard the above comparison between $\mathscr{E}(\Phi)$ and $\mathscr{E}(\Phi')$ as being in the same spirit as hyperbolicity.

The same method applies also to *semihyperbolicity* (cf. Ehrenpreis [5]), that is, where we do not require extension of a solution $w \in \mathscr{E}(q; a)$ to all x, but only to $x_1 > -a, \ldots, x_q > -a$.

See Problem 8.3.

VIII.4. Quasihyperbolicity

The heat equation (that is, $r = l = 1$, $D = \partial/\partial x_1 - \partial^2/\partial x_2^2$) has the following interesting property which can be proved by an analysis of the Cauchy problem on the time axis (that is, $x_2 = 0$): Let f be indefinitely differentiable on $|x_1| \leqslant a$ with $Df = 0$. Then there is a $g \in \mathscr{E}$ such that $g = f$ on $|x_1| \leqslant a$ and $Dg = 0$ in all of R. (However, g is *not* uniquely determined by f and a.) To prove this we want to extend the Cauchy data of f on $x_2 = 0$ from $|x_1| \leqslant a$ to the whole axis. In order to do this we must know three things:

1. Characterize the possible Cauchy data of solutions in $|x_1| \leqslant a$ and for solutions on all of R.

If follows easily from the method of proof of Theorem 5.26 that the Cauchy data consist of all pairs of functions (f_1, f_2) of x_1 (for $|x_1| \leqslant a$ and for all x_1, respectively) which satisfy inequalities of the form

$$|d^m f_j(x_1)/dx_1^m| \leqslant A\varepsilon^m (2m)! \tag{8.8}$$

for every $\varepsilon > 0$ uniformly in $|x_1| \leqslant a$ (respectively, uniformly on compact sets in which case A can depend on the compact set).

2. Prove that functions f_j defined on $|x_1| \leqslant a$ and satisfying such inequalities can be extended to functions g_j on the whole x_1 axis which satisfy the same type of inequalities.

This is done in Chapter XIII, Theorem 13.16.

3. If we denote by g the solution of the Cauchy problem with Cauchy data (g_1, g_2), then $g = f$ on $|x_1| \leqslant a$.

This follows from the Holmgren uniqueness theorem (see Section IX.9). We are led to the following general $(r, l \geqslant 1)$

DEFINITION. We say that $\boxed{\mathrm{D}}$ is *quasihyperbolic* in $x_1, \ldots, x_q$ if there is an $a > 0$ such that any solution $\vec{\mathrm{f}} \in \mathscr{E}^l(q; \bar{a})$ of $\boxed{\mathrm{D}}\vec{\mathrm{f}} = 0$ coincides on $|x_1| \leqslant a, \ldots, |x_q| \leqslant a$ with a $\vec{\mathrm{g}} \in \mathscr{E}^l$ which satisfies $\boxed{\mathrm{D}}\vec{\mathrm{g}} = 0$ in all of R.

Here $\mathscr{E}^l(q; \bar{a})$ is the space of indefinitely differentiable functions on $|x_1| \leqslant a, \ldots, |x_q| \leqslant a$. For $\vec{\mathrm{f}} \in \mathscr{E}^l(q; \bar{a})$ we write $\boxed{\mathrm{D}}\vec{\mathrm{f}} = 0$ to mean $\boxed{\mathrm{D}}\vec{\mathrm{f}} = 0$ on the interior of this strip.

Observe that there are two important differences between hyperbolicity and quasihyperbolicity:

1. In hyperbolicity we require extension from the open strip $|x_1| < a, \ldots, |x_q| < a$, while in quasihyperbolicity we require extension from the closed strip $|x_1| \leqslant a, \ldots, |x_q| \leqslant a$.
2. In hyperbolicity we require uniqueness of the extension while in quasihyperbolicity we do not.

Thus the operator which is identically zero is quasihyperbolic (for any q) but is not hyperbolic.

The requirements 1 and 2 are not independent. Using the technique of Chapter IX it is possible to show that, assuming quasihyperbolicity, either of the two requirements for hyperbolicity implies the other. For example, suppose the uniqueness condition 2 holds. In order to extend f from the open strip $|x_1| < a, \ldots, |x_q| < a$, we extend f to $\tilde{f}$ from the closed strips $|x_1| \leqslant ja/(j+1), \ldots, |x_q| \leqslant ja/(j+1)$ for each j, to all of R. By uniqueness, $\tilde{f}$ is independent of j, so $\tilde{f} = f$ on each of these strips, hence on $|x_1| < a, \ldots, |x_q| < a$.

The proof that Condition 1 implies Condition 2 is more complicated and involves an analysis of the multiplicity variety $\vec{\mathfrak{V}}$.

See Problem 8.4 (Exercise)

It is also possible to define a different type of hyperbolicity, say *metahyperbolicity*, by requiring that any $\vec{\mathrm{f}} \in \mathscr{E}^l(q; \bar{a})$ which can be extended to a solution in $\mathscr{E}^l(q; a')$ for some $a' > a$ can be extended to a solution in all of R.

See Problem 8.5.

Using Theorem 8.1 as in the proof of Theorem 8.5 we deduce

THEOREM 8.8. *$\boxed{\mathrm{D}}$ is quasihyperbolic in $x_1, \ldots, x_q$ if there is a collection $\{\sigma_j^i\}$ which is $\mathscr{E}(q; \bar{a})$ sufficient for $\vec{\mathfrak{V}}$ and a $b > 0$ so that for $t = 1, \ldots, q$*

$$|\mathrm{Im}\ z_\lambda| \leqslant b(1 + |\mathrm{Im}\ z_{q+1}| + \cdots + |\mathrm{Im}\ z_n|) \tag{8.9}$$

whenever z lies in some σ_j^i.

(The concept of sufficiency is defined in Chapter VII.)

Let us illustrate Theorem 8.8 by the heat equation $\mathfrak{V} = (V, \text{identity})$ where V is the set $iz_1 + z_2^2 = 0$. Thus, D is quasihyperbolic in x_1 if there is a $b > 0$ so that the set of $(z_1, z_2) \in V$ with

(8.10) $|\text{Im } z_1| \leqslant b(1 + |\text{Im } z_2|)$

is $\mathscr{E}(1; \bar{a})$ sufficient for V. (8.10) is

$$|\text{Im } (iz_2^2)| \leqslant b(1 + |\text{Im } z_2|)$$

or

(8.11) $|(\text{Re } z_2)^2 - (\text{Im } z_2)^2| \leqslant b(1 + |\text{Im } z_2|).$

Now, (8.11) is certainly satisfied if $(\text{Re } z_2)^2 - (\text{Im } z_2)^2 = 0$, that is, $\text{Re } z_2 = \pm \text{Im } z_2$. We denote by σ the set of points $(-iz_2^2, z_2)$ with $\text{Re } z_2 = \pm \text{Im } z_2$; we claim that σ is $\mathscr{E}(1; \bar{a})$ sufficient for V for any a.

To prove this, we consider the following way of identifying functions on V with functions of z_2: For any G defined on V we define $\nu(G)(z_2) = G(iz_2^2, z_2)$. Given any H defined on the z_2 plane, we set $\xi(H)(iz_2^2, z_2) = H(z_2)$. Clearly, $\nu\xi = \text{identity} = \xi\nu$.

We find easily that $\nu[\mathbf{E}'(1; \bar{a})(V)]$ is the space of entire functions H of z_2 which satisfy

$$|H(z_2)| \leqslant \text{const } (1 + |z_2|)^N \exp (a|\text{Re } (z_2^2)| + N|\text{Im } z_2|)$$

for some N. The functions of an AU structure for $\nu[\mathbf{E}'(1; \bar{a})(V)]$ must dominate all

$$(1 + |z_2|)^N \exp(a |\text{Re } (z_2^2)| + N |\text{Im } z_2|).$$

That the set $\text{Re } z_2 = \pm \text{Im } z_2$ is sufficient for any a can be deduced by an harmonic majorant argument as in the proof of Theorem 5.5. $\square$

Remarks

Remark 8.1.   See page 200.

Though we shall not go into the details here, under certain circumstances there is a "duality" between the methods (a) and (c), which may be thought of as a generalization of the Serre duality theorem (see Serre [2]). (See Section XI.2 for an example.)

Remark 8.2.   See page 204.

In Ehrenpreis [4] several variations of this question are studied in case $l = r = 1$.

Remark 8.3.   See page 205.

In case $r = l = 1$, Theorem 8.3 is due to Petrowsky [1]. Theorem 8.4 for $r = l = 1$ was proved by Hörmander [1] and (independently) by the author in a form suitable for convolution equations (see Ehrenpreis [4]). The case $r, l > 1$ can be reduced to $r = l = 1$ by a result of Lech [1] (see Hörmander [4] where this is carried out).

Remark 8.4. See page 205.
Theorems 8.3 and 8.4 can be proved by the construction of a suitable fundamental solution (see Section VI.2 and Ehrenpreis [4]).

Remark 8.5. See page 207.
It is to be noted that the conditions for ellipticity and hyperbolicity are opposite, the former asserting the largeness of Im z on V and the latter its smallness.

Remark 8.6. See page 209.

Problems

PROBLEM 8.1. (Exercise) See page 205.
Carry out the details of the lacunary series arguments in the proofs of Theorems 8.3, 8.4, and 8.5.

PROBLEM 8.2. See page 209.

PROBLEM 8.3. See page 210.
Study the analogs for $r > 1$ of the cones of hyperbolic time directions which occur in the classical theory (see e.g. Gårding [1]). In particular, is semihyperbolicity the same as hyperbolicity? (In case $r = 1$ the fact that the two concepts are the same is due to Gårding [1].)

PROBLEM 8.4. (Exercise) See page 211.
Carry out the details.

PROBLEM 8.5. See page 211.
Is quasihyperbolicity the same as metahyperbolicity?

CHAPTER IX

General Theory of Cauchy's Problem

Summary

Let $\vec{\mathrm{D}}=(D_1, \ldots, D_r)$, let $\vec{\mathrm{P}}=(P_1, \ldots, P_r)$ be the Fourier transform of its adjoint, and let $\mathfrak{V}$ be a multiplicity variety associated to $\vec{\mathrm{P}}$ by the results of Chapter IV. Let $\mathscr{W}$ be an LAU space, and denote by ${}_{\vec{\mathrm{D}}}\mathscr{W}^r$ the image in $\mathscr{W}^r$ of the map $f \to \vec{\mathrm{D}}f=(D_1 f, \ldots, D_r f)$ for $f \in \mathscr{W}$; ${}_{\vec{\mathrm{D}}}\mathscr{W}^r$ is described in Chapter VI. Call $d=$ dimension $\mathfrak{V}$. For $j=0, 1, \ldots, d$ let T^j be a plane through the origin of dimension j. Suppose the restriction maps $f \to f|_{T^j}$ are defined for all j and map $\mathscr{W}$ onto an AU space which we denote by ${}^j\mathscr{W}$ or $\mathscr{W}(T^j)$. For each j we choose a sequence of linear constant coefficient differential operators q_{ij} for $i=1, 2, \ldots, l^j$, the Fourier transforms of their adjoints being written Q_{ij}. Then the *Cauchy problem* consists of determining conditions under which the map

$$\gamma : f \to (D_1 f, \ldots, D_r f;\ q_{10} f|_{T^0}, \ldots, q_{l^d,d} f|_{T^d})$$

is a topological isomorphism of $\mathscr{W}$ onto ${}_{\vec{\mathrm{D}}}\mathscr{W} \oplus ({}^0\mathscr{W})^{l^0} \oplus \cdots \oplus ({}^d\mathscr{W})^{l^d}$. If this is the case we say the Cauchy problem is *well posed*.

Let us denote by γ' the adjoint of γ and by $\hat{\gamma}'$ the Fourier transform of γ'. We use the variables ξ^j on T^j and t^{n-j} orthogonal to T^j. The dual variables under Fourier transform are denoted by w^j and by s^{n-j}, respectively. Then we prove: *A necessary and sufficient condition that the Cauchy problem be well-posed is that*

$$\hat{\gamma}'_{\mathfrak{V}} : (L_{10}(w^0), \ldots, L_{l^d,d}(w^d)) \to \text{restriction to } \mathfrak{V} \text{ of } \sum Q_{ij}(s^{n-j}) L_{ij}(w^j)$$

be a topological isomorphism of $({}^0\mathbf{W}')^{l^0} \oplus \cdots \oplus ({}^d\mathbf{W}')^{l^d}$ *onto* $\mathbf{W}'(\mathfrak{V})$.

In case $r=1$, $\gamma'_{\mathfrak{V}}$ is related to the Lagrange interpolation formula. In the general case $r \geqslant 1$, $\hat{\gamma}'_{\mathfrak{V}}$ is related to an extension of the Lagrange interpolation formula.

For general $\mathscr{W}$ there may not exist any well-posed Cauchy problem. For a class of spaces called *CK spaces* there always exist well-posed

Cauchy problems for $\vec{\mathrm{D}}$. This may be regarded as a generalization of the Cauchy-Kowalewski theorem, hence, the terminology "*CK space.*" $\mathscr{W}$ is a *CK* space if it is LAU and if

1. There exists an AU structure K consisting only of functions of $|z|$ which are monotonically increasing.
2. For any $F \in \mathbf{W}'$ the Taylor series for F at any point converges to F in the topology of $\mathbf{W}'$.
3. For any $a > 0$ and any $k \in K$ there is a $k' \in K$ with $k'(a|z|) \leqslant k(|z|)$ for all z.

For examples of *CK* spaces we consider $x = (x_1, \ldots, x_n)$ as complex variables. The spaces of entire functions, of functions analytic in a neighborhood of the origin, and of entire functions of finite order provide examples of *CK* spaces.

We prove: *Let $\mathscr{W}$ be a CK space. Then there exists a well-posed Cauchy problem for $\vec{\mathrm{D}}$.*

When we have a well-posed Cauchy problem for $\vec{\mathrm{D}}$ we call $(T^0, \ldots, T^d)$ a *noncharacteristic for* $\vec{\mathrm{D}}$. We think of T_j as being the empty set if there are no q_{ij}. T^d is called a *principal noncharacteristic for* $\vec{\mathrm{D}}$ if there exists a *noncharacteristic for $\vec{\mathrm{D}}$ of the form $(T^0, \ldots, T^{d-1}, T^d)$. A necessary and sufficient condition that T^d be a principal noncharacteristic for $\vec{\mathrm{D}}$ is the existence of a $c > 0$ so that for $(w^d, s^{n-d}) \in \mathfrak{V}$* (the multiplicity variety associated to the Fourier transform $\vec{\mathrm{P}}$ of $\vec{\mathrm{D}}$ by Theorem 4.1) *we have*

$$|s^{n-d}| \leqslant c(1 + |w^d|).$$

Given that T^d is a principal noncharacteristic for $\vec{\mathrm{D}}$ we can determine a new system ${}^{d-1}\vec{\mathrm{D}}$ so that T^{d-1} is a principal noncharacteristic of ${}^{d-1}\vec{\mathrm{D}}$ if and only if there exists a noncharacteristic $(T^0, \ldots, T^{d-2}, T^{d-1}, T^d)$ of $\vec{\mathrm{D}}$. ${}^{d-1}\vec{\mathrm{D}}$ depends on T^d as well as on $\vec{\mathrm{D}}$. Proceeding in this way we construct a *derived sequence* $(\vec{\mathrm{D}} = {}^{d}\vec{\mathrm{D}}, {}^{d-1}\vec{\mathrm{D}}, \ldots, {}^{0}\vec{\mathrm{D}})$ of $\vec{\mathrm{D}}$ and the problem of constructing noncharacteristics can be reduced to that of constructing principal noncharacteristics which is solved by the above.

We call $\vec{\mathrm{D}}$ *Cauchy hyperbolic* (CH) if there exists a well-posed Cauchy problem for the space $\mathscr{E}$ of indefinitely differentiable functions (see Section IX.3). Let T^d be a principal noncharacteristic for $\vec{\mathrm{D}}$. We say that $\vec{\mathrm{D}}$ is *principally Cauchy hyperbolic* (PCH) if the set of $(q_{1d}f|_{T^d}, \ldots, q_{l^d,d}f|_{T^d})$ for $f \in \mathscr{E}$, $Df = 0$, is equal to $(\mathscr{E}(T^d))^{l^d}$. Then we have: *A necessary and sufficient condition that $\vec{\mathrm{D}}$ be* PCH *for the principal noncharacteristic*

T^d *is that* $\vec{D}$ *be hyperbolic in* t^{n-d} (in the sense of Chapter VIII). Thus, the problem of CH is reduced to hyperbolicity and to derived sequences.

In Section IX.4 we discuss the initial value problem instead of the Cauchy problem.

We show in Section IX.5 how certain well-posed Cauchy problems lead to $n-d$ parameter commutative groups. This happens if each $T^j \subset T^d$. The group is the group of transformations of the Cauchy data on $t^{n-d}=0$ onto the "same" Cauchy data on $t^{n-d}=$ const. In case it is an "initial value problem" which is well-posed instead of the Cauchy problem we get an $n-d$ parameter commutative semigroup. We study the infinitesimal generators of these groups and semigroups.

Next, in Section IX.6, we study the image of γ (see above) on $\mathscr{E}$. We also generalize the result that (for $n=1$) every $f \in \mathscr{E}$ is of the form $f^+ + f^-$ where $f^+(f^-)$ is analytic in the upper (lower) half of the complex plane and indefinitely differentiable in the closure of this half plane: Suppose for simplicity that $r=1$, and that $D=D_1$ is of degree m in $t=t^1$, and that the coefficient of $\partial^m/\partial t^m$ in D is a constant. *Given any* $(f_0, \ldots, f_{m-1}) \in \mathscr{E}(\xi^{n-1})$ *there exist* $f^+ \in \mathscr{E}(t \geqslant 0)$ and $f^- \in \mathscr{E}(t \leqslant 0)$ such that $Df^{\pm}=0$ *and*

$$f_i(\xi) = \partial^i f^+(0, \xi)/\partial t^i + \partial^i f^-(0, \xi)/\partial t^i$$

for all i.

Using a *parametrix* which is a modification of a fundamental solution (see Chapter VI) we show in Section IX.7 how it is possible to solve the Cauchy problem for some nonlinear T^j.

In Section IX.8 we make some remarks on domain of dependence and lacunas for the Cauchy problem.

The problem we treat in Section IX.9 is that of the uniqueness of the Cauchy problem. In case $r=1$, Holmgren proved that if $T=T^d=T^{n-1}$ is a noncharacteristic for D, for $f \in \mathscr{E}$, $Df=0$ and $(q_{i,n-1}f)(0, \xi)=0$ for all i imply $f=0$. We prove a similar result for $r>1$.

Täcklind [1] proved a uniqueness theorem for the heat operator $D=\partial/\partial t - \partial^2/\partial \xi^2$. This result is not included in Holmgren's because $t=0$ is characteristic for D. Täcklind's result is as follows: *Let* $h(\xi)$ *be a positive function. Consider the subspace* M_h of $\mathscr{E}$ *consisting of those* f *which satisfy* $|f(t, \xi)| \leqslant \exp(c|\xi|h(|\xi|))$ *for some* c *uniformly on compact sets in* t. *(*c *may depend on the compact set.) A necessary and sufficient condition that no function except* 0 *in* M_h *can vanish on* $t=0$ *is*

$$\int d\xi/\bar{h}(\xi) = \infty.$$

Here $\bar{h}$ *is the largest nondecreasing minorant of* h.

We prove an extension of this result to arbitrary $\vec{D}$. Our proof is essentially the same as our proof of the generalization of Holmgren's theorem mentioned above.

Our next uniqueness theorem is an extension of the fact that a solution of an elliptic equation cannot vanish together with all its derivatives at a point. Courant and Hilbert [1] (Vol. II, pp. 427–430) and John [1] have extended this result for $r = 1$ by showing that there may exist subvarieties $\tilde{T}$ of $T^{n-1} = T$ of dimension $< n - 1$ such that $f \in \mathscr{E}$, $Df = 0$, f and all its derivatives vanish on $\tilde{T}$, imply $f = 0$. We prove a similar result for $r > 1$.

These results are capable of a further modification. Instead of requiring that f and all its derivatives vanish on $\tilde{T}$ we could require that $f \to 0$ very rapidly in a suitable neighborhood of $\tilde{T}$. In Section IX.10 we determine conditions under which such a rapid decrease implies that $f = 0$.

A by-product of our investigation is related to the following problem. Let $f \not\equiv 0$ be harmonic and bounded in $t = t^1 > 0$. Can $\lim_{t \to 0^+} \partial f(t, \xi)/\partial \xi_j = 0$ for $j = 1, \ldots, n - 1$ for ξ in a set of positive measure? We show in Section IX.11 that f can be continued to be analytic and bounded in a domain $\Omega \subset C$ which is independent of f and is maximal for this analytic continuation property. Ω contains a linear set Ω_1 of real dimension $n - 1$ in its boundary with the property that the vanishing (in a limit sense) of f and its first derivatives on a set of positive (real) $n - 1$ dimensional measure of Ω_1 implies $f = 0$.

In Section IX.12 we point out how to extend the results of this chapter to the case when $\vec{\mathrm{D}}$ is replaced by $\boxed{\mathrm{D}}$ which is an $r \times l$ matrix of linear constant coefficient partial differential operators.

IX.1. Formulations of the Problems

The classical Cauchy problem can be stated as follows: Let

$$D = \frac{\partial^m}{\partial x_1^m} + L_1\left(\frac{\partial}{\partial x_2}, \ldots, \frac{\partial}{\partial x_n}\right)\frac{\partial^{m-1}}{\partial x_1^{m-1}} + \cdots + L_m\left(\frac{\partial}{\partial x_2}, \ldots, \frac{\partial}{\partial x_n}\right),$$

where the L_j are polynomials. Let g be a function of x and let $f_0, \ldots, f_{m-1}$ be functions of $x_2, \ldots, x_m$. Determine f so that

$$Df = g$$

and

$$\left(\frac{\partial^j f}{\partial x_1^j}\right)(0, x_2, \ldots, x_n) = f_j(x_2, \ldots, x_n) \qquad \text{for} \quad j = 0, 1, \ldots, m - 1.$$

(Of course, the problem is not yet precisely stated because we have not stated what kind of functions f, g, f_j are admitted.)

There is a mapping associated with the Cauchy problem, namely,

$$(9.1)\qquad \gamma : f \to \left[Df;\ f(0, x_2, \ldots, x_n), \ldots, \left(\frac{\partial^{m-1}}{\partial x_1^{m-1}} f\right)(0, x_2, \ldots, x_n)\right].$$

Our point of view will be the study of γ by means of the Fourier transform of its adjoint. Before doing this we shall generalize the problem so as to apply to overdetermined systems. We shall consider at first only the case of one unknown function. We shall see that the Cauchy data can no longer be given on a single linear subspace, but we need several linear subspaces of different dimensions.

For $j = 0, 1, \ldots, n-1$ let T^j be a linear subspace of R through the origin of dimension j. For each j we introduce coordinates ξ^j on T^j and t^{n-j} orthogonal to T^j. The dual variables under Fourier transform are denoted by w^j, s^{n-j}, respectively. For each j we are given l^j constant coefficient differential operators q_{ij} ($\not\equiv 0$) in the variables t^{n-j}. The Fourier transform Q_{ij} of the adjoint of q_{ij} is a polynomial in s^{n-j}. For a function or distribution f on R we denote by $q_{ij}f|_{T^j}$ the restriction of $q_{ij}f$ to T^j. The assumption that $0 \in T^j$ for all j is made just for convenience. In Section IX.5 we shall use affine planes.

Let $\mathscr{W}$ be an LAU space of functions or distributions. For each j let ${}^j\mathscr{W}$ be an AU space on T^j. We assume that ${}^j\mathbf{W}'$ is exactly the space of restrictions of $\mathbf{W}'$ to the linear variety $\hat{T}^j$ defined by $s^{n-j} = 0$, that is ${}^j\mathbf{W}' = \mathbf{W}'(\hat{T}^j)$. We require, in fact that for $G \in {}^j\mathbf{W}'$ the extension $\tilde{G}(s^{n-j}, w^j) = G(w^j)$ belong to $\mathbf{W}'$ and the map $G \to \tilde{G}$ be a topological isomorphism of ${}^j\mathbf{W}'$ into $\mathbf{W}'$. In this case we can define the *restriction map* $g' \to g|_{T^j}$ for $g \in \mathscr{W}$, namely

$${}^jg' \cdot (g\,|_{T^j}) = g' \cdot g$$

for any ${}^jg' \in {}^j\mathscr{W}'$, where $\hat{g}'$ (the Fourier transform of g') is the extension of ${}^j\hat{g}'$ to C defined to be constant in s^j. The above shows that $g\,|_{T^j} \in {}^j\mathscr{W}$.

Let $\vec{\mathrm{D}} = (D_1, \ldots, D_r)$ be a system of linear constant coefficient partial differential operators; the Fourier transforms of their adjoints are $P_1, \ldots, P_r$. We denote by $\mathscr{W}_{\vec{\mathrm{D}}}$ the set of $g \in \mathscr{W}$ with $D_j g = 0$ for $j = 1, 2, \ldots, r$.

General Cauchy problem. Is the map

$$(9.2)\qquad \gamma : f \to (D_1 f, \ldots, D_r f;\ q_{10} f\,|_{T^0}, \ldots, q_{l^{n-1}, n-1} f\,|_{T^{n-1}})$$

a topological ismorphism of $\mathscr{W}$ onto ${}_{\vec{\mathrm{D}}}\mathscr{W}^r \oplus {}^0\mathscr{W}^{l^0} \oplus \cdots \oplus {}^{n-1}\mathscr{W}^{l^{n-1}}$, where ${}_{\vec{\mathrm{D}}}\mathscr{W}^r$ is the subspace of $\mathscr{W}^r$ (=r-fold direct sum of $\mathscr{W}$ with itself) which satisfies the compatibility conditions (see Theorem 6.1)? If this is the case we say that the Cauchy problem is *well-posed for* $\mathscr{W}$.

See Remark 9.1.

We compute the adjoint γ' of γ. This maps ${}_{\vec{D}}\mathscr{W}'^r \oplus {}^0\mathscr{W}'^{l^0} \oplus \cdots \to \mathscr{W}'$. Given ${}^1S, \ldots, {}^rS \in \mathscr{W}'$, $S_{ij} \in {}^j\mathscr{W}'^{r^j}$ we have

$$(9.3) \qquad \gamma'({}^1S, \ldots, {}^rS; S_{10}, \ldots) \cdot f = \sum {}^iS \cdot D_i f + \sum S_{ij} \cdot q_{ij} f|_{T^j}$$
$$= [\sum D_i' {}^iS + \sum q_{ij}' S_{ij}|^{T^j}] \cdot f,$$

where $S_{ij}|^{T^j}$ is the extension of S_{ij} to R (the adjoint of restriction to T^j).

Denote by $\hat{\gamma}'$ the Fourier transform of γ'. Then the above leads to the *Equivalent formulation of the general Cauchy problem.* Is the map

$$(9.4) \qquad \hat{\gamma}' : [{}^1L, \ldots, {}^rL; L_{10}(w^0), \ldots, L_{l^{n-1}, n-1}(w^{n-1})]$$
$$\to \sum P_i(z) {}^iL(z) + \sum Q_{ij}(s^{n-j}) L_{ij}(w^j)$$

a topological isomorphism onto $\mathbf{W}'$? [Actually this formulation is slightly stronger than the one given in (9.2), but the two formulations coincide for reflexive spaces which comprise the most interesting examples. In any case, we shall concern ourselves with (9.4).]

Let $\vec{\mathrm{P}}$ be the row vector $(P_1, \ldots, P_r)$ acting on the column space $\mathbf{W}'^r$. Denote by $\mathfrak{B}$ the multiplicity variety associated with the ideal generated by the P_i by Theorems 4.1 and 4.2. Let d be the complex dimension of $\mathfrak{B}$, that is, the maximum of the complex dimensions of the varieties of $\mathfrak{B}$.

LEMMA 9.1. *A necessary and sufficient condition that the Cauchy problem be well-posed is that the mapping*

$$(9.5) \qquad \hat{\gamma}'_{\mathfrak{B}} : [L_{10}(w^0), \ldots, L_{l^{n-1}, n-1}(w^{n-1})]$$
$$\to \text{restriction to } \mathfrak{B} \text{ of } \sum Q_{ij}(s^{n-j}) L_{ij}(w^j)$$

be a topological isomorphism onto $\mathbf{W}'(\mathfrak{B})$.

PROOF. Suppose first that $\hat{\gamma}'$ is a topological isomorphism. Then $\hat{\gamma}'_{\mathfrak{B}}$ is a continuous linear map. It is onto by Theorem 4.2. It is also one-to-one by Theorem 4.2 because every function which is zero on $\mathfrak{B}$ belongs to the ideal generated by the P_i, hence is not of the form $\sum Q_{ij} L_{ij}$ because $\hat{\gamma}'$ is one-to-one. Finally, $\hat{\gamma}'$ is topological. For, suppose $F \in \mathbf{W}'(\mathfrak{B})$, $F \to 0$. Then by Theorem 4.2 we can find H whose restriction to $\mathfrak{B}$ is F such that $H \to 0$ in $\mathbf{W}'$. We write $H = \hat{\gamma}'({}^iL; L_{ij})$ so ${}^iL \to 0$ and $L_{ij} \to 0$. Since $\hat{\gamma}'_{\mathfrak{B}}(L_{ij}) = F$, it follows that $\hat{\gamma}'_{\mathfrak{B}}$ is topological.

Conversely, suppose $\hat{\gamma}'_{\mathfrak{B}}$ is a topological isomorphism. Now $\hat{\gamma}'$ is clearly continuous. Moreover, $\hat{\gamma}'$ is one-to-one. For if $\hat{\gamma}'({}^iL; L_{ij}) = 0$, then $\hat{\gamma}'_{\mathfrak{B}}(L_{ij}) =$ restriction to $\mathfrak{B}$ of $\sum Q_{ij} L_{ij}$ is certainly 0. By the one-to-oneness of $\hat{\gamma}'_{\mathfrak{B}}$ this means that each $L_{ij} = 0$. Hence, $\sum P_i {}^iL = 0$, so $({}^iL)$ belongs to the module of relations of $\vec{\mathrm{P}}$. Thus, by the definition of ${}_{\vec{D}}\mathscr{W}'^r$, $({}^iL)$ is zero in $({}_{\vec{D}}\mathbf{W}^r)'$.

Next $\hat{\gamma}'$ is onto. For we can write any $H \in \mathbf{W}'$ as $\hat{\gamma}'(0; L_{ij}) + H^0$, where H^0 is zero on $\mathfrak{B}$, because $\hat{\gamma}'_{\mathfrak{B}}$ is onto. By Theorem 4.2 we may write $H^0 = \sum P_i\, {}^iL$ so $\hat{\gamma}'$ is onto.

Finally, $\hat{\gamma}'$ is topological. For, if $\hat{\gamma}'({}^iL; L_{ij}) \to 0$ in $\mathbf{W}'$, then certainly $\hat{\gamma}'_{\mathfrak{B}}(L_{ij}) \to 0$ in $\mathbf{W}'(\mathfrak{B})$. Thus, $L_{ij} \to 0$ because $\hat{\gamma}'_{\mathfrak{B}}$ is topological. Hence, $\sum P_i\, {}^iL \to 0$ in $\mathbf{W}'$ which implies (see Theorems 4.2 and 4.8) that $({}^iL) \to 0$ in $({}_{\vec{\mathrm{D}}}\mathbf{W}^r)'$.

Lemma 9.1 is therefore proved. ▯

Let d_0 be the largest number such that not all q_{id_0} are identically zero. Lemma 9.1 states that in the case of a well-posed Cauchy problem all functions in $\mathbf{W}'(\mathfrak{B})$ can be expressed as the restriction to $\mathfrak{B}$ of a finite sum of functions each of which is the product of a fixed function (namely, Q_{ij}) by a variable function depending on not more than d_0 variables w^j. Using the definition of the dimension of an algebraic variety (see e.g., Weil [1], Samuel [1]) we see easily that $d_0 = d$ if $\mathbf{W}'$ is dense in the space of entire functions, that is

LEMMA 9.2. *For a well-posed Cauchy problem,* $q_{ij} = 0$ *for* $j > d = \dim \mathfrak{B}$, *but not all* $q_{id} = 0$ *if* $\mathbf{W}'$ *is dense in the space of entire functions.*

We say that the set of planes $\{T^j\}$ is *regular for* $\vec{\mathrm{D}}$ if for each j we can find q_{ij} not all zero so that the Cauchy problem is well-posed for the space of formal power series. (We omit all T^j for which the corresponding $l^j = 0$.) By Chapter V, Example 8 and the above, this is equivalent to saying that every polynomial F can be uniquely expressed, mod $\vec{\mathrm{P}}\mathscr{P}^r$, in the form

$$F(z) = \sum Q_{ij}(s^{n-j}) L_{ij}(w^j).$$

Here the Q_{ij} are fixed polynomials and the L_{ij} are variable polynomials. ($\mathscr{P}$ is the ring of polynomials.)

In general, it is very difficult to say anything definitive about the general Cauchy problem as we have posed it. One of the sources of great difficulty is the fact that T^j may not be contained in T^{j+1}. We shall therefore make the

Assumption. $T^j \subset T^{j+1}$ for all j.

We assume that the coordinate systems on $\hat{T}^j$ are so chosen that $w^j_p = w^d_p$ for $p = 1, 2, \ldots, j$.

We shall have some occasion to consider the situation in which the assumption does not hold. That case will be called the *Goursat Problem* rather than the Cauchy Problem (see Section X.7).

We shall show in the next section that for many interesting spaces well-posed Cauchy problems exist for any $\vec{\mathrm{D}}$.

Suppose we are given $\vec{\mathrm{D}}$ and a well-posed Cauchy problem for the space of formal power series, and let $d = \dim \mathfrak{B}$. We use $\mathscr{P}/\vec{\mathrm{P}}\mathscr{P}^r$ to construct a vector space over the field $\mathscr{R}(\hat{T}^d)$ of rational functions on $\hat{T}^d$ which we again denote by $\mathscr{P}/\vec{\mathrm{P}}\mathscr{P}^r$; this is the quotient of $\mathscr{R}(\hat{T}^d)\mathscr{P}$ by $\mathscr{R}(\hat{T}^d)\vec{\mathrm{P}}\mathscr{P}^r$.

THEOREM 9.3. *The Q_{id} form a basis for this vector space. In particular, for all well-posed Cauchy problems for $\vec{\mathrm{D}}$ with T^d fixed, the number l^d is constant.*

PROOF. The second part of Proposition 9.3 is an immediate consequence of the first.

We note first that the Q_{id} are linearly independent over $\mathscr{R}(\hat{T}^d)$. For, if we had a nontrivial relation

$$\sum R_i(w^d)Q_{id}(s^{n-d}) + \sum U_i(w^d)\,{}^iL(z)P_i(z) = 0$$

with R_i, $U_i \in \mathscr{R}(\hat{T}^d)$ and ${}^iL$ polynomials, then multiplication by a suitable polynomial $R(w^d) \neq 0$ would yield a similar relation in which R_i, $U_i \in \mathscr{P}(\hat{T}^d)$. This is impossible because $\hat{\gamma}'$ is one-to-one for a well-posed Cauchy problem.

To prove that the Q_{id} span $\mathscr{P}/\vec{\mathrm{P}}\mathscr{P}^r$ we observe that the Q_{ij} for all i, j span $\mathscr{P}/\vec{\mathrm{P}}\mathscr{P}^r$ as follows from the fact that $\hat{\gamma}'$ is onto for a well-posed problem and, by our assumption, $T^j \subset T^d$ for all j. Thus, it suffices to express each $Q_{i_0 j_0}$ with $j_0 < d$ in terms of Q_{id} and P_i. Using the fact that $\hat{\gamma}'$ is onto we can write

$$w_d^d Q_{i_0 j_0}(s^{n-j_0}) = \sum P_i(z)\,{}^iL^0(z) + \sum Q_{ij}(s^{n-j})L^0_{ij}(w^j).$$

Since the coefficient $L^0_{i_0 j_0}$ of $Q_{i_0 j_0}$ on the right side of this formula does not involve w_d^d, this allows us to express $R_{i_0 j_0}(w^d)Q_{i_0 j_0}(s^{n-j_0})$ for some nontrivial polynomial $R_{i_0 j_0}(w^d)$ as a linear combination of the Q_{ij} with $(i, j) \neq (i_0, j_0)$ with coefficients in $\mathscr{P}(\hat{T}^j)$, and the $P_i(z)$ with coefficients in $\mathscr{P}$.

Now, let $j_1 < d$ and let $(i_1, j_1) \neq (i_0, j_0)$. Then we can write again

$$w_d^d Q_{i_1 j_1}(s^{n-j_1}) = \sum P_i(z)\,{}^iL^1(z) + \sum Q_{ij}(s^{n-j})L^1_{ij}(w^j).$$

Multiplying by $R_{i_0 j_0}$ gives

$$\begin{aligned}
&[w_d^d - L^1_{i_1 j_1}(w^{j_1})]R_{i_0 j_0}(w^d)Q_{i_1 j_1}(s^{n-j_1}) \\
&\quad = \sum R_{i_0 j_0}(w^d)P_i(z)\,{}^iL^1(z) + \sum_{(i,j)\neq(i_1,j_1)} R_{i_0 j_0}(w^d)Q_{ij}(s^{n-j})L^1_{ij}(w^j) \\
&\quad = \sum P_i(z)\,{}^i\tilde{L}(z) + \sum_{(i,j)\neq(i_0,j_0)} Q_{ij}(s^{n-j})\tilde{L}_{ij}(w^d)
\end{aligned}$$

by the above expression for $R_{i_0 j_0}(w^d)Q_{i_0 j_0}(s^{n-j_0})$. In the right side of the above formula, $\tilde{L}_{i_1 j_1}$ is a polynomial in w^{j_1} only, hence does not involve w_d^d.

On the other hand, $w_d^d - L^1_{i_1j_1}(w^{j_1})$ involves w_d^d in a nontrivial way since w^{j_1} does not involve w_d^d. Thus, as in the case of $Q_{i_0j_0}$, we may transpose the term on the right side involving $Q_{i_1j_1}$ and obtain a nontrivial polynomial $R_{i_1j_1}(w^d)$ such that $R_{i_1j_1}(w^d)Q_{i_1j_1}(s^{n-j_1})$ can be expressed in terms of the Q_{ij} for $(i, j) \neq (i_0, j_0)$ and $(i, j) \neq (i_1, j_1)$ with coefficients which are sums of polynomials in $\mathscr{P}(\hat{T}^j)$ with products of polynomials in $\mathscr{P}(\hat{T}^j)$ by $R_{i_0j_0}(w^d)$, and the $P_i(z)$ with coefficients in $\mathscr{P}$.

We claim that we can find that, if we order the (i, j) with $j < d$, say by writing (i_m, j_m), then for each m we can find a polynomial $R_{i_mj_m}(w^d)$ so that $R_{i_mj_m}(w^d)Q_{i_mj_m}(s^{n-j_m})$ can be expressed in terms of the Q_{ij} for $(i, j) \neq (i_0, j_0), (i_1, j_1), \ldots, (i_m, j_m)$ with coefficients in the ring generated by $\mathscr{P}(\hat{T}^j)$ and $R_{i_0j_0}, \ldots, R_{i_{m-1}j_{m-1}}$, but not involving any powers $(R_{i_qj_q})^a$ with a sufficiently large, say $a > a_0$, and $P_i(z)$ with coefficients in $\mathscr{P}$. To see this, suppose $m > 1$ and the result has been established for $0, 1, \ldots, m-1$. For any b we can write

$$(w_d^d)^b Q_{i_mj_m}(s^{n-j_m}) = \sum P_i(z)\, {}^iL^m(z) + \sum Q_{ij}(s^{n-j}) L^m_{ij}(w^j).$$

Multiplying by $R_{i_0j_0} R_{i_1j_1} \cdots R_{i_{m-1}j_{m-1}}$ and using our induction hypothesis, we arrive at an expression

$$(w_d^d)^b R_{i_0j_0}(w^d) R_{i_1j_1}(w^d) \cdots R_{i_{m-1}j_{m-1}}(w^d) Q_{i_mj_m}(s^{n-j_m})$$
$$= \sum P_i(z)\, {}^i\tilde{L}(z) + \sum{}' Q_{ij}(s^{n-j}) \tilde{L}_{ij}(w^d).$$

Here $\sum'$ means we sum over $(i, j) \neq (i_0, j_0), \ldots, (i_{m-1}, j_{m-1})$. The ${}^i\tilde{L}(z)$ belong to $\mathscr{P}$ and the $\tilde{L}_{ij}$ are sums of products of polynomials in $\mathscr{P}(\hat{T}^j)$ by products of $1, R_{i_0j_0}, \ldots, R_{i_{m-1}j_{m-1}}$ with the condition that no power of any $R_{i_qj_q}$ greater than the a_0 should appear. In particular the power of w_d^d that appears in $\tilde{L}_{ij}$ can be bounded independently of b. This means that $\tilde{L}_{i_mj_m}$ is different from

$$(w_d^d)^b R_{i_0j_0}(w^d) \cdots R_{i_{m-1}j_{m-1}}(w^d)$$

for b large enough. Thus, our assertion regarding the expression for $Q_{i_mj_m}$ is proved if we set

$$R_{i_mj_m}(w^d) = (w_d^d)^b R_{i_0j_0}(w^d) \cdots R_{i_{m-1}\,m-1}(w^d) - \tilde{L}_{ij}(w^d)$$

for some large b.

Theorem 9.3 now follows easily.

See Remark 9.2.

Using the technique of the proof of Theorem 9.3 we find

THEOREM 9.4. *For any fixed* j_0 *the* Q_{ij_0} *form a basis for*

$$\mathscr{P}\Big/\Big[\vec{\mathrm{P}}\mathscr{P}^r + \sum_{j>j_0} Q_{ij}\,\mathscr{P}(\hat{T}^j)\Big]$$

which is considered a vector space over $\mathscr{R}(\hat{T}^{j_0})$ *as in Theorem* 9.3. *In particular, if* T^{j_0} *and the* T^j *for* $j > j_0$ *and the* Q_{ij} *for* $j > j_0$ *are given, then* l^{j_0} *is determined.*

To show our ignorance of the Cauchy problem, we shall pose two problems which we cannot answer:

PROBLEM 9.1.
Let $\mathscr{W}$ be an LAU space for which $\mathbf{W}'$ is dense in the space of entire functions. Suppose we are given a well-posed Cauchy problem for $\mathscr{W}$. Is this well-posed for the space of formal power series?

If the answer to Problem 1 is "yes" then we might expect that, for a well-posed Cauchy problem for $\mathscr{W}$, if we write any $F \in \mathbf{W}'$ in accordance with (9.4)

$$F(z) = \sum P_i(z)\,{}^iL(z) + \sum Q_{ij}(s^{n-j})L_{ij}(w^j),$$

the values ${}^iL(z)$ and $L_{ij}(w^j)$ should be expressible in terms of certain values $F(z)$ by means of suitable "algebraic expressions," for example, like those in (2.15). In some sense these should be generalizations of the Lagrange interpolation formula.

PROBLEM 9.2
Is l^d equal to the "generic" number m_0 of points of $\mathfrak{V}$ lying over a point in T^d?

By the "generic" number of points in $\mathfrak{V}$ lying over a point in $\hat{T}^d$ is meant the following: For each fixed $w^d \in \hat{T}^d$, denote by $m(w^d)$ the number of points of the form (s^{n-d}, w^d) belonging to $\mathfrak{V}$. Here, if $\mathfrak{V} = (V_1, \partial_{11}, \ldots, \partial_{1p_1}; \ldots; V_a, \partial_{a1}, \ldots, \partial_{ap_a})$, then for each j such that $(s^{n-d}, w^d) \in V_j$ we count its multiplicity as p_j. Using techniques of algebraic geometry, we could show easily that the set of w^d for which $m(w^d)$ takes a fixed value forms a Z-variety. Thus, there is a number m_0 such that the set of w^d with $m(w^d) = m_0$ forms a Z-variety whose closure is all of $\hat{T}^d$. The number m_0 is the generic number of Problem 2.

We can show that the answer to Problem 2 is "yes" in the following situation: Suppose that $\mathfrak{V}$ is an irreducible variety V (no multiplicities) so $\vec{\mathrm{P}}$ defines a prime ideal in the ring $\mathscr{P}$ of polynomials. To prove our assertion we note that the field of rational functions on V, $\mathscr{R}(V)$, which is the field of fractions of $\mathscr{P}/\vec{\mathrm{P}}\mathscr{P}^r$, is an extension of the complex numbers of transcendence degree d. It is a simple consequence of the fact that the map (9.5) is one-to-one that $w_1^d, \ldots, w_d^d$ are independent transcendental elements of $\mathscr{R}(V)$. We claim that the generic number m_0 of points of V lying over a point in $\hat{T}^d$ is just the degree m of the algebraic extension $\mathscr{R}(V)$ of $C(w_1^d, \ldots, w_d^d)$.

To see this, let $a \in \mathscr{R}(V)$ generate this field over $C(w_1^d, \ldots, w_d^d)$. Thus, a satisfies an irreducible equation $M(a) = 0$ of degree m over $C(w_1^d, \ldots, w_d^d)$. By multiplying a by a suitable polynomial in $w_1^d, \ldots, w_d^d$, we may assume the leading coefficient of M is one. Now, if we fix the values of w^d at a d-tuple ${}_0w^d$ of complex numbers, the corresponding equation ${}_0M(a) = 0$ will have m distinct roots for almost all ${}_0w^d$, for otherwise the discriminant of M would be identically zero, so M would not be reducible.

The powers of $s_1^{n-d}, \ldots, s_{n-d}^{n-d}$ generate $\mathscr{R}(V)$ over $C(w_1^d, \ldots, w_d^d)$. For, the powers of the s_i^{n-j} for $j \leqslant d$ generate by the assumption that the Cauchy Problem is well-posed, and the s_i^{n-j} for $j < d$, $i < n - d$ can be assumed to be included among the w_k^d because of our assumption that $T^j \subset T^d$. This means we can express the s_j^{n-d} rationally in terms of a and vice versa. Since we can express the s_j^{n-d} rationally in terms of a and since a takes m values for each generic value ${}_0w^d$ of w^d, it follows that $m_0 \leqslant m$. Since we can express a rationally in terms of the s_j^{n-d}, it follows similarly that $m \leqslant m_0$. Thus, $m = m_0$.

This means that any basis for $\mathscr{R}(V)$ over $C(w^d)$ will have m_0 generators. Since $C(w^d)$ is the same as $\mathscr{R}(\hat{T}^d)$, our assertion follows from Theorem 9.3.

It does not seem unreasonable that the proof in the above special case could be suitably modified so as to handle the general case. The main difficulty is that we do not know how to modify the step in the proof where we have used the fact that an algebraic extension of finite degree of a field of characteristic zero can be generated by a single element.

IX.2. Generalization of the Cauchy-Kowalewski Theorem

We are now going to derive an extension of the Cauchy-Kowalewski theorem to our situation. We shall call $\mathscr{W}$ a CK *space* if it is LAU and if

1. An AU structure K can be formed of functions k which depend only on $|z|$ and are monotonically increasing.
2. For any $F \in \mathbf{W}'$ the Taylor series for F at any point converges in the topology of $\mathbf{W}'$.
3. For any $a > 0$ and any $k \in K$ there is a $k' \in K$ with $k'(a|z|) \leqslant k(|z|)$.

Examples of CK spaces are the space of entire functions, the space of local analytic functions, the space of entire functions of finite order (see Chapter V. Example 1).

THEOREM 9.5. *Let $\mathscr{W}$ be a* CK *space. Then there exists a well-posed Cauchy problem for $\mathscr{W}$, that is, it is possible to choose linear subspaces T^j and differential operators q_{ij} so that the corresponding Cauchy problem is well-posed for $\mathscr{W}$. The T^j and q_{ij} depend on $\vec{\mathrm{D}}$ but not on $\mathscr{W}$.*

This theorem, in view of Lemma 9.1 is a global analog of Theorems 2.5 and 3.2, since it gives a global parametrization of $\mathbf{W}'/\vec{\mathbf{P}}\mathbf{W}'^r$. In order to explain the proof we introduce

DEFINITION. Let $P(z)$ be a polynomial. We say that the hyperplane $z_1 = 0$ is *noncharacteristic for* P if the degree of P in z_1 is equal to its degree.

Actually, it is more proper to speak of characteristic directions than characteristic hyperplanes. However, the two concepts can be used interchangeably because we pass from a hyperplane to its normal.

LEMMA 9.6. *A necessary and sufficient condition that $z_1 = 0$ be noncharacteristic for P is: There is a constant $a > 0$ so that if $P(z_1, \ldots, z_n) = 0$, then*

$$|z_1| \leqslant a(1 + |z_2| + \cdots + |z_n|). \tag{9.6}$$

If 9.6 holds for all $z_2, \ldots, z_n$ except for a subvariety, then $z_1 = 0$ is noncharacteristic.

PROOF. The necessity is a consequence of the fact that the polynomial in one variable $a_0\zeta^m + a_1\zeta^{m-1} + \cdots + a_m$ has all its zeros in the circle $|\zeta| \leqslant \sum |a_j/a_0|^{1/j}$ (see Marden [1], p. 98, Example 5).

The sufficiency follows from the fact that (see Marden [1] p. 98, Example 7) there exists a root of this polynomial with

$$|\zeta| \geqslant (1/m) \sum |a_j/a_0 C(m, j)|^{1/j}. \ \square$$

In the proof of Lemma 3.7 we have remarked

LEMMA 9.7. *For any polynomial P, almost all hyper-planes are noncharacteristic, that is, for a Z open set of Λ in $GL(n, C)$ (complex general linear group) z_1 is noncharacteristic for $P(\Lambda z)$.*

PROOF OF THEOREM 9.5. We follow step by step the proof of Theorem 2.5 or 3.2 except that we are now in a global situation. The coefficients L_{ij} that occur in those theorems are obtained by successive applications of the Lagrange interpolation formula. We may assume by Lemma 9.7 that all these Lagrange interpolation formulas involve noncharacteristic directions. Then by Lemma 9.6, and the definitions of CK space and $L_{ij}(w^j)$, and the Lagrange interpolation formula (note that the denominators that occur can easily be handled by Theorem 1.4), we see that the $L_{ij}(w^j)$ are in the restriction of $\mathbf{W}'$ to $\hat{T}^j$.

For the same reason, the L_{ij} depend continuously on the $H \in \mathbf{W}'(\mathfrak{V})$ of which they are the coefficients. By construction they determine H uniquely. On the other hand, it is clear that

$$\sum Q_{ij} L_{ij} \to \text{restriction to } \mathfrak{V} \text{ of } \sum Q_{ij} L_{ij}$$

is continuous. Thus $\hat{\gamma}'_{\mathfrak{V}}$ is a topological isomorphism. Our result now follows from Lemma 9.1 □

See Remark 9.3.

DEFINITION. We call a system $\{T^j\}$ a *noncharacteristic for* $\vec{\mathrm{D}}$ if for each j we can find q_{ij} not all zero so that the Cauchy problem is well-posed for it for every CK space. We call T^d (where $d = \dim \mathfrak{V}$) a *principal noncharacteristic plane for* $\vec{\mathrm{D}}$.

We wish now to find conditions for a plane in order that it be a principal noncharacteristic. Note by Lemma 9.2 that no T^j for $j > d$ occurs.

THEOREM 9.8. *A necessary and sufficient condition for T^d to be a principal noncharacteristic for $\vec{\mathrm{D}}$ is the existence of a constant c so that for $(s^{n-d}, w^d) \in \mathfrak{V}$ we have*

$$|s^{n-d}| \leqslant c(1 + |w^d|). \tag{9.7}$$

PROOF. We show first the necessity of (9.7). For this we take for $\mathscr{W}$ the space $\mathscr{H}$ of entire functions (see Chapter V, Example 1). If (9.7) fails then after making a suitable complex change of coordinates we may assume that we can find points (s^{n-d}, w^d) in $\mathfrak{V}$ so that for no c is

$$|s_1^{n-d}| \leqslant c(1 + |w^d|) \tag{9.8}$$

and, moreover,

$$Rs_1^{n-d} \geqslant \tfrac{1}{2}(1 + |s_1^{n-d}|). \tag{9.9}$$

For, (9.8) can clearly be arranged by a change of variables. We can now make a rotation in the complex s_1^{n-d} plane to insure that (after taking a subsequence) the arguments of the s_1^{n-d} coordinates of the points converge to zero. The projection of $\mathfrak{V}$ on the complex s_1^{n-d} plane is a Zariski open set of a subvariety (see Samuel [1]). Because (9.8) fails, this must be the s_1^{n-d} plane except for a finite number of points, since the only subvarieties of a complex plane are the whole plane and a finite set of points.

We claim that $\exp(s_1^{n-d})$ cannot be written on $\mathfrak{V}$ in the form $\sum Q_{ij}(s^{n-j}) L_{ij}(w^j)$ for any Q_{ij}, L_{ij} with Q_{ij}, polynomials and $L_{ij} \in \mathbf{H}'$. For on $\mathfrak{V}$ we have

$$\left|\sum Q_{ij}(s^{n-j}) L_{ij}(w^j)\right| \leqslant a(1 + |z|)^a \exp(|aw^d|) \tag{9.10}$$

for some a because the L_{id} are in $\mathbf{H}'$, hence are of exponential type (see Chapter V, Example 1). But for some large $|s_1^{n-d}|$ the right side of (9.10) is $<|\exp(s_1^{n-d})|$ because of (9.8) and (9.9). Thus, the necessity of (9.7) is proved.

For the sufficiency we must again use some algebraic geometry: Call V the union of all the varieties of $\mathfrak{V}$ and denote by I the idéal of all polynomials which vanish on V. Denote by $\mathscr{P}$, $\mathscr{R}$, $\mathscr{R}(V)$, $\mathscr{R}(\hat{T}^d)$ the ring of polynomials, the rational functions, the rational functions on V, and the rational functions on $\hat{T}^d$, respectively. It is clear from (9.7) that for each $w^d \in \hat{T}^d$ there are only finitely many s^{n-d} with $(s^{n-d}, w^d) \in V$.

Thus, dim $V = d$ and $\mathscr{R}(V)$ is an algebraic extension of $\mathscr{R}(\hat{T}^d)$.

In particular, each s_p^{n-d} is algebraic over $\mathscr{R}(\hat{T}^d)$. Now the projection of V on $\{s_1^{n-d}=0 \ldots, s_{p-1}^{n-d}=0, s_{p+1}^{n-d}=0, \ldots, s_{n-d}^{n-d}=0\}$ is well known to be a bunch of Z-varieties all of whose dimensions are $\leqslant d$ and some of which have dimension $=d$ (see Samuel [1], Weil [1]). Let $\tilde{V}^p$ be a variety of dimension d in (s_p^{n-d}, w^d) space containing this bunch. Then $\tilde{V}^p$ is a hypersurface, so on $\tilde{V}^p$ we can write

$$A_p^0(w^d)(s_p^{n-d})^{m_p} + \cdots + A_p^{m_p}(w^d) = 0, \tag{9.11}$$

where the A_p^i are polynomials. We now apply (9.7) and the second part of Lemma 9.6 to deduce that A_p^0 is a nonzero constant and degree $A_p^i \leqslant i$. We shall assume in the following that $A_p^0 = 1$.

Next we note, by Hilbert's Nullstellensatz (see Van der Waerden [1]), that a sufficiently high power of the left side of (9.11) belongs to the ideal generated by the P_i. Thus, we have relations

$$(s_p^{n-d})^{l_p m_p} = B_p(s_p^{n-d}, w^d) + C_p(s_p^{n-d}, w^d), \tag{9.12}$$

where B_p is of degree $<l_p m_p$ in s_p^{n-d} and C_p is the l_p power of the left side of (9.11) and so belongs to the ideal generated by the P_i.

LEMMA 9.9. *Let $F(s^{n-d}, w^d) \in \mathbf{W}'$, where $\mathscr{W}$ is a CK space. We use (9.12) and the relations derived from it for $(s_p^{n-d})^j$ for $j > l_p m_p$ (by multiplication by s_p^{n-d} and successive use of (9.12) to reduce all the powers of s_p^{n-d} above $l_p m_p - 1$). Thus we write*

$$F(s^{n-d}, w^d) = {}^1F(s^{n-d}, w^d) + {}^2F(s^{n-d}, w^d)C_p(s_p^{n-d}, w^d), \tag{9.13}$$

where ${}^1F$ contains s_p^{n-d} only with powers $<l_p m_p$. Then ${}^1F$ and ${}^2F \in \mathbf{W}'$; they depend continuously on $F \in \mathbf{W}'$.

PROOF OF LEMMA 9.9. This is essentially the same as the proof of the local analog which occurs in the proof of Theorem 2.8. We could also use, as in the proof of Theorem 9.5, a method based on the Lagrange interpolation formula. ☐

By successive application of the above procedure we can reduce the powers of all s_p^{n-d} to finitely many. That is, we can write every $F \in \mathbf{W}'$ in the form

$$F(s^{n-d}, w^d) = \sum (s_1^{n-d})^{i_1} \cdots (s_{n-d}^{n-d})^{i_{n-d}} F_{i_1 \cdots i_{n-d}}(w^d) + G(s^{n-d}, w^d), \tag{9.14}$$

where $F_{i_1 \cdots i_{n-d}}(w^d) \in \mathbf{W}'(\hat{T}^d)$, G vanishes on $\mathfrak{B}$, and the sum is finite. This representation is, of course, not necessarily unique because the $(s_1^{n-d})^{i_1} \cdots (s_{n-d}^{n-d})^{i_{n-d}}$ are not necessarily linearly independent over the ideal $\vec{\mathrm{P}}\mathscr{P}^r$ generated by the P_i. We denote by Q_{id} a maximal set of these monomials such that there is no relation

$$\sum R_i(w^d) Q_{id}(s^{n-d}) + \sum S_i P_i = 0, \tag{9.15}$$

where R_i are polynomials (not all zero) in w^d and S_i are polynomials in (s^{n-d}, w^d). Then for each other monomial Q there are polynomials $R(w^d) \not\equiv 0$, $R_i(w^d)$ and $S_i(s^{n-d}, w^d)$ so that

$$RQ + \sum R_i Q_{id} + \sum S_i P_i = 0. \tag{9.16}$$

We consider (9.16) for all those monomials Q of the form

$$(s_1^{n-d})^{i_1} \cdots (s_{n-d}^{n-d})^{i_{n-d}}$$

which are needed in the right side of (9.14). For each such Q, call $\Delta(Q)$ the ideal of all possible $R(w^d)$. By $\Delta^{d'}$ we denote the ideal in the ring of polynomials in z which is the least common multiple of all the $\Delta(Q)$. (The integer d' will be described below; note that d' is a superscript, not a power.)

It is a well-known result that the assumption that (9.15) does not hold for polynomials implies that (9.15) cannot hold even for entire functions $R_i(w^d)$, $S_i(s^{n-d}, w^d)$ (see e.g., Serre [1]). Thus, by Lemma 9.9 for every $F \in \mathbf{W}'$, $R \in \Delta^{d'}$, RF can be represented by

$$RF = \sum L_{id} Q_{id} + \sum P_i \, {}^iL, \tag{9.17}$$

where $L_{id} \in \mathbf{W}'(\hat{T}^d)$ and ${}^iL \in \mathbf{W}'$ depend uniquely and continuously on F.

We denote by $\delta^{d'}$ the adjoint of the inverse Fourier transform of $\Delta^{d'}$, by ${}^{d'}\vec{\mathrm{D}}$ the system $(D_1, \ldots, D_r, \delta^{d'})$, and by ${}^{d'}\vec{\mathrm{P}} = (P_1, \ldots, P_r, \Delta^{d'})$. (Here we have used the notations $\Delta^{d'}$ and $\delta^{d'}$ to denote both the ideals and suitable sets of generators.) Since $\Delta^{d'}$ is an ideal generated by polynomials in w^d which is not identically zero, the dimension of the variety ${}^{d'}V$ of zeros of P_i, $\Delta^{d'}$ is $d' < d$ because, as we have already noted, (9.7) implies that the number of points above any $w^d \in \hat{T}^d$ is finite. Call ${}^{d'}\mathfrak{B}$ the multiplicity variety associated with ${}^{d'}\vec{\mathrm{P}}$.

We have in (9.17) represented $\Delta^{d'}\mathbf{W}'^{r'}$ modulo $\vec{\mathrm{P}}$ in terms of the $L_{id}Q_{id}$. In order to represent all of $\mathbf{W}'$ we must consider the quotient $\mathbf{W}'/{}^{d'}\vec{\mathbf{P}}\mathbf{W}'^{r'+r}$. We are thus led to study the quotient space $\mathscr{W}/{}^{d'}\vec{\mathrm{D}}\mathscr{W}^{r'+r}$, where r' is the number of generators of $\Delta^{d'}$. By Theorem 9.5 we can find a principal noncharacteristic $T^{d'} \subset T^d$ for the intersection of $\Delta^{d'}$ with the ring of polynomials in w^d. If we denote by $w^{d'}$ the variables on $\hat{T}^{d'}$ and $s^{d-d'}$, the "orthogonal" variables in $\hat{T}^d$, then by the proof of the necessity of (9.7) we will have

$$|s^{d-d'}| \leqslant c(1+|w^{d'}|),$$

whenever $(s^{d-d'}, w^{d'})$ is a zero of $\Delta^{d'}$. In view of (9.7) this means that, if we denote by $s^{n-d'}$ the "orthogonal" variables to $w^{d'}$ in C^n, then

$$|s^{n-d'}| \leqslant c(1+|w^{d'}|)$$

for $(s^{n-d'}, w^{d'}) \in {}^{d'}V$, that is, $\hat{T}^{d'}$ satisfies the analog of (9.7) for the variety ${}^{d'}V$.

We can now proceed as for $\hat{T}^d$ to find monomials $(s_1^{n-d'})^{i_1} \ldots (s_{n-d'}^{n-d'})^{i_{n-d'}}$ in finite number so that every $F \in \mathbf{W}'$ can be written in the form

$$F(s^{n-d'}, w^{d'}) = \sum {}^{i_1 \ldots i_{n-d'}}F(w^{d'})(s_1^{n-d'})^{i_1} \cdots (s_{n-d'}^{n-d'})^{i_{n-d'}} \tag{9.18}$$
$$+ \sum {}^iF(s^{n-d'}, w^{d'})P_i + \vec{\mathrm{G}}(s^{n-d'}, w^{d'})\Delta^{d'},$$

where ${}^{i_1 \ldots i_{n-d'}}F \in \mathbf{W}'(\hat{T}^{d'})$, ${}^iF \in \mathbf{W}'$, $\vec{\mathrm{G}} \in \mathbf{W}'^{r'}$. We can now use (9.17) to express $\Delta^{d'}\vec{\mathrm{G}}$ uniquely in terms of the $Q_{id}(s^{n-d})$ and the P_i.

We denote by $Q_{id'}$ a maximal set of the monomials $(s_1^{n-d'})^{i_1} \cdots (s_{n-d'}^{n-d'})^{i_{n-d'}}$ such that there is no relation

$$\sum R_{id}(w^d)Q_{id}(s^{n-d}) + \sum R_{id'}(w^{d'})Q_{id'}(s^{n-d'}) + \sum S_i(z)P_i(z) = 0, \tag{9.19}$$

where the R_{id}, $R_{id'}$, S_i are polynomials and not all $R_{id'}$ are identically zero. (Hence (see, e.g., Serre [1]), there is no such relation for R_{id}, $R_{id'}$, S_i entire.) Then for every other one of the monomials Q there are polynomials $R(w^{d'}) \not\equiv 0$, $R_{id'}(w^{d'})$, $R_{id}(w^d)$, and $S_i(z)$ so that

$$RQ + \sum R_{id'}Q_{id'} + \sum R_{id}Q_{id} + \sum S_iP_i = 0. \tag{9.20}$$

For each Q we denote by $\Delta(Q)$ the ideal of all such R and denote by $\Delta^{d''}$ the ideal in $\mathscr{P}$ which is the least common multiple of all such $\Delta(Q)$. Then by (9.18) and (9.17) for $F \in \mathbf{W}'$, $R \in \Delta^{d''}$, RF can be expressed in the form

$$RF = \sum L_{id}(w^d)Q_{id}(s^{n-d}) + \sum L_{id'}(w^{d'})Q_{id'}(s^{n-d'}) + \sum P_i\,{}^iL, \tag{9.21}$$

where the L_{id}, $L_{id'}$, ${}^iL$ depend uniquely and continuously on F.

We now define the systems ${}^{d''}\vec{\mathrm{D}} = (D_1, \ldots, D_r, \delta^{d'}, \delta^{d''})$ and ${}^{d''}\vec{\mathrm{P}} = (P_1, \ldots, P_r, \Delta^{d'}, \Delta^{d''})$, where $\delta^{d''}$ is the adjoint of the inverse Fourier transform of $\Delta^{d''}$. We proceed in this way until we arrive at the ideal $\Delta^0 = \mathscr{P}$ and we are finished. □

See Remark 9.4.

DEFINITION. A sequence of systems $(\vec{\mathrm{D}}, {}^{d'}\vec{\mathrm{D}}, {}^{d''}\vec{\mathrm{D}}, \ldots, {}^{0}\vec{\mathrm{D}})$ of the type constructed in the proof of Theorem 9.8 is called a *derived sequence* of $\vec{\mathrm{D}}$.

More precisely, a derived sequence is defined as follows: Let T^d be a principal noncharacteristic for $\vec{\mathrm{D}}$. Let $s_1^{n-d}, \ldots, s_{n-d}^{n-d}$ be coordinates for a complement of $\hat{T}^d$. Let $\{(s_1^{n-d})^{i_1} \cdots (s_{n-d}^{n-d})^{i_{n-d}}\} = M$ be a finite set of monomials such that every $F \in \mathbf{W}'$ can be written in the form (9.14) with $F_{i_1 \cdots i_{n-d}}(w^d) \in \mathbf{W}'(\hat{T}^d)$ and $G = 0$ on $\mathfrak{B}$. Let $\{Q_{id}(s^{n-d})\}$ be a maximum set of monomials in M such that no relation of the type (9.15) with R_i polynomials in w^d (not all zero) and S_i polynomials in (s^{n-d}, w^d) is possible. Then for each $Q \in M$, we can write an expression of the type (9.16) where $R(w^d) \not\equiv 0$, $R_i(w^d)$, $S_i(s^{n-d}, w^d)$ are polynomials. For each $Q \in M$ denote by $\Delta(Q)$ the ideal of all R which can occur in (9.16). We denote by $\Delta^{d'}$ the ideal which is the least common multiple in the ring of polynomials in z of all the $\Delta(Q)$.

DEFINITION. $\Delta^{d'}$ is called the *different* of the basis $\{Q_{id}\}$.

We see easily that $\Delta^{d'}$ is the largest polynomial ideal which, modulo the ideal generated by the P_i, is contained in the set of sums

$$\sum Q_{id}(s^{n-d}) L_{id}(w^d)$$

where L_{id} are polynomials. Thus, our definition is in comformity with the customary usage in algebraic number theory (compare Hecke [1]).

We define ${}^{d'}\vec{\mathrm{D}}$ as the system $(D_1, \ldots, D_r, \delta^{d'})$, where $\delta^{d'}$ is the adjoint of the inverse Fourier transform of $\Delta^{d'}$. Then ${}^{d''}\vec{\mathrm{D}}$ bears the same relation to ${}^{d'}\vec{\mathrm{D}}$ as ${}^{d'}\vec{\mathrm{D}}$ does to $\vec{\mathrm{D}}$, etc.

See Remark 9.5 and Problems 9.3 and 9.4.

The method of proof of Theorem 9.8 yields

THEOREM 9.10. *Suppose we have a Cauchy problem which is obtained from a derived system as in the proof of Theorem 9.8. For each j consider the map*

$${}^j\gamma : f \to (D_1 f, \ldots, D_r f; q_{1j} f|_{T^j}, \ldots, q_{l^j j}|_{T^j}, \ldots, q_{1d} f|_{T^d}, \ldots, q_{l^d d} f|_{T^d}).$$

Then f is determined by ${}^j\gamma$ modulo an element of the kernel of ${}^{j'}\vec{\mathrm{D}}$, where j' is the largest integer $< j$ for which ${}^{j'}\vec{\mathrm{D}}$ is defined.

IX.3. Hyperbolic Systems

We wish now to give an analogous treatment for the space $\mathscr{E}$ (see Chapter V, Example 5) instead of CK spaces. Naturally, it is possible to treat other spaces by the same method.

DEFINITION. The system $\vec{\mathrm{D}}f = g$ is called *Cauchy hyperbolic* (abbreviated CH), if the Cauchy problem is well-posed for the space $\mathscr{E}$, more precisely, if there exists a Cauchy problem which is well-posed for $\mathscr{E}$ (see Chapter V, Example 5). The system is called *principally Cauchy hyperbolic* (PCH) if there exists a well-posed Cauchy problem for $\mathscr{H}$ (the space of entire functions) such that T^d is a principal noncharacteristic for $\vec{\mathrm{D}}$ and the set

$$(q_{1d}f\,|_{T^d}, \ldots, q_{l^d d}f\,|_{T^d}) \tag{9.22}$$

for $f \in \mathscr{E}$, $\vec{\mathrm{D}}f = 0$, consists of all of $\mathscr{E}(T^d) \oplus \cdots \oplus \mathscr{E}(T^d)$.

THEOREM 9.11. *A necessary and sufficient condition for $\vec{\mathrm{D}}$ to be* PCH *with principal noncharacteristic T^d is that $\vec{\mathrm{D}}$ should be hyperbolic (see Chapter* VIII) *with space variables being those on T^d.*

PROOF. This is proved by essentially the same method as that used in the proof of Theorem 9.8. The use of inequality (9.7) is replaced by that of (8.6) which is the inequality equivalent to hyperbolicity.

Combining Theorem 9.11 with the proof of Theorem 9.8 we deduce

THEOREM 9.12. *Suppose we have a Cauchy problem which is obtained from a derived system as in the proof of Theorem* 9.8. *A necessary and sufficient condition for $\vec{\mathrm{D}}$ to be Cauchy hyperbolic with noncharacteristic $(T^d \ldots, T^0)$ is that each derived system ${}^j\vec{\mathrm{D}}$ be hyperbolic with space variables T^j.*

As in the case $r = 1$, it would be of interest to study cones of time-like directions.

IX.4. Parabolic Systems

Without going into the details, let us explain how the above extends to parabolic systems. First we must modify the Cauchy problem to obtain the *initial value* problem. Suppose we are given linear spaces $T^0, T^1, \ldots, T^d$ with corresponding differential operators q_{ij}, as before. We assume for simplicity that each $T^j \subset T^d$. We assume $d = \dim\, \mathfrak{V}$

but we do not assume that $(T^0, \ldots, T^d)$ is noncharacteristic. We assume, however, the restriction properties of $\mathscr{W}$ on each T^j as before.

We use variables ξ^d on T^d and t^d for the others. We write $t^d \geqslant 0$ to mean all the components of t^d are $\geqslant 0$. By $\mathscr{W}^+$ we denote the restrictions of the elements of $\mathscr{W}$ to $t^d \geqslant 0$. The initial value problem is said to be *well-posed* if

$$(9.23) \qquad \gamma : f \to \gamma f$$

is a topological isomorphism of $\mathscr{W}^+$ onto ${}_{\vec{\mathrm{D}}}\mathscr{W}^{+r} \oplus {}^0\mathscr{W}^{l^0} \oplus \cdots \oplus {}^d\mathscr{W}^{l^d}$. $\vec{\mathrm{D}}$ is called *Cauchy parabolic* (CP) if the initial value problem is well posed for the space $\mathscr{O}_M$ (see Schwartz [1]) of indefinitely differentiable functions g such that all derivatives of g are polynomially increasing at infinity.

It is easy to derive the conditions for parabolicity. We could, of course, consider systems which are parabolic in some variables and hyperbolic in others.

IX.5. Groups and Semigroups Associated with Cauchy and Initial Value Problems

In the classical theory of Cauchy's problem for one unknown function and one operator, there is associated a one-parameter group, namely, the group which transforms the Cauchy data from $t = 0$ to another t. For parabolic systems we get a corresponding semigroup for the initial value problem. We wish to generalize this to the case of a general $\vec{\mathrm{D}}$ (still $l = 1$).

Let us assume we are given a well-posed Cauchy problem for the space $\mathscr{W}$. Suppose that $T^j \subset T^d$ for all j and suppose for simplicity that $0 \in T^j$ for all j. We shall denote as usual by ξ^d the variables on T^d and by t^{n-d} the orthogonal ones; the dual (Fourier transform) variables are (s^{n-d}, w^d). For any t_0^{n-d} and any j we denote by $T^j_{t_0^{n-d}}$ the (affine) linear subspace of $t^{n-d} = t_0^{n-d}$ defined as the set of those points (t_0^{n-d}, ξ^d) for which $(0, \xi^d) \in T^j$, that is, $T^j_{t_0^{n-d}}$ is obtained from T^j by a translation in which the t^{n-d} coordinate is changed from 0 to t_0^{n-d}. It is clear that $T^j_0 = T^j$.

We shall assume in this section that translation in the t^{n-d} direction is a topological isomorphism of $\mathscr{W}$ onto itself.

LEMMA 9.13. *For any t_0^{n-d} the Cauchy problem for $\{(T^j_{t_0^{n-d}}, q_{ij})\}$ is well-posed.*

PROOF. This is an immediate consequence of the fact that the D_j commute with translation and our assumption that translation in t^{n-d} is a topological isomorphism of W. ∎

We shall write for simplicity t for t^{n-d}.

Let us suppose we are given a well-posed Cauchy problem as above. For any t^0, t^1 we denote by $\tau(t^1, t^0)$ the transformation which maps Cauchy data on $t = t^0$ into the Cauchy data on $t = t^1$, which give rise to the same solution as the original data on $t = t^0$. (This exists and is unique by Lemma 9.13.) We denote by $\tau(t) = \tau(t, 0)$. We regard $\tau(t^0)$ as a mapping of the space of Cauchy data on $t = 0$ onto itself by means of the identification of the space of Cauchy data on $t = t^0$ with that on $t = 0$ obtained by translation by t^0.

THEOREM 9.14. *The transformations $\tau(t)$ form an $(n-d)$-parameter commutative group of topological isomorphisms of the Cauchy data on $t = 0$ onto itself.*

PROOF. First we observe that

$$\tau(t^2, t^1)\tau(t^1, t^0) = \tau(t^2, t^0). \tag{9.24}$$

This is an immediate consequence of the uniqueness of the solution of Cauchy's problem. Similarly, $\tau(t^1, t^0)$ is the inverse of $\tau(t^0, t^1)$. Moreover, we claim that if we identify the Cauchy data on $t = t^0, t^1$ with that on $t = 0$ by means of translation then

$$\tau(t^1, t^0) = \tau(t^1 - t^0). \tag{9.25}$$

This is an immediate consequence of the fact that the D_j commute with translation.

We claim that

$$\begin{aligned} \tau(t^1 + t^2) &= \tau(t^1)\tau(t^2) \\ &= \tau(t^2)\tau(t^1). \end{aligned} \tag{9.26}$$

For the left side is

$$\begin{aligned} \tau(t^1 + t^2, 0) &= \tau(t^1 + t^2, t^1)\tau(t^1, 0) \\ &= \tau(t^2, 0)\tau(t^1, 0) \end{aligned}$$

by (9.24) and (9.25). Similarly for the second equation in (9.26). Since $\tau(0)$ is clearly the identity, we are finished. □

We can give an interesting interpretation to the infinitesimal generators ∂_j of the group. These are, of course, just the transformations defined by

(9.28) ∂_j: Cauchy data $\rightarrow$ solution f

$\rightarrow$ Cauchy data corresponding to $\partial f/\partial t_j$.

To compute the adjoint of this map let us denote by $\mathscr{C}\mathscr{W}$ the Cauchy data of $\mathscr{W}$, so $\mathscr{C}\mathscr{W}$ is just the direct sum ${}^0\mathscr{W}^{l^0} \oplus \cdots \oplus {}^d\mathscr{W}^{l^d}$. We denote by η

the map of $\mathscr{W} \to \mathscr{C}\mathscr{W}$ defined by (compare (9.2))

$$\eta f = (q_{10} f \,|\, _{T^0}, \; \ldots, \; q_{l^d d} f \,|\, _{T^d}).$$

We note that an element C in the dual of the space of Cauchy data is uniquely determined by its values on the Cauchy data of solutions f of $\vec{\mathrm{D}} f = 0$. Thus, the adjoint of ∂_j can be written as ∂_j', where

$$\partial_j' C \cdot \eta f = C \cdot \eta(\partial f / \partial t_j) \tag{9.29}$$

for any f with $\vec{\mathrm{D}} f = 0$. (9.29) shows that, as elements of $\mathscr{W}_{\vec{\mathrm{D}}}'$ (dual of the space of $f \in \mathscr{W}$ with $\vec{\mathrm{D}} f = 0$)

$$\eta' \partial_j' C = -\partial(\eta' C)/\partial t_j .$$

We want to compute $\partial_j' C$; C is a $\sum l^i$-tuple of distributions, where for each i there are l^i components with support on T^i. By Theorem 7.1 we may compute $\partial_j' C$ by restricting our considerations in (9.29) to those f which are suitable exponentials and certain derivatives of them which are in $\mathscr{W}_{\vec{\mathrm{D}}}$. For example, if $f(t, \xi) = \exp(it \cdot s + i\xi \cdot w)$ with $(s, w) \in V$, and only the ik component C_{ik} of C is different from zero, then (9.29) becomes

$$\sum_{l,m} (\partial_j' C)^\wedge{}_{lm}(s, w) Q_{ml}(s) = i\hat{C}_{ik}(s, w) s_j Q_{ik}(s), \tag{9.30}$$

since by definition of $Q_{lm}(s)$, the Cauchy data of $\exp(it \cdot s + i\xi \cdot w)$ is just $\{Q_{lm}(s) \exp(i\xi \cdot w)\}$. Here, as usual, $(\quad)^\wedge$ refers to the Fourier transform. (We have omitted superscripts over s, w.)

In the particular case where $C_{ik} = \delta$ is the Dirac δ mass at the origin, (9.30) becomes

$$s_j Q_{ik}(s) = \sum_{l,m} Q_{lm}(s) \, {}_j L_{lm}^{ik}(w). \tag{9.31}$$

If we use a derivative of an exponential for f, then we find that (9.30) for $(s, w) \in \mathfrak{V}$ is necessary and sufficient for (9.29). Thus the ${}_j L_{lm}^{ik}(w^m)$ can be constructed explicitly by the methods of the proof of Theorems 2.5 or 3.2. They can be expressed by iterated use of Lagrange interpolation which implies by Theorem 1.4 that they are of polynomial growth.

Thus, ${}_j L_{lm}^{ik}$ is a polynomial. If we denote by ${}_j\boxed{\mathrm{L}}$ the matrix $({}_j L_{lm}^{ik})$ then we have shown

THEOREM 9.15. *The matrix* ${}_j\boxed{\mathrm{L}}$ *of polynomials "represents" the Fourier transform of* ∂_j' *in the sense that it describes the action of* ∂_j' *on vectors whose components consist of zeros and multiples of the Dirac* δ *measure.*

See Remarks 9.7, 9.8 and 9.9.

In the case of parabolic systems the same method leads to semigroups instead of groups. Since the modifications to handle parabolicity are obvious, we shall leave them to the reader.

See Problem 9.5.

IX.6. Structure of Cauchy Data

We consider first the case $r = 1$, namely, a single equation $Df = 0$. Then $d = n - 1$ and we write T for T^{n-1}, ξ for ξ^{n-1}, etc. We shall *not* assume that $t = 0$ is noncharacteristic, but we shall assume that $t = 0$ is nondegenerate, that is, that

$$D = \partial^m/\partial t^m + D_0(\partial/\partial\xi)\partial^{m-1}/\partial t^{m-1} + \cdots + D_m(\partial/\partial\xi). \tag{9.32}$$

The problem is to determine necessary and sufficient conditions on $f_0, \ldots, f_{m-1} \in \mathscr{W}(T)$ in order that there exist an $f \in \mathscr{W}$ with $Df = 0$ and $\partial^j f(0, \xi)/\partial t^j = f_j(\xi)$. We say in this case that $(f_0, \ldots, f_{m-1})$ is *admissible Cauchy data* (for $\mathscr{W}$, D). Of course, if the Cauchy problem is well-posed, then there are no conditions on the f_j. On the other hand, if $t = 0$ is characteristic then (see Hörmander [1]) f need not be unique.

It follows easily from properties of AU spaces that the polynomials belong to $\mathscr{W}(T)$ and are dense in $\mathscr{W}(T)$ if we assume that $\mathscr{W}(T)$ is reflexive. Moreover, we can choose the f_j to be polynomials and still have a solution $f \in \mathscr{W}$. Thus, the set of admissible Cauchy data is dense in $(\mathscr{W}(T))^m$.

We can give necessary and sufficient conditions for $f_0, \ldots, f_{m-1}$ to be Cauchy data by means of Theorem 7.1 or Theorem 7.2. For simplicity we use Theorem 7.1 and obtain (in the notation of that theorem)

THEOREM 9.16. *A necessary and sufficient condition that* $(f_0, \ldots, f_{m-1})$ *should be Cauchy data is that there exist bounded measures* μ_j on V_j *and a* $k \in K$ *so that*

$$f_l(\xi) = \sum_{j=1}^{p} \int \left\{\frac{\partial^l}{\partial t^l} [\partial_j \exp(iz \cdot x)]\right\}_{t=0} d\mu_j(z)/k(z). \tag{9.33}$$

This result is somewhat complicated and we should like a simpler formulation. The difficulty is that the conditions on the f_l are not independent, that is, the fact that for each l there exists an $f^l \in \mathscr{W}$ with $Df^l = 0$ and $\partial^l f^l(0, \xi)/\partial t^l = f_l(\xi)$ does not imply that $(f_0, \ldots, f_{m-1})$ is Cauchy data. The simplest example of such a phenomenon is when D is the product of an elliptic and a hyperbolic operator. Less trivial examples can be constructed using Theorem 9.17 below. We may, therefore, ask when an f_l has the above property

THEOREM 9.17. *A necessary and sufficient condition that (for i fixed) there exists an $f \in \mathscr{W}$ with $Df = 0$ and $\partial^i f(0, \xi)/\partial t^i = a(\xi)$ is that a admits a representation*

$$a(\xi) = \int \exp (iw \cdot \xi)\, dv(w)/r(w). \tag{9.34}$$

Here v is a bounded measure and r is a continuous positive function satisfying

$$r(w) \geqslant k[s_q(w), w] \tag{9.35}$$

for some k in an AU structure for $\mathscr{W}$ and for some q. (q may depend on w.) The $s_q(w)$ are the roots of $P(s, w) = 0$.

PROOF. For simplicity we shall consider only the case $i = 0$ as the other cases are treated similarly. Suppose first that there exists an f with $Df = 0$ and $f(0, \xi) = a(\xi)$. Then by Theorem 9.16 we can write

$$a(\xi) = \sum_j \int \{\partial_j \exp (iz \cdot x)\}_{t=0}\, d\mu_j(z)/k(z). \tag{9.36}$$

We note that the ∂_j are differential operators with polynomial coefficients; thus $\partial_j \exp (iz \cdot x)$ is of the form

$$\partial_j \exp (iz \cdot x) = \sum b_{j\alpha}(x) c_{j\alpha}(z) \exp (iz \cdot x),$$

where the $b_{j\alpha}$ and the $c_{j\alpha}$ are polynomials. We write, formally, $d\mu_j(s_\beta(w), w)$ for $d\mu_j(z)$ on V_j. Then (9.36) becomes (formally)

$$a(\xi) = \sum_{j,\alpha} \int [b_{j\alpha}(0, \xi) \exp (iw \cdot \xi)] \sum_\beta c_{j\alpha}(s_\beta(w), w)\, d\mu_j(s_\beta(w), w)/k(s_\beta(w), w). \tag{9.37}$$

The functions $c_{j\alpha}(s_\beta(w), w)$ are polynomially increasing in w and the "measures" $d\mu_j(s_\beta(w), w)$ are of total bounded variation in w. It is easy to justify this formal argument and it follows that we can write

$$a(\xi) = \sum_{j,\alpha} \int b_{j\alpha}(0, \xi) \exp (iw \cdot \xi)\, dv_{j\alpha}(w)/r'(w), \tag{9.38}$$

where r' satisfies (9.35) and the $v_{j\alpha}$ are bounded measures. But this gives

$$a(\xi) = \sum_{j,\alpha} \int [b_{j\alpha}(0, -i\partial/\partial w) \exp (iw \cdot \xi)]\, dv_{j\alpha}(w)/r'(w) \tag{9.39}$$

from which our result follows easily from Cauchy's integral formula (which allows us to replace differentiation in w by integration).

Conversely, suppose a is of the form (9.34). We want to construct μ_j and k so that (9.36) holds. It simplifies our writing somewhat to observe that for the multiplicity variety $\mathfrak{V}$ each of the varieties that make up $\mathfrak{V}$ occurs with the identity operator in $\mathfrak{V}$. (Of course, it may occur with other operators as well.) For, in the present situation, $l = r = 1$, so we can appeal to Theorem 3.8 and we see that all varieties in $\mathfrak{V}$ occur with the identity operator. When constructing the μ_j we shall make them zero if ∂_j is not the identity. Thus, we shall assume that each ∂_j is the identity. Here $\mathfrak{V} = (V_1, \partial_1; \cdots; V_p, \partial_p)$.

We choose k satisfying (9.35). Then for each w we choose some $q = q(w)$ so that (9.35) holds. We construct μ (formally) by

$$\mu(s_\beta(w), w) = \begin{cases} v(w)k(s_q(w), w)/r(w) & \text{for } \beta = q \\ 0 & \text{for } \beta \neq q. \end{cases} \tag{9.40}$$

Then we find, as in (9.37) that if we define f by

$$f(x) = \int \exp(iz \cdot x)\, d\mu(z)/k(z), \tag{9.41}$$

then

$$\begin{aligned} f(0, \xi) &= \int \exp(iw \cdot \xi) \sum_\beta d\mu(s_\beta(w), w)/k(s_\beta(w), w) \\ &= \int \exp(iw \cdot \xi)\, dv(w)/r(w) \\ &= a(\xi) \end{aligned} \tag{9.42}$$

which is the desired result.

It is easy to justify the above formal procedure by a slight modification of the definition of μ in (9.40) and working with μ and v as measures instead of functions. We leave the details to the reader. ▯

It is proved in Ehrenpreis [10] that every indefinitely differentiable function $a(x)$ $(n = 1)$ can be written in the form

$$a(\xi) = f^+(\xi) + f^-(\xi),$$

where f^+ (f^-) is analytic in the open upper (lower) half-plane and indefinitely differentiable in the closed half-plane. We can recast this result in the following manner: We weaken the Cauchy (or initial value) problem for $\partial/\partial t + i\partial/\partial\xi$ and ask when a given $a(\xi)$ can be written as the sum of the Cauchy data of solutions in the half-planes $t > 0$ and $t < 0$. The path to generalization is clear: Suppose first that $r = 1$. When can every $(f_0(\xi), \ldots, f_{m-1}(\xi))$ be expressed as the sum of Cauchy data of f^+ and f^- where $Df^+ = 0$ $(Df^- = 0)$ in the half space $t > 0$ $(t < 0)$?

THEOREM 9.18. *Given any* $(f_0, \ldots, f_{m-1}) \in (\mathscr{W}(\xi))^m$ *there exist* $f^+ \in \mathscr{W}(t \geqslant 0)$ *and* $f^- \in \mathscr{W}(t \leqslant 0)$ *with* $Df^{\pm} = 0$ *and*

$$f_l(\xi) = \partial^l f^+(0, \xi)/\partial t^l + \partial^l f^-(0, \xi)/\partial t^l \tag{9.43}$$

for all l.

PROOF. The method is somewhat similar to the proof of the sufficiency of (9.34) in Theorem 9.17 but is more complicated. We write each f_l in the form

$$f_l(\xi) = \int \exp(iw \cdot \xi)\, dv_l(w)/k_l(w), \tag{9.44}$$

where v_l are bounded measures and k_l belong to an AU structure for $\mathscr{W}(t = 0)$. We want to construct bounded measures $\mu_j^{\pm}$ on V_j and $k_j^{\pm}$ in an AU structure for $\mathscr{W}$ so that, on defining

$$f^{\pm}(x) = \sum \int [\partial_j \exp(iz \cdot x)]\, d\mu_j^{\pm}(z)/k_j^{\pm}(z) \tag{9.45}$$

we have (9.43).

In order to make the ideas clear, let us assume first that all ∂_j are the identity so we may write $\mu^{\pm}$ for $\mu_j^{\pm}$. The proof of (9.37) shows that we should try to make, for every l,

$$\sum_{\beta} s_{\beta}^l(w)[\mu^+(s_{\beta}(w), w)/k^+(s_{\beta}(w), w) + \mu^-(s_{\beta}(w), w)/k^-(s_{\beta}(w), w)] = v_l(w)/k_l(w). \tag{9.46}$$

In order to understand the construction we note that we should try to make $\mu^{\pm}/k^{\pm}$ very small when $\operatorname{Im} s_{\beta}(w) \lessgtr 0$ while trying to keep the $\mu^{\pm}/k^{\pm}$ about the same size as v/k for other $z \in \mathfrak{V}$. For then the integrals in (9.45) would converge in the appropriate half-plane in the same way as the integrals (9.44). (Convergence is, in fact, helped by the fact that the factor exp(its) is small.) The reason for this is that $f^{\pm}$ is to be a solution only in the upper (lower) half-plane. Thus the integrals on the right side of (9.45) need converge only in this half-plane. The reason they may not converge in the other half-plane is that $\mu^{\pm}/k^{\pm}$ can be large in $\operatorname{Im} s_{\beta} \gtrless 0$. The fact that they can be large there is needed to satisfy (9.46). The simplest way of making the $\mu^{\pm}/k^{\pm}$ small in the appropriate half-plane is to make them zero. There is still the case of real $s_{\beta}(w)$ but we note that the complement of a neighborhood of this set is sufficient for $\mathscr{W}$ (see Chapter VII), so we may assume that no $s_{\beta}(w)$ is real since real $s_{\beta}(w)$ need not be used in the integral representation. (We are assuming that $n > 0$; if $n = 0$, Theorem 9.18 is trivial.) We are thus led to the system of equations

$$\text{(9.47)}\quad \text{(a)}\quad \sum s^l_\beta(w)[\mu^+(s_\beta(w), w)/k^+(s_\beta(w), w) + \mu^-(s_\beta(w), w)/k^-(s_\beta(w), w)] = v_l(w)/k_l(w)$$

$$\text{(b)}\quad \mu^+(s_\beta(w), w) = 0 \qquad \text{for } Is_\beta(w) < 0$$

$$\mu^-(s_\beta(w), w) = 0 \qquad \text{for } Is_\beta(w) > 0$$

for the unknown quantities

$$\mu^+(s_\beta(w), w)/k^+(s_\beta(w), w) \quad \text{and} \quad \mu^-(s_\beta(w), w)/k^-(s_\beta(w), w).$$

For fixed w, because of (9.47b), the system (9.47a) is a system of m equations in m unknowns. The complement of a neighborhood of the set of w where the determinant of the coefficients (Vandermonde determinant) is zero is again $\mathscr{W}$-sufficient so we may assume the determinant is always different from zero (in fact, even $\geqslant 1$ in absolute value). Thus, (9.47) admits a unique solution.

By Cramer's rule we deduce from (9.47) and Theorem 1.4 (which is needed to handle the denominators that occur) the following bounds for $\mu^\pm/k^\pm$:

$$\text{(9.48)}\qquad |\mu^\pm(s_\beta(w), w)/k^\pm(s_\beta(w), w)| \leqslant c(1 + \max |s_\beta(w)|)^c(\max |v_l(w)/k_l(w)|).$$

This shows that, except for polynomial factors—which, of course, are ignorable for LAU spaces—$\mu^\pm(s_\beta(w), w)/k^\pm(s_\beta(w), w)$ are zero in appropriate regions and behave in the rest like the $|v_l(w)/k_l(w)|$. But we note that in the integrals (9.45) if $t > 0$ ($t < 0$), then in the part of the integral where $Is > 0$ ($Is < 0$) we are aided by a factor $\exp(-t|Is|)$. Thus, the integrals in these regions converge better than the integrals in (9.44), so that $f^\pm \in \mathscr{W}^\pm$.

In case the ∂_j cannot all be chosen to be the identity, then we know by the method of their construction that they can always be chosen to be of the form $\partial^{\alpha_j}/\partial s^{\alpha_j}$ (compare Theorem 2.8). Then formally we shift the operators ∂_j to $\mu^\pm/k^\pm$ in (9.45). Thus, (9.47a) becomes

$$\text{(9.49)}\qquad \sum s^l_\beta(w)(\partial^{\alpha_j}/\partial s^{\alpha_j})[\mu^+_j(s_\beta(w), w)/k^+(s_\beta(w), w) + \mu^-_j(s_\beta(w), w)/k^-(s_\beta(w), w)] = v_l(w)/k_l(w).$$

The system (9.49) and (9.47b) can be solved because it corresponds to the Lagrange interpolation formula with multiple points (compare Lemma 2.12). We obtain estimates similar to those in (9.48).

It is a simple matter to rigorize all these formal calculations. ▯

See Remark 9.10.

IX.7. The Fundamental Solution and Nonlinear Initial Surfaces

In the classical theory of partial differential equations much importance is attached to the fundamental solutions, that is, distributions f such that $Df = \delta$ (Dirac measure). In particular, suitable fundamental solutions are used to solve boundary value problems, in particular Cauchy's problem. We shall write $\vec{\mathrm{D}} = (D_1, \ldots, D_r)$. As in Section VI.2 we call $\vec{\mathrm{F}} = (F_1, \ldots, F_r)$, where each $F_j \in \mathscr{W}$, *a fundamental solution for* $\vec{\mathrm{D}}$ if $\sum D_j F_j = \delta$. If we have instead $\sum D_j F_j = \delta_a$ we call F a *fundamental solution at* a.

Thus, the existence of a fundamental solution in $\mathscr{W}$ implies that $\delta \in \mathscr{W}$.

Actually, as is also apparent in Chapter XII, the fundamental solution is usually not too useful, but it is the *parametrix*, which is a suitable approximation to the fundamental solution, that is important. To see what this is for the Cauchy problem, let us suppose for simplicity that $\mathscr{W}$ is a space of functions and that $\mathscr{W}'$ contains δ_a for any a. Suppose we are given a well-posed Cauchy problem for $\vec{\mathrm{D}}$. Then we can write by (9.3)

$$\delta_a = \sum D_i'\,{}^iS^a + \sum q_{ij}' S_{ij}^a, \tag{9.50}$$

where the S_{ij}^a have their supports on T^j. Thus, we see that $({}^iS^0)$ is "approximately" a fundamental solution at zero, the "error" being measured by $\sum q_{ij}' S_{ij}^0$. Our control over the error term is that we know its support. [Actually, it is D_i' rather than D_i that appears in (9.50). Thus $({}^iS^0)$ is "approximately" a fundamental solution for $(D_1', \ldots, D_r')$ rather than for $\vec{\mathrm{D}}$. The reason for the change from D_i to D_i' is that (9.50) is the analog of (8.2)].

In case $r = 1$ it is easily seen that the terms S^a and S_{ij}^a on the right side of (9.50) are uniquely determined by a. This is, of course, false for $r > 1$ because we can add to $({}^1S^a, \ldots, {}^rS^a)$ any element of the module of relations of $\vec{\mathrm{D}}'$.

As in Section IX.2, let us assume that all the T^j are contained in $T^d = T$. We write t for t^{n-d}, ξ for ξ^d, etc. As above, we write $t > t'$ if $t \neq t'$ and $t_j \geqslant t_j'$ for all j. As in our discussion of groups related to the Cauchy problem, we shall consider the Cauchy problem with the T^j shifted to $T_{t^0}^j$ which is the translation from $t = 0$ to $t = t^0$. Then, in view of Lemma 9.13 we may write, instead of (9.50), for any a, t

$$\delta_a = \sum D_i'\,{}_{t^0}^{\;i}S^a + \sum q_{ij}'\,{}_{t^0}S_{ij}^a \tag{9.51}$$

where support ${}_{t^0}S_{ij}^a$ is on $T_{t^0}^j$.

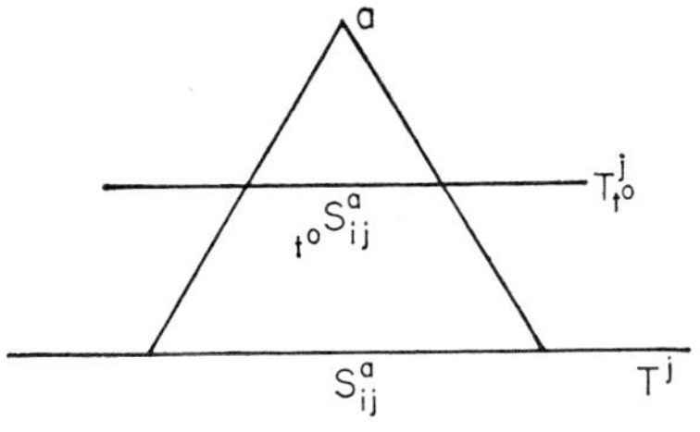

FIGURE 14

When $r = 1$ we can "piece" together the $_{t^0}^{\,i}S^a$ to obtain a fundamental solution, namely the limit of $_{t^0}^{\,i}S^a$ as $t^0 \to -\infty$ (see Fig. 14). This is possible because of the uniqueness of (9.51). In case $r > 1$ we need a different procedure.

We shall consider only the case of a Cauchy hyperbolic system $\vec{D}$, although much of what we say applies to more general situations. Thus, we shall assume in the rest of this section that $\mathscr{W} = \mathscr{E}$ is the space of indefinitely differentiable functions. Then each $_{t^0}S^a_{ij}$ is of compact support on T^i. It is an immediate consequence of the definition of "well-posed Cauchy Problem" that they are uniquely determined by a, t^0. Thus in (9.51) we want to make a "best" choice of the $_{t^0}^{\,i}S^a$. A "good" choice is afforded by the following result:

LEMMA 9.19. *Let a, t^0 be fixed. Let $A(a, t^0)$ be the convex hull of a and the union of the supports of the $_{t^0}S^a_{ij}$. Then we can choose the $_{t^0}^{\,i}S^a$ to have their supports in A.*

PROOF. We rewrite (9.51) as

$$\sum D'_i \, {}_{t^0}^{\,i}S^a = \delta_a - \sum q_{ij} \, {}_{t^0}S^a_{ij} \tag{9.52}$$

and we regard (9.52) as an equation in terms of distributions on $A(a, t^0)$ for the unknowns $_{t^0}^{\,i}S^a$. By Theorem 6.1 and Chapter V, Example 5 (which applies because we need only check bounds relevant to convex polyhedra) this can be solved if the right side of (9.52) satisfies certain necessary conditions. But these necessary conditions are independent of $A(a, t^0)$ and, since there exist solutions $_{t^0}^{\,i}S^a \in \mathscr{E}'(R^n)$, it follows that they hold. The lemma is therefore proved. □

The possibility of relating the $_{t^0}^{\,i}S^a$ for varying a and t^0 is

LEMMA 9.20. *For any a, t^0, i, j we have*

$$_{t^0}S^a_{ij} = S^{a-t^0}_{ij}. \tag{9.53}$$

Moreover, for an appropriate choice of $_{t^0}^{\,i}S^a$ we have

$$_{t^0}^{\,i}S^a = {}^iS^{a-t^0}, \tag{9.54}$$

where $a - t^0$ *is defined by identifying* t^0 *with* $(t^0, 0) \in R^n$.

PROOF. (9.53) is proved in the proof of (9.25). (9.54) then follows immediately. ▯

For a more precise way of relating the ${}_{t^0}^{\,i}S^a$ for fixed a and varying t^0 we need

LEMMA 9.21. *We can choose the* ${}_{t^0}^{\,i}S^a$ *[considered as a distribution in* $x = (t, \xi)$*] to be indefinitely differentiable in t relative to* ξ *for* $x \neq a$ *and for* x *off* $t = t^0$.

Here by a distribution being indefinitely differentiable in t relative to ξ (see Ehrenpreis [4]) we mean that its convolution with any $h \in \mathscr{D}(\xi)$ has the property that all its t derivatives are distributions of bounded order.

PROOF. Except for $x = a$ or x on $t = t^0$, we have $\sum D_i' \, {}_{t^0}^{\,i}S^a = 0$. Thus, what we need is an "elliptic" property of $\vec{\mathrm{D}}$, that is, a property which guarantees regularity of solutions of homogeneous equations. But $t = t^0$ is a principal noncharacteristic so the result can be proved by the methods of Ehrenpreis [4] or of Chapter VIII. ▯

It follows from Lemma 9.21 that if $a = (t^2, \xi)$, then for $t^0 < t^1 < t^2$ and any $l = 1, 2, \ldots, n - d$, we can multiply each ${}_{t^0}^{\,i}S^a$ by $\chi(t_l \geqslant t_l^1)$ which we denote also by $\chi^+(t_l^1)$, the characteristic function of $t_l \geqslant t_l^1$. We should like to say that, for a suitable l,

$$\sum D_i'[\chi^+(t_l^1)\,{}_{t^0}^{\,i}S^a] = \delta_a - \sum q_{ij}\,{}_{t^1}S_{ij}^a. \tag{9.55}$$

Unfortunately, this cannot be true in general because the ${}_{t^0}^{\,i}S^a$ are not uniquely determined. Thus we are led to formulate

PROBLEM 9.6.

Can we make a "coherent" choice of ${}_{t^0}^{\,i}S^a$ so that (9.55) holds?

If this can be done, then

$$\lim_{t^0 \to (-\infty, \ldots, -\infty)} {}_{t^0}^{\,i}S^a$$

would be a fundamental solution which would be a "good" one for the Cauchy problem, namely, we could use it as in the classical case of $r = 1$ to solve Cauchy problems in some cases when the Cauchy data is given on some nonlinear surfaces (see the remarks following the proof of Proposition 8.6). The limit is to be taken in the topology of $\mathscr{D}'$. Even when the above limit does not exist we can use the ${}_{t^0}^{\,i}S^a$ to solve the Cauchy Problem with data given on some nonlinear surfaces.

If the answer to Problem 9.6 is in the affirmative then the fundamental solution we construct would have its support contained in the cone with vertex at a whose base is contained in a d-dimensional linear variety.

The existence of such fundamental solutions can be proved using the type of construction mentioned in Section VI.2 combined with the methods of Ehrenpreis [4] (compare also Section XI.2).

We can now define the general concept of a spacelike surface for $\vec{D}$.

DEFINITION. An indefinitely differentiable surface T of dimension d in R^n is called *spacelike* if each x^0 in T has a neighborhood $N(x^0)$ in R so that for every $a \in N(x^0)$, $a \notin T$, we can find distributions of compact support ${}_{x^0}^{\ i}S^a$ on R so that

$$\delta_a - \sum D_i' \, {}_{x^0}^{\ i}S^a \text{ has support } \subset T. \tag{9.56}$$

From (9.56) it follows that for any f with $\vec{D}f = 0$ we can express $f(a)$ in terms of values of derivatives of f on T.

PROBLEM 9.7. (CONJECTURE).

T is spacelike if and only if for each a near T there is a compact subset A_a of T such that there exists a fundamental solution at a whose support is the union of the infinite rays (half lines) which start from a and pass through A_a, and such that this fundamental solution has a restriction to T.

It is clear that any T satisfying the conditions of the conjecture is spacelike. Moreover by use of the theory of hyperbolic operators for $r = 1$, we can establish the converse in that case, so the conjecture is true for $r = 1$. But for $r > 1$ we have not investigated it nor have we investigated the question of which surfaces T have the property of the existence of fundamental solutions with the properties mentioned in the conjecture.

See Remark 9.11.

IX.8. Domain of Dependence and Lacunas

We assume again that $\vec{D}$ is Cauchy hyperbolic. We assume we are given a well-posed Cauchy problem (on linear initial surfaces T^j).

DEFINITION. For $a \in R$, the union of the supports of S^a_{ij} is called the *domain of dependence of* a. For a set $A \subset \cup T^j$ the *sphere of influence* of A is the set of all a whose domain of dependence intersects A.

We observe that the Fourier transforms of the S^a_{ij} are given by explicit formulas involving iterations of Lagrange interpolation. From this we can obtain bounds on the convex hull of the supports of the S^a_{ij}, because we can determine the exponential type of their Fourier transforms. In special cases it should be possible to use the method of Petrowsky [2] to obtain more precise information on the support of S_{ij}^a, in particular, to determine if lacunas exist, that is, the support of S_{ij}^a is the union of disjoint parts some of which have dimension $<j$.

IX.9. Uniqueness of the Cauchy Problem

Historically, the first important uniqueness result in the direction we are aiming is due to Holmgren [1] who proved that, for $r = 1$, any $f \in \mathscr{E}$ with $Df = 0$ is determined by its Cauchy data on any noncharacteristic hypersurface. This result has been extended to certain characteristic cases, the most complete results being due to Täcklind [1] who considered the heat equation and certain similar equations. In the characteristic case we must impose growth conditions on f in order to guarantee uniqueness and Täcklind determined the appropriate growth conditions. We shall derive the extension of the Holmgren-Täcklind results to $r \geqslant 1$. Even for $r = 1$ our results are new in the characteristic case, except for those equations which were treated by Täcklind. (His methods seem essentially restricted to parabolic equations of a special kind.)

We illustrate first how our method can be used heuristically (for $r = 1$) to prove Holmgren's theorem. Let $f \in \mathscr{E}$, $Df = 0$, and f and all its derivatives vanish on $t = 0$ which is noncharacteristic. To conclude that $f = 0$, we must show that $U \cdot f = 0$ for a dense set of $U \in \mathscr{E}'$. It suffices to produce a dense set of $U \in \mathscr{E}'$ of the form $U_1 + D'U_2$ where U_1 has its support on T; thus, for suitable U, we want to solve the equation

$$U = U_1 + D'U_2 . \tag{9.57}$$

Note that (9.57) is of the form (8.2). Thus our method for proving uniqueness is a parametrix method.

The Fourier transform of (9.57) can be written in the form

$$F(s, w) = \sum s^i F_i(w) + P(s, w)G(s, w), \tag{9.58}$$

where by our often used method we may assume that $i < m =$ degree P. Of course, if D is hyperbolic then (9.58) has a solution $F_i \in \mathbf{E}'$, $G \in \mathbf{E}'$ for any $F \in \mathbf{E}'$. But for D nonhyperbolic we still want to produce a dense set of F for which (9.58) has a solution.

We can solve (9.58) uniquely for entire functions F_i, G by power series methods as in the proof of Theorem 2.8. Then it suffices to show that $F_i \in \mathbf{E}'$. We know that $F_i(w)$ can be expressed in terms of the values $F(s^j(w), w)$, where $P(s^j(w), w) = 0$ and polynomial factors which play an unimportant role. The fact that $t = 0$ is noncharacteristic means (see

Lemma 9.6) that $|s^j(w)| \leqslant \text{const } (1 + |w|)$. Thus, Holmgren's uniqueness theorem is a consequence of

LEMMA 9.22. *For any $c > 0$ there exists a dense set of $F \in \mathbf{E}'$ which in the "angular region"*

$$|s| \leqslant c(1 + |w|) \tag{9.59}$$

satisfies

$$F(s, w) = 0(1 + |s| + |w|)^{c'} \exp (c'|\text{Im } w|). \tag{9.60}$$

PROOF. In order to understand the construction, let us consider first case $n = 2$. Let $\eta^{\pm}$ be the part of the parabola

$$t = \pm(1 - \xi_1^2)$$

for which $t \gtrless 0$. For any polynomial A let $U_A^{\pm}$ denote the distribution of compact support which is A multiplied by the mass $d\xi_1$ on $\eta^{\pm}$. It is clear that the linear combinations of the $U_A^{\pm}$, and their translations in the ξ direction, and their dilations, that is, the distributions obtained by mapping $x \to bx$ for $b > 0$, are dense in $\mathscr{E}'$. We claim that $F_A^{\pm}$ which is the Fourier transform of $U_A^{\pm}$ satisfies (9.60) in the region (9.59) for a suitable c which can be chosen independent of A. The same then clearly holds for the Fourier transform of the ξ translates, and the dilations of the $U_A^{\pm}$.

Since the Fourier transform of multiplication by A is application of a linear partial differential operator with constant coefficients, it suffices, by Cauchy's formula, to prove our contention for $A = 1$. We shall prove the result for F_1^+ as the result for F_1^- is similar. We write F for F_1^+.

By definition,

$$\begin{aligned} F(s, w) &= \int_{\eta^+} \exp (its + i\xi_1 \cdot w_1) d\xi_1 \\ &= \int_{|\xi_1| \leq 1} \exp [is(1 - \xi_1^2) + i\xi_1 \cdot w_1] d\xi_1. \end{aligned}$$

For simplicity we may assume that $\text{Re } w_1 \geqslant 0$. We may also assume that actually

$$|s| \leqslant c(1 + \text{Re } w_1)$$

since if (9.59) holds and $|\text{Im } w_1| \geqslant |\text{Re } w_1|$, then (9.60) is clear since $F \in \mathbf{E}'$. We shall assume that c is small.

Now, F clearly satisfies (9.60) in $\text{Im } s \geqslant 0$ so we may assume that $\text{Im } s < 0$. We shift the contour in the complex ξ_1 plane to the semicircle $|\xi_1| = 1$, $\text{Im } \xi_1 > 0$. We claim that on the shifted contour

$$\text{Im } [s(1 - \xi_1^2) + \xi_1 \cdot w_1] \geqslant -c''(|\text{Im } w_1| + 1)$$

for a suitable c''; this clearly implies that F satisfies (9.60). We write $\xi_1 = e^{i\theta}$. The left side of the above inequality is

$$(\operatorname{Im} s)(1 - \cos 2\theta) - (\operatorname{Re} s) \sin 2\theta + (\operatorname{Im} w_1) \cos \theta + (\operatorname{Re} w_1) \sin \theta.$$

Now, the terms involving $\operatorname{Im} w_1$ give no trouble. The term $(\operatorname{Re} s) \sin 2\theta$ is certainly dominated by the term $(\operatorname{Re} w_1) \sin \theta$ since, by the above, $|s| \leqslant c(1 + \operatorname{Re} w_1)$ for some small c. Since $\operatorname{Re} w_1 \geqslant 0$ and $\sin \theta \geqslant 0$, these terms present no problems. As for the term involving $\operatorname{Im} s$, we have

$$(\operatorname{Im} s)(1 - \cos 2\theta) = 2(\operatorname{Im} s) \sin^2 \theta$$

which can again be absorbed in $(\operatorname{Re} w_1) \sin \theta$.

Thus our contention

$$\operatorname{Im} [s(1 - \xi_1^2) + \xi_1 \cdot w_1] \geqslant -c''(|\operatorname{Im} w_1| + 1)$$

is established. This proves the lemma for small c. The result for arbitrary c is obtained by apply dilatations in the ξ_1 direction. This completes the proof of Theorem 9.22 in case $n = 2$.

In case $n > 2$, we define $\eta^{\pm}$ as the parabolic region

$$t = \pm(1 - \xi_1^2 - \cdots - \xi_{n-1}^2), \qquad \sum \xi_j^2 \leqslant 1.$$

For any polynomial A we define $U_A^{\pm}$ as the distribution which is A multiplied by $d\xi_1 \cdots d\xi_{n-1}$ on $\eta^{\pm}$. As above we want to prove that F, which is the Fourier transform of U_1^{+}, satisfies (9.60) in the region (9.59).

We write

$$F(s, w) = \int_{\eta^+} \exp\,(its + i\xi \cdot \operatorname{Re} w + i\,\xi \cdot \operatorname{Im} w)\, dS.$$

As in the case $n = 2$ treated above, the terms involving $\operatorname{Im} w$ will present no difficulty. Since the integrand is invariant under rotations in ξ, $F(s, w)$ is invariant under rotations in w. We may assume, therefore, that $\operatorname{Re} w_2 = \cdots = \operatorname{Re} w_{n-1} = 0$. Thus

$$\begin{aligned} F(s, w) &= \int_{\eta^+} \exp\,(its + i\xi_1 \operatorname{Re} w_1 + i\xi \cdot \operatorname{Im} w) d\xi_1 \cdots d\xi_{n-1} \\ &= \int d\xi_2 \cdots d\xi_{n-1} \int \exp\,(its + i\xi_1 \operatorname{Re} w_1 + i\xi \cdot \operatorname{Im} w)\, d\xi_1. \end{aligned}$$

The ξ_1 integral is estimated as in the case $n = 2$ treated above. This yields easily the desired estimate for F. ▯

The proof of Lemma 9.22 yields

LEMMA 9.22*. *For any $c > 0$ there exists a dense set of $G \in \mathbf{E}'(t \geqslant 0)$ (or $G \in \mathbf{E}'(t \leqslant 0)$) which satisfies* (9.60) *in* (9.59).

We also have

LEMMA 9.22**. *Let us write* $z = (a, b)$, *where* $a = (z_1, \ldots, z_p)$ *and* $b = (z_{p+1}, \ldots, z_n)$. *Then, for any* $c > 0$ *there is a dense set of* $F \in \mathbf{E}'$ *which in the "angular region"*

$$|a| \leqslant c(1 + |b|) \tag{9.59}**$$

satisfies

$$|F(z)| \leqslant c'(1 + |z|)^{c'} \exp(c'|\mathrm{Im}\, b|). \tag{9.60}**$$

Lemma 9.22** is proven by applying the construction used in the proof of Lemma 9.22 to linear varieties of dimension $n - p + 1$ which contain the $n - p$ dimensional plane defined by $a = 0$. We leave the details to the reader.

There are, of course, analogous results for $\mathbf{E}'(\Omega)$, where Ω is a half-space or a convex set of a special kind.

It should be observed that (9.60) is a necessary condition that F should be of the form (9.58) for *all* P whose zeros satisfy (9.59).

Next let us examine the case when $t = 0$ is characteristic (still for $r = 1$). Then it is proved by Hörmander [1] that the obvious analog of Holmgren's uniqueness theorem cannot hold, namely, there exist solutions $f \in \mathscr{E}$ of $Df = 0$ which vanish together with all their derivatives on $t = 0$. Thus, we must consider a more restricted class of f in order to obtain uniqueness, We shall, therefore, assume that $f \in \mathscr{E}(\Phi)$, where $\mathscr{E}(\Phi)$ is a space of rapidly increasing functions (see Chapter V, Example 7). We wish to determine the conditions on Φ in order that uniqueness should hold for D. We suppose now that Φ is a function of the ξ variables and that $\mathscr{E}(\Phi)$ has no growth restrictions in the t variable. Thus, our notation differs slightly from that of Chapter V, Example 7. For simplicity we shall assume that $\Phi(\xi) = \varphi(\xi_1) + \cdots + \varphi(\xi_{n-1})$, where φ is even, though the general case can probably be handled by using the Oka embedding method of Section IV.5. We assume that D is of the form (9.32).

Let us assume that all the zeros of P lie in the region

$$|s| \leqslant c(1 + |w|^{\alpha}), \tag{9.61}$$

where $\alpha \geqslant 1$. Then again the problem is that of determining conditions on Φ in order that the set of $F \in \mathbf{E}'(\Phi)$ which satisfy the analog of (9.60) in (9.61) should be dense in $\mathbf{E}'(\Phi)$. This seems to be quite difficult, so we make the following

PROBLEM 9.8. (CONJECTURE).
A necessary and sufficient condition that the set of $F \in \mathbf{E}'(\Phi)$ which satisfies for some b, c'

$$F(s, w) = 0[(1 + |s| + |w|)^{c'} \exp(\Psi(b\, \mathrm{Im}\, w))] \tag{9.62}$$

in the region (9.61) be dense in the space $\mathbf{E}'(\Phi)$ is that

$$\int \frac{\Psi(\operatorname{Im} w)}{1+|\operatorname{Im} w|^{\alpha+1}}\, d \operatorname{Im} w = \infty. \tag{9.63}$$

Here Ψ is the conjugate of Φ in the sense of Chapter V, Example 7.

Although we cannot establish our conjecture, nevertheless we can prove (by somewhat different methods which will be explained below)

THEOREM 9.23. (9.63) *is a necessary and sufficient condition for the uniqueness of the Cauchy problem for D in the space $\mathscr{E}(\Phi)$.*

See Remark 9.12.

We wish now to explain how our method extends to over-determined systems ($r > 1$). In order to do this, we must first decide on which sequences (T^d, q_{id}, T^{d-1}, $q_{i,d-1}, \ldots$) will be admitted as candidates. We recall the definition of Section IX.1: The sequence (T^d, q_{id}, T^{d-1}, $q_{i,d-1}, \ldots$) will be called *regular* if the corresponding Cauchy problem is well-posed for the space of formal power series. T^d is called a *principal regular characteristic* for $\vec{\mathrm{D}}$.

Exactly as in the Cauchy-Kowalewski case treated above, we can define the derived sequence: $\vec{\mathrm{D}} = {}^d\vec{\mathrm{D}}$, ${}^{d-1}\vec{\mathrm{D}}, \ldots, {}^0\vec{\mathrm{D}}$. For Cauchy problems arising from derived sequences, the proof of Theorem 9.8 yields easily the fact that $\hat{\gamma}'$ is a topological isomorphism on the space of entire functions, that is, the Cauchy problem is well-posed for the space $\mathbf{H}'$ of entire functions of exponential type.

By analyzing the proof of Theorem 9.8 we find

LEMMA 9.24. *Suppose we construct a well-posed Cauchy problem for $\mathbf{H}'$ by the method of derived sequences. Let F be an entire function and write*

$$F(z) = \sum F_{ij}(w^j) Q_{ij}(s^{n-j}) \bmod (P_1, \ldots, P_r). \tag{9.64}$$

Then the $F_{id}(w^d)$ can be expressed in terms of $F(z)$ in "Lagrange-like" formulas, that is,

$$F_{id}(w^d) = \sum_l \frac{R_{id}((s^{n-d})_l, w^d)(\partial_l F)((s^{n-d})_l, w^d)}{\Delta_{id}((s^{n-d})_l, w^d)}, \tag{9.65}$$

where ∂_l are differential operators with algebraic coefficients, $((s^{n-d})_l, w^d)$ are the points on $\mathfrak{V}$ above w^d, R_{id} and Δ_{id} are polynomials with $\Delta_{id} \not\equiv 0$. Here (9.65) *holds for those w^d for which $\Delta_{id} \neq 0$.*

We now define orders (c_j, α_j) for the system $\vec{\mathrm{D}}$: α_j is the smallest exponent so that

$$|s^{n-j}| = 0(1 + |w^j|)^{\alpha} \tag{9.66}$$

for all points in ${}^j\mathfrak{V}$, where ${}^j\mathfrak{V}$ is the multiplicity variety corresponding to ${}^j\vec{\mathrm{D}}$. Then c_j is the smallest number for which

$$|s^{n-j}| \leqslant c_j(1 + |w^j|)^{\alpha_j} \tag{9.67}$$

for $(s^{n-j}, w^j) \in {}^j\mathfrak{V}$ sufficiently large. It is easily seen from the algebraic nature of the ${}^j\mathfrak{V}$ that α_j is rational.

From Lemma 9.24 we deduce easily

LEMMA 9.25. *Under the same hypothesis as in Lemma 9.24, suppose there exist $c_j' > c_j$ so that the set of $F \in \mathbf{E}'(\Phi)$ which for each j, for some b, c'', satisfy*

$$F(s^{n-j}, w^j) = 0(1 + |s^{n-j}| + |w^j|)^{c''} \exp(\Psi(b \operatorname{Im} w^j)) \tag{9.68}$$

in the region

$$|s^{n-j}| \leqslant c_j'(1 + |w^j|)^{\alpha_j} \tag{9.69}$$

is dense in $\mathbf{E}'(\Phi)$. Then there is uniqueness for the Cauchy problem in $\mathscr{E}(\Phi)$

We cannot decide when there is density, but we make the

PROBLEM 9.9. (CONJECTURE)

A necessary and sufficient condition that the set of $F \in \mathbf{E}'(\Phi)$ which satisfy (9.68) in the region (9.69) be dense in $\mathbf{E}'(\Phi)$ is (9.63) for $\alpha = \max \alpha_j$.

Although we cannot prove the conjecture, we shall prove below

THEOREM 9.26. *A sufficient condition that every $f \in \mathscr{E}_{\vec{\mathrm{D}}}(\Phi)$ which vanishes together with all its derivatives on $t^{n-d} = 0$ be identically zero is*

$$\int \frac{\Psi(w)}{1 + |\operatorname{Im} w|^{\alpha_d + 1}}\, dw = \infty. \tag{9.70}$$

In general, Theorem 9.26 is weaker than the result we would get on combining Lemma 9.25 with our conjecture. For if our conjecture is true, then we would establish that for $f \in \mathscr{E}(\Phi)$, $\vec{\mathrm{D}}f = 0$, the vanishing of certain t^{n-d} derivatives of f on T^d together with the vanishing of certain t^{n-d-1} derivatives on T^{d-1}, etc., would imply that $f \equiv 0$. Since each $T^j \subset T^d$, this is weaker than requiring that all t^{n-d} derivatives of f vanish on T^d.

We shall also establish a converse to Theorem 9.26 under some special conditions.

We shall not prove this theorem yet, as it will be a consequence of a more general result proved below.

Our next generalization of the uniqueness problem had its beginning with a result of Courant-Hilbert [1], Vol. II, pp. 427–430 which was extended by John in [1] to cover a great number of situations for $r = 1$.

These results show that on T^{n-1} there may exist a subvariety $\tilde{T}$ of dimension $<n-1$ so that the vanishing of certain Cauchy data together with all their ξ derivatives on $\tilde{T}$ implies their vanishing on T^{n-1}.

The prototype of this situation is the case where D is analytic elliptic in which case we may take for $\tilde{T}$ a point. This shows us what we should expect, namely, in order for $\tilde{T}$ to reduce to a point (say the origin), we need to know not only that T^{n-1} is noncharacteristic but that *all* ρT^{n-1} obtained by applying to T^{n-1} a (real) rotation are noncharacteristic. This means that there is a $c > 0$ so that $\mathfrak{V}$ is contained in the region consisting of all $z \in C$ for which

$$|(\rho z)_n| \leqslant c(1 + |(\rho z)_1| + \cdots + |(\rho z)_{n-1}|)$$

for all ρ in the (real) rotation group. It could be shown (and this is important for what follows) that this is the same as saying that $\mathfrak{V}$ satisfies the conditions for analytic ellipticity of Chapter VIII. John has shown in [1] how to modify this in the nonanalytic elliptic case. Namely, if $\tilde{T}$ has dimension $\geqslant 0$, then we want the ρ to go over those rotations which preserve $\tilde{T}$.

The extension of John's results to the case when some ρT^{n-1} are characteristic can be made. But rather than explain this, we shall pass directly to the case of overdetermined systems. We shall suppose that we are given a regular Cauchy problem $(T^d, q_{id}, \ldots, T^0, q_{i0})$. Suppose we are given for each j a linear subspace $\tilde{T}^j$ of T^j, and a subsequence $\{q_{i'j}\}$ of the q_{ij}. (It is permitted that this subsequence be empty.) Then we ask

PROBLEM 9.10.

Under what conditions do the conditions $f \in \mathscr{W}$, $\vec{D}f = 0$, all ξ^j derivatives of $q_{i'j}f = 0$ on $\tilde{T}^j$ for all i', j imply $q_{ij}f = 0$ on T^j for all i, j?

We shall consider this problem only in case all the subsequences $\{q_{i'j}\}$ are empty except one and that one consists only of the identity, though the general case can probably be handled by similar methods. Slightly generalized this is

PROBLEM 9.11.

Let T be a linear subspace of R and $\tilde{T}$ a linear subspace of T. Under what conditions does the vanishing of a solution f of $\vec{D}f = 0$ with all its T derivatives on $\tilde{T}$ imply $f = 0$ on T?

If $\tilde{T}$ is a principal regular characteristic and $T = R$, then this is the problem dealt with in Theorem 9.26.

As usual we shall use coordinates ξ on T, (t, ξ) on R, with dual variables (s, w). We shall write $\xi = (\beta, \gamma)$, where γ is the coordinate on $\tilde{T}$. The dual variables will be denoted by u, v. See Fig. 15.

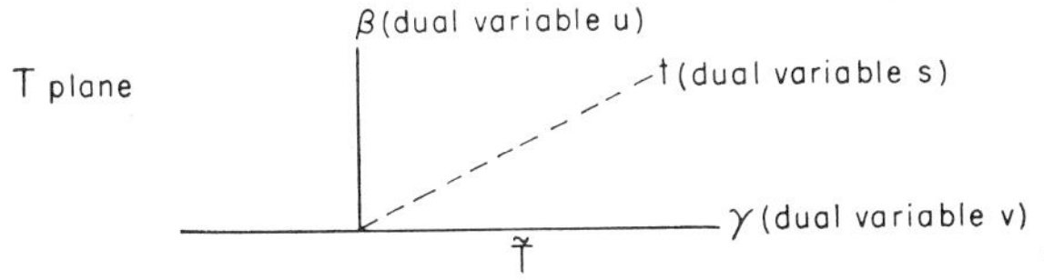

FIGURE 15

As mentioned above, if $\vec{\mathrm{D}}$ is analytic elliptic then f must be analytic so the vanishing of all ξ derivatives on any $\tilde{T}$ (even if $\tilde{T}$ is a point) imply $f=0$ on T. In his treatment of the case $r=1$ John in [1] showed that certain integrals of f were analytic in suitable variables from which the vanishing of f on T could be concluded. We shall show that a similar situation persists in general. We consider first the case of the space $\mathscr{E}$.

Let y be a complex parameter. Let $A \subset B$ be regions in w-space. Suppose we can find a function $H(y, w)$ with the following properties:

(a) $H(y, w)$ is an analytic function of the complex variables y, w.
(b) $\delta H(0, w)$ is a finite sum of terms, each of which is a polynomial in u times an element of $\mathbf{E}'(\tilde{T})$ for δ any derivative in y.
(c) For some fixed point y^0 with $|y^0|=1$ we have $H(y^0, w) \in \mathbf{E}'(T)$.
(d) For $|y| \leqslant 1$ and $w \in A$ we have

$$|H(y, w)| \leqslant c(1+|w|)^c \exp(c|\operatorname{Im} w|). \tag{9.71}$$

Here c should be independent of y in $|y| \leqslant 1$.

(e) For $|y| \leqslant 1$ for all $w \in B$ we have

$$|H(y, w)| \leqslant c \exp(c|u| + c|\operatorname{Im} w|). \tag{9.72}$$

Again c is independent of y in $|y| \leqslant 1$.

Any such H will be called $(\mathscr{E}; A, B)$ *admissible*.

If the set of $(\mathscr{E}; A, B)$ admissible functions has the property that the set of linear combinations of $H(y^0, w)$ is dense in $\mathbf{E}'(T)$, then we say that (A, B) is *analytically closed* for $\mathscr{E}$. (Of course, this depends on $\tilde{T}$ and T.) In case B is all of T we say that A is *analytically closed* for $\mathscr{E}$ if the above holds.

Let $\mathfrak{V}$ be the multiplicity variety for $\vec{\mathrm{D}}$ and let V be the union of the varieties of $\mathfrak{V}$. Let $c(A)$ be the cylinder of base A, that is, the set of (s, w) with $w \in A$. We say that $\mathfrak{V}$ is *analytic in* β *outside* $c(A)$ *if*

$$|\operatorname{Im} z| \geqslant c'(1+|\operatorname{Re} u|) \quad \text{for} \quad z \in V, z \notin c(A). \tag{9.73}$$

LEMMA 9.27. *Let B be a region such that the interior of $c(B)$ contains the set of z whose distance from V is $\leqslant 1$. Suppose (A, B) is analytically closed* *for* $\mathscr{E}$ *and* $\mathfrak{B}$ *is analytic in* β *outside* $c(A)$. *If* $f \in \mathscr{E}$, $\vec{D}f = 0$, f *and all* ξ *derivatives vanish on* $\tilde{T}$, *then* $f = 0$ *on* T.

PROOF. Let H be admissible. We use Theorem 7.1 to represent f (in that theorem we used w in place of f). By the definition of the integral in (7.1) we have

$$h(y^0, x) \cdot f(x) = \sum \int (\partial_l H)(y^0, z) \, d\mu_l(z)/k(z),$$

where $H(y^0, z)$ is the Fourier transform in x of $h(y^0, x)$. (Here $H(y, w)$ is extended to $H(y, z)$ by making it constant in s.) By the definition of analytically closed (see especially property (b) of ($\mathscr{E}$; A, B) admissible) and the fact that f and all ξ derivatives vanish on $\tilde{T}$, we must show that for μ_l bounded measures on V_l and k in an AU structure for $\mathscr{E}$ we have

$$\sum \int (\partial_l H)(y^0, z) \, d\mu_l/k = 0, \tag{9.74}$$

provided that

$$\sum \int (\delta \partial_l H)(0, z) \, d\mu_l/k = 0 \tag{9.75}$$

for all δ which are differential operators in y. It suffices to show that each integral

$$\int (\partial_l H)(y, z) \, d\mu/k \tag{9.76}$$

is analytic in $|y| < 1$ and continuous in $|y| \leqslant 1$.

We can clearly write $\mu_l = \mu_l^1 + \mu_l^2$, where μ_l^1 has its support on the closure of $c(A)$ and μ_l^2 has its support on the closure of its complement. Using (9.71) and (9.72), and the description of the AU structure for $\mathscr{E}$ given in Theorem 5.19, we see easily that

$$\int (\partial_l H)(y, z) \, d\mu_l^i/k \tag{9.77}$$

converges uniformly in $|y| \leqslant 1$, hence has the desired analyticity for $i = 1, 2$. Thus, (9.76) has the desired analyticity so the lemma is proved. □

We wish to generalize the above concepts to an arbitrary LAU space $\mathscr{W}$. We make the assumption here and in the following that if k is in an AU structure for $\mathscr{W}$, so is $\sqrt{k}$. Let $b(u)$ be a continuous positive function. We shall say that $H(y, w)$ ($\mathscr{W}$; A, B) *is analytically admissible relative to* b if

1. $H(y, w)$ is an analytic function of the complex variables y, w.
2. $\delta H(0, w)$ is a finite sum of terms each of which is polynomial in u times an element of $\mathbf{W}'(\tilde{T})$ for δ any derivative in y.

3. $H(y^0, w) \in \mathbf{W}'(T)$ for some y^0 with $|y^0| = 1$.
4. For any k in an AU structure for $\mathscr{W}(T)$ we have $|H(y, w)|/k(w)$ uniformly bounded in $|y| \leqslant 1$ and $w \in A$.
5. For any k in an AU structure for $\mathscr{W}(T)$ we have

$$(9.78) \qquad \sup_{|y| \leqslant 1, \, w \in B} \frac{|H(y, w)|}{k(u, v) b(u)} < \infty.$$

If the set of $(\mathscr{W}; A, B)$ analytically admissible functions relative to a fixed b has the property that the linear combinations of the $H(y^0, w)$ are dense in $\mathbf{W}'(T)$, then we say that (A, B) is *analytically closed* for $\mathscr{W}$ relative to b.

Using the above notation, we say that a set L in C is *well behaved* in β outside $c(A)$ (relative to b) if, given any k, there is a c so that

$$(9.79) \qquad b(u) \leqslant ck(z)$$

for all $z = (s, u, v) \in L$, $z \notin c(A)$.

An easy modification of the proof of Lemma 9.27 gives

LEMMA 9.28. *Let B be a region as in Lemma 9.27. Suppose (A, B) is analytically closed for $\mathscr{W}$ relative to b and $\mathfrak{V}$ is well behaved in β outside $c(A)$. If $f \in \mathscr{W}$, $\vec{\mathrm{D}}f = 0$, f and all ξ derivatives vanish on $\tilde{T}$, then $f = 0$ on T.*

We can go one step further. Namely, we do not need analyticity in y of the integrals (9.76) but only quasianalyticity. Then, let $M = \{M_j\}$ be a fixed quasianalytic class as in Section V.6. We say that $H(y, w)$, which is defined and indefinitely differentiable for real y, is $(\mathscr{W}; A, B)$ *quasianalytically admissible relative to b* if it satisfies Conditions 2 and 3 for analytic admissibility and

1'. For fixed y, $H(y, w)$ is analytic in w and for fixed w it belongs to $\mathscr{E}_M$ in y.
4'. For any k in an AU structure for $\mathscr{W}(T)$ the functions $H(y, w)/k(w)$ are uniformly bounded in $\mathscr{E}_M(|y| \leqslant 1)$ for $w \in A$.
5'. For any k in an AU structure for $\mathscr{W}(T)$ and any $c > 0$ we have

$$(9.80) \qquad \sup_{j, \, w \in B, \, |y| \leqslant 1} \frac{|\delta^{(j)} H(y, w)|/c^{|j|+1} M_j}{k(w) b(u)} < \infty.$$

Here, $\delta^{(j)} = \partial^{j_1 + \cdots + j_a}/\partial y_1^{j_1} \cdots \partial y_a^{j_a}$.

If the set of $(\mathscr{W}; A, B)$ quasianalytically admissible functions relative to a fixed b has the property that, the linear combinations of the $H(y^0, w)$ are dense in $\mathbf{W}'(T)$, then we say that (A, B) is *quasianalytically closed* for $\mathscr{W}$ (relative to b).

Of course, it should be borne in mind that wherever the expression "quasianalytic" appears it really means "quasianalytic relative to M." We can now imitate the proof of Lemma 9.27 replacing analyticity by quasianalyticity to obtain

LEMMA 9.29. *Let B be a region as in Lemma 9.27. Suppose (A, B) is quasianalytically closed for $\mathscr{W}$ relative to b and $\mathfrak{B}$ is well behaved in β outside $c(A)$. If $f \in \mathscr{W}$, $\vec{\mathrm{D}}f = 0$, f and all ξ derivatives vanish on $\tilde{T}$, then $f = 0$ on T.*

We are now in a position to state the main uniqueness theorem. This result contains Theorems 9.23 and 9.26 if $T = R$ and $\tilde{T}$ is a principal regular characteristic. We shall now assume that Φ is of the form $\varphi(\gamma_1) + \varphi(\gamma_2) + \cdots + \varphi(\gamma_m)$, and the space $\mathscr{E}(\Phi)$ will have no growth conditions in β, t.

THEOREM 9.30. *Let $\tilde{T} \subset T$ be any linear subvarieties of R. A sufficient condition that $f \in \mathscr{E}(\Phi)$, $\vec{\mathrm{D}}f = 0$, f and all ξ derivatives $= 0$ on $\tilde{T}$ imply $f = 0$ on T is: There exists an $\alpha(\geqslant 1)$ and a $c > 0$ so that if we denote by A the region*

$$|u| \leqslant c(1 + |v|)^\alpha, \tag{9.81}$$

then $\mathfrak{B}$ is analytic in β outside $c(A)$ and, moreover, (9.63) holds. Here, if $\alpha = 1$, the space $\mathscr{E}(\Phi)$ is to be replaced by $\mathscr{E}$.

PROOF. There are two cases to distinguish, namely, $\alpha = 1$ and $\alpha > 1$. As the latter is more difficult we shall present it in detail and we shall then outline the modifications for the case $\alpha = 1$.

As mentioned above, we are considering the case in which $\Psi(v) = \psi(v_1) + \psi(v_2) + \cdots + \psi(v_m)$. By Fubini's theorem, (9.63) implies

$$\int \frac{\psi(\mathrm{Im}\, v_1)}{1 + |\mathrm{Im}\, v_1|^{\alpha+1}}\, d(\mathrm{Im}\, v_1) = \infty. \tag{9.63*}$$

We wish now to make several "regularizations" of ψ. We shall not give the detailed construction of these regularizations as the reader who is not interested in the details can assume that we are dealing only with the special class of ψ for which such a regularization is possible.

Regularization 1. There exists a convex minorant ψ' of ψ for which (9.63*) (for ψ' in place of ψ) holds and such that the set of $F \in \mathbf{E}'(\Phi)(\tilde{T})$ which satisfy

$$|F(v)| \leqslant c \exp\left[-\Psi'(b\ \mathrm{Re}\ v) + \Psi(b'\ \mathrm{Im}\ v)\right] \tag{9.82}$$

for some b, b', c, is dense in $\mathbf{E}'(\Phi)(\tilde{T})$. Here

$$\Psi'(\mathrm{Re}\ v) = \psi'(\mathrm{Re}\ v_1) + \cdots + \psi'(\mathrm{Re}\ v_m).$$

Outline of proof. The first point to notice is that it is sufficient to prove our assertion for ψ, ψ' in place of Ψ, Ψ' (i.e., $m = 1$). For, if we know the result for $m = 1$, then for any γ we can approximate $\exp(i\gamma \cdot v)$ in the topology of $\mathbf{E}'(\Phi)(\tilde{T})$ by functions $F(v) = F_1(v_1)F_2(v_2) \cdots F_m(v_m)$ which satisfy (9.82). Namely, we just let $F_j(v_j)$ approximate $\exp(i\gamma_j v_j)$ which is possible by the case $m = 1$. Now, the linear combinations of the $\exp(i\gamma \cdot v)$ for varying (real) γ are dense in $\mathbf{E}'(\Phi)(\tilde{T})$ because the inverse Fourier transform of $\exp(i\gamma \cdot v)$ is just δ_γ and no $f \in \mathscr{E}(\Phi)(\tilde{T})$ can be orthogonal to all δ_γ except $f \equiv 0$.

We now assume $m = 1$. We claim that the set of real v for which $\psi(v) \leqslant |v|$ is compact. To see this we recall (see Chapter V, Example 7) that ψ is convex, non-negative, and $\psi(0) = 0$. Thus, if $\psi(v_0) \leqslant |v_0|$ then also $\psi(v) \leqslant |v|$ for $v \leqslant v_0$. But, clearly,

$$\int_1^{v_0} \frac{v}{1 + v^{\alpha+1}}\, dv$$

is bounded independently of v_0 ($\geqslant 1$). This is in contradiction with (9.63*) if the set of v_0 such that $\psi(v_0) \leqslant |v_0|$ is unbounded, so our assertion is established.

We try to write $F(v)$ satisfying (9.82) in the form

$$F(v) = \prod \frac{\sin (v/d_j)}{(v/d_j)}$$

By choosing the sequences $\{d_j\}$ suitably we can show the existence of such an $F \not\equiv 0$ satisfying (9.82) for a minorant ψ' of ψ which satisfies (9.63*). We needed our above assertion that $\psi(v) > |v|$ except for a compact set of v in order to be sure that

$$\frac{\sin (v/d_j)}{(v/d_j)} \in \mathbf{E}'(\Phi)(\tilde{T}).$$

The details of the construction follow the general lines of Chapter V; we leave them for an exercise for the reader.

Once we have one nontrivial F satisfying (9.82) it follows that the set of F satisfying (9.82) is dense in $\mathbf{E}'(\Phi)(\tilde{T})$ if we know that for any a there is a b such that $\psi'(v + a) \leqslant b\psi'(v)$. That ψ' can be chosen with this additional property follows from our construction and the fact that the same is assumed about ψ (see 5.126). Thus, each real translate of a multiple of F by $\exp(i\gamma v)$ for real γ also satisfies (9.82). We claim that the linear combinations of the real translates of multiples of F by $\exp(i\gamma v)$ are dense in $\mathbf{E}'(\Phi)(\tilde{T})$. By Fourier transform this is the same as saying that the linear combinations of $\exp(i\gamma v) f(\gamma + \gamma')$ for varying v and γ' are dense in $\mathscr{E}'(\Phi)(\tilde{T})$, where f is the inverse Fourier transform of F.

The linear combinations of the exp $(i\gamma v)$ are dense in the space $\mathcal{O}_M(\tilde{T})$ of Schwartz [1] which consists of all indefinitely differentiable functions $h(\gamma)$ such that h and all its derivatives are of polynomial growth. We can approximate any indefinitely differentiable function of compact support in the topology of $\mathcal{O}_M(\tilde{T})$ by linear combinations of exp $(i\gamma v)$. Since $g \to gf$ is a continuous map of $\mathcal{O}_M \to \mathcal{E}'(\Phi)$, it follows easily that we can approximate any $h \in \mathcal{D}(\tilde{T})$ in the topology of $\mathcal{E}'(\Phi)(\tilde{T})$ by linear combinations of $\exp(i\gamma v) f(\gamma + \gamma')$. Since $D(\tilde{T})$ is dense in $E'(\Phi)(\tilde{T})$, our claim is established. Thus, Regularization 1 is possible. □

Regularization 2. We can choose a minorant ψ_1 of $\psi(b''v)$ for some $b'' > 0$ so that Regularization 1 is satisfied (for ψ_1 in place of ψ') and, in addition, we can write ψ_1 in the form

$$\psi_1(v) = \sum a_j v^{2j}$$

with non-negative a_j.

PROOF. We can apply Regularization 1 to ψ' in place of ψ. Let ψ'' bear the same relation to ψ' as ψ' does to ψ. The analog of (9.82) for $m = 1$ becomes

$$|F(v)| \leqslant c \exp[-\psi''(b \operatorname{Re} v) + \psi'(b' \operatorname{Im} v)]. \tag{9.82$'$}$$

In the construction described in Regularization 1, the functions $F(v)$ are products of

$$\frac{\sin(v/d_j)}{(v/d_j)}$$

for suitable d_j. Given any such F, we set

$$\psi_1(v) = \log F(iv).$$

Then ψ_1 is a minorant of $\psi'(b'v)$, hence is a minorant of $\psi(b'v)$. Since the Taylor expansion at the origin of $(\sin iv)/iv$ has non-negative coefficients, we can write

$$\psi_1(v) = \sum a_j v^{2j}$$

with non-negative a_j. Since ψ_1 is a minorant of $\psi'(b'v)$, we can replace ψ' by ψ_1 in (9.82). Thus, Regularization 2 is completed. □

Regularization 3. Define

$$\lambda_1(v) = \exp \psi_1(|v|^{1/\alpha}) \tag{9.83}$$

for ψ_1 as in Regularization 2 and v real. Define the sequence

$$m' = \{m_j'\}$$

by

$$m'_j = \max |v|^j/\lambda_1(v),$$

and set $\lambda'(v) = \sum |v|^j/m'_j$ as in Chapter V, Example 6. Then $\mathscr{E}_{m'}$, is quasi-analytic, that is,

$$\int \frac{\log \lambda'(v)}{1+v^2}\, dv = \infty.$$

PROOF. By our construction of ψ_1 we know that (9.63*) is satisfied for ψ_1 so that

$$\int \frac{\log \lambda_1(v)}{1+v^2}\, dv = \infty.$$

By Regularization 2 we may write

$$\log \lambda_1(v) = \sum a_j |v|^{2j/\alpha}$$

with $a_j \geqslant 0$ for all j. A simple modification of an argument of Carleman (in his case the exponents $2j/\alpha$ are integers) (see Carleman [1]) shows that $\mathscr{E}_{m'}$ is quasianalytic so that (see Chapter V, Example 6)

$$\int \frac{\log \lambda'(v)}{1+v^2}\, dv = \infty. \quad \square$$

Regularization 4. We may replace the sequence m' by a majorant $m = \{m_j\}$ so that $\mathscr{E}_m$ is still quasianalytic and for $\lambda(v) = \sum |v|^j/m_j$ we have

$$\lambda(v) \leqslant \exp(|v|).$$

PROOF. It is clear from the definition of m'_j that $(m'_j)^{1/j} \leqslant (m'_{j+1})^{1/j+1}$ for all j large enough. Carleman's condition for quasianalyticity (see Carleman [1]) is

$$\sum (m'_j)^{-1/j} = \infty.$$

Define $m_j = \max(m'_j, j!)$. We claim first that

$$\sum (m_j)^{-1/j} = \infty.$$

To see this, we note first that we may assume that $j! > m'_j$ for infinitely many j, for otherwise our assertion is trivial. Thus, there exists a sequence $\{j_\nu\}$ with $j_{\nu+1} \geqslant 2j_\nu$ for which

$$m'_{j_\nu} < j_\nu!.$$

Since $(m'_j)^{1/j}$ is monotonic, this implies that

$$(m'_j)^{1/j} \leqslant (j_\nu!)^{1/(j_\nu)} \quad \text{for } j \leqslant j_\nu.$$

By Stirling's formula this implies that

$$\sum_{j=j_\nu}^{j_{\nu+1}} (m_j)^{-1/j} \geqslant (j_{\nu+1} - j_\nu)[(j_{\nu+1})!]^{-1/(j_{\nu+1})}$$

$$\geqslant \tfrac{1}{2} j_{\nu+1} \cdot (j_{\nu+1})^{-1}(c/2)$$

$$= c/4.$$

Thus, the series $\sum m_j^{-1/j}$ diverges which means that $\mathscr{E}_m$ is quasianalytic. This proves Regularization 4. □

PROOF OF THEOREM 9.30. Now we define M as the multisequence $M_{j_1 j_2 \cdots j_a} = m_{j_1} m_{j_2} \cdots m_{j_a}$. We define $\mathscr{E}_M$ as the space of functions of $u = (u_1, u_2, \ldots, u_a)$ defined by this multisequence as in Chapter V, Example 6. By the above $\mathscr{E}_M$ is quasianalytic.

In order to apply Lemma 9.29 we want to prove that A is quasianalytically closed for $\mathscr{E}(\Phi)$ relative to $\exp(|u|)$. For any F satisfying (9.82) we define

$$H(y, w) = \exp(i\tau y \cdot u) F(v). \tag{9.84}$$

Here τ is a real scalar. Since the estimates we shall need depend in a simple manner on τ we shall assume $\tau = 1$. However we need all real τ in (9.84) in order to know the density of the linear combinations of $H(y_0, w)$.

Since by hypothesis, $\mathfrak{B}$ is well behaved in $\mathfrak{z}$ outside $c(A)$ relative to $\exp(|u|)$, we must show that $H(y, w)$ is $(\mathscr{E}(\Phi); A, T)$ quasianalytically admissible relative to $\exp(|u|)$. Conditions 1′, 2, and 3 are immediate. To verify Condition 5′, we note that the left side of (9.80) for $c = 1$ is by (9.84)

$$\leqslant \sup_{j,u,v} \frac{|u_1|^{j_1}|u_2|^{j_2} \cdots |u_a|^{j_a}|F(v)|}{m_{j_1} m_{j_2} \cdots m_{j_a} k(v) \exp(|u|)}$$

$$\leqslant \sup_v \frac{|F(v)|}{k(v)} \sup_{j,u} \frac{|u_1|^{j_1}|u_2|^{j_2} \cdots |u_a|^{j_a}}{m_{j_1} m_{j_2} \cdots m_{j_a} \lambda(u_1)\lambda(u_2) \cdots \lambda(u_a)}$$

(by Regularization 4)

$$\leqslant \sup_v \frac{|F(v)|}{k(v)}$$

(by definition of λ)

$$< \infty$$

(by (9.82)).

It remains to verify Condition 4′. It suffices to show that for varying $(u, v) \in A$ the functions

$$H(y, w)/k(v)$$

are uniformly bounded in $\mathscr{E}_M(|y| \leqslant 1)$, Thus, we must estimate

$$\max_{|y| \leqslant 1} |\delta^{(j)} H(y, w)|/k(v) M_j = |u_1|^{j_1} \cdots |u_a|^{j_a} |F(v)|/k(v) m_{j_1} \cdots m_{j_a}$$
$$\leqslant \lambda(u_1) \cdots \lambda(u_a) |F(v)|/k(v)$$

by the definition of λ. Using (9.82) (with v replaced by $c_1 v$) and the definition of k, it suffices to show that

$$\lambda(u_1) \cdots \lambda(u_a) \leqslant \exp[\Psi_1(bc_1 \operatorname{Re} v) + \Psi(bc_1 \operatorname{Im} v)] \tag{9.85}$$

for suitable b, c_1 when (u, v) satisfies (9.81). By (9.83) this is the same as

$$\psi_1(|u_1|^{1/\alpha}) + \psi_1(|u_2|^{1/\alpha}) + \cdots + \psi_1(|u_a|^{1/\alpha}) \leqslant \Psi_1(bc_1 \operatorname{Re} v) + \Psi(bc_1 \operatorname{Im} v). \tag{9.86}$$

By (9.81), and the fact that ψ_1 is monotonic, (9.86) is a consequence of

$$\Psi_1(c'v) \leqslant \Psi_1(bc_1 \operatorname{Re} v) + \Psi(bc_1 \operatorname{Im} v), \tag{9.87}$$

where c' depends on c_1. Since $\psi_1 \leqslant \psi$ and since c_1 is arbitrary, (9.87) is proved. Hence, the sufficiency of the conditions in Theorem 9.30 is established in case $\alpha > 1$.

For $\alpha = 1$ we define $H(y, w)$ for y complex variables by

$$H(y, w) = G(y_1 u_1, \ldots, y_a u_a, v). \tag{9.88}$$

Here $G(w) \in \mathbf{E}'(T)$ satisfies

$$|G(u, v)| \leqslant c(1 + |w|)^c \exp(c|\operatorname{Im} v)|) \tag{9.89}$$

in $|u| \leqslant c'(1 + |v|)$. The existence and density properties of such G follow from Lemma 9.22**. It is now a simple matter to show that Lemma 9.27 applies and we have our result.

We have thus proved Theorem 9.30. ▯

We wish to derive a (strong) converse to Theorem 9.30.

Let $\alpha \geqslant 1$ be chosen so that there exists a c so that for A defined by (9.81) we have $\mathfrak{B}$ analytic in β outside $c(A)$. We shall assume that α is chosen to be minimal with this property. It can be shown (in fact by the methods described below) that if any α exists then a minimal α exists. We shall assume that α exists. In case $\alpha = 1$ we have nothing to prove, so we assume $\alpha > 1$.

We shall consider curves Γ on V which arise as the intersection of V with a complex linear variety $\tilde{\Gamma}$ of dimension $n - d + 1$. We consider those $\tilde{\Gamma}$ for which Γ is a curve which is a Riemann surface covering a complex line Λ in the $\{v_j\}$ space (i.e., $u = 0$, $s = 0$), which is normalized so that Λ is the v_1 axis. We may think of each z_l on Γ as an algebraic function of v_1 to which we may apply the Puiseux expansion (see Lemma 1.1). We may also speak of the various "branches" of Γ.

We wish to examine the possible Puiseux expansions of the v_i, u_i, s_i, on Γ in terms of v_1. First of all, let us examine an expansion of u_j:

$$u_j = a_j v_1^{b_j} + \text{lower order terms}$$

with $a_j \neq 0$. If each $b_j < \alpha$ then Γ would be contained in $c(A)$, with A defined by (9.81) even if we replaced α in (9.81) by $\max b_j$. If the same were true for every choice of $\tilde{\Gamma}$ we could easily show (by a "compactness" argument) that α did not have its required minumum property. Thus, we may assume that some b_j, say b_1, is $\geqslant \alpha$. This "compactness" argument, as well as the one needed below, are best formulated in terms of the points at infinity on V in projective space. We leave this formulation to the reader.

If actually $b_1 > \alpha$ then for large v_1 the part of Γ on which $u_1 \sim a_1 v_1^{b_1}$ would lie outside $c(A)$ and so would lie in the region analytic in β. If for every $\tilde{\Gamma}$ and every branch of Γ we could find a j so that the corresponding $b_j > \alpha$ then a "compactness" argument would show that there is a $c' > 0$ so that

$$|\operatorname{Im} z| \geqslant c'(1 + |\operatorname{Re} u|)$$

for all $z \in V$. By a simple modification of the proof of Theorem 8.3 this implies that every solution $f \in \mathscr{E}$ of $\vec{D}f = 0$ would be analytic in β so we would have uniqueness. We shall, henceforth, exclude this case.

Thus we are led to the situation in which there is a branch with

$$u_1 = a_1 v_1^{\alpha} + \text{lower order terms} \tag{9.90}$$

and each $b_j \leqslant \alpha$ on this branch. We normalize so that this branch contains the positive u_1 axis. The same argument as the one we used to show (9.90) shows also that we may assume that the branch of Γ which we are considering lies outside the set where V is analytic in β. This means first that for those j with $b_j = \alpha$ we must have a_j/a_1 real. In fact, for any l, write the Puiseux expansion

$$z_l = a^l v_1^{b^l} + \text{lower order terms} \tag{9.91}$$

with $a^l \neq 0$. Then to avoid (9.73) on the positive real u_1 axis, if $b^l \geqslant \alpha$ we require by (9.90) that $a^l/(a_1)^{b^l/\alpha}$ be real. Here the branch of $a_1^{b^l/\alpha}$ chosen is in conformity to the inverse formula to (9.90) on the positive u_1 axis. A similar requirement is to be put on those lower order terms in the expansion of z_l which involve v_1^b with $b \geqslant \alpha$.

Now consider the case when $z_l = v_j$, $j \neq 1$. Then we claim that all the coefficients of v_1^b with $b > 1$ will be zero. For, if not, then we could write all the Puiseux expansions in terms of this v_j instead of v_1, and we would arrive at a smaller α. Moreover, the coefficient of v_1 is real.

We define ζ as the part of the curve in the v_1 plane which corresponds to the positive real u_1 axis on our branch of Γ on which v_1 is large enough so that each z_l on this branch is a regular function of v_1 in a neighborhood of ζ.

Next, as in the proof of Theorem 9.30, we define

$$\lambda_1(v_1) = \exp \psi(|v_1|^{1/\alpha}). \tag{9.92}$$

Assumption.

$$\int \frac{\log \lambda_1(v_1)}{1+|v_1|^2}\, dv < \infty. \tag{9.93}$$

Again we may perform some regularization procedures until we construct a majorant λ of λ_1 that that λ still satisfies (9.93) and $\lambda(v_1)$ is of the form

$$\lambda(v_1) = \sum |v_1|^j/m_j,$$

where for $m = \{m_j\}$ the space $\mathscr{E}_m(\gamma_1)$ (see Chapter V, Example 6 and Chapter XIII) is nonquasianalytic.

We can be a little more precise: If we look at the complex u_1 plane, then there are only a finite number of points u_1^* at which any of the functions $z_l(u_1)$ (z_l on Γ) will be branched. We let η be a path which lies in the upper half u_1 plane and surrounds all these points and agrees with real u_1 for large $|u_1|$ (see Fig. 16). Here all z_l when considered as functions of u_1 are extended into the upper half u_1 plane above η by direct analytic continuation from the values on $\eta \cap$ positive u_1 axis.

By the standard theory of quasianalyticity (see e.g., Paley and Wiener [1]) we can construct a function $L(u_1)$ holomorphic in the upper half u_1 plane which satisfies, for every $b > 0$,

$$L(u_1) = 0(\lambda^{-1}(b \operatorname{Re} u_1) \exp(-\operatorname{Im} u_1)). \tag{9.94}$$

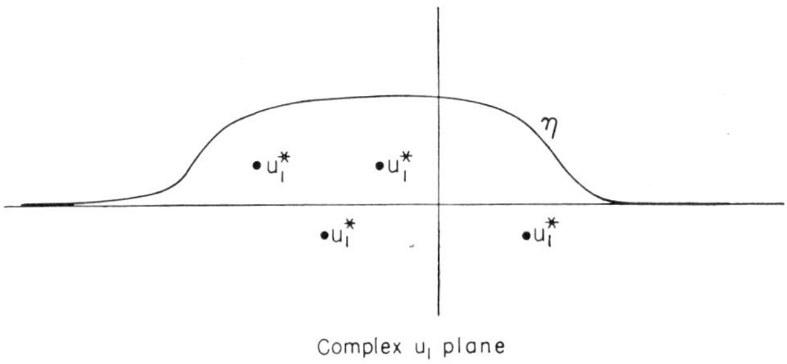

FIGURE 16

Again, as in the proof of Theorem 9.30, given any $\alpha_1 < 1$ we may regularize λ so that in addition to the previous properties we have

(9.95) $$\limsup_{u_1 \text{ real}} \exp(|u_1|^{\alpha_1})/\lambda(u_1) = 0.$$

Thus, using (9.94) and Cauchy's formula we deduce (by shifting the contour to infinity)

(9.96) $$\int_\eta L(u_1)G(u_1)\,du_1 = 0$$

whenever G is holomorphic on the exterior of η in the upper half u_1 plane, has continuous boundary values on η, and such that

(9.97) $$G(u_1) = 0\,(\exp(|u_1|^{\alpha_1})) \qquad \text{for} \quad \alpha < 1.$$

We define

(9.98) $$f(x) = \int_\eta \exp(ix \cdot z)L(u_1)\,du_1.$$

Since $L(u_1) = 0(\lambda(bu_1)^{-1})$ on η, and λ is a majorant of

$$\lambda_1(v) = \exp \psi(|v_1|^{1/\alpha}),$$

it follows from (9.90) that, on η, we have for every $b > 0$,

$$L(u_1) = 0[\exp(-\psi(|bv_1|))].$$

By our construction (see (9.91) and following) the Im v_j for $j \neq 1$ are $0(|\text{Im } v_1| + |v_1|^{\alpha'})$ for some $\alpha' < 1$ on Γ so that, by (9.95),

$$L(u_1) = 0[\exp(-\psi(b \text{ Im } v_1) + \cdots + \psi\,(b \text{ Im } v_m))]$$

on η. Except for polynomial factors, which can be handled by a slight modification of the definition of L, this would mean that the measure

$$L(u_1)\,du_1(v_1) \qquad \text{on} \quad \eta$$

(η considered as a subset of Γ) would belong to the dual of $\mathbf{E}'(\Phi)$ by the explicit description of $\mathbf{E}'(\Phi)$ given in Chapter V, Example 7, if we knew that the z_l were real on η near infinity. However, the z_l need not be real near infinity on η. By the remarks following (9.91) we find easily: There exists an $\alpha_0 < 1$ so that, on η for every l,

(9.99) $$|\text{Im } z_l| = 0(1 + |u_1|)^{\alpha_0}.$$

By (9.99), (9.95), and the above it follows that $L(u_1)\,du_1$ on η defines an element of $\mathbf{E}(\Phi)$ so that $f \in \mathscr{E}(\Phi)$. Moreover, by Theorem 7.1, $\vec{\mathbf{D}}f = 0$.

We claim that $f \not\equiv 0$ on T. To see this we evaluate f when $\gamma = t = 0$

and $\beta_j = 0$ for $j \neq 1$. We find from (9.98)

$$f(\beta_1, 0) = \int_\eta \exp(i\beta_1 u_1) L(u_1)\, du_1.$$

Now, by construction, $L(u_1)$ is holomorphic and $\not\equiv 0$ in the open upper half plane and continuous on the boundary. Thus, we may deform the contour to the real axis in which case the above is an ordinary Fourier integral so $f(\beta_1, 0) \not\equiv 0$.

We now apply (9.96) to f. If ∂ is any differential operator with constant coefficients and x is chosen so that the β and t coordinates are zero (write $x = (0, \gamma, 0)$), then by (9.98)

$$(\partial f)(x) = \int_\eta \exp(i\gamma \cdot v) L(u_1) P(z)\, du_1, \tag{9.100}$$

where P is a polynomial. By our construction of η, each z_l is regular in the exterior of η.

In the exterior of η in the upper half plane, we have for any fixed γ

$$\exp(i\gamma \cdot v) = 0(|u_1|^{1/\alpha})$$

as is seen from the formulas for the Puiseux expansion of the v_j discussed in (9.91) and following. Thus, we can apply (9.96) to

$$G(u_1) = \exp(i\gamma \cdot v) P(z).$$

It follows that $\partial f \equiv 0$ on $\tilde{T}$.

THEOREM 9.31. *(a) Suppose that there exist α, c so that $\mathfrak{B}$ is analytic in β outside $c(A)$, where A is defined by* (9.81). *Suppose that for every such α*

$$\int \frac{\psi(\operatorname{Im} v_1)}{1 + |\operatorname{Im} v_1|^{\alpha+1}}\, d(\operatorname{Im} v_1) < \infty. \tag{9.101}$$

Suppose, moreover, there is a branch on which (9.90) *holds and each $b_j \leqslant \alpha$. Then there exists a function $f \in \mathscr{E}(\Phi)$, with $\vec{D}f = 0$, f and all its derivatives identically zero on $\tilde{T}$, and $f \not\equiv 0$ on T.*

(b) If there do not exist α, c for which $\mathfrak{B}$ is analytic in β outside $c(A)$, then for any Φ for which (9.101) *holds for some α we find an $f \in \mathscr{E}(\Phi)$ with $\vec{D}f = 0$, f and all its derivatives identically zero on $\tilde{T}$, and $f \not\equiv 0$ on T.*

PROOF. Statement (*a*) has already been proved. To prove (*b*) we observe easily that if there do not exist α, c for which $\mathfrak{B}$ is analytic outside $c(A)$, then there exists a finite point v^0 such that V has points (v, u, s) v arbitrarily close to v^0 and $|u|$ arbitrarily large. (In the language of projective geometry, V has a point (v^0, u, s) with some coordinate of u infinite.) We can now argue in a manner similar to the proof of (*a*) to obtain (*b*). $\square$

See Remarks 9.13, 9.14, and 9.15.

EXAMPLE. For the wave equation

$$\frac{\partial^2}{\partial\gamma^2}-\frac{\partial^2}{\partial\beta^2}=\frac{\partial^2}{\partial t^2}$$

the corresponding polynomial is $v^2-u^2-s^2$. We see easily that Theorem 9.30 applies with $\alpha=1$. (This case is treated in John [1].)

We wish to make some remarks on another aspect of these uniqueness theorems. For simplicity we shall consider only the case $r=1$. Suppose T^{n-1} is noncharacteristic. Then Holmgren's uniqueness theorem, upon Fourier transform, says that the functions of the form

$$\sum_{j<m} F_j(w)s^j$$

are dense in $\mathbf{E}'(\mathfrak{B})$, where m is the order of D. If we want to approximate a given $G(z)$ on $\mathfrak{B}$ then the Lagrange interpolation formula gives

$$G(z)=\sum_{j<m} G_j(w)s^j$$

on $\mathfrak{B}$, where the G_j are entire functions which in general no longer belong to $\mathbf{E}'$. However, since T^{n-1} is noncharacteristic the Lagrange interpolation formula shows that the G_j are entire functions of exponential type. Thus, we want to approximate $\{G_j\}$ by $\{F_j\}$, where the F_j are in $\mathbf{E}'$. The natural topology of the $\{G_j\}$, that is, the topology that makes $\{G_j\}\to\sum G_j(w,s)|\mathfrak{B}$ a topological isomorphism onto $\mathbf{E}'(\mathfrak{B})$ is complicated (compare Theorem 9.16). Nevertheless, we know that approximation is possible. The essential reason for this is the fact that the G_j are of exponential type. The underlying principle seems to be: The functions in $\mathbf{E}'$ will be dense in any "reasonable" subspace of $\mathbf{H}'$ which contains $\mathbf{E}'$. A similar property holds for vector functions.

When $t=0$ is characteristic then the G_j cease to be of exponential type, though they remain of finite order as can be seen from their expressions by means of the Lagrange interpolation formula. Thus, to approximate $\{G_j\}$ by $\{F_j\}$ we want to approximate functions of larger than exponential type by functions of exponential type. This problem is akin to the one discussed in Chapter XIII (Bernstein's problem) which is related to the question of approximation of entire functions of exponential type by polynomials. In certain norms it is possible and in some it is not. This is the general situation when we want to approximate functions of one order by functions of a lower order (order being understood in terms of bounds on the maximum modulus). Thus, for our analog of Tacklind's theorem we make the norms on the G_j weak enough so that approximation is possible.

A similar situation prevails in the strengthened uniqueness theorems which culminate in Theorem 9.30. Here, for example, it is a question of approximating say $G(u, v)$ by $F(v)Q(u)$, where the Q are polynomials and $F \in \mathbf{E}'$. (Here $G(u, v) \in \mathbf{E}'$ is the Fourier transform of a distribution in $\mathscr{E}'(T)$ and the topology on the G is that given by considering $\mathscr{E}'(T)$ as acting on the restrictions to T of f with $Df = 0$.) In order to be able to perform this approximation we must know that the norms on G are, for fixed v, "weak" as $|u| \to \infty$. This is the content of the assumption that $\mathfrak{V}$ is analytic outside of a suitable set.

PROBLEM 9.12.
Study these and similar approximation problems from a more direct point of view.

Some further discussion will be made in Chapter XI.

IX.10 More Refined Uniqueness Results

The type of uniqueness questions we shall deal with here stems from an attempt to extend to systems $\vec{\mathrm{D}}f = 0$ as much as possible of the uniqueness properties of analytic functions. Most of the deeper uniqueness properties of analytic functions are intimately related to the subharmonicity of $\log |f|$. Unfortunately, we know of no analog of this for general systems. Nevertheless, we shall be able to extend to arbitrary systems many of the properties of analytic functions which depend on subharmonicity, or the algebra structure, or composition. A notable exception is an analog of Picard's theorem and its refinements. This seems to be due to our inability to formulate any analog of the Poisson-Jensen formula, although some form of Nevanlinna's second inequality may be possible.

The reason for this possibility is that, for functions of one complex variable, the predecessor to Nevanlinna theory is Liouville's theorem which states that a bounded entire function is a constant. If we think of a meromorphic function as a holomorphic mapping of part of the Riemann sphere S into S, then we can reformulate Liouville's theorem as follows: A holomorphic map f of S *minus* a point p into S whose image omits a neighborhood of infinity is a constant. Note that, since $f(s)$ omits a neighborhood of infinity, we can use the theorem of removable singularities to continue f to a holomorphic map on all of S whose image omits a neighborhood of infinity. The constancy of f now follows from the maximum modulus theorem.

If we want to abstract the above set-up to general differential equations, then we are led to the following situation: We are given two compact manifolds, M and N, and a differential operator D which acts on indefinitely differentiable maps of M into N. D can be defined, for example, in terms of local coordinates. If f is an indefinitely differentiable map of M into N, then we allow the possibility that Df is a more general type of object, say a cross section of a suitable vector bundle (e.g., D could be the ordinary d on M).

Then the first problem we must study is that of removable singularities for the kernel of D (cf. Sections XI.2 and XI.3).

For example, Picard's theorem states that, if $n = 1$, and if f is a holomorphic map of a disk minus the center p, into S such that there is a neighborhood U of p for which the image $f(U)$ omits at least three points on S, then f can be continued to be holomorphic on all of the disk. (Since we are considering maps into S, a function is considered to be regular at a pole.)

To obtain Nevanlinna's second theorem, the above would have to be quantified.

As a substitute for the maximum modulus theorem, we require that the kernel of D should consist of open maps.

See Problem 9.13.

The first property we shall be concerned with is the fact that an analytic function which is bounded in the upper half-plane cannot approach zero too rapidly on the real axis at infinity (see Paley-Wiener [1]). We shall see in Chapter XIII that (roughly speaking) this property is equivalent to an approximation property of polynomials or of entire functions of fixed exponential type which are in $\mathbf{E}'$. If we map the upper half-plane into the strip $|\text{Im } x| \leqslant 1$, then the property is that an analytic function which is bounded in the strip cannot approach zero too rapidly on the boundary as $\text{Re } x \to \pm\infty$. Again this is dual to the possibility of suitably approximating continuous functions on the boundary by suitable functions which are analytic in the strip. At first this might seem paradoxical, for Holmgren's theorem for the Cauchy-Riemann equations tells us that the values of an analytic function on $\text{Im } x = -1$ determine those in the strip. Yet we can approximate continuous functions on $\text{Im } x = \pm 1$ independently. The point is, of course, that the values on $\text{Im } x = +1$ do *not* depend continuously on the values on $\text{Im } x = -1$, except in very special topologies. (This is best seen on mapping the strip $|\text{Im } x| \leqslant 1$ onto the unit disk. A sequence of bounded analytic functions on the disk may converge to zero over an arc of the boundary without converging to zero.) Actually we can see that approximation of continuous functions on $\text{Im } x = \pm 1$ is not possible in some cases even when the values on $\text{Im } x = +1$ do not depend continuously on the values on $\text{Im } x = -1$. This will be made precise below.

We may regard the above as a sharpening of the problems dealt with in Section IX.9 for the following reason: In Section IX.9 we could have studied the question of whether a function f which is analytic in the open upper half-plane, indefinitely differentiable on the boundary, and satisfying a suitable growth condition could vanish with all its derivatives at the origin. In this section we are concerned with how fast $f(x)$ can tend to zero at the origin. This type of problem stands in close relationship with the quasianalytic problems of Turan and Mandelbrojt mentioned in Section XIII.1.

We want to generalize the above to an arbitrary $\vec{D}$. We assume first that $r = 1$. Let f satisfy $Df = 0$ in $0 \leqslant t \leqslant 1$.

PROBLEM 9.14.

How fast can f and all its derivatives approach zero uniformly in $0 \leqslant t \leqslant 1$ as $|\xi| \to \infty$?

It is clear that the problem depends on D. For D hyperbolic with respect to t we see that f could be of compact support in ξ. That hyperbolicity of D (or more precisely, the "weak" hyperbolicity of some irreducible factor of D) is the only way in which such an f could exist, is proved by John [2] if $t = 0$ is noncharacteristic. We shall see below that a similar situation prevails in general.

Another observation that is of importance is the following: We know that, for $r = 1$, there exist solutions of $Df = 0$ which vanish on one side of any characteristic. Thus, if D is not elliptic we know that f can actually *equal* zero in "half" of the strip $0 \leqslant t \leqslant 1$. Yet we shall see that f cannot be too small for *all* ξ with $|\xi| \to \infty$.

In case $r > 1$ we can formulate the problem in a similar fashion. We require that there be given a noncharacteristic or regular characteristic $(T^d, T^{d-1}, \ldots T^0)$ where all $T^j \subset T^{j+1}$. Then our problem can be formulated: Let $\mathscr{W}$ be an AU space for which $\mathscr{W}'$ is a space of functions on T^d. When is it true that any $f \in \mathscr{E}$ with $Df = 0$ and such that the restriction of f and all of its derivatives to all planes $t = t^0$, where $0 \leqslant t_j^0 \leqslant 1$ for all j, belong to $\mathscr{W}'$ must be zero? (We have written t for t^d.)

See Remarks 9.16 and 9.17.

Our method of solution of the problem is very similar to that used by John [2], except that we need to refine some of his arguments.

Our study of this problem is distinguished from our other methods in that we take the Fourier transform in *certain* variables rather than in all. Actually, we could take the Fourier transform in all variables but the method would be somewhat more complicated.

Let $\mathfrak{U}'$ denote the space of functions or distributions on R^n which is identified with the space of indefinitely differentiable maps of t into $\mathscr{W}'$. Щ′ is given the natural topology. By **Щ**′ we denote the "spatial" Fourier transform of Щ′ which is the space of indefinitely differentiable maps of t into $\mathbf{W}'$, the Fourier transform of an element of Щ′ being defined in the obvious way:

$$S(t) \to \widehat{S}(t) = \widehat{S(t)} \quad \text{for } S(t) \in Щ'.$$

We shall assume that the space $\mathscr{D}$ (see Chapter V, Example 4) is contained in Щ (the dual of Щ′) and is dense in Щ. We shall also make the

Assumption. The given noncharacteristic (or regular characteristic) consists of T^d only.

Of course, Щ′ is *not* AU but we want to prove an analog of our main theorems for it. As mentioned above, there are two ways to proceed: Take the Fourier transform in all variables, or take the spatial Fourier transform. We explain first the former method as it is a good way to "guess" the result.

Let $D_1 f = D_2 f = \cdots = D_r f = 0$, $f \in Щ'$. Suppose we want to represent f as a Fourier integral. Then we can do this in the manner of Chapter VII and the support of the "Fourier transform" of f would be the whole complex variety $\mathfrak{V}$ associated to $D_1 \ldots, D_r$. However, we should expect much more, for f is small on the surfaces $t = \text{const}$. Thus, the restriction of f to $t = \text{const}$ can be represented as a classical Fourier transform. This means that the support of this transform is contained in the *real* w space. Thus, we might expect that f could be represented as a Fourier integral over the part of $\mathfrak{V}$ with w real. This is indeed the case and will be proved below. The proof in fact will be, to a large extent, independent of the method of Chapters II, III, and IV.

Let $f(t) \in Щ'$. As in Section IX.5 we denote by $\eta_{t_0} f$ the set of restrictions of the $q_{id} f$ to the plane $t = t_0$. We want to give the condition on $\eta_{t_0} f$ in order that $D_j f = 0$ for $j = 1, 2, \ldots, r$. For each fixed $t = t_0$ we denote by $\hat{f}(t_0)$ the collection of Fourier transforms of $\eta_{t_0} f$. We can consider $\hat{f}(t)$ and $\eta_t f$ as vector-valued functions of t so it makes sense to form t derivatives.

LEMMA 9.32. *Let* ${}_j\boxed{\mathrm{L}}$ *be the Fourier transform of the adjoint of* ${}_j\boxed{\mathrm{N}}$ *(see Theorem* 9.15). *A necessary condition that* $D_1 f = \cdots = D_r f = 0$ *is* $i\partial(\eta_t f)/\partial t_j = {}_j\boxed{\mathrm{N}}\,\partial_t f$ *or, what is the same thing,*

$$i\,\frac{\partial \hat{f}}{\partial t_j} = {}_j\boxed{\mathrm{L}}\,\hat{f}.$$

If $\vec{\mathrm{D}} f = 0$, *then also*

$$\hat{f}(t) = \exp\left(-i\,{}_1\boxed{\mathrm{L}}\,t_1\right) \cdots \exp\left(-i\,{}_{n-d}\boxed{\mathrm{L}}\,t_{n-d}\right)\hat{f}(0).$$

PROOF. Suppose first that $\vec{D}f = 0$. The statement that

$$i\partial(\eta_t f)/\partial t_j = {}_j\boxed{N}\eta_t f$$

means that for any ξ and any l we have

$$i\left[\frac{\partial}{\partial t_j}(q_{ld}f)\right](t,\xi) = \sum_i ({}_jN_{id}^{ld}q_{id}f)(t,\xi)$$

because $\partial/\partial t_j$ and ${}_jN_{ld}^{id}$ commute with the q_{id}. This is the same as

$$i\left(\frac{\partial}{\partial t_j}\right)' q'_{ld}S = \sum_{i,k} {}_j(N_{id}^{ld})' q'_{id}S$$

on solutions of $\vec{D}f = 0$ for any $S \in Ш$. By our assumption, the space $\mathscr{D}$ is dense in $Ш$, so it suffices in the above to restrict our attention to $S \in \mathscr{D}$. We now take the Fourier transform in ξ and the result follows from (9.31).

The formula relating $\hat{f}(t)$ to $\hat{f}(0)$ is an immediate consequence of this. ▯

Let us denote by M the map

$$M: \hat{f}(t) \to \sum Q_{ld}(s)\hat{f}_{ld}(t,w).$$

Then we have by (9.31)

LEMMA 9.33.

$$s_j[M\hat{f}(t)](s,w) = M[{}_j\boxed{L}\hat{f}(t)](s,w) \quad \text{for } (s,w) \in \mathfrak{B}$$

so that

$$\exp(it_j s_j)[M\hat{f}(t)](s,w) = M[\exp(i\,{}_j\boxed{L}t_j)\hat{f}(t)](s,w) \quad \text{for } (s,w) \in \mathfrak{B}.$$

We can now explain how to obtain our results. We use the expression for $\hat{f}(t)$ in terms of $\hat{f}(0)$ given by Lemma 9.32 in Lemma 9.33; it follows that

$$M[\hat{f}(0)] = \exp(it_j s_j)M[\hat{f}(t_j)] \quad \text{on } \mathfrak{B}, \tag{9.102}$$

where we have written $\hat{f}(t_j)$ for $\hat{f}(0, \ldots, t_j, \ldots, 0)$. We now assume that $\mathscr{W}$ is a space $\mathscr{E}(\Phi)$ (see Chapter V). Then we know that each component of $\hat{f}(t)$ is majorized by $(1 + |w|^a)\exp(\Psi(a \operatorname{Im} w))$, where Ψ is the conjugate of Φ. Thus, we have

$$|M[\hat{f}(t)](s,w)| \leqslant \text{const}\,(1 + |s| + |w|)^a \exp(\Psi(a \operatorname{Im} w)), \tag{9.103}$$

where a and the const can be chosen independent of t as long as t is in a compact set. Combining (9.103) with (9.102) gives

$$\begin{aligned}(9.104)\qquad &|M[\hat{f}(0)](s,w)| \\ &\leqslant \text{const}\min[1, |\exp is_j|](1 + |s| + |w|)^a \exp(\Psi(a \operatorname{Im} w))\end{aligned}$$

on $\mathfrak{B}$ for any j.

Suppose first that dim $\mathfrak{B} = d = 1$. We use the Puiseux expansion for s_j in terms of w on $\mathfrak{B}$. We get expressions of the form

$$s_j(w) = b_j w^{\alpha_j}(1 + \gamma_j^1 w^{-\beta_j} + \gamma_j^2 w^{-2\beta_j} + \cdots), \tag{9.105}$$

where α_j, β_j are positive rationals and where $b_j \neq 0$ and the series converges for $|w|$ sufficiently large. We exclude two cases: $\alpha_j > 1$ and ($\alpha_j = 1$; b_j real) as these are the "weakly" hyperbolic cases, that is, the cases for which the results of Section VIII.3 apply if the space $\mathscr{E}$ of indefinitely differentiable functions is replaced by a non-quasianalytic space $\mathscr{E}_A$ (see Chapter V, Example 6). Here A will depend on $\vec{\mathrm{D}}$ and can be chosen as a product of sequences $\{(cj)!\}$ for suitable $c > 1$.

Suppose first that $\alpha_j = 1$, b_j is complex. Then we derive immediately

$$|\exp is_j| \leqslant \exp[(\operatorname{Im} b_j)(\operatorname{Re} w) + \gamma(|w|^\beta + |\operatorname{Im} w|)], \tag{9.106}$$

where $\beta < 1$ and γ is some positive constant. Combining this with (9.104) gives

$$\begin{aligned} &|M[\hat{f}(0)](s, w)| \\ &\leqslant \mathrm{Const}(1 + |w|)^a \exp[(\operatorname{Im} b_j)(\operatorname{Re} w) + \gamma(|w|^\beta + |\operatorname{Im} w|) + \Psi(a \operatorname{Im} w)]. \end{aligned} \tag{9.107}$$

Since $\operatorname{Im} b_j \neq 0$ the terms $(1 + |w|)^a$ and $|w|^\beta$ are easily seen to be ignorable. Also, since $\Psi(\operatorname{Im} w) \geqslant |\operatorname{Im} w|$, we can rewrite (9.107) as

$$|M[\hat{f}(0)](s, w)| \leqslant \mathrm{const} \exp[-b_j' \operatorname{Re} w + \Psi(a \operatorname{Im} w)], \tag{9.108}$$

where b_j' is real and different from zero. Now we use the principle of harmonic majorant. (9.108) for $b_j' \operatorname{Re} w > 0$ and (9.104) for $b_j' \operatorname{Re} w < 0$ imply $M[\hat{f}(0)] = 0$ on the part of $\mathfrak{B}$ on which s_j has the expansion (9.105) for some j if for every $p > 0$

$$\int \Phi(\xi) \exp(-p|\xi|)\, d\xi = \infty. \tag{9.109}$$

To see this in detail, suppose for simplicity that $b_j' > 0$. In a half-plane where $\operatorname{Im} w \geqslant c$, all $s_k(w)$ are regular single-valued functions of w, hence so is $M[\hat{f}(0)]$. In this half-plane we have (9.108) for $\operatorname{Re} w > 0$ and, by (9.104), modulo unimportant terms (that is, terms whose growth is small enough so they do not affect the present argument)

$$|M[\hat{f}(0)](s, w)| \leqslant \mathrm{const} \exp[\Psi(a \operatorname{Im} w)]$$

for $\operatorname{Re} w < 0$. We now take the Fourier transform, that is, we form

$$\tilde{f}(\xi) = \int_{\operatorname{Im} w = c} M[\hat{f}(0)](s, w) \exp(i\xi w)\, dw.$$

The above inequalities imply that $\tilde{f}$ is analytic in the strip

$$S\colon 0 > \operatorname{Im} \xi > -b_j'.$$

Moreover f is bounded in closed substrips of S and, in S,

$$\tilde{f}(\xi) = 0[\exp(-\Phi(a' \operatorname{Re} \xi))]$$

as $\operatorname{Re} \xi \to \infty$ for some $a' > 0$. Using the principle of harmonic majorant we see that (9.109) implies $\tilde{f} \equiv 0$. (For those unfamiliar with harmonic majorant for a strip, use an exponential map to reduce the question to a half-plane.) Thus, if V is irreducible, (9.109) implies $M[\hat{f}(0)] = 0$ on $\mathfrak{V}$. It follows easily from the definitions that $\hat{f}(0)$ is itself zero.

Next suppose that $\alpha_j > 1$. We shall assume that $b_j = ie^{i\varphi}$ as the general case will involve merely a change of notation. We write $w = |w|e^{i\theta}$. Then we have

$$|\exp(-is_j)| \leqslant \exp[|w|^{\alpha_j} \cos(\alpha_j \theta + \varphi) + \gamma |w|^{\beta}],$$

where $\gamma > 0$, $\beta < \alpha_j$. Combining this with (9.104) gives

$$|M[\hat{f}(0)](s, w)| \tag{9.110}$$

$$\leqslant \text{const} \exp\{\min[0, |w|^{\alpha_j} \cos(\alpha_j \theta + \varphi) + \gamma |w|^{\beta}] + \Psi(a \operatorname{Im} w)\},$$

where we have again ignored the factor $(1 + |s| + |w|)^a$, which is clearly permissible.

Using the minimum modulus theorem (see Titchmarsh [2]), we see that (9.110) implies the vanishing of $M[\hat{f}(0)]$, and hence of $\hat{f}(0)$ if V is irreducible, if

$$\Psi(w) = o(|w|^{\alpha_j}). \tag{9.111}$$

THEOREM 9.34. *Let V be a union of irreducible varieties V^p each of which is zerodimensional or satisfies the following condition: There exists an intersection of V^p with a complex linear affine variety U^p such that $U^p \cap V^p$ is of complex dimension* 1 *and such that* Ψ *satisfies the appropriate condition* (9.109), (9.111) *above for some j. Then if $f \in \mathfrak{Z}'(\Phi)$, $D_1 f = \cdots = D_r f = 0$, then $f = 0$. Here by $\mathfrak{Z}'(\Phi)$ we mean* $\text{Щ}'$ for $\mathscr{W} = \mathscr{E}(\Phi)$.

We have proved the theorem in case $\dim V^p = 1$ for all p. The case of $\dim V^p = 0$ is trivial. The higher dimensional case can be reduced to the case of dimension 1 exactly as in John [2].

See Remark 9.18.

IX.11. The Zeros of Solutions of Elliptic Systems

We wish now to prove the analog of the theorem that a bounded analytic function f in the unit disk cannot have "too many" zeros in the closed disk. This means that the Blaschke product formed by the zeros inside the disk must converge, that is, $\sum \log |z^k| > -\infty$, where z^k are the zeros of f, and f can have zero radial limits only on a set of measure zero. If we use the upper half-plane instead of the unit disk then the "radial limit" is replaced by "vertical limit" and the interior condition by $\sum (\text{Im } z^k)/(1+|z^k|^2) < \infty$.

For solutions of general systems of differential equations we shall study a slightly different problem, though it is likely a suitable modification of our methods would lead to the solution of the exact analog.

We denote by $\mathscr{S}$ the space of L. Schwartz of indefinitely differentiable functions of ξ each of whose derivatives decreases more rapidly than $(1+|\xi|)^{-a}$ for any a as $|\xi| \to \infty$. By $\mathscr{B}^+\mathscr{S}$ we denote the space of indefinitely differentiable maps of $t \geqslant 0$ into $\mathscr{S}$, where $t \geqslant 0$ means $t_j \geqslant 0$ for all j, which together with their derivatives are uniformly bounded in $t \geqslant 0$. The spatial Fourier transform of $\mathscr{B}^+\mathscr{S}$ is again $\mathscr{B}^+\mathscr{S}$.

DEFINITION. By a *Cauchy zero* of a function

$$f \in \mathscr{B}^+\mathscr{S} \text{ with } D_1 f = \cdots D_r f = 0$$

we mean a point (t, ξ) such that $(q_{ij} f)(t, \xi) = 0$ for all i, j.

Of course, the definition depends on the Cauchy problem considered, but this will be assumed fixed.

We want to find conditions on a sequence $\{x^k\}$ of points in $t \geqslant 0$ in order that every $f \in \mathscr{B}^+\mathscr{S}$ with $D_1 f = \cdots = D_r f = 0$ which has Cauchy zeros at the points x^k must be identically zero. If the system is not elliptic, then we know that there exist solutions f which can vanish in a half-plane. This remark in conjunction with Theorem 9.34 shows that an appropriate condition on $\{x^k\}$ would be both a type of density and also some sort of "large distribution" in all directions (or at least in many directions) as $\xi \to \infty$. This problem is undoubtedly very interesting and we shall make some remarks about it later, though our methods are too weak to solve it. Our main results will be confined to elliptic systems when only a density condition is relevant.

Let $\mathfrak{V}$ be a multiplicity variety associated with $(D_1, \ldots, D_r)$. We assume that $\mathfrak{V}$ does not contain any varieties which are contained in linear varieties of the form $w =$ constant. We define $\mathscr{B}_0^+\mathscr{S}(\mathfrak{V})$ as follows: First we define $\mathscr{B}_0^+\mathscr{S}(V)$ when V is a variety of dimension d which is not contained in $w =$ constant. We denote by $\tilde{V}$ the intersection of V with the part of C^n where w is real. By an *indefinitely differentiable* function on $\tilde{V}$ we mean a function which can be extended to be indefinitely differentiable in C. It is clear that w is a set of local coordinates almost everywhere. We define $\mathscr{B}_0^+\mathscr{S}(V)$ as the space of all indefinitely differentiable functions f on $\tilde{V}$ which are zero unless $\text{Im } s > 0$ and such that all w

derivatives approach zero faster than $(1+|w|+|s|)^{-j}$ for all j as $w \to \infty$ through values at which w is a local parameter.

The definition of $\mathscr{B}_0^+\mathscr{S}(\mathfrak{V})$ for any multiplicity variety $\mathfrak{V}$ is now clear, namely, if $\mathfrak{V} = (V_1, \partial_1; \ldots; V_l \partial_l)$, then an element of $\mathscr{B}_0^+\mathscr{S}(\mathfrak{V})$ is an l-tuple $\vec{\mathrm{f}} = (f_1, \ldots, f_l)$ with

$$\vec{\mathrm{f}} \in \sum_{\oplus} \mathscr{B}_0^+\mathscr{S}(V_j)$$

and such that for each point $z \in \cup V_j$ there is a neighborhood N in C and an indefinitely differentiable function f on this neighborhood such that $\partial_j f = f_j$ on $V_j \cap N$.

In case $r = 1$ and D is a homogeneous real elliptic operator then $\tilde{V}$ contains a cone. In particular, if D is the Laplacian Δ then $\tilde{V}$ contains the positive light cone. We shall denote by V^* the "positive light cone," that is, $s^2 = \xi_1^2 + \cdots + \xi_d^2$, $s > 0$. Our uniqueness results will apply only to systems which are sufficiently like Δ (see Remark 9.21), so we shall restrict ourselves to this case. Let $f \in \mathscr{B}_0^+\mathscr{S}(V^*)$ which means that f is an indefinitely differentiable function on the positive light cone whose derivatives decrease at infinity faster than the reciprocal of any polynomial, and f and all is derivatives are zero at the origin. We form

$$\tilde{F}(t, \xi) = \int \exp i(ts + \xi_1 w_1 + \cdots + \xi_d w_d) f(s, w)\, dw,$$

where $s = +(w_1^2 + \cdots + w_d^2)^{1/2}$. We set $F(t, \xi) = \tilde{F}(it, \xi)$. It is clear that $\tilde{F}$ can be extended to be an analytic function of t in $\operatorname{Im} t > 0$ which is bounded in $\operatorname{Im} t \geq 0$, all real ξ. In fact, we see easily that $F \in \mathscr{B}^+\mathscr{S}$. In addition, $\Delta F = 0$.

Conversely, any $F \in \mathscr{B}^+\mathscr{S}$ which satisfies $\Delta F = 0$ has such an integral representation. To see this, let $G(t, w)$ be the spacial Fourier transform of F. Then G clearly satisfies the (ordinary) differential equation

$$\frac{\partial^2 G}{\partial t^2} = (w_1^2 + \cdots + w_d^2) G.$$

Thus, for each w, $G(t, w)$ is of the form

$$G(t, w) = a_+ \exp\left(+(w^2)^{1/2} t\right) + a_- \exp\left(-(w^2)^{1/2} t\right)$$

for suitable $a_\pm(w)$. We must have $a_+ = 0$ since $F \in \mathscr{B}^+\mathscr{S}$. Set

$$f(s, w) = a_-(w) \quad \text{for } s = +(w^2)^{1/2}.$$

We claim that F has the desired integral representation

$$F(t, \xi) = \int f((w^2)^{1/2}, w) \exp\left[-t(w^2)^{1/2} + i(\xi_1 w_1 + \cdots + \xi_d w_d)\right] dw.$$

The right side is

$$\int a_-(w) \exp[-t(w^2)^{1/2}] \exp(i\xi \cdot w)\, dw = \int G(t, w) \exp(i\xi \cdot w)\, dw = F(t, w)$$

because G and F are Fourier transforms.

The fact that $f \in \mathscr{B}^+\mathscr{S}\,(V^*)$ is easily established, so we have proved that the above integral representation characterizes the $F \in \mathscr{B}^+\mathscr{S}$ which satisfy $\Delta F = 0$. Even more is true: Let us take any plane of support A of the positive light cone $\tilde{V}$. We denote by s_A the normal coordinate to A which is positive in $\tilde{V}$ and by w_A suitable coordinates on A. We denote by t_A, ξ_A dual variables. We may assume that (s_A, w_A) are obtained from (s, w) by applying a linear transformation. Then again $\tilde{F}$ can be continued analytically to that portion of complex t, ξ space corresponding to $\operatorname{Im} t_A > 0$, ξ_A real. In fact, the extension is bounded in $\operatorname{Im} t_A \geqslant 0$, ξ_A real, and the bound can be chosen independent of A.

In this manner we see easily that F has been continued to be analytic in an open set U of complex t, ξ space and F can be extended to be continuous and bounded in $cl(U)$ which is the closure of U. U can be seen to be the tube domain whose base is a cone, that is, U is the set of complex (t, ξ) for which $\operatorname{Im} t$, $\operatorname{Re} \xi$ are unrestricted and $\operatorname{Re} t > 0$, $(\operatorname{Re} t)^2 > (\operatorname{Im} \xi_1)^2 + \cdots + ((\operatorname{Im} \xi_d)^2$. We note that the original half-space t real, $t \geqslant 0$, ξ real, is contained in $cl(U)$ and $t = 0$, ξ real is contained in the boundary of $cl(U)$. We should like to say that F is determined by its Cauchy zeros on a "small" subset of this part of the boundary. It is easily seen by reflection that F could not have Cauchy zeros on an open set of $t = 0$, ξ real without vanishing identically.

PROBLEM 9.15.

Can "open set' be replaced by "set of positive measure" in the above ?

This problem is related to a classical problem on Riesz conjugate functions (see Zygmund [1]). Our methods give no information on this problem except to transform it into the question of when linear combinations of the form

$$\sum (a_j \exp(i\alpha^j \cdot w) + b_j s \exp(i\alpha^j \cdot w))$$

are dense in the dual of $\mathscr{B}_0^+\mathscr{S}(V^*)$, where α^j belong to the set in question.

Although we cannot say anything about the above problem, we can say something about the determination of F on other parts of the boundary of $cl(U)$. We shall consider, in particular, the subset of $cl(U)$ contained in t real, ξ imaginary, $\xi_1 = 0$.

DEFINITION. By a *Cauchy zero* of F on this set we shall mean a point (t, ξ) with $F(t, \xi) = \partial F(\xi, t)/\partial \xi_1 = 0$.

The reason for this definition is that in the space t real, ξ imaginary, F satisfies a hyperbolic equation (the wave equation) and we are considering the Cauchy problem for this on $\xi_1 = 0$.

THEOREM 9.35. *Let $\{\alpha^j\}$ be a set on $\xi_1 = 0$, t real, ξ imaginary, of Cauchy zeros of an F which is not identically zero. Then the Lebesgue (real) $n-1$ dimension measure of $\{\alpha^j\}$ is zero.*

PROOF. By the above integral representation for F, the theorem is equivalent to the statement that if the measure of $\{\alpha^j\}$ is positive then the linear combinations of $\exp(i\tilde{\alpha}^j \cdot z)$ and $w_1 \exp(i\tilde{\alpha}^j \cdot z)$ are dense in the dual of $\mathscr{B}_0^+\mathscr{S}(V^*)$. Here $\tilde{\alpha}^j$ is obtained from α^j by multiplying the t coordinate by i.

We denote by $rV^*(w_1 = 0)$ the positive light cone on $w_1 = 0$ together with its interior. By $\mathscr{B}_{00}^+\mathscr{S}(rV^*)$ we denote the space of functions on $w_1 = 0$ which are zero outside rV^*, indefinitely differentiable on rV^*, have all derivatives vanish at the origin, and have all derivatives tend to zero at infinity faster than $(1 + |s| + |w|)^{-a}$ for any a.

LEMMA 9.36. *We can write $\mathscr{B}_0^+\mathscr{S}(V^*)$ as the direct sum of two copies of $\mathscr{B}_{00}^+\mathscr{S}(rV^*)$ in such a way that if $f \leftrightarrow (f_1, f_2)$ then for all j*

$$(9.112)\qquad [a_j \exp(i\tilde{\alpha}^j \cdot z) + b_j w_1 \exp(i\tilde{\alpha}^j \cdot z)] \cdot f$$
$$= a_j \exp[i\tilde{\alpha}^j \cdot (s, w_2, \ldots, w_d)] \cdot f_1$$
$$+ b_j \exp[i\tilde{\alpha}^j \cdot (s, w_2, \ldots, w_d)] \cdot f_2.$$

PROOF. We make $(f_1, f_2) \in \mathscr{B}_{00}^+\mathscr{S}(rV^*)$ correspond to the restriction of $f_1 + w_1 f_2$ to V^*. Clearly, this restriction belongs to $\mathscr{B}_0^+\mathscr{S}(V^*)$. Conversely given any $f \in \mathscr{B}_0^+\mathscr{S}(V^*)$, define (f_1, f_2) by

$$(9.113)\qquad f_1(s, w_2, \ldots, w_d) = \left[f\left(s, +\left(s^2 - \sum_{j>1} w_j^2\right)^{1/2}, w_2, \ldots, w_d\right)\right.$$
$$\left. + f\left(s, -\left(s^2 - \sum w_j^2\right)^{1/2}, w_2, \ldots, w_d\right)\right] \Big/ 2$$

$$f_2(s, w_2, \ldots, w_d) = \left[f\left(s, +\left(s^2 - \sum w_j^2\right)^{1/2}, w_2, \ldots, w_d\right)\right.$$
$$\left. - f\left(s, -\left(s^2 - \sum w_j^2\right)^{1/2}, w_2, \ldots, w_d\right)\right] \Big/ 2\left(s^2 - \sum w_j^2\right)^{1/2}.$$

It is clear that $f_1, f_2 \in \mathscr{B}_{00}^+\mathscr{S}(rV^*)$ and that $f_1 + w_1 f_2 = f$ on V^*. Moreover, the correspondence $f \leftrightarrow (f_1, f_2)$ is easily seen to be a topological isomorphism. Finally, since the w_1 component of α^j is zero, (9.112) follows from the definitions. [One should be careful to note precisely how functions are identified with distributions in $\mathscr{B}_0^+\mathscr{S}(V^*)$ and $\mathscr{B}_{00}^+\mathscr{S}(rV^*)$.]

Remark 9.19.
(9.112) is nothing but another expression for our often used adjoint of the Cauchy problem.

Proof of Theorem, continued. In view of Lemma 9.36, the theorem will be proved if we can show that the linear combinations of $\exp(i\tilde{\alpha}^j \cdot (s, w_2, \ldots, w_d))$ are dense in the dual of $\mathscr{B}_{00}^+\mathscr{S}(r\tilde{V})$ if the measure of $\{\alpha^j\}$ is positive. We note, however, that the Fourier transform G of any $g \in \mathscr{B}_{00}^+\mathscr{S}(r\tilde{V})$ will be extendable to an analytic function of d complex variables in some region Y. Y is the tube domain with base $r\tilde{V}$, that is, Y is the set of complex $(t, \xi_2, \ldots, \xi_d)$ whose real parts are unrestricted and for which

$$\operatorname{Im} t > 0, \ (\operatorname{Im} t)^2 > (\operatorname{Im} \xi_2)^2 + \cdots + (\operatorname{Im} \xi_d)^2.$$

Moreover, G is bounded as we approach the imaginary part of the boundary. In addition, $G(\tilde{\alpha}^j) = 0$ for all $\tilde{\alpha}^j$ if g is orthogonal to all $\exp(i\tilde{\alpha}^j \cdot (s, w_2, \ldots, w_d))$. This means that $G = 0$ if the measure of $\{\alpha^j\}$ is positive, since it can easily be shown by induction that a nonzero holomorphic function cannot vanish on a set of positive measure of the base of a tube. (See Zygmund [1] for the case $n = 1$.) The theorem is thus proved. ▯

See Remarks 9.20–9.24.

IX.12. General Systems

Up to now we have considered only systems for one unknown function. There is no difficulty in extending all the above to general systems. The only difference that arises is that in the Cauchy problem we may have to allow some of the T^j to be of dimension n. For example, let $r = 1$, $l = n = 3$, $\vec{\mathrm{D}} = (\partial/\partial x_1, \partial/\partial x_2, \partial/\partial x_3)$. Then $\vec{\mathrm{D}}\vec{\mathrm{f}}$ is just div $\vec{\mathrm{f}}$, so $\vec{\mathrm{D}}\vec{\mathrm{f}} = 0$ mean $\vec{\mathrm{f}} = \operatorname{curl} g$. Here g is an arbitrary function of three variables, so we need to use functions of three variables in the parametrization of solutions of $\vec{\mathrm{D}}\vec{\mathrm{f}} = 0$.

Remarks

Remark 9.1. See page 218.
There is another way to state the definition of Cauchy problem being well-posed which is in the spirit of the method of orthogonal projection. Actually, what we say

applies to any general boundary value problem (see Chapter X). Suppose the Cauchy problem is well-posed and let $f \in \mathscr{W}$. Then there exists a unique solution $h \in \mathscr{W}$ of $Dh = 0$ for which h and f have the same Cauchy data. Thus, $f - h = h' \in \mathscr{W}$ has zero Cauchy data. This means that every $f \in \mathscr{W}$ can be written uniquely in the form

$$f = h + h',$$

where $Dh = 0$ and h' has vanishing Cauchy data. Thus, $\mathscr{W}$ is the direct sum

$$\mathscr{W} = \mathscr{W}_{\vec{\mathrm{D}}} + \mathring{\mathscr{W}},$$

where $\mathring{\mathscr{W}}$ is the subspace with zero Cauchy data. If we think of $\mathscr{W}_{\vec{\mathrm{D}}}$ as being parametrized by its Cauchy data, we can identify $\mathscr{W}'_{\vec{\mathrm{D}}}$ with the dual of the space of Cauchy data so that $\mathring{\mathscr{W}}$ becomes the subspace of $\mathscr{W}$ which is "orthogonal" to $\mathscr{W}'_{\vec{\mathrm{D}}}$. The Fourier transform arguments below show that $\mathring{\mathscr{W}}'$ can be identified with the subspace of $\mathscr{W}'$ generated (as a module) by the D'_j. This is, of course, just the orthogonal space to $\mathscr{W}_{\vec{\mathrm{D}}}$.

It is easy to see that if we have a direct sum splitting $\mathscr{W} = \mathscr{W}_{\vec{\mathrm{D}}} + \mathring{\mathscr{W}}$ then the Cauchy problem is well-posed if the map of $\mathscr{W}$ into the space of Cauchy data is onto.

Remark 9.2. See page 222.

One might be tempted to guess that if $\vec{\mathrm{D}}$ and a nondegenerate set of planes $(T^d, \ldots, T^0)$ are given then the l^j are determined. That this is not the case can be seen from the example: $r = 1$, $n = 2$, $D = \partial^2/\partial x_1^2 - \partial^2/\partial x_2^2$. Using the fact that every solution f of $Df = 0$ can be written in the form $g(x_1 + x_2) + h(x_1 - x_2)$, we see easily that for any m we have the well-posed problems for the space of formal power series (also for the spaces of entire functions, indefinitely differentiable functions, etc.):

T^1 is the plane $x_1 = 0$, T^0 is the origin; $l^1 = 2$, $l^0 = 2m$; $q_{11} = \partial^m/\partial x_1^m$, $q_{21} = \partial^{m+1}/\partial x_1^{m+1}$. For m odd, $q_{10} =$ identity, $q_{20} = \partial/\partial x_2, \ldots, q_{m+1,0} = \partial^m/\partial x_2^m$, $q_{m+2,0} = \partial/\partial x_1, \ldots, q_{2m,0} = \partial^{m-1}/\partial x_1 \partial x_2^{m-2}$, while for m even, $q_{10} =$ identity, $q_{20} = \partial/\partial x_2, \ldots,$ $q_{m0} = \partial^{m-1}/\partial x_2^{m-1}$, $q_{m+1,0} = \partial/\partial x_1, \ldots, q_{2m,0} = \partial^m/\partial x_1 \partial x_2^{m-1}$.

Remark 9.3. See page 226.

We could have used a pure power series argument, avoiding the Lagrange interpolation formula. This is carried out in Ehrenpreis, Guillemin, and Sternberg [1].

Remark 9.4. See page 230.

We could easily eliminate the use of Theorem 9.5 in the proof of Theorem 9.8. In fact, we could use the proof of Theorem 9.8 to give another proof of Theorem 9.5.

Remark 9.5. See page 230.

In the proof of Theorem 9.8 we have given a method of constructing noncharacteristics $\{T^j\}$ in terms of derived sequences. The natural question to ask is: How can we construct all noncharacteristics? Since Theorem 9.8 tells us all possible T^d, we may rephrase our question to: Given T^d, what are all possible d' and T^d so there exists a noncharacteristic of the form $T^d \supset T^{d'} \supset \cdots$. By the remark following the proof of Theorem 9.3, d' is *not* determined by T^d and $\vec{\mathrm{D}}$. Thus, we may ask Problems 9.3 and 9.4 (see below).

Remark 9.6. See page 242.

We had previously assumed (only to simplify the notation) that the spaces on which the Cauchy data are given go through the origin, but it is clear that all the above applies without this restriction.

Remark 9.7. See page 234.

It cannot be said that in general ${}_j\boxed{\mathrm{L}}$ describes the action of ∂_j completely, for the components of C may belong to spaces of distributions of different dimensions.

However, if there is only one $T^i = T$, say, then we see easily from the definitions that ${}_j\boxed{\mathrm{L}}$ describes the action of ∂_j completely, namely, ∂_j commutes with convolution by elements of $\mathscr{W}'(T)$. In the general case (when there is more than one T^j) the situation is much more complicated as is evidenced by Theorem 2.5.

Remark 9.8. See page 234.

It follows from (9.31) or rather the general formula (9.30) that the infinitesimal generators give a method of expressing multiples of s_j in terms of the basis of functions Q_{lm}.

Remark 9.9. See page 234.

We may regard Theorem 9.15 as an extension of the classical procedure of passing from an ordinary differential equation of order >1 to a system of first order equations.

Remark 9.10. See page 239.

Although we have not done so, it should be possible to extend the results of Theorems 9.17 and 9.18 to general systems. The analog of Theorem 9.18 for the Cauchy-Riemann system in several variables was proved in Ehrenpreis [10].

Remark 9.11. See page 243.

One might think that by allowing nonlinear T we are studying some over-determined systems with variable coefficients. For $r = 1$ we see by a change of variables that we can reduce the problem to the case of linear T but with coefficients varying only in t. In a sense the constancy in t is not too essential. It is used to apply the Lagrange interpolation formula, but this can usually be replaced by the solution of an ordinary differential equation in t if we take Fourier transforms in ξ and not in t.

Remark 9.12. See page 248.

For the heat equation and certain other parabolic equations Theorem 9.23 is due to Täcklind [1]. A weaker result for general D is proved by Gelfand and Šilov [1].

Remark 9.13. See page 264.

We have assumed in the above that we are considering solutions in the whole x space. The same methods would apply to many other convex regions, for example to $t \geqslant 0$. For a single equation and $\alpha = 1$ this is studied in John [1].

Remark 9.14. See page 264.

We have considered only those functions $\Phi(\gamma)$ of the form $\varphi(\gamma_1) + \cdots + \varphi(\gamma_m)$, where φ is even. Presumably our methods could be modified to treat Φ of the form $\varphi_1(\gamma_1) + \cdots + \varphi_m(\gamma_m)$, where the φ_j are no longer assumed to be even. Then we would have to use the more general notion of quasianalyticity studied in Chapter XIII. By using the Oka embedding of Section IV.5 and taking limits we could probably treat arbitrarily convex Φ.

Remark 9.15. See page 264.

Theorem 9.30 shows that under suitable conditions the restrictions of $f \in \mathscr{E}_{\vec{\mathrm{D}}}(\Phi)$ to T have certain quasianalytic properties. It would be interesting to derive them from the explicit structure of these restrictions as discussed in IX.6 above.

Remark 9.16. See page 267.

We could make the problem somewhat more precise if instead of asking for *all* derivatives of f to belong to $\mathscr{W}'$ we asked that the Cauchy data belong to $\mathscr{W}'$ and to analogous spaces on T^j for $j < d$. We suspect that our methods apply also to this refinement.

Remark 9.17. See page 267.

Instead of requiring that the restriction of f to $t = t^0$ belong to $\mathscr{W}'$ for all t^0 with $0 \leqslant t_j^0 \leqslant 1$, we could require it for *certain* t^0 with weaker conditions for the other t^0. Then a Phragmén-Lindelöf principle would imply that the restrictions belong to $\mathscr{W}'$ for all such t^0. We shall not discuss this further.

Remark 9.18. See page 271.

Presumably the conditions in Theorem 9.34 are best possible, but we have not studied this question.

Remark 9.19. See page 276.

Remarks 9.20–9.24. See page 276.

Remark 9.20. In case $n = 2$ our theorem reduces to the classical result that a harmonic function in $\mathscr{B}^+\mathscr{S}$ cannot have Cauchy zeros on $t = 0$ on a set of positive measure. This follows easily if we write such a harmonic function $h = h_1(\xi + it) + h_2(\xi - it)$, where h_1 and h_2 are analytic. Then $t = 0$, ξ real means the arguments in h_1 and h_2 are both real which is also the case if $\xi = 0$, t imaginary. A similar remark applies to the derivatives.

Remark 9.21. The analog of Theorem 9.35 can be proved for elliptic systems whenever we have an analog of Lemma 9.36 (that is, a suitable Cauchy problem) in which the spaces $\mathscr{B}^+_{00}\mathscr{S}(r\tilde{V})$ are replaced by spaces on "small sets," that is, for which we know the Fourier transform is analytic in enough variables.

Remark 9.22. Although F itself is analytic in $d + 1$ complex variables, Theorem 9.36 does not follow directly because the relevant property of $\{\alpha^j\}$ is its (real) d-dimensional measure.

Remark 9.23. We could prove a similar type of result for Cauchy zeros in some other parts of $cl(U)$. For example, if we wanted to get bounds on the number of Cauchy zeros of F on the (complex) part of $cl(U)$, where $\xi_1 = 0$, then we would be led to a density problem as above except that the α^j would now be permitted to be suitable *complex* points. Again Lemma 9.36 and the following would enable us to show that F could not have Cauchy zeros at the α^j, unless there exists a suitable analytic function of d complex variables which vanishes there. The result is, of course, interesting only if the α^j are contained in the boundary of $cl(U)$.

Remark 9.24. In the above proof we have parametrized harmonic functions in n variables by pairs of functions which are complex analytic in $n - 1$ variables.

Problems

PROBLEM 9.1. See page 223.

Let $\mathscr{W}$ be an LAU space for which $\mathbf{W}'$ is dense in the space of entire functions. Suppose we are given a well posed Cauchy problem for $\mathscr{W}$. Is this well-posed for the space of formal power series?

PROBLEM 9.2. See page 223.

Is l^d equal to the "generic" number m_0 of points of $\mathfrak{V}$ lying over a point in T^d?

PROBLEM 9.3.

Given $\vec{\mathrm{D}}$, T^d and the Q_{id}, is d' determined?

Theorem 9.4 shows that given $\vec{\mathrm{D}}$, T^d, the Q_{id}, and $T^{d'}$, then $l^{d'}$ is determined. In view of Theorem 9.4 a determination of the $T^{d'}$ is dependent on

PROBLEM 9.4.

For which $T^{d'}$ is

$$\mathscr{P}/[\vec{\mathrm{P}}\mathscr{P}^r + \sum Q_{id}\,\mathscr{P}(\hat{T}^d)]$$

a vector space of finite nonzero dimension over $\mathscr{R}\,(\hat{T}^{d'})$?

If the set

$$M^d = [\vec{\mathrm{P}}\mathscr{P}^r + \sum Q_{id}\mathscr{P}(\hat{T}^d)]$$

were an ideal, then in view of Theorem 9.3 (applied to this ideal instead of $\vec{\mathrm{P}}\mathscr{P}$), Theorem 9.5 would characterize all possible $T^{d'}$. Thus, we must extend Theorem 9.5 so as to cover the case when $\vec{\mathrm{P}}\mathscr{P}^r$ is replaced by M^d. Of course, the notion of principal noncharacteristic must be suitably modified, since $\mathfrak{B}$ and (9.7) are apparently not meaningful for M^d if it is not an ideal, but rather a module over $\mathscr{P}(\hat{T}^d)$ of a special form (though infinitely generated). In particular, we must find the analogs of the reasoning which led to (9.11) and (9.12). We do not know how to proceed.

PROBLEM 9.5. See page 235

There is one special case which is of particular interest, namely, the case in which there is only one $T^i = T$, say, and in which the ${}_j\,\boxed{\mathrm{L}}$ are symmetric matrices. We refer to this as a *symmetric hyperbolic* (or *parabolic*) *system*. It bears many resemblances to the usual symmetric systems. Can our results be extended to apply to symmetric hyperbolic systems with variable coefficients ?

PROBLEM 9.6. See page 242.

Can we make a "coherent" choice of ${}_{t0}^{\,i}S^a$ so that (9.55) holds ?

PROBLEM 9.7. (CONJECTURE). See page 243.

T is spacelike if and only if for each a near T there is a compact subset A_a of T such that there exists a fundamental solution at a whose support is the union of the infinite rays (half lines) which start from a and pass through A_a, and such that this fundamental solution has a restriction to T.

PROBLEM 9.8. (CONJECTURE). See page 247.

A necessary and sufficient condition that the set of $F \in \mathbf{E}'(\Phi)$ which satisfy for some b, c'

$$F(s, w) = 0[(1 + |s| + |w|)^{c'} \exp(\Psi(b \operatorname{Im} w))]$$

in the region $|s| \leqslant c(1 + |w|^a)$ be dense in the space $\mathbf{E}'(\Phi)$ is that

$$\int \frac{\Psi(\operatorname{Im} w)}{1 + |\operatorname{Im} w|^{a+1}}\, dw = \infty.$$

Here Ψ is the conjugate of Φ in the sense of Chapter V, Example 7.

PROBLEM 9.9 (CONJECTURE). See page 249

A necessary and sufficient condition that the set of $F \in \mathbf{E}'(\Phi)$ which satisfy (9.68) in the region (9.69) be dense in $\mathbf{E}'(\Phi)$ is (9.63) for $\alpha = \max \alpha_j$.

PROBLEM 9.10. See page 250.

PROBLEM 9.11. See page 250.

PROBLEM 9.12. See page 265.

Study these and similar approximation problems from a more direct point of view.

Problem 9.13. See page 266.

Carry out the details of the above program. Find an extension of Nevanlinna's second inequality.

Problem 9.14. See page 267.

How fast can f and all its derivatives approach zero uniformly on $0 \leqslant t \leqslant 1$ as $|\xi| \to \infty$?

Problem 9.15. See page 274.

Can "open set" be replaced by "set of positive measure" in the above ?

This problem is related to a classical problem on Riesz conjugate functions (see Zygmund [1]). Our methods give no information on this problem except to transform it into the question of when linear combinations of the form

$$\sum (a_j \exp(\alpha^j \cdot w) + b_j s \exp(\alpha^j \cdot w))$$

are dense in the dual of $\mathscr{B}_0^+ \mathscr{S}(\tilde{V})$, where the α^j belong to the set in question.

CHAPTER X

Balayage and General Boundary Value Problems

Summary

Roughly speaking, *balayage* is a method of choosing representatives from a quotient space of a space of functions or distributions. Let Ω be the closed unit cube in $R = R^n$ and let $\vec{\mathrm{D}} = (D_1, \ldots, D_r)$. Let $B_1, \ldots, B_h$ be parts of Ω; for each j let δ_j be a linear partial differential operator with constant coefficients. We say that $(B_1, \delta_1; \ldots; B_h, \delta_h)$ define a *parametrization problem* for $\vec{\mathrm{D}}$. We say that the parametrization problem is *well-posed* if the map

$$\alpha : f \to (D_1 f, \ldots, D_r f; \delta_1 f | B_1, \ldots, \delta_h f | B_h)$$

is a topological isomorphism from $\mathscr{W}$ onto ${}_{\vec{\mathrm{D}}}\mathscr{W} \oplus \mathscr{W}_1 \tilde{\oplus} \cdots \tilde{\oplus} \mathscr{W}_h$. Here $\mathscr{W}$ (respectively $\mathscr{W}_j$) is an AU space of functions or distributions on Ω (respectively B_j), and $\tilde{\oplus}$ indicates that where the B_j intersect, certain compatibility conditions are to be imposed. ${}_{\vec{\mathrm{D}}}\mathscr{W}$ is the subspace of $\mathscr{W}^r$ which satisfies the same differential relations as the D_j (see Chapter VI). (If Ω were replaced by C, and B_j were linear varieties passing through the origin, then this would be the Cauchy problem.)

In Section X.1 we show that, by duality, the parametrization problem becomes the balayage problem. By Fourier transform these become interpolation problems.

In Section X.2 we study the example of the Laplacian on the square from our point of view.

In Sections X.3 and X.4 we study the Dirichlet problem, which is the parametrization problem where the B_j are faces of Ω. Under suitable conditions of symmetry we can give necessary and sufficient conditions in order that the Dirichlet problem be well-posed.

Using a method based on the Schwarz alternating procedure for the Dirichlet problem for the Laplacian, we show in Section X.5 how to extend some of our results to general convex polyhedra instead of the unit cube.

In Section X.6 we extend the notion of a well-posed problem so as to allow α to have a kernel and to have a closed range (instead of being onto). We can formulate various notions of *index* for these problems.

We discuss a generalization of the Dirichlet problem in Section X.7 which is in the direction of the Wiener-Hopf problem. A generalization from a different point of view is dealt with in Section X.8.

In Section X.9 we explain how to extend the results to $l > 1$.

X.1. Balayage, General Boundary Value Problems, and Interpolation

In his study of harmonic functions and the Newtonian potential, Poincaré introduced the concept of balayage (sweeping out). Roughly speaking this may be formulated as follows: Given any compact set A in R and any measure μ of compact support, there is a measure μ_A whose support lies on A and whose Newtonian potential, that is,

$$\begin{cases} \int |x-y|^{2-n}\, d\mu_A(y) & \text{for } n > 2 \\ \int \log|x-y|\, d\mu_A(y) & \text{for } n = 2 \end{cases}$$

is equal to that of μ almost everywhere on A.

In 1959–1960 Beurling discussed a different type of "balayage" problems in his lectures at the Institute for Advanced Study (unpublished): Let $\{\lambda_j\}$ be a sequence of real numbers. Under what conditions is every function f, whose Fourier transform is a bounded measure, equal on the interval $[-1, 1]$ to a function $\sum a_j \exp(ix\lambda_j)$, where $\sum |a_j| < \infty$.

The similarity in the two examples is the following: We are given some object (μ or f) for which we want a suitable representative which agrees with it in some property (having equal potential on A or being equal on $[-1, 1]$, respectively). "To be equal in a given property" is, roughly speaking, the same as "to belong to the same coset in some quotient space." We shall see that in this manner we can formulate a very general balayage problem which contains the above examples and is related to general boundary value problems.

In order to make this clear, let us consider the case in which A is the boundary of a (nice) domain S and support $\mu \subset S$. Let us denote the (Newtonian) potential of the mass ν by U^ν. If h is harmonic and "not too large" on S, we can write h in the form U^ν for some ν whose support is contained in A. Actually ν may be a distribution (see Kellogg [1], p. 218) but we shall ignore this in our formal argument.

Thus, by well-known potential theory arguments (see Brelot [1], p. 78) we have

$$(10.1)\qquad \begin{aligned}\int h\,d\mu &= \int U^{\nu}\,d\mu\\ &= \int U^{\mu}\,d\nu\\ &= \int U^{\mu_A}\,d\nu\\ &= \int h\,d\mu_A\end{aligned}$$

because support $\nu \subset A$. This means that $\mu = \mu_A$ on the space of harmonic functions on S. Since this space of harmonic functions is a closed subspace of the space of continuous functions on S, if we use the duality between subspace and quotient space, we see that μ and μ_A belong to the same coset in the space of measures on S modulo the subspace of measures which are zero on harmonic functions.

A similar situation prevails in the problem considered by Beurling: We regard two functions whose Fourier transforms are bounded measures as being equal if they agree on $[-1, 1]$. Then in each equivalence class we seek a representative of the form $\sum a_j \exp(i\lambda_j x)$ with $\sum |a_j| < \infty$.

The *general balayage problem* can be formulated as follows: Let $\mathscr{U}$ be a topological vector space of functions or distributions on a region Ω and let $\mathscr{U}_1$ be a closed subspace. Let Ω_2 be a subset of Ω and let $\mathscr{U}_2$ be a space of functions or distributions on Ω_2 which is contained in $\mathscr{U}$ in some natural way. Then we say that we can perform *balayage* from $\mathscr{U}/\mathscr{U}_1$ on Ω_2 (or, more precisely, on $\mathscr{U}_2$) if each element of $\mathscr{U}/\mathscr{U}_1$ has a representative in $\mathscr{U}_2$, that is, if the map $\mathscr{U}_2 \to \mathscr{U}/\mathscr{U}_1$ is onto.

Thus, in the example in potential theory, $\mathscr{U}$ is the space of measures on S, $\mathscr{U}_1$ the subspace of those which are zero on harmonic functions, and $\mathscr{U}_2$ the space of measures on A. In Beurling's example, $\mathscr{U}$ is the space of bounded measures on $R(n = 1)$, $\mathscr{U}_1$ the subspace of those whose Fourier transform vanishes on $[-1, 1]$, and $\mathscr{U}_2$ the subspace of those whose support is on $\{\lambda_j\}$.

Another example occurs in the theory of sufficient sets in AU spaces (see Chapters I, V, and VII). Let $\mathscr{W}$ be an AU space and $K = \{k\}$ an AU structure. Let $\mathscr{U}$ be the space of measures on C of the form μ/k, where μ is a bounded measure and $k \in K$. Let $\mathscr{U}_1$ be the subspace of those μ/k which are zero on $\mathbf{W}'$ and let $\mathscr{U}_2$ be the subspace of those which have their support on σ where σ is $\mathscr{W}$ sufficient. The sufficiency of σ means that balayage is possible.

Let us also mention that our quotient structure theorem (see Theorem 4.1) is also a type of balayage property, since we are representing elements of $\mathbf{W}'/\vec{P}\mathbf{W}'^r$ by functions on $\mathfrak{B}$. This is not, strictly speaking, an example of our general balayage problem because $\mathbf{W}'(\mathfrak{B})$ is not a subspace of $\mathbf{W}'$. We could easily modify our definitions to allow this also to be a case of balayage but we shall not do this.

We wish to show the relation of the above to boundary value problems. For the Laplace equation this is the classical relation of Poincaré's balayage to the Dirichlet problem. Let $\vec{D} = (D_1, \ldots, D_r)$ and let Ω be a domain in R.

DEFINITION. Let $B_1, \ldots, B_h$ be parts of the boundary of Ω. (They may overlap.) For each j we associate a linear partial differential operator δ_j with constant coefficients in R such that δ_j is not a differential operator tangent to B_j. Then we say that $(B_1, \delta_1; \ldots; B_h, \delta_h)$ define a *boundary value problem* for $\vec{D}$. If we assume that the B_j are contained in Ω but not necessarily in its boundary then we speak of the above as a *parametrization problem* for $\vec{D}$.

Let $\mathscr{W}$ be a space of functions or distributions on Ω such that $\vec{D}$ is defined on $\mathscr{W}$ and the restriction of $\delta_j w$ to B_j is defined for each $w \in \mathscr{W}$ and each j. Let $\mathscr{W}_j$ be a space of functions or distributions on B_j such that the above restriction maps are continuous from $\mathscr{W}$ into $\mathscr{W}_j$. Then we say that the parametrization problem is *well-posed* (for $\mathscr{W}; \mathscr{W}_1 \ldots, \mathscr{W}_h$) if the mapping

$$\alpha : f \to (D_1 f, \ldots, D_r f; \delta_1 f | B_1, \ldots, \delta_h f | B_h) \tag{10.2}$$

(where $\delta_j f | B_j$ denotes the restriction of $\delta_j f$ to B_j) is a topological isomorphism of $\mathscr{W}$ onto ${}_{\vec{D}}\mathscr{W} \oplus \mathscr{W}_1 \tilde{\oplus} \cdots \tilde{\oplus} \mathscr{W}_h$. Here ${}_{\vec{D}}\mathscr{W}$ is the subspace of the r-fold direct sum of $\mathscr{W}$ with itself which satisfies the compatibility conditions (cf. Chapter VI) and we have written $\tilde{\oplus}$ to indicate that where the B_j intersect, compatibility conditions are to be satisfied.

In case the B_j are linear varieties passing through the origin which are either equal or of different dimension, the parametrization problem becomes the Cauchy problem discussed in Chapter IX. In Ehrenpreis [5] we treat an analog of the Cauchy problem for a single hyperbolic operator $(r = 1)$ in which the B_j are parallel hyperplanes.

In what follows we shall assume that $\mathscr{W}$ is LAU and metrizable, and that the spaces $\mathscr{W}$ and $\mathscr{W}_j$ are reflexive.

LEMMA 10.1. *Suppose that $f \to (\delta_1 f | B_1, \ldots, \delta_h f | B_h)$ is a topological isomorphism of $\mathscr{W}_{\vec{D}}$ onto $\mathscr{W}_1 \tilde{\oplus} \cdots \tilde{\oplus} \mathscr{W}_h$. Then the parametrization problem is well-posed.*

Here $\mathscr{W}_{\vec{D}}$ is the subspace of $\mathscr{W}$ consisting of those f with $\vec{D}f = 0$.

PROOF. Our hypothesis implies that α is one-to-one; it is easily seen that α is onto. It is continuous by definition. Suppose $\alpha f \to 0$. Let $h_i = D_i f$

so $h = (h_1, \ldots, h_r) \in \vec{\mathrm{D}}\mathscr{W}$ and $h \to 0$. Thus, by Theorem 6.4 and the localizability of $\mathscr{W}$, there are $g \in \mathscr{W}$ such that $\vec{\mathrm{D}}g = h$ and $g \to 0$ in the topology of $\mathscr{W}$. But then $\delta_j g \mid B_j \to 0$ for each j, because the restriction maps are continuous. Hence, by our hypothesis, $f - g$ (which is in $\mathscr{W}_{\vec{\mathrm{D}}}$) must approach zero. Thus, our result is proved. ▯

We can now show the relation of parametrization to balayage. Let α' denote the adjoint of α. In order to decide whether the parametrization problem is well-posed, it is sufficient, by Lemma 10.1, to study the restriction of α' to $(\mathscr{W}_1 \tilde{\oplus} \cdots \tilde{\oplus} \mathscr{W}_h)'$. (We shall therefore denote this restriction again by α') α' is a mapping into $\mathscr{W}'_{\vec{\mathrm{D}}}$. Now suppose that the parametrization problem is well-posed. Then we can write, for any $T \in \mathscr{W}'_{\vec{\mathrm{D}}}$,

$$(\alpha')^{-1}T = (S_1, \ldots, S_h), \tag{10.3}$$

where $S_j \in \mathscr{W}'_j$. Thus, S_j is a distribution whose support is on B_j. Note that $T \in \mathscr{W}'_{\vec{\mathrm{D}}}$ is represented by a coset in $\mathscr{W}'$ of the module in $\mathscr{W}'$ generated by the D'_i. Thus, (10.3) is a balayage of this quotient space onto the "parameter space" $\{B_j\}$. In case each B_j is contained in the boundary of Ω (10.3) represents the balayage of the dual of the space of solutions onto the boundary of Ω. As such it represents the generalization of Poincaré's balayage, which is relevant to the Dirichlet problem for the Laplacian. (In this formal argument we have not been too concerned with the distinction between $\oplus$ and $\tilde{\oplus}$.)

Let us denote by $\hat{\alpha}'$ the Fourier transform of α'. Then we see easily that (10.3) becomes

$$\hat{T} = \sum \hat{\delta}_j \hat{S}_j \text{ as elements of } \mathbf{W}'_{\vec{\mathrm{D}}}. \tag{10.4}$$

The $\hat{\delta}_j$ are polynomials. Now we use our quotient structure theorem (Theorem 4.1). This asserts that $\hat{T}$ is completely determined as an element of $\mathbf{W}'_{\vec{\mathrm{D}}}$ by its restriction to $\mathfrak{V}$. Thus (10.4), becomes

$$\hat{T} = \sum \hat{\delta}_j \hat{S}_j \quad \text{on } \mathfrak{V}.$$

This leads to another interpretation of (10.3). $\hat{T}$ is an arbitrary element of $\mathbf{W}'(\mathfrak{V})$. Thus, (10.5) says that every element of $\mathbf{W}'(\mathfrak{V})$ can be uniquely (modulo some compatibility conditions) expressed in the form $\sum \hat{\delta}_j \hat{S}_j$, where $\hat{S}_j \in \mathbf{W}'_j$. In the case of the Cauchy problem, we saw that the possibility of such an expression was determined by the Lagrange interpolation formula. We shall call the possibility of writing an element in $\mathbf{W}'(\mathfrak{V})$ in the form $\sum \hat{\delta}_j \hat{S}_j$ as the *general interpolation problem*. Thus, we have shown the equivalence of parametrization, balayage, and interpolation.

It is interesting to note that the determination of some well-posed parametrizations will proceed in two steps.

1. Use of some classical interpolation formulas to obtain the result in "simple" cases.
2. Use of a limiting argument to obtain more complex results.

Now, step 1 is similar in spirit to Beurling's method while step 2 is similar in spirit to Poincaré's.

X.2. An Example

In order to explain the general method for rectangular regions, let us begin with an example. Let $r = 1$, $n = 2$, $D = \partial^2/\partial x^2 + \partial^2/dy^2$. Let Ω be the rectangle $|x| \leqslant 1$, $|y| \leqslant \lambda$, where $\lambda > 0$ is fixed. Let B_j be the (closed) sides of Ω as shown below (see Fig. 17):

For each j let δ_j be the identity. Thus, our boundary value problem is just the classical Dirichlet problem.

$\mathscr{W}$ is the space $\mathscr{E}(\Omega)$ of indefinitely differentiable functions on Ω and $\mathscr{W}_j$ is the space $\mathscr{E}(B_j)$ of indefinitely differentiable functions on B_j.

To fix the ideas, suppose first that we want to perform balayage of δ on the boundary of Ω. Thus, we want to write

$$\delta = S_1 + S_2 + S_3 + S_4 \quad \text{mod } D'\mathscr{E}'(\Omega)$$

where the S_j belong to $\mathscr{E}'(B_j)$. We note that the Fourier transform of S_1 can be written in the form $\exp(iz)F_1(s)$, where $F_1(s)$ is an entire function of exponential type λ, which is slowly increasing on the real axis. [We have written (z, s) for the dual variables to (x, y).] This is a consequence of the Paley-Wiener theorem (see Chapter V, Example 5). A similar expression holds for the other S_j. Thus, the Fourier transform of the above expression for δ becomes.

$$(10.6) \qquad 1 = e^{iz}F_1(s) + e^{-i\lambda s}F_2(z) + e^{-iz}F_3(s) + e^{i\lambda s}F_4(z) \ (\text{mod } (s^2 + z^2)),$$

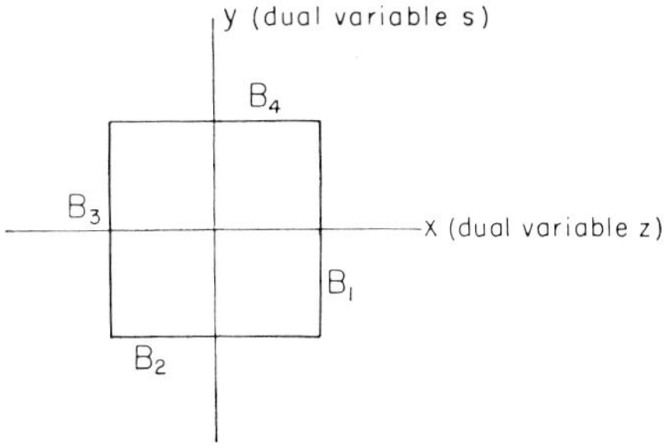

FIGURE 17

where we have written "mod $(s^2 + z^2)$" to mean "modulo the module generated by $s^2 + z^2$ in the Fourier transform space $\mathbf{E}'(\Omega)$." By the results of Chapter IV, (10.6) means

$$1 = e^{iz}F_1(s) + e^{-i\lambda s}F_2(z) + e^{-iz}F_3(s) + e^{i\lambda s}F_4(z) \quad \text{on } V, \tag{10.7}$$

where V is the variety $s = \pm iz$. This gives two equations

$$\begin{aligned} 1 &= e^{iz}F_1(iz) + e^{\lambda z}F_2(z) + e^{-iz}F_3(iz) + e^{-\lambda z}F_4(z) \\ 1 &= e^{iz}F_1(-iz) + e^{-\lambda z}F_2(z) + e^{-iz}F_3(-iz) + e^{\lambda z}F_4(z). \end{aligned} \tag{10.8}$$

This might cause one to wonder, for we know that the Dirichlet problem is well-posed so the *two* equations (10.8) should determine the *four* functions F_j (essentially) uniquely! The reason for this is that the F_j are restricted greatly at infinity and this causes (10.8) to have a unique solution. This will be seen in the following.

In order to see how to solve (10.8) let us make the following observation which is of the greatest importance: Since δ is invariant under $x \to -x$ or $y \to -y$, we should expect that $F_1 = F_3$ and $F_2 = F_4$. If we make this assumption then (10.8) becomes

$$1 = 2F_1(iz) \cos z + 2F_2(z) \cosh \lambda z. \tag{10.9}$$

It may seem that we have gained nothing, for we have replaced the two equations (10.8) in four unknowns by one equation in two unknowns. Yet this equation determines F_1 and F_2 as follows: For $z = (-i\pi/\lambda)(j + \frac{1}{2})$, (10.9) becomes

$$1 = 2F_1\left(\frac{\pi}{\lambda}\left(j + \frac{1}{2}\right)\right) \cosh \frac{\pi}{\lambda}\left(j + \frac{1}{2}\right). \tag{10.10}$$

Thus, F_1 is determined at the points $(\pi/\lambda)(j + \frac{1}{2})$. Similarly, setting $z = \pi(j + \frac{1}{2})$ gives

$$1 = 2F_2(\pi(j + \tfrac{1}{2})) \cosh \lambda\pi(j + \tfrac{1}{2}) \tag{10.11}$$

which determines F_2 at the points $\pi(j + \frac{1}{2})$. However, (see Chapter V, Example 5) F_1, F_2 are entire functions of exponential type λ, 1, respectively, which are polynomially increasing on the real axis. Hence, by a well-known result (see e.g., Boas [1]) they are determined (if they exist) by (10.10) and (10.11), respectively, except that we can add $Q_1(z) \cos \lambda z$ or $Q_2(z) \cos z$, respectively, where Q_1 are arbitrarily polynomials, that is, polynomials times $\cos z$ are the only entire functions L of exponential type 1 which are of polynomial growth on R and vanish at the points $\pi(j + \frac{1}{2})$. (This can be seen by applying Liouville's theorem to $L/\cos z$.) Since the Fourier transform of $Q(z) \cos z$ has its support at the points ± 1, the nonuniqueness terms give effects only at the corners of Ω. Since these are easily handled, we can state that the F_j are completely determined by the necessary conditions (10.10) and (10.11). It follows from the remarks below that for these F_j (10.9) is satisfied for all z, i.e., (10.10) and (10.11) are sufficient.

The above method works when we replace δ by any distribution on $\mathscr{E}'(\Omega)$ which is invariant under $x \to -x$ and $y \to -y$. A similar method works if we replace invariance by changing sign (i.e., odd instead of even). We allow the group X of symmetries of Ω to act on the distributions on Ω in the obvious manner. This defines a representation of X on $\mathscr{E}'(\Omega)$. We decompose this representation into its irreducible parts. For a distribution belonging to any irreducible part we can proceed exactly as in the case of distributions invariant under X. Thus, we can perform balayage of an arbitrary distribution in $\mathscr{E}'(\Omega)$ and show that the Dirichlet problem is well-posed for Ω.

The above method is subject to an interesting interpretation. Let $G(s, z) \in \mathbf{E}'(\Omega)(V)$. Then the above says that we can write

$$G(s, z) = e^{iz}H_1(s) + e^{-i\lambda s}H_2(z) + e^{-iz}H_3(s) + e^{i\lambda s}H_4(z) \tag{10.12}$$

on V for suitable H_1, $H_3 \in \mathbf{E}'[-\lambda, \lambda]$ and H_2, $H_4 \in \mathbf{E}'[-1, 1]$. The determination of the H_j by the above procedure would involve the values $G(s, z)$ for

$$z = \pi\left(j + \frac{1}{2}\right),\ i\frac{\pi}{\lambda}\left(j + \frac{1}{2}\right),\ \pi j,\ i\frac{\pi}{\lambda}j. \tag{10.13}$$

This can be understood as follows: We note that V consists of two planes, namely, $s = \pm iz$. If, for example, we look at $s = iz$, then we see that the functions G satisfy

$$|G(iz, z)| = 0(1 + |z|)^a \exp\left(|\text{Im } z| + \lambda\,|\text{Re } z|\right). \tag{10.14}$$

An analytic function of z satisfying the inequality (10.14) is determined, except for certain simple terms which correspond to the above nonuniqueness terms, by its values at the points $z = \pi(j + \frac{1}{2})$, $z = i(\pi/\lambda)j$. The reason we need twice as many points in (10.13) is that we want a suitable "separation" property to hold (see below).

By modifying the class of functions considered we could eliminate the simple nonuniqueness terms. This can be done in many ways, for example, we could eliminate the polynomial factor in (10.14) and replace 0 by o. In this case we could show that the above points form a *sufficient set* in the sense described in Chapter VII. Thus, as in Chapter VII, every harmonic function on Ω belonging to a suitable space can be represented as a convergent series of these exponentials.

We can now note the remarkable similarity between this last statement and the statement of Beurling's balayage problem described at the beginning of this chapter. Indeed, in the problem treated by Beurling, balayage on the set $\{\lambda_j\}$ is possible if and only if the points λ_j form a sufficient set for the topology of the dual of the space of functions which have absolutely convergent Fourier integrals, two such functions being identified if they agree on $[-1, 1]$.

We recapitulate the above: Boundary problems (of a suitable kind) for rectangles lead to questions of sufficient sets on V which are related to balayage problems in the sense of Beurling. Thus, two balayage problems which apparently have no connection, namely, Poincaré's and Beurling's are actually closely related.

We could try now to reverse the procedure. Can we start with a sufficient set on V (of a suitable type) and find a related boundary value problem, or at least a parametrization problem? Such an interpretation is indeed possible in many cases. Suppose, for example, that the sufficient set has the property that of the points $(z^j, s^j) \in V$ at least one of z^j, s^j is real. Consider those (z^j, s^j) for which z^j is real. Then on the boundary of Ω, on B_2 say, $\exp(ixz^j + iys^j) = \exp(-is^j)\exp(ixz^j)$. (We shall assume from now on that $\lambda = 1$.) Suppose that we expand every harmonic function in $\mathscr{E}(\Omega)$ in a series

$$h(x, y) = \sum a_j \exp i(xz^j + ys^j). \tag{10.15}$$

Then on B_2 we would have

$$h(x, -1) = \sum a_j \exp(-is^j) \exp(ixz^j). \tag{10.16}$$

Thus, if every function f in $\mathscr{E}(B_2)$ could be expanded in a series

$$f(x) = \sum b_k \exp(ixz^k), \tag{10.17}$$

the sum being over only those z^k which are real, then we would know that $h \to h(x, -1)$ maps *onto* $\mathscr{E}(B_2)$. A similar result holds for any B_l. However, we are not finished. For to know that α is onto means not only that we can map onto each B_l but that we can do this simultaneously for all l. It seems that the reason we required above [following (10.14)] twice as many points as would appear necessary is because of this "simultaneity" requirement. Moreover, to know that the boundary value problem is well-posed we must know in addition that α is one-to-one and in fact topological.

If we do not assume that one of (z^j, s^j) is real, then we can still proceed as above except that we need an expansion of the form (10.17), where the z^j are no longer real. It should be noted that for z^j real, (10.17) is a modification of Beurling's balayage problem, so in the case of complex z^j we are led to a generalization of Beurling's problem. Actually we can think of the problem of sufficient sets on a general $\mathfrak{V}$ as a generalization of Beurling's balayage problem to complex exponentials which lie on multiplicity varieties.

We wish now to show how to carry out the above procedure of passage from sufficient sets to the Dirichlet problem for the Laplacian on a rectangle, as it is quite instructive in understanding the apparently mysterious role played by symmetry in our first method.

We shall consider the general case when neither z^j nor s^j is assumed real. For each j let $f_j \in \mathscr{E}(B_j)$. Then, in order to satisfy (10.15), we must solve four equations

(10.18)
$$\text{(a)}\quad f_1(y) = \sum a_j \exp(iz^j)\exp(iys^j)$$
$$\text{(b)}\quad f_2(x) = \sum a_j \exp(-is^j)\exp(ixz^j)$$
$$\text{(c)}\quad f_3(y) = \sum a_j \exp(-iz^j)\exp(iys^j)$$
$$\text{(d)}\quad f_4(x) = \sum a_j \exp(is^j)\exp(ixz^j).$$

In order to understand the meaning of equations (10.18) let us suppose that we had a "perfect" set $\{(z^j, s^j)\}$, that is, each h has a unique representation (10.15). (This is possible only for very special spaces $\mathscr{W}$, but can be satisfied approximately in general.) Then to solve (10.18) uniquely we would want to know that there are in some sense as many points (z^j, s^j) as would be needed to balayage uniquely four functions, namely, a single function in each $\mathscr{E}(B_j)$. In the specific case treated above the points (z^j, s^j) are $[\pi(j+\frac{1}{2}), \pm i\pi(j+\frac{1}{2})]$, and $[i\pi(j+\frac{1}{2}), \pm\pi(j+\frac{1}{2})]$, and also $(\pi j, \pm i\pi j)$, and $(i\pi j, \pm\pi j)$ for all integers j. If we realize that each function in $\mathscr{E}(B_l)$ can, roughly speaking, be balayaged uniquely on the points $\pi(j+\frac{1}{2})$, we see that we have apparently twice the requisite number of points.

There is a further remark of interest in this connection. V consists of two complex planes so $\mathbf{E}'(\Omega)(V)$ is essentially the direct sum (there is a compatibility condition at the origin) of $\mathbf{E}'(\Omega)(V^+)$ and $\mathbf{E}'(\Omega)(V^-)$, where $V^\pm$ is $s = \pm iz$. The spaces $\mathbf{E}'(\Omega)(V^+)$ and $\mathbf{E}'(B_2)$ are both spaces of entire functions of exponential type one, but those in $\mathbf{E}'(\Omega)(V^+)$ can grow fast on the real axis, while those in $\mathbf{E}'(B_2)$ are small on the real axis. Thus, we see again that the space $\mathbf{E}'(\Omega)(V^+)$ is larger than $\mathbf{E}'(B_2)$.

Returning to the system (10.18), we see that even in the best possible situation we must solve infinitely many equations in infinitely many unknowns. This is of the same order of magnitude of difficulty as solving an integral equation—of course, integral equations methods are not new in boundary value problems.

However, in certain cases this system of equations can be "separated."

Suppose $f_1 = f_3$ and $f_2 = f_4$. Then, using (10.18), we have

$$(10.19) \qquad \begin{aligned} &\text{(a)}\ f_1(y) = \sum a_j \cos z^j \exp(iys^j) \\ &\text{(b)}\ f_2(x) = \sum a_j \cos s^j \exp(ixz^j). \end{aligned}$$

We can now understand the above example when the points (z^j, s^j) are of the form $(\pi(j+\frac{1}{2}), \pm i\pi(j+\frac{1}{2}))$ or $(i\pi(j+\frac{1}{2}), \pm\pi(j+\frac{1}{2}))$. For then, half the terms in (10.19a) are zero and the other half in (10.19b) are zero so the equations for the a_j are separated, that is, each a_j appears in only one equation. Similar results hold if $f_1 = -f_3$ or $f_2 = -f_4$.

It is important to note that for this separation we need to know that Ω is a rectangle. Suppose, for example, that the boundary of Ω were a regular polygon which is not a rectangle. Now, the above method depends on the fact that we can construct simple trigonometric polynomials which are harmonic and vanish on all of the boundary of Ω, except for a pair of opposite sides. Since a trigonometric polynomial is real analytic, this would imply that it would vanish on the union of the infinite extensions of the other sides of the polygons. This contains the boundary of a compact set so by the maximum property the trigonometric polynomial must be identically zero.

The above remark does not apply, for example, to $(\partial^2/\partial x^2 + \partial^2/\partial y^2)^j$ for $j > 1$, so it might be possible to solve the Dirichlet problem for these operators in finite terms on other regions by our method. We shall, however, not consider this here.

See Problem 10.1.

We wish to point out exactly how the symmetry is used. We note that we can derive (10.19) from (10.18) if we assume that $f_1 = f_3$, $f_2 = f_4$. The only way we could know this in advance would be by knowing that h in (10.15) is symmetric. In general we decompose h according to the characters of the symmetry of the rectangle and we find equations like (10.19) (where cos may be replaced by sin).

Finally, we must solve (10.19). If we use for (z^j, s^j) the points of the form $(\pm i\pi(j+\frac{1}{2}), \pi(j+\frac{1}{2}))$ and $(\pi(j+\frac{1}{2}), \pm i\pi(j+\frac{1}{2}))$, then (10.19) becomes

$$(10.20) \qquad \begin{aligned} &(a)\ f_1(y) = \sum (b_j^+ + b_j^-) \cosh \pi(j+\tfrac{1}{2}) \exp i\pi(j+\tfrac{1}{2})y \\ &\text{(b)}\ f_2(x) = \sum (c_j^+ + c_j^-) \cosh \pi(j+\tfrac{1}{2}) \exp i\pi(j+\tfrac{1}{2})x, \end{aligned}$$

where we have written $b_j^\pm$, $c_j^\pm$ for the coefficient of $\exp i(z^l x + s^l y)$, when $(z^l, s^l) = [\pm i\pi(j+\frac{1}{2}), \pi(j+\frac{1}{2})], [\pi(j+\frac{1}{2}), \pm i\pi(j+\frac{1}{2})]$, respectively. The right side of (10.20) is an arbitrary periodic function of y (or x) with period two. (10.20) prescribes the Fourier coefficients by giving the value of this periodic function on $[-1, 1]$. There are some difficulties at the endpoints which we shall not go into. Thus, (10.20) determines $b_j^+ + b_j^-$

and $c_j^+ + c_j^-$, respectively. Note, however, that $b_j^+ = b_j^-$ and $c_j^+ = c_j^-$ if h is invariant under $x \to -x$ and $y \to -y$. Thus, $b_j^\pm$ and $c_j^\pm$ are determined.

If, for example, $h(x, -y) = -h(x, y) = h(-x, y)$, then we would expand h in terms of other exponentials (z^l, s^l), namely, $(\pm i\pi j, \pi j)$ and $(\pi j, \pm i\pi j)$ for j integral. A similar computation would apply.

Finally, we realize that these computations are possible, because when a given harmonic function h is expanded according to the irreducible representations of the group of symmetries of the rectangle, each of the terms in the expansion is still harmonic, because the Laplacian commutes with the group of symmetries.

It should be remarked that the method which has (10.15) as its starting point is essentially dual to the method which starts with (10.6).

X.3. The General Case

We pass now to the case of a general system and $n \geqslant 2$. We shall assume that $\vec{\mathrm{D}}$ is *symmetric*, that is, that a multiplicity variety $\mathfrak{V}$ can be chosen which is invariant under $z_i \to -z_i$ for any i. We assume also that if $\mathfrak{V}$ has any components of dimension zero they do not lie on any of the coordinate planes $z_i = 0$. Let d be the dimension of $\mathfrak{V}$.

We shall assume for simplicity that all the differential operators of $\mathfrak{V}$ are equal to the identity. It is only a technical matter to ameliorate this assumption.

Let p be any permutation of $1, 2, \ldots, n$. By T_p^m we denote the linear space defined by $x_{p(m+1)} = \cdots = x_{p(n)} = 0$. For each $m \leqslant d$ and each p we choose linear constant coefficient differential operators q_{jp}^m on R with Fourier transforms Q_{jp}^m such that

1. For any j, p, m, i, changing z_i into $-z_i$ leaves Q_{jp}^m invariant.
2. For any p let $z_{p(1)}, \ldots, z_{p(d)}$ be fixed integers or half integers. Then the Q_{jp}^d "parametrize" the part of $\mathfrak{V}$ above $z_{p(1)}, \ldots, z_{p(d)}$. That is, let G be any function in $\mathscr{H}(\mathfrak{V})$ which is invariant under $z_{p(l)} \to -z_{p(l)}$ for any $l > d$. Then we can write

$$G(z_{p(1)}, \ldots, z_{p(d)}, z_{p(d+1)}, \ldots, z_{p(n)}) = \sum L_{jp}^d(z_{p(1)}, \ldots, z_{p(d)}) Q_{jp}^d(z) \quad \text{on } \mathfrak{V} \text{ above } (z_{p(1)}, \ldots, z_{p(d)}), \tag{10.21}$$

where the L_{jp}^d are uniquely determined by G and depend continuously on G in the sense that

$$|L_{jp}^d(z_{p(1)}, \ldots, z_{p(d)})| \leqslant c(1+|z|)^c \max_{\substack{z \in \mathfrak{V} \\ z \text{ above} \\ (z_{p(1)}, \ldots, z_{p(d)})}} |G(z)|. \tag{10.22}$$

3. Suppose conditions on the $Q_{jp}^{m'}$ have been explained for

$$m' = m + 1, \ldots, d.$$

Then we shall give conditions on the Q_{jp}^{m}. Let $z_{p(1)}, \ldots, z_{p(m)}$ be fixed integers or half integers. Then the Q_{jp}^{m} "parametrize" the part of $\mathfrak{B}$ above $z_{p(1)}, \ldots, z_{p(m)}$. That is, let G be any function in $\mathscr{H}(\mathfrak{B})$ which is invariant under $z_{p(i)} \to -z_{p(i)}$ for $i > m$. Then we can write

(10.23)

$$G(z_{p(1)}, \ldots, z_{p(m)}, z_{p(m+1)}, \ldots, z_{p(n)}) = \sum L_{jp}^{m}(z_{p(1)}, \ldots, z_{p(m)}) Q_{jp}^{m}(z)$$
$$+ \sum_{\substack{d \geqslant m' > m \\ p'}} M_{jp'}^{m'}(z_{p(1)}, \ldots, z_{p(m)}, z_{p'(m+1)}, \ldots, z_{p'(m')}) Q_{jp'}^{m'}(z) \quad \text{on} \quad \mathfrak{B},$$

where the sum over p' is over all p' such that $p'(1) = p(1), \ldots, p'(m) = p(m)$. Here the $M_{jp'}^{m'}$ are any entire functions of

$$z_{p'(1)}, \ldots, z_{p'(m')}.$$

We require that the L_{jp}^{m} should be uniquely determined by G (though the $M_{jp'}^{m'}$ need not be). Moreover, they should depend continuously on B in the following sense: Suppose that

$$G(z_{p(1)}, \ldots, z_{p(n)})$$

can be extended to be a polynomial in $z_{p(m+1)}, z_{p(n)}$, which we again denote by G satisfying

$$|G(z_{p(1)}, \ldots, z_{p(m)}, z_{p(m+1)}, \ldots, z_{p(n)})| \leqslant c'(1 + |z|)^{c'}. \tag{10.24}$$

Then the $M_{jp'}^{m'}$ can be chosen to be polynomials in $z_{p'(m+1)}, \ldots, z_{p'(m')}$. Moreover, we can find a c which depends on c' such that

$$|L_{jp}^{m}(z_{p(1)}, \ldots, z_{p(m)})| \leqslant c(1 + |z|)^{c} \tag{10.25}$$
$$|M_{jp'}^{m'}(z_{p'(1)}, \ldots, z_{p'(m')})| \leqslant c(1 + |z|)^{c}$$

for some choice of $M_{jp'}^{m'}$.

We wish to explain how our assumptions differ from the type of conclusion we want (see Theorem 10.2). The type of parametrization required in Condition 2 is related to the Cauchy problem (compare Chapter IX). In this problem we parametrize the solutions of $\vec{\mathrm{D}}f = 0$ by their values on *whole* linear subvarieties (in this case T_p^d). In Theorem 10.2 we parametrize the solutions by their values on *parts* of linear varieties. Condition 3 is related to the Goursat problem (see Section X.7). Our assumptions are related to parametrizing solutions of $\vec{D}f = 0$ by their values on the *whole* linear varieties $T_{p'}^{m'}$ for $m' \geqslant m$.

In applications, it is often possible to use the explicit formulas for the Cauchy and Goursat problems, for example, the explicit formula for the Cauchy problem in terms of the Lagrange interpolation formulas as given in Chapter IX, to verify Conditions 2 and 3. We shall sometimes refer to Condition 3 as the *corner condition*.

See Remark 10.1.

X.4. The Dirichlet Problem

Let Ω be the unit cube. Call $\tilde{T}_p^m$ the intersection of T_p^m with Ω.

DEFINITION. By the *Dirichlet problem for* $\vec{D}$ *on* Ω we mean the boundary value problem in which we prescribe the q_{jp}^m on each face which is obtained from $\tilde{T}_p^m$ by translating each coordinate $x_{p(m+1)}, \ldots, x_{p(n)}$ by ± 1. The associated function spaces are $\mathscr{E}(\Omega)$ and $\mathscr{E}(\tilde{T}_p^m)$. In order for this to have meaning we require that $q_{jp}^m = q_{jp'}^m$, if $p(1) = p'(1), \ldots, p(m) = p'(m)$.

We seek conditions in order that the Dirichlet problem be well-posed. Let $S \in \mathscr{E}'(\Omega)$ and suppose that S is invariant under each transformation $x_i \to -x_i$ (the case of changing sign under some $x_i \to -x_i$ is handled similarly). Then, as in the above example of the Laplacian on the square, we try to find a representation

$$S = \sum (q_{jp}^d)' \, {}^t S_{jp}^d \mod \vec{D}'\mathscr{E}'(\Omega). \tag{10.26}$$

where t is a vector with $n - d$ components of the form $t = (\pm 1, \pm 1, \ldots, \pm 1)$ and ${}^t S_{jp}^d$ has its support on $\{(x_{p(d+1)}, \ldots, x_{p(n)}) = t\} \cap \Omega$. Note that ${}^t S_{jp}^d = {}^t S_{jp'}^d$ if the sets $(p(1), \ldots, p(d))$ and $(p'(1), \ldots, p'(d))$ are the same (though their order may be different) because ${}^t S_{jp}^d$ depends on T_p^d and on t but not on the order of writing the coordinates on T_p^d. Actually, we cannot hope that (10.26) can hold exactly since terms coming from the lower dimensional faces of Ω must appear. However, we try to solve (10.26) "as well as possible." The "error" term will then be taken care of by the lower dimensional faces.

By symmetry we seek a representation (10.26) in which for fixed j, p all the ${}^t S_{jp}^d$ are obtained by translation from a fixed S_{jp}^d whose support is on T_p^d. Thus, the Fourier transform of (10.26) becomes (the calculations given below are correct modulo a constant factor which we shall ignore)

$$F(z) = \sum Q_{jp}^d(z_{p(1)}, \ldots, z_{p(n)}) F_{jp}^d(z_{p(1)}, \ldots, z_{p(d)}) \cos z_{p(d+1)} \cdots \cos z_{p(n)} \quad \text{on } \mathfrak{B}. \tag{10.27}$$

Let p_0 be a fixed permutation and let

$$z_{p_0(1)} = \pi(j_1 + \tfrac{1}{2}), \ldots, z_{p_0(d)} = \pi(j_d + \tfrac{1}{2}),$$

where the j_i are integers. Then all the terms in the right side of (10.27) are zero except the one corresponding to p_0, because $F^d_{jp} = F^d_{jp'}$ if the sets $(p(1), \ldots, p(d))$ and $(p'(1), \ldots, p'(d))$ are the same, and we obtain

$$F(z) = \sum_j Q^d_{jp_0}(z_{p_0(1)}, \ldots, z_{p_0(n)}) \tag{10.28}$$

$$F^d_{jp_0}(\pi(j_1 + \tfrac{1}{2}), \ldots, \pi(j_d + \tfrac{1}{2})) \cos z_{p_0(d+1)} \cdots \cos z_{p_0(n)}$$

for all z in $\mathfrak{V}$ above $z_{p_0(1)} = \pi(j_1 + \frac{1}{2}), \ldots, z_{p_0(d)} = \pi(j_d + \frac{1}{2})$. We now use (10.21) which allows us to solve for $F^d_{jp_0}(\pi(j_1 + \frac{1}{2}), \ldots, \pi(j_d + \frac{1}{2}))$ in terms of $F(z)$. Then, as in the case of the Laplacian, we can interpolate the values of $F^d_{jp_0}$ and solve our problem. In order to do this we must first know that the values $F^d_{jp_0}(\pi(j_1 + \frac{1}{2}), \ldots, \pi(j_d + \frac{1}{2}))$ are slowly (polynomially) increasing. (In fact they may not even be defined.) To insure this we introduce the

Diophantine condition. There exists an $a > 0$ so that for any p we have

$$\begin{aligned} |\cos z_{p(i)}| &\geqslant (a + |j_1| + \cdots + |j_d|)^{-a} \\ |\sin z_{p(i)}| &\geqslant (a + |j_1| + \cdots + |j_d|)^{-a} \end{aligned} \tag{10.29}$$

for $i = d + 1, \ldots, n$, wherever $(\pi j_1, \ldots, \pi j_d, z_{p(d+1)}, \ldots, z_{p(n)}) \in \mathfrak{V}$. Here $j_1, \ldots, j_d$ are half-integers or integers. We do not require the condition (10.29) on $\sin z_{p(i)}$ if $z_{p(i)} = 0$.

More generally, let $0 < m \leqslant d$ and let $j_1, \ldots, j_m$ be integers or half-integers. Let us consider the part of $\mathfrak{V}$ which intersects the plane $z_{p(1)} = \pi j_1, \ldots, z_{p(m)} = \pi j_m$. Then on all components of that intersection which are of dimension zero, we require

$$\begin{aligned} |\cos z_{p(i)}| &\geqslant (a + |j_1| + \cdots + |j_d|)^{-a} \\ |\sin z_{p(i)}| &\geqslant (a + |j_1| + \cdots + |j_d|)^{-a} \end{aligned} \tag{10.29*}$$

for $i = m + 1, \ldots, n$, whenever $(\pi j_1, \ldots, \pi j_m, z_{p(m+1)}, \ldots, z_{p(n)}) \in \mathfrak{V}$ is such a component of dimension zero, except for the condition on $\sin z_{p(i)}$ if $z_{p(i)} = 0$.

For those components which are of dimension >0, we require that each point $z'_{p(m+1)}, \ldots, z'_{p(n)}$, on a component be within a distance $a \log (a + |z'_{p(m+1)}| + \cdots + |z'_{p(n)}|)$ from a point $(z_{p(m+1)}, \ldots, z_{p(n)})$ in the same component at which (10.29*) holds. (It is easily seen that this condition is of importance only if these components of intersection are linear.)

Remark 10.2. The reason we do not require any condition on $\sin z_{p(i)}$ if $z_{p(i)} = 0$ is that the term $\sin z_{p(i)}$ is used when we are dealing with odd functions, which all vanish at the origin, so there is no problem in dividing by $\sin z_{p(i)}$.

(10.29) can be stated in another form:

$$\min_{\substack{j \text{ integer or} \\ \text{half-integer}}} |z_{p(i)} - \pi j| \geqslant (a + |j_1| + \cdots + |j_d|)^{-a}. \tag{10.30}$$

The equivalence of (10.29) and (10.30) is readily verified. (10.30) implies

$$|\cos z_{p(i)}| \geqslant (a + |j_1| + \cdots + |j_d|)^{-a} \exp(|\text{Im } z_{p(i)}|) \tag{10.31}$$

$$|\sin z_{p(i)}| \geqslant (a + |j_1| + \cdots + |j_d|)^{-a} \exp(|\text{Im } z_{p(i)}|).$$

THEOREM 10.2. *A necessary and sufficient condition that the Dirichlet problem be well-posed is that the Diophantine condition should hold.*

Remark 10.3. In case $\vec{\text{D}}$ is the single operator $\partial^2/\partial x_1^2 - \lambda^2\partial^2/\partial x_2^2$ the Diophantine condition becomes: λ is irrational and there exists an $a > 0$ so that the inequality

$$|l\lambda - m| < (al)^{-a} \tag{10.32}$$

can hold for only a finite number of integer pairs l, m. The equivalence of (10.32) with the fact that the Dirichlet problem is well-posed seems to be known to experts. It is a curiosity that, by Liouville's theorem, the set of λ for which the Dirichlet problem is well-posed contains all irrational algebraic numbers. It is known (see Cassels [1]) that the complement is uncountable and of measure zero.

PROOF OF THEOREM 10.2. We prove the sufficiency first. Since $F \in \mathscr{E}'(\Omega)$, we have (see Chapter V, Example 5)

$$|F(z)| \leqslant b(1 + |z|)^b \exp(|\text{Im } z_1| + \cdots + |\text{Im } z_n|). \tag{10.33}$$

Hence, by (10.28) and (10.31), this gives

(10.34)

$$\left|\sum_j Q^d_{jp_0}(z_{p_0(1)}, \ldots, z_{p_0(n)}) F^d_{jp_0}[\pi(j_1 + \tfrac{1}{2}), \ldots, \pi(j_d + \tfrac{1}{2})\right| \leqslant b'(1 + |z|)^{b'}.$$

Combining this with (10.22) gives

$$|F^d_{jp_0}(\pi(j_1 + \tfrac{1}{2}), \ldots, \pi(j_d + \tfrac{1}{2}))| \leqslant (b'' + |j_1| + \cdots + |j_d|)^{b''}. \tag{10.35}$$

Thus, $F^d_{j\,p_0}$ is slowly increasing on the half-integers times π.

We need

LEMMA 10.3. *Given any slowly increasing sequence σ on the integers times π there exists a function τ in $\mathbf{E}'(\tilde{T}^d)$ which interpolates it. τ can be chosen to depend continuously on σ in the sense that if σ stays in a bounded set in its natural topology, τ can be chosen in a bounded set in $\mathbf{E}'(\tilde{T}^d)$. τ is unique* *modulo addition of a function of the form*

$$N_d(z_1, \ldots, z_d) = N^1(z_1, \ldots, z_d) \sin z_1 + \cdots + N^d(z_1, \ldots, z_d) \sin z_d,$$

where each N^i belongs to the tensor product of $\mathbf{E}'(\tilde{T}_i^{d-1})$ and the polynomials in z_i. Here $\tilde{T}_i^{d-1}$ is the intersection of $z_i = 0$ with $\tilde{T}^d$.

Remark 10.4. The case of half-integers is obtained by translation.

PROOF OF LEMMA 10.3. For $d = 1$ the result is classical (see Boas [1], Paley and Wiener [1]). The idea of the proof is as follows: Let $\sigma = \{c_j\}$ and choose m so large that $c_j = 0(1 + j^{2m})$. Then the function $\tau_1 \in \mathbf{E}'(\tilde{T}^1)$ equal to

$$\tau_1(z) = \sum \frac{c_j}{(1 + j^{2m+2})} \frac{\sin z}{(z - \pi j)}$$

satisfies $\tau_1(\pi j) = c_j/(1 + j^{2m+2})$. Thus, $\tau(z) = (1 + z^{2m+2})\tau_1(z)$ interpolates σ. The continuity properties are seen from a simple modification of the above.

Another way of looking at this is as follows: By dividing by a polynomial we may assume that $\sigma \in l^2$ (square summable sequences). Thus, there exists a periodic function h whose Fourier coefficients are just σ. We multiply h by the characteristic function of T^d and obtain a function h' whose Fourier transform τ clearly interpolates σ. (This is just a simple case of the Poisson summation formula.) The continuity of τ as a function of σ is clear.

For the uniqueness, suppose that $N \in \mathbf{E}'(\tilde{T}^d)$ is zero at π times all lattice points; call n the Fourier transform of N. Then the Poisson summation formula (see Schwartz ([1]) shows that n must become zero when we identify the opposite sides of the cube $\tilde{T}^d$ so as to form a torus. This can only happen if the support of n is on the boundary of $\tilde{T}^d$ and if n is odd. This is exactly the stated structure of N. Lemma 10.3 is therefore proved. □

Proof of Theorem 10.2 *continued.* Thus, we have constructed F^d_{jp} with the property that (10.27) holds whenever any d of the z_i are π times half-integers.

We wish to go further and get an expression for F which holds whenever $d - 1$ of the z^l are π times half-integers. To do this we must use the $(d - 1)$-dimensional faces of Ω. Call F^d the difference between the left and right sides of (10.27). Then we want to write

$$F^d(z) = \sum Q^{d-1}_{jp}(z) F^{d-1}_{jp}(z_{p(1)}, \ldots, z_{p(d-1)}) \cos z_{p(d)} \cdots \cos z_{p(n)} \text{ on } \mathfrak{B} \tag{10.36}$$

As above, let p_0 be a fixed permutation, and let

$$z_{p_0(1)} = \pi(j_1 + \tfrac{1}{2}), \ldots, z_{p_0(d-1)} = \pi(j_{d-1} + \tfrac{1}{2}),$$

where the j_i are integers. Then all the terms on the right side of (10.36) vanish except the one corresponding to p_0 and we obtain

$$(10.37)\quad F^d(z) = \sum Q^{d-1}_{jp_0}(z)F^{d-1}_{jp_0}(\pi(j_1+\tfrac{1}{2}), \ldots, \pi(j_{d-1}+\tfrac{1}{2}))\cos z_{p_0(d)} \cdots \cos z_{p_0(n)}$$

for all z in $\mathfrak{V}$ above $z_{p_0(1)} = \pi(j_1+\frac{1}{2}), \ldots, z_{p_0(d-1)} = \pi(j_{d-1}+\frac{1}{2})$.

We shall actually proceed as follows: we use (10.37), or rather, a slight modification of it, to define $F^{d-1}_{jp_0}(\pi(j_1+\frac{1}{2}), \ldots, \pi(j_{d-1}+\frac{1}{2}))$. We shall then obtain an interpolation of these values of $F^{d-1}_{jp_0}$ and verify (10.36).

We now apply the Diophantine condition. We want to show that $F^d(z)/\cos z_{p_0(d)} \cdots \cos z_{p_0(n)}$ is of polynomial growth on the part of $\mathfrak{V}$ above $z_{p_0(1)} = \pi(j_1+\frac{1}{2}), \ldots, z_{p_0(d-1)} = \pi(j_{d-1}+\frac{1}{2})$. We note first that $F^d(z)$ vanishes at those points of $\mathfrak{V}$ d of whose coordinates are π times half-integers. Thus, $F^d(z)/\cos z_{p_0(d)} \cdots \cos z_{p_0(n)}$ is regular on the part of $\mathfrak{V}$ above $z_{p_0(1)} = \pi(j_1+\frac{1}{2}), \ldots, z_{p_0(d-1)} = \pi(j_{d-1}+\frac{1}{2})$, because by (10.29) at most one of $z_{p_0(d)}, \ldots, z_{p_0(n)}$ can be a half-integer on this part. For those components of this part of $\mathfrak{V}$ which are of dimension zero the polynomial growth follows from the Diophantine condition as in (10.34). For those components which are of dimension >0 we use the Diophantine condition to show that the desired inequalities hold at certain points. Then we apply a maximum–minimum modulus type of argument as in Theorem 1.4 to obtain the desired result for all points. Since $F^d(z)/\cos z_{p_0(d)} \cdots \cos z_{p_0(n)}$ is slowly increasing, we can use Example 8 of Chapter V and Theorem 4.1 to extend this function from $\mathscr{V}$ to a polynomial in $z_{p_0(d)}, \ldots, z_{p_0(n)}$.

Now we use Condition 3 on the $Q^{d-1}_{jp_0}$. This says we can write

$$(10.38)\quad F^d(z)/\cos z_{p_0(d)} \cdots \cos z_{p_0(n)} = \sum Q^{d-1}_{jp_0}(z)F^{d-1}_{jp_0}(\pi(j_1+\tfrac{1}{2}), \ldots, \pi(j_{d-1}+\tfrac{1}{2})) + \sum M^d_{jp'}(\pi(j_1+\tfrac{1}{2}), \ldots, \pi(j_{d-1}+\tfrac{1}{2}), z_{p'(d)})Q^d_{jp'}(z)$$

on $\mathfrak{V}$ where the $M^d_{jp'}$ are polynomials in $z_{p'(d)}$. We note that $p'(d)$ is one of the numbers $p_0(d), \ldots, p_0(n)$, and that $\cos z_{p_0(d)} \cdots \cos z_{p_0(n)} = \cos z_{p'(d)} \cdots \cos z_{p'(n)}$ for any p' in the above sum. We can use Lemma 10.3 to interpolate the $F^{d-1}_{jp_0}$ to a function in $\mathbf{E}'(x_{p_0(d)} = 0, \ldots, x_{p_0(n)} = 0) \cap \mathbf{E}'(\Omega)$ and the $M^d_{jp'}$ to a finite linear combination of monomials in $z_{p'(d)}$ times functions in $\mathbf{E}'(x_{p_0(d)} = 0, \ldots, x_{p_0(n)} = 0) \cap \mathbf{E}'(\Omega)$. We denote these extensions again by $F^{d-1}_{jp_0}$ and $M^d_{jp_0}$.

We now examine

$$(10.39)\quad F(z) - \sum Q^d_{jp}(z)F^d_{jp}(z_{p(1)}, \ldots, z_{p(d)})\cos z_{p(d+1)} \cdots \cos z_{p(n)} - \sum Q^{d-1}_{jp_0}(z)F^{d-1}_{jp_0}(z_{p_0(1)}, \ldots, z_{p_0(d-1)})\cos z_{p_0(d)} \cdots \cos z_{p_0(n)} - \sum M^d_{jp'}(z_{p'(1)}, \ldots, z_{p'(d)})Q^d_{jp'}(z)\cos z_{p'(d)} \cdots \cos z_{p'(n)}.$$

The above construction shows that this function vanishes on $\mathfrak{V}$, whenever $z_{p_0(1)}, \ldots, z_{p_0(d-1)}$ is π times a half-integer. We rewrite (10.39) as

$$(10.40)\quad F(z) - \sum_{j,p'} Q^d_{jp'}(z)[F^d_{jp'}(z_{p'(1)}, \ldots, z_{p'(d)})$$
$$+ M^d_{jp'}(z_{p'(1)}, \ldots, z_{p'(d)}) \cos z_{p'(d)}] \cos z_{p'(d+1)} \cdots \cos z_{p'(n)}$$
$$- \sum_{\substack{j \\ p \neq \text{some } p'}} Q^d_{jp}(z) F^d_{jp}(z_{p(1)}, \ldots, z_{p(d)}) \cos z_{p(d+1)} \cdots \cos z_{p(n)}$$
$$- \sum_j Q^{d-1}_{jp_0}(z) F^{d-1}_{jp_0}(z_{p_0(1)}, \ldots, z_{p_0(d-1)}) \cos z_{p_0(d)} \cdots \cos z_{p_0(n)}.$$

This shows that we need to use $M^d_{jp'} \cos z_{p'(d)}$ as a "correction" to $F^d_{jp'}$ in order to get vanishing when $z_{p_0(1)}, \ldots, z_{p_0(d-1)}$ are π times half-integers. Certainly, $M^d_{jp'} \cos z_{p'(d)} \in \mathbf{E}'(x_{p'(d+1)} = 0, \ldots, x_{p'(n)} = 0) \cap \mathbf{E}'(\Omega)$. It is clear that (10.40) vanishes whenever any d of the $z_1, \ldots, z_n$ are π times half-integers. Thus, we may continue the above construction for all permutations p_0. We finally obtain functions $'F^d_{jp} \in \mathbf{E}'(x_{p(d+1)} = 0, \ldots, x_{p(n)} = 0) \cap \mathbf{E}'(\Omega)$ and $F^{d-1}_{jp_0} \in \mathbf{E}'(x_{p_0(d)} = 0, \ldots, x_{p_0(n)} = 0) \cap \mathbf{E}'(\Omega)$, so that

$$(10.41)\quad F(z) - \sum_{jp} Q^d_{jp}(z)\, 'F^d_{jp}(z_{p(1)}, \ldots, z_{p(d)}) \cos z_{p(d+1)} \cdots \cos z_{p(n)}$$
$$- \sum_{jp_0} Q^{d-1}_{jp_0} F^{d-1}_{jp_0}(z_{p_0(1)}, \ldots, z_{p_0(d-1)}) \cos z_{p_0(d)} \cdots \cos z_{p_0(n)}$$

vanishes whenever $d - 1$ of the z_l are π times half-integers.

We can now continue the process. We obtain functions

$$F^m_{jp}(z_{p(1)}, \ldots, z_{p(m)}) \in \mathbf{E}'(x_{p(m+1)} = 0, \ldots, x_{p(n)} = 0) \cap \mathbf{E}'(\Omega)$$

for $0 < m \leqslant d$, so that

$$(10.42)\quad F(z) = \sum_{j,p,m} Q^m_{jp}(z) F^m_{jp}(z_{p(1)}, \ldots, z_{p(m)}) \cos z_{p(m+1)} \cdots \cos z_{p(n)}$$

on $\mathfrak{V}$ whenever any z_i is π times a half-integer. Also, the F^m_{jp} can be assumed invariant under $z_{p(i)} \to -z_{p(i)}$ for $1 \leqslant i \leqslant m$.

We need

LEMMA 10.4. *Suppose $\mathfrak{V}$ satisfies the Diophantine condition. Let $G \in \mathbf{E}'(\Omega)(\mathfrak{V})$ vanish whenever any z_i is π times a half-integer. Then G is, on $\mathfrak{V}$, a polynomial times* $\cos z_1 \cdots \cos z_n$.

PROOF OF LEMMA 10.4. $G/\cos z_1 \cdots \cos z_n$ is certainly regular on $\mathfrak{V}$. The Diophantine condition in the form (10.31), together with the fact that

$$|G(z)| \leqslant c(1 + |z|)^c \exp(|\operatorname{Im} z_1| + \cdots + |\operatorname{Im} z_n|) \quad \text{on } \mathfrak{V}$$

(see Chapter V, Example 5) implies immediately that this function is of polynomial growth on $\mathfrak{V}$. By Chapter V, Example 8 this means that $G/\cos z_1 \cdots \cos z_n$ is the restriction of a polynomial to $\mathfrak{V}$ which is the desired result. □

Now we use again our "corner conditions" for Q^0_{jp}, and we have

$$(10.43)\quad F(z) = \sum_{jpm} Q^m_{jp}(z) F^m_{jp}(z_{p(1)}, \ldots, z_{p(m)}) \cos z_{p(m+1)} \cdots \cos z_{p(n)}$$

the sum over m going from 0 to d. This is the desired expression for F.

We wish now to describe the lack of uniqueness in the F^m_{jp} in (10.43). To see this we go through the step-by-step construction. The $F^d_{jp}(z_{p(1)} \cdots, z_{p(d)})$ are uniquely determined when $z_{p(1)}, \ldots, z_{p(d)}$ are π times half-integers. Thus, Lemma 10.3 shows that F^d_{jp} is unique up to addition of a function of the form

$$(10.44)\quad N^d_{jp}(z_{p(1)}, \ldots, z_{p(d)}) = N^{d1}_{jp}(z_{p(1)}, \ldots, z_{p(d)}) \cos z_{p(1)} + \cdots + N^{dd}_{jp}(z_{p(1)}, \ldots, z_{p(d)}) \cos z_{p(d)}.$$

In the construction of the $F^{d-1}_{jp_0}$ certain conditions are placed on the N^d_{jp}. A little reflection shows that the only possible N^d_{jp} that could occur result from the following: Let a $(d-1)$-dimensional face Ω^{d-1} of Ω be written as the intersection of two d-dimensional faces Ω^d_1 and Ω^d_2. Then certain distributions A on R which have their supports on Ω^{d-1} could be expressed in the forms

$$(10.45)\quad A = \sum_j q^d_{jp_1} B^1_j = \sum_j q^d_{jp_2} B^2_j,$$

where $B^l_j \in \mathscr{E}'(\Omega^d_l)$, and where Ω^d_l is defined by $p_l(d+1) = \cdots = p_l(n) = 0$ for $l = 1, 2$. It is at this point that the assumption that the $B^{m'}_{jp'}$ in Condition 3 can be chosen to be polynomials in suitable variables is used. We call the relations so obtained, and those obtained similarly for lower-dimensional faces, the *trivial relations*. Then our method of construction leads readily to the fact that *the F^m_{jp} are uniquely determined modulo the trivial relations.*

Next we observe that the continuity conditions in Conditions 2 and 3 on the Q^m_{jp} and the continuity result in Lemma 10.3 easily enable us to show that if F lies in a bounded set in $\mathbf{E}'(\Omega)(\mathfrak{V})$, then the F^m_{jp} could be chosen to lie in a bounded set in $\mathbf{E}'(x_{p(m+1)} = 0, \ldots, x_{p(n)} = 0) \cap \mathbf{E}'(\Omega)$. Moreover, the map $F \to (F^m_{jp})$ is clearly linear into the direct sum of the various $\mathbf{E}'(x_{p(m+1)} = 0, \ldots, x_{p(n)} = 0) \cap \mathbf{E}'(\Omega)$ modulo the trivial relations. Thus, the map is sequentially continuous. If we knew that the space $\mathbf{E}'(\Omega)(\mathfrak{V})$ is bornologic, that is, any convex circled set which swallows every bounded set is a neighborhood of zero, (which we suspect to be true) we would know that the map is continuous. Actually the continuity can be proved by going through the above construction with some care but we shall omit the details.

Thus, the sufficiency of the Diophantine condition for the well-posedness of the Dirichlet problem is proved.

The necessity of the condition is seen as follows: Suppose first that it is (10.29) which is violated. This can happen in two ways.

Case 1. For some half-integers or integers $j_1, \ldots, j_d$ we have $\cos z_{p(j)} = 0$ or $\sin z_{p(j)} = 0$ for some $j > d$. Suppose for simplicity that all j_l are half-integers and it is $\cos z_{p(n)} = 0$. (The other cases are treated similarly.) Then the $F^d_{jp_0}$ are undefined at $(\pi j_1, \ldots, \pi j_d)$ as is seen from (10.28). This means that the interpolating function is not unique. The lack of uniqueness cannot be compensated for by the F^m_{jp} with $m < d$ because of the nature of our assumptions. Thus, the mapping $F \to (F^m_{jp})$ described above is *not* one-to-one onto the direct sum of the $\mathbf{E}'(x_{p(m+1)} = 0, \ldots, x_{p(n)} = 0) \cap \mathbf{E}'(\Omega)$ modulo the trivial relation. In fact, it is not even defined on those F for which $F(\pi j_1, \ldots, \pi j_d, z_{p(1)}, \ldots, z_{p(n)} \neq 0$ for any points in $\mathfrak{V}$ above $(\pi j_l, \ldots, \pi j_d)$.

It is clear why the case $z_{p(j)} = 0$ causes no difficulty.

Case 2. In (10.29) no $\cos z_{p(j)} = 0$ and no $\sin z_{p(j)} = 0$, but (10.29) does not hold. Let $\{{}^s z\}$ be a sequence of points with ${}^s z_{p(l)} = j_l \pi$ for $1 \leqslant l \leqslant d$ so that, say $|\cos {}^s z_{p(n)}| \leqslant (s + |j_1| + \cdots + |j_d|)^{-s}$. It is easy to see that we can find a function $F \in \mathbf{E}'(\Omega)$ such that $|F({}^s z)| \geqslant \exp(|\mathrm{Im}\ {}^s z_{p(d+1)}| + \cdots + |\mathrm{Im}\ {}^s z_{p(n)}|)$. Then (10.28) shows that the $F^d_{jp_0}(\pi j_1, \ldots, \pi j_d)$ cannot be polynomially increasing. We leave the details of this and the following to the reader.

In case it is (10.29*) which does not hold, we modify the above construction by the methods of Ehrenpreis [4] and we obtain our result.

Theorem 10.2 is thus completely proved. □

Remark 10.5. We could easily modify the above to obtain the solution to "mixed" boundary value problems for Ω. By this we mean that for certain sets of opposite faces of Ω of dimension $\leqslant d$ we give all the data on one of the pairs of sides instead of on both. (Here by a "set" of opposite faces we mean those obtained from one of them by translation.) Then, for this side we give Cauchy data which means that in the analogs of Conditions 1, 2, and 3 on the q^m_{jp} we drop the symmetry properties. However, our method applies to the case when at most *one* pair of opposite faces of dimensions $d, d-1, \ldots, 0$ is replaced by one of those in the set. For the general case, see our discussion of the Goursat problem below. An example of this is the well-known problem for the heat equation $\partial/\partial x_1 = \partial^2/\partial x_2^2$, where the function is prescribed on the faces of $\Omega : x_1 = -1$ and $x_2 = \pm 1$.

X.5. Convex Polyhedra

It may be possible to use the Oka embedding (see Section V.5) to study boundary value problems on convex polyhedra. We have not studied this and, instead, we use a different method.

There are certain cases in which we could extend the above method from a cube Ω to more general convex polyhedra. First we could find well-posed problems for an image of Ω by an affine transformation A for the system $A\vec{\mathrm{D}}$ obtained from $\vec{\mathrm{D}}$ by the affine transformation. Thus, if we have well-posed problems for $\vec{\mathrm{D}}$ and $A^{-1}\vec{\mathrm{D}}$ on the cube we get well-posed problems for $\vec{\mathrm{D}}$ on Ω and $A\Omega$.

We say that Ω and $A\Omega$ *intersect properly* if $\Omega \cap A\Omega \neq \phi$ and, whenever a face of either meets the interior of the union, then the interior of this face is contained in the interior of the other. We say that the *Dirichlet conditions are compatible* if they are the same on any face of the intersection. For example, see Fig. 18.

We wish to formulate an analog of the Schwarz alternating procedure, which is a modification of Poincaré's balayage method mentioned at the beginning of this chapter which will allow us to find a well-posed problem on $\Omega \cup A\Omega$. (Note that $\Omega \cup A\Omega$ is a convex polyhedron under our assumption of proper intersection.) Let $S \in \mathscr{E}'(\Omega \cup A\Omega)$. By using a partition of unity, we may suppose $S \in \mathscr{E}'(\Omega)$. Using the well-posed problem for Ω, we perform "balayage" of S on the boundary of Ω. Then there will be a part, say ${}_1S$, on the part of the boundary of Ω which lies in the interior of $A\Omega$. (The rest will lie on the boundary of $\Omega \cup A\Omega$.) Using the well-posed problem for $A\Omega$, we perform balayage of ${}_1S$ on the boundary of $A\Omega$. We then obtain a part ${}_2S$ on the part of the boundary of $A\Omega$

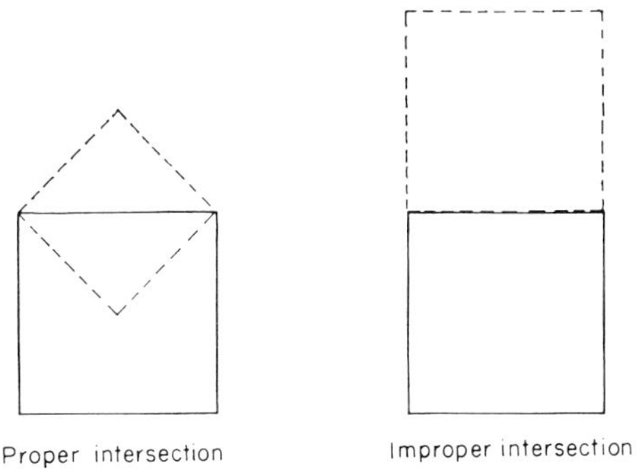

FIGURE 18

which lies in Ω. We proceed in this way, and obtain a sequence ${}_1S$, ${}_2S$, $\cdots$ which we hope converges to zero, while the parts on the boundary of $\Omega \cup A\Omega$ converge.

We shall give a condition which is easily seen to lead to convergence of the process and hence to the fact that the boundary value problem is well-posed on $\Omega \cup A\Omega$. We denote by $bd^*(\Omega)$ the part of the boundary of Ω contained in the interior of $A\Omega$; $bd^*(A\Omega)$ is defined similarly. For each integer a we define the norm $\|U\|_a$ for $U \in \mathscr{E}'(\Omega \cup A\Omega)$ by ($\hat{U}$ is the Fourier transform of U)

$$\|U\|_a = \sup_{z \text{ real}} |\hat{U}(z)|/(1+|z|)^a. \tag{10.46}$$

(Of course, this may be infinite.) We denote by $\mathscr{E}'(\Omega \cup A\Omega)_a$ the subspace of $\mathscr{E}'(\Omega \cup A\Omega)$ with finite a norm. The spaces $\mathscr{E}'(\Omega)_a$, $\mathscr{E}'(A\Omega)_a$ are defined similarly. Then we require that for some infinite sequence of a there exist a' which $\to \infty$ and numbers $b_a < 1$ so that if $U \in \mathscr{E}'(bd^*\Omega)_a$, then when we perform balayage of U on the boundary of $A\Omega$ we can choose the components $U^* \in \mathscr{E}'(bd^*A\Omega)_{a'}$ and U_λ in $\mathscr{E}'$ (face $A\Omega)_{a'}$ in such a way that

$$\|U^*\|_{a'} \leqslant b_a \|U\|_a, \qquad \|U_\lambda\|_{a'} \leqslant b_a \|U\|_a. \tag{10.47}$$

Conversely, given ${}^1U \in \mathscr{E}'(bd^*A\Omega)_a$, we can choose the components ${}^1U^*$ and ${}^1U_\lambda$ of the balayage of ${}^1U$ on the boundary of Ω so that

$$\|{}^1U^*\|_{a'} \leqslant b_a \|{}^1U\|_a, \qquad \|{}^1U_\lambda\|_{a'} \leqslant b_a \|{}^1U\|_a. \tag{10.48}$$

We refer to the above as the *alternating hypothesis*.

THEOREM 10.5. *If the alternating hypothesis holds, then the boundary value problem for $\Omega \cup A\Omega$ is well-posed.*

PROOF. By the results of Section X.1 we need to prove that for any $S \in \mathscr{E}'(\Omega \cup A\Omega)$ we can perform balayage of S on the boundary of $\Omega \cup A\Omega$, that is, write $S = T + D_1U_1 + \cdots + D_rU_r$, where T, $U_j \in \mathscr{E}'(\Omega \cup A\Omega)$ and the support of T is contained in the boundary of $\Omega \cup A\Omega$, and T is of a suitable form which is defined by the boundary value problem. The above construction shows how to achieve this. (T is the result of the balayage of S on the boundary of $\Omega \cup A\Omega$.) ▯

Remark 10.6. We have used the fact that $\Omega \cup A\Omega$ is convex, since in the proof of the equivalence of well-posed boundary value problems with balayage (see Lemma 10.1 and following) we have had to apply Theorem 4.1 with $\mathscr{W} = \mathscr{E}(\Omega \cup A\Omega)$. By Chapter V, Example 5 this will be LAU if $\Omega \cup A\Omega$ is convex. (It is clearly not even AU if $\Omega \cup A\Omega$ is not convex.) However, the proof of the existence of balayage did not require explicitly that $\Omega \cup A\Omega$ be convex, though the convexity is a consequence of the fact that Ω and $A\Omega$ intersect properly.

In many cases one could decide by our method of proof of Theorem 10.2

whether or not the alternating hypothesis holds. It is also possible to weaken the hypothesis slightly in several ways and still obtain convergence.

Of course, we could iterate the alternating process and obtain well-posed boundary value problems for convex polyhedra.

X.6. The Fredholm Alternative

We wish now to extend Theorem 10.2 so as to permit a certain non-uniqueness in boundary value or parametrization problems. In view of its relation to the classical problem, we shall call this the Fredholm alternative.

Suppose we are given a boundary value problem such that there is a subspace N of solutions in a space $\mathscr{W}$ which have zero boundary data. Then we want to say what it means for the boundary value problem to be well-posed "modulo N."

Before doing this we must recall that in the classical theory, when there is a nontrivial N it is usually also the case that the image of α (in (10.2)) is not everything. Call I the image of α.

DEFINITION. We say that the Fredholm alternative holds if I is closed and if α defines a topological isomorphism of $\mathscr{W}/N$ onto I.

Let us denote by $d(\alpha)$ the dimension of N and by $d'(\alpha)$ the dimension of the co-kernel of α, that is, the dimension of the possible image space modulo I.

DEFINITION. If $d(\alpha)$ or $d'(\alpha)$ is finite, we define the *index* of α as $d(\alpha) - d'(\alpha)$.

We denote by $\mathscr{W}'_N$ the subspace of $\mathscr{W}'$ which vanishes on N. Thus, $\mathscr{W}'_N$ is the dual of $\mathscr{W}/N$. Denote by $\vec{\mathscr{W}}$ the possible image space of α, that is, $\vec{\mathscr{W}}$ is the direct sum ${}_D\mathscr{W} \oplus \mathscr{W}_1 \oplus \cdots \oplus \mathscr{W}_h$ modulo certain obvious relations. Denote by $\vec{\mathscr{W}}'_I$ the subspace of $\vec{\mathscr{W}}'$ which vanishes on I so $\vec{\mathscr{W}}'/\vec{\mathscr{W}}'_I$ is the dual of I. Thus, the adjoint α' of α maps $\vec{\mathscr{W}}'$ onto $\mathscr{W}'_N$ and has kernel $\vec{\mathscr{W}}'_I$ if the Fredholm alternative holds.

LEMMA 10.6. *A necessary and sufficient condition that the Fredholm alternative hold is that there should exist a closed subspace $\vec{\mathscr{W}}'_{\otimes}$ of $\vec{\mathscr{W}}'$ so that α' is a continuous open mapping of $\vec{\mathscr{W}}'$ onto $\mathscr{W}'_N$ with kernel $\vec{\mathscr{W}}'_{\otimes}$. The image I of α is then the subspace of $\vec{\mathscr{W}}$ which is annihilated by $\vec{\mathscr{W}}'_{\otimes}$.*

PROOF. If the Fredholm alternative holds, then set $\vec{\mathscr{W}}'_{\otimes} = \vec{\mathscr{W}}'_I$ and the result is obvious.

Conversely, suppose α' is continuous and open with kernel $\vec{\mathscr{W}}'_{\otimes}$. Define I as the set annihilated by $\vec{\mathscr{W}}'_{\otimes}$. Then I is clearly closed. Let $a \in I$. We want to find a $b \in \mathscr{W}$ with $\alpha b = a$, that is, for any $S \in \vec{\mathscr{W}}'$ we want

$$S \cdot \alpha b = S \cdot a \tag{10.49}$$

$$\alpha' S \cdot b = S \cdot a. \tag{10.50}$$

Thus, we must know (using the Hahn-Banach theorem) that if $\alpha' S \to 0$, then $S \cdot a \to 0$. Our assumptions imply that $S \to 0$ in $\vec{\mathscr{W}}'$ modulo $\vec{\mathscr{W}}'_{\S}$ so that $S \cdot a \to 0$. (Recall our assumption that $\mathscr{W}$ and the $\mathscr{W}_j$ are reflexive.)

Conversely, if there is a b satisfying $\alpha b = a$, then by (10.50), since the kernel of α' is $\vec{\mathscr{W}}'_{\S}$, we must have $\vec{\mathscr{W}}'_{\S} \cdot a = 0$ so that $a \in I$. Thus, I is exactly the image of α.

The fact that α is topological is easily seen by duality. ▯

Now let Ω be the unit cube. We wish to extend Theorem 10.2 so as to obtain necessary and sufficient conditions in order that the Fredholm alternative holds. With Lemma 10.6 in mind, we examine (10.28) to see what is the kernel of $\hat{\alpha}'$. We can get elements in the kernel if for some integers or half-integers $(j_1, \ldots, j_d)$ the following degeneracy occurs:

Degeneracy 1. The $Q^d_{jp}(\pi j_1 + \frac{1}{2}), \ldots, \pi(j_d + \frac{1}{2}), z_{p(d+1)}, \ldots, z_{p(n)})$ do not uniquely parametrize the even functions on the part of $\mathfrak{B}$ above $z_{p(1)} = \pi(j_1 + \frac{1}{2}), \ldots, z_{p(d)} = \pi(j_d + \frac{1}{2})$ (see (10.21)), that is, the $F^d_{jp_0}$ $(\pi(j_1 + \frac{1}{2}), \ldots, \pi(j_d + \frac{1}{2}))$ which exist satisfying (10.28) are not unique.

There are two kinds of degeneracies which are relevant towards decreasing the image of $\hat{\alpha}'$ (and increasing N):

Degeneracy 2. For some $l > d$ we have $\cos z_{p(l)} = 0$.

Degeneracy 3. The $Q^d_{jp_0}(\pi(j_1 + \frac{1}{2}), \ldots, \pi(j_d + \frac{1}{2}), z_{p(d+1)}, \ldots, z_{p(n)})$ do not parametrize the even functions on the part of $\mathfrak{B}$ above $z_{p(1)} = \pi(j_1 + \frac{1}{2}), \ldots, z_{p(d)} = \pi(j_d + \frac{1}{2})$. That is, not for all F do solutions $F^d_{jp_0}(\pi(j_1 + \frac{1}{2}), \ldots, \pi(j_d + \frac{1}{2}))$ of (10.21) exist.

For Degeneracies 1, 2, and 3 we define degeneracy degrees $d'(\pi(j_1 + \frac{1}{2}), \ldots, \pi(j_d + \frac{1}{2}))$ and $d(\pi(j_1 + \frac{1}{2}), \ldots, \pi(j_d + \frac{1}{2}))$ as follows: $d(\pi(j_1 + \frac{1}{2}), \ldots, \pi(j_d + \frac{1}{2}))$ is the dimension of the space representing lack of uniqueness (Degeneracy 1), and d' is the dimension of the space representing lack of existence (Degeneracies 2 and 3), that is, the codimension of the image of the right side of (10.28) in the space of possible values. Of course, in the definitions of d' and d, symmetry must be taken into account.

We define the *local d index* $i(\pi(j_1 + \frac{1}{2}), \ldots, \pi(j_d + \frac{1}{2}))$ as the difference $d(\pi(j_1 + \frac{1}{2}), \ldots, \pi(j_d + \frac{1}{2})) - d'(\pi(j_1 + \frac{1}{2}), \ldots, \pi(j_d + \frac{1}{2}))$.

We define the *total d index* i_d as the sum of all local indices when all except a finite number are different from zero.

We could now do the same thing in relation to the construction of the F^m_{jp} for $0 \leq m \leq d$. We obtain local m indices and global m indices i_m.

Finally, we define the *grand total index* i as

$$i = i_1 + i_2 + \cdots + i_d . \tag{10.51}$$

THEOREM 10.7. *Let Ω be the unit cube. Suppose the Q^m_{jp} satisfy Condition 1 of the Dirichlet problem. Suppose instead of Condition 2 they satisfy*

2′. *In the notation of Condition 2, we can write some G in the form* (10.21). *The L^d_{jp} are not uniquely determined but if they exist then they can be chosen to satisfy* (10.22).

3′. *In the notation of Condition 3, some G can be written in the form* (10.23). *The L^m_{jp} and the $L^{m'}_{jp'}$ are not uniquely determined by G, but, if they exist, then we can choose them satisfying* (10.25).

Suppose, moreover, that the Diophantine condition is weakened so that (10.29) *and* (10.29*) *are required to hold only when the left side is not zero; we make a similar modification in the condition on the components of dimension* >0.

Then the Fredholm alternative holds. If the index exists, then the grand total index exists and is equal to it.

The proof of Theorem 10.7 follows along the same lines as that of Theorem 10.2; we shall omit the details.

We can also derive an extension of Theorem 10.5 to the case of the Fredholm alternative. We use the notation preceding Theorem 10.5. Suppose we are given Fredholm alternative problems for Ω and $A\Omega$ with the compatibility conditions on the faces of the boundary of $\Omega \cup A\Omega$.

DEFINITION. We say the Fredholm alternative data is *compatible* if every boundary distribution $\sum (q^m_{jp})' \, {}^tS^m_{jp}$ on the closure of $bd^*\Omega$ is orthogonal to the null space N_A on $A\Omega$ and similarly if the roles of Ω and $A\Omega$ are reversed.

Let $S \in \mathscr{E}'(\Omega \cup A\Omega)$ be of the form $S' + S''$, where $S' \in \mathscr{E}'(\Omega)$ is zero on N and $S'' \in \mathscr{E}'(A\Omega)$ is zero on N_A. If we try to carry out the alternating process as in the Dirichlet problem we realize that we must make a choice at each stage of balayage. This leads to the

DEFINITION. We say that the *Fredholm alternating* hypothesis holds if there exists an infinite sequence $\{a\}$ and a corresponding sequence $\{a'\}$, where $a' \to \infty$ and numbers $b_a < 1$ so that if $U \in \mathscr{E}'(bd^*\Omega)_a$ is of the form $\sum (q^m_{jp})' \, {}^tS^m_{jp}$, then when we balayage U on the boundary of $A\Omega$ we can choose the components U^* so that (10.47) holds. Similarly, when the roles of Ω and $A\Omega$ are interchanged.

In analogy to Theorem 10.5 we obtain

THEOREM 10.8. *If the Fredholm alternating hypothesis holds then the Fredholm alternative holds for $\Omega \cup A\Omega$.*

Of course, we could iterate the process and obtain results for convex polyhedra.

An interesting example of the Fredholm alternative is the $\bar{\partial}$ Neuman problem studied by Spencer, Garabedian, Morrey, and Kohn (see Kohn [1].) In its simplest form for functions on the unit square Ω in R^2 this is the Dirichlet problem for the differential operator Δ (the Laplacian) with boundary conditions $\partial/\partial x_1 + i\partial/\partial x_2$ on each face and no corner conditions. It is easily seen that Theorem 10.7 applies. The index is not defined but the grand total index is zero. A similar result holds in R^n for n even.

It would be interesting to carry through the procedure of Theorem 10.8 to see for which domains our method leads to the solvability of the $\bar{\partial}$ Neuman problem (that is, the Fredholm alternative holds). Kohn in [1] has shown that this holds for strongly pseudoconvex domains with smooth boundary (so his results do *not* apply to polyhedra).

There is an interesting problem in case the Fredholm alternative holds and d and d' are finite. Consider, for example, the Dirichlet problem on the unit square, for $n = 2$, $r = 1$, $D_\lambda = \partial^2/\partial x_1^2 + \partial^2/\partial x_2^2 + \lambda$. We are thus studying the eigenvalue problem for the Laplacian D_0 on the space of functions on the unit square with vanishing Dirichlet data. We write, more precisely, d_λ for the dimension of N. It is easily seen that $d_\lambda \neq 0$ if and only if λ can be represented as a sum of the squares

$$\lambda = \pi^2(m_1^2 + m_2^2),$$

where m_1 and m_2 are integers or half-integers.

Much work in number theory (see Walfisz [1]) has been devoted to the problem of determining the number of such λ with $|\lambda| \leqslant R$. This is clearly equivalent to the *lattice point problem*, that is, the problem of determining the number of points with integral coordinates which lie in a given circle. The best known results show that the number $\nu(R)$ of such λ with $|\lambda| \leqslant R$ satisfies

$$\nu(R) = 4R/\pi + 0(R^p),$$

where p is somewhat $< \frac{1}{3}$. The best possible (and conjectured result) is that any $p > \frac{1}{4}$ works.

Much work has been done towards generalizing this result by considering other elliptic operators instead of the Laplacian. However, we feel that these methods cannot lead to an understanding of how to improve p. The reason is that the lattice points are best described by *two* differential operators, namely, $\partial/\partial x_1$, $\partial/\partial x_2$ rather than by the Laplacian D_0. Thus the solutions of

$$\partial f/\partial x_1 = m_1 f, \qquad \partial f/\partial x_2 = m_2 f$$

with f satisfying *periodic* boundary conditions on the unit square are just $\exp(im_1x_1 + im_2x_2)$, m_1, m_2 being π times integers. It seems to us that passing to D_0 loses too much of the structure to determine p accurately. In fact, the proofs of the best estimates for p use the two-dimensional structure.

PROBLEM 10.2.

Study analogous problems for other systems with $r > 1$.

X.7. The Wiener-Hopf and Goursat Problems

In Ehrenpreis [5] we discussed the relation of a certain type of Wiener-Hopf problem to the Cauchy problem. We shall now discuss a more refined type of Wiener-Hopf problem.

We work formally first. For $n = 1$ the Wiener-Hopf problem can be stated as follows: Let k' be a function. Then we want to find functions f' which satisfy

$$\int_0^\infty k'(x - x')f'(x')\,dx' = f'(x) \qquad \text{for } x > 0. \tag{10.52}$$

If we define f to be f' for $x > 0$ and 0 for $x \leqslant 0$, then (10.52) is equivalent to

$$\begin{aligned}(k' * f)(x) &= f(x) \qquad \text{for } x > 0\\ f(x) &= 0 \qquad \text{for } x \leqslant 0.\end{aligned} \tag{10.53}$$

Next let k be $k' - \delta$. Then (10.53) becomes

$$\begin{aligned}(k * f)(x) &= 0 \quad \text{for } x > 0.\\ f(x) &= 0 \quad \text{for } x \leqslant 0.\end{aligned} \tag{10.54}$$

Let us assume that k', hence k, is of compact support. Then we generalize (10.54) by replacing k by an arbitrary $S \in \mathscr{E}'$. Thus, we look for $f \in \mathscr{E}$ with

$$\begin{aligned}(S * f)(x) &= 0 \quad \text{for } x > 0\\ f(x) &= 0 \quad \text{for } x \leqslant 0.\end{aligned} \tag{10.55}$$

Suppose $[a, b]$ is the smallest interval containing the support of S. Then to solve (10.55), it is sufficient to solve

$$\begin{aligned}(S * f)(x) &= 0 \quad \text{for } x > 0\\ f(x) &= 0 \quad \text{for } -b \leqslant x \leqslant 0\end{aligned} \tag{10.56}$$

if $b \geqslant 0$ while the second condition is vacuous if $b < 0$.

Assume that $0 \in [a, b]$ (which is certainly the case if $S = k$, since $k = k' - \delta$ and k' is a function). Then (10.56) is a special case of the initial value problem

$$(10.57) \qquad (S * f)(x) = g(x) \quad \text{for } x > 0$$
$$f(x) = h(x) \quad \text{for } -b \leqslant x \leqslant 0,$$

namely, (10.56) arises from (10.57) if $g = 0$ and $h = 0$ for $-b \leqslant x \leqslant 0$.

In Ehrenpreis [5] an analogous problem is treated for hyperbolic S for $n > 0$. ($S \in \mathscr{E}'$ is called hyperbolic in x_n if $S * \mathscr{E} = \mathscr{E}$ and if for c sufficiently large, any $l(x)$ which is defined and indefinitely differentiable on $|x_n| < c$ and satisfies $(S * l)(x) = 0$ for $|x_n| < 1$ possesses a unique extension $m \in \mathscr{E}$ with $(S * m)(x) = 0$ for all x. Compare Section VIII.3.) The problem considered there is concerned with the case when we want to solve

$$(10.58) \qquad (S * f)(x) = g(x)$$

in a half-plane with f having suitable initial data. It has, however, long been realized that the true generalization of (10.52) is

$$(10.59) \qquad \int_0^\infty \cdots \int_0^\infty k'(x - x') f'(x') \, dx_1 \cdots dx_n = f'(x)$$
$$\text{for } x_1 > 0, \ldots, x_n > 0.$$

The above procedure would therefore lead us to consider the equation

$$(10.60) \qquad (S * f)(x) = g(x) \quad \text{for } x_1 > 0, \ldots, x_n > 0$$

with f having suitable "initial data" near the boundary of this domain. We denote $x_1 > 0, \ldots, x_n > 0$ by R^+.

The simplest case of (10.60) is that in which S is a constant coefficient partial differential operator. More generally, we could consider the boundary value problem

$$(10.61) \qquad D_j f = g_j \quad \text{on } R^+$$
$$\delta_i f = h_i \quad \text{on } B_i^+,$$

where B_i^+ are faces of R^+. It is clear that this problem bears a similarity to the boundary value problem for the cube Ω discussed at the beginning of this chapter, except that the boundary of R^+ is like half the boundary of Ω. (See Remark 10.4 on mixed boundary value problems at the end of X.4.)

We shall refer to (10.61) as the *Wiener-Hopf problem* for D.

For $n = 1$ the Wiener-Hopf problem is the initial value problem. This is related in an obvious way to the Cauchy problem. If we denote by

B_i the linear space containing B_i^+ then the parametrization problem

$$\text{(10.62)} \qquad \begin{aligned} D_j f &= g_j \quad \text{on } R \\ \delta_i f &= h_i \quad \text{on } B_i \end{aligned}$$

will be called the *Goursat problem*. (Goursat considered the case when there is only one D_j.)

The Goursat and Wiener-Hopf problems lead, by Fourier transform, to the problem of parametrizing the functions in $\mathbf{W}'(\mathfrak{V})$ by suitable functions which are polynomials times functions which depend on the dual variables to B_i. This problem is almost always *not* solvable in closed algebraic form (as is the Cauchy problem by means of the Lagrange interpolation formula) but depends on power series methods. (See, however, Section X.8 below.)

This is easily seen even from the simple example $D = (\partial^2/\partial x_1\, \partial x_2) + 1$, $B_i = x_i$ axis for $i = 1, 2$, $\delta_i = \text{identity}$ for $i = 1, 2$. Then we must write any $F \in \mathbf{W}'(z_1 z_2 = 1)$ in the form $H_1(z_1) + H_2(z_2)$. We can easily solve for H_1 and H_2 by power series methods but a closed algebraic form seems impossible, because, unlike the situation in the Cauchy problem, the expression $F = H_1 + H_2$ on $z_1 z_2 = 1$ is highly nonunique unless F, H_1 and H_2 are assumed entire. [The nonuniqueness is seen as follows: We can make $H_1(z_1) + H_2(z_2) = 0$ on $z_1 z_2 = 1$ by setting $H_2(z) = -H_1(1/z)$ where H_1 is of compact support.]

In the application needed in Theorem 10.2 we need only interpolate polynomials so the power series method should be useful in examples. In some cases we can use the fundamental solution (see Section IX.7) to obtain explicit formulas. (See also the following section.)

X.8. The Newman-Shapiro Parametrization Problem

Newman and Shapiro [1] have considered the following uniqueness problem: Let $r = 1$.
For

$$D = \sum a_{j_1, \ldots, j_n} \partial^j / \partial x_1^{j_1} \ldots \partial x_n^{j_n},$$

we set

$$\tilde{P} = \sum \bar{a}_{j_1, \ldots, j_n} x_1^{j_1}, \ldots, x_n^{j_n}.$$

Does there exist an entire function f with $Df = 0$ and $f = 0$ on the set V of zeros (with multiplicity) of $\tilde{P}$? Newman and Shapiro succeeded in proving uniqueness in case f is assumed to be quadratically integrable with respect to $\exp(-|x|^2)\,dx$. They also showed that if D is "almost homogeneous," then uniqueness holds for any f.

The formal reason for suspecting uniqueness is the following: Suppose f is a polynomial. The assumption that f vanishes on V implies that $f = \tilde{P}g$ for some polynomial g. Thus

$$0 = Df = D\tilde{P}g$$

so

$$(D_1 D)(\tilde{P}g) = 0,$$

where D_1 is a linear constant coefficient operator such that $g = \tilde{P}_1$. Now it is easily seen by direct computation that for any $D_2 \neq 0$, we have

$$(D_2 \tilde{P}_2)(0) > 0.$$

This gives the desired result if we set $D_2 = DD_1$.

It is easy to generalize the problem in many ways. For example, we could start with two operators D_1, D_2 of the same degree (say m) and ask whether there exists an entire function f such that $D_1 f = 0$ and f divisible by $\tilde{P}_2$. The particular case $D_2 = \partial^m/\partial x_n^m$ corresponds to the Cauchy problem. For $\tilde{P}_2$ a product of m linear functions, we obtain the Goursat problem.

In case D_1 and D_2 are hyperbolic, then there would exist a C^∞ formulation of the Newman-Shapiro problem.

Naturally the above extends to $r > 1$.

Instead of considering the uniqueness problem we could consider the corresponding parametrization problem. An approach to this problem which is successful in some cases is the following: Suppose first that $r = 1$ and $D_1 = D_2 = D$ is a product of m distinct linear factors ${}_jD$. Then, by Theorem 7.1 (or else, directly) any solution f of $Df = 0$ can be written as a sum

$$f = \sum f_j,$$

where for each j we have ${}_jDf_j = 0$. Let us set ${}_jD = \sum a_{ij}\,\partial/\partial x_i$ and suppose that $a_{nj} \neq 0$ for all j. Then we may write for each j

$$f_j = g_j(x_1 - a_{1j}x_n/a_{nj}, \ldots, x_{n-1} - a_{n-1,j}x_n/a_{nj}),$$

where g_j are arbitrary entire functions of $n-1$ variables. For the parametrization problem, we want to prescribe f uniquely by its restriction to the variety V of zeros of

$$\prod \Big(\sum_i \bar{a}_{ij} x_j\Big) = 0.$$

If we fix $x_1, \ldots, x_{n-1}$, then there are, in general, m distinct x_n such that $(x_1, \ldots, x_{n-1}, x_n) \in V$. These are given by $-\sum \bar{a}_{ij} x_i / \bar{a}_{nj}$. Thus, the parametrization problem leads us to the study of the system of linear equations

$$(10.63) \quad \sum_j g_j \left[x_1 + \sum_{i=1}^{n-1} a_{1j} \bar{a}_{ie} x_i / a_{nj} \bar{a}_{ne}, \ldots, x_{n-1} + \sum_{i=1}^{n-1} a_{n-1,j} \bar{a}_{ie} x_i / a_{nj} \bar{a}_{ne} \right] = h_e(x_1, \ldots, x_{n-1})$$

Here h_e are given entire functions of $n - 1$ complex variables [for (10.63) is just the prescription of f on V].

Equations (10.63) play the same role in the present situation as the Lagrange interpolation formula plays for the Cauchy problem (see Chapter IX). Though (10.63) seems formidable as it stands, it can be greatly simplified by performing a suitable transformation. In order to see what this is, let us consider first the case $n = 2$. Then (10.63) is just (writing x for x_1)

$$(10.64) \qquad \sum_j g_j[(1 + a_{1j} \bar{a}_{1e} / a_{2j} \bar{a}_{2e}) x] = h_e(x).$$

We operate formally first. If we denote (formally) by $\hat{g}$, $\hat{h}_e$ the respective Mellin transforms of g_j, h_e then the fact that the Mellin transform is defined by the characters of the *multiplicative* group yields

$$(10.65) \qquad \sum_j (1 + a_{1j} \bar{a}_{1e} / a_{2j} \bar{a}_{2e})^{-s} \hat{g}_j(s) = \hat{h}_e(s).$$

We must now know that the determinant of $(1 + a_{1j} \bar{a}_{1e} / a_{2j} \bar{a}_{2e})^{-s}$ is different from zero. If we call $\lambda_j = a_{1j} / a_{2j}$, then we must show that

$$(10.66) \qquad \det\, (1 + \lambda_j \bar{\lambda}_e)^{-s} \neq 0.$$

(This determinant replaces the Vandermonde determinant which was important for the Cauchy problem.) For $s = -1$, (10.66) follows easily from the fact that the matrix $(\lambda_j \bar{\lambda}_e)$ is positive definite. For general s, the proof involves some difficulties and we shall omit it.

In the above we took a "formal" Mellin transform. By studying the adjoint of our parametrization problem, it is easy to make everything rigorous.

In case $n > 2$ we must replace the Mellin transform by a more complicated transform.

The case $r = 1$, and $D_1 = D_2 = D$ is *not* a product of linear factors is much more difficult. However, if we want to study the uniqueness question for entire solutions, then there is this approach: Since an entire function is determined by its values on an arbitrarily small open set, we need expressions of the form (10.5) only for $S = \delta_a$, a in some small open set. We now use the fact that, in general, V is "close to" linear.

Thus we try to find expressions of the form (10.5) for $S = \delta_a$ in which we expect that the support of S_j would be on the part of V which is close to linear. Now the idea of Section IX.7 indicates that a "good" choice of a would be such that we could find a fundamental solution at a whose support intersects V in a part of V which is close to linear.

Once this is done, then we could try to solve (10.5) by using the above idea for the case when D is a product of distinct linear factors combined with a "perturbation." This method can be carried through when $n = 2$ by use of the Puiseux expansion (Theorem 1.1).

X.9. General Systems

As in the case of the Cauchy problem we can extend all the above to general systems, that is, in which there may be more than one unknown function. The only difference is that we may have to give some of the boundary data on the whole of Ω.

Remarks

Remark 10.1. See page 295.

The case in which some of the Q_{jp}^m change sign under some of the symmetries $z_i \to -z_i$ can *not* be treated by the same methods. It is easy to see what the modified conditions corresponding to Conditions 2 and 3 should be, but the proof given below does not carry over to this case.

Remark 10.2. See page 297.

Remark 10.3. See page 297.

In case $\vec{D}$ is the single operator $\partial^2/\partial x_1^2 - \lambda^2\partial^2/\partial x_2^2$ the Diophantine condition becomes: λ is irrational and there exists an $a > 0$ so that the inequality

$$|l\lambda - m| < (al)^{-a}$$

can hold for only a finite number of integer pairs l, m. The equivalence of the above inequality with the fact that the Dirichlet problem is well-posed seems to be known to experts. It is a curiosity that, by Liouville's theorem, the set of λ for which the Dirichlet problem is well-posed contains all irrational algebraic numbers. It is known (see Cassels [1]) that the complement is uncountable and of measure zero.

Remark 10.4. See page 298.

Remark 10.5. See page 302.

We could easily modify the above to obtain the solution to "mixed" boundary value problems for Ω. By this we mean that for certain sets of opposite faces of Ω of dimension $< d$ we give all the data on one of the pairs of sides instead of on both. (Here, by a "set" of opposite faces we mean those obtained from one of them by translation.) Then for this side we give Cauchy data which means that in the analogs of Conditions 1, 2, and 3 on the q_{jp}^m we drop the symmetry properties. However, our method applies to the case when at most *one* pair of opposite faces of dimensions

d, $d-1$, ..., 0 is replaced by one of those in the set. For the general case see our discussion of the Goursat problem in Section X.7. An example of this is the well-known problem for the heat equation $\partial/\partial x_1 = \partial^2/\partial x_2^2$, where the function is prescribed on the faces of $\Omega : x_1 = -1$, and $x_2 = \pm 1$.

Remark 10.6. See page 304.

We have used the fact that $\Omega \cup A\Omega$ is convex, since in the proof of the equivalence of well-posed boundary value problems with balayage (see Lemma 10.1 and following) we have had to apply Theorem 4.1 with $\mathscr{W} = \mathscr{E}(\Omega \cup A\Omega)$. By Chapter V, Example 5 this will be LAU if $\Omega \cup A\Omega$ is convex. (It is clearly not even AU if $\Omega \cup A\Omega$ is not convex.) However, the proof of the existence of balayage did not require explicity that $\Omega \cup A\Omega$ be convex, though the convexity is a consequence of the fact that Ω and $A\Omega$ intersect properly.

Problems

PROBLEM 10.1. See page 292.

By use of the Oka embedding (Section V.5) can we extend the results to convex polyhedra?

(In Section X.5 we shall explain a different method.)

PROBLEM 10.2. See page 309.

CHAPTER XI

Miscellanea

Summary

We extend some of the main results of Chapter IV to the case when $\boxed{\mathrm{D}}$ is a matrix of convolution operators with kernels of compact support. (In Chapter IV we assumed that $\boxed{\mathrm{D}}$ was a matrix of constant coefficient partial differential operators.) This extension can be made only for very special $\boxed{\mathrm{D}}$ which are called *slowly decreasing*. We discuss an extension of the uniqueness results of Chapter IX to some convolution systems.

In Section XI.2 we discuss problems of singularities of solutions of $\vec{\mathrm{D}}f = 0$. These are of two kinds:

1. f is given to be a solution of $\vec{\mathrm{D}}f = 0$ on some $\Omega_0 \subset R$. Can we extend f to be a solution on a larger set? In case Ω_0 is the set $|x_1| < b, \ldots, |x_q| < b, x_{q+1}, \ldots, x_n$ arbitrary, this is just the hyperbolicity problem dealt with in Chapter VIII. In the case in which Ω_0 is a "shell," that is, $\Omega_0 = \Omega_2 - \Omega_1$, where Ω_1 and Ω_2 are open sets such that the closure of Ω_1 is contained in Ω_2, we can find necessary and sufficient conditions. This result generalizes the classical theorem of Hartogs which is the case $n = 2m$,

$$\vec{\mathrm{D}} = (\partial/\partial x_1 + i\partial/\partial x_{m+1}, \ldots, \partial/\partial x_m + i\partial/\partial x_n).$$

   More general cases are considered.
2. f is given to be a solution of $\vec{\mathrm{D}}f = 0$ on all of R and f is "regular" on some $\Omega_0 \subset R$. Is f regular on all of R?

In Section XI.3 we try to find those domains A such that we can solve $\boxed{\mathrm{D}}\vec{\mathrm{f}} = \vec{\mathrm{g}}$ for $\vec{\mathrm{f}} \in \mathscr{E}^l(A)$, for every $\vec{\mathrm{g}} \in \mathscr{E}^r(A)$ which satisfies the necessary differential relations (see Chapter VI). For A a convex polyhedron we showed in Chapter VI that this is possible for every $\boxed{\mathrm{D}}$. However, if A is not convex, the solvability of $\boxed{\mathrm{D}}\vec{\mathrm{f}} = \vec{\mathrm{g}}$ depends on a relation between A and $\boxed{D}$. The method is based on the study of singularities given in Section XI.2.

In Section XI.4 we show how our theory can be used to study some of the classical special functions such as the Bessel function and the Legendre function. A relation to group representations is discussed.

We explain how to use generalized Fourier analysis to study the Cauchy problem for variable partial differential equations in Section XI.5. Other approaches to the study of variable coefficient equations are discussed.

XI.1. Extension to Convolution Systems

We should like to know to what extent the results of Chapter IV can be extended to the case when the D_{ij} are convolution operators. Thus, the P_{ij} are analytic functions of a suitable kind. For simplicity we shall discuss only the case of $\mathscr{W} = \mathscr{E}$; the general case is handled by similar methods.

In Section I.3 we introduced the concept of a slowly decreasing function. In Ehrenpreis [4] we proved the following result characterizing those $F \in \mathbf{E}'$ which are slowly decreasing for $\mathscr{E}$: $F \in \mathbf{E}'$ is slowly decreasing for $\mathscr{E}$ if there exists an $a > 0$ so that to each real z there is another real z' with

$$|z - z'| \leqslant a \log (1 + |z|) \tag{11.1}$$

$$|F(z')| \geqslant (a + |z|)^{-a}.$$

(We shall consider here slowly decreasing only for $\mathscr{E}$ so that we shall often not say "for $\mathscr{E}$.")

Let F be the Fourier transform of $S \in \mathscr{E}'$. It was shown in Ehrenpreis [4] that F is slowly decreasing if and only if $S * \mathscr{E} = \mathscr{E}$. Moreover in case $n = 1$ we showed that every solution of $S * f = 0$, $f \in \mathscr{E}$ could be represented as a sum of exponential polynomial solutions, where the series converges in the topology of $\mathscr{E}$ upon a proper choice of grouping of terms.

We wish to extend the above, as much as we can, to general convolution systems. We shall also derive an analog of the above expansion of f in case $n > 1$.

For any open set B in C we call $\mathscr{O}(B)$ the ring of holomorphic functions on B.

DEFINITION. Let A be a module in $\mathbf{E}'^l$ (the l-fold sum of $\mathbf{E}'$ with itself) over $\mathbf{E}'$. We say that A is *slowly decreasing* if there exist finitely many elements $\vec{F}_1, \ldots, \vec{F}_r$ of A which generate A such that for every $b > 0$ there is an $a > 0$ such that the following property holds: Let $z \in C$, and $\vec{H} = (H^1, \ldots, H^l)$ where the H^j belong to

$$\mathcal{O}\{|z' - z| < a[\log(|z| + 1) + |\text{Im } z| + 1]\}.$$

Suppose $\vec{H}$ belongs to the module generated by the $\vec{F}_p$ in the ring

$$\mathcal{O}\{|z' - z| < a[\log(|z| + 1) + |\text{Im } z| + 1]\}$$

and each $|H^j| \leq 1$ on the set $|z' - z| < a[\log(|z| + 1) + |\text{Im } z| + 1]$. Then on $|z' - z| < b$ we can write

$$\vec{H} = \sum G_p \vec{F}_p, \tag{11.2}$$

where each $G_p \in \mathcal{O}(|z' - z| < b)$, and on $|z' - z| < b$ we have

$$|G_p(z)| \leq a(1 + |z|)^a \exp[a(|\text{Im } z| + 1)]. \tag{11.3}$$

We call the generators $\vec{F}_p$ *slowly decreasing generators* for A.

We say that A is *properly slowly decreasing* if there exist slowly decreasing generators whose module of relations over $\mathcal{O}(C)$ is generated over $\mathcal{O}(C)$ by a module A_1, which is slowly decreasing. Similarly, for the module of relations of a set of slowly decreasing generators of A_1, etc., up to the $2n$th module of relations. The reason for putting conditions on the successive modules of relations up to the $2n$th is that this is what is needed to extend the proof of Theorem 4.1 (see Theorem 11.1 below).

In Chapter III (see Theorem 3.18) we showed that if A is generated by polynomials, then A is properly slowly decreasing. (In this case the term $\log|z|$ is superfluous.)

The minimum modulus methods of Ehrenpreis [4] show that if $l = r = 1$ then the above concept of slowly decreasing is the same as our previous one. (Compare the argument in the proof of Theorem 1.4 which is the case $r = l = 1$, A generated by a polynomial.)

DEFINITION. Let $c > 0$, and let B be a subset of C. By the $(\log c|\ |, c|\text{Im}\ |)$ *neighborhood of* B we mean the set $N(B; \log c|\ |, c|\text{Im}\ |)$ of all points $z \in C$ for which there is a $z' \in B$ with $|z - z'| < \log c(1 + |z|) + c|\text{Im } z|$. By $\mathbf{E}'(N(B; \log c|\ |, c|\text{Im}\ |))$ we mean the space of functions which are holomorphic on $N(B; \log c|\ |, c|\text{Im}\ |)$ and which satisfy the same growth conditions there as the functions of $\mathbf{E}'$. The topology is given by the seminorms

$$\|F\|_k = \max_{z \in N} |F(z)|/k(z)$$

for k is an AU structure for $\mathscr{E}$.

We have the following extension of Theorem 4.1:

THEOREM 11.1. *Let A be a properly slowly decreasing ideal and let V denote the variety of common zeros of the functions of A. Then there exists a $c > 0$ so that the map*

$$\rho_N : H \dashrightarrow H \,|\, N(V; \log c|\ \ |, c|\text{Im}\ \ |) \tag{11.4}$$

is a topological isomorphism of

$$\mathbf{E}'/A \qquad \textit{onto} \qquad \mathbf{E}'(N(V; \log c|\ \ |, c|\text{Im}\ \ |))/A_c,$$

where A_c is the ideal generated by A in the ring

$$\mathbf{E}'(N(V; \log c|\ \ |, c|\text{Im}\ \ |).$$

The proof proceeds along the same lines as the proof of Theorem 4.1 with one major change: To define cochains, nice cochains, etc., we must consider cubes, $\gamma(\alpha; \beta(\alpha))$, where $\beta(\alpha)$ may increase like

$$c' \log (1 + |\alpha|) + c'\,|\text{Im}\ \alpha|.$$

This affects also the analog of the part of the proof of Theorem 4.6 with a odd in which case the strips used in the proof of Theorem 4.6 are to be replaced by strips bounded by suitable curves rather than by straight lines. We shall omit the details, as they are straightforward.

We could also formulate and prove an analog of Theorem 4.2. Rather than do this, we shall content ourselves to state two of the consequences which follow as in the case of polynomial modules. They are the analogs of Theorems 6.1 and 6.2, and 7.1, respectively.

THEOREM 11.2. *Let $\boxed{\mathrm{D}} = (D_{ij})$ be an $r \times l$ matrix of distributions in $\mathscr{E}'$. Suppose that the columns of the Fourier transform of the adjoint of $\boxed{\mathrm{D}}$ are slowly decreasing generators of a properly slowly decreasing module. Then the image of $\boxed{\mathrm{D}}$ in $\mathscr{E}^r$ consists exactly of those $\vec{\mathrm{g}} = (g_i) \in \mathscr{E}^r$ which satisfy $\sum \partial_i g_i = 0$ whenever (∂_i) belongs to the module of relations of the rows of $\boxed{\mathrm{D}}$ in $\mathscr{E}'$. The image of the adjoint of $\boxed{\mathrm{D}}$ on $\mathscr{E}'^r$ is closed in $\mathscr{E}'^l$.*

As an example of Theorem 11.2 let us consider the case $n = 1$, $l = 1$, $r = 2$ [so we write $\vec{\mathrm{D}} = (D_1, D_2)$] where the Fourier transform of D_1 is $(\sin z)/z$ and the Fourier transform of D_2 is $\sin \alpha z$, where α is irrational. It follows from the theory of mean periodic functions that 1 belongs to the closure of the ideal in $\mathbf{E}'$ generated by $(\sin z)/z$ and $\sin \alpha z$, since these functions have no common zero (see e.g., Schwartz [4]). But under suitable Diophantine conditions on α, it is easily seen that 1 cannot belong to the ideal generated by $(\sin z)/z$, $\sin \alpha z$ in $\mathbf{E}'$. For, if

$$1 = (F(z) \sin z)/z + G(z) \sin \alpha z,$$

then $G(z)$ must be $1/\sin \alpha z$, whenever $z = \pi j$, j an integer $\neq 0$. If α is well approximated by rationals, for example, if for any c we can find rationals p/q such that $|\alpha - p/q| < 1/q^c$ then

$$|\sin \alpha\pi q| \leqslant |\sin \pi p| + q^{-c} = q^{-c}.$$

This is incompatible with the fact that $G(\pi q) = 0(1 + q)^a$ for some a, since $G \in \mathbf{E}'$ (see Chapter V, Example 5). Thus, the ideal is not closed so that by Theorem 11.2, $(\sin z)/z$ and $\sin \alpha z$ are *not* slowly decreasing generators of an ideal.

The difficulty in using Theorem 11.2 is that it is usually hard to find information about the module of relations of the columns of the Fourier transform of $\boxed{\mathbf{D}}'$. In some cases, however, we can find enough elements in the module of relations for our purposes. For example, suppose that $P_i(z) \in \mathbf{E}'$ satisfy

$$\sum_{i=1}^{r} |P_i(z)| \geqslant (a + |z|)^{-a} \exp(-a\,|\text{Im } z|)$$

for some $a > 0$. We claim that 1 belongs to the ideal generated by the P_i in $\mathbf{E}'$.

To see this, we want to use the methods of Chapters III and IV to construct $\vec{\text{H}} = (H_i)$ such that $\sum P_i H_i = 1$. Note that there is a simple solution to the semilocal problem. Near each point z^0, for at least one choice of i, say $i(z^0)$, we have

$$|P_i(z)| \geqslant (a + |z|)^{-a} \exp(-a\,|\text{Im } z|)/r.$$

Near this z^0 we choose $\vec{\text{H}}(z^0; z)$ to be the vector whose ith component is $1/P_i(z)$ for this particular $i = i(z^0)$ and whose other components are zero. We have $\vec{\text{P}}\vec{\text{H}} = 1$ near z^0.

Now, $P_i(z) \in \mathbf{E}'$ so its derivatives are bounded by a polynomial times $\exp(a'\,|\text{Im } z|)$ for suitable a'. Thus, $|P_i(z)|$ cannot decrease too fast, that is, for $i = i(z^0)$

$$|P_i(z)| \geqslant (a + |z|)^{-a} \exp(-a\,|\text{Im } z|)/2r$$

as long as z belongs to $N(z^0)$ defined by

$$|z - z^0| \leqslant (b + |z|)^{-b} \exp(-b\,|\text{Im } z|)$$

for suitable b. Thus, $\vec{\text{H}}$ is not too large on sets which are not too small.

We can now use the method of Chapter IV to piece these semilocal $\vec{\text{H}}$ together. Although we do not know (directly) how to get the whole module of relations of $\vec{\text{P}}$, we do know that $\vec{\text{K}}_{ij}(z)$ which is the vector whose ith component is $P_j(z)$ and whose jth component is $-P_i(z)$ belongs to the module of relations. This is all that is needed in the first step of the piecing together process of Chapter IV, that is, the proof of (4.23) for $a = 0$. (A similar method applies to the other steps.) For, when we want to piece together $\vec{\mathrm{H}}(z^0; z)$ and $\vec{\mathrm{H}}(z^{0'}; z)$, we note that $\vec{\mathrm{H}}(z^0; z) - \vec{\mathrm{H}}(z^{0'}; z)$ is the vector whose $i(z^0)$ component is $[P_{i(z^0)}(z)]^{-1}$, whose $i(z^{0'})$ component is $-[P_{i(z^{0'})}(z)]^{-1}$, and whose other components are all zero. (If $i(z^0) = i(z^{0'})$ there is no problem.) Thus, on $N(z^0) \cap N(z^{0'})$

$$\vec{\mathrm{H}}(z^0; z) - \vec{\mathrm{H}}(z^{0'}; z) = \vec{\mathrm{K}}_{i(z^0), i(z^{0'})}/P_{i(z^0)}(z)P_{i(z^{0'})}(z).$$

This means that, for our purposes, we do not need to know anything about the module of relations of $\vec{\mathrm{P}}$ except the $\vec{\mathrm{K}}_{ij}$. The method of Chapter IV now yields the result that 1 belongs to the ideal generated by the P_i in $\mathbf{E}'$. We leave the details to the reader. By a slight modification of the above, we can treat the case when the P_i have a finite number of common zeros. More generally, if the multiplicity variety defined by the P_j coincides with an algebraic multiplicity variety $\mathfrak{V}$ and if

$$\sum_i |P_i(z)| \geqslant (a + |z|)^{-a} \exp(-a|\mathrm{Im}\, z|)$$

away from $\mathfrak{V}$, then the P_i generate the same ideal as the polynomials Q_j which have $\mathfrak{V}$ as a multiplicity variety.

It should be noted that the corona problem discussed by Carleson [2] for $n = 1$ is in the same spirit. Carleson proved that a necessary and sufficient condition that the ideal generated by the bounded analytic functions in the unit disk $P_1(z), \ldots, P_r(z)$ be the whole ring is

$$\sum_i |P_i(z)| \geqslant \delta > 0$$

for any z.

An analogous result for $n = 1$ for the space $\mathscr{H}$ was proved by Kelleher and Taylor (see Kelleher and Taylor [1]) by a different method. (Our method applies also to $\mathscr{H}$.) Hörmander [7] has shown how to apply his method for proving vanishing of cohomology with bounds to this problem.

In general, we do not know of any "good" conditions to guarantee that a set of $\vec{\mathrm{F}}_j$ are slowly decreasing generators of a properly slowly decreasing module. The only sufficient conditions we know are

1. The $\vec{\mathrm{F}}_j$ are vectors of polynomials.
2. There is only one $\vec{\mathrm{F}}_j$ and it is a scalar ($l = r = 1$), which is slowly decreasing.

PROBLEM 11.1. (Even for $n = 1$).

Give necessary and sufficient conditions on the $\vec{\mathrm{F}}_j$ if they are exponential polynomials in order that they be slowly decreasing generators of a properly slowly decreasing module. (Note that in Ehrenpreis [2] it is proved that for $r = l = 1$, every exponential polynomial is slowly decreasing.)

CONJECTURE.

If F_j are exponential polynomials for which the exponentials have algebraic frequencies then they are slowly decreasing generators of a properly slowly decreasing ideal.

THEOREM 11.3. *Let the* F_j *be slowly decreasing generators of a properly slowly decreasing ideal* $(l = 1)$. *There exists a* $c > 0$ *so that to each solution* $f \in \mathscr{E}$ *of* $D_j f = 0$ *for* $j = 1, \ldots, r$ *there is a bounded measure* μ *on* $N(V; \log c|\ |, \exp c|\ |)$ *and a function* k *of an analytic uniform structure for* $\mathscr{E}$ *so that*

$$f(x) = \int \exp(iz \cdot x)\, d\mu(z)/k(z). \tag{11.5}$$

While Theorem 11.3 is weaker than Theorem 7.1, it (together with its analogs for spaces other than $\mathscr{E}$) can be applied to yield extensions of some of the results of Chapter VIII.

We can regard Theorem 11.3 as giving a weak analog of the case $n = r = l = 1$ mentioned at the beginning of this chapter. In that case (if F is slowly decreasing) every solution can be represented as a convergent series of groups of exponential polynomial solutions. (For the meaning of grouping, see Section XII.9.) In this connection we remark that in Ehrenpreis and Malliavin [1] the necessary and sufficient conditions (still for $n = r = l = 1$) in order that no grouping should be needed are determined.

Although we cannot solve the general problem of deciding when an ideal is slowly decreasing, we can give one sufficient condition for solvability of the equations $D_j f = g_j$, where g_j are given in $\mathscr{E}$ and $f \in \mathscr{E}$ is to be found. (Of course the g_j must satisfy suitable compatibility conditions.) Suppose $D_1, \ldots, D_{r-1}$ are linear differential operators with constant coefficients and D_r is an element of $\mathscr{E}'$ acting by convolution. Then by Theorem 6.1 we can solve $D_1 f^0 = g_1, \ldots, D_{r-1} f^0 = g_{r-1}$. By replacing f by $f - f^0$, we may assume $g_1, \ldots, g_{r-1}$ are all zero. We ask the question: When can we solve $D_r f = g$, $D_1 f = \cdots = D_{r-1} f = 0$ provided that $D_1 g = \cdots = D_{r-1} g = 0$?

DEFINITION. Let A be an irreducible variety and let $F \in \mathbf{E}'$. We say that F is *slowly decreasing on* A if there is a $c > 0$ so that

$$|F(z)| \geq c(1 + |z|)^{-c} \exp(-c\,|\text{Im } z|) \tag{11.6}$$

for z in a sufficient set on A (for $\mathscr{E}$).

THEOREM 11.4. *Let* $\mathfrak{V}^0$ *be a multiplicity variety for* $D_1, \ldots, D_{r-1}$ *and call* P_r *the Fourier transform of the adjoint of* D_r. *Suppose* P_r *is slowly decreasing on every irreducible component of every variety in* $\mathfrak{B}^0$. *Then we can always solve (in* $\mathscr{E}$*) the system* $D_j f = g_j$ *for* $j = 1, \ldots, r$, *provided that* $g_1, \ldots, g_{r-1}$ *satisfy the compatability relations for* $D_1, \ldots, D_{r-1}$ *and also* $D_j g_r = D_r g_j$ *for* $j = 1, \ldots, r-1$.

This theorem follows from Theorem 4.1 and the Hahn-Banach theorem. The main point is that the condition that P_r is slowly decreasing on every irreducible component of every variety in $\mathfrak{B}^0$ enables one to show, using Theorem 4.1, that if $S \in \mathscr{E}'/(D_1', \ldots, D_{r-1}')\mathscr{E}'$ and $D_r'S \to 0$ in the topology of $\mathscr{E}'/(D_1', \ldots, D_{r-1}')\mathscr{E}'$, then also $S \to 0$ in the topology of $\mathscr{E}'/(D_1', \ldots, D_{r-1}')\mathscr{E}'$. The Hahn-Banach theorem then allows us to solve the equations. We shall omit the details.

We can prove that every $f \in \mathscr{E}$ which satisfies $D_j f = 0$ for all j is a *limit* of linear combinations of exponential polynomial solutions. (This is a weakened version of Theorem 7.1.) For the same type of argument can now be used to show that every $S \in \mathscr{E}'$ which vanishes on all such exponential polynomials belongs to the ideal generated by the D_j'; the result then follows from the Hahn-Banach theorem.

See Remark 11.1.

We wish to make some remarks on uniqueness questions for Cauchy-like problems for convolution systems. Let us begin with the example $l = r = 1$, $n = 2$, $Df(t, \xi) = \partial f(t, \xi)/\partial t - f(t, \xi + 1)$. We note first that there exist solutions of $Df = 0$ which vanish together with all their derivatives for $t \leqslant 0$. For we may begin by constructing f arbitrarily in $0 \leqslant \xi \leqslant 1$ subject only to the condition that $\partial f(t, 0)/\partial t = f(t, 1)$. Then we extend f to all $\xi \geqslant 1$ by using the equation $Df = 0$. Similarly, we extend $\partial f/\partial t$ to $-1 \leqslant \xi < 1$, and we obtain f (assuming, for example, that $f \to 0$ rapidly in $0 \leqslant \xi \leqslant 1$ as $t \to -\infty$) by forming

$$\int_{-\infty}^{t} \frac{\partial f}{\partial t}\, dt.$$

Continuing in this way, we construct f for all t, ξ. If we had assumed that $f = 0$ for $0 \leqslant \xi \leqslant 1$, $t \leqslant 0$, then our construction shows that $f = 0$ whenever $t \leqslant 0$. Thus, the Cauchy problem has no uniqueness on $t = 0$, that is, $t = 0$ is characteristic. (It should be noted that f could not vanish for $0 \leqslant \xi \leqslant 1$ without vanishing identically in $\xi \geqslant 0$. However such an f need not vanish for $\xi < 0$; it can, for example, be a nonzero constant in t in $-1 < \xi < 0$, this constant being an indefinitely differentiable function of ξ.) Thus, unlike the situation for differential equations, there is a difference in uniqueness properties for $\xi > 0$ and $\xi < 0$.

We observe that for m a positive integer and any t we have

$$f(t, m) = \partial^m f(t, 0)/\partial t^m. \tag{11.7}$$

Thus, if $f(t, 0) \not\equiv 0$ and $f(t, 0) = 0$ for $t \leqslant 0$, then (11.7) shows that f cannot be very small as $\xi \to \infty$. (Compare Section IX.9 and Chapter V, Example 6.)

In order to see the relation between this example and the methods of Section IX.9, we examine the Fourier transform P of the adjoint of D. This is $[-is\text{–}\exp(iw)]$. Call V the set of zeros of P; we regard V as a covering of the s plane. The real s axis is covered by the set where $\exp(iw)$ is imaginary, that is, $Rw = \pi(j + \frac{1}{2})$, where j is an integer. The strip between two such lines covers either the upper or lower half of the complex s plane. This means that we have an angle 0 at infinity in the w plane between the positive and negative parts of the real s axis on a "branch" of V (that is an inverse image in V under projection on the s plane of "most" of the upper or lower half s plane.) The relation between this and the proof of Theorem 9.30 is as follows: The variable (s, w) above correspond to the variables (u, v), respectively, of that theorem. By (9.90) the angle mentioned above corresponds to π/α. Thus, the above situation is analogous to one in which $\alpha = \infty$. This is, of course, in conformity with the fact that we do not have uniqueness.

The general situation seems to be roughly the following (again in the notation of Theorem 9.30): Let Λ be a line in the real u plane; we write $\Lambda = \Lambda^+ \cup \Lambda^-$ where $\Lambda^\pm$ are half-lines. We form the projections of V on the u and v spaces and denote by $\text{proj}_u^{-1}(\Lambda^\pm)$ the inverse of $\Lambda^\pm$ under the projection on the u space. Let $\nu^\pm$ be two curves (real dimension 1) in v space for which there exists an irreducible component V^0 of V such that $\text{proj}_v[\text{proj}_u^{-1} \Lambda^\pm \cap V^0] \supset \nu^\pm$. We want to know how far apart (at infinity) ν^+ and ν^- can be. (We may think of $\nu^\pm$ as being an algebroid correspondence of $\Lambda^\pm$ by means of V^0.) This will determine (roughly) the relative sizes of u and v on a "branch" of V which as above is essentially related to α. See Fig. 19.

To understand this last point, observe that α is defined (see (9.81)) in terms of the relative sizes of u and v on V. In the algebraic case treated in

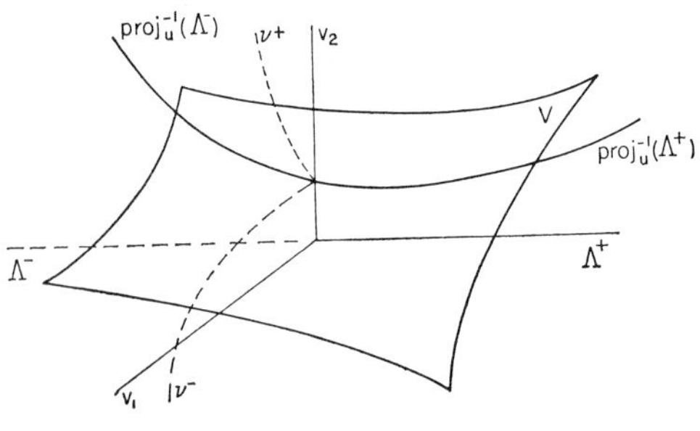

FIGURE 19

Chapter IX, on suitable algebraic curves contained in V the Puiseux expansion shows that $u \sim cv^{\alpha}$ (compare (9.90)). Thus with Λ as above, the angle at infinity between ν^+ and ν^- is π/α.

See Problem 11.2.

XI.2. Singularities

The first topic we shall deal with has its origin in the results of Hartogs that the singularities of a holomorphic function of several variables must obey certain conditions. For example, a holomorphic function defined in a "shell" domain can be extended to its interior. It was Bochner [1] who recognized that this result could be put within the framework of overdetermined systems of partial differential equations. In Ehrenpreis [20] we generalized Bochner's results (which were proved by him for special elliptic systems) to general overdetermined systems for one unknown function. (It is easy to extend the results to general systems.) We begin by repeating the results and proof in Ehrenpreis [20].

Let Ω_0 be a shell in $R(n > 1)$. By this we mean there exist open relatively compact sets Ω_1, Ω_2 such that $\Omega_0 = \Omega_2 - \Omega_1$, and closure $\Omega_1 \subset \Omega_2$.

Call Γ_j the boundary of Ω_j and let $N'(\Gamma_1)$ be a small neighborhood of Γ_1 in Ω_2. Let f be indefinitely differentiable on $\Omega_0 \cup N'(\Gamma_1)$, and suppose $\vec{D}f = 0$ on $\Omega_0 \cup N'(\Gamma_1)$. See Fig. 20.

Let g be defined and C^∞ on Ω_2 such that $g = f$ on the union of Ω_0 with a small neighborhood $N(\Gamma_1)$ in Ω_2 whose closure is contained in $N'(\Gamma_1)$. Call $\Omega_3 = \Omega_1 - N(\Gamma_1)$. Thus, $\vec{D}g = 0$ on $\Omega_2 - \Omega_3$. We set $g_j = D_j g$ so g_j are C^∞ and have their supports in the closure of Ω_3; in particular, the g_j are of compact support. For any j, k

$$D_k g_j = D_j g_k \tag{11.8}$$

since both sides are equal to $D_k D_j g$ in Ω_3 and zero outside.

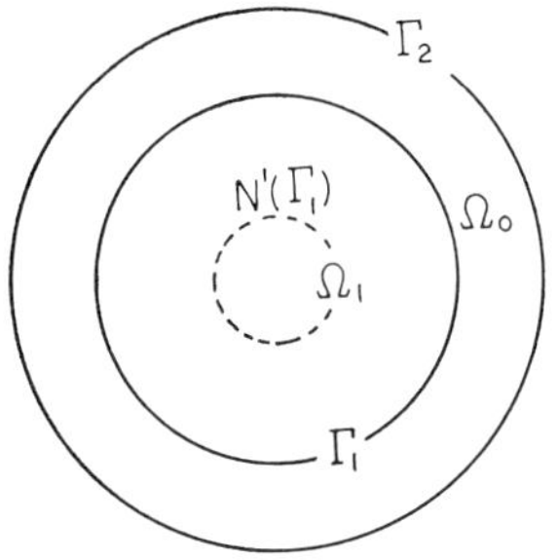

FIGURE 20

Call G_k the Fourier transform of g_k. Then (11.8) implies

$$P_k G_j = P_j G_k. \tag{11.9}$$

(strictly speaking, in order for the notation convention given in Example 3 of Chapter V to hold, G_k is the Fourier transform of g_k thought of as an element of $\mathscr{E}$).

We introduce

Assumption (α). The polynomials $\{P_j\}$ have no common factor.

By (11.9) $H = G_k/P_k$, which is independent of k, has its singularities on the common zeros of the P_k. By assumption (α) this variety is of complex codimension at least 2, so by a classical theorem (see Bochner and Martin [1]) H is entire. Using the usual inequalities on polynomials (see Theorem 1.4) and using Chapter V, Examples 3 and 5, H is the Fourier transform of h which is C^∞ and has its support on K, the convex hull of Ω_3. Moreover, we clearly have $D_k h = g_k$ for all k.

Assumption (β). For any $N(\Gamma_1)$ the following unique continuation property holds: If a is C^∞ on R, and $a = 0$ outside K, and $\vec{\mathrm{D}}a = 0$ outside Ω_3 then $a = 0$ outside of some compact subset Ω_4 of Ω_1.

Remark 11.2. A simple argument shows that the existence of Ω_4 depending on a is equivalent to the existence of an Ω_4 independent of a.

Using this we may assume $h = 0$ outside Ω_4.

Now set $\tilde{f} = g - h$. Then clearly

(a) $\tilde{f}$ is C^∞ on Ω_2
(b) $\tilde{f} = g$ on $\Omega_2 - \Omega_4$
(c) $D_j \tilde{f} = D_j g - D_j h = g_j - g_j = 0$ on Ω_2.

Thus, $\tilde{f}$ is an extension of f. $\tilde{f}$ is unique, for if there were another extension $\tilde{f}_1$ then $\tilde{f} - \tilde{f}_1$ would be of compact support and $\vec{\mathrm{D}}(\tilde{f} - \tilde{f}_1) = 0$. Since the Fourier transform of $\tilde{f} - \tilde{f}_1$ is entire, and its product with P_j is zero, this implies $\tilde{f} - \tilde{f}_1 = 0$.

It follows from the definition of $\tilde{f}$ that $f = \tilde{f}$ on $\Omega_2 - \Omega_3$.

THEOREM 11.5. *Assumptions* (α) *and* (β) *are necessary and sufficient in order that any f which is C^∞ on the union of Ω_0 with some neighborhood $N'(\Gamma_1)$ and satisfies $\vec{\mathrm{D}}f = 0$ on $\Omega_0 \cup N'(\Gamma_1)$ possesses a unique C^∞ extension $\tilde{f}$ to Ω_2 with $\vec{\mathrm{D}}\tilde{f} = 0$ on Ω_2, $\tilde{f} = f$ on the union of Ω_0 with some neighborhood of Γ_1 in Ω_1.*

PROOF. We have already demonstrated the sufficiency so we proceed to the necessity.

Necessity of (α). Suppose the P_j have a common factor, say P, whose inverse Fourier transform is D. Now, by Theorem 6.1, D maps $\mathscr{D}'$ onto

$\mathscr{D}'$. In particular, there exists a fundamental solution for D, that is, an $S \in \mathscr{D}'$ such that $DS = \delta_a$, where δ_a is the unit mass at the point $a \in$ interior Ω_1. Let $m \in \mathscr{D}$ have support so small that the support of $m * \delta_a$ is still contained in the interior of Ω_1. Assume also that P does not divide the Fourier transform M of m.

The restriction f of $S * m$ to the union of Ω_0 with some neighborhood $N'(\Gamma_1)$ satisfies $D(S * m) = m * \delta_a = 0$ on $\Omega_0 \cup N'(\Gamma_1)$, hence $\vec{D}(S * m) = 0$ on $\Omega_0 \cup N'(\Gamma_1)$. We claim there is no $\tilde{f}$ which is indefinitely differentiable on Ω_2 with $\vec{D}\tilde{f} = 0$ on Ω_2 and $\tilde{f} = f$ on Ω_0. For such an $\tilde{f}$ would have the property that $l = \tilde{f} - S * m$ would be zero on Ω_0 (hence would be in $\mathscr{D}$) and $\vec{D}l = -\vec{D}(S * m)$.

We claim that $-\vec{D}(S * m)$ is not of the form $\vec{D}l$ for $l \in \mathscr{D}$. To see this, write $D_j = D\partial_j$ for each j. Then we must have

$$\partial_j\, Dl = -\partial_j m * \delta_a$$

which means that

$$\partial_j(Dl + m * \delta_a) = 0.$$

Since $Dl + m * \delta_a$ is of compact support, this implies, as in the proof of the first part of Theorem 11.5, that $Dl + m * \delta_a = 0$. By Fourier transform this implies

$$PL = -M \exp(ia \cdot \quad),$$

(where L is the Fourier transform of l) which means that P divides M contrary to our assumptions. Thus, the necessity of (α) is proved. ▯

Necessity of (β). If (β) fails, then there is an $N(\Gamma_1)$ and an indefinitely differentiable function a such that

1. Support $a \subset K$.
2. $\vec{D}a = 0$ outside Ω_3.
3. The support of a is not contained in a compact subset of Ω_1.

Now, define the function b on $\Omega_2 - \Omega_3$ as the restriction of a to $\Omega_2 - \Omega_3$. By condition 2 we have $\vec{D}b = 0$ on $\Omega_2 - \Omega_3$.

We claim that for no neighborhood $N'(\Gamma_1)$ in Ω_2 does there exist a $\tilde{b} \in C^\infty(\Omega_2)$ with $\vec{D}\tilde{b} = 0$ on Ω_2 and $\tilde{b} = b$ on $\Omega_0 \cup N'(\Gamma_1)$. For, we could extend $\tilde{b}$ to be C^∞ on all of R^n by defining $\tilde{b} = a$ outside Ω_2. Thus, $\tilde{b}$ would be of compact support and $\vec{D}\tilde{b} = 0$ everywhere. Thus, $\tilde{b} \equiv 0$. This means that $b \equiv 0$ on $\Omega_0 \cup N'(\Gamma_1)$ which contradicts condition 3. Thus, (β) is necessary. ▯

Remark 11.3. A similar result holds for distribution solutions. See Remark 11.4 at the end of the chapter.

Remark 11.5. In case $\vec{D}$ is analytic elliptic, that is, all solutions of $\vec{D}f = 0$ are real analytic, then, as is easily seen, assumption (β) will be satisfied if and only if the complement of Ω_1 is connected, that is Ω_1 is simply connected. When applied to the Cauchy-Riemann equations for holomorphic functions of several complex variables our Theorem 11.5 is just Hartogs' original result on analytic continuation.
See Problem 11.3 (Exercise)

For later purposes, we wish to derive a modification of Theorem 11.5 in which we extend "approximate solutions" from Ω_0 to Ω_2. We shall consider only the case of convex Ω_1, Ω_2.

THEOREM 11.5*. *Let the notation be as in Theorem* 11.5. *Suppose* Ω_1 *and* Ω_2 *are convex and assumption* (α) *holds. Let* B *be a set of* C^∞ *functions on the union of* Ω_0 *with some neighborhood* $N'(\Gamma_1)$ *such that* $\vec{D}B$ is *bounded in the topology of* $\mathscr{E}^r(\Omega_0 \cup N'(\Gamma_1))$.
Then for each $f \in B$ *we can find an* $\tilde{f} \in \mathscr{E}(\Omega_2)$ *such that*

(*a*) $\{f - \tilde{f}\}$ *is bounded in the topology of* $\mathscr{E}(\Omega_0 \cup N(\Gamma_1))$ *for some neighborhood* $N(\Gamma_1)$.

(*b*) $\{\mathrm{D}\tilde{f}\}$ *is bounded in the topology of* $\mathscr{E}^r(\Omega_2)$.

PROOF. We proceed as in the proof of Theorem 11.5. We may assume that Ω_3 is a convex polyhedron. By multiplying the g_j by a function in $\mathscr{D}(\Omega_2)$ which is one on $\bar{\Omega}_3$, we can assume that $g_j \in \mathscr{D}(\Omega_2)$ and, instead of (11.8), that

$$\{D_k g_j - D_j g_k\} \qquad \text{is bounded in } \mathscr{D}(\Omega_4) \tag{11.8*}$$

because, by assumption, $\{g_j\}$ is bounded on $\Omega_0 \cup N'(\Gamma_1)$. Here Ω_4 is a convex polyhedron containing $\bar{\Omega}_3$ and $\bar{\Omega}_4 \subset \Omega_1$.
From (11.8*) we deduce

$$\{P_k G_j - P_j G_k\} \qquad \text{is bounded in } \mathbf{D}(\Omega_4). \tag{11.9*}$$

We *cannot* now use the fact that the singularities of a meromorphic function must have complex codimension one. Instead we examine the consequences of (11.9*) on the multiplicity varieties $\mathfrak{V}_k$ corresponding, by Theorem 4.2, to each of the P_k. By assumption (α), given k and any variety V_{km} in $\mathfrak{V}_k$, there is a j so that P_j does not vanish identically on V_{km}. Thus, by (11.9*), G_k is bounded in the topology of $\mathbf{D}(\Omega_4)(V_{km})$ outside the subset of V_{km} where $|P_j| \leqslant 1$. We noted in Chapter VII (see Problem 7.1) that this subset is removable. Thus G_k is bounded in $\mathbf{D}(\Omega_4)(V_{km})$.

We conclude that each G_k is bounded in the topology of $\mathbf{D}(\Omega_4)(\mathfrak{V}_k)$. By Theorem 4.1 and Examples 3 or 4 of Chapter V, we can find $\tilde{G}_k \in \mathbf{D}(\Omega_4)$ such that $\tilde{G}_k$ is bounded in $\mathbf{D}(\Omega_4)$ and $G_k - \tilde{G}_k = P_k H_k$ for some $H_k \in \mathbf{D}(\Omega_4)$.

We may assume that $D_1 \neq 0$; set $H = H_1$. Call h the inverse Fourier transform of H. Then the above shows that, for any k,

$$D_k h - g_k = (D_k h_k - g_k) + (D_k h - D_k h_k)$$
$$= -\tilde{g}_k + (D_k h - D_k h_k).$$

On the other hand, by (11.9*) and the definitions of h, h_k

$$D_1 D_k h - D_1 D_k h_k = D_k D_1 h_1 - D_1 D_k h_k$$
$$\sim D_k g_1 - D_1 g_k$$
$$\sim 0.$$

Here we have used the symbol $\sim$ to indicate equality modulo bounded terms. It is noted in Theorem 6.5 that $m \to D_1 m$ is a topological isomorphism on $\mathbf{D}(\Omega_4)$. Thus $D_k h \sim D_k h_k$ so $D_k h - g_k \sim 0$. It is easy to see that $\tilde{f} = g - h$ satisfies the conclusion of Theorem 11.5*. (Actually $\tilde{f} = f$ on $\Omega_0 \cup N(\Gamma_1)$ which is stronger than (a).) ▯

The method used to prove Theorem 11.5 is Method c described at the beginning of Chapter VIII. In the language of sheaves, we showed how to reduce the extension problem to the problem of proving that $H^1_*(\Omega_1, \mathscr{S}) = 0$ (see, e.g., Ehrenpreis [14]) where $\mathscr{S}$ is the sheaf of germs of solutions of $\vec{\mathrm{D}}f = 0$ and H^1_* is the first cohomology group with compact support. Readers unfamiliar with cohomology theory can think of $H^1_*(\Omega_1, \mathscr{S})$ as the quotient of the kernel of $\boxed{\mathrm{D}}^1$ on $\mathscr{D}^r(\Omega_1)$ by the image of $\vec{\mathrm{D}}$ on $\mathscr{D}(\Omega_1)$. Here $\boxed{\mathrm{D}}^1$ is an $r_1 \times r$ matrix whose rows generate the module of relations of the D_j. (In the proof of Theorem 11.5, $\vec{\mathrm{D}}g$ defined an element of $H^1_*(\Omega, \mathscr{S})$ which we wanted to show was zero in this group; we showed that $\vec{\mathrm{D}}g = \vec{\mathrm{D}}h$ when $h \in \mathscr{D}(\Omega_1)$.)

Let us see what happens if we apply Method a. Let $\vec{\mathrm{D}}f = 0$ on $\Omega_0 \cup N'(\Gamma_1)$ and suppose we want to extend f to a solution on Ω_2. Let $S \in \mathscr{E}'(\Omega_0)$ and suppose $S \to S_0$ in the topology of $\mathscr{E}'(\Omega_2)$, where S_0 vanishes on the kernel of $\vec{\mathrm{D}}$ in $\mathscr{E}(\Omega_2)$. From the fact that S_0 vanishes on the kernel of $\vec{\mathrm{D}}$ in $\mathscr{E}(\Omega_2)$ we see from the Hahn-Banach theorem that S_0 belongs to the closure of the image of $\vec{\mathrm{D}}'$ on $\mathscr{E}'^r(\Omega_2)$. Suppose for simplicity that actually $S_0 = \vec{\mathrm{D}}'\vec{\mathrm{T}}_0$, where $\vec{\mathrm{T}}_0 \in \mathscr{E}'^r(\Omega_2)$.

Now, $\vec{\mathrm{D}}'\vec{\mathrm{T}}_0 = 0$ on Ω_1. Suppose that ${}^*H^1(\Omega_1, \mathscr{S}_1) = 0$ where $\mathscr{S}_1$ is the sheaf of germs of distribution solutions of $(\boxed{\mathrm{D}}^1)'$, and where we can interpret ${}^*H^1(\Omega_1, \mathscr{S}_1)$ as the quotient of the kernel of $\vec{\mathrm{D}}'$ by $(\boxed{\mathrm{D}}^1)'\mathscr{D}'^{r_1}(\Omega_1)$. Then we can write, $\vec{\mathrm{T}}_0 = (\boxed{\mathrm{D}}^1)'\vec{\mathrm{T}}_1$, where $\vec{\mathrm{T}}_1 \in \mathscr{D}'^{r_1}(\Omega_1)$. Let h be an indefinitely differentiable function whose support is contained in a compact subset of Ω_1 and such that $h = 1$ on a neighborhood of $\Omega_1 - N'(\Gamma_1)$.

Then,

$$\vec{T}_2 = \vec{T}_0 - (\boxed{\vec{D}}^1)' h \vec{T}_1 \in \mathscr{E}'^r(\Omega_0 \cup N'(\Gamma_1)) \tag{11.10}$$

and $\vec{D}'\vec{T}_2 = \vec{D}'\vec{T}_0 = S_0$. Since $\vec{D}f = 0$ on $\Omega_0 \cup N'(\Gamma_1)$, we have $S_0 \cdot f = \vec{D}'\vec{T}_2 \cdot f = \vec{T}_2 \cdot \vec{D}f = 0$ which proves that the restriction of f to $\Omega_0 \cup N'(\Gamma_1)$ can be extended to a solution on Ω_2.

Thus, we see the relationship between the vanishing of $H^1_*(\Omega_1, \mathscr{S})$ and $^*H^1(\Omega_1, \mathscr{S}_1)$. This is similar to the Serre duality theorem (see Serre [2] and Remark 8.1).

Note that if Ω_1 and Ω_2 are convex, then the above use of Method a applies by the results of Chapter VI to prove this case of Theorem 11.5.

Theorem 11.5 may be regarded as saying that certain compact sets of singularities are "removable." When we speak of a "compact set of singularities" we mean that the set Ω_1 into which we want to extend the given solution f has compact closure in R and the boundary of Ω_1 is contained in the closure of the set on which f is given as a solution. The removability of noncompact sets of singularities poses a very difficult problem. Some results can be obtained by the method of proof of Theorem 11.5 (see Problem 11.3). However, these are incomplete and are not illuminating as they do not bring into play enough structure of the specific $\vec{D}$ we are dealing with. For this reason we shall follow another approach which, in fact, gives all the results obtainable from the previous method. The results we obtain will be used in Section XI.3 for the study of solution of inhomogeneous systems on domains which are not convex. Questions about removable singularities seem to be closely related to an extension of Nevanlinna theory to partial differential equations (see the remarks at the beginning of Section IX.1).

In Section VIII.3 we discussed a removable singularities (extension) problem related to hyperbolic systems. Since hyperbolic operators possess special kinds of fundamental solutions, we might suspect that there is a relation between fundamental solutions and extension problems. That this is the case will be seen in the following. It should be noted that the type of extension problem treated in Theorem 11.5 is related to the hyperbolicity problem discussed in Section VIII.3 rather than the problem of quasi-hyperbolicity discussed in Section VIII.4. The reason for this is the unicity requirement (β). As in Section VIII.4, we can use the unicity condition to prove that if (α) and (β) hold, then if f is defined and indefinitely differentiable on the *open* ring: Ω_2 — closure Ω_1 and $\vec{D}f = 0$ there then f possesses a unique C^∞ extension $\tilde{f}$ to all of Ω_2 satisfying $\vec{D}\tilde{f} = 0$ on Ω_2, $\tilde{f} = f$ on Ω_2 — closure Ω_1.

We shall begin by discussing a special type of local extension which is obtained by use of a suitable fundamental solution. We shall then treat the question in more detail.

Let Λ be a linear variety, let A be a domain, and let $a \in A$. We say that a can be *bitten out of A by Λ* if for $0 \leqslant b \leqslant 1$ we can find a set of translates Λ_b of Λ, which depend continuously on b, and on each of them a domain B_b with smooth boundary ∂B_b, depending continuously on b such that

1. $B_0 \subset$ complement of A.
2. For each b, $\partial B_b \subset$ complement of A.
3. For some b_0 we have $a \in B_{b_0}$.

For example, if A is compact, all points in A can be bitten out by any Λ.

We define a similar concept locally in the neighborhood of a boundary point of A; the local analog of bite is called *nibble*. A more classical terminology would be that A is *strictly convex* with respect to Λ. We shall discuss below this condition in terms of the derivatives of the function defining the boundary of A.

In the definition of nibbling, let b^0 be the smallest value of b for which B_b meets A, so that B_{b^0} is tangent to A, say at the point a^0 (see Fig. 21). Then we say that a^0 can be *just nibbled out of A by Λ*.

Let $\vec{D} = (D_1, \ldots, D_r)$ as usual and let L be a linear variety passing through the origin. We call L a *mouth* for $\vec{D}$ if there exist $e_1, \ldots, e_r$ which are distributions on R having support on L such that

$$\sum D'_j e_j = \delta.$$

Thus $\vec{e}$ is a fundamental solution (see Section VI.2) for the system obtained by replacing D_j by D'_j. In particular, L will be a mouth for $\vec{D}$ if there exists an operator ∂ in the ideal generated by the D'_j in the ring of linear constant coefficient operators such that ∂ depends only on derivatives in directions along L. For, by Theorem 6.1 (applied to L in place of R^n) there exists a distribution T on L with $\partial T = \delta$ (in the sense of distributions on L). If

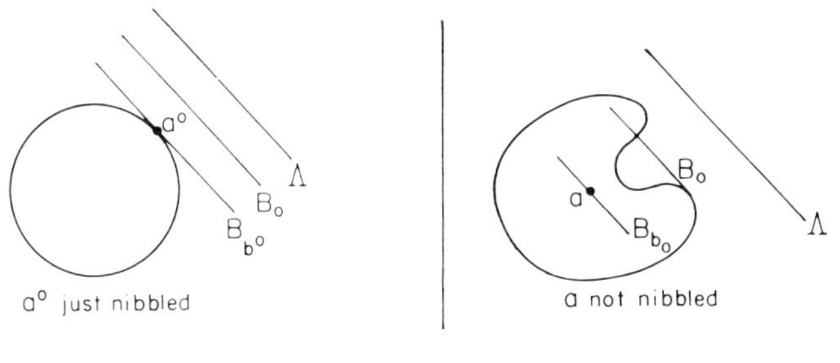

FIGURE 21

we write $\partial = \sum D'_j \partial_j$, then $(\partial_1(\delta_L \times T), \ldots, \partial_r(\delta_L \times T))$ is clearly a fundamental solution for $\vec{\mathrm{D}}'$ with support on L. Here we have denoted by $\delta_L \times T$ the extension of T to R^n which is the unit mass at the origin in the coordinates orthogonal to L. We shall make the following conjecture:

PROBLEM 11.4.

L is a mouth for D if and only if there is an operator ∂ in the ideal generated by the D'_j in the ring of linear constant coefficient operators such that ∂ depends only on derivatives in directions along L.

We suspect also that the existence of such a ∂ is necessary and sufficient for the existence of distributions $S_j \neq 0$ which are of the form $S_j = T_j \times \delta_L$, with T_j a distribution on L and δ_L the unit mass at the origin in the directions "orthogonal" to L such that

$$\sum D'_j S_j = 0.$$

We can verify the conjectures in many special cases by the following method: Suppose $\vec{\mathrm{S}}$ is a fundamental solution for $\vec{\mathrm{D}}'$ with support on L. Use coordinates ξ on L and t orthogonal to L. Assume for simplicity that $\vec{\mathrm{S}}$ is a distribution of finite order (see Chapter 5, Example 3). Then we can write each S_j in the form

$$S_j = \sum S^i_j \times \eta_i,$$

where η_i are differentiations in t with $\eta_0 = \delta$ and S^i_j are distributions of finite order on L. Write each D'_j in the form

$$D'_j = \sum \lambda^i_j \times \eta_i,$$

where λ^i_j are differential operators in ξ. The equation

$$\sum D'_j S_j = \delta$$

becomes

$$\sum_j \lambda^0_j S^0_j = \delta \tag{11.11a}$$

$$\sum \lambda^i_j S^{i'}_j = 0, \tag{11.11b}$$

where the second sum is over all j and all pairs i, i' for which $\eta_i \eta_{i'}$ is some fixed differentiation $\neq \delta$.

Our first conjecture asserts that if these equations have a solution $\{S^i_j\} \in \mathscr{D}'_F$ then there is a solution to (11.11b) with each S^i_j having its support at the origin. In many cases this can be verified by applying Theorem 6.1 which shows that the possibility of solving (11.11a) and (11.11b) with $S^i_j \in \mathscr{D}'_F$ can be translated into an algebraic question.

The second conjecture can be reduced to an algebraic question by applying Theorem 7.3 to solutions of $\sum D_i' S_i = 0$.

The concepts of biting, nibbling, and mouth can be extended to the case when Λ, L are no longer assumed to be linear varieties, but rather smooth (C^∞) local manifolds with boundary. For L to be a mouth at some point $p \in L$ we require the existence of a local fundamental solution at p, that is, there must exist $S_1^p, \ldots, S_r^p$ with support on L such that $\sum D_j' S_j^p = \delta_p$ near p; here δ_p is the unit mass at p. In this case, we say that a^0 can be just nibbled by the mouth L, if, in addition to the requirement in the case of L linear, we require $a^0 = \tau p$, where τ is that translation which, when applied to L, makes B_{b^0} tangent to A.

For the case where L is not linear we cannot make a conjecture analogous to the one above as the example of the wave operator in an odd number of dimensions >1 shows. Namely, there exists a fundamental solution with support on a cone although D cannot be expressed in terms of derivatives in directions along the cone.

For the converse we must also take care. Suppose there exists a linear differential operator ∂ with *variable coefficients* of the form

$$\partial = \sum \lambda_j D_j',$$

where $\lambda_j(x)$ are differential operators with indefinitely differentiable coefficients such that ∂ depends only on L, that is, we can choose coordinates y in the neighborhood in R^n of any $p \in L$ in such a way that L can be defined locally by equations $y_1 = 0, \ldots, y_q = 0$ and such that, at each $y \in L$ we can write ∂ as a differential operator in $\partial/\partial y_{q+1}, \ldots, \partial/\partial y_n$. We denote this differential operator by $\partial_L(y, \partial/\partial y)$, which we may think of as a differential operator on L. However, $\partial_L(y, \partial/\partial y)$ may *not have a fundamental solution*. An example of this phenomenon occurs for $n = 4$, $r = 2$, $D_1 = \partial/\partial x_1 + i\partial/\partial x_2$, $D_2 = \partial/\partial x_3 + i\partial/\partial x_4$, and L the boundary of a strictly convex domain. This example is due to H. Lewy [1]. It provided the first example of a differential equation (even with real analytic coefficients) with no fundamental solution.

EXAMPLE 1. $r = n$, $D_j = \partial/\partial x_j$. Then every manifold is a mouth.

EXAMPLE 2. $n = 2m$, $r = m$, $D_j = \partial/\partial x_j + i\partial/\partial x_{m+j}$. Then a manifold L is a mouth if and only if the tangent plane at each point contains a complex analytic line. In particular, L will be a mouth if $\dim_R L > m$.

For our local analog of Theorem 11.5, we need a uniqueness property which plays the role of assumption (β) of Theorem 11.5. The uniqueness property always holds if $\vec{\mathrm{D}}$ is analytic elliptic (see Section VIII.2) that is, if all distribution solutions of $\vec{\mathrm{D}}f = 0$ are real analytic functions.

DEFINITION. Let A be a domain, a a boundary point of A which can be just nibbled by the mouth L for $\vec{\mathrm{D}}$. Let $\tilde{N}$ be a neighborhood of a in R. By the *part of $\tilde{N}$ above a* (relative to L) we mean the points a' in $\tilde{N}$ which can be obtained by translating a to a' in such a way that, in the translation process, some fixed neighborhood B of a in L does not meet A.

We say that $\vec{\mathrm{D}}$ has the *L uniqueness property at a* if, as above, a can be just nibbled by L, and if whenever f is an indefinitely differentiable function on the union of the part U of $\tilde{N}$ above a with a neighborhood U_1 of the complement of A in $\tilde{N}$ near a such that $\vec{\mathrm{D}}f = 0$ on $U \cup U_1$ and $f = 0$ on U, then $f = 0$ on U_2 where U_2 is a neighborhood of the complement of A near a. U_2 depends on $\tilde{N}$ and U_1 but not on f. In addition, we require that if f is small on U and $\vec{\mathrm{D}}f$ is small on $U \cup U_1$, then f is small on U_2. (Here by "small" we mean "small in the topology of indefinitely differentiable functions on the set in question.") More precisely, there exist arbitrarily small neighborboods $\tilde{N}$ of a with this property.

We say that $\vec{\mathrm{D}}$ has the *strong L uniqueness property at a* if, given a_1 close to a and N_2 a small neighborhood of a_1 in A, there exists a neighborhood N_1 of a_1 which is small (in the sense that its diameter shrinks to zero as the diameter of N_2 shrinks to zero) such that if $U_1 \supset N_1$ then $U_2 \supset N_2$. See Fig. 22.

The question arises as to how to decide when L uniqueness and strong L uniqueness hold. The general procedure is based on the following idea: Let p be a point near a, with $p \notin U$. Suppose there exists a mouth L_p for $\vec{\mathrm{D}}$ at p which "enters U," that is, L_p meets the closure of U in a compact set which bounds an open subset $\tilde{L}_p$ of L_p containing p (see Fig. 23). Suppose, moreover, that $\tilde{L}_p$ is contained in an open subset of $U \cup U_1$. Then, if $\vec{\mathrm{D}}f$ is small on $U \cup U_1$ and f is small on U, it follows that f is small at p. (A similar method applies for showing $f(p) = 0$.)

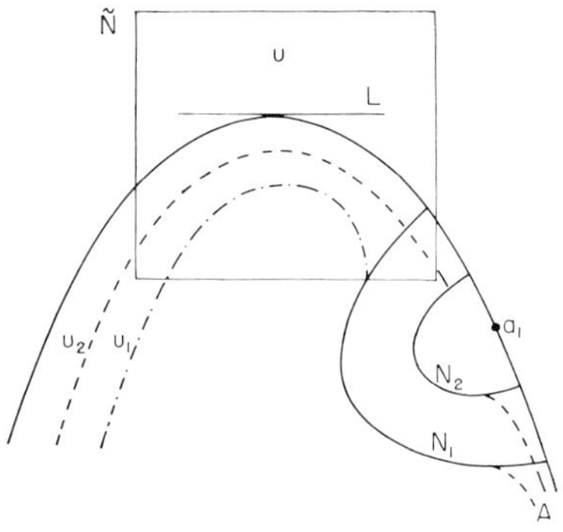

FIGURE 22

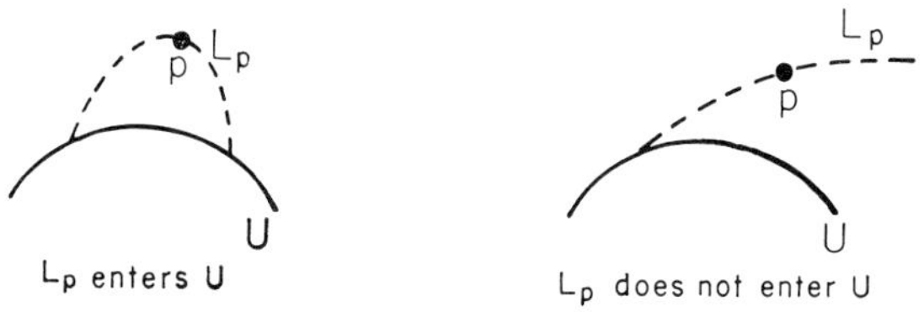

FIGURE 23

This fact is verified as follows: By the definition of a mouth, we can write $\delta_p = \sum D'_j e_j^p$, where e_j^p are distributions with support on L_p. We now cut off the e_j^p by multiplying by a function η which is C^∞ on L_p, with $\eta \equiv 1$ in $\tilde{L}_p$ and η changes from being $\equiv 1$ to $\equiv 0$ rapidly near the boundary of L_p (see the proof of Theorem 11.6 below for more details).

Thus,

$$\begin{aligned} f(p) &= (f *' \delta_p)(0) \\ &= (f *' \zeta_p)(0) + \sum [(D_j f) *' (\eta e_j^p)](0). \end{aligned}$$

Here

$$\sum D'_j[\eta e_j^p] = \delta_p - \zeta_p$$

and ζ_p has its support in U. The term $(f *' \zeta_p)(0)$ is small because f is small on U and the term $\sum [(D_j f) *' (\eta e_j^p)](0)$ is small because $D_j f$ is small on $U \cup U_1$.

We conclude that f is small at p. Thus, using fundamental solutions we derive sets larger than U on which f is small. Continuing in this manner we can often verify L uniqueness and strong L uniqueness at a.

It may be possible to use the technique of proof of Theorem 9.30 to derive the L uniqueness property, but we have not attempted this.

THEOREM 11.6. *Let A be a (local) domain with a smooth boundary. Let L be a mouth for $\vec{\mathrm{D}}$ at a; suppose that a can be just nibbled by L, and suppose that $\vec{\mathrm{D}}$ has the L uniqueness property at a. Let f be an indefinitely differentiable function on the union of a connected neighborhood N of a in the complement of A with a neighborhood N' of the intersection of the boundary of A near a with the closure of N. Suppose $\vec{\mathrm{D}}f = 0$ on $N \cup N'$. Then there exists a neighborhood $\tilde{N}$ of a in R and a function $\tilde{f} \in \mathscr{E}(\tilde{N})$ with $\vec{\mathrm{D}}\tilde{f} = 0$ on $\tilde{N}$ and*

$$\tilde{f} = f \qquad \text{on } \tilde{N} \cap (N \cup N'').$$

Here N'' is a neighborhood of the intersection of the boundary of A near a with $\tilde{N}$. $\tilde{N}$ depends on $\vec{\mathrm{D}}$ and N but not on f or N'. N'' may depend on N'.

PROOF. After performing a translation we may assume that $a = 0$ and that L has the following properties: There exists a domain $B \subset L$ with a smooth boundary such that $0 \in B$ and boundary $B \subset N$. Moreover there is at least one direction of translating B so as (locally) to move B into N. (See Fig. 24.)

Let $e_1, \ldots, e_r$ be distributions with support on L satisfying (locally) $\sum D'_j e_j = \delta$ on B. For each x, denote by $\tau_x B$ or $\tau_x L$ the respective translates of B and L be x, that is, $\tau_x B = \{b + x\}_{b \in B}$, $\tau_x L = \{l + x\}_{l \in L}$. If x is near zero, the $\tau_x B$ meets N and its whole boundary is contained in N. Thus there is an indefinitely differentiable function η_x on $\tau_x B$ which is equal to one near x, and changes from being identically one to identically zero on the strict interior of the intersection of $\tau_x B$ with N (that is, somewhat away from the boundary of the intersection).

As usual, let $\tau_x e_j$ denote the translate of e_j by x. Thus, the support of $\tau_x e_j$ is on $\tau_x L$. It follows that

$$\sum D'_j[\eta_x(\tau_x e_j)] = \delta_x - \zeta_x,$$

where ζ_x is an indefinitely differentiable function on $\tau_x L$ (perhaps multiplied by differential operators in directions orthogonal to $\tau_x L$) having its support in the strict interior of $N \cap \tau_x B$.

Thus, for $x \in N$ near zero, we have

$$\begin{aligned} f(x) &= (f *' \delta_x)(0) \\ &= (f *' \zeta_x)(0) \end{aligned}$$

because $D_j f = 0$ on N. Here $*'$ denotes the adjoint of $*$, that is, formally,

$$(g *' h)(t) = \int g(t')h(t + t')\, dt'$$

(see Example 3 of Chapter V). In our case, f is defined only locally so care must be taken that the functions or distributions with which we convolve

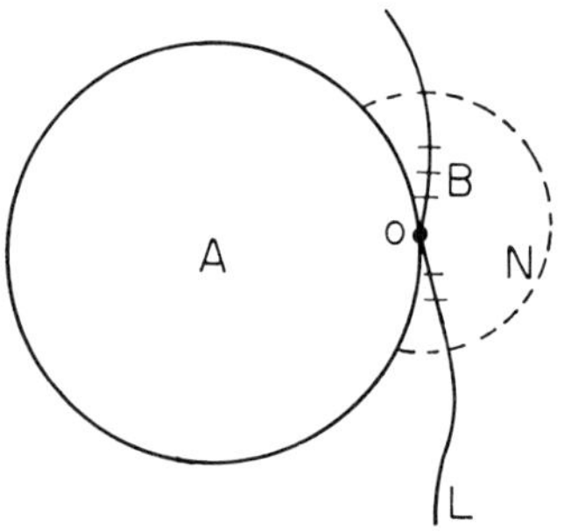

FIGURE 24

f have sufficiently small supports. (We could also "cut off" f by multiplying f by a suitable function in D. This would lead to the same results.) We shall use some commutativity and associativity properties of convolution; these are readily verified for the cases to which we shall apply them. In fact, for y close to x, we can write

$$f(y) = (f *' \delta_x)(x - y) \tag{11.12}$$
$$= (f *' \zeta_x)(x - y).$$

(11.12) holds for all y for which the support of $\eta_x(\tau_x e_j)$ lies in the interior of $N + x - y$. For, in order to know that (11.12) holds, we need $\{f *' D_j'[\eta_x(\tau_x e_j)]\}(x - y) = 0$. In order for this adjoint convolution to be meaningful and in order to commute the term D_j' so as to write

$$\{f *' D_j'[\eta_x(\tau_x e_j)]\}(x - y) = \{D_j f *' [\eta_x(\tau_x e_j)]\}(x - y) = 0$$

(since $D_j f = 0$ on N) we must know that the support of $\eta_x(\tau_x e_j)$ lies in the interior of $N + x - y$. This is seen from the formal calculation

$$\{f *' D_j'[\eta_x(\tau_x e_j)]\}(x - y) = \int f(t')\{D_i'[\eta_x(\tau_x e_j)]\}(t' + x - y)\, dt'$$
$$= \int (D_j f)(t')[\eta_x(\tau_x e_j)](t' + x - y)\, dt'.$$

In order for the integrals to be meaningful and in order that the last integral be zero, we want to know that only values of $t' \in N$ enter in the integral. Now, only values of t' for which $t' + x - y \in$ support $\eta_x(\tau_x e_j)$ enter. Thus we want

$$t' \in \text{support } \eta_x(\tau_x e_j) - x + y$$

to imply

$$t' \in N.$$

This is just our condition

$$\text{support } \eta_x(\tau_x e_j) \subset N + x - y.$$

On the other hand, by applying D_j to the right side of (11.12) we see that it defines a solution of $\vec{\mathrm{D}}$ as long as the support of ζ_x lies in the interior of $N + x - y$ since $\vec{\mathrm{D}}f = 0$ on N. (The change from D_j' to D_j occurs because $-y$ appears on the right side of (11.12).) Since the support of ζ_x is much smaller than the support of $\eta_x(\tau_x e_j)$, the set of y for which the right side of (11.12) is a solution of $\vec{\mathrm{D}}$ may contain points not in N, that is, points of A. In fact, it is geometrically clear that we can start with $x \in N$ very close to 0 and construct η_x so that we can "move" B

by translation so that x moves into $y \in A$ but the translates of the support of ζ_x stay in N. (See Fig. 25.)

In this way we have extended f to a solution $\tilde{f}$ on a full neighborhood of zero. It is clear that $f = \tilde{f}$ above a. It follows from the definition of the L uniqueness property that $f = \tilde{f}$ on $\tilde{N} \cap (N \cup N'')$. □

It is clear from (11.12) that the extension is unique. We could continue the process and extend f further. There is no difficulty in the uniqueness of the extension as long as we extend to a simply connected neighborhood of 0.

Remark 11.6.

Using Proposition 8.6 we can show Theorem 11.6 contains Theorem 8.5. We can also show that Theorem 11.6 implies Theorem 11.5.

An interesting problem is the question of which domains A are *domains of existence* for $\vec{D}$, that is, domains A for which there exists an $f \in \mathscr{E}(A)$ such that $\vec{D}f = 0$ but for no domain $A' \supset A$, $A' \neq A$ is there a $g \in \mathscr{E}(A')$ with $\vec{D}g = 0$ and $g = f$ on A. One of the most celebrated theorems of the theory of functions of several complex variables (Oka's theorem) asserts that, for the Cauchy-Riemann system, domains on which we can solve inhomogeneous equations (cf. Section XI.3) are the same as domains of existence; they are called *domains of holomorphy*. However, as one sees easily from the example $\partial/\partial x_1$, these two concepts do not always coincide. Nevertheless the proof given in Section XI.3 that on certain domains we can solve inhomogeneous equations involves the use of continuation for a related system. The precise relation between the two concepts is still unclear.

Now, Theorems 11.5 and 11.6 give us ways of showing that a domain is not a domain of existence.

PROBLEM 11.5 (Conjecture)

If no point in the boundary of the complement of A can be just nibbled by a mouth for $\vec{D}$ then A is a domain of existence for $\vec{D}$.

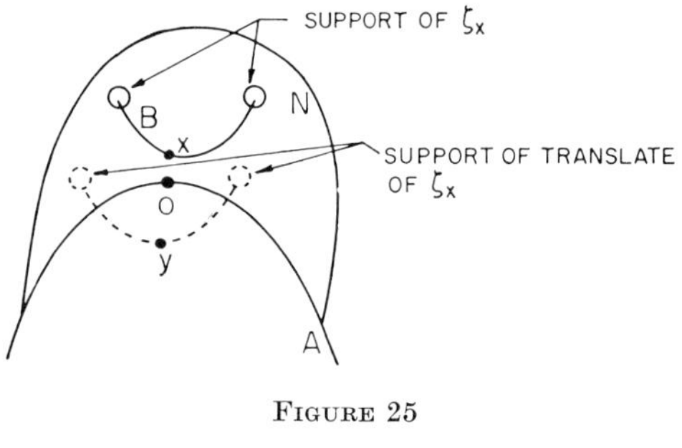

FIGURE 25

Although the boundary of A is the same as the boundary of its complement, the question of whether a point can be just nibbled depends on whether we consider it as a boundary point of A or of *complement* A.

Let M be a submanifold of A, that is, M is a set of common zeros of a finite set of functions in $\mathscr{E}(A)$. We say that M is a $\vec{\mathrm{D}}$-*submanifold* (of A) if every function f in $\mathscr{E}(M)$ which satisfies the equations induced by $\vec{\mathrm{D}}$ on M (that is, $\partial f = 0$ whenever ∂ is a differential operator with variable coefficients on M of the form $\sum \partial_j D_j$, where ∂_j are operators with variable coefficients) can be extended to a function $\tilde{f} \in \mathscr{E}(\tilde{M})$, where $\tilde{M}$ is a neighborhood of M, such that $\vec{\mathrm{D}}\tilde{f} = 0$ on $\tilde{M}$.

For example, if $\vec{\mathrm{D}}$ is the Cauchy-Riemann system, then a $\vec{\mathrm{D}}$-submanifold is any (complex) analytic subvariety, while if $\vec{\mathrm{D}}$ is hyperbolic then a $\vec{\mathrm{D}}$-submanifold is spacelike (see Section VIII.3). By using the methods of Section VIII.3 or the method of proof of Theorem 9.17, we could characterize all linear $\vec{\mathrm{D}}$-submanifolds.

PROBLEM 11.6 (Conjecture)

If A is a domain of existence for $\vec{\mathrm{D}}$ and M is a $\vec{\mathrm{D}}$-submanifold of A whose boundary is contained in the boundary of A, then any $f \in \mathscr{E}(M)$ which satisfies the equations induced by $\vec{\mathrm{D}}$ on M can be extended to a solution on all of A.

For the Cauchy-Riemann system, a domain of existence is a domain of holomorphy. The conjecture is true in this case by virtue of a theorem of Oka (see Gunning and Rossi [1] and Bers [1]).

We make the conjecture for the following reason: We like to think of extension properties as "hyperbolic" properties of $\vec{\mathrm{D}}$. Thus, roughly speaking domains of existence and maximal sets into which f can be extended should have a structure determined by the mouths of $\vec{\mathrm{D}}$, that is, if extension is possible at all, it should be possible to make the extension using mouths, as in the proof of Theorem 11.6.

We wish now to discuss the local extension problem in more detail. Let A be a domain with a smooth boundary and suppose that $0 \in \partial A$. We shall assume that the boundary of A is defined near 0 by a C^∞ function φ such that A is defined locally by $\varphi < 0$. We shall assume that $\det(\varphi_{x_i x_j}(0)) \neq 0$; we denote the matrix $(\varphi_{x_i x_j})$ by Φ. We normalize so that $x_1 = 0$ is the tangent plane to A at 0 and, locally, A contains the negative x_1 axis.

Case 1. *A is concave at* 0.

See Fig. 26.

Suppose f is (locally) a C^∞ function on the complement of the closure of A satisfying $\vec{\mathrm{D}}f = 0$ there. Can we extend f to a C^∞ function $\tilde{f}$ on a full

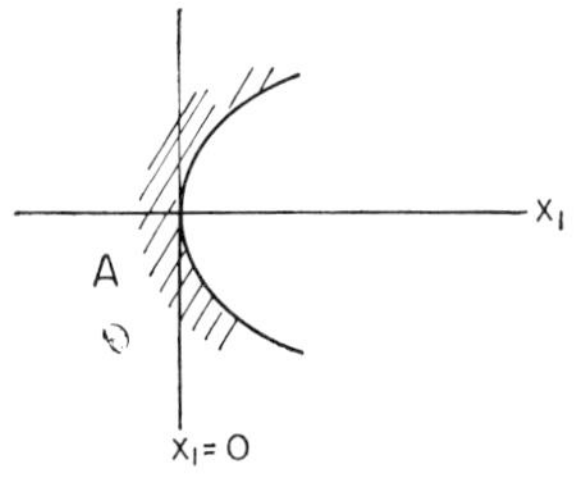

FIGURE 26

neighborhood of 0 satisfying $\vec{D}f = 0$? If this can be done then certainly any f_1 which is C^∞ in $x_1 > 0$ satisfying $\vec{D}f_1 = 0$ there possesses an extension. In Section VIII.4 (see Theorem 8.5 and statement 1 following the definition of quasihyperbolic; actually we need a local analog of the result which is proved in the same manner) we showed that such an extension is possible if and only if $\vec{D}$ is hyperbolic with time variable x_1. Using Proposition 8.6 and the classical theory of hyperbolic operators, (see Gårding [1]) it follows that there exists a fundamental solution e for some hyperbolic operator ∂ in the ideal generated by D_j in the ring of linear constant coefficient partial differential operators such that the support of e is contained in a cone of the form

$$x_1 \geqslant 0, \quad (|x_2| + \cdots + |x_n|) \leqslant a\,|x_1| \tag{11.13}$$

for some $a > 0$.

Conversely, if such a fundamental solution exists then the proof of Theorem 11.6 or the remarks following the proof of Proposition 8.6 show that extension is possible. (There is a change from D_j to D_j' in Theorem 11.6 which is easily seen to be unimportant.) Thus in Case 1, extension is possible if and only if $\vec{D}$ is hyperbolic with time variable x_1. (The uniqueness requirement of Theorem 11.6 holds because of the remarks preceeding the statement of Theorem 11.6.)

Case 2. *A is convex at* 0.
See Fig. 27.

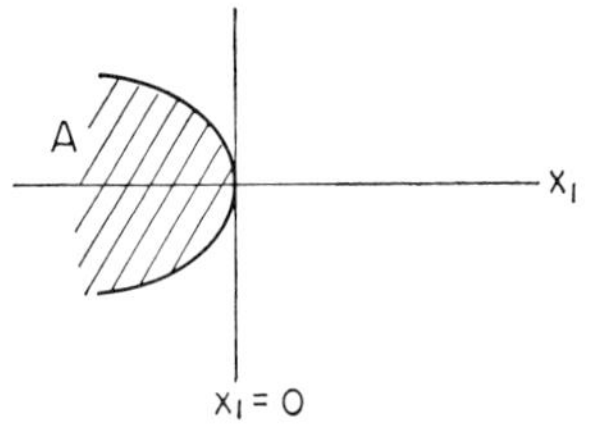

FIGURE 27

In this case, in order to apply Theorem 11.6 we need to know that, for any $a > 0$, $\vec{\mathrm{D}}$ has a fundamental solution whose support lies outside the cone

$$(11.14) \qquad x_1 < 0, \quad (|x_2| + \cdots + |x_n|) < a\,|x_1|.$$

In particular this will be the case if $\vec{\mathrm{D}}$ has a fundamental solution whose support lies in the half-plane $x_1 \leqslant 0$.

Note that if we are not searching for a completely local theory, that is, we consider solutions in some fixed neighborhood of 0 exterior to A then we do not require (11.14) for all a, but only for $a < a_0$, where a_0 depends on the given neighborhood.

If $r > 1$ and the polynomials P_j have no common factor, then a simple algebraic argument shows that for almost all (that is, all except a proper subvariety) complex linear hyperplanes Λ of C^n there exists a nontrivial P_Λ in the ideal generated by the P_j such that P_Λ depends only on the variables on Λ. This means that for almost all real linear hyperplanes Ξ of R^n we can find a nontrivial constant coefficient D_Ξ in the ideal generated by the D_j depending only on the coordinates of Ξ. Given any a we can certainly find a Ξ with this property which does not meet the cone (11.14). Since D_Ξ, thought of as a differential operator on Ξ, has a fundamental solution e_Ξ (see Chapter VI) D_Ξ, thought of as a differential operator on R^n has $e_\Xi \times \delta_{\Xi^\perp}$ as a fundamental solution. Here $\delta_{\Xi^\perp}$ is the δ function in the direction orthogonal to Ξ. Thus, if $r > 1$ and the P_j have no common factor, extension is possible if the uniqueness requirement of Theorem 11.6 holds.

If $r = 1$ or if the D_j have a nontrivial common factor, let D by the greatest common factor of the D_j. If D is hyperbolic with space variable x_1, then (see Case 1 above) there certainly exists a fundamental solution for D with support outside (11.14). The operators D_j/D have no common factor, so for almost all Ξ there exists a fundamental solution for $\vec{\mathrm{D}}/D = (D_1/D, \ldots, D_r/D)$ whose support lies on Ξ.

Now, D is hyperbolic with space variables on $x_1 = 0$. If Ξ is close enough to $x_1 = 0$ then D is also hyperbolic with space variables along Ξ. This can be seen in two ways:

1. As noted in Section VIII.3, the condition for hyperbolicity is equivalent to the existence of a fundamental solution with support in a cone around the time axis. It is clear that this condition is invariant under small variations of the time axis.
2. It is a simple matter to show that Condition (8.6) is preserved under linear transformations close to the identity.

Now, let ∂_Ξ be an operator in the ideal generated by the D_j/D which depends only on the coordinates on Ξ. Call $D_\Xi = D\partial_\Xi$. We claim that, for Ξ close to $x_1 = 0$, D_Ξ has a fundamental solution whose support is contained in the half-plane $\Xi^\perp \geqslant 0$. Here $\Xi^\perp$ is the coordinate on the orthogonal complement to Ξ oriented by continuous transformation from x_1.

Let E_Ξ be a fundamental solution for ∂_Ξ and let E be a fundamental solution for D whose support lies in a cone. Then the solution of

$$D_\Xi e = \delta$$

is given by

$$e = E_\Xi * E.$$

To see this, note first that the convolution $E_\Xi * E$ is well defined because of the nature of the supports of E_Ξ and E. Formally, (see Chapter V, Example 3 for the notational convention)

$$(E_\Xi * E)(x) = \int E_\Xi(x - y)E(y)\, dy.$$

The integration is over a compact set, namely the set where the cone which is the support of E meets the translate of Ξ by x. (Since $x - y \in \Xi$ means $y \in \Xi + x$.) For the same reason, it is clear that the integral vanishes when $\Xi^\perp < 0$. Also we have

$$D_\Xi e = D\partial_\Xi E_\Xi * E = \delta.$$

It is easy to justify the formal arguments and obtain our assertion.

Thus, in case D is hyperbolic then Theorem 11.6 shows there is extension. (Again, the difference between D and D' is unimportant.)

It is possible for D to have a fundamental solution with support outside any (11.14) without D being hyperbolic. For example, D could depend only on the variables on $x_1 = 0$. It is possible to show that D has such fundamental solutions if either

1. D depends only on the variables along $x_1 = 0$.
2. For fixed real $z_2, \ldots, z_n$, the imaginary parts of the zeros of $P(z_1, z_2, \ldots, z_n) = 0$ are bounded.
3. D is hyperbolic with time variables x_1.

By Theorem 8.5 we know that Condition 3 is stronger than Condition 2. (The operator $i\partial/\partial x_1 - \partial^2/\partial x_2^2$ satisfies Condition 2 but not 3.) As noted above, in Case 3 there exist suitable fundamental solutions for $\vec{\mathrm{D}}$. This need no longer be true if Condition 1 or 2 hold, unless there exists a non-trivial operator $D_{x_1=0}$ in the ideal generated by the D_j/D which depends only on the coordinates on $x_1 = 0$.

PROBLEM 11.7. (Exercise)

(a) Using the technique of construction of fundamental solutions given in the proof of Theorem 11.10, show the existence of a fundamental solution for D with support in $x_1 \geqslant 0$ if Condition 2 holds.

(b) Show in Cases 1 and 2 that if there exists a nontrivial $D_{x_1=0}$ as above then $\vec{\mathrm{D}}$ has a fundamental solution with support in $x_1 \leqslant 0$.

(c) Give an example where Condition 1 or 2 holds, no $D_{x_1=0}$ exists, and $\vec{\mathrm{D}}$ does not have a fundamental solution with support outside every region (11.14).

We conclude that if Condition 1, 2, or 3 holds, and if $D_{x_1=0}$ exists in case it is Condition 1 or 2 that holds, and if the uniqueness requirement of Theorem 11.6 is met, then the extension property will hold.

Case 3. *General case.*

We have assumed that the quadratic form $\sum \varphi_{ij}(0)x_i x_j$ is nondegenerate. We perform a change of variables so it is of the form $\Phi(x) = \sum a_i x_i^2$, so that

$$\varphi(x) = x_1 + \sum a_i x_i^2 + \text{higher order terms.}$$

We assume all a_i are $\neq 0$, that $a_1, \ldots, a_p > 0$ and $a_{p+1}, \ldots, a_n < 0$. For $p = 0$, 1 we are in Case 1, for $p = n$ we are in Case 2. We now consider arbitrary p. It should be noted that the linear transformation necessary to diagonalize Φ may modify $\vec{\mathrm{D}}$ somewhat. For example, if $\vec{\mathrm{D}}$ is the Cauchy-Riemann system, then the linear transformation may destroy this fact (unless it is complex linear).

Cases 1 and 2 suggest that we should look for a fundamental solution (actually, for the system $\vec{\mathrm{D}}'$ obtained by replacing each D_j by D_j'; we shall ignore the trivial differences between $\vec{\mathrm{D}}$ and $\vec{\mathrm{D}}'$ as they only lead to differences in sign) whose support lies in a set of a suitable kind. According to Theorem 11.6, "suitable kind" means "meeting A only at 0." It is easily seen that a_1 is unimportant since x_1 dominates $a_1 x_1^2$ locally.

Now, taking Case 2 into account, we see that we might try to find a fundamental solution for $\vec{\mathrm{D}}$ which lies in a linear subvariety, contained in the complement of A, or a halfspace of dimension $\leqslant n$ which is contained in the complement of A. On the other hand, Case 1 suggests that we look for a fundamental solution whose support lies in a suitable cone.

The two viewpoints can be consolidated as follows: For each line Λ through the origin in the hyperplane $x_1 = 0$, consider the plane Λ_1 determined by Λ and the x_1 axis. The intersection of Λ_1 with A is one of three types: [type (c) is only infinitesimal]

(a) A parabolic concave region.
(b) A parabolic convex region.
(c) The half plane $x_1 < 0$.

The three types correspond to the three situations in which $\sum a_i x_i^2$ is

(a′) Positive on Λ.
(b′) Negative on Λ.
(c′) 0 on Λ.

In situations (a) and (b) we can use the arguments in Cases 1 and 2 discussed above in order to obtain conditions on the variety V of common zeros of the P_j so that, restricted to Λ_1, we can find a suitable fundamental solution. Situation (c) is treated in a similar manner. The conditions on the variety V of common zeros of the P_j will be in terms of the intersection of V with the complex subspace Λ_1^C of C which is dual to Λ_1. (Since we have a chosen quadratic form, namely $\sum z_i^2$ in C, we can identify subspaces of C with quotient spaces.)

PROBLEM 11.8 (Conjecture)

(a) Show that if the intersection of V with each Λ_1^C satisfies the conditions required for the existence of a suitable fundamental solution, then $\vec{\mathrm{D}}$ has a fundamental solution with support outside $-A$.
(b) Determine when strong uniqueness holds.

We shall discuss the converse problem at the end of this section.

A word of caution. When discussing Cases 1 and 2 we tried to find conditions for the existence of fundamental solutions whose supports lie in cones or half-planes. Sometimes this may fail, but we can find a fundamental solution whose support lies in a suitable parabolic region.

EXAMPLE. Let $n = 2m$, $\vec{\mathrm{D}} = (\partial/\partial x_1 + i\partial/\partial x_2 \ldots,\ \partial/\partial x_{n-1} + i\partial/\partial x_n)$. The theory of functions of several complex variables (see, e.g., Bers [1]) shows that it is of interest to consider domains for which the *Levi form*

$$\sum \frac{\partial^2 \varphi(0)}{\partial \zeta_i\, \partial \bar{\zeta}_j}\, \lambda_i \bar{\lambda}_j$$

is positive definite on the set of λ satisfying $\sum \varphi_{\zeta_i}(0)\lambda_i = 0$. Here $\partial/\partial\zeta_i = \partial/\partial x_{2i-1} - i\partial/\partial x_{2i}$. Under our normalization for φ this means that

$$\sum_{i=2}^{m} (a_{2i-1} + a_{2i})\, |\lambda_i|^2 > 0,$$

that is,

$$a_{2i-1} + a_{2i} > 0 \quad \text{for} \quad i = 2, \ldots, m.$$

If A has the property that the Levi form is positive definite at each boundary point, then A is called *strongly pseudoconvex*. If, at each boundary point of A, at least q of the eigenvalues of the Levi form are positive then A is called *strongly q pseudoconvex*. (Thus strong $m-1$ pseudoconvexity equals strong pseudoconvexity.)

The importance of the Levi form for us is the following: Suppose A is strongly 1 convex at 0. Then, for the normalization of φ which we have given, $a_{2i-1}+a_{2i}>0$ for some $i=2,\ldots,m$; we suppose this $i=2$, that is, $a_3+a_4>0$. If both a_3 and a_4 are >0 then the ζ_2 plane (that is, the plane determined by the x_3 and x_4 axes) would meet the closure of A locally only at 0. For, on the ζ_2 plane near 0 we have $\varphi(x)=a_3x_3^2+a_4x_4^2>0$ except for $x_3=x_4=0$. Since A is defined locally by $\varphi<0$, the ζ_2 plane meets the closure of A only at 0. Now the operator $\partial/\partial\bar{\zeta}_2$ is one of the components of $\vec{\mathrm{D}}$. Thus if a_3, $a_4>0$ we would have a fundamental solution of the type required by Theorem 11.6.

In case not both a_3, $a_4>0$, suppose $a_4<0$. Now $a_3+a_4>0$, so $a_3>-a_4$. Consider the "parabola"

$$\zeta_1=-a'\zeta_2^2, \quad \zeta_3=\zeta_4=\cdots=\zeta_m=0,$$

where $|a_4|<a'<a_3$. On the parabola we have,

$$x_1=-a'(x_3^2-x_4^2), \quad x_2=-2a'x_3x_4.$$

Thus, up to second order terms, on the parabola,

$$\varphi(x)=a'(x_4^2-x_3^2)+a_3x_3^2+a_4x_4^2.$$

Hence, since $|a_4|<a'<a_3$, we have $\varphi(x)>0$ if $x\neq 0$. Thus the parabola meets the closure of A only at 0.

Since the parabola is a complex analytic curve, it is readily verified that there exists an operator (namely, complex differentiation on this curve) which is a linear combination of the $\partial/\partial\bar{\zeta}_j$ with complex analytic coefficients and has a fundamental solution with support on the parabola. Thus the parabola is a mouth so the continuation property of Theorem 11.6 holds.

There is no difficulty in providing a similar treatment in case φ is not in the given normal form.

If should be noted that, in this example, we are using a parabolic curve $\zeta_1=-a'\zeta_2^2$ instead of the linear varieties considered in Problem 11.8.

Theorem 11.5 and Remark 11.3 show that a truly overdetermined system is characterized by the property that there do not exist distributions which are solutions except at one point. For the Cauchy-Riemann equations for holomorphic functions in several complex variables much more is true: Singularities on any variety of real codimension >2 are removable, and only on complex analytic subvarieties of real codimension 2 can we have nonremovable singularities.

If we note that complex analytic curves (that is, complex dimension one) are mouths for the Cauchy-Riemann system, we see that the reason we cannot apply Theorem 11.6 to remove singularities from a complex analytic subvariety of complex codimension one is the *preservation of analytic intersection number*: If in C^n we are given two (local) complex analytic varieties V_1, V_2 of complementary dimension which meet at a point, they cannot be separated by a small deformation. (In fact, the definition of algebraic intersection multiplicity of Weil [1] is based essentially on this principle.) This leads us to make the following

DEFINITION. Let M be a local manifold in $R = R^n$ through 0. We say that M has the property of *preservation of* $\vec{\mathrm{D}}$ *intersection* if, whenever L is a mouth for $\vec{\mathrm{D}}$ which meets M only at 0, every small translate of L meets M near 0. (See Fig. 28.)

Theorem 11.6 (and the remarks on uniqueness preceeding it) then implies immediately:

THEOREM 11.7. *Let M be a local manifold passing through 0. If M does not have the property of preservation of $\vec{\mathrm{D}}$ intersection, then every C^∞ function f, defined in a connected component N of a neighborhood of 0 in the complement of M and satisfying $\vec{\mathrm{D}}f = 0$ on N, possesses an extension g to a full neighborhood $\tilde{N}$ of 0 which satisfies $\vec{\mathrm{D}}g = 0$ on $\tilde{N}$.*

COROLLARY. *Suppose there is an operator ∂ in the ideal generated by the D_j which depends only on the variables of a linear variety L of codimension $s \geqslant 1$. Then every local manifold M of dimension $< s$ which has the property that some translate of M meets L in a single point is removable.*

See Remark 11.7.

All above results are in the direction of proving that certain manifolds are removable. To state results in the converse direction, let us write $x = (t, \xi)$ with dual variables $z = (s, w)$. (Here t and ξ may be several variables.)

THEOREM 11.8. *Suppose for some fixed w_0 the $P_j(s, w_0)$ have a common factor and are not identically zero. Assume that the following uniqueness property holds for $\vec{\mathrm{D}}$: If S is a distribution such that $S = 0$ on $|t| > a$ and*

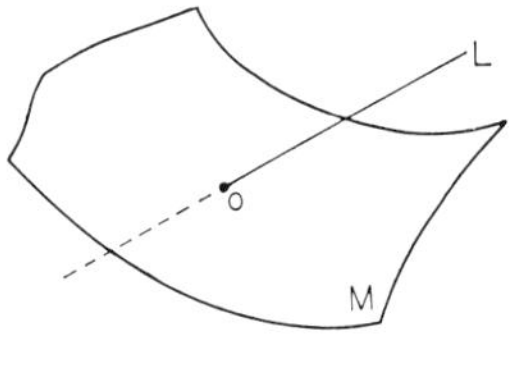

FIGURE 28

$\vec{\mathrm{D}}S = 0$ on $|t| > a_0$ (with $a_0 < a$), then $S = 0$ on $|t| > a_0$. Suppose also that for some i the operator D_i, when written as a sum of derivatives, contains a nontrivial term involving t derivatives only whose order is equal to the order of D_i in t. Then there is a distribution T of the form $T^0(t)\exp(i\xi \cdot w_0)$ such that $\vec{\mathrm{D}}T = 0$ off $t = 0$ but $t = 0$ is not removable, that is, for no $b > 0$ does there exist a distribution $\tilde{T}$ on R with $\vec{\mathrm{D}}\tilde{T} = 0$ on R and $\tilde{T} = T$ on $|t| > b$.

PROOF. Let $P(s)$ be a common factor of the $P_j(s, w_0)$. Let ∂' be the Fourier transform of P and let $T^0(t)$ be a fundamental solution for ∂. Call $T(t, \xi) = T^0(t)\exp(i\xi \cdot w_0)$. Then for each j (see Chapter V, Example 3 for notations)

$$\begin{aligned} D_j T &= D_j T^0 \exp(iw_0 \cdot \quad) \\ &= D_j^1(w_0)T \end{aligned}$$

where $D_j^1(w_0)$ is a linear partial differential operator in t, the Fourier transform in t of its adjoint being $P_j(s, w_0)$. Since $D_j^1(w_0)$ is of the form $D_j^2\partial$, where D_j^2 is a linear coefficient operator in t, we have

$$D_j T = D_j^2\, \partial T = D_j^2\, \delta(t) \times \exp(iw_0 \cdot \quad)$$

which has its support on $t = 0$ and is not $\equiv 0$ for some j.

We claim $t = 0$ is not removable. For if there were a $\tilde{T}$ with $\vec{\mathrm{D}}\tilde{T} = 0$ on R and $\tilde{T} = T$ on $|t| > b$ we would have

$$D_j(T - \tilde{T}) = D_j T = D_j^2\, \delta(t) \times \exp(iw_0 \cdot \quad) \tag{11.15}$$

for all j and

$$T - \tilde{T} = 0 \quad \text{for } |t| > b. \tag{11.16}$$

By hypothesis this implies that

$$T - \tilde{T} = 0 \quad \text{for } |t| > 0. \tag{11.17}$$

Since $T - \tilde{T}$ has its support on $t = 0$, we see that in a neighborhood of the origin, $T - \tilde{T}$ is a sum of terms of the form $w \times w_1$, where w_1 is a distribution near the origin on $t = 0$ and w is a distribution in t whose support is the origin {that is, a finite sum of derivatives of $\delta_{t=0}$ (see Schwartz [1])}. Thus, $D_i(T - \tilde{T})$ will be a sum of the same form. But by our assumptions, this sum will contain a nonzero term of the form $\tilde{\partial}\delta_{t=0} \times w_2$, where $\tilde{\partial}$ is a differentiation in t of order equal to the maximum order of w plus the order of D_i in t. This order is $\geqslant$ order of D_i in t. Note, however, that the order of D_j^2 in t is less than the order of D_j in t. This contradicts (11.15) for $j = i$ so our assertion is established. ∎

Remark 11.8. If $\vec{\mathrm{D}}$ consists of the single operator $\partial/\partial t - \partial^2/\partial\xi^2$, then in Chapter VIII (see the beginning of Section VIII.4) we showed that $t = 0$ is removable in the sense of Theorem 11.8. Thus, the unique continuation hypothesis in Theorem 11.8 is essential.

Even if the hypothesis of the Corollary to Theorem 11.7 does not apply to a linear variety T, it is possible that "mild" singularities on T may be removable. The prototype of this is the classical theorem of Riemann that a function f of a single complex variable x which is holomorphic in the neighborhood N of a point x_0 except possibly at x_0 but which is bounded at x_0 can be redefined at x_0 so as to be holomorphic there. The situation is studied in detail in Bochner and Martin [1]. Our methods allow us to treat many cases but since cur results are incomplete we shall omit them.

The Corollary to Theorem 11.7 shows that varieties of small dimension are removable for singularities of solutions of $\vec{D}f = 0$. Thus there exists a minimal dimension for nonremovable singularities. It seems natural to call the distributions f with $\vec{D}f = 0$ except possibly on a variety of minimal dimension the *meromorphic solutions of* $\vec{D}$. In case $r = 1$ (that is, $\vec{D}$ consists of single operator) we see that for any open relatively compact set $\overline{\Omega}$ the meromorphic solutions with singularities off $\overline{\Omega}$ are dense (in a suitable topology) in the solutions of $Df = 0$ on $\overline{\Omega}$. For if g is a C^∞ function on R which vanishes outside Ω and if g is orthogonal to all meromorphic solutions, then g is orthogonal to all solutions in the whole space (entire solutions). In particular, g is orthogonal to exponential polynomial solutions. It follows by Fourier transform that $g = D'k$, where $k \in \mathscr{D}$. Moreover, if $g \cdot h_a = 0$ whenever $Dh_a = \delta_a$, $a \notin \overline{\Omega}$, we have

$$\begin{aligned} k \cdot \delta_a &= k \cdot Dh_a \\ &= D'k \cdot h_a \\ &= g \cdot h_a \\ &= 0. \end{aligned} \tag{11.18}$$

Thus, k vanishes outside Ω. By duality this gives

THEOREM 11.9. *Let $r = 1$ and let Ω be any relatively compact open set with a smooth boundary. The meromorphic solutions of D with singularities off $\overline{\Omega}$ are dense in the topology of $\mathscr{D}'(\overline{\Omega})$ in the solutions on $\overline{\Omega}$ of $Df = 0$.*

In case D is analytic elliptic, we do not need all meromorphic solutions with singularities off $\overline{\Omega}$. It suffices, by the above proof and unique continuation, to allow singularities at one point of each component of $R - \overline{\Omega}$. This result (and its proof) is due to Malgrange [1].

It would be of interest to extend the above result to $r > 1$. Some condition on Ω is necessary as the case of holomorphic functions indicates. Our conjecture (assuming Ω is maximal in the sense that there is no domain containing Ω into which all solutions can be continued) is: For any $x \notin \Omega$ there exists a variety L which is the singular set of a meromorphic solution of D which passes through x and does not meet Ω. An example of Wermer [1] shows that even for elliptic systems it is not sufficient to know that in each component of the complement of $\overline{\Omega}$ there exists a singular set of meromorphic solution. (Wermer does not show this completely, but H. Rossi has pointed out that it follows from Wermer's results.) Our conjecture can be verified for holomorphic functions.

In this connection it is of interest to note that the proof of the converse Theorem 11.8 gives a method of constructing some meromorphic solutions of $\vec{\mathrm{D}}$, namely, distributions of the form $T^0 \exp(iw_0)$ and hence suitable integrals of such over the variety of possible w_0.

See Problem 11.9.

There is another class of "special" solutions of $\vec{\mathrm{D}}$ which one might use as an alternative definition of "meromorphic solutions." To understand this, consider first the case $r = 1$. A fundamental solution f of D can be expressed formally by

$$f(x) = \int_{z \text{ real}} \exp(ix \cdot z)\, dz/P(z). \tag{11.19}$$

If we formally shift the contour in the integral, we can use the residue calculus to express f in terms of integrals on $\mathfrak{V}$. The expressions for $f(x)$ may differ for different x. For example, if all $x_j > 0$ we would shift the contour to $\operatorname{Im} z_1 = \infty, \ldots, \operatorname{Im} z_n = \infty$ while if all $x_j < 0$ we shift to $\operatorname{Im} z_1 = -\infty, \ldots, \operatorname{Im} z_n = -\infty$.

Thus a fundamental solution of D is expressed in terms of the residues on $\mathfrak{V}$ of a meromorphic function whose denominator is P. For $r > 1$, we might consider functions expressed in terms of a suitable type of residue on $\mathfrak{V}$ of meromorphic "objects" whose "polar set" is suitably restricted. Giving a meaning to this "residue calculus" seems to be of interest in itself; we shall outline an approach.

To understand the idea, let us consider first the case $n = 1$. Let I be the ideal generated by z^2 so it corresponds to the multiplicity variety $\mathfrak{V} = (0, \text{identity}; 0, d/dz)$. Let f be a meromorphic function having a pole at the origin. Then we define the residue of f modulo I as follows: Let $(z^2)^m f$ be regular at 0. Then

$$\operatorname{res}_I f = \left[\frac{1}{(2m-2)!}\,\frac{d^{2m-2}(z^{2m}f)}{dz^{2m-2}}(0),\ \frac{1}{(2m-1)!}\,\frac{d^{2m-1}(z^{2m}f)}{dz^{2m-1}}(0)\right]. \tag{11.20}$$

Note that this definition does not depend on m.

For $n > 0$, let I be a polynomial ideal; let P be an element of I, and let $\mathfrak{B}$ be a multiplicity variety corresponding to I. Let f be a meromorphic function having the property that Pf is entire. Then for a suitable differential operator δ with polynomial coefficients we define

$$\text{res}_I f = \delta(Pf)\,|_{\mathfrak{B}}\,.$$

For example, let $n = 2$, let I be the ideal with generators (z_1, z_2) so $\mathfrak{B} = \{0\}$, let $P = z_1 z_2$ and let $\delta =$ identity. For any f having simple poles on $\{z_1 = 0\} \cup \{z_2 = 0\}$ we have

$$\text{res}_I f = (z_1 z_2\, f)(0).$$

This is the classical formulation of the residue calculus (see Leray [1]).

Instead of considering meromorphic functions f we could consider vectors $\vec{\mathrm{f}}$ of meromorphic functions whose common polar set is suitably restricted. It would take us too far afield to enter into a detailed discussion of this.

Leray has pointed out another method of constructing solutions of equations which have singularities on low dimensional sets. Let $F(x, w)$ be a rational function of x and w. Let γ be a cycle in w space. Consider

$$J(x) = \int_\gamma F(x, w)\, dw.$$

By direct calculation, one can often show that J satisfies $\vec{\mathrm{D}}J = 0$ for a suitable $\vec{\mathrm{D}}$ depending on F. Moreover, many properties of the singularities of J can be determined from the integral expression for J and from the fact that $\vec{\mathrm{D}}J = 0$. (It should be noted that J is analytic in "most" of the complex x space so, by classical theory, its singularities must be contained in the bicharacteristics of each D_j. Moreover, the form of the integral places other restrictions on the possible positions of singularities of J.)

In Section VI.2 we introduced the concept of semifundamental solution. As explained there, the semifundamental solution plays a similar role for the case $l > 1$ as the fundamental solution does for $l = 1$. In case $l > 1$ we define a mouth for $\boxed{\mathrm{D}}$ at p to be a local manifold L with boundary containing p such that there are solutions $\vec{\mathrm{e}}_\lambda$ of $\boxed{\mathrm{D}}'\,\vec{\mathrm{e}}_\lambda = 0$ near p with $\vec{\mathrm{e}}_\lambda \in \mathscr{E}^r$ and support $\vec{\mathrm{e}}_\lambda \subset L_\lambda$. The L_λ are arbitrarily close to L. We require further that for any small neighborhood N_1 of p there is a function $\beta \in \mathscr{D}$ which is near one p and zero outside N_1 so that the linear combinations of $\{(\beta\vec{\mathrm{e}}_\lambda)\}$ are dense in $\mathscr{E}'^r$ near p. There might be some confusion in case

$l = 1$ as to which concept of mouth we are using. However, it will always be clear from the context.

We can now extend Theorem 11.6 to the case $l > 1$. The uniqueness requirement is replaced by a density requirement on linear combinations of suitable $\beta\vec{e}_\lambda$. The proof of the result for $l > 1$ goes along the lines of the proof of Theorem 11.6 except that we use the method of extension by means of the semifundamental solution. There are three points to notice:

1. We are in a "semilocal" situation rather than the global situation of Section VI.2. Thus, as in the proof of Theorem 11.6, care must be taken as to which translates can be used and that all expressions are defined.
2. In (6.24) (or rather, its generalization from curl to arbitrary $\boxed{\mathrm{D}}$) we used the fact that $\vec{\mathrm{g}}$ satisfies the compatibility conditions on all of R, or, at least, on a convex set containing the set on which we want to define $\vec{\tilde{\mathrm{f}}}$. Note that in our case $\vec{\mathrm{g}} = 0$ so the compatibility holds on all of R; we do *not* use the fact that $\vec{\tilde{\mathrm{f}}}$ is defined on all of R.
3. We are trying to solve

$$\boxed{\mathrm{D}}\vec{\tilde{\mathrm{f}}} = 0, \quad \vec{\tilde{\mathrm{f}}} = \vec{\mathrm{f}} \quad \text{on} \quad N \cup N''.$$

   Let μ be a cube containing the supports of all $(\beta\vec{e}_\lambda)$. According to Section VI.2 we know that $\vec{\tilde{\mathrm{f}}}$ must take the value 0 on the set M. By the above, $\vec{\tilde{\mathrm{f}}} = \vec{\mathrm{f}}$ on the set $\{\delta_x\}_{x \in N \cup N''}$. These conditions might conflict if some δ_x, $x \in N \cup N''$ is the limit of elements of M. However, it is easily seen from the definitions that there is no possible conflict if the distance from x to the boundary of N' is greater than twice the diameter of μ. This accounts for the fact that we can only hope to make $\vec{\tilde{\mathrm{f}}} = \vec{\mathrm{f}}$ on $\tilde{N} \cap (N \cup N'')$ rather than on $\tilde{N} \cap (N \cup N')$.

PROBLEM 11.10 (Exercise)

Carry out the details. Extend Theorem 11.5* to the present situation.

We turn now to another question about singularities. This had its origin in the result discovered by John [3] and by Malgrange [2] (for $r = 1$): Let ψ be the convex hull of a strictly convex hypersurface λ. (That is, λ is the part of the boundary of ψ at which ψ is strictly convex.) Let f be a distribution on ψ which is an indefinitely differentiable function in the neighborhood of λ. Suppose $Df = 0$. Then $f \in \mathscr{E}(\psi)$. This result is similar to Holmgren's uniqueness theorem discussed in Section

IX.9, except that there "indefinitely differentiable function" is replaced by "holomorphic function" and so it has been called the "unique continuation of singularities." (See also Hörmander [3].)

However, Holmgren's theorem applies to the case when ψ is a half-space, and when λ is a hyperplane. We wish to derive the analog of the unique continuation of singularities for this case. We do not assume $r=1$. We use variables (t, ξ) in R^n with t a variable in R^1. As usual the dual variables are (s, w).

THEOREM 11.10. *A sufficient condition that any $f \in \mathscr{D}'$ which satisfies $\vec{\mathrm{D}}f = 0$ and $f \in \mathscr{E}(t<0)$ should be in $\mathscr{E}$ is: Call V the variety of common zeros of the P_j. We can write $V = V_1 \cup V_2$ where, for $(s, w) \in V_1$ we have*

$$|\mathrm{Im}\ s| \leqslant c(1 + |\mathrm{Im}\ w| + \log(1 + |s| + |w|)) \tag{11.21}$$

for some c, while for $(s, w) \in V_2$ we have for any c'

$$|\mathrm{Im}\ s| + |\mathrm{Im}\ w| \geqslant c' \log(1 + |s| + |w|) \tag{11.22}$$

whenever $|s| + |w|$ is sufficiently large (depending on c').

Our conditions mean that V can be split into a "hyperbolic" and an "elliptic" part (see Chapter VIII). Note, however, that we may not be able to write V as a union $V = V_1 \cup V_2$ with V_1, V_2 as above, where V_1 and V_2 are *algebraic varieties*. This is easily seen in case $r=1$, $n=2$, $(\partial^2/\partial t^2 + \partial^2/\partial \xi^2)(\partial^2/\partial t^2 - \partial^2/\partial \xi^2) + 1$ which satisfies the hypothesis of Theorem 11.10 but for which V is irreducible.

PROOF. First we claim that it is sufficient to prove that for any $d>0$ there exists a parametrix which is a d times differentiable function outside a forward cone. By this we mean that there exist $a', a'' > 1$ independent of d, and distributions $e_j \in \mathscr{D}'$ depending on d so that

$$\sum D_j e_j = \delta + \zeta. \tag{11.23}$$

The e_j are C^d functions outside of

$$t \geqslant |\xi|/a'' - a' \tag{11.24}$$

and $\zeta \in \mathscr{E}$. Although the e_j depend on d, they are distributions of uniformly bounded order on any compact set K of R, that is, on K we can write

$$e_j = \sum D_{jl}\, \mu_l,$$

where μ_l are measures on K and D_{jl} are differential operators with constant coefficients whose orders are bounded independently of d (although the bound may depend on K).

That this implies the sufficiency of our condition is seen as follows: Let h be an indefinitely differentiable function of compact support such that $h = 1$ on $\|x\| \leqslant a$ and the support of h is contained in $\|x\| \leqslant a + 1$. Then $\sum D_j he_j = \delta + \zeta$ on $\|x\| \leqslant a$ and $\sum D_j he_j \in \mathscr{E}'$. Moreover, we can write $\sum D_j he_j = \delta + \zeta_1$, where $\zeta_1 \in \mathscr{E}'$ is a d times differentiable function on the union of $\|x\| \leqslant a$ with the exterior of $t \geqslant |\xi|/a'' - a'$. In fact, the only place where ζ_1 may not be a d times differentiable function is the intersection of $t \geqslant |\xi|/a'' - a'$ with the shell $a \leqslant \|x\| \leqslant a + 1$.

Thus, we may write $\zeta_1 = \zeta_2 + \zeta_3$, where ζ_2 is a d times differentiable function whose support is contained in $\|x\| \leqslant a + 1$ and the support of ζ_3 is contained in the intersection of $t \geqslant |\xi|/a'' - a'$ with $a - 1 \leqslant \|x\| \leqslant a + 1$. (Actually, a' may have to be increased slightly for this to be true.) Suppose $a > 2a' + 1$. A point (t, ξ) in the support of ζ_3 satisfies

$$t \geqslant -a' + (a-1)/2na''.$$

This is true if $|\xi| \geqslant (a-1)/2n$ since $t \geqslant |\xi|/a'' - a'$. If not, then since $\|x\| \geqslant a - 1$, we must have $|t| \geqslant (2n-1)(a-1)/2n$. Thus either

$$t \geqslant (2n-1)(a-1)/2n \geqslant (a-1)/2n \geqslant -a' + (a-1)/2na''$$

or

$$t \leqslant -(2n-1)(a-1)/2n < -a'.$$

The latter is impossible since $t \geqslant -a'$. Thus our assertion is established. (See Fig. 29).

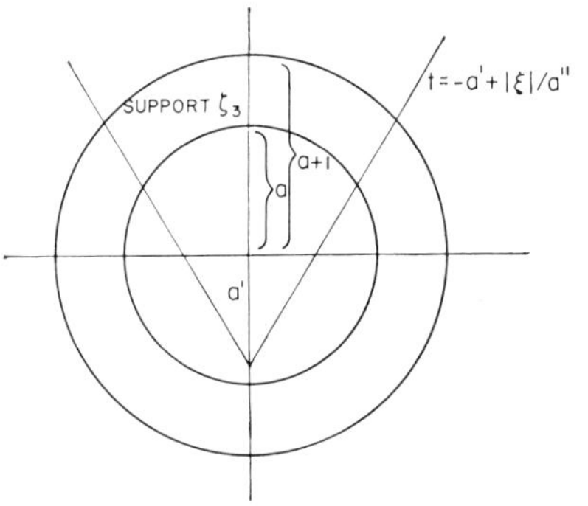

FIGURE 29

Suppose $f \in \mathscr{D}'$, $\vec{\mathrm{D}}f = 0$, and $f \in \mathscr{E}(t < 0)$. Then (see our conventions regarding convolution in Chapter V, Example 3)

$$f = f * \delta = -f * \zeta_1$$
$$= -f * \zeta_2 - f * \zeta_3 .$$

Let b be a positive integer. Suppose we want to prove that f is a b times differentiable function on $\|x\| < b$. Note that if $a > b + 1$ and if $b + a' - (a-1)/2na'' < 0$ then, by our above remarks on the t coordinates of the points in the support of ζ_3, the values of $f * \zeta_3$ on $\|x\| \leqslant b$ depend only on the values of f in $t < 0$. Thus, $f * \zeta_3$ is an indefinitely differentiable function on $\|x\| < b$. On the other hand, since the support of ζ_2 is contained in $\|x\| \leqslant a + 1$, the values of $f * \zeta_2$ on $\|x\| < b$ will depend only on the values of f on $\|x\| < a + b + 1$. Since f is a distribution, it will be of finite order m on $\|x\| < a + b + 1$, that is, we can write f there in the form $\sum \partial_i \mu_i$, where μ_i are measures and ∂_i differentiations of order $\leqslant m$. Thus $f * \zeta_2$ will be a function which is $d - m$ times differentiable on $\|x\| < b$, which gives our assertion that f is a b times differentiable function on $\|x\| < b$ if $d = m + b$. Note that a' and a'' are independent of d so a can be chosen to depend only on b and not on d; thus, m depends only on b and not on d.

The same argument shows that if, instead of assuming the e_j are d times differentiable functions outside of the cone, we assume the e_j are d times differentiable in ξ outside of the cone, then we would conclude that f is indefinitely differentiable in ξ in all of R. By this we mean that, near every point p outside of the cone, there is an $m > 0$ so that if ψ is a continuous function whose support is in a small set containing p, and ψ is an m times differentiable function of t, then $\partial e_j * \psi$ is continuous near p whenever ∂ is a constant coefficient differential operator in the ξ variables of order $\leqslant d$. To state this in another way, each ∂e_j with ∂ as before can be expressed near p in the form

$$\partial e_j = \sum D_{jl} \mu_l ,$$

where μ_l are measures and D_{jl} are differential operators in t of order $\leqslant m$.

Note that the conditions given in (11.21) and (11.22) are invariant under small real rotations of R. That is, if we apply a small rotation ρ to x and let it act on z by adjoint action ρ', then the new $s' = \rho' s$, $w' = \rho' w$ on V_1 and V_2 satisfy (11.21) and (11.22) (with slightly different constants). More precisely, if we think of ρ' as a matrix close to the identity, then s' is the first component of $\rho'(s, w)$ and w' are the other components.

It is clear that (11.22), for (s', w') replacing (s, w), is satisfied for $(s', w') \in V_2$. To verify the corresponding statement for V_1, let $(s', w') \in V_1$, say

$(s', w') = \rho'(s, w)$. Then, since ρ' is real,

$$|\text{Im } s'| = |\rho'(\text{Im } s, \text{Im } w)| \leqslant (1+\varepsilon)|\text{Im } s| + \varepsilon|\text{Im } w|$$

since ρ' is close to the identity. Similarly,

$$\begin{aligned} |\text{Im } w| &\leqslant \varepsilon|\text{Im } s'| + (1+\varepsilon)|\text{Im } w'| \\ |w| &\leqslant \varepsilon|s'| + (1+\varepsilon)|w'| \\ |s| &\leqslant (1+\varepsilon)|s'| + \varepsilon|w'|. \end{aligned}$$

Combining this with (11.21) gives

$$\begin{aligned} |\text{Im } s'| &\leqslant (1+2\varepsilon)c(1 + |\text{Im } w| + \log(1+|s|+|w|)) \\ &\leqslant (1+2\varepsilon)(1+2\varepsilon)c(1+|\text{Im } w'| + \log(1+|s'|+|w'|)) \\ &\quad + (1+2\varepsilon)\varepsilon c|\text{Im } s'|. \end{aligned}$$

As long as

$$\varepsilon(1+2\varepsilon)c < 1$$

we have an inequality like (11.21) with c replaced by

$$c(1+2\varepsilon)^2[1 - \varepsilon c(1+2\varepsilon)]^{-1}.$$

It is not difficult to show that if f is indefinitely differentiable in $\rho(\xi)$ for all ρ close to the identity then $f \in \mathscr{E}(R)$ (see Schwartz [1]). Thus it remains to construct e_j which are d times differentiable functions in ξ outside a cone. For then the same argument as that given above would show that f is indefinitely differentiable in $\rho(\xi)$ for all ρ close to the identity; hence $f \in \mathscr{E}$.

To construct e_j we suppose first that $r = 1$. In this case we write e for e_1. The idea of the proof is as follows: We shall construct e by writing

$$e(x) = \int_{\Gamma_1} \exp(ix \cdot z)\, dz/P(z), \tag{11.25}$$

where Γ_1 lies in the part of C slightly away from that defined by (11.21). Γ_1 will be chosen so that $|P|$ is not too small and so that we can use Cauchy's formula to shift the contour from Γ_1 to $\Gamma_2(x)$, where again $|P|$ is not too small on $\Gamma_2(x)$ but $\Gamma_2(x)$ lies in the part of C close to that satisfying (11.22). Using the fact that, for suitable x, $\exp(ix \cdot z)$ and its x derivatives are small on $\Gamma_2(x)$, we shall be able to deduce the required regularity of e.

It is important to understand that we start with the integral over Γ_1 in order to know that the integral defines a distribution. For this reason we need to know that Γ_1 is close to the real axis. The Paley-Wiener-Schwartz theorem (see Chapter V, Example 3) and (11.21) show that the integral defining e will converge in the space of distributions. However, $\Gamma_2(x)$ depends on x. For any x_0 the integral $\int_{\Gamma_2(x_0)} \exp(ix \cdot z)\, dz/P(z)$

will still converge in the space $\mathscr{D}'$ but, *a priori*, as c' increases this distribution will become more and more irregular for certain x. For this reason we must start with a fixed Γ_1 which explains why we need the splitting of V into V_1 and V_2.

We give now the details of the proof. We shall suppose for simplicity that ξ(and hence w) is a single variable; the general case is handled by a simple modification. From the hypothesis (11.21) it follows that the curves Γ'_w in the complex s plane defined (for w complex) by

$$\text{(11.26)} \qquad \operatorname{Im} s = -4(c+1)(1+|\operatorname{Im} w| + \log(1+|s|+|w|))$$

will have the property that (Γ'_w, w) will be at a distance at least 1 from V_1. Moreover, by (11.22) for any $c' > 0$, if

$$\text{(11.27)} \qquad |\operatorname{Im} w| \leqslant (c')(1+\log(1+|w|)),$$

then, except for a bounded set of s, w (which may depend on c'), the curves Γ''_w defined by

$$\text{(11.28)} \qquad \operatorname{Im} s = -4(c')(1+|\operatorname{Im} w| + \log(1+|s|+|w|))$$

will be such that (Γ''_w, w) will be at a distance at least 1 from V_2. If (11.27) holds then, except for a bounded set of s, w, both (Γ'_w, w) and (Γ''_w, w) will be at a distance at least 1 from all of V.

Let us suppose that for fixed c' the exceptional bounded set of s, w is contained in $|s| \leqslant c''$, $|w| \leqslant c''$. Now if c' is large enough, there will exist fixed points s_0, s_1 with $\operatorname{Re} s_0 = c''$, $\operatorname{Re} s_1 = -c''$ which lie between Γ'_w and Γ''_w for all w with $|w| \leqslant c''$. We define the curves γ'_w, γ''_w for $|w| \geqslant c''$ to be just the curves Γ'_w, Γ''_w, respectively. For $|w| < c''$ they are defined to lie below the real s axis, except for $|\operatorname{Re} s| < c''$, and to be obtained by joining s_0 and s_1 to Γ'_w and Γ''_w, respectively, by vertical lines and then continuing along Γ'_w and Γ''_w, respectively (see Fig. 30).

We define e by

$$\text{(11.29)} \qquad e(t, \xi) = \int_{w \text{ real}} e^{i\xi w}\, dw \int_{\gamma'_w} e^{its}\, ds/P(s, w).$$

We claim that the integral is convergent in the space of distributions. First observe that by the method of proof of Theorem 1.4 there is a $b > 0$ such that $|P(s, w)| \geqslant (b+|s|+|w|)^{-b}$ whenever (s, w) is at a distance $\geqslant 1$ from V. In particular, this inequality applies to those values of (s, w) that appear in the integral in (11.29). Thus, by the Paley-Wiener-Schwartz theorem (see Chapter V, Example 3) the map

$$g \to \int_{w \text{ real}} dw \int_{\gamma'_w} G(s, w)\, ds/P(s, w),$$

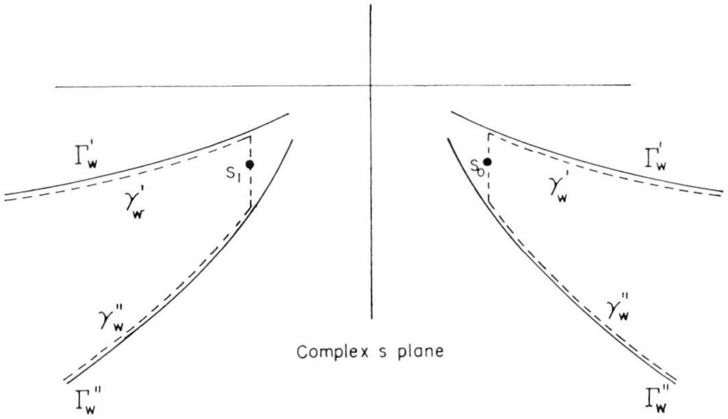

FIGURE 30

where G is the Fourier transform of $g \in \mathscr{D}$, is well defined on $\mathscr{D}$ and will be continuous on $\mathscr{D}$ by Lemma 5.8 (since by Schwartz [1] sequential continuity of linear functions on $\mathscr{D}$ implies they are in $\mathscr{D}'$). By definition this means that the integral converges in the topology of $\mathscr{D}'$.

Note that e depends on c' because c'' does, so we could write more precisely $e^{c'}$. The above argument shows that all the $e^{c'}$ are of bounded order on any compact set of R.

We write $e = e^1 + e^2$, where

$$e^1 = \int_{\substack{w \text{ real} \\ |w| \geqslant c''}} e^{i\xi w}\, dw \int_{\gamma'_w} e^{its}\, ds/P(s, w)$$

$$e^2 = \int_{\substack{w \text{ real} \\ |w| \leqslant c''}} e^{i\xi w}\, dw \int_{\gamma'_w} e^{its}\, ds/P(s, w).$$

We shall discuss e^1 only, as e^2 can be treated in a similar but simpler fashion.

We want to shift the contour into the part of complex w, s space, where Im w, Im s are large so as to be able to use the fact that exp $(ix \cdot z)$ is small for suitable x if Im w, Im s are large. To do this we note that for any fixed w_0 satisfying (11.27) and $|w_0| \geqslant c''$, for any w_1 close to w_0 the integrals

$$\int_{\gamma'_w} e^{its}\, ds/P(s, w) \tag{11.30}$$

are independent of w_1 since by the continuity of the roots of a polynomial there will be no zeros of P between γ'_{w_1} and γ'_{w_0}. [This also follows from the definitions and (11.21) and (11.22).] This means that

$$\int_{\gamma'_w} e^{its}\, ds/P(s, w) \tag{11.31}$$

depends holomorphically on w. We want to use Cauchy's formula to write

$$e^1(t, \xi) = \int_{\gamma_\pm} e^{i\xi w}\, dw \int_{\gamma'_w} e^{its}\, ds/P(s, w), \tag{11.32}$$

$$e^{\,1}(t, \xi) = \int_{\gamma_\pm} e^{i\xi w}\, dw \int_{\gamma''_w} e^{its}\, ds/P(s, w), \tag{11.33}$$

where $\gamma_\pm$ is a curve lying over the real w axis except for $|w| < c''$; $\gamma_\pm$ has endpoints at $w = \pm c''$ (the endpoints are the same for both γ_+ and γ_-); for w large, $\gamma_\pm$ is of the form

$$\operatorname{Im} w = \pm(c'/4)(1 + \log (1 + |w|)); \tag{11.34}$$

all $w \in \gamma_\pm$ satisfy (11.27); and the length of arc on $\gamma_\pm$ over any segment of the real axis of length l is $\leqslant 2(l + 1)^2$. Exactly as in the argument following (11.29) we see that the integrals on the right side of (11.32) and (11.33) are convergent in the space of distributions.

We construct $\gamma_\pm$ as follows: (We shall show how to construct γ_+, as the construction of γ_- is similar.) For w_0 real and large, we try to use Cauchy's formula to replace the integral

$$\int_{\substack{w \text{ real} \\ |w - w_0| \leqslant 1}} e^{i\xi w}\, dw \int_{\gamma'_w} e^{its}\, ds/P(s, w)$$

by

$$\int_{\lambda_{w_0}} e^{iw\xi}\, dw \int_{\gamma'_w} e^{its}\, ds/P(s, w).$$

Here λ_{w_0} is the curve consisting of the part of (11.34) lying above w real, $|w - w_0| \leqslant 1$ together with the vertical segments joining the endpoints of this arc to the real axis.

Denote by Λ_{w_0} the area enclosed by λ_{w_0} and the part of the real w axis lying between $w_0 - 1$ and $w_0 + 1$. Suppose we can show that for any $w_1 \in \Lambda_{w_0}$ for all points s lying between γ'_{w_1} and γ'_{w_0}, the point (s, w) is at a distance at least 1 from V. By the reasoning of the proof of Theorem 1.4 it follows that $|P(s, w_1)| \geqslant (b + |s| + |w_1|)^{-b}$ for some b for all such (s, w). Hence, by the Paley-Wiener-Schwartz theorem (see Chapter V,

Example 3) for any $f \in \mathscr{D}$ we can apply Cauchy's formula to deduce

$$\int_{\substack{w \text{ real} \\ |w-w_0| \leqslant 1}} dw \int_{\gamma'_w} F(s, w)\, ds/P(s, w) = \int_{\substack{w \text{ real} \\ |w-w_0| \leqslant 1}} dw \int_{\gamma'_{w_0}} F(s, w)\, ds/P(s, w)$$

$$= \int_{\lambda_{w_0}} dw \int_{\gamma'_{w_0}} F(s, w)\, ds/P(s, w)$$

$$= \int_{\lambda_{w_0}} dw \int_{\gamma'_w} F(s, w)\, ds/P(s, w).$$

Here F is the Fourier transform of f. By definition of the integral this means that

$$\int_{\substack{w \text{ real} \\ |w-w_0| \leqslant 1}} e^{i\xi w}\, dw \int_{\gamma'_w} e^{its}\, ds/P(s, w) = \int_{\lambda_{w_0}} e^{i\xi w}\, dw \int_{\gamma'_w} e^{its}\, ds/P(s, w).$$

Of course, the above argument can easily be modified: We do not need to know that for any $w_1 \in \Lambda_{w_0}$, s between γ'_{w_1} and γ'_{γ_0} the point (s, w_1) is at a distance at least 1 from V. It suffices to pass from w real, $|w - w_0| \leqslant 1$ to λ_{w_0} by stages in which, say, we increase the imaginary part one unit at a time. In order to accomplish this, we need to know that if w_0 is sufficiently large (not necessarily real) and satisfies (11.27) then for

$$s \in \bigcup_{|w-w_0| \leqslant 1} \gamma'_w, \quad |w - w_0| \leqslant 1$$

the point (s, w) is at a distance $\geqslant 1$ from V. It is clear from the definition of γ'_w that this can be arranged (perhaps by adjusting c'' slightly).

We have shown how to prove (11.32). Using the same type of argument we can verify (11.33).

The essential point of this contour changing is that in all the stages of change we stay between the regions defined by (11.21) and (11.22) so the zeros of P do not cause any trouble. Thus we keep $|\text{Im } s|$ larger than the right side of (11.21) and $|\text{Im } s| + |\text{Im } w|$ smaller than the right side of (11.22).

We can now use (11.32) and (11.33) to show that for any $d > 0$ we can choose c' large enough so that e^1 is a d times differentiable function outside of a cone. It is to be understood that we use γ_+ for $\xi > 0$, γ_- for $\xi < 0$.

If $t < -1$, then on γ''_w we will have

$$(11.35) \qquad |e^{its}| \leqslant e^{\text{Im } s} \leqslant (1 + |s| + |w|)^{-4c'} e^{-4c'|\text{Im } w|}$$

for $|s|$ large. Thus, the integral in (11.33) together with its derivatives of order $\leqslant 4c' - 2$ will converge uniformly on compact sets of $t < -1$ which means that e^1 is a function in $t < -1$ which is differentiable of order at least $4c' - 3$.

For $t \geqslant -1$ we must use a different method, since the term e^{its} in the right side of (11.33) or (11.32) will certainly not help us if $t > 0$. In this case we use (11.32). Since w belongs to $\gamma_\pm$, the factor $\exp(i\xi w)$ is small. In fact, as c' increases we can differentiate it as many times as we want with respect to ξ and it is still small [see (11.34)]. On the other hand, by (11.26), $\operatorname{Im} s$ is not too large on γ'_w. Thus e^1 can be made to be d times differentiable.

More precisely, for large $|s|$ we have on γ'_w

$$(11.36)\qquad |e^{its}| \leqslant e^{|t|\,|\operatorname{Im} s|}$$

$$\leqslant (1 + |s| + |w|)^{4|t|(c+1)}\, e^{4|t|(c+1)(1+|\operatorname{Im} w|)} \quad (\text{by } (11.26))$$

$$(11.37)\qquad \leqslant (1 + |s|)^{4|t|(c+1)}\, (1 + |w|)^{4|t|(c+1)}\, e^{4|t|(c+1)(1+|\operatorname{Im} w|)}.$$

On the other hand, for $w \in \gamma_\pm$, $|w|$ large, we have

$$(11.38)\qquad |e^{i\xi w}| = e^{-|\xi|\,|\operatorname{Im} w|}$$

$$(11.39)\qquad \leqslant (1 + |w|)^{-c'|\xi|/4} \quad (\text{by } (11.34)).$$

Combining (11.37) with (11.39) we have by (11.34)

$$(11.40)\qquad |e^{its+i\xi w}| \leqslant (1 + |s|)^{4|t|(c+1)} \times (1 + |w|)^{4|t|(c+1)+4|t|(c+1)c'/4-c'|\xi|/4} \times e^{4|t|(c+1)(1+c'/4)}.$$

As long as

$$(11.41)\qquad |\xi| \geqslant 4\,|t|(c+1) + 1$$

the right side of (11.40) $\to 0$ as $|\xi| \to \infty$ faster than $|\xi|^{-d}$ if c' is chosen large enough (depending on d). On the other hand, in s the growth is like

$$(11.42)\qquad (1 + |s|)^{4|t|(c+1)}$$

which is bounded by a fixed power of $(1 + |s|)$ for t in a compact set.

It follows easily from this that e^1 is d times differentiable in ξ in (11.41). This, together with (11.35) and following, shows that e^1 is d times differentiable in ξ outside a forward cone.

As we have mentioned, a similar method applies to e^2. Thus e is a d times differentiable function outside of a cone.

Finally, we note that, from the definition (11.29) of e we have

$$(11.43)\qquad De(t, \xi) = \int_{w \text{ real}} e^{i\xi w}\, dw \int_{\gamma'_w} e^{its}\, ds.$$

A simple application of Cauchy's formula shows that this differs from

$$\delta = \int_{w \text{ real}} e^{i\xi w}\, dw \int_{s \text{ real}} e^{its}\, ds$$

by an integral over a compact set in the (s, w) space. Such an integral is easily seen to converge uniformly on compact sets of the complex (t, ξ) space and is thus an entire function in (t, ξ). Thus, our assertion that e is a parametrix is proved. As noted at the beginning of the proof, this is what is needed to establish Theorem 11.10 in case $r = 1$.

Next we pass to the case $r > 1$. Here there is great difficulty because the usual way of constructing fundamental solutions for $r = 1$ as the Fourier transform of the reciprocal of a polynomial does not make sense for $r > 1$. What we must do is find r functions which play, for $r > 1$, the same role that $1/P$ plays for $r = 1$.

From our hypotheses (11.21) and (11.22), if Im w satisfies (11.27), then, for c, c' as before, except for a bounded set of s, w, there will be no points of V such that

$$(11.44)\quad 4c(1 + |\text{Im } w| + \log(1 + |s| + |w|)) \leqslant |\text{Im } s| \leqslant (c' + 1)(1 + |\text{Im } w| + \log(1 + |s| + |w|)).$$

Using the techniques of Chapters III and IV we find that in a slightly smaller region we can find holomorphic functions $G^1, \ldots, G^r$ which are of growth $0(1 + |s| + |w|)^m$ for m independent of c, c' such that

$$\sum P^j G^j = 1.$$

Namely, Theorem 3.18 shows how to construct the G^j semilocally, that is, in a disk $|z - z^0| \leqslant 1$ if z^0 is sufficiently far from V. We cannot apply Theorem 4.2 to "piece together" the semilocally defined G^j, because the piecing-together method of the proof of Theorem 4.2 applies to the whole space C^n, whereas in our present situation we want to apply the method to the complement in C^n of a suitable neighborhood of V. (Note that by Chapter V, Example 8 the ring of polynomials is PLAU.) However it is possible to modify the proof of Theorem 4.2 so as to obtain global G^j which are holomorphic and of polynomial growth on a set which is large enough for us to perform all the desired changes of contour that we shall need. We shall omit the details as they are essentially straightforward.

We now define e_j by

$$(11.45)\qquad e_j = \int_{w \text{ real}} e^{i\xi w}\, dw \int_{\gamma'_w} G^j(s, w) e^{its}\, ds.$$

We can now proceed as before and we obtain our result. ▯

See Remark 11.9.

Problem 11.11 (Exercise) Carry out the details.

There is also an analog of Theorem 11.10 for distributions of finite order:

THEOREM 11.11. *Suppose we can write* $V = V_1 \cup V_2$, *where for* $(s, w) \in V_1$ *and any* $\varepsilon > 0$ *we have*

$$(11.46) \qquad |\operatorname{Im} s| \leqslant \varepsilon(1 + |\operatorname{Im} w| + \log(1 + |s| + |w|))$$

except for a bounded set of (s, w) *(depending on* ε*), while for* $(s, w) \in V_2$ *we have*

$$(11.47) \qquad |\operatorname{Im} s| + |\operatorname{Im} w| \geqslant c \log(1 + |s| + |w|)$$

for some c *except on a bounded set. Then every distribution* f *of finite order which satisfies* $Df = 0$ *and is such that for any* $l > 0$ *there is an* $l' > 0$ *so that* f *is an* l *times differentiable function in* $t < -l'$ *must be on* $\mathscr{E}$.

PROOF. The proof is similar to that of the previous theorem except we now must be careful to see that all the e_j are of some finite order independent of x. □

See Remark 11.10.

By similar methods we can prove

THEOREM 11.12. *A sufficient condition that any* $f \in \mathscr{E}$ *which satisfies* $Df = 0$ *and is real analytic in* $\operatorname{Re} t < 0$ *should be real analytic in all* t, ξ *is: We can write* $V = V_1 \cup V_2$, *where for* $(s, w) \in V_1$ *we have*

$$(11.48) \qquad |\operatorname{Im} s| \leqslant c(1 + |\operatorname{Im} w| + (|s| + |w|)^{\alpha})$$

for some c *and some* $\alpha < 1$, *while for* $(s, w) \in V_2$

$$(11.49) \qquad |\operatorname{Im} s| + |\operatorname{Im} w| \geqslant c_1(|s| + |w|)$$

for some c_1 *for* $|s| + |w|$ *sufficiently large.*

The conditions in Theorem 11.10 are not necessary. This can be seen from the example: $r = 1$, $n = 3$, $D = (1 + i\varepsilon)\partial^2/\partial t^2 + \partial^2/\partial\xi_1^2 - \partial^2/\partial\xi_2^2$. It is seen in this case that for $\varepsilon > 0$ the hypotheses of Theorem 11.10 are not satisfied. However, a result of Hörmander ([3], p. 216, Theorem 8.8.1) shows that the conclusion of Theorem 11.10 holds in this case.

Some partial results concerning the converse of Theorem 11.10 can be obtained using a construction which was first used by Zerner and then generalized by Hörmander (see Hörmander [3], Chapter VIII). The usual method of lacunary series cannot be used to prove the necessity because the AU intersection (see Chapter XII below)

$$\mathscr{D}' \overset{a}{\cap} \mathscr{E}(t<0) = \mathscr{E},$$

as can easily be seen from the expressions for the AU structures of $\mathscr{D}'$ and $\mathscr{E}$ given in Chapter V, Examples 4 and 5. For the usual method of lacunary series we would attempt to find a function, or distribution $f(x)$ of the form

$$f(x) = \sum a^j \exp(ix \cdot z^j), \tag{11.50}$$

where $\{z^j\}$ is lacunary, such that

(a) Each $z^j \in V$ (so $\vec{\mathrm{D}}f = 0$).
(b) The series converges in the topology of $\mathscr{D}'$.
(c) $f \in \mathscr{E}(t<0)$.
(d) $f \notin \mathscr{E}$.

Such a construction is impossible by the results of Chapter XII, since $\mathscr{D}' \overset{a}{\cap} \mathscr{E}(t<0) = \mathscr{E}$ which means that any f satisfying (a), (b), and (c) is in $\mathscr{E}$.

However, the method of lacunary series yields easily

THEOREM 11.13. *Suppose the hypothesis of Theorem* 11.11 *is violated. Then there is an* $f \in \mathscr{D}'$ *with* $\vec{\mathrm{D}}f = 0$ *and such that for any* $l > 0$ *there is an* $l' > 0$ *so that* f *is an* l *times differentiable function in* $t < -l'$ *(or in* $t > l'$*), but* $f \notin \mathscr{E}$.

Although the method of lacunary series cannot be applied directly, there is a modification which can be applied to many examples:

In order to understand the idea, let us try for $n = 1$ to construct a distribution F by Fourier analysis which has prescribed regularity properties in $x > 0$ and in $x < 0$. (These regularity conditions may differ in $x > 0$ and $x < 0$.) Let Γ_j be the contours $\operatorname{Im} s = \varphi_j(\operatorname{Re} s)$ for $\operatorname{Re} s > 0$. The functions φ_j will be prescribed later. For $j = 1, 2$ let $\{s_{ij}\}$ be a lacunary sequence of points on Γ_j. We define the function $f(s)$ by

$$f(s) = s^{-a} \sum_{i,j} \lambda_{ij}(s - s_{ij})^{-1}. \tag{11.51}$$

The λ_{ij} and a will be prescribed later. We require that they should be bounded so f is a meromorphic function with poles at the points s_{ij}, 0.

Let Γ be a contour of the form $\operatorname{Im} s = \varphi(\operatorname{Re} s)$ defined for $\operatorname{Re} s$ sufficiently large and lying between the Γ_j. We shall assume that $|\varphi(\alpha)| < \log \alpha$ and that the element of length on Γ is $\leqslant$const $(1 + |s|)^{a'}$. We now require that $a > a' + 4$. The integral

$$F(t) = \int_\Gamma \exp(its) f(s)\, ds \tag{11.52}$$

defines F as a distribution on R for, if t stays in a bounded set, we will have $|\exp(its)| \leqslant (m + |s|)^m$ for some m depending on this bounded set. Thus, for any $g \in \mathscr{D}$ the integral $\int F(t)g(t)\,dt$ which is defined as $\int_\Gamma G(s)f(s)\,ds$, where G is the Fourier transform of g, will exist. It is clear that $g \to \int F(t)g(t)\,dt$ is continuous on $\mathscr{D}$. (See Fig. 31.)

Assume now that $|\varphi_j(\alpha)| \leqslant \alpha$ for all j, α. Then for $t > 0 (t < 0)$ we can shift the contour in (11.52) to $\Gamma_\pm$ which coincides with $\operatorname{Im} s = \pm 2 \operatorname{Re} s$ for $\operatorname{Re} s$ sufficiently large, of course, taking into account the residues at the poles of f. From (11.51) and (11.52) we have

$$(11.53)\quad F(t) = \begin{cases} \sum \lambda_{i1}(s_{i1})^{-a} \exp(its_{i1}) + \int_{\Gamma+} \exp(its)f(s)\,ds & \text{for } t > 0 \\ \sum \lambda_{i2}(s_{i2})^{-a} \exp(its_{i2}) + \int_{\Gamma-} \exp(its)f(s)\,ds & \text{for } t < 0. \end{cases}$$

Now $\int_{\Gamma_\pm} \exp(its)f(s)\,ds$ is clearly real analytic in $t > 0 (t < 0)$. Thus, the regularity properties (weaker than real analyticity) of F in $t > 0$ and $t < 0$ depend on the sequences s_{ij} and λ_{ij}. These regularity properties can be completely controlled because of the lacunary nature of the s_{ij}. By the results of Chapter XII the regularity of F in $t > 0$ depends only on the rate of growth of $\{\lambda_{i1}\}$, while the regularity in $t < 0$ depends only on the rate of growth of $\{\lambda_{i2}\}$. Since these sequences are independent, we can prescribe regularity conditions on F in $t > 0$ and in $t < 0$, as we desire.

Of course, in many cases, we may not have to shift the contour as far as $\operatorname{Im} s = \pm 2 \operatorname{Re} s$ for s large. We must be able to shift far enough so that the shifted integrals will have as much regularity as is needed.

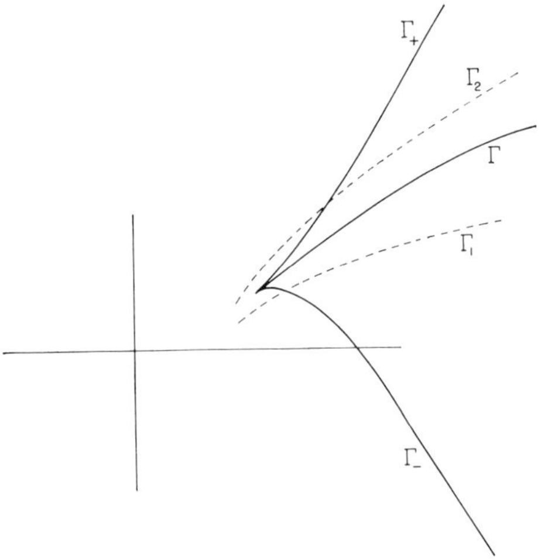

FIGURE 31

The application of this construction to the converses of Theorems 11.10, 11.11, and 11.12 is as follows: We try to find a complex analytic curve Ψ in V which lies above the portion of the complex s plane between curve Γ_- in $\operatorname{Im} s < 0$ and Γ_+ in $\operatorname{Im} s > 0$. If $w_j = \psi_j(s)$ denotes the coordinates of the points in Ψ above the point s, certain conditions will be imposed on the ψ_j below.

We can now "lift" the above construction to the variety V by defining, in analogy to (11.52),

$$F(t, \xi_1, \ldots, \xi_{n-1}) = \int_{(\Gamma,\psi(\Gamma))} \exp(its + i\xi \cdot w) f(s)\, ds, \tag{11.54}$$

where f is again defined by (11.51), and where by $(\Gamma, \psi(\Gamma))$ we mean $s \in \Gamma$, $w = \psi(s)$. We can now shift the contour as before. In analogy to (11.53) we obtain

$$F(t) = \begin{cases} \sum \lambda_{i1}(s_{i1})^{-a} \exp[its_{i1} + i\xi \cdot \psi(s_{i1})] \\ \qquad + \displaystyle\int_{(\Gamma_+,\psi(\Gamma_+))} \exp[its + i\xi \cdot \psi(s)] f(s)\, ds & \text{for } t > 0 \\ \sum \lambda_{i2}(s_{i2})^{-a} \exp[its_{i2} + i\xi \cdot \psi(s_{i2})] \\ \qquad + \displaystyle\int_{(\Gamma_-,\psi(\Gamma_-))} \exp[its + i\xi \cdot \psi(s)] f(s)\, ds & \text{for } t < 0. \end{cases} \tag{11.55}$$

The following conditions must be placed on ψ in order that the construction can work

1. For $s \in \Gamma_\pm$, we have $|\operatorname{Im}(\psi_j(s))|/|\operatorname{Im} s| \to 0$, as $s \to \infty$ for any j.
2. We have control over the imaginary parts of the points $\psi(s_{ij})$.

Condition 1 is needed in order that the regularity of the integrals in the right side of (11.55) should be controlled by whether t is >0 or <0, that is, the terms $\exp[i\xi \cdot \psi(s)]$ do not interfere.

Condition 2 is needed in order to know that the regularity of the series on the right side of (11.55) is controlled by the $\lambda_{i1}, \lambda_{i2}$, that is, the terms $\exp[i\xi \cdot \psi(s_{i1})]$, $\exp[i\xi \cdot \psi(s_{i2})]$ do not cause too much disturbance. This could happen if $|\operatorname{Im} \psi(s_{ij})|$ were very large for then the series could be regular for $|\xi|$ small but not for all $|\xi|$ large unless we make the λ_{ij} so small that we would force F to be more regular than we had originally intended.

We also need another condition on $\Gamma_\pm$:

3. For $s \in \Gamma_\pm$, $|\operatorname{Im} s|$ should be large (in a suitable sense) compared to $|s| + |w|$.

Condition 3 is also needed to guarantee the regularity of the integrals on the right side of (11.55)

EXAMPLE. Let $t=0$ be a timelike surface for the wave operator, that is, $r=1$ and $D=\partial^2/\partial t^2-\partial^2/\partial\xi_1^2+\partial^2/\partial\xi_2^2+\cdots+\partial^2/\partial\xi_{n-1}^2$. (Note the change from the usual notation.) We see easily that the hypotheses of Theorems 11.10 and 11.11 do not hold. Examples showing that the conclusions of Theorem 11.10 and 11.11 do not hold were constructed by Zerner [1]. We should like to apply our method to the converse of Theorem 11.12

Unfortunately, we do not have a complete proof; our method gives a partial answer. We may clearly restrict our considerations to the case $n=3$. We shall first construct an $F\in\mathscr{E}$ with $DF=0$ such that F is real analytic in $\xi_1+\xi_2>0$ but is not real analytic everywhere. Note that $\xi_1+\xi_2=0$ is characteristic so by a result of Hörmander [1] such an F exists and can even be chosen to be 0 in $\xi_1+\xi_2>0$. However, we shall construct an F which is bounded. Such an F could not be 0 in $\xi_1+\xi_2>0$ by the results of Chapter IX (see Theorem 9.30).

We introduce coordinates $\lambda_1=\frac{1}{2}(\xi_1+\xi_2)$, $\lambda_2=\frac{1}{2}(\xi_1-\xi_2)$ with dual coordinates $\mu_1=(w_1+w_2)$, $\mu_2=(w_1-w_2)$. Then $s^2-w_1^2+w_2^2=s^2-\mu_1\mu_2$. We now write, in analogy to (11.54),

$$F(t,\xi_1,\xi_2)=\int\exp\,(its+i\lambda_1\mu_1+i\lambda_2\,\mu_2)f(\mu_1)\,d\mu_1, \tag{11.56}$$

where the integral is over the positive real μ_1 axis. Here we set $\mu_2=\alpha(\mu_1)$ and $s=[\mu_1\alpha(\mu_1)]^{1/2}$, where α is holomorphic in the right half-plane and α is real on the real axis. We chose $\alpha\equiv 1$. Our method shows that for a suitable choice of f we can make F real analytic in $\lambda_1>0$, but not in $\lambda_1<0$. We leave the details to the reader.

It may be possible that by a suitable use of the method of Zerner [1] we could use such F to construct a solution of the wave equation which belongs to $\mathscr{E}$ which is real analytic in $t>0$ but not in $t<0$.

It is interesting to observe why our method does not yield directly the existence of such an F. Our construction requires that Im s be large on $\Gamma_\pm$. By condition 1, Im μ_1 and Im μ_2 must be small on $\psi(\Gamma_\pm)$. Since $s^2=\mu_1\mu_2$, this can only be the case if $\Gamma_\pm$ are essentially arg $s=\pm\pi/2$, and say on arg $s=+\pi/2$, μ_1 is close to the negative real axis and μ_2 close to the positive real axis. By Condition 2, $\psi(s)$ is essentially real on the real axis. But then $s\to\mu_1(s)$ essentially maps the region $0\leqslant\arg s\leqslant\pi/2$ onto $0\leqslant\arg\mu_1\leqslant\pi$. Thus $|\mu_1(s)|$ and hence $|\psi(s)|$ must be of the order $|s|^2$ on arg $s=\pi/2$ which contradicts Condition 3.

We leave the precise details to the reader.

The same type of construction can be used to show, in many cases, that the local boundary conditions on A described following the proof of Theorem 11.6 lead to necessary and sufficient conditions for the extension property of Theorem 11.6 to hold. The simplest type of construction would give an example of a function for $n = 1$ which is C^∞ in $|x| > 1$ but has singularities at $x = \pm 1$. This can be constructed in a manner similar to the above construction of F with prescribed regularity in $x > 0$ and $x < 0$. We can then "lift the construction" to V as above and construct examples of solutions of $\vec{D}F = 0$ with "prescribed" essential singularities. We shall not enter into the details.

See Remark 11.11.
See Problem 11.12.

XI.3. General Domains

Let $\boxed{D}$ be an $r \times l$ matrix of linear partial differential operators with constant coefficients. In Chapter VI we showed that if A is an open convex set then we can always solve the system

$$\boxed{D}\vec{f} = \vec{g}$$

for $\vec{f} \in \mathscr{E}^l(A)$ if $\vec{g} \in {}_{\boxed{D}}\mathscr{E}^r(A)$. Here ${}_{\boxed{D}}\mathscr{E}^r(A)$ consists of all functions in $\mathscr{E}^r(A)$ which satisfy the same differential relations as the rows of $\boxed{D}$.

We wish to extend this result to nonconvex A. Our method goes as follows: By the Hahn-Banach theorem, $\boxed{D}$ is *onto* if the inverse of its adjoint $\boxed{D}'$ is continuous. Now $\boxed{D}'$: ${}_{\boxed{D}}\mathscr{E}'^r(A) \to \mathscr{E}'^l(A)$. Roughly speaking, a map is continuous if it takes bounded sets into bounded sets. For a set B in $\mathscr{E}'(A)$ to be bounded, two requirements must be met (see Chapter V, Example 5 for more details on the space $\mathscr{E}$):

(a) All $S \in B$ have their supports in a fixed compact set K of A.
(b) The $S \in B$ satisfy certain inequalities on K.

Thus, the proof of the continuity of the inverse of $\boxed{D}'$ can be subdivided into two parts:

(c) If $\{\boxed{D}'\vec{S}\}$ have their supports in a fixed compact set then we can choose $\{\vec{S}_1\}$ so that $\vec{S}_1 \in \mathscr{E}'^r$, $\vec{S} = \vec{S}_1$ as an element of ${}_{\boxed{D}}\mathscr{E}'^r$, and $\{\vec{S}_1\}$ have their supports in a fixed compact set.
(d) If $\{\boxed{D}'\vec{S}\}$ is bounded in the topology of $\mathscr{E}'^l(A)$ then we can choose $\{\vec{S}_1\}$ satisfying the conclusions of Condition (c) and so that they satisfy certain inequalities.

In case $r = l = 1$ (so ${}_{\boxed{D}}\mathscr{E}'^r = \mathscr{E}'$) Malgrange [1] showed that it suffices to prove Condition (c) to obtain the fact that $\boxed{D}$ is onto. If, in addition, $n = 2$, it is fairly easy to describe Malgrange's condition in terms of the geometry of A: Every characteristic line for D (see Chapter IX) meets A in an open interval.

See Remark 11.12.

Observe that if A is not convex, the solvability of $\boxed{D}\vec{f} = g$ depends on a relation between A and $\boxed{D}$; for some A, $\boxed{D}$ it is solvable while for others it is not.

The situation for r, $l > 1$ seems to be much more complicated. In case $n = 3$ we have the classical equations

$$\text{grad}\, f = \vec{g}, \quad \text{curl f} = \vec{g}$$

It is well-known that the possibility of solving these equations depends on the topological nature of A. Thus we can solve grad $f = \vec{g}$ for any $\vec{g}$ satisfying the necessary differential relation curl $\vec{g} = 0$ if and only if $H^1(A, C)$, the first cohomology group of A with complex coefficients, vanishes. We can solve curl $\vec{f} = \vec{g}$ for all $\vec{g}$ satisfying div $\vec{g} = 0$ if and only if $H^2(A, C) = 0$.

Notice that the possibility of solving the equations grad $f = \vec{g}$, curl $\vec{f} = \vec{g}$, for all $\vec{g}$ satisfying the respective compatibility conditions curl $\vec{g} = 0$, div $\vec{g} = 0$, depends only on global properties of A. An opposite situation is met if we try to solve the complex analogs of these equations: Suppose $n = 6$. Write

$$\partial_j = \partial/\partial x_{2j-1} + i\partial/\partial x_{2j}, \quad j = 1, 2, 3.$$

Write

$$\text{grad}^C f = (\partial_1 f, \partial_2 f, \partial_3 f)$$

$$\text{curl}^C \vec{f} = (\partial_3 f_2 - \partial_2 f_3, \partial_1 f_3 - \partial_3 f_1, \partial_2 f_1 - \partial_1 f_2)$$

$$\text{div}^C \vec{f} = \partial_1 f_1 + \partial_2 f_2 + \partial_3 f_3 .$$

The theory of functions of several complex variables (see Bers [1]) shows that the possibility of solving the equations grad^C $f = \vec{g}$, curl^C $\vec{f} = \vec{g}$, for all $\vec{g}$ satisfying the respective compatibility conditions curl^C $\vec{g} = 0$, div^C $\vec{g} = 0$, depends on local conditions on the boundary of A. For example, if A is pseudoconvex (see Section XI.2) then we can always solve the equations. [Actually, as will be apparent from what follows, this result is somewhat misleading. The apparent discrepancy between the two types of conditions, namely the global conditions on A for say grad $f = \vec{g}$ and the local conditions for grad^C $f = \vec{g}$, is not so great if we realize that the local conditions on the boundary of A imply topological restrictions on A. For example, Thom (unpublished) showed, by means of Morse theory, that if A is a pseudoconvex domain in C^n then $H^j(A, Z)$, the jth cohomology group with integer coefficients, vanishes for $j > n$. Also, if A is a strictly pseudoconvex domain in a complex manifold, there can still be topological obstructions to solving $\mathrm{grad}^C f = \vec{\mathrm{g}}$, $\mathrm{curl}^C \vec{\mathrm{f}} = \vec{\mathrm{g}}$, etc.]

We shall first treat a special case. Our proof contains most of the ideas of the proof of the general result to be given below. The proof is modeled after the treatment of the Cauchy-Riemann system in Ehrenpreis [15].

THEOREM 11.14. *Let A be a domain with smooth boundary. Suppose the module of relations of the rows of $\boxed{\mathrm{D}}$ has one generator which we denote by $\vec{\mathrm{D}}_1'$, so $\vec{\mathrm{D}}_1'$ is a row vector. Assume that $\boxed{\mathrm{D}}'$ generates the module of relations of the rows of $\vec{\mathrm{D}}_1$. Suppose there exists a C^∞ real-valued function φ on R so that*

1. *A is defined by $0 \leqslant \varphi(x) < 1$.*
2. *$\varphi(x) = 0$ at a single point of A.*
3. *For any t with $0 < t < 1$, the domain A_t defined by $\varphi(x) < t$ has the property that each boundary point can be just nibbled by a mouth for $\vec{\mathrm{D}}_1$ and the strong uniqueness property for that mouth (see Section XI.2) holds at each boundary point.*
4. *There is no point in the closure of A at which all derivatives of φ vanish.*

Then the system $\boxed{\mathrm{D}}\vec{\mathrm{f}} = \vec{\mathrm{g}}$ has a solution $\vec{\mathrm{f}} \in \mathscr{E}^l(A)$ for any $\vec{\mathrm{g}} \in \mathscr{E}^r(A)$ satisfying $\vec{\mathrm{D}}_1'\vec{\mathrm{g}} = 0$.

PROOF. We wish to verify Condition (c) and (d) described on p. 367. We begin with (c). Suppose $\{\boxed{\mathrm{D}}'\vec{\mathrm{S}}\}$ have their supports in a fixed compact set K of A. Let a be a boundary point of A. Then in the neighborhood of a we have $\boxed{\mathrm{D}}'\mathrm{S} = 0$. Our first task is to replace $\vec{\mathrm{S}}$ by $\vec{\mathrm{S}}_1 = \vec{\mathrm{S}} + \vec{\mathrm{D}}_1 T$, where $T \in \mathscr{E}'(A)$ in such a way that the support of S_1 does not meet some fixed neighborhood of a. (Note that $\vec{\mathrm{S}}_1 = \vec{\mathrm{S}}$ as an element of ${}_{\boxed{\mathrm{D}}}\mathscr{E}'^r(A)$ since ${}_{\boxed{\mathrm{D}}}\mathscr{E}^r(A)$ consists of all $\vec{\mathrm{g}} \in \mathscr{E}^r(A)$ with $\vec{\mathrm{D}}_1'\vec{\mathrm{g}} = 0$.) This being done, we can continue the process and remove a fixed neighborhood of the boundary of A from the support of $\vec{\mathrm{S}}$.

There are two ways to construct T:

Method 1. Since $\boxed{\mathrm{D}}'S = 0$ near a, there is a cube $\tilde{N}$, center a, which does not meet K. Let N be the complement of A in $\tilde{N}$. By our assumption and the results of Chapter VI and Example 3 of Chapter V, we can find $V \in \mathscr{D}_F'(\tilde{N})$ satisfying $\vec{\mathrm{D}}_1 V = \vec{\mathrm{S}}$ on $\tilde{N}$. Now, $\mathrm{D}_1 V = 0$ on a neighborhood N_1 of N in $\tilde{N}$ because $\vec{\mathrm{S}}$ is of compact support in A. Thus, by our hypothesis and Theorem 11.6 there exists a $\tilde{V}$ which is a distribution of finite order on the union of N_2 (which is a suitable neighborhood of N in $\tilde{N}$)

with a neighborhood M of a in R and satisfies $\vec{\mathrm{D}}_1 \tilde{V} = 0$ on $N_2 \cup M$, and $\tilde{V} = V$ on N_2. (Actually Theorem 11.6 is stated for indefinitely differentiable functions, but there is no difficulty in extending the result to distributions.) Here N_2 depends on $\vec{\mathrm{S}}$ but M depends only on $\tilde{N}$, hence K, and $\vec{\mathrm{D}}_1$. (See Fig. 32.)

Let α be a C^∞ function satisfying

$$\alpha = \begin{cases} 0 & \text{outside } M \cup N_2 \\ 1 & \text{on a neighborhood } m \\ & \text{of } a \text{ in } M \cup N_2 . \end{cases} \tag{11.57}$$

We set

$$\vec{\mathrm{S}}_1 = \vec{\mathrm{S}} - \vec{\mathrm{D}}_1[\alpha(V - \tilde{V})]. \tag{11.58}$$

Then $\vec{\mathrm{S}}_1 = \vec{\mathrm{S}}$ as an element of ${}_{\boxed{\mathrm{D}}}\mathscr{E}'^{r}(A)$ since $\alpha(V - \tilde{V}) \in \mathscr{E}'(A)$. Moreover, the construction shows that the support of S_1 does not meet m. Thus we have "nibbled" m out of the support of $\vec{\mathrm{S}}$.

Method 2. We try to solve $\vec{\mathrm{D}}_1 U = \vec{\mathrm{S}}$ in a neighborhood M of a in R with $U = 0$ on some neighborhood N_3 of the boundary of A near a. Here N_3 might depend on $\vec{\mathrm{S}}$ but M should not. However, N_3 must contain the intersection of M with the boundary of A. Note that this is very similar to a Cauchy problem.

We use the notation of the proof of Theorem 11.6, except that now for x near a, which is again normalized to be 0, η_x "cuts off" B so that ζ_x has its support outside A. Then we define, in analogy to (11.12)

$$U(y) = \sum \{S_j *' [\eta_x(\tau_x e_j)]\}(x - y). \tag{11.59}$$

Thus, for any i,

$$\begin{aligned} (D_i U)(y) &= \sum \{D_i S_j *' [\eta_x(\tau_x e_j)]\}(x - y) \\ &= \sum \{D_j S_i *' [\eta_x(\tau_x e_j)]\}(x - y) \\ &= \{S_i *' \sum D_j'[\eta_x(\tau_x e_j)]\}(x - y) \\ &= S_i(y). \end{aligned} \tag{11.60}$$

(Note that since $-y$ appears on the right side of (11.60) there is a change from D_i' to D_i.) Here we have used the fact that $\vec{\mathrm{S}}$ satisfies the compatibility relation $D_i S_j = D_j S_i$, that

$$\sum D_j'[\eta_x(\tau_x e_j)] = \delta_x - \zeta_x$$

and that the support of ζ_x lies outside $A + x - y$. As in the proof of Theorem 11.6, there is no difficulty in applying the commutativity and

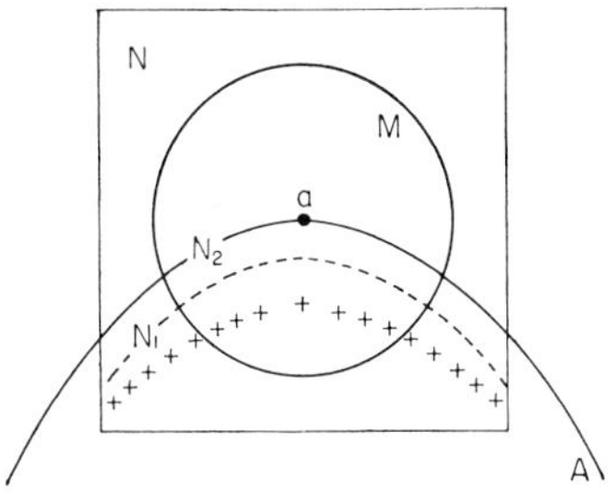

FIGURE 32

associativity of convolution because of the nature of the supports (see Schwartz [1]). We have set $\vec{\mathrm{D}}_1 = (D_1, \ldots, D_r)$.

It is interesting to observe that we do not need the full compatibility conditions on $\vec{\mathrm{S}}$ but only $D_i S_j = D_j S_i$. The reason for this is that it can be shown that $D_i S_j = D_j S_i$ imply the full compatibility conditions for $\vec{\mathrm{S}}$ on M since $\vec{\mathrm{S}} = 0$ on N.

It is clear from (11.59) that $U = 0$ above a in a neighborhood M of a in R. Using the uniqueness property we conclude that $U = 0$ on a neighborhood of the intersection of M (or perhaps a slightly smaller neighborhood of a) with the boundary of A. Thus $\vec{\mathrm{D}}_1 U = \vec{\mathrm{S}}$ on M and $U = 0$ on N_3 which is what we desired. We can now, as in Method 1, multiply U by a suitable C^∞ function α' and "nibble" a neighborhood m of a out of the support of $\vec{\mathrm{S}}$, that is, we replace $\vec{\mathrm{S}}$ by

$$\vec{\mathrm{S}}_1 = \vec{\mathrm{S}} - \vec{\mathrm{D}}_1(\alpha' U) \tag{11.61}$$

and the support of $\vec{\mathrm{S}}_1$ does not meet m.

It should be noted that it is an easy consequence of the definition of U that U vanishes on some neighborhood of a in R. However, we need the vanishing on N_3 which contains the intersection of M with the boundary of A.

From either Method 1 or 2, we see that if the support of $\vec{\mathrm{S}}$ were given as not meeting some neighborhood m' of a piece of the boundary near a, then the above construction and the strong uniqueness condition preserve this vanishing on m' (or slightly smaller) and produce vanishing on $m \cup m'$ (perhaps made slightly smaller). We see this as follows: Let us treat Method 1 and leave Method 2 to the reader. See Fig. 33. In the notation of Method 1, we have

$$\vec{\mathrm{S}}_1 = \vec{\mathrm{S}} - \vec{\mathrm{D}}_1[\alpha(V - \tilde{V})].$$

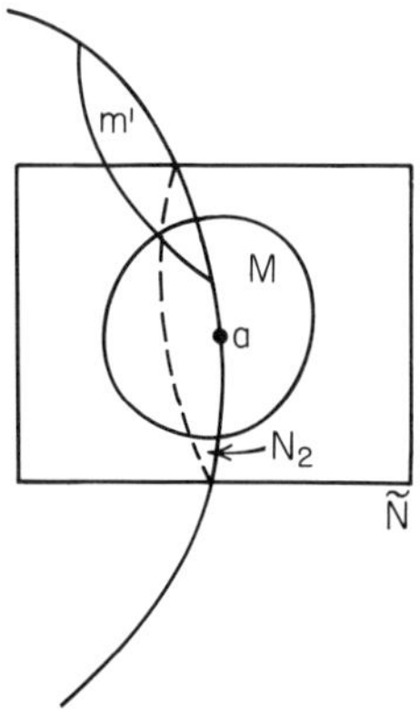

FIGURE 33

By our construction, the definition of α can be modified to obtain the vanishing of $\vec{S}_1$ on a slightly smaller set than the one on which $\tilde{V}$ is defined with $\vec{D}_1\tilde{V} = 0$. Now, $\vec{D}_1\tilde{V} = 0$ on $N_2 \cup M$. Moreover, $\vec{D}_1 V = \vec{S}_1$ on $\tilde{N}$, so $\vec{D}_1 V = 0$ on $(\tilde{N} \cap m') \cup N_1$. We want to use V to continue $\tilde{V}$ from $N_2 \cup M$ into

$$(N_2 \cup M) \cup (\tilde{N} \cap m').$$

To know that this can be done we must verify that

$$V = \tilde{V} \quad \text{on} \quad (\tilde{N} \cap m') \cap (N_2 \cup M).$$

But, $V = \tilde{V}$ on N_2 and $\vec{D}_1\tilde{V} = 0$ on $M \cup N_2$ while $\vec{D}_1 V = 0$ on $\tilde{N} \cap m'$. The strong uniqueness property shows, therefore, that $V = \tilde{V}$ on a (possibly slightly smaller set than) $(\tilde{N} \cap m') \cap (N_2 \cup M)$. This set is large enough to enable us to define $\tilde{V}$ in almost all of m' to be equal to V. As mentioned above, this gives the vanishing of $\vec{S}_1$ on essentially all of $m \cup m'$.

Thus we may iterate the process, that is, starting with $\vec{S}_1$ we can nibble more of the support. Since the boundary of A is compact, after a finite set of nibbles we can be sure that the support of $\vec{S}$ lies in some fixed compact set. This completes the proof of (c). ☐

Proof of (d). We should now like to show that $\vec{S}$ can be chosen to be small on A. After having performed the nibbling in (c), we can assume that there is a t_0 with $0 < t_0 < 1$ such that $\vec{S}$ has its support in A_{t_0}, that is, $\vec{S} = 0$ outside A_{t_0}, and $\boxed{\mathrm{D}}'\vec{S}$ is small on A_{t_0}. Since $\vec{S} = 0$ outside A_{t_0}, $\vec{S}$ is small outside A_{t_0}.

We want to nibble A_{t_0} and increase the size of the set on which $\vec{S}$ is small. For this purpose we shall use a modification of Method 2; we leave to the reader the task of showing how to apply Method 1 instead.

Let a be a boundary point of A_{t_0}. We define U as in (11.59). The argument in (11.60) shows that $D_i U$ is close to S_i. The reason for this is that we do not have $D_i S_j = D_j S_i$ but rather $D_i S_j - D_j S_i$ is small. We can use U as before to nibble the set on which $\vec{S}$ is small, that is, $\vec{S} - \vec{D}_1 U$ is small on a neighborhood m of a in A_{t_0} and is zero outside A_{t_0}. Moreover, U is zero outside A_{t_0}. For this reason we do not introduce any large terms when applying (11.61), because α' changes from the constant 0 to the constant 1 only where U is zero.

As in the proof of (c) there is no interference in continuing the process and nibbling a neighbohood of the boundary of A_{t_0}. We then arrive at the following situation:

(α) Support $\vec{S} \subset A_{t_0}$.

(β) $\vec{S}$ is small outside A_{t_1} where $0 < t_1 < t_0$.

We can now continue the process of increasing the set on which $\vec{S}$ is small. We are in a slightly different situation in that (β) replaces the fact that $\vec{S} = 0$ outside A_{t_0} which was used before. This is, however, of no consequence for the argument.

Thus we can find a sequence $t_0 > t_1 > \cdots > 0$ such that we have (α) and

(β') $\vec{S}$ is small outside A_{t_j}.

Does the process end? Using Condition 4 on φ and the discussion following the proof of Theorem 11.6 of the local boundary behavior in case the hypotheses of Theorem 11.6 are satisfied, it is possible to estimate the t_j. In particular, we can show that $t_j \to 0$.

PROBLEM 11.13 (Exercise)

Carry out the details.

Because of hypothesis 2 we arrive at the situation in which A_{t_j} is contained in some convex polyhedron $\tilde{A}$ whose closure is contained in A. We can now apply Theorem 11.5*. In the notation of Theorem 11.5* we may assume that

$$\bar{A}_{t_j} \subset \Omega_1 - N'(\Gamma_1)$$
$$\bar{\Omega}_2 \subset A.$$

Moreover, $\vec{S}$ is bounded outside $\Omega_1 - N'(\Gamma_1)$ and $\boxed{\mathrm{D}}'\vec{S}$ is bounded on Ω_2. (By this we mean "bounded in the topologies of the spaces $\mathscr{D}_F'^r$ on the respective sets.")

By multiplying $\vec{S}$ by a function in $\mathscr{D}(\Omega_2)$ we may assume that $\vec{S} \in \mathscr{E}'^r(\Omega_2)$. Let e_1 be a fundamental solution for D_1. Call $f = e_1 * S_1$. Thus $f \in \mathscr{D}'_F$ and

$D_1 f = S_1$. Moreover, for any j, since S_1 is of compact support,

$$\begin{aligned} D_j f &= D_j e_1 * S_1 \\ &= e_1 * D_j S_1 \\ &\sim e_1 * D_1 S_j \\ &= S_j . \end{aligned}$$

Here we have used the symbol $\sim$ to indicate equality modulo a bounded set.

Since $\vec{\mathrm{S}}$ is bounded on $\Omega_0 \cup N'(\Gamma_1)$, f satisfies the hypothesis of Theorem 11.5*, except that f is a distribution instead of being a C^∞, function. This does not affect the proof of Theorem 11.5* and we conclude the existence of $\tilde{f} \in \mathscr{D}'_F\,(\Omega_2)$ such that

(a) $f \sim \tilde{f}$ on $\Omega_0 \cup N(\Gamma_1)$.

(b) $\{\vec{\mathrm{D}}_1 \tilde{f}\}$ is bounded in the topology of $\mathscr{D}'^r_F(\Omega_2)$.

We now multiply $f - \tilde{f}$ by a function α in $\mathscr{D}(\Omega_2)$ which is one on Ω_1. Then

$$\vec{\mathrm{S}} - \vec{\mathrm{D}}_1[\alpha(f - \tilde{f})] = \vec{\mathrm{T}}$$

can be used to replace $\vec{\mathrm{S}}$ and $\vec{\mathrm{T}}$ is small on all of Ω_2. Also $\vec{\mathrm{T}} = \vec{\mathrm{S}}$ outside Ω_2 so there is no interference with $\vec{\mathrm{S}}$ outside Ω_2. The proof of Theorem 11.14 is complete. ▯

We leave as an exercise for the reader the proof of the extension of Theorem 11.14 to the general case (using a semifundamental solution instead of a fundamental solution):

THEOREM 11.15. *Let the notations be as in Theorem* 11.14 *except that we no longer assume that* $\boxed{\mathrm{D}}_1$ *is a column vector. Assume that* $\boxed{\mathrm{D}}'$ *generates the module of relations of the rows of* $\boxed{\mathrm{D}}_1$. *Suppose Statements* 1, 2, 3, *and* 4 *of Theorem* 11.14 *hold. Then the system* $\boxed{\mathrm{D}}f = \vec{\mathrm{g}}$ *has a solution* $\vec{\mathrm{f}} \in \mathscr{E}^l\,(A)$ *for any* $\vec{\mathrm{g}} \in \mathscr{E}^r\,(A)$ *satisfying* $\boxed{\mathrm{D}}'_1\vec{\mathrm{g}} = 0$.

See Remarks 11.13 and 11.14 and Problem 11.14.

We wish now to discuss what happens when we ameliorate the hypotheses of Theorems 11.14 and 11.15.

(a) *Suppose that* $\boxed{\mathrm{D}}'$ *does not generate the module of relations of the rows of* $\boxed{\mathrm{D}}_1$. The simplest example of this occurs when $r = l = 1$, for then $\boxed{\mathrm{D}}_1 = 0$. In this case, in the proof of Theorem 11.14, we cannot nibble out the support of $\vec{\mathrm{S}}$ by replacing $\vec{\mathrm{S}}$ by $\vec{\mathrm{S}}_1 = \vec{\mathrm{S}} + \vec{\mathrm{D}}_1 T$ because $\boxed{\mathrm{D}}'\vec{\mathrm{S}} = 0$ does *not* imply that we can locally write $\vec{\mathrm{S}}$ in the form $\vec{\mathrm{D}}_1 V$ (as the example $r = l = 1$ shows). However, it is fairly easy to show that if $\boxed{\mathrm{D}}'\vec{\mathrm{S}} = 0$

then we can locally write $\vec{S} = \vec{D}_1 V + \vec{S}_2$ where $\vec{S}_2$ satisfies a *determined* or *overdetermined* system $\boxed{D}_2$ (see Section VI.1 for definitions). For $\vec{S}_2$ we can use the uniqueness results of Section IX.9; we do not need any nibbling. It is thus possible to eliminate the assumption that $\boxed{D}'$ generates the module of relations of the rows of $\boxed{D}_1$. To the hypothesis that each point of the boundary of A_t can be just nibbled by a mouth for $\boxed{D}_1$ we must add a condition guaranteeing uniqueness of the Cauchy problem locally near each boundary point of A_t. The problem essentially reduces to a combination of Theorems 11.14 and 11.15 with the case of determined systems. We leave the details to the reader.

(b) *Suppose the set A_0 is not a point.* This will happen, in particular, if A is not homologically trivial. In this case we cannot conclude that we can always solve $\boxed{D}\vec{f} = \vec{g}$ for all $\vec{g}$ satisfying $\boxed{D}'_1\vec{g} = 0$ as is shown by the example: $n = 2$, A is the annulus $1 < x_1^2 + x_2^2 < 2$, $l = 1$, $r = 2$, $\vec{D} = \text{grad}$.

In this case the proofs of Theorems 11.14 and 11.15 show that we can reduce the set on which $\vec{S}$ may be large to an arbitrarily small neighborhood of A_0. In some cases we can use Theorem 11.7 to nibble A_0 down to a point and hence conclude as in the proofs of Theorems 11.14 and 11.15.

PROBLEM 11.15 (Exercise)

Show that this is the case for the Cauchy-Riemann system. Conclude that if A is strictly q pseudoconvex then $H^j(A, \mathcal{S}) = 0$ for $j \geqslant n/2 - q$. Here $\mathcal{S}$ is the sheaf of germs of holomorphic functions (see Bers [1]).

In some cases we cannot nibble as the above example grad shows. The general situation seems to be quite complicated. However, if there is some kind of "ellipticity," for example if the system $\boxed{D}\vec{f} = \vec{g}$ arises from the study of a cohomology group $H^j(A, \mathcal{S})$, where $\mathcal{S}$ is the sheaf of germs of solutions of an analytic elliptic system and $j > 0$, then the situation seems to be simpler. Malgrange in [5] showed that the nibbling method implies, in many cases, that $H^j(A, \mathcal{S})$ is finite dimensional, hence by Serre duality (Serre [2]) is equal to the cohomology with compact supports $H^{j'}_*(A, \mathcal{S}')$ for suitable j', $\mathcal{S}'$.

This and other evidence suggests that in the elliptic case and even in some nonelliptic cases the situation is as follows:

PROBLEM 11.16 (Conjecture)

We can find a finite collection Ψ of mouths ψ for $\boxed{D}_1$ contained in a compact subset of A. These mouths are cycles in the topological sense. The elements of $H^{j'}_*(A, \mathcal{S}')$ can be represented by solutions $\vec{T}$ of $\boxed{D}'\vec{T} = 0$ with support $\vec{T} \subset$ some ψ and no two such $\vec{T}$ define the same element of $H^{j'}_*(A, \mathcal{S}')$.

The nibbling method used in the proof of Theorem 11.14 shows how to "nibble" the support of "obstructions" to solving $\boxed{D}\vec{f} = \vec{g}$. This nibbling might be thought of as some refinement of the Morse theory (see Milnor [1]) which, roughly speaking, could be used to give a proof of the conjecture for systems arising out of the usual differential operator on differential forms. (Of course, in that case, our conjecture is, in view of de Rham's theorem, just the duality between homology and cohomology.) Our conjecture, if true, would thus lead to a "homology" theory for partial differential equations. In the case of the $\bar{\partial}$ operator on complex manifolds (see Bers (1], Gunning and Rossi [1] our conjecture asserts the known fact that, if A is a strictly pseudoconvex domain in a complex manifold, then the cohomology groups $H^i(A, \mathscr{S})$, where $\mathscr{S}$ is the sheaf of germs of holomorphic functions, are "supported" on compactly embedded complex submanifolds of A.

(c) *Instead of supposing that each boundary point of A can be just nibbled by a mouth for* $\boxed{\text{D}}_1$, *suppose that each boundary point a can either be just nibbled by a mouth for A or else there is a mouth which is contained in the boundary of A near a.*

In this case the possibility of nibbling depends on global phenomena. For example, it depends on whether the mouths which lie in the boundary of A eventually enter A or leave A (see Fig. 34). If the mouth leaves then we can still apply the nibbling method. If the mouth enters completely

Mouth eventually completely enters A   Mouth eventually leaves A

FIGURE 34

then the quotient of the space of $\vec{\text{g}}$ satisfying the compatibility conditions by the set of $\boxed{\text{D}}\vec{\text{f}}$ will usually be infinite. The reason for this is that we can displace the mouth slightly so that it is totally inside A. We can then find an $\vec{\text{S}} \in \mathscr{E}'^r(A)$ with support on the mouth so that $\boxed{\text{D}}'\vec{\text{S}} = 0$ except on a compact set of A. Since the support of $\vec{\text{S}}$ is arbitrarily close to the boundary of A, there is a chance that (a) might fail. (Though, as is clear from the system grad $f = \vec{\text{g}}$, (a) need not fail.)

This type of reasoning accounts for the existence of pseudoconvex domains A in manifolds which are not strictly pseudoconvex and for which $H^j(A, \mathscr{S})$ can be infinite dimensional ($\mathscr{S}$ = sheaf of germs of holomorphic functions). It also explains why, if A is a pseudoconvex subset of C^n, then $H^j(A, \mathscr{S}) = 0$ for $j \geqq 1$.

PROBLEM 11.17.

Give a general result regarding the possibility of solving $\boxed{\mathrm{D}}\vec{\mathrm{f}} = \vec{\mathrm{g}}$ in this case.

See Problems 11.18, 11.19, and 11.20.

It should be pointed out that the nibbling used in the proofs of Theorems 11.14 and 11.15, when using Method 1 of proof, is more related to quasi-hyperbolicity than to hyperbolicity (see Sections VIII.3 and VIII.4). The reason for this is that, in the notation of the proof of Theorem 11.14, we are allowed to diminish N_1 to N_2 and we do not care about uniqueness of the extension. As pointed out following the proof of Theorem 11.6, the proof of Theorem 11.6 is more "hyperbolic" than "quasihyperbolic" in character because extensions obtained by use of fundamental solutions are generally unique. However, the compatibility conditions needed to continue the process and nibble the whole boundary of A in Theorems 11.14 and 11.15 seem to require a uniqueness property.

In order to show a converse direction, that is, lack of local extension implies that the conclusion of Theorem 11.14 and 11.15 do not hold, we content ourselves with the following result for the case $l = 1$ and for $\vec{\mathrm{D}}_1$ analytic elliptic (that is, $\vec{\mathrm{D}}_1 h = 0$ implies h is analytic):

THEOREM 11.16. *Let the notation be as in the previous theorem. We assume that A has a smooth boundary and that $\vec{\mathrm{D}}_1$ is analytic elliptic. Let p be a boundary point of A. Suppose there exist neighborhoods N_b of p in A for $0 \leqslant b \leqslant 1$ with smooth boundaries and functions $k_b \in \mathscr{E}(A)$ for $b \leqslant 1$ such that $\{k_b\}$ is bounded in the topology of $\mathscr{E}(A)$ and*

1. *$N_b \subset N_{b'}$ for $b > b'$.*
2. *The boundaries of N_b all have the same intersection $\partial N_b \cap \partial A$ with the boundary of A. Moreover, there exists a fixed neighborhood M of $\partial N_b \cap \partial A$ in A such that the intersection of ∂N_b with M is independent of b. Moreover, $N_b \cup M$ is connected.*
3. *$\vec{\mathrm{D}}_1 k_b = 0$ on $N_b \cup M$ but $k_b \not\equiv 0$ on $N_b \cup M$ for $b \leqslant 1$.*
4. *There exist points $p_b \in N_b$ with $p_b \to p$ as $b \to 1$ from below such that we cannot extend k_p to a solution of $\vec{\mathrm{D}}_1$ in the neighborhood of p_b.*

Then we cannot always solve $\boxed{\mathrm{D}}\vec{\mathrm{f}} = \vec{\mathrm{g}}$.

See Fig. 35.

PROOF. It suffices, by applying some properties of topological vector spaces, to produce a set $B \subset \mathscr{E}'^{r}(A)$ with $\boxed{\mathrm{D}}'B$ bounded but B not bounded in the topology of $\mathscr{E}'^{r}(A)/\vec{\mathrm{D}}_1\mathscr{E}'(A)$

Let h be a function in $\mathscr{E}(A)$ which is 0 outside $N_0 \cup M$ and is 1 on $N_1 - M$ except possibly for a small neighborhood of the part of the boundary of N_1 which does not meet M. For $b \leqslant 1$ define

$$\vec{\mathrm{S}}_b = h\vec{\mathrm{D}}_1 k_b\,.$$

Since $\vec{\mathrm{D}}_1 k_b = 0$ on M for all b, we see that $\vec{\mathrm{S}}_b = 0$ on M. Thus

$$\boxed{\mathrm{D}}'\vec{\mathrm{S}}_b = 0 \quad \text{on } M \cup N_1$$

which imples that $\boxed{\mathrm{D}}'\vec{\mathrm{S}}_b = 0$ except on the fixed compact set $N_0 - N_1$. Thus by multiplying k_b by sufficiently small constants if necessary, $B = \{\boxed{\mathrm{D}}'\vec{\mathrm{S}}_b\}$ is a bounded set.

We claim that $\{\vec{\mathrm{S}}_b\}$ is not bounded in the topology of $\mathscr{E}'^{r}(A)/\vec{\mathrm{D}}_1\mathscr{E}'(A)$. In fact, we cannot find $T_b \in \mathscr{E}'(A)$ such that all $\vec{\mathrm{S}}_b + \vec{\mathrm{D}}_1 T_b$ have fixed compact support. For, suppose there is some neighborhood N of p on A and $T_b \in \mathscr{E}'(A)$ such that $\vec{\mathrm{S}}_b + \vec{\mathrm{D}}_1 T_b = 0$ on N. By shrinking N if necessary, we may assume that $h = 1$ on N and that N is connected. Thus, on N, $\vec{\mathrm{S}}_b = \vec{\mathrm{D}}_1 k_b$ so

$$\mathrm{D}_1(T_b + k_b) = 0 \quad \text{on } N. \tag{11.62}$$

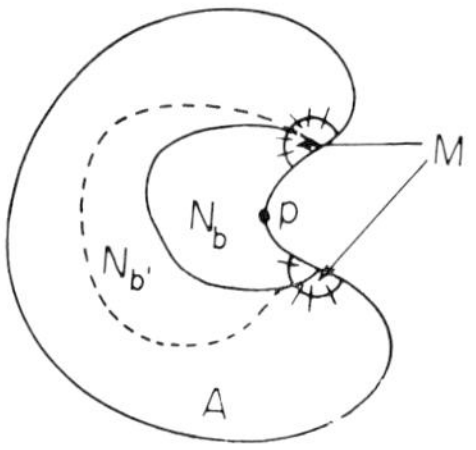

FIGURE 35

Since $T_b = 0$ in a neighborhood of ∂A, it follows from Condition 3 and the fact that $\vec{\mathrm{D}}_1$ is analytic elliptic that

$$T_b = 0 \quad \text{on } N_b \cap N.$$

Thus,

$$k_b + T_b = k_b \quad \text{on } N_b \cap N. \tag{11.63}$$

Hence, by (11.62), $k_b + T_b$ provides an extension of k_b to all of N (in particular, to a neighborhood of p_b if b is large enough) satisfying the given equation. In view of Condition 4 this gives our result. □

See Remark 11.15.

Theorems 11.14, 11.15, and 11.16 together with the remarks at the end of Section XI.2 can often be used to give "almost necessary and sufficient conditions" on the boundary of A for the solvability of $\boxed{\mathrm{D}}\vec{\mathrm{f}} = \vec{\mathrm{g}}$. In any case, the method seems to yield the "best" geometric conditions on the boundary of A.

There is another possible approach to the problem of determining domains A on which we can solve inhomogeneous problems. Suppose that A is a polyhedron. We decompose A into convex polyhedra A_j. On each convex polyhedron we can solve inhomogenous problems by Theorems 6.1 and Example 5 of Chapter 5. The problem becomes the study or compatibility conditions on the intersections of the A_j. At present we have not been able to obtain very many results by this approach.

XI.4. Special Functions and Group Representations

Most of the classical special functions arise from the study of special types of solutions or eigenfunctions of the Laplacian. Usually the special solutions come from separation of variables or from group invariance properties. We shall consider the latter approach.

As the simplest illustration of our method, let us construct rotationally invariant solutions of Laplace's equation $\Delta f = 0$ in the x_1, x_2 plane. It is clear that the only such solution in the whole plane is the constant; if we allow a singularity at the origin then we may take also $f = \log |\zeta|$, where $\zeta = x_1 + ix_2$. We wish to construct these functions by a general method.

By Theorem 7.1, we may write

$$f(x_1, x_2) = \int_{z_1 = \pm iz_2} \exp(ix_1 z_1 + ix_2 z_2)\, d\mu(z_1, z_2)$$

for a suitable measure μ. (We assume $f \in \mathscr{E}$. We deviate somewhat from the notation of Theorem 7.1 by absorbing the factor k in the measure.) We shall denote a plane rotation by

$$l_\theta = \begin{pmatrix} \cos\theta & -\sin\theta \\ \sin\theta & \cos\theta \end{pmatrix}$$

and we write $(l_\theta f)(x_1, x_2) = f(l_\theta(x_1, x_2))$, where (x_1, x_2) is thought of as a column vector. It is clear that

$$\begin{aligned} (l_\theta f)(x_1, x_2) &= \int \exp\left[i(x_1, x_2)\cdot l'_\theta(z_1, z_2)\right] d\mu(z_1, z_2) \\ &= \int \exp\left(ix_1 z_1 + ix_2 z_2\right) d\mu(l_\theta(z_1, z_2)), \end{aligned}$$

where l'_θ is the adjoint of l_θ. Thus the simplest way of choosing μ to make f invariant is as follows: Fix some point $z^0 \in \{z_1^2 + z_2^2 = 0\}$. Denote by Lz^0 the orbit of z^0 under the rotation group L. (Note that $Lz^0 \subset \{z_1^2 + z_2^2 = 0\}$.) Choose μ as the rotationally invariant measure on Lz^0.

A simple example is $z^0 = (1, i)$. Then the orbit of z^0 is

$$\begin{aligned} l'_\theta z^0 &= \begin{pmatrix} \cos\theta & \sin\theta \\ -\sin\theta & \cos\theta \end{pmatrix}\begin{pmatrix} 1 \\ i \end{pmatrix} \\ &= \begin{pmatrix} \cos\theta + i\sin\theta \\ -\sin\theta + i\cos\theta \end{pmatrix} \\ &= e^{i\theta}\begin{pmatrix} 1 \\ i \end{pmatrix}. \end{aligned}$$

Our choice of $d\mu$ is just $d\theta$ on the orbit. We thus obtain

$$\begin{aligned} f(x_1, x_2) &= \int \exp\left[ie^{i\theta}(x_1 + ix_2)\right] d\theta \\ &= \int \exp\left(ie^{i\theta}\zeta\right) d\theta \\ &= \int_{|\omega|=1} \exp(\omega\zeta)\, d\omega/\omega \\ &= 1. \end{aligned}$$

Here ω is a complex variable. The final step is by Cauchy's formula. We are, of course, ignoring factors of $2\pi i$.)

A similar analysis shows that if we started with any z^0 we would get $f = \text{const}$. Thus we cannot obtain the solution $\log|\zeta|$ using orbits of L. In order to obtain the solution $f = \log|\zeta|$ we make the following observavation: If f is invariant under L and f has some analyticity properties then f is also invariant under L^C which is the complexification of L (that is, allow θ to be an arbitrary complex number). We therefore try to choose μ

as the invariant measure of an orbit $L^C z^0$. A simple argument shows that such a μ is too large to yield a meaningful f (at least if the integral for f is to be interpreted in any "reasonable" sense). However, just as invariance for analytic functions under L is equivalent to invariance under L^C, so it is equivalent to invariance under any real group L^R whose complexification is L^C. Since L^C is isomorphic to the multiplicative group of complex numbers, the only choices for L^R are L and a real multiplicative half-line (which corresponds to θ pure imaginary). We take for L^R this latter choice.

The orbit of $z^0 = (-i, 1)$ is now, by the above

$$l'_\theta(-i, 1) = -ie^{-\theta}(1, i),$$

where $\theta \in (-\infty, \infty)$. Thus

$$\begin{aligned} f(x_1, x_2) &= \int_{-\infty}^{\infty} \exp\,[e^{-\theta}(x_1 + ix_2)]\, d\theta \\ &= \int_{-\infty}^{\infty} \exp\,[e^{-\theta}\zeta]\, d\theta \\ &= \int_0^{\infty} e^{a\zeta}\, da/a. \end{aligned}$$

The integral clearly does not exist in the usual sense; in general there is trouble at both limits of integration. Formally one might think that f is a constant because da/a should be invariant under a multiplicative change of variables. We shall see that this is not so.

For simplicity let ζ be real and negative. We interpret the above integral as a Cauchy principal value. Since the integrand is positive this does not help matters but, if we "subtract an infinite constant" we get an interesting result: Write

$$f_\varepsilon(\zeta) = \int_\varepsilon^{\infty} e^{a\zeta}\, da/a.$$

Then

$$\begin{aligned} f'_\varepsilon(\zeta) &= \int_\varepsilon^{\infty} e^{a\zeta}\, da \\ &= -e^{\varepsilon\zeta}/\zeta \end{aligned}$$

since $\zeta < 0$. Thus, $\lim f'_\varepsilon(\zeta) = -1/\zeta$ so, formally, $f(\zeta) = -\log\zeta + \text{const}$. Of course the constant $= \infty$.

In order to get a more meaningful result, we want to subtract two such f.

If we choose the orbit of $(-i, -1)$ we obtain the function

$$\tilde{f}(\zeta) = \int_{-\infty}^{\infty} \exp\,[e^{\theta}(x_1 - ix_2)]\, d\theta$$

$$= \int_{0}^{\infty} e^{a\bar{\zeta}}\, da/a.$$

Now for ζ real, $\tilde{f} = f$. Suppose instead that ζ is pure imaginary, say $\zeta = it$. We observe that for any $\varepsilon > 0$

$$\int_{\varepsilon}^{\infty} e^{iat}\, da/a$$

exists in the space $\mathscr{D}'$ of distributions (see Chapter 5, Example 5). Again, clearly

$$\int_{0}^{\infty} (e^{iat} - e^{-iat}) da/a = 2i \int_{0}^{\infty} \sin at\, da/a$$

exists in the space $\mathscr{D}'$.

We claim that

$$f(it) - \tilde{f}(it) = iH(t) + \text{const},$$

where H is the Heaviside function

$$H(t) = \begin{cases} 1 & \text{for } t > 0 \\ 0 & \text{for } t < 0, \end{cases}$$

where the const is now finite. This follows by integrating the relation

$$\int_{0}^{\infty} \cos at\, da = \frac{1}{2} \int_{-\infty}^{\infty} \exp\,(iat)\, da$$

$$= \frac{1}{2}\,\delta.$$

The last identity is standard (see Schwartz [1]).

This suggests that instead of looking for a way of defining $f(\zeta)$ and $\tilde{f}(\zeta)$ precisely, we try to make sense of some linear combinations, and hope to obtain $\log|\zeta| +$ (finite) const from some combination. To accomplish this, we observe that the fundamental solution to Δ is $\log|\zeta|$, that is, $\Delta \log|\zeta| = \delta$ By Fourier transform we have formally)

$$\log|\zeta| = -\int_{z\text{ real}} e^{ix_1z_1 + ix_2z_2}\, dz_1 dz_2/(z_1^2 + z_2^2).$$

Now, for $x_2 > 0$ we can shift the z_2 contour to $+i\infty$. We obtain by residues formally

$$\log |\zeta| = \frac{1}{2} \int_{z_1>0} e^{ix_1z_1 - x_2z_1}\, dz_1/z_1 + \frac{1}{2} \int_{z_1<0} e^{ix_1z_1 + x_2z_1}\, dz_1/z_1$$

$$= \frac{1}{2} \int_{z_1>0} (e^{iz_1\zeta} + e^{-iz_1\bar{\zeta}})\, dz_1/z_1.$$

This agrees with our heuristic calculation of f and $\tilde{f}$ up to sign; the sign, of course, has no group theoretic meaning. Note that the factor 2π that usually occurs in Cauchy's formula is absent because of our normalization of the measure dz (see Chapter V, Example 3).

This suggests that the true way to interpret f and $\tilde{f}$ is to find some rational function of z_1, z_2 whose "residue" on $z_1^2 + z_2^2$ is the integrand in f. This is somewhat similar in principle to Theorem 4.1. (Compare also the remarks on meromorphic solutions in Section XI.2.) It would take us too far afield here to go into further details. We shall content ourselves with giving two more examples.

EXAMPLE 2. *Bessel Functions.* Here we study solutions in the plane of $\Delta f + f = 0$ which are invariant under L. We can write

$$f(x_1, x_2) = \int_{z_1^2+z_2^2=1} \exp\,[ix_1z_1 + ix_2z_2]\, d\mu(z_1, z_2).$$

The simplest choice of μ for an f invariant under L is the invariant measure of the orbit under L of (1, 0). This yields

$$f(x_1, x_2) = \int \exp\,[ix_1 \cos\theta - ix_2 \sin\theta]\, d\theta$$

$$= \int e^{i|\zeta| \cos\theta}\, d\theta$$

by a simple change of variables. This is the standard formula for the Bessel function denoted by $J_0(|\zeta|)$ (see Bateman [1], Vol. 2).

The orbit under L^R of (1, 0) leads to

$$K_0(|\zeta|) = \int_{-\infty}^{\infty} e^{-\theta \cosh\theta}\, d\theta.$$

EXAMPLE 3. *Legendre Functions.* We study solutions in R^3 of the wave equation

$$\Box f \equiv \partial^2 f/\partial x_1^2 - \partial^2 f/\partial x_2^2 - \partial^2 f/\partial x_3^2 = 0$$

which are invariant under the group M of rotations about the x_1 axis. We impose on f a condition of *homogeneity*:

$$f(tx_1, tx_2, tx_3) = t^{-s} f(x_1, x_2, x_3)$$

for some complex s (fixed) for all real $t > 0$.

The interest of such functions arises from the theory of group representations as follows: Denote by G the three-dimensional Lorentz group, that is, the group of linear transformations of R^3 leaving $x_1^2 - x_2^2 - x_3^2$ invariant. We obtain a representation of G on functions h on R^3 by setting, for $g \in G$,

$$(gh)(x) = h(g^{-1}(x)).$$

Now, each $g \in G$ commutes with $\square$ so G acts on the subspace of solutions of $\square h = 0$. G also commutes with scalar multiplication so it acts on the subspace of h which satisfy a homogeneity condition. If we "decompose" this representation space according to the subgroup M of G, we look for functions like f above and also similar functions where "invariance under M" is replaced by "transform according to a character of M." (Technically it is better to replace R^3 by the forward light cone: $x_1 > 0$, $x_1^2 - x_2^2 - x_3^2 > 0$.)

We shall work purely formally. If $\square F = 0$, we can write

$$F(x_1, x_2, x_3) = \int_{z_1^2 = z_2^2 + z_3^2} \exp\,(ix_1 z_1 + ix_2 z_2 + ix_3 z_3)\, d\mu(z_1, z_2, z_3).$$

The simplest choice of μ to make F invariant under M is the invariant measure of the orbit $(1, 0, 1)$. We represent elements of M as

$$\begin{pmatrix} 1 & 0 & 0 \\ 0 & \cos\theta & -\sin\theta \\ 0 & \sin\theta & \cos\theta \end{pmatrix}$$

Thus

$$F(x_1, x_2, x_3) = \int \exp\,(ix_1 + ix_2 \sin\theta + ix_3 \cos\theta)\, d\theta.$$

To impose the homogeneity condition we form the Mellin transform

$$\begin{aligned} f(x_1, x_2, x_3) &= \int F(tx_1, tx_2, tx_3) t^s\, dt/t \\ &= \iint \exp\,(itx_1 + itx_2 \sin\theta + itx_3 \cos\theta) t^s\, d\theta\, dt. \end{aligned}$$

We change the order of integration. Using

$$\Gamma(s) = \int e^{-a} a^s\, da/a$$

we have

$$\int e^{-a\lambda} a^s\, da/a = \lambda^{-s}\Gamma(s).$$

Thus formally,

$$f(x_1, x_2, x_3) = \int [-(ix_1 + ix_2 \sin \theta + ix_3 \cos \theta)]^{-s} \, d\theta.$$

This, in suitable coordinates, is just the Legendre function P_{-s} (see Bateman [1], Vol. 1). If we replace M by M^R we obtain the associated Legendre function Q_{-s}.

See Remark 11.16.

EXAMPLE 4. *Semi-simple Lie Groups.*

Let G be a real semi-simple Lie group. We think of G acting on its Lie algebra $\mathfrak{g}$ by adjoint representation. It is known that there exist r algebraically independent polynomials on $\mathfrak{g}$ which are invariant under this action of G (r is the rank of G). Thus we are led to generalization of Example 3 in which we study the simultaneous solutions of r constant coefficient operators. It would take us too far afield to go into the details; they will appear elsewhere.

See Remark 11.17.

XI.5. Variable Coefficient Equations

We wish to give some idea of an application of the theory of general Fourier analysis to variable coefficient linear partial differential equations in Ehrenpreis [21]. In that paper, we show how to prove the Cauchy-Kowaleswki theorem and some results on the Cauchy problem for linear hyperbolic equations (see Chapter IX). We shall give only the idea of the method. We shall give below two different approaches to variable coefficient problems.

Let D be an operator with variable coefficients, say

$$D = \frac{\partial^m}{\partial t} + D_1(t, \xi, \partial/\partial\xi) \frac{\partial^{m-1}}{\partial t^{m-1}} + \cdots + D_m(t, \xi, \partial/\partial\xi),$$

where D_j have coefficients in $\mathscr{H}$ (see Chapter V, Example 1). As in Chapter IX, the Cauchy problem is related to the map

$$\gamma : f \to [Df; f(0, \xi), \partial f(0, \xi)/\partial t, \ldots, \partial^{m-1} f(0, \xi)/\partial t^{m-1}]$$

of $\mathscr{H}$ into $\mathscr{H} \oplus \mathscr{H}_0^m$, where $\mathscr{H}_0$ is the space of entire functions of ξ. To show that γ is a topological isomorphism is the same as showing that γ' (the adjoint of γ) is, which is the same as showing that $\hat{\gamma}'$ (the Fourier transform of γ') is. We shall show that $\hat{\gamma}'$ is one-to-one if $n = 1$ (that is, D

is an ordinary differential operator) and the coefficients of D are polynomials. The passage to the case when the coefficients are in $\mathscr{H}$ is accomplished by use of Theorem 1.5. The passage to $n > 1$ and to the proof that $\hat{\gamma}'$ is a topological isomorphism involves only technical difficulties; the interested reader should consult Ehrenpreis [21].

To show $\hat{\gamma}'$ is one-to-one we must show that it is impossible to have, for $G \in \mathbf{H}'$,

$$\hat{D}'G + \sum_{j=0}^{m-1} (is)^j c_j = 0. \tag{11.64}$$

Here c_j are constants and

$$\hat{D}'G = (is)^m\, G + \sum \lambda_{lj}(s^l G), \tag{11.65}$$

where λ_{lj} are ordinary differential operators with constant coefficients. At first sight it seems as though we have not gained anything because (11.64) is again a differential equation. However, G is now an entire function of *exponential type*; this is crucial for our method.

By assumption, there is an $a' > 0$ so that for, say $|s| \geqslant R$, we have

$$|G(s)| \leqslant \exp\,(a'\,|s|). \tag{11.66}$$

For simplicity we may assume $a' = 1$. From (11.66) and Cauchy's formula we have for $|s| \leqslant R - 1$

$$|\lambda_{lj}(s^l G)| \leqslant M R^{m-1} \exp\,(R) \tag{11.67}$$

for all l, j for a suitable $M > 0$. (11.64) and (11.65) give

$$|s|^m\,|G| \leqslant \sum_{l<m} |\lambda_{lj}(s^l G)| + \sum_{j<m} |s|^j\,|c_j|. \tag{11.68}$$

Combining this with (11.67) yields for $|s| = R - 1$ (hence for $|s| \leqslant R - 1$)

$$|G(s)| \leqslant M' R^{m-1} \exp\,(R)/(R-1)^m + m \max |c_j|/(R-1), \tag{11.69}$$

where $M' = M$ times the number of λ_{lj}. Since the c_j are constants, say

$$m \max |c_j| = M''$$

we can deduce from (11.69)

$$|G(s)| \leqslant M_1 R^{m-1} \exp\,(R)/(R-1)^m + M_1/(R-1), \tag{11.70}$$

where $M_1 = \max\,(M', M'')$.

Note that (11.70) is an improvement over our original assumption (11.66) because of the factor $R^{m-1}/(R-1)^m$.

We can now iterate the process: Instead of (11.67) we deduce for $|s| \leqslant R - 2$

$$(11.71)\quad |\lambda_{lj}(s^l G)| \leqslant M M_1 R^{m-1}(R-1)^{m-1} \exp(R)/(R-1)^m + M M_1 (R-1)^{m-1}/(R-1).$$

Combining (11.68) and (11.71) yields

$$(11.72)\quad |G(s)| \leqslant M_1^2 R^{m-1}(R-1)^{m-1} \exp(R)/(R-1)^m (R-2)^m + M_1^2 (R-1)^{m-1}/(R-1)(R-2)^m + M_1/(R-2).$$

We iterate k times and we have for $|s| \leqslant R - k - 1$

$$(11.73)\quad \begin{aligned} |G(s)| \leqslant\, & M_1^k R^{m-1} \exp(R)/(R-1)(R-2)\cdots(R-k)^m \\ & + M_1^k (R-1)^{m-1}/(R-1)(R-2)\cdots(R-k)^m \\ & + M_1^{k-1}(R-2)^{m-1}/(R-2)\cdots(R-k)^m \\ & + \cdots + M_1/(R-k). \end{aligned}$$

For k close to $R/2$, R large, the term $(R-1)(R-2)\cdots(R-k)$ in (11.73) is close to $\Gamma(R)/\Gamma(R/2)$ which is larger than the numerator. Similarly for the other terms. Since R is arbitrarily large we find easily that $G \equiv 0$. ▯

We wish to explain two other approaches to variable coefficient problems from the point of view of complex Fourier analysis. We hope to study them in detail elsewhere.

1. The main point of the approach of Chapters VI–X is that we can take the Fourier transform to "replace" problems involving linear partial differential equations with constant coefficients by algebraic problems. If D is a linear partial differential equation with polynomial coefficients, then the Fourier transform $\hat{D}$ of D is again of the same form. However, if D has linear coefficients then $\hat{D}$ is a linear partial differential equation of first order. As such, problems involving $\hat{D}$ can be solved, classically, by reducing them to ordinary differential equations. If D has polynomial coefficients it is often possible to take a more complicated "Fourier-like" transform in which $\exp(ix \cdot z)$ is replaced by $\exp(Z(x))$, where Z runs through all polynomials if a given degree ($\geqslant 1$) in such a way as to "reduce" problems involving D to ordinary differential equations.

2. This approach is based on trying to interpret all the arguments used to prove Theorem 4.2 directly in the space $\mathscr{W}$ without using Fourier transform, and then trying to apply this method to variable coefficient equations. What is the meaning of the passage from semilocal to global in Chapter IV? Roughly speaking this means we partition $\mathscr{W}$ into a sum

$$(11.74)\qquad \mathscr{W} = \sum W_{\alpha,\beta},$$

where α runs through all lattice points in C^n, and $\beta > 1$, and $W_{\alpha,\beta}$ consists of all $w \in \mathscr{W}$ which are entire functions of exponential type satisfying

$$(11.75) \qquad |w(x) \exp(-ix \cdot \alpha)| \leqslant A \exp(\beta \|x\|)$$

for all complex x. This corresponds to the condition that w can be represented as the Fourier transform of a measure with support on the cube center α, side 2β.

Of course, the sum in (11.74) is not direct because the cubes intersect; convergence of the sums in (11.74) must be understood on terms of the seminorms k. The lack of directness is what accounts for the cohomology problems of Chapter IV.

In Chapters II and III we studied the local and semilocal problems. One of the essential tools we employed is the Lagrange interpolation formula. In Chapter IX we saw that the Lagrange interpolation formula is intimately related to the Cauchy-Kowalewski theorem, which makes sense for partial differential equations with *analytic* coefficients. Thus there is the possibility of extending our results to equations with analytic coefficients. It would take us too far afield to enter into more details.

Remarks

Remark 11.1. See page 323.

These results indicate that Theorem 4.1 is more powerful than Theorems 6.1 and 7.1, since we do not see how to derive them from these theorems. However, Theorem 11.4 could be derived from Theorem 7.2.

Remark 11.2. See page 326.

Remark 11.3. See page 327.

A similar result holds for distribution solutions.

Remark 11.4. See page 327.

The same method applies to systems of convolution equations if we assume some relative slowly decreasing property (see Ehrenpreis [4] and Section XI.1.)

Remark 11.5. See page 328.

In case $\vec{D}$ is analytic elliptic, that is, all solutions of $\vec{D}f = 0$ are real analytic, then, as is easily seen, assumption (β) will be satisfied if and only if the complement of Ω_1 is connected, that is, Ω_1 is simply connected. When applied to the Cauchy-Riemann equations for holomorphic functions of several complex variables, our Theorem 11.5 is just Hartog's original result on analytic continuation.

Remark 11.6. See page 338.

Remark 11.7. See page 346.

If we tried to derive Theorems 11.6 and 11.7 by Methods a or c of Section VIII.1 we would be led to the problem of proving analogs of Theorem 6.1 on certain domains Ω for spaces of infinitely differentiable functions which are zero in the neighborhood of certain parts of the boundary. Unfortunately, even for the simplest Ω we do not know of any direct approach and our only method of handling this problem is to use Theorems 11.6 and 11.7.

Remark 11.8. See page 347.

If $\vec{\mathrm{D}}$ consists of the single operator $\partial/\partial t - \partial^2/\partial\xi^2$, we showed in Chapter VIII (see the beginning of Section VIII.4) that $t = 0$ is removable in the sense of Theorem 11.8. Thus, the unique continuation hypothesis in Theorem 11.8 is essential.

Remark 11.9. See page 362.

Essentially the same proof would apply to the case $r = 1$, D replaced by a convolution operator by a distribution S of compact support such that $\hat{S}$ is slowly decreasing (see Section XI.1). The type of modifications of the proof of Theorem 11.10 that are necessary to handle this situation can be found in Ehrenpreis [5].

Theorem 11.10 shows that regularity in a half-plane implies, under suitable conditions, regularity in the whole space. Actually, the above proof shows that we did not have to assume that f (in Theorem 11.10) is C^∞ in all of $t < 0$ in order to conclude that f is C^∞ in $t > 0$; we needed to assume only that f is C^∞ in a thin strip $-\varepsilon < t < \varepsilon/2$. In view of the ideas of Chapter IX, we should expect that, if $r > 0$, we should be able, for suitable $\vec{\mathrm{D}}$, to deduce that f is C^∞ in all of R by assuming it is C^∞ in the neighborhood of a noncharacteristic. We have not investigated this question.

Remark 11.10. See page 362.

We suspect that the hypotheses in Theorems 11.10 and 11.11 are the same. However, as stated, these assertions apply (for $r = 1$) to certain invertible convolution operators (compare Ehrenpreis [4], [5]).

Remark 11.11. See page 367.

It is not too surprising that a method based on (11.54) will often fail to produce examples of the converses of Theorems 11.10, 11.11, and 11.12. For the hypotheses of these theorems mean that V can be split into a "hyperbolic" part V_1 and an "elliptic" part V_2. These can intersect only where Im w is large. Thus, points with $|\text{Im } s|$ small can be joined on V to points with $|\text{Im } s|$ large only through points on which $|\text{Im } w|$ is large. In the opposite case they can be connected even through small values of $|\text{Im } w|$. However, (11.54) tries to use a complex analytic curve Ψ (one complex dimension) for a connecting device. As our discussion of the failure of (11.54) shows, a precise meaning to the connected concept, which would lead to necessary and sufficient conditions in these theorems, may be very subtle. The final result would probably necessitate the introduction of a suitable concept of *analytic connectivity* which would modify the usual concept of connectivity by adding certain complex analytic conditions.

Remark 11.12. See page 368.

This result was found independently by Hörmander and the author (unpublished). Hörmander and Malgrange gave a partial extension to $l = r = 1$, $n > 2$ (see Hörmander [3], pp. 91–92, Theorems 3.7.3 and 3.7.4).

Remark 11.13. See page 374.

In the above proofs we nibbled A to a small convex set contained in A. It is possible to obtain similar results by starting with a suitable convex set A^c containing A, using the known results (see Chapter VI) for A^c, and then nibbling from A^c to A. We leave the precise formulation to the reader.

Remark 11.14. See page 374.

In Theorem 11.15 we used the semifundamental solution to prove Theorem 11.15 rather than a fundamental solution. It is sometimes possible to use, instead, a suitable fundamental solution. Consider the system

$$\text{grad}^C \vec{\mathrm{f}} = \vec{\mathrm{g}}.$$

In order to understand the idea, suppose that $0 \in \partial A$ and 0 can be just nibbled by the complex $\zeta_1 = x_1 + ix_2$ plane: $x_3 = x_4 = x_5 = x_6 = 0$. Suppose, as in the proof of Theorem 11.14 that we want to nibble a neighborhood of 0 out of the support of $\vec{S}$. We try to solve the equation

$$\text{curl}^C \vec{U} = \vec{S}$$

on M with $\vec{U} = 0$ on N_3. (Notation is as in the proof of Theorem 11.14.)

As in the case of curl treated in Section VI.2, a fundamental solution for curl^C is

$$e_1 = (0, 0, E_3); \quad e_2 = E_3; \quad e_3 = 0,$$

where E_3 is a fundamental solution for $\partial/\partial\bar{\zeta}_3$ which we can take to be the product of $1/\zeta_3$ on the ζ_3 plane by the ∂ function in x_1, x_2, x_3, x_4. If we apply this fundamental solution as in (11.59) we find that $\vec{U}$ is only zero "above" the ζ_3 plane rather than on a full neighborhood of a in the boundary of A near a (more precisely, above a small neighborhood of the ζ_3 plane). (See Fig. 36.)

The reason for the difficulty is that the mouth in question is tangent to A only at a. If, however, we use a "moving coordinate system" for which each boundary point of A near a can be just nibbled by the ζ_3 plane, then we can conclude that $\vec{U}$ vanishes on N_3. In order for this argument, based on moving coordinate systems, to work we need to know that A is strongly pseudoconvex.

See Problem 11.14.

Remark 11.15. See page 379.

As noted by Serre, the possibility of solving $\boxed{D}_1\vec{f} = \vec{g}$ implies, by Serre's duality theorem, that we have global extensions as in Theorem 11.5 (see Ehrenpreis [15]). Theorem 11.16 shows that we also have the local extension property.

Remark 11.16. See page 385.

The passage between L and L^R or M and M^R explains why the integral representations for the pairs of classical functions involve the same integrand; only the path of integration in the complex plane is altered.

Remark 11.17. See page 385.

The above can be used in some cases for matrix groups defined over finite fields k and other fields. The main point to observe is that we think of an expression (for z fixed)

$$\exp(ix_1z_1 + ix_2z_2 + ix_3z_3)$$

as an additive character on $k \oplus k \oplus k$. We then think of expressions such as

$$F(x_1, x_2, x_3) = \int_{z_1^2 = z_2^2 + z_3^2} \exp(ix_1z_1 + ix_2z_2 + ix_3z_3)\,d\mu$$

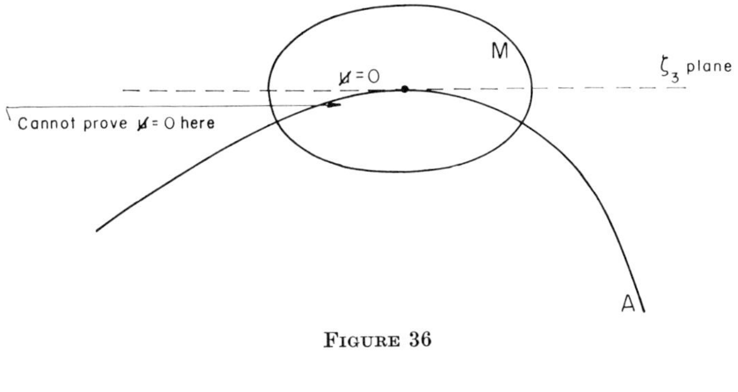

FIGURE 36

as defining what we mean by a solution of $\square F = 0$. We shall not go into the details here. We content ourselves with the remark that the analog of K (see Example 2) for finite fields involves Kloosterman sums.

Problems

Problem 11.1 See page 321.

(Even for $n = 1$.) Give necessary and sufficient conditions on the $\vec{F}_j$ if they are exponential polynomials in order that they be slowly decreasing generators of a properly slowly decreasing module. (Note that in Ehrenpreis [2] it is proved that for $r = l = 1$, every exponential polynomial is slowly decreasing.)

Conjecture. If F_j are exponential polynomials for which the exponentials have algebraic frequencies, then they are slowly decreasing generators of a properly slowly decreasing ideal.

Problem 11.2 (Exercise) See page 325.

Carry out the details.

Problem 11.3 (Exercise) See page 328.

(a) The simplest case of Theorem 11.5 occurs when Ω_1 and Ω_2 are convex. Prove an analogous result for this case when Ω_0 is a "cap," rather than a shell, that is, $0 \in \Omega_1$ and Ω_0 is the part of $\Omega_2 - \Omega_1$ which lies in the half-plane $x_1 \geqslant 0$. [In this case, in the notation of the proof of Theorem 11.5, we have to multiply $\vec{D}g$ by a suitable function, say u, to make $u\vec{D}g$ of compact support. Then $u\vec{D}g$ does not satisfy (11.8) near $x_1 = 0$, but we can subtract a (vector) function $\vec{u}'$ of support near $x_1 = 0$ so that $\vec{g}' = u\vec{D}g - \vec{u}'$ satisfies (11.8); the existence of such a $\vec{u}'$ follows from Theorem 6.5. We can now continue as before.]

Use this method as in the proof of Theorem 11.5 to derive analogous results in case Ω_1 and Ω_2 are not convex.

(b) Extend the results of Theorem 11.5 and (a) above to the case $l > 1$, that is solutions of $\boxed{D}\vec{f} = 0$. In this case the analog of (11.8) is $\boxed{D}^1\vec{g} = 0$, where $\boxed{D}^1$ is an $r_1 \times r$ matrix generating the module of relations of the rows of $\boxed{D}$. The existence of h [following (11.9)] means in our situation that, if $\boxed{D}^1\vec{g} = 0$, then $\vec{g} = \boxed{D}\vec{h}$ where support $\vec{h}$ is contained in the convex hull of support $\vec{g}$. For this we must (by Theorem 6.5) know the following property:

(Ext). $\boxed{D}$ generates the right module of relations for $\boxed{D}^1$. The condition (Ext) generalizes (α) to $l > 1$.

The type of extension problem dealt with in (a) and (b) is found in Ehrenpreis [15] for the Cauchy-Riemann and related systems. Theorem 11.5 is due to Ehrenpreis [20]. Malgrange, in Leray [2], extended those results to the Cases (a) and (b) above whenever Ω_1 and Ω_2 are convex. The condition (Ext) is due to Malgrange.

Problem 11.4 (Conjecture) See page 332.

L is a mouth for $\vec{D}$ if and only if there is an operator ∂ in the ideal generated by the D_j' in the ring of linear constant coefficient operators such that ∂ depends only on derivatives in directions along L.

We suspect also that the existence of such a ∂ is necessary and sufficient for the existence of distributions S_j which are of the form $S_j = T_j \times \delta_L$ with T_j a distribution on L and δ_L the unit mass at the origin in the directions "orthogonal" to L such that

$$\sum D_j' S_j = 0.$$

The conjecture can be verified for $r = 2$.

PROBLEM 11.5 (Conjecture) See page 338.

If no point in the boundary of the complement of A can be just nibbled by a mouth for $\vec{\mathrm{D}}$ then A is a domain of existence for $\vec{\mathrm{D}}$.

Although the boundary of A is the same as the boundary of its complement, the question of whether a point can be just nibbled depends on whether we consider it as a boundary point of A or of the complement of A.

PROBLEM 11.6 (Conjecture) See page 339.

If A is a domain of existence for $\vec{\mathrm{D}}$ and M is a $\vec{\mathrm{D}}$-submanifold of A whose boundary is contained in the boundary of A, then any $f \in \mathscr{E}(M)$ which satisfies the equations induced by $\vec{\mathrm{D}}$ on M can be extended to a solution on all of A.

PROBLEM 11.7 (Exercise) See page 343.

PROBLEM 11.8 (Exercise) See page 344.

PROBLEM 11.9 See page 349.

Can we construct in this manner all meromorphic solutions of $\vec{\mathrm{D}}$ (or at least those with a suitable growth condition) which have their singularities on L?

In case $\vec{\mathrm{D}}$ is the Cauchy-Riemann system in several complex variables we see easily that this is the case.

PROBLEM 11.10 (Exercise) See page 351.

PROBLEM 11.11 (Exercise) See page 362.

PROBLEM 11.12 See page 367.

Modify Theorems 11.10, 11.11, and 11.12 to obtain necessary and sufficient conditions.

PROBLEM 11.13 (Exercise) See page 373.

PROBLEM 11.14 (Exercise) See Remark 11.14 and page 374.

Derive Theorem 11.15 using a fundamental solution instead of a semifundamental solution. (The concept of mouth may have to be modified.)

PROBLEM 11.15 (Exercise) See page 375.

PROBLEM 11.16 (Conjecture) See page 375.

PROBLEM 11.17 See page 377.

PROBLEM 11.18 See page 377.

In the above we studied solutions of $\boxed{\mathrm{D}}\,\vec{\mathrm{f}} = \vec{\mathrm{g}}$ in $\mathscr{E}(A)$. Study the possibility of solving the system in other spaces, for example, $\mathscr{D}'(A)$ and $\mathscr{E}(\bar{A})$, the space of C^∞ functions on the closure $\bar{A}$ of A. Study also analogous problems for spaces defined by finiteness of certain integrals. Also, if $n = 2m$ and we think of $R^n = C^m$, study the possibility of solution of $\boxed{D}\vec{f} = \vec{g}$ in the space $\mathscr{H}(A)$ of holomorphic functions on A. Here we suppose that A is pseudoconvex and the operators in $\boxed{D}$ depend only on the $\partial/\partial z_j$.

In this connection it should be pointed out that J. J. Kohn [1] showed the possibility of solution in $\mathscr{E}(\bar{A})$ when A is strictly peudoconvex and $\boxed{D}$ is derived from the Cauchy-Riemann system. (This follows also from our methods.)

Problem 11.19 See page 377.

Study how the domains for which we can always solve $\boxed{D}\vec{f} = \vec{g}$ vary when the coefficients of $\boxed{D}$ are slightly perturbed.

Problem 11.20 See page 377.

In the case the hypothesis of Theorems 11.14 and 11.15 do not hold, what is the "obstruction" in a given $\vec{g}$ to solving $\boxed{D}\vec{f} = \vec{g}$?

CHAPTER XII

Lacunary Series. Refined Comparison Theorems

Summary

Let x be a complex variable. Let $\{c^j\}$ be a sequence of complex numbers and let $\{a^j\}$ be a sequence of real numbers. The classical Fabry gap theorem (see Levinson [1]) asserts: Suppose the series

$$f(x) = \sum c^j \exp(ia^j x)$$

converges uniformly on compact subsets of the strip $|\text{Im } x| < 1$ to a function $f(x)$. Suppose f can be continued to be analytic in the neighborhood of some point x^0 on the boundary of the strip. Then, if the sequence $\{a^j\}$ is "lacunary", f can be continued to be analytic in a strip containing x^0, and the series $\sum c^j \exp(ia^j x)$ converges uniformly on compact subsets of this larger strip.

We put the Fabry gap theorem in the general framework of AU spaces.

Let $\mathscr{W}$ and $\mathscr{W}_1$ be AU spaces with AU structures $K = \{k\}$ and $K_1 = \{k_1\}$, respectively. We define K_2 as the family of functions k_2 on C for which there exist $k \in K$, $k_1 \in K_1$ so that

$$k_2(z) = \max [k(z), k_1(z)].$$

Then K_2 becomes in a natural way an AU structure for a space $\mathscr{W}_2$ which we call the AU *intersection* of $\mathscr{W}$ and $\mathscr{W}_1$ and which we denote by $\mathscr{W} \overset{a}{\cap} \mathscr{W}_1$. For example, if Ω and Ω_1 are two open convex sets in R with nonempty intersection and if Ω_2 denotes the convex hull of $\Omega \cup \Omega_1$, then $\mathscr{E}(\Omega) \overset{a}{\cap} \mathscr{E}(\Omega_1) = \mathscr{E}(\Omega_2)$.

We think of the function $\exp(ia^j x)$ as the solution of the differential equation $df^j/dx - ia^j f^j = 0$. Suppose that $\mathscr{W}$ and $\mathscr{W}_1$ are LAU. Then, in view of the results of Chapter VII, we generalize the above as follows: Let $\vec{\mathrm{D}}^j = (\mathrm{D}_1^j, \ldots, D_{s_j}^j)$. Then we consider series of the form

$$f = \sum f^j,$$

where $\vec{\mathrm{D}}^j f^j = 0$.

Let $\mathscr{W}_3$ be an AU space with $\mathscr{W} \overset{a}{\cap} \mathscr{W}_1 \subset \mathscr{W}_3 \subset \mathscr{W}$. We say that the $(\mathscr{W}, \mathscr{W}_1)$ *density of* $\{\vec{D}^j\}$ is $\leqslant \mathscr{W}_3$ whenever the following holds: Let $\{f^j\}$ be a sequence of elements of $\mathscr{W}_3$ such that $\vec{D}^j f^j = 0$ for each j. If the series $\sum f^j$ converges in the topology of $\mathscr{W}$ to f which is in $\mathscr{W} \cap \mathscr{W}_1$, then f must be in $\mathscr{W}_3$ and the series converges in the topology of $\mathscr{W}_3$. If we can take $\mathscr{W}_3 = \mathscr{W} \overset{a}{\cap} \mathscr{W}_1$, then we say that $\{\vec{D}^j\}$ is $(\mathscr{W}, \mathscr{W}_1)$ *lacunary*.

Our density provides a generalization and refinement of the Pólya maximal density (see Pólya [1], Levinson [1]).

Our main tool in the computation of the $(\mathscr{W}, \mathscr{W}_1)$ density is the notion of a $(\mathscr{W}, \mathscr{W}_1)$ *parametrix* which is an analog of the parametrix used for elliptic partial differential equations. The most important part of our general computation gives a geometric criterion for the $(\mathscr{W}, \mathscr{W}_1)$ density of $\{\vec{D}^j\}$ to be $\leqslant \mathscr{W}_3$: Let V^j be the algebraic variety of common zeros of $P_1^j, \ldots, P_{s_j}^j$. Then certain regions of C should meet few V^j. This geometric condition reduces to that of Pólya when we specialize to his situation.

The following result is proved (see Theorem 12.9 and following): *There exists a $B > 0$ with the following property: Let $f(s) = \sum c_n n^{-s}$ where the series converges in some half-plane. Suppose that f can be extended to an entire function satisfying*

$$f(s) = 0(\exp(\beta |s| \log |s|).$$

for some $\beta < 1$. Then

$$c_n = 0(\exp(-\exp Bn)).$$

The example $f(s) = (1 - 2^{1-s})\zeta(s)$, where $\zeta(s)$ is the Riemann ζ-function, shows that we could not have $\beta \geqslant 1$ in the above theorem. In some sense our result shows that the Dirichlet series for $\zeta(s)$ has a "maximum possible cancellation."

XII.1. Formulation of the Problem

The first example of the type of result we wish to discuss is the classical result that the Taylor series about x_0 of an analytic function converges uniformly on compact sets of the largest disk centered at x_0 in which f is analytic. To put things another way, if we know (x is a complex variable)

1. The series $\sum c^j x^j$ converges uniformly on compact sets of $|x| < 1$, say to $f(x)$.

(Note that x^j is the jth power of x but j is a superscript in c^j, and also in a^j below.)

2. $f(x)$ is analytic in $|x| < 1 + \delta$ for some $\delta > 0$.

Then we can conclude

4. The series converges uniformly on compact sets of $|x| < 1 + \delta$.

(The reason there is no "3" will appear presently.)

There have been many sharpenings of this result in the direction of weakening Condition 2, which is only possible if we place conditions on the c^j. An important result of the type we are interested in is the celebrated Fabry gap theorem.

Suppose 1 holds as above and also

2′. $f(x)$ is analytic in some neighborhood of the point 1, say in $|x - 1| < \delta$.

3′. $c^j = 0$ except for j belonging to a subsequence $\{a^j\}$ with the property that the number of a^j such that $a^j \leqslant N$ is $o(N)$ as $N \to \infty$.

Then the conclusion 4 holds as before.

Before studying the structure of these theorems more carefully, let us make an exponential change of variables, i.e., replace x by $\exp(ix)$. Then the results are:

(I) Suppose

(a) $\sum c^j \exp(ijx)$ converges uniformly on compact sets of $\operatorname{Im} x > 0$ to $f(x)$, say. (Here $\operatorname{Im} x$ denotes the imaginary part of x.)

(b) $f(x)$ is analytic in $\operatorname{Im} x > -\delta$ for some $\delta > 0$.

Then

(d) The series converges uniformly on compact sets of $\operatorname{Im} x > -\delta$.

The Fabry gap theorem becomes

(II) Suppose (a) as in (I) and

(b′) $f(x)$ analytic in $|x| < \delta$.

(c′) $c^j = 0$ except for j in a subsequence $\{a_j\}$ as in 3′.

Then (d) follows.

Now (a) and (d) can easily be formulated in terms of growth conditions on the c^j, namely, (a) is equivalent to

$$c^j = 0(\exp(\varepsilon j)) \quad \text{for any } \varepsilon > 0,$$

while (d) is equivalent to

$$c^j = 0\,(\exp(-\delta + \varepsilon)j) \qquad \text{for any } \varepsilon > 0.$$

On the other hand, (b) [or (b′)] is not, *a priori*, a condition on the growth of the c^j. *However*, there is a growth condition on the c^j which is "naturally" associated to (b) [or (b′)], namely, that growth condition on the c^j which would force (b) [or (b′)]. This is just

$$c^j = 0\,(\exp(-\delta + \varepsilon)j) \qquad \text{for any } \varepsilon > 0.$$

Thus the growth condition for the conclusion (d) is the same as the growth condition related to (b) or (b′). In order to generalize this, we must decide on what types of growth conditions we shall consider. At present, we shall study those which are related to analytically uniform spaces.

Let us put the above in a more general framework. Denote by $\mathscr{W}$ the space of functions which are analytic in $|\text{Im } x| < 1$ and by $\mathscr{W}_1$ the space of functions which are analytic in some neighborhood of x^0. (The topology of $\mathscr{W}_1$ is the inductive limit topology of the spaces of functions analytic on fixed neighborhoods of the origin. The reader unfamiliar with inductive limits need not concern himself with this as we shall never use it.) Then the theorem states that if the series $\sum c^j \exp(ia^j x)$ converges to f in the topology of $\mathscr{W}$, and if $\{a^j\}$ is "lacunary," and if $f \in \mathscr{W}_1$, then the series converges in the topology of $\mathscr{W}_1$. Moreover, f belongs to some space, which we denote by $\mathscr{W} \overset{a}{\cap} \mathscr{W}_1$ which is smaller than the ordinary intersection of $\mathscr{W}$ and $\mathscr{W}_1$. (In our case $\mathscr{W} \overset{a}{\cap} \mathscr{W}_1$ is the space of functions analytic in some strip containing $|\text{Im } x| < 1$ and x^0.)

Our generalization is not yet complete, because we want to extend the concepts of exponential sums. The function $\exp(ia^j x)$ can be defined as the solution of the ordinary differential equation $dg/dx = ia^j g$. Thus, it would be natural, in a general theory, to study sums of solutions of systems of linear constant-coefficient partial differential equations. In view of Theorem 7.1 this means we are considering sums of integrals of exponentials over algebraic varieties.

Actually, we could extend the concept of exponential sum even further. Namely, we can regard (formally) a sum

$$\sum c^j \exp(ia^j x) = f(x)$$

as a solution of the convolution equation $S * f = 0$, where the Fourier transform of S is just $\prod[1 - z/a^j]$ (perhaps with some convergence factors added). Thus, a suitable generalization should involve solutions of systems of convolution equations for suitable kernels. However, our methods are not powerful enough to treat this situation.

See Problem 12.1.

From this point of view, it is interesting to note the relation between the work of this chapter and some of our previous results. For $n = 1$, it is remarked in Chapter XI following Theorem 11.3 that, under suitable conditions on S, every solution $f \in \mathscr{E}$ of $S * f = 0$ can be written in the form

$$f(x) = \sum c^j \exp(ia^j x),$$

where the series converges in the topology of $\mathscr{E}$. In fact, this means that actually

$$c^j = 0((1 + |a^j|)^{-l} \exp(-l|\text{Im } a^j|))$$

for any l. Thus we obtain a conclusion of the form (d) from a set of hypotheses which are different from (a), (b′), and (c′).

Actually, the difference is not so great as it seems. The "suitable conditions on S" referred to above are in the same spirit as lacunary conditions on the sequence $\{a^j\}$ (see Ehrenpreis and Malliavin [1]). However, there is no assumption corresponding to (a) which, as has already been remarked, corresponds to a growth condition on the c^j. (This will be made clear in the sequel.) In fact, by examining $S \in \mathscr{E}'$ whose Fourier transform has zeros at a^j with a^j, a^{j+1} very close together for an infinite number of j, we see that no estimate is possible on the size of the c^j without an assumption on S (cf. Section XII.11).

We see, therefore, that the problem of lacunary series differs from the problem of expanding solutions of $S * f = 0$ in that the latter problem has nothing corresponding to (a). (The assumption that $f \in \mathscr{E}$ corresponds to (b) or (b′).) For this reason we like to think of the problem of lacunary series as a *relative* problem because we begin with an estimate on the size of the c^j and try to derive a better estimate, while the problem of expansion of f as a solution of $S * f = 0$ is an *absolute* problem.

Tauberian theorems bear a close relationship to questions involving lacunary (gap) series. One of the most important classical Tauberian theorems is due to Littlewood: Let $\{c^j\}$ be a sequence satisfying $c^j = 0(1/j)$. Suppose

$$f(x) = \sum c^j x^j \to s$$

as $x \to 1$ through real values. Then the series $\sum c^j$ is convergent and $\sum c^j = s$.

Let us consider the structure of the theorem from a general point of view: Let $\mathscr{U}$ be the space of functions $f(x) = \sum c^j x^j$ with $c^j = 0(1/j)$. Let $\mathscr{U}_1$ be the space of functions analytic in $|x| < 1$ having a radial limit as $x \to 1$. Our hypothesis is that $f \in \mathscr{U} \cap \mathscr{U}_1$. Moreover, just as in the above case of gap series, the space $\mathscr{U}$ determines a growth condition on the c^j. However, in the present case the spaces $\mathscr{U}$ and $\mathscr{U}_1$ are *not* AU (similarly after an exponential change of variables). Moreover, we seek a slightly different type of conclusion now from the previous case, namely we want $\sum c^j$ convergent (which is not as before exactly a growth condition on the c^j) and we want to compute $\sum c^j$.

We can rephrase the type of conclusion we are seeking as follows: Determine the dual of $\mathscr{U} \cap \mathscr{U}_1$. More precisely, we see easily that $\mathscr{U}'$ is dense in $(\mathscr{U} \cap \mathscr{U}_1)'$, so each $S \in (\mathscr{U} \cap \mathscr{U}_1)'$ is a limit of elements of $\mathscr{U}'$. Can we say which sequences of elements of $\mathscr{U}'$ converge in the topology of $(\mathscr{U} \cap \mathscr{U}_1)'$ and what the limit is when they converge? In the Littlewood theorem we want the limit of χ_t to exist, where

$$\chi_t \cdot f = \sum_{j=0}^{t} c^j.$$

The speed of convergence in $(\mathscr{U} \cap \mathscr{U}_1)'$ can also be used to determine error terms.

The gap series problem can be formulated similarly. Namely, call $\mathscr{W}_0$ the space of series $f(x) = \sum c^j \exp(ia^j x)$ which converge in $\mathscr{W}$. Then we want to know that the dual of $\mathscr{W}_0 \cap \mathscr{W}_1$ is the same as the dual of $\mathscr{W}_0 \overset{a}{\cap} \mathscr{W}_1$.

We can state the problems in another form: For each x we may think of $\sum c^j x^j$ or $\sum c^j \exp(ia^j x)$ as an average of the sequence $\{c^j\}$. Thus a behavior of $f(x)$ (e.g., $f \in \mathscr{W}_1$) tells a behavior of these averages. The general problem is to determine more about $\{c^j\}$ from the knowledge of certain averages. In the gap series problems we usually want to know that c^j is small as $j \to \infty$. In the Tauberian problems we usually want to find other averages of $\{c^j\}$ (in Littlewood's theorem, $\sum c^j$). We can, of course, also ask for error terms in these averages.

Before going on, let us consider another example of a Tauberian theorem, namely Cantor's uniqueness theorem for trigonometric series as sharpened by Rajchman and Verblunsky. (We now allow c^j to be defined for positive and negative integers j; f is defined as $\sum_{j \geqq 0} c^j x^j + \sum_{j>0} c^{-j} \bar{x}^j$, so f is an harmonic function.) This asserts that if $c^j = o(j)$ or, more generally, if $f(x) = o((1-|x|)^{-2})$ as $|x| \to 1$, and if $f(|x|e^{i\theta}) \to 0$ as $|x| \to 1$ for any θ, then $c^j \equiv 0$. Thus in Cantor's theorem we require information about *all* radial limits (in Littlewood s theorem only *one* radial limit enters).

There are, in general, three methods in Tauberian theorems:

(a) *Toeplitz method.* Given that certain kernels, that is, averages in $\mathscr{W}'$ converge in $(\mathscr{W} \cap \mathscr{W}_1)'$ we can apply *positive* kernels to the original ones and obtain further convergent sequences in $(\mathscr{W} \cap \mathscr{W}_1)'$ (see Zygmund [1]). For example, in the proof of Cantor's theorem we use the fact that if $f(|x|\, e^{i\theta_0}) \to 0$ for θ_0 fixed then the lower generalized second derivative of the function $F(\theta) = \sum c^j e^{ij\theta}/j^2$ is ≤ 0 at θ_0. The lower generalized second derivative of F is, by definition,

$$\varliminf_{h \to 0} \frac{F(\theta+h) - 2F(\theta) + F(\theta-h)}{4h^2}.$$

Now,

$$e^{ij(\theta+h)} - 2e^{ij\theta} + e^{ij(\theta-h)} = -e^{ij\theta}(4\sin^2 jh/2),$$

so the method is Toeplitz since $\sin^2 jh/2 \geqslant 0$. If the series $\sum c^j \exp(ij\theta)$ converges then we deduce that the generalized second derivative of F is zero.

(b) *Linear combination of translates.* Suppose $y \in [0, 1]$ and

$$\sum_j k(y, j)c^j$$

converges as $y \to 1$. Here k is some kernel. If g_l are any continuous maps of $[0, 1]$ into itself taking $1 \to 1$ then we can replace $k(y, j)$ above by

$$\sum_{\text{finite}} \alpha_l\, k(g_l(y), j).$$

Hence we can also replace it by an element of the closure of $\sum \alpha_l\, k(g_l(y), j)$ in the dual topology of possible $\{c^j\}$, that is, in $\mathscr{U}'$.

For example, in Wiener's proof of the Littlewood theorem we use Wiener's Tauberian theorem to approximate all l_1 kernels (acting by a type of convolution on the sequence $\{c^j\}$), assuming only that $c^j = o(1)$. (This is done after summation by parts, so c^j is replaced by the partial sums $S(j)$.) If in addition $\{S(j)\}$ satisfies an "equicontinuity" condition, that is,

$$S((1+\varepsilon)j) - S(j) = 0(1),$$

which is a consequence of $c^j = O(1/j)$, then we can approximate δ by l_1 kernels and obtain the result.

Karamata and Littlewood use explicit approximations to δ in their proofs so an error term can be extracted from their methods.

(c) *Special points or variational methods.* When we have information about many averages then we might try to find special averages amongst them to deduce further information. For example, by Method (a) above, to complete the proof of Cantor's theorem we need Schwarz's theorem: If F is continuous on [a, b] and the generalized second derivative of F is zero, then F is constant on [a, b]. To prove this we perform a linear transformation and assume $F(a) = F(b) = 0$ and $a > 0$. Assume $F > 0$ at some points of [a, b]. For small $\eta > 0$, $F_\eta = F - \eta(x-a)(b-x)$ still takes positive values in $[a, b]$; we look at a maximum point θ_0. Since the generalized second derivative of F is zero, we find easily that the generalized second derivative of F_η at θ_0 is $2\eta\theta_0 > 0$. This is easily seen to be incompatible with the fact that F_η has a maximum at θ_0.

In what follows we shall use different methods to deal with lacunary series. It is to be hoped that if they are combined with Methods (a), (b), and (c), then we could obtain deeper results perhaps beyond the framework of AU spaces.

See Problem 12.2.

We can also regard lacunary series in the framework of the removable singularities results of Section XI.2. For example, the Fabry gap theorem says that if hypotheses (a), (b'), and (c') hold then any singularities of f in the region $\operatorname{Im} x > -\delta$, $\operatorname{Im} x \leqslant 0$, $|x| \geqslant \delta$ are removable. There is a sort of unifying principle in the hypothesis (c') of the Fabry theorem and the hypothesis (α) of Theorem 11.5: They both assert that f can be represented as the Fourier integral of a measure whose support is "very thin." (In the case of Theorem 11.5, by Theorem 7.1 the support of the measure is on a complex variety of complex codimension $\geqslant 2$.)

We shall now make precise definitions. We shall place a condition on all AU structures with which we deal in this chapter: Let $\mathscr{W}$ be AU. Then all AU structures for $\mathscr{W}$ with which we shall concern ourselves will have the property that if $K = \{k\}$ and $K' = \{k'\}$ are any two such AU structures, then for any k there is an $a > 0$ and a k' so that $k'(z) \leqslant ak(z)$ for all $z \in C$. We place a similar condition if $\mathscr{W}$ is a subspace of $\mathscr{W}_1$ and the topology of $\mathscr{W}$ is stronger than the induced topology. Another condition will be placed on the AU structures K (see "*Assumption on* $\mathscr{W}$" preceding Proposition 12.2 below).

Let $\mathscr{W}$ and $\mathscr{W}_1$ br AU spaces with AU structures $K = \{k\}$ and $K_1 = \{k_1\}$, respectively. Let K_2 be the family of functions on C of the form

$$k_2(z) = \max\,[k(z),\, k_1(z)] \tag{12.1}$$

for any $k \in K$, $k_1 \in K_1$. We define $\mathbf{W}_2'$ as the space of entire functions $G(z)$, such that

$$|G(z)|/k_2(z) \to 0 \quad \text{as } |z| \to \infty \tag{12.2}$$

for any $k_2 \in K_2$.

LEMMA 12.1. *The space* $\mathbf{W}_2'$ *depends only on* $\mathscr{W}$ *and* $\mathscr{W}_1$, *that is, it is independent of the choice of* AU *structures.*

PROOF. Let K', K_1' be different AU structures. By definition, given any $k \in K$, $k_1 \in K_1$ there are $k' \in K'$ and $k_1' \in K_1'$ and an $a > 0$ so that $k'(z) \leqslant ak(z)$ and $k_1'(z) \leqslant ak_1(z)$ for all $z \in C$. Thus,

$$k_2'(z) = \max\,[k'(z),\, k_1'(z)] \leqslant ak_2(z)$$

which gives the result. ▯

Given any $\hat{w}_2 \in \mathbf{W}_2$ (the dual of $\mathbf{W}_2'$), we define its Fourier transform by representing $\hat{w}_2$ in the form $d\mu/k_2$, where μ is a bounded measure on C

and $k_2 \in K_2$. (The possibility of such a representation follows from the Hahn-Banach theorem; cf. Chapter I. Actually we have to use a finite sum of such expressions, but we shall be a little sloppy and ignore this fact.) Then we see easily that the integral

$$w_2 = \int \exp\,(iz \cdot \quad)\, d\mu(z)/k_2(z) \tag{12.3}$$

converges in the topology of $\mathscr{W}$ to w, say, and in the topology of $\mathscr{W}_1$ to w_1, hence, by the definition given in Chapter VIII, it defines an element $(w, w_1) \in \mathscr{W} \cap \mathscr{W}_1$. The space $\mathscr{W}_2$ of w_2 defined by (12.3), which can be thought of as a subspace of either $\mathscr{W}$ or $\mathscr{W}_1$, is given the topology to make the Fourier transform an isomorphism. It is readily verified that $\mathscr{W}_2$ is AU with AU structure K_2. Moreover, $\mathscr{W}_2$ is the largest AU space contained in $\mathscr{W} \cap \mathscr{W}_1$.

DEFINITION. We call $\mathscr{W}_2$ the AU *intersection* of $\mathscr{W}$ and $\mathscr{W}_1$ and denote it by $\mathscr{W} \overset{a}{\cap} \mathscr{W}_1$. The AU intersection of several AU spaces is defined similarly.

We can define the AU union in an analogous way. This is denoted by $\mathscr{W} \overset{a}{\cup} \mathscr{W}_1$.

EXAMPLE A. If $\mathscr{W}$ is a subspace of $\mathscr{W}_1$ with the topology of $\mathscr{W}$ stronger than that induced from $\mathscr{W}_1$ then $\mathscr{W} \overset{a}{\cap} \mathscr{W}_1 = \mathscr{W}$.

EXAMPLE B. Let Ω and Ω_1 be open convex sets in R. Let $\mathscr{W} = \mathscr{E}(\Omega)$, the space of indefinitely differentiable functions on Ω, and $\mathscr{W}_1 = \mathscr{E}(\Omega_1)$. Then $\mathscr{W} \overset{a}{\cap} \mathscr{W}_1 = \mathscr{E}(\Omega_2)$, where Ω_2 is the convex hull of Ω and Ω_1. This follows easily from the explicit description of the spaces $\mathbf{E}'(\Omega)$, $\mathbf{E}'(\Omega_1)$, and $\mathbf{E}'(\Omega_2)$ (see Chapter V, Example 5).

EXAMPLE C. Let Ω and Ω_1 be convex polyhedra in $C^m = R^n$, where $n = 2m$. Let $\mathscr{H}(\Omega)$, $\mathscr{H}(\Omega_1)$ be the spaces of analytic functions on Ω and Ω_1 respectively. Thus, $\mathscr{H}(\Omega)$ is the kernel of the Cauchy-Riemann system

$$\vec{\mathrm{D}} = (\partial/\partial x_1 + i\partial/\partial x_{m+1}, \ldots, \partial/\partial x_m + i\partial/\partial x_{2m})$$

on $\mathscr{E}(\Omega)$ with a similar property for $\mathscr{H}(\Omega_1)$. Using Theorem 4.1, we see that $\mathscr{H}(\Omega)$, $\mathscr{H}(\Omega_1)$ are AU (see Theorem 5.21 and its proof). Using the result of Example B, it follows that

$$\mathscr{H}(\Omega) \overset{a}{\cap} \mathscr{H}(\Omega_1) = \mathscr{H}(\Omega_2), \tag{12.4}$$

where Ω_2 is the convex hull of Ω and Ω_1.

EXAMPLE D. Let Ω again be an open convex set in R and set $\mathscr{W} = \mathscr{E}(\Omega)$. Let x^0 be a point in the boundary of Ω such that the plane of support L of Ω at x^0 intersects the boundary of Ω in the closure of an open set Γ of L containing x^0 in its interior. Let γ be an open convex neighborhood of x^0 in Γ and set $\mathscr{W}_1 = \mathscr{E}(\gamma)$. Then

$$\mathscr{E}(\Omega) \overset{a}{\cap} \mathscr{E}(\gamma) = \mathscr{E}(\Omega \cup \gamma). \tag{12.5}$$

This can be proved by the methods of Chapter V, Example 5, which show how to describe the AU structures of $\mathscr{E}(\Omega)$, $\mathscr{E}(\gamma)$, and $\mathscr{E}(\Omega \cup \gamma)$.

The general problem of lacunary series can now be formulated:

MAIN PROBLEM. Let $\{\vec{\mathrm{D}}^j\}$ be a sequence of systems of linear constant coefficient partial differential operators. Let $\mathscr{W}$ and $\mathscr{W}_1$ be AU spaces and let $\mathscr{W}_2$ be an AU subspace of $\mathscr{W}$ containing $\mathscr{W} \overset{a}{\cap} \mathscr{W}_1$ such that the induced topologies of $\mathscr{W}$ on $\mathscr{W}_2$ and of $\mathscr{W}_2$ on $\mathscr{W} \overset{a}{\cap} \mathscr{W}_1$ are weaker than the respective topologies of these spaces. Under what conditions is the following true: Given any sequence $\{f^j\}$ with $f^j \in \mathscr{W} \cap \mathscr{W}_1$, $\vec{\mathrm{D}}^j f^j = 0$ such that the series $\sum f^j$ converges in the topology of $\mathscr{W}$ and

$$f = \sum f^j \tag{12.6}$$

$f \in \mathscr{W} \cap \mathscr{W}_1$; then f must be in $\mathscr{W}_2$ and the series $\sum f^j$ must converge in the topology of $\mathscr{W}_2$?

The Main Problem is a special case of the following: Let $\mathscr{M}_i$ be AU spaces and for each i let ${}_i\mathscr{M}$ be a closed subspace of $\mathscr{M}_i$. Let $\mathscr{M}$ be an AU space containing $\mathscr{M}_1 \overset{a}{\cap} \mathscr{M}_2 \overset{a}{\cap} \cdots \overset{a}{\cap} \mathscr{M}_t$, and such that the topology of the AU intersection is stronger than the induced topology. Let $\mathscr{N}$ be a closed subspace of $\mathscr{M}$. Under what condition is it true that ${}_1\mathscr{M} \cap {}_2\mathscr{M} \cap \cdots \cap {}_t\mathscr{M} \subset \mathscr{N}$? We call this problem the General Comparison Problem. It is a refinement of the comparison problem dealt with in Chapter VIII, because $\cap \mathscr{M}_i$ may not be AU.

The Main Problem corresponds to $t = 2$, $\mathscr{W} = \mathscr{M}_1$, $\mathscr{W}_1 = \mathscr{M}_2$, $\mathscr{W}_2 = \mathscr{M} = \mathscr{N}$, and ${}_i\mathscr{M}$ is the subspace of $\mathscr{M}_i$ formed by sums of the form (12.6) which converge in the topology of $\mathscr{M}_1$.

We have already met examples of the General Comparison Problem in Section XI.2.

The General Comparison Problem is an example of an Extension Problem so Methods (a), (b), or (c) of Chapter VIII could be attempted. In particular, Method (a) takes an interesting form which we should like to describe:

We denote by ${}_i\mathscr{M}^\perp$ the set of elements in $\mathscr{M}_i'$ which vanish on ${}_i\mathscr{M}$. Then to show that $\cap\, {}_i\mathscr{M} \subset \mathscr{N}$ we must show that if $S \in (\sum \mathscr{M}_i') \cap \mathscr{M}'$ and

$S \to T$ in the topology of $\mathscr{M}'$, where $T \in \mathscr{N}^{\perp}$, then we can write S in the form $\sum (S_i + {}_iS)$. where $S_i \in \mathscr{M}'_i$, $S_i \to 0$ in the topology of $\mathscr{M}'_i$ and ${}_iS \in {}_i\mathscr{M}^{\perp}$.

Note that, as in Chapter VIII, $\sum \mathscr{M}'_i$ is defined as the inverse Fourier transform of the set of sums $\sum \hat{S}_i$ where $\hat{S}_i \in \mathbf{M}'_i$. This makes sense because the $\mathscr{M}_i$ are AU.

Let us consider the example in which $\mathscr{M}$, $\mathscr{M}_i$ are LAU and $\mathscr{N}$, ${}_i\mathscr{M}$ are defined as the kernels of a system $\vec{\mathrm{D}}$ of linear constant coefficient differential operators on $\mathscr{M}$ and $\mathscr{M}_i$ respectively. We shall assume that $\mathbf{M}'_i \subset \mathbf{M}'$ for all i. This is the case in particular if $\mathscr{M}$ is the AU intersection of the $\mathscr{M}_i$. Then $\mathscr{N}^{\perp} = \vec{\mathrm{D}}'\mathscr{M}'^r$ and ${}_i\mathscr{M}^{\perp} = \vec{\mathrm{D}}'\mathscr{M}'^r_i$ by the results of Chapter VI. Suppose

(α) *Cutting up condition on* C: $\mathscr{M}' \subset \sum \mathscr{M}'_i$.

Then in the above we may assume that $T = 0$. Let $\mathfrak{V}$ be the multiplicity variety corresponding to $\vec{\mathrm{D}}$. Suppose we have also

(β) *Cutting up condition on* $\mathfrak{V}$: We can find a map (not necessarily linear)

$$\lambda : \mathbf{M}'(\mathfrak{V}) \to \sum \mathbf{M}'_i(\mathfrak{V})$$

which is continuous at zero and such that, for $F \in \mathbf{M}'(\mathfrak{V})$ we have $F = \sum (\lambda F)_i$.

We claim that if (α) and (β) hold, then $\cap {}_i\mathscr{M} \subset \mathscr{N}$. Suppose $S \to 0$ in the topology of $\mathscr{M}'$. If F denotes the Fourier transform of S then certainly the restriction of F to $\mathfrak{V} \to 0$ in the topology of $\mathbf{M}'(\mathfrak{V})$. By ($\beta$) we can write $F = \sum F_i$ on $\mathfrak{V}$, where $F_i \in \mathbf{M}'_i(\mathfrak{V})$ and $F_i \to 0$ in the topology of $\mathbf{M}'_i(\mathfrak{V})$. By Theorem 4.2 we may even assume that $F_i \to 0$ in the topology of $\mathbf{M}'_i$. Since $F = \sum F_i$ on $\mathfrak{V}$ we can write

$$F - \sum F_i = \vec{\mathrm{P}} \cdot \vec{\mathrm{H}},$$

where $\vec{\mathrm{H}} \in \mathbf{M}'^r$ and $\vec{\mathrm{P}}$ is the Fourier transform of $\vec{\mathrm{D}}'$. Using (α) we can write

$$\vec{\mathrm{H}} = \sum {}_i\vec{\mathrm{H}},$$

where ${}_i\vec{\mathrm{H}} \in \mathbf{M}'^r_i$.

Now, let S_i denote the inverse Fourier transform of F_i and ${}_i\vec{\mathrm{T}}$ the Fourier transform of ${}_i\vec{\mathrm{H}}$. Then the above yields

$$S = \sum (S_i + \vec{\mathrm{D}}'{}_i\vec{\mathrm{T}}).$$

Since $S_i \to 0$ in the topology of $\mathscr{M}'_i$ and $\vec{\mathrm{D}}'{}_i\vec{\mathrm{T}} = {}_iS \in {}_i\mathscr{M}^{\perp}$ we have our result. ▯

We want to explain why we call (α) and (β) "cutting up conditions." Let $F \in \mathbf{M}'$. Then we know that

$$F(z) = 0(k(z))$$

for any k in an AU structure for $\mathscr{M}$. Since $\mathscr{M}$ contains the AU intersection of the $\mathscr{M}_i$ and the topology of the AU intersection is stronger than the induced topology, for each k we can find k_i in an AU structure for $\mathscr{M}_i$ such that

$$k(z) \leq \max\,[k_1(z), \ldots, k_t(z)]$$

and conversely, given k_i in an AU structure for $\mathscr{M}_i$ there is a k in an AU structure for $\mathscr{M}$ satisfying the above inequality.

Thus the growth conditions for functions F in $\mathbf{M}'$ are stronger than the maximum of the growth conditions for the spaces $\mathbf{M}'_i$. Call C_i the portion of C where the growth conditions for $\mathbf{M}'$ are stronger than those for $\mathbf{M}'_i$. (We assume this is meaningful which is the case in most applications.) The above suggests that, in general, $\cup\, C_i = C$. Given $F \in \mathbf{M}'$ we try to write

$$F(z) = \sum F_i(z),$$

where each $F_i \in \mathbf{M}'_i$ and F_i is about the size of F on C_i and F_i is smaller than F off C_i.

Another way to look at this is that the above inequality on k allows us to write

$$F(z) = \sum F'_i(z),$$

where the F'_i are *continuous* functions satisfying the requisite growth conditions of $\mathbf{M}'_i$. (The F'_i are obtained by "cutting up" F.) We now want to modify the F'_i and make them analytic and still satisfy the requisite growth conditions. That this is possible in spirit is part of the general "Oka principle": What can be done with continuous functions can be done with analytic functions in spaces where suitable cohomology groups vanish (i.e., the arguments of Chapter IV hold).

The cutting up condition (β) is to be understood similarly. If $F \in \mathbf{M}'(\mathfrak{B})$ then we try to write

$$F = \sum F_i \quad \text{on } \mathfrak{B},$$

where each $F_i \in \mathbf{M}'_i(\mathfrak{B})$ and $F \to (F_i)$ is continuous at zero. This problem is of essentially the same nature as (α) except that we require continuity and C is replaced by $\mathfrak{B}$.

We wish to explain a technique for dealing with cutting up problems. This technique has some success, for example it is used in Section XIII.1 in a slightly different form:

The main difficulty is that the space $\cap \mathcal{M}_i$ is not AU. In order to overcome this difficulty we proceed as follows: Suppose $\mathcal{M}_i$ consists of functions or distributions of x. Let us denote by X the variable $X = (x, x, \ldots, x)$ (t-fold) with Z the dual variable to X. We write $X = (x^1, x^2, \ldots, x^t)$, $Z = (z^1, z^2, \ldots, z^t)$. We denote by M the tensor product of the $\mathcal{M}_i$. Thus M is the AU space of functions or distributions of X with AU structure given by the set of

$$k(Z) = k_1(z^1)k_2(z^2) \cdots k_t(z^t),$$

where $\{k_i\}$ form an AU structure for $\mathcal{M}_i$.

Now, we can "embed" $\cap \mathcal{M}_i$ into M by sending

$$f \to f(x^1)f(x^2) \cdots f(x^t) = F(X).$$

Thus Theorem 1.5 gives us a Fourier representation of F hence, if f is a function with $f(0) = 1$, of $f(x) = F(x, 0, \ldots, 0)$.

There are sometimes other embeddings of $\cap \mathcal{M}_i$ into M. For example, in Section XIII.1 we shall use an embedding of the form

$$f \to f(x^1 + x^2 + \cdots + x^t).$$

In general it is very difficult to verify the cutting up condition. For this reason our technique in dealing with the General Comparison Problem will be a combination of Method (b) of Chapter VIII with a "dual cutting up" as described in Section VIII.1 and in the next section under the name of "parametrix."

Remark 12.1. We could generalize the main problem by not requiring that $\mathcal{W}_1$ be a topological vector space but rather using the generalized extension described in Section VIII.1. The methods of this chapter could be modified so as to treat such problems.

The Main Problem falls naturally into two parts, Problem A and Problem B:

PROBLEM A. Let $\vec{\mathrm{D}}$ be a system of constant coefficient linear partial differential operators. Let $\mathcal{W}$, $\mathcal{W}_1$ and $\mathcal{W}_2$ be AU spaces. Under what conditions is it true that $f \in \mathcal{W} \cap \mathcal{W}_1$, $\vec{\mathrm{D}}f = 0$ imply $f \in \mathcal{W}_2$?

Problem A is essentially the case $t = 2$ of the example of the General Comparison Problem discussed above. We shall not discuss Problem A any further in this chapter. Problem B will be formulated below.

XII.2. ($\mathcal{W}$, $\mathcal{W}_1$) Density

DEFINITION. Let $\mathcal{W}$, $\mathcal{W}_1$, $\mathcal{W}_2$ be AU spaces with $\mathcal{W} \overset{a}{\cap} \mathcal{W}_1 \subset \mathcal{W}_2 \subset \mathcal{W}$, where, as above, "$\subset$" means "contained in with topology stronger than that induced." We say that the $(\mathscr{W}, \mathscr{W}_1)$ *density of* $\{\vec{D}^j\}$ *is* $\leqslant \mathscr{W}_2$ whenever the following holds: Let $\{f^j\}$ be a sequence of elements of $\mathscr{W}_2$ such that $\vec{D}^j f^j = 0$, and $\sum f^j$ converges in the topology of $\mathscr{W}$ to f which is in $\mathscr{W} \cap \mathscr{W}_1$; then $f \in \mathscr{W}_2$ and the series $\sum f^j$ converges in the topology of $\mathscr{W}_2$. In case $\mathscr{W}_2 = \mathscr{W} \overset{a}{\cap} \mathscr{W}_1$ we say that $\{\vec{D}^j\}$ is $(\mathscr{W}, \mathscr{W}_1)$ *lacunary.*

See Remark 12.2.

PROBLEM B. Determine the $(\mathscr{W}, \mathscr{W}_1)$ density of $\{\vec{D}^j\}$.

In what follows, we shall consider only those AU spaces $\mathscr{W}$ such that differentiation is continuous on $\mathbf{W}'$. Thus, $\mathbf{W}$ contains all $\partial\delta_a$ for $a \in C$, for ∂ any linear constant coefficient partial differential operator. The Fourier transform of $\partial\delta_a$ is an arbitrary polynomial times $\exp(ia \cdot x)$. Thus, any AU space contains all exponential polynomials. Hence, for $n = 1$ Problem A is trivially solvable.

For an illustration of Problem B, we let $n = 1$, $D^j = d/dx - ia^j$ with a^j real. We let $\mathscr{W}$ be the space of functions which are analytic in $|\text{Im } x| < a$, and for x^0 a point on the boundary of this strip, we let $\mathscr{W}_1$ be the space of functions analytic in $|\text{Im } x^0 - \text{Im } x| < b$, $|\text{Re } x^0 - \text{Re } x| < b'$. Then the classical Fabry gap theorem can be suitably modified (see, e.g., Dienes [1], p. 373) to show that a necessary and sufficient condition in order that $\{\vec{D}^j\}$ be $(\mathscr{W}, \mathscr{W}_1)$ lacunary is that the number of j with $|a^j| < N$ be $o(N)$ and that the a^j do not come too close together. Pólya [1] has extended the Fabry gap theorem so as to be able to compute the $(\mathscr{W}, \mathscr{W}_1)$ density in some special cases, namely, he considers the situation in which $\mathscr{W}_2$ is the space of analytic functions on $a' < \text{Im } x < a''$, where $a' \leqslant -a$ and $a'' \geqslant a$. The amount that a'' must be less than $a + \max(0, b - a + \text{Im } x^0)$ or a' greater than $-a - \max(0, b - a - \text{Im } x^0)$ in order that the $(\mathscr{W}, \mathscr{W}_1)$ density of $\{a^j\}$ be $\leqslant \mathscr{W}_2$ is given by the Pólya density of $\{a^j\}$. We may thus regard our density as a generalization and refinement of the Pólya density (see Example C above).

The most important tool in the computation of the $(\mathscr{W}, \mathscr{W}_1)$ density of $\{\vec{D}^j\}$ is the notion of $(\mathscr{W}, \mathscr{W}_1)$ *parametrix* (see Chapter VIII).

In our present problem of determining the $(\mathscr{W}, \mathscr{W}_1)$ density of $\{\vec{D}^j\}$ we apply the parametrix method as follows: For each j let $f^j \in \mathscr{W} \overset{a}{\cap} \mathscr{W}_1$ satisfy $\vec{D}^j f^j = 0$, and let (formally) $f = \sum f^j$. We let $\mathscr{U}$ be the space of all such sequences $\{f^j\}$ for which the series converges in the topology of $\mathscr{W}$; let $\mathscr{V}_2$ be the subset of those sequences for which the series $\sum f^j$ converges in the topology of $\mathscr{W}_2$, and let $\mathscr{V}_1$ be the subspace for which the series converges in the topology of $\mathscr{W}$ to f which is in $\mathscr{W}_1$. Let $\mathscr{U}_1 = \mathscr{W}$ and $\mathscr{U}_2 = \mathscr{W}_1$. L_1 is the map $L_1\{f^j\} = \sum f^j = f$, where f is considered

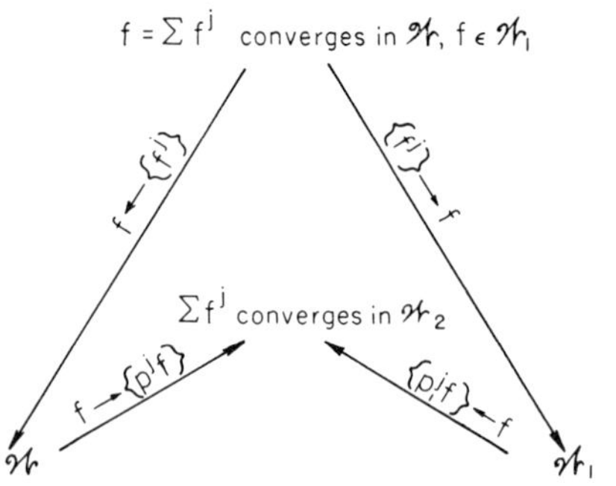

FIGURE 37

as an element of $\mathscr{W}$ and L_2 is the same map, except that f is now considered as an element of $\mathscr{W}_1$. Thus, we seek an $(L_1, L_2; \mathscr{V}_2)$ parametrix.

To construct the parametrix, we first want to write for each j

$$f^j = p^j \cdot f + p_1^j \cdot f, \tag{12.7}$$

where p^j are continuous linear maps of $\mathscr{W} \to \mathscr{W}_2$ and p_1^j of $\mathscr{W}_1 \to \mathscr{W}_2$. Then we want to show that the series $\sum p^j$ and $\sum p_1^j$ converge in a suitable sense. The sums of these series will be M_1 and M_2, respectively. See Fig. 37.

To clarify the construction of the p^j, p_1^j let us consider first the case $n = 1$, $D^j = d/dx - ia^j$ (a^j complex). Then $f^j(x) = c^j \exp(ia^j x)$. Let A^j be a suitable neighborhood of a^j in C. Let $G^j(z)$ be a function in $\mathbf{W}' \cap \mathbf{W}_1'$ satisfying

$$\begin{aligned} &G^j(a^j) = 1 \\ &G^j(a^{j'}) = 0 \quad \text{for } a^{j'} \in A^j \text{ and } j \neq j'. \end{aligned} \tag{12.8}$$

Let p_1^j be the map $f \to (\hat{G}^j \cdot f) \exp(ia^j x)$, where $\hat{G}^j$ is the inverse Fourier transform of G^j, and let $p^j f = f^j - p_1^j f$.

Heuristically we expect the following: Since the series $\sum c^j \exp(ia^j x)$ converges in the topology of $\mathscr{W}$ the c^j are "small" for j large. Moreover, because $\sum f^j$ converges in $\mathscr{W}$, we can apply $\hat{G}^j$ to the series, term by term. Using (12.8) this gives

$$\hat{G}^j \cdot f = c^j + \sum_{a^{j'} \notin A^j} c^{j'} G^j(a^{j'}). \tag{12.9}$$

If A^j is large enough, the right side of (12.9) will turn out to be close to c^j, because the sum on the right side of (12.9) will be small, since it is the tail of a convergent series. If we can show that the G^j are small in the topology of $\mathbf{W}_1'$, then the left side of (12.9) will be small so the c^j will be small.

To proceed rigorously, let us rewrite (12.9) as

$$(12.10)\quad c^j \exp(ia^j \cdot \quad) = (\hat{G}^j \cdot f) \exp(ia^j \cdot \quad) - \left[\sum_{a^{j'} \notin A^j} c^{j'} G^j(a^{j'})\right] \exp(ia^j \cdot \quad).$$

Rather than find conditions for the series $\sum c^j \exp(ia^j \cdot \quad)$ to converge in $\mathscr{W}_2$, we shall investigate when the sequence $\{c^j \exp(ia^j \cdot \quad)\}$ is bounded in the topology of $\mathscr{W}_2$. Since $\sum c^j \exp(ia^j \cdot \quad)$ converges in $\mathscr{W}$, we know that $\{c^j \exp(ia^j \cdot \quad)\}$ is bounded in $\mathscr{W}$. We now make

Assumption on $\mathscr{W}$. If $\{c^j \exp(ia^j \cdot \quad)\}$ is bounded in the topology of $\mathscr{W}$, then for some k in an AU structure for $\mathscr{W}$ we have $c^j = 0(1/k(a^j))$. Moreover, $\mathscr{W}'$ is reflexive.

This assumption can be verified for all the spaces we deal with, and we shall make it for the remainder of this chapter without explicitly stating so. This assumption implies a condition on the AU structure $\{k\}$ which we use. For, usually, we could remove a neighborhood of the points a^j in C and be left with a sufficient set (see Section I.3) for $\mathscr{W}$. Thus we can find AU structures for $\mathscr{W}$ for which the functions are infinite on $\{a^j\}$. The above assumption implies that this is not the case and, in fact, the AU structure is not trivial on $\{a^j\}$.

The method of verification of this assumption is as follows: Since $\{c^j \exp(ia^j \cdot \quad)\}$ is bounded, there is a neighborhood N of zero in $\mathscr{W}'$ such that for $S \in N$ we have

$$|c^j\hat{S}(a^j)| = |S \cdot c^j \exp(ia^j \cdot \quad)| \leqslant 1.$$

Since $\mathscr{W}$ is AU, there is a k in an AU structure for $\mathscr{W}$ such that N consists of all $S \in \mathscr{W}$ with $|\hat{S}(z)| \leqslant k(z)$ for all z.

Suppose k could be chosen so that for each j there is an $S \in N$ such that $|\hat{S}(a^j)| \to k'(a^j)$ for some k' (independent of j) in an AU structure for $\mathscr{W}$. (This is easily verified for all the examples of Chapter V.) Then the above shows that

$$|c^j| \leqslant 1/|\hat{S}(a^j)| \to 1/k'(a^j)$$

which is the desired result.

PROPOSITION 12.2. *Suppose for any k in an AU structure for $\mathscr{W}$ we can choose sequences $\{A^j\}$ and $\{G^j\}$ as above and a k_2 in an AU structure for $\mathscr{W}_2$ so that the conditions $c^j = 0(1/k(a^j))$ imply*

1. *$\{k_2(a^j)G^j\}$ is bounded in the topology of $\mathscr{W}_1$.*
2. $\sum_{a^{j'} \notin A^j} c^{j'} G^j(a^{j'}) = 0(1/k_2(a^j))$.

Then the sequence $\{c^j \exp(ia^j \cdot \quad)\}$ is bounded in the topology of $\mathscr{W}_2$.

PROOF. By (12.9) we deduce that $c^j = 0(1/k_2(a^j))$. Now, for any $h \in \mathscr{W}'_2$, $h \cdot c^j \exp(ia^j \cdot \quad) = c^j H(a^j)$, where H is the Fourier transform of h. Thus, $\{c^j \exp(ia^j \cdot \quad)\}$ is uniformly bounded on the set of h for which $H(z) = 0(k_2(z))$. Since $\mathscr{W}$ is AU, this set is a neighborhood of zero in $\mathscr{W}'_2$. Thus, $\{c^j \exp(ia^j \cdot \quad)\}$ is bounded in $\mathscr{W}_2$. ☐

Note that hypotheses 1 and 2 imply $M = \{p^j\}$ and $M_1 = \{p_1^j\}$ are maps of the set of f into the space of sequences $\{f^j\}$ which are bounded in the topology of $\mathscr{W}_2$, that is, the parametrix for this problem exists.

As we shall see below, Proposition 12.2 is sufficient for many interesting examples.

XII.3. Analytic ($\mathscr{W}$, $\mathscr{W}_1$) Density

Let us now pass to the general case. We assume $\mathscr{W}$, $\mathscr{W}_1$, $\mathscr{W}_2$ are LAU. Write $\mathfrak{B}^j = (V_1^j, \partial_1^j; \ \ldots; \ V_r^j, \partial_r^j)$, where $\mathfrak{B}^j$ is the multiplicity variety associated by Theorem 4.2 to $\vec{\mathrm{P}}^j$. For each j suppose we can find a neighborhood A^j of V^j and a function $G^j \in \mathbf{W}' \cap \mathbf{W}'_1$ such that

1. $G^j = 1$ on $\mathfrak{B}^j$, that is, for any entire function F on C and any i we have $\partial_i^j(FG^j) = \partial_i^j F$ on V_i^j.
2. For any $j' \neq j$ if $\mathfrak{B}^{j'}$ meets A^j, then $G^j = 0$ on $\mathfrak{B}^{j'}$, that is, for any entire function F on C and any i we have $\partial_i^{j'}(FG^j) = 0$ on $V_i^{j'}$.
3. For any k in an AU structure for $\mathscr{W}$ there is a k' in an AU structure for $\mathscr{W}$ so that for any j, j' if $\mathfrak{B}^{j'}$ does not meet A^j and if $L \in \mathbf{W}'$ is $\leqslant k'(z)$ on $\mathfrak{B}^{j'}$, then there exists an $L' \in \mathbf{W}'$ which is $\leqslant k(z)$ in all of C and $L = L'$ on $\mathfrak{B}^{j'}$.
4. For any k_1 in an AU structure for $\mathscr{W}_1$ there is a k_2 in an AU structure for $\mathscr{W}_2$ such that, uniformly in j.

(12.11) $$|G^j(z)|/k_1(z) = 0(1/k_2(z)),$$

   that is,

$$|G^j(z)|k_2(z) \leqslant \text{const } k_1(z).$$

5. For any k in an AU structure for $\mathscr{W}$ we can find a k_2 in an AU structure for $\mathscr{W}_2$ such that, uniformly in j

(12.12) $$|G^j(z)|/k(z) = 0(1/k_2(z)),$$

   that is,

$$|G^j(z)|k_2(z) \leqslant \text{const } k(z)$$

   for z in any $\mathfrak{B}^{j'}$ which does not meet A^j.
6. G^j defines, by multiplication, a continuous map of $\mathbf{W}'$ into $\mathbf{W}'$.

DEFINITION. If the $\{G^j\}$ and $\{A^j\}$ exist as above, then we say that the *analytic* $(\mathscr{W}, \mathscr{W}_1)$ *density of* $\{\vec{\mathrm{D}}^j\}$ *is* $\leqslant \mathscr{W}_2$.

The reason for this terminology is that Conditions 1, 2, 3, 4, 5, and 6 are conditions for the existence of analytic functions with prescribed properties.

THEOREM 12.3. *Let* $\{f^j\}$ *be a sequence in* $\mathscr{W} \overset{a}{\cap} \mathscr{W}_1$ *such that for each* j *we have* $\vec{\mathrm{D}}^j f^j = 0$. *Suppose the series*

$$f = \sum_j f^j$$

is convergent in $\mathscr{W}$ *and that* $f \in \mathscr{W}_1$. *Suppose that the analytic* $(\mathscr{W}, \mathscr{W}_1)$ *density of* $\{\vec{\mathrm{D}}^j\}$ *is* $\leqslant \mathscr{W}_2$. *Then the sequence* $\{f^j\}$ *is bounded in* $\mathscr{W}_2$.

We recall that the convergence of $\sum f^j$ in the topology of $\mathscr{W}$ means that for any $S \in \mathscr{W}'$ the series $\sum |S \cdot f^j|$ is convergent, and these series converge uniformly for S in any bounded set of $\mathscr{W}'$.

See Remark 12.3.

PROOF. Let $\hat{G}^j$ denote the inverse Fourier transform of G^j. By (12.11) and the definition of AU, multiplication by G^j defines a continuous linear map of $\mathbf{W}_2'$ into $\mathbf{W}_1'$. Hence in an obvious way, convolution by $\hat{G}^j$ defines a continuous linear map of $\mathbf{W}_2'$ into $\mathbf{W}_1'$, hence its adjoint $*'$ is a continuous linear map of $\mathscr{W}_1$ into $\mathscr{W}_2$. By 6. above, convolution by $\hat{G}^j$ defines also a continuous linear map of $\mathscr{W}'$ into $\mathscr{W}'$.

For each j we may write

$$\hat{G}^j *' f^j = \hat{G}^j *' f - \sum_{j' \neq j} \hat{G}^j *' f^{j'} \tag{12.13}$$

because $\sum f^{j'}$ converges in $\mathscr{W}$. For any $H \in \mathbf{W}'$ we have

$$\hat{H} \cdot (\hat{G}^j *' f^j) = (\hat{H} * \hat{G}^j) \cdot f^j. \tag{12.14}$$

Since the Fourier transform of $\hat{H} * \hat{G}^j$ is HG^j which has the same restriction to $\mathfrak{V}^j$ as H does, it follows from Theorem 4.2 that

$$\hat{H} \cdot (\hat{G}^j *' f^j) = \hat{H} \cdot f^j, \tag{12.15}$$

that is,

$$\hat{G}^j *' f^j = f^j. \tag{12.16}$$

Similar reasoning shows

$$\hat{G}^j *' f^{j'} = 0 \quad \text{if } \mathfrak{V}^{j'} \text{ meets } A^j. \tag{12.17}$$

Thus, (12.13) becomes

$$f^j = -\sum\nolimits^{j'} \hat{G}^j *' f^{j'} + \hat{G}^j *' f, \tag{12.18}$$

where $\sum^{j'}$ means the sum over those $\mathfrak{B}^{j'}$ which do not meet A^j.

Let p^j be the map $f \to -\sum^{j'} \hat{G}^j *' f^{j'}$ and let p_1^j be the map $f \to \hat{G}^j *' f$, that is, $f^j = p^j f + p_1^j f$. To prove Theorem 12.3 we show that $\{p^j f\}$ and $\{p_1^j f\}$ are bounded in $\mathscr{W}_2$. We begin with $p_1^j f$: For any $H \in \mathbf{W}_2'$ we have

$$\hat{H} \cdot (\hat{G}^j *' f) = (\hat{H} * \hat{G}^j) \cdot f. \tag{12.19}$$

Now, f is bounded on a neighborhood of zero in $\mathscr{W}_1'$. Thus, there exists a k_1 in an AU structure for $\mathscr{W}_1$ so that f is bounded on the set of $\hat{J} \in \mathscr{W}_1'$ for which $|J(z)| \leqslant k_1(z)$. By Condition 4 above [see (12.11)] there is a k_2 in an AU structure for $\mathscr{W}_2$ so that the conditions $H \in \mathbf{W}_2'$, $|H(z)| \leqslant k_2(z)$ imply $|H(z)G^j(z)| = 0(k_1(z))$ uniformly in j. Thus, by (12.19) $\{\hat{G}^j *' f\}$ is uniformly bounded for all j on a neighborhood of zero in $\mathscr{W}_2'$, so by the definition of the topology of $\mathscr{W}_2$, $\{\hat{G}^j * f\}$ is bounded in $\mathscr{W}_2$, that is, $\{p_1^j f\}$ is bounded in $\mathscr{W}_2$.

Next we note that for $H \in \mathbf{W}'$ we have

$$\hat{H} \cdot p^j f = -\sum\nolimits^{j'} (\hat{H} * \hat{G}^j) \cdot f^{j'}. \tag{12.20}$$

Since $\sum f^{j'}$ is convergent in $\mathscr{W}$, there is a k in an AU structure for $\mathscr{W}$ so that if $J \in \mathbf{W}'$ and $|J(z)| \leqslant k(z)$, then $|\sum \hat{J} \cdot f^{j'}| \leqslant 1$ (since we can write $\hat{J} \cdot \sum f^{j'} = \sum \hat{J} \cdot f^{j'}$). By Condition 5 above [see (12.12)], given any k in an AU structure for $\mathscr{W}$, there is a k_2 in an AU structure for $\mathscr{W}_2$ (independent of j) so that if $H \in \mathbf{W}'$ satisfies $|H(z)| \leqslant k_2(z)$ for z in any $\mathfrak{B}^{j'}$ which does not meet A^j, then also $|G^j(z)H(z)| \leqslant k(z)$ for z in any $\mathfrak{B}^{j'}$ which does not meet A^j. Since $\vec{\mathrm{D}}^{j'} f^{j'} = 0$, the value $(\hat{H} * \hat{G}^j) \cdot f^{j'}$ depends (by Theorem 4.2) only on the restriction of HG^j to $\mathfrak{B}^{j'}$. By Condition 3 we may therefore assume that $|G^j(z)H(z)| \leqslant k(z)$ for all $z \in C$.

All the above shows that for $H \in \mathbf{W}'$, $|H(z)| \leqslant k_2(z)$ for all z we have

$$\left|\sum\nolimits^{j'} (\hat{H} * G^j) \cdot f^{j'}\right| \leqslant 1. \tag{12.21}$$

Combining this with (12.20) shows readily that $\{p^j f\}$ is uniformly bounded in j on a neighborhood of zero in $\mathscr{W}_2'$, hence, $\{p^j f\}$ is bounded in $\mathscr{W}_2$. This completes the proof of Theorem 12.3. ☐

DEFINITION. Let $\{j_t\}$ be a subsequence of $\{j\}$. Suppose that, in the definition of analytic $(\mathscr{W}, \mathscr{W}_1)$ density of $\{D^j\}$ we assume the existence of the G^j, A^j satisfying Conditions 1, 2, 3, 4, 5, and 6 hold only for $j \in \{j_t\}$ and for all j' [in 2, 3, and 5]. Then we say that the *analytic* $(\mathscr{W}, \mathscr{W}_1)$ *density of* $\{\vec{\mathrm{D}}^{j_t}\}$ *relative to* $\{\vec{\mathrm{D}}^j\}$ is $\leqslant \mathscr{W}_2$.

The proof of Theorem 12.3 actually shows

THEOREM 12.4. *Let* $\{f^j\}$ *be a sequence in* $\mathscr{W} \overset{a}{\cap} \mathscr{W}_1$ *such that* $\vec{\mathrm{D}}^j f^j = 0$ *for each* j. *Suppose the series* $f = \sum f^j$ *is convergent in* $\mathscr{W}$ *and that* $f \in \mathscr{W}_1$. *Let* $\{j_t\}$ *be a subsequence such that the analytic* $(\mathscr{W}, \mathscr{W}_1\}$ *density of* $\{\vec{\mathrm{D}}^{j_t}\}$ *relative to* $\{\vec{\mathrm{D}}^j\}$ *is* $\leqslant \mathscr{W}_2$. *Then the subsequence* $\{f^{j_t}\}$ *is bounded in* $\mathscr{W}_2$.

See Remark 12.4.

XII.4. Geometric Density

In general, when the sequence $\{\vec{\mathrm{D}}^j\}$ and the spaces $\mathscr{W}$, $\mathscr{W}_1$ are given, we should like to find some "reasonable" method of determining the $(\mathscr{W}, \mathscr{W}_1)$ density. Thus, the usual conditions in Fabry's theorem ($D^j = d/dx - ia^j$) are given in terms of the distribution of the a^j; it is usually possible to decide in any given situation whether these conditions are satisfied. In general, we should like some condition relating the geometric distribution of the $\mathfrak{B}^j$ to the analytic $(\mathscr{W}, \mathscr{W}_1)$ density.

In order to explain the meaning of this "geometric density," let us scrutinize the properties of the functions G^j and the sets A^j. Property 1 says that $G^j = 1$ on $\mathfrak{B}^j$, that is, G^j and 1 have the same restrictions to $\mathfrak{B}^j$. This means that $1 - G$ belongs to the ideal generated by $\mathrm{P}^j = (P_1^j, \ldots, P_r^j)$ which is the Fourier transform of $\vec{\mathrm{D}}'^j$.

Condition 2 says that G^j has zero restriction to $\mathfrak{B}^{j'}$ if $j' \neq j$ and $\mathfrak{B}^{j'}$ meets A^j. Hence, for such j', G^j belongs to the ideal generated by the $P_i^{j'}$. Thus, for each such j' we want to construct a function $G^{jj'}$ so that

$$G^{jj'} = 1 + \sum_i N_i^j P_i^j = \sum_i Q_i^{jj'} P_i^{j'} \tag{12.22}$$

for suitable N_i^j and $Q_i^{jj'}$ which will be prescribed below.

Now suppose that for each fixed j' for which $\mathfrak{B}^{j'}$ meets A^j we have chosen a $G^{jj'}$ satisfying (12.22). Then we could find a G^j which works for all such j' by setting (if the number of j' is finite)

$$G^j = \prod_{j'} G^{jj'}, \tag{12.23}$$

as is clear from (12.22).

We could now multiply G^j by a suitable function which is 1 on $\mathfrak{B}^j$ [so (1) and (2) will still hold] in order to satisfy (4), (5), and (6).

Although the above method is quite general, it seems quite difficult to go further with the general theory. We shall henceforth consider only the case when the underlying varieties of each $\mathfrak{B}^j$ are of dimension 0.

See Problem 12.3.

We shall assume first that $n = 1$ and that each $\mathfrak{B}^j = (a^j$, identity; $a^j, d/dz; \ldots; a^j, d^{e^j-1}/dz^{e^j-1})$. Let $a^{j'} \in A^j$. Thus, by (12.22) we must have

$$G^{jj'}(z) = 1 + N^j(z)(z - a^j)^{e^j} = Q^{jj'}(z)(z - a^{j'})^{e^{j'}} \tag{12.24}$$

for entire functions N^j and $Q^{jj'}$. For fixed j, j' a "simple" solution $G^{jj'}$ of (12.24) is given by

$$G^{jj'}(z) = [1 - (z - a^j)^{e^j}/(a^{j'} - a^j)^{e^j}]^{e^{j'}}. \tag{12.25}$$

Thus, using (12.23) a possible choice $'G^j$ for G^j is

$$'G^j(z) = \prod_{j'} [1 - (z - a^j)^{e^j}/(a^{j'} - a^j)^{e^j}]^{e^{j'}}. \tag{12.26}$$

We can modify $'G^j$ by multiplication by a suitable function $''G^j$ which differs from the constant 1 by a function which has a zero of order at least e^j at a^j. We shall choose $''G^j \in \mathbf{W}' \cap \mathbf{W}'_1$, so that

$$G^j = {'G^j}\, {''G^j} \tag{12.27}$$

will be in $\mathbf{W}' \cap \mathbf{W}'_1$. Of course, we want to choose a $''G^j$ so as to improve our estimates.

Still for $n = 1$, we may consider the general

$$\mathfrak{B}^j = \{a^{j_1}, \text{identity}; \ldots; a^{j_1}, d^{e^j_1-1}/dz^{e^j_1-1}; \ldots; a^{j_r}, d^{e^j_r-1}/dz^{e^j_r-1}\}.$$

Then we modify (12.25) to

$$G^{jj'}(z) = \prod_{i'} \left\{1 - \prod_i [(z - a^{j_i})/(a^{j'_{i'}} - a^{j_i})]^{e^j_i}\right\}^{e^{j'}_{i'}}. \tag{12.28}$$

For $n = 1$ a better choice for $'G^j$ will be discussed in Section XII.9.

In order to make the appropriate modifications for $n > 1$, since we assume that the underlying varieties of each $\mathfrak{B}^j$ are of dimension 0, we may write

$$\mathfrak{B}^j = (a^{j_1}, \partial^{j_{11}}; \ldots; a^{j_1}, \partial^{j_{1s_1}}; \ldots; a^{j_r}, \partial^{j_{rs_r}}).$$

For each j_i we can find suitable positive integers $e_1^{j_i}, \ldots, e_n^{j_i}$ so that the ideal generated by the Fourier transform of the $D_i^{\prime j}$ (j fixed) contains the functions

$$(z_1 - a_1^{j_i})^{e_1^{j_i}}, \ldots, (z_n - a_n^{j_i})^{e_n^{j_i}}. \tag{12.29}$$

Then (12.28) generalizes to

$$G^{jj'}(z) = \prod_{i'} \left\{1 - \prod_i [(z_t - a_t^{j_i})/(a_t^{j'_{i'}} - a_t^{j_i})]^{e_t^{j_i}}\right\}^{e_{t i'}^{j'}}. \tag{12.30}$$

for any t. Of course, the appropriate choice of t will depend on j_i and $j'_{i'}$. (Thus $G^{jj'}$ may not depend only on one fixed z_t.) We could modify (12.30) somewhat by allowing a linear change of variables in C which does not change the spaces $\mathbf{W}'$ and $\mathbf{W}_1'$.

Finally, we define

$$'G^j = \prod_{j'} G^{jj'} \quad \text{and} \quad G^j = {'G^j} \, {''G^j}.$$

where $''G^j \in \mathbf{W}' \cap \mathbf{W}_1'$ differs from 1 by an element of the ideal generated by the functions of (12.29).

We wish now to show how to verify Conditions 1, 2, 3, 4, 5, and 6 on the G^j in the examples to which we shall apply our theory. For this we are led to

DEFINITION. For each j let A^j be a neighborhood of $\mathfrak{B}^j$. Let $\{u^j\}$, $\{v_1^j\}$, and $\{v_2^j\}$ be sequences of non-negative numbers. We say that the *geometric density of* $\{\mathfrak{B}^j\}$ *is* $\leqslant \{(A^j, u^j, v_1^j, v_2^j)\}$ if

(a) For each $\mathfrak{B}^{j'}$ that meets A^j we can choose t (depending on $j'_{i'}$ and j_i) so that

$$\prod_{i,j',i'} |a_t^{j'_{i'}} - a_t^{j_i}|^{e_t^{j_i} e_t^{j'_{i'}}} \geqslant v_1^j + v_2^j(\tfrac{1}{2}\tilde{u}^j)^{\tilde{u}^j}, \tag{12.31}$$

where $\tilde{u}^j = \sum_{i,j',i'} e_t^{j_i} e_t^{j'_{i'}}$.

(b) It is possible to choose the t satisfying (12.31) in such a way that

$$\tilde{u}^j \leqslant u^j. \tag{12.32}$$

(c) Call s^j the diameter of A^j. If $\mathfrak{B}^{j'}$ meets A^j, then for any i, i', t we have

$$|a_t^{j'_{i'}} - a_t^{j_i}| \leqslant cs^j,$$

where c is a constant independent of j, j', i, i', t.

Before exploring the consequences of the geometric density of $\{\mathfrak{B}^j\} \leqslant \{(A^j, u^j, v_1^j, v_2^j)\}$, let us see what this concept means in case $n = 1$, and each $\mathfrak{B}^j$ is of the form (a^j, identity), a^j real, and A^j is an interval on the real axis containing a^j. Condition (b) states that the number of $a^{j'}$ in A^j is not too large ($\leqslant u^j$). Thus, (b) gives our intuitive concept of a sequence of real numbers a^j being "not too dense," namely, that the number of $a^{j'}$ near a given a^j (that is, the number in A^j) is small (that is, $\leqslant u^j$). Condition (a) is a "uniform discreteness" condition.

We shall see below that, in general, an inequality on a suitable geometric density will imply an inequality on an analytic density, hence, by Theorem 12.3 on a density. Call s^j the diameter of A^j. We can now make our estimates: We have, using $'G^j = \prod_{j'} G^{jj'}$ and (12.30)),

$$(12.33) \quad |'G^j| = \left\{ \prod_{i,j',i'} |a_t^{j'_{i'}} - a_t^{j_i}|^{-e_t^{j_i} e_t^{j'_{i'}}} \right\}$$

$$\times \left\{ \prod_{i,j',i'} |(a_t^{j'_{i'}} - a_t^{j_i})^{e_t^{j_i}} - (z_t - a_t^{j_i})^{e_t^{j_i}}|^{e_t^{j'_{i'}}} \right\}.$$

Since a finite number of terms play no role, we may assume each $|a^{j_i}| > 1$. Then, using (12.31), (12.32), and the fact that $a^m + b^m \leqslant (a+b)^m$ for $a, b \geqslant 0$ and $m \geqslant 1$, we have

$$(12.34) \qquad |'G^j(z)| \leqslant \{v_1^j + v_2^j(\tilde{u}^j)^{\tilde{u}^j}\}^{-1} c_1^{\tilde{u}^j} (s^j + |z_t - a_t^{j_i}|)^{\tilde{u}^j}$$

for any fixed i, where c_1 is an appropriate constant. Thus, we have shown

THEOREM 12.5. *If the geometric density of* $\{\vec{\mathrm{D}}^j\}$ *is* $\leqslant \{(A^j, u^j, v_1^j, v_2^j)\}$, *then* (12.34) *holds.*

We can use Theorem 12.5 for many examples. It should be noted that the fact that $\mathscr{W}$ is localizeable was used, essentially, in (12.15). In the examples discussed below, $\mathfrak{B}^j$ is reduced to a point and it is easily seen that the assumption of localizeability can be dispensed with.

XII.5. Example 1. The Classical Fabry Gap Theorem

Here $n = 1$ and x is a complex variable; $\mathscr{W}$ is the space of functions analytic say in $|\mathrm{Im}\, x - 1| < 1$. For simplicity of notation, let us take for $\mathscr{W}_1$ the space of functions analytic in $|\mathrm{Im}\, x| < \delta$, $|\mathrm{Re}\, x| < b$, where $\delta, b > 0$. Let $D^j = d/dx - ia^j$, where a^j, is real and suppose $|a^j - a^{j'}| \geqslant 1$ for all j, j'. Let A^j be the open interval $(\frac{1}{2}a^j, \frac{3}{2}a^j)$. Let $1 > \varepsilon > 0$ and let $\{u^j\}$ be a sequence with $u^j \leqslant \varepsilon |a^j|$ for $|a^j|$ sufficiently large. Let $v_1^j = 0$ and let $v_2^j = \exp(-\varepsilon |a^j|)$ for $|a^j|$ large.

By Stirling's formula, Condition (a) is already implied by (b). Thus, the geometric density of $\{D_j\} \leqslant \{(A^j, u^j, v_1^j, v_2^j)\}$ is essentially equivalent to the classical density of $\{a^j\}$ being $\leqslant \varepsilon$ (see Levinson [1]).

The spaces $\mathscr{W}$ and $\mathscr{W}_1$ are AU.

The fact that $\mathscr{W}$ is AU results from an application of Theorem 4.1 for $D = \partial/\partial\bar{x}$ to Chapter V, Example 5, since $\mathscr{W}$ is the kernel of $\partial/\partial\bar{x}$ on $\mathscr{E}(|\mathrm{Im}\, x - 1)| < 1$. (Compare Theorem 5.21 which is a slightly weaker statement). The fact that $\mathscr{W}$ is AU is proved in the same manner.) $\mathscr{W}_1$ is AU for the same reason. An AU structure for $\mathscr{W}$ is given by all functions $k(z)$ which dominate all $\exp(\alpha\, \mathrm{Re}\, z + \beta\, |\mathrm{Im}\, z|)$ for all β and for $-2 < \alpha < 0$.

An AU structure for $\mathscr{W}_1$ consists of all $k_1(z)$ which dominate all $\exp[\xi|\operatorname{Re} z| + \eta|\operatorname{Im} z|]$ for $|\xi| < \delta$ and $|\eta| < b$.

From (12.34) we have, for $|z - a^j| \leqslant \frac{1}{2}|a^j|$ by our assumptions on v_1^j and v_2^j and (12.32)

$$|'G^j(z)| \leqslant \exp(\varepsilon|a^j|)(\tilde{u}^j)^{-\tilde{u}^j} c_2^{\tilde{u}^j} |a^j|^{\tilde{u}^j}. \tag{12.35}$$

By differentiating the function $\log[(a/x)^x]$ we see that it is increasing for $0 < x < a/e$. Thus, if $\varepsilon < 1/e$, for j sufficiently large we can replace $\tilde{u}^j$ by u^j in (12.35). This gives

$$(|a^j|/\tilde{u}^j)^{\tilde{u}^j} \leqslant \varepsilon^{-\varepsilon|a^j|}.$$

Combining this with (12.35) gives, for $|z - a^j| \leqslant \frac{1}{2}|a^j|$,

$$|'G^j(z)| \leqslant \exp(c_3\varepsilon|a^j|). \tag{12.36}$$

Note that c_3 is of the order of $-\log\varepsilon$.

Again using (12.34), we have for $|z - a^j| \geqslant \frac{1}{2}|a^j|$

$$\begin{aligned} |'G^j(z)| &\leqslant \exp(c_4\varepsilon|a^j|)|\tilde{u}^j|^{-\tilde{u}^j}(|a^j| + |z|)^{\tilde{u}^j} \\ &\leqslant \exp(c_4\varepsilon|a^j|)[\varepsilon^{-1}(1 + |z|/|a^j|)]^{\varepsilon|a^j|} \\ &\leqslant \exp(c_5\varepsilon|a^j|)(1 + |z|/|a^j|)^{\varepsilon|a^j|} \\ &\leqslant \exp(c_5\varepsilon|a^j|)\exp(\varepsilon|z|). \end{aligned} \tag{12.37}$$

Note that the constants are of the order $-\log\varepsilon$.

Next we define $''G^j(z)$ by

$$''G^j(z) = [\exp(-\varepsilon_1(z - a^j))]\{[\sin\varepsilon_2(z - a^j)/l^j]/[\varepsilon_2(z - a^j)/l^j]\}^{l^j}, \tag{12.38}$$

where the ε_i, l^j will be fixed later. For ε_1, ε_2 small enough $''G^j$ is certainly in $\mathbf{W}' \cap \mathbf{W}_1'$.

Let us note that

$$\left|\frac{\sin z}{z}\right| \leqslant e^{-1}\exp(|\operatorname{Im} z|) \qquad |z| \geqslant e. \tag{12.39}$$

Moreover, $|\sin z| \leqslant |z|$ for real z while on the imaginary axis we find by power series that $|\sin z|/|z| \leqslant \exp(|\operatorname{Im} z|)$. Thus, by the Phragmén-Lindelöf theorem (see Titchmarsh [2]), or else directly,

$$\left|\frac{\sin z}{z}\right| \leqslant \exp(|\operatorname{Im} z|) \quad \text{for all } z. \tag{12.40}$$

Combining this with (12.38) gives

$$|''G^j(z)| \leqslant |\exp[\varepsilon_1|a^j| - \varepsilon_1 z + \varepsilon_2|\operatorname{Im} z|]|. \tag{12.41}$$

We now set $G^j(z) = {}'G^j(z)''G^j(z)$. By (12.41) and (12.36) we obtain for $|z - a^j| \leqslant \frac{1}{2}|a^j|$

$$|G^j(z)| \leqslant |\exp[(c_3\varepsilon + \varepsilon_1)|a^j| - \varepsilon_1 z + \varepsilon_2|\mathrm{Im}\, z|]|. \tag{12.42}$$

We now choose l^j as the greatest integer $\leqslant \varepsilon_3|a^j|$, where ε_3 will be defined below. Then by (12.38) and (12.39) we have

$$|''G^j(z)| \leqslant e\,|\exp(-\varepsilon_1 z + \varepsilon_1|a^j| - \varepsilon_3|a^j| + \varepsilon_2|\mathrm{Im}\, z|)| \tag{12.43}$$

for $|z - a^j| \geqslant \varepsilon_3 e|a^j|/\varepsilon_2$. Suppose that $\varepsilon_3 = \varepsilon_2/2e$. Then we can combine (12.43) with (12.37) to obtain for $|z - a^j| \geqslant \frac{1}{2}|a^j|$

$$|G^j(z)| \leqslant e|\exp[c_5\varepsilon + \varepsilon_1 - \varepsilon_3)|a^j| + \varepsilon|z| - \varepsilon_1 z + \varepsilon_2|\mathrm{Im}\, z|]|. \tag{12.44}$$

Now let $\mathscr{W}_2$ be the space of functions holomorphic on $|\mathrm{Im}\, x| < \delta - \varepsilon_4$, $|\mathrm{Re}\, x| < b - \varepsilon_5$, where $0 < \varepsilon_4 < \delta$ and $0 < \varepsilon_5 < b$. We claim that for ε small enough the analytic $(\mathscr{W}, \mathscr{W}_1)$ density of $\{D^j\}$ is $\leqslant \mathscr{W}_2$. Note that $\mathscr{W}_2$ is AU with AU structure consisting of all continuous positive $k_2(z)$ which dominate all $\exp(\xi|\mathrm{Re}\, z| + \eta|\mathrm{Im}\, z|)$ for $|\xi| < \delta - \varepsilon_4$ and $|\eta| < b - \varepsilon_5$. (The proof is the same as for $\mathscr{W}_1$.)

To verify (12.11) it clearly suffices to show that for all z

$$G^j(z) = 0(\exp(\varepsilon_4|\mathrm{Re}\, z| + \varepsilon_5|\mathrm{Im}\, z|)).$$

Here 0 is to be understood uniformly in j. For $|z - a^j| \leqslant \frac{1}{2}|a^j|$, this follows from (12.42) if $2(c_3\varepsilon + \varepsilon_1) + \varepsilon_1 < \varepsilon_4$ and $\varepsilon_2 < \varepsilon_5$. (Note that $|z - a^j| \leqslant \frac{1}{2}|a^j|$ implies $|a^j| \leqslant 2|\mathrm{Re}\, z|$.) For $|z - a^j| \geqslant \frac{1}{2}|a^j|$, the derived inequality follows from (12.44) if ε_3 is chosen larger than $c_5\varepsilon + \varepsilon_1$, if $\varepsilon + \varepsilon_1 < \varepsilon_4$, and $\varepsilon + \varepsilon_2 < \varepsilon_5$.

To verify (12.12) it suffices to show that

$$G^j = 0\{\exp[-(\delta - \varepsilon_4)z]\} \quad \text{for } z \geqslant 0,\ |z - a^j| \geqslant \tfrac{1}{2}|a^j| \tag{12.45a}$$

$$G^j = 0\{\exp[(2 - (\delta - \varepsilon_4))|z|]\} \quad \text{for } z \leqslant 0,\ |z - a^j| \geqslant \tfrac{1}{2}|a^j| \tag{12.45b}$$

as follows from our description of the AU structures for $\mathscr{W}$, $\mathscr{W}_2$. To verify (12.45a) we use (12.44). We choose $\varepsilon_3 > c_5\varepsilon + \varepsilon_1$ and $\varepsilon_1 = (\delta - \varepsilon_4) + \varepsilon$. Then by (12.44), (12.45b) is satisfied if $\delta - \varepsilon_4 + 2\varepsilon \leqslant 2 - (\delta - \varepsilon_4)$, that is, $2(\delta - \varepsilon_4 + \varepsilon) \leqslant 2$. It is easily seen that for any $\delta < 1$ and any ε_4 with $\varepsilon_4 < \delta < \frac{4}{3}\varepsilon_4$, for ε small enough, we can choose ε_1, ε_2 and ε_5 so that all the above requirements are met. The point is, that if ε is small and ε_4 is close to δ, ε_1 is small. If, in addition, ε_5 is large and ε_2 is small, then all the inequalities are satisfied.

Conditions 3 and 6 for analytic density are readily verified.

Thus, by Theorem 12.4 we have the following sharpening of the Fabry gap theorem:

THEOREM 12.6. *Let $\{a^j\}$ be a sequence of real numbers and let $\{a^{j_t}\}$ be a subsequence. Suppose there exists an $\eta > 0$ so that for every j_t and every $\varepsilon > 0$ the number of a^j in $|z - a^{j_t}| < \eta |a^{j_t}|$ is $\leqslant \varepsilon |a^{j_t}|$ for $|a^{j_t}|$ sufficiently large (depending on ε) and that these a^j satisfy $|a^j - a^{j'}| \geqslant l > 0$. Suppose the series*

$$f(x) = \sum c^j \exp (ia^j)x$$

converges absolutely uniformly on compact subsets of an open strip S and f can be continued to be analytic in a neighborhood of a boundary point x^0. Then the series

$$g(x) = \sum c^{j_t} \exp (ia^{j_t}x)$$

converges in a neighborhood of x^0 and hence in an open strip containing S and x^0.

See Remarks 12.5 and 12.6.

XII.6. Example 2. Analog of the Fabry Gap Theorem for $\mathscr{E}$

Here $n = 1$ and x is a complex variable; $\mathscr{W}$ is the space of functions analytic in $|\text{Im } x - 1| < 1$ which have boundary values on $\text{Im } x = 0$ which are distributions. For $\mathscr{W}_1$ we take the space of indefinitely differentiable functions on $\text{Im } x = 0, |\text{Re } x| < \delta$. Let $D^j = d/dx - ia^j$, where a^j is real. We suppose again that $|a^j - a^{j'}| \geqslant 1$ for any j, j'. Let A^j_N be the open interval $(a^j - N \log |a^j|, a^j + N \log |a^j|)$ for $|a^j|$ sufficiently large. Let $\{u^j\}$ be a sequence with $|u^j| \leqslant \varepsilon \log |a^j|$ for $|a^j|$ sufficiently large. Let $v^j_1 = 0$ and let $v^j_2 = \exp (-\varepsilon \log |a^j|)$ for $|a^j|$ large.

By Stirling's formula, Condition (a) for geometric density is already implied by (b). The statement that the geometric density of

$$\{D^j\} \leqslant \{(A^j, u^j, v^j_1, v^j_2)\}$$

is a refinement of the statement that the classical density of $\{a^j\}$ is $\leqslant \varepsilon/N$.

As in Example 1 above, $\mathscr{W}$ is AU. (Similar spaces are discussed in detail in XIII.2.) $\mathscr{W}_1$ is AU by Chapter V, Example 5. An AU structure for $\mathscr{W}_1$ consists of all $k_1(z)$ which dominate all

$$(1 + |z|)^\alpha \exp (\beta |\text{Im } z|)$$

for all α and for $\beta < \delta$. Rather than describe the complicated AU structure for $\mathscr{W}$ again, we need only its restriction to the real axis. The reason that we need only the AU structure on the real axis is that in the Conditions 1–5 for analytic density the only place the AU structure for $\mathscr{W}$ enters is Condition 5. Since (12.12) is to be satisfied only for $z = a^{j'}$, we need only know $k(a^{j'})$. In our example $a^{j'}$ is real (cf. Remark 12.3). Condition 6 seems to involve the AU structure for $\mathscr{W}$. However, Condition 6 can be readily verified by Fourier transform.

On the real axis the AU structure consists of functions $k(z) = (1 + |z|)^{-\gamma}$ for some γ for Re $z > 0$ and $k(z)$ dominates all exp $(-\eta z)$ for all $\eta < 2$ for Re $z < 0$.

By (12.34) we have for $|z - a^j| \leqslant N \log |a^j|$

$$(12.46) \qquad |'G^j(z)| \leqslant \exp(\varepsilon \log |a^j|)(\tilde{u}^j)^{-\tilde{u}^j} c_2^{\tilde{u}^j} (N \log |a^j|)^{\tilde{u}^j}.$$

As in Example 1, (12.36), we find for $|a^j|$ large enough, if ε is small

$$(12.47) \qquad |'G^j(z)| \leqslant \exp[(c_3 \varepsilon + \varepsilon \log (N/\varepsilon)) \log |a^j|].$$

For $|z - a^j| \geqslant N \log |a^j|$ we find from (12.34)

$$(12.48) \qquad |'G^j(z)| \leqslant \exp[c_4 \varepsilon \log |a^j|](\tilde{u}^j)^{-\tilde{u}^j} |z - a^j|^{\tilde{u}^j}$$
$$\leqslant \exp[(c_4 \varepsilon + \varepsilon \log (1/\varepsilon)) \log |a^j|] \, |(z - a^j)/\log |a^j||^{\varepsilon \log |a^j|}.$$

Let $''g(z) = ''g(\varepsilon_1; z)$ be an entire function of z of exponential type ε_1 such that $''g(0) = 1$ and

$$(12.49) \qquad |''g(\varepsilon_1; z)| \leqslant c_5 \exp[\varepsilon_1 |\text{Im } z| - |\text{Re } z|^{1/2}],$$

where $c_5 = c_5(\varepsilon_1)$. It is well known that such functions exist (see Chapter V, Example 6).

We define

$$(12.50) \qquad ''G^j(z) = ''g(z - a^j)\{[(\sin \varepsilon_2 (z - a^j)/l^j]/[\varepsilon_2 (z - a^j)/l^j]\}^{l^j},$$

where l^j is the greatest integer in $\varepsilon_3 \log |a^j|$, where ε_3 will be prescribed below. By (12.50), (12.40), and (12.49) we have for all z

$$(12.51) \qquad |''G^j(z)| \leqslant c_5 \exp[(\varepsilon_1 + \varepsilon_2)|\text{Im } z| - |\text{Re } (z - a^j)|^{1/2}].$$

For all z we can use (12.50), (12.49), and the fact that $|\sin z| \leqslant \exp(|\text{Im } z|)$ to obtain

$$(12.52) \qquad |''G^j(z)| \leqslant c_5 \exp[(\varepsilon_1 + \varepsilon_2)|\text{Im } z| - |\text{Re } (z - a^j)|^{1/2}$$
$$+ \varepsilon_3 \log \left(\frac{\varepsilon_3}{\varepsilon_2}\right) \log |a^j|] \, |z - a^j|^{-\varepsilon_3 \log |a^j|} (\log |a^j|)^{\varepsilon_3 \log |a^j|}.$$

Combining (12.51) with (12.47), we have for $|z - a^j| \leqslant N \log |a^j|$

$$(12.53) \qquad |G^j(z)| \leqslant c_5 \exp[(c_3 \varepsilon + \varepsilon \log (N/\varepsilon)) \log |a^j| + (\varepsilon_1 + \varepsilon_2)|\text{Im } z|].$$

Here we have written $G^j = {'G^j} \, {''G^j}$. For $|z - a^j| \geqslant N \log |a^j|$ we combine

(12.52) with (12.48) to obtain

$$(12.54)\quad |G^j(z)| \leqslant c_5 \exp \{[c_4\,\varepsilon + \varepsilon \log(1/\varepsilon) + \varepsilon_3 \log(\varepsilon_3/\varepsilon_2)] \log |a^j| - |\mathrm{Re}\,(z - a^j)|^{1/2} + (\varepsilon_1 + \varepsilon_2)|\mathrm{Im}\, z|\} \times |z - a^j|^{(\varepsilon-\varepsilon_3)\log|a^j|}(\log |a^j|)^{(\varepsilon_3-\varepsilon)\log|a^j|}.$$

Let $\mathscr{W}_2$ be the space of indefinitely differentiable functions on $\mathrm{Im}\, x = 0$. $|\mathrm{Re}\, x| < \delta - \varepsilon_4$. By Chapter V, Example 5 $\mathscr{W}_2$ is AU. An AU structure consists of all continuous positive k_2 which dominate all

$$(1 + |z|)^{\alpha} \exp(\beta\,|\mathrm{Im}\, z|)$$

for α arbitrary and $\beta < \delta - \varepsilon_4$. We claim that if ε can be chosen arbitrarily small (and hence N arbitrarily large) for $|a^j|$ large enough, for suitable ε_4, the analytic $(\mathscr{W}, \mathscr{W}_1)$ density of $\{D^j\}$ is $\leqslant \mathscr{W}_2$. Again, Condition (3) for analytic density is trivial, so we need verify (12.11) and (12.12).

In the present case, (12.11) will be satisfied if

$$|G^j(z)| \leqslant \text{const}\,(1 + |z|)^p \exp(\varepsilon_4\,|\mathrm{Im}\, z|)$$

for some p. By (12.53) we can verify this for $|z - a^j| \leqslant N \log |a^j|$ if $\varepsilon_1 + \varepsilon_2 \leqslant \varepsilon_4$ because, in this interval, $|a^j| \leqslant 2\,|z|$ if z is large enough. For $N \log |a^j| \leqslant |z - a^j| \leqslant 4\,|a^j|$ we use (12.54). Suppose $\varepsilon_3 > \varepsilon$, $\varepsilon_3 = \varepsilon_2$. Then, if ε is small enough, (12.54) gives for such z, using the fact that we have $|z - a^j|/\log|a^j| \geqslant N$,

$$(12.55)\quad |G^j(z)| \leqslant c_5 \exp[(\varepsilon' - (\varepsilon_3 - \varepsilon) \log N) \log |a^j| + (\varepsilon_1 + \varepsilon_2)\,|\mathrm{Im}\, z|].$$

where ε' is arbitrarily small and N is arbitrarily large. For such values of z this gives, since $|a^j| \geqslant |z|/5$,

$$(12.56)\quad |G^j(z)| \leqslant c_6(1 + |z|)^{-\varepsilon_3 \log N/5} \exp[(\varepsilon_1 + \varepsilon_2)\,|\mathrm{Im}\, z|].$$

Thus, (12.11) holds for these values of z. For $|z - a^j| \geqslant 4|a^j|$ we again use (12.54) to obtain, since $\log |a^j| \leqslant \varepsilon'\,|\mathrm{Re}\,(z - a^j)| + \varepsilon'\,|\mathrm{Im}\, z|$ for ε' arbitrarily small, we can find an arbitrarily small ε'' so that

$$(12.57)\quad |G^j(z)| \leqslant c_7 \exp[-(1 - \varepsilon'')\,|\mathrm{Re}\,(z - a^j)|^{1/2} + (\varepsilon_1 + \varepsilon_2 + \varepsilon')\,|\mathrm{Im}\, z|] \leqslant c_7 \exp[-\tfrac{1}{2}\,|\mathrm{Re}\, z|^{1/2} + (\varepsilon_1 + \varepsilon_2 + \varepsilon')\,|\mathrm{Im}\, z|],$$

so (12.11) holds for these z also. Finally, (12.56) and (12.57) show that (12.12) holds. Thus, using Theorem 12.4 we have

THEOREM 12.7. *Let $\{a^j\}$ be a sequence of real numbers and let $\{a^{j_t}\}$ be a subsequence. Suppose for every j_t and for every $\varepsilon > 0$ the number of a^j in $|z - a^{j_t}| < \log |a^{j_t}|$ is $\leqslant \varepsilon \log |a^{j_t}|$ for $|a^{j_t}|$ sufficiently large and that these a^j*

satisfy $|a^j - a^{j'}| \geqslant l > 0$. *Suppose the series*

$$f(x) = \sum c^j \exp (ia^j x)$$

is convergent in the space $\mathscr{W}$ *and that* f *is indefinitely differentiable in a neighborhood of some real boundary point of the strip. Then the series*

$$g(x) = \sum c^{j_t} \exp (ia^{j_t} x)$$

converges in the topology of the space of functions which are analytic in $|\text{Im } x - 1| < 1$ *and have indefinitely differentiable boundary values on* $\text{Im } x = 0$. *In particular,* $c^{j_t} = 0(|a^{j_t}|^{-T})$ *for every* $T > 0$.

See Remark 12.7.

XII.7. Example 3. Relation to the Riemann ζ Function

Here, again $n = 1$. $\mathscr{W}$ is the space of functions which are analytic in $\text{Im } x < -1$. $\mathscr{W}_1$ is the space of entire functions $f(x)$ which are $0(\exp (\varepsilon |x| \log |x|))$ for all $\varepsilon > 0$. $\mathscr{W}$ and $\mathscr{W}_1$ are given "natural" topologies (compare Chapter V). Again $D^j = d/dx - ia^j$, where a^j is real and negative. For our "uniform discreteness" condition we assume that $|a^j - a^{j'}| \geqslant \exp (-2 |a^j|)$ for any j, j' if $|a^j|$ is large enough. (We shall derive below [see (12.59)] an inequality which shows how to replace this uniform discreteness condition by a prescription of v_1^j and v_2^j.) A^j is the interval $(10a^j, 0)$. Let $\{u^j\}$ be a sequence with $u^j < \exp (\delta |a^j|)$ for some $\delta > 0$.

The spaces $\mathscr{W}$ and $\mathscr{W}_1$ are AU. An AU structure for $\mathscr{W}_1$ consists of all continuous positive functions which dominate $\exp (\exp b|z|)$ for all $b > 0$. $\mathbf{W}'$ consists of all entire functions $L(z)$ which satisfy an inequality of the form

$$|L(z)| = \begin{cases} 0[\exp ((1+\varepsilon) \text{ Re } z + b|\text{Im } z|)] & \text{for some } \varepsilon, b > 0 \quad \text{for Re } z \leqslant 0 \\ 0[\exp (b \text{ Re } z + b\,|\text{Im } z|)] & \text{for some } b > 0 \quad \text{for Re } z \geqslant 0. \end{cases}$$

The AU structure is defined in the usual way, that is, all continuous positive functions which dominate these bounds.

The proof of the assertion for $\mathscr{W}_1$ can be obtained by applying Theorem 4.1. for the differential operator $\partial/\partial \bar{x}$ to the results of Chapter V, Example 7. Now, $\mathscr{W}_1$ is the kernel of $\partial/\partial \bar{x} = \partial/\partial x_1 + i\partial/\partial x_2$ on the space $\mathscr{E}(\Phi)$ for

$$\Phi(x_1, x_2) = [x_1 \log (1 + |x_1|) + x_2 \log (1 + |x_2|)].$$

The results for $\mathscr{W}_1$ now follow by the observation that the conjugate of $x_1 \log x_1$ is $\exp (y_1 - 1)$.

The result for $\mathscr{W}$ is essentially contained in the results of Chapter V, Example 5.

An example of a sequence which has geometric density $\leqslant \{(A^j, u^j, v_1^j, v_2^j)\}$ is $\{-\log j\}$. Thus, though any classical density for this sequence is infinite, ours is not. We have chosen our normalization so that the function $\zeta_1(x) = (1 - 2^{1-ix})\zeta(ix)$ has the property that the Dirichlet series representing it converges in the topology of $\mathscr{W}$, but this function would have "just failed" to belong to $\mathscr{W}_1$ if we would have changed the definition slightly be replacing the condition "all $\varepsilon > 0$" by "some $\beta < 1$." (Here, $\zeta(x)$ is the Riemann zeta function.) The significance of this will appear later.

As usual, we define

$$'G^j(z) = \prod_{\substack{a^{j'} \in A^j \\ a^{j'} \neq a^j}} [1 - (z - a^j)/(a^{j'} - a^j)] = \prod (a^{j'} - z)/(a^{j'} - a^j). \tag{12.58}$$

Now the distance between any two $a^{j'}$ in A^j is $\geqslant \exp(-20\,|a^j|)$. It follows easily that

$$\prod |a^{j'} - a^j| \geqslant \exp(-20\,|a^j|\,\tilde{u}^j)[(\tilde{u}^j/2)\,!]^2 \geqslant \exp[\tilde{u}^j(\log \tilde{u}^j/2 - 20\,|a^j| - 1)] \tag{12.59}$$

by Stirling's formula.

For $|z - a^j| \leqslant 2\,|a^j|$ we have by the definition of A^j, $|a^{j'} - z| \leqslant 11\,|a^j|$, so

$$\prod |a^{j'} - z| \leqslant |11a^j|^{\tilde{u}^j} = \exp(\tilde{u}^j \log |11a^j|). \tag{12.60}$$

For $|z - a^j| \geqslant 2\,|a^j|$ we have $|a^j| \leqslant |z|$. Thus $|a^{j'} - z| \leqslant 11\,|z|$, so

$$\prod |a^{j'} - z| \leqslant |11z|^{\tilde{u}^j} = \exp(\tilde{u}^j \log |11z|). \tag{12.61}$$

(12.58), (12.59), and (12.60) give for $|z - a^j| \leqslant 2\,|a^j|$

$$|'G^j(z)| \leqslant \exp[\tilde{u}^j(20\,|a^j| + \log 11\,|a^j| - \log \tilde{u}^j/2 + 1)] \leqslant \exp(40\,|a^j| \exp \delta\,|a^j|) \tag{12.62}$$

for $|a^j|$ sufficiently large.

(12.58), (12.59), and (12.61) imply for $|z - a^j| \geqslant 2\,|a^j|$ (so $|z| \geqslant |a^j|$)

$$|'G^j(z)| \leqslant \exp[\tilde{u}^j(\log |11z| + 20\,|a^j| - \log \tilde{u}^j/2 + 1)] \leqslant \exp[(\log |11z| + 30\,|a^j|) \exp \delta\,|a^j|)] \leqslant \exp[2 \log |z| \exp \delta'\,|a^j|] \tag{12.63}$$

for any $\delta' > \delta$, for $|a^j|$ sufficiently large.

We have

$$\text{(12.64)} \qquad \operatorname{Re}(\cosh z) = \tfrac{1}{2}\operatorname{Re}(e^{x+iy} + e^{-x-iy}) = \tfrac{1}{2}[e^x \cos y + e^{-x}\cos y] = \cosh x \cos y.$$

Now let us define

$$\text{(12.65)} \qquad {}''G^j(z) = \{[\sinh(z-a^j)]/(z-a^j)\}^{l^j} \times \exp\{-2l^j(z-a^j) - \cosh[c(z-a^j)] + 1\}.$$

It is clear that $''G^j \in \mathbf{W}_1'$, but $''G^j$ is certainly *not* in $\mathbf{W}'$. However, the a^j are real, and the topology of $\mathbf{W}'$ enters into our considerations only insofar as estimates near the a^j are concerned; more precisely, the topology of $\mathbf{W}'$ is used to determine bounds on the c^j in $\sum c^j \exp(ia^jx)$. For this reason, we can replace the space $\mathbf{W}'$ by any space which has the same growth conditions on the set $\{a^j\}$; in particular, any space which has the same growth conditions on the negative real axis. It is clear from (12.65) that $''G^j$ satisfies the growth conditions for $\mathbf{W}'$ on the negative real axis if $c > 0$ (cf. Remark 12.3).

For all z we have by (12.40), (12.64), and (12.65)

$$\text{(12.66)} \qquad |''G^j(z)| \leqslant \exp\{l^j|\operatorname{Re}(z-a^j)| - 2l^j \operatorname{Re}(z-a^j) + \exp[c|\operatorname{Re}(z-a^j)|] + 1\}.$$

In particular, if $|z-a^j| \leqslant \frac{1}{2}|a^j|$ so that $|a^j| \leqslant 2|z|$

$$|''G^j(z)| \leqslant \exp[6l^j|z| + \exp(c|z|) + 1].$$

For any real z we have by (12.65) and (12.40)

$$\text{(12.67)} \qquad |''G^j(z)| \leqslant |\exp\{l^j|z-a^j| - 2l^j\operatorname{Re}(z-a^j) - \tfrac{1}{2}\exp c|z-a^j| + 1\}|.$$

Let $\mathscr{W}_2$ be the space of entire functions which are $0(\exp(\beta|x|\log|x|))$ for all $\beta > \beta_1$ (for a suitable $\beta_1 > 0$ which will *not* be allowed to be taken $=0$). We could modify the proof of Chapter V, Example 7 to show that $\mathscr{W}_2$ is LAU. W_2' consists of all entire functions $H(z)$ which are $0\{\exp[\exp(c'|z|)]\}$ for some $c' < 1/\beta_1$. An AU structure consists of all continuous positive functions which dominate all $\exp[\exp(c'|z|)]$ for $c' < 1/\beta_1$.

See Remark 12.8.

Let $G^j = {}'G^j\,''G^j$ as usual. We wish to verify (12.11) and (12.12), as the other conditions are easily demonstrated. We consider first (12.12) for which it is sufficient to consider real z. From the expressions for the AU structures of $\mathscr{W}$ and $\mathscr{W}_2$ it suffices to show that

(12.68) $$|G^j(z)| \leqslant \text{const} \exp [-\exp (|z|/\beta_1)]$$

for z real, $z \leqslant 0$, $z \notin A^j$, because there is no condition for z real, $z \geqslant 0$, since all $a^{j'}$ are negative.

For z real, $z < 0$, $z \notin A^j$, we have $|z - a^j| \geqslant 9\,|a^j|$. We can therefore use (12.63) in conjunction with (12.67). Since $|z| \leqslant |a^j| + |z - a^j| \leqslant 2\,|z - a^j|$, the term $\exp [-\frac{1}{2} \exp c\,|z - a^j|]$ dominates the terms in the right side of (12.63) if $c > \delta/18$. It also dominates the term $l^j\,|z - a^j| - 2l^j \operatorname{Re} (z - a^j)$ in the right side of (12.67) if $l^j < \exp 4c\,|a^j|$.

Hence the term $\exp [-\frac{1}{2} \exp c\,|z - a^j|]$ dominates all the terms in G^j. since $|z| \leqslant 2\,|z - a^j|$ this implies (12.68) if $c/4 > 1/\beta_1$.

We now verify (12.11) for which, because of the AU structures for $\mathscr{W}$, $\mathscr{W}_1$, it suffices to prove that

(12.69) $$|G^j(z)| \leqslant \text{const} \exp [\exp \beta\,|z|]$$

for some $\beta > 0$.

For $|z| \geqslant \frac{1}{2}\,|a^j|$ we derive (12.69) from (12.66) and (12.62) or (12.63), depending on whether $|z - a^j| \leqslant 2\,|a^j|$ or $|z - a^j| \geqslant 2\,|a^j|$, since $|z - a^j| \leqslant |z| + |a^j| \leqslant 3\,|z|$. For this we require that β be large enough so that

$$\max (2\delta, 3c) < \beta$$

and

$$l^j \leqslant \exp (|a^j|\,\beta'/3)$$

for some $\beta' < \beta$.

Finally, for $|z| \leqslant \frac{1}{2}\,|a^j|$ we have $|\operatorname{Re} (z - a^j)| = \operatorname{Re} (z - a^j) \geqslant \frac{1}{2}\,|a^j| > 1$ if j is large enough and also $|z - a^j| \leqslant \frac{3}{2}\,|a^j|$. From (12.66) we derive

$$|''G^j(z)| \leqslant \exp [-l^j + \exp (\tfrac{3}{2}c\,|a^j|)].$$

We take $l^j =$ greatest integer in $4 \exp (\frac{3}{2}c\,|a^j|)$, which, combined with (12.62), gives the result if $c > \delta'$. (Note that the conditions on β, c, and δ are not contradictory.)

We can thus apply Theorems 12.3 or 12.4 to our situation. We content ourselves with the application of Theorem 12.3.

THEOREM 12.8. *Let $\{a^j\}$ be a discrete sequence of real negative numbers with the properties (for $|a^j|$ large enough): $|a^j - a^{j'}| \geqslant \exp (-2\,|a^j|)$ for any j, j', and the number of $a^{j'}$ in $(10a^j, 0)$ is $\leqslant \exp (\delta\,|a^j|)$ for some $\delta > 0$. Suppose the series*

$$f(x) = \sum c^j \exp (ia^j x)$$

is absolutely convergent in Im < -1, *uniformly on compact subsets of this* *strip. Suppose $f(x)$ can be continued to be an entire function which satisfies*

$$|f(x)| = 0(\exp(\beta|x|\log|x|)) \tag{12.70}$$

for any $\beta > 0$. Then $c^j = 0(\exp(-\exp(B|a^j|)))$ for any $B > 0$.

It is easy to modify the proof of Theorem 12.8 to require (12.70) for some sufficiently small β (depending on δ). The conclusion is then that $c^j = 0(\exp(-\exp(B|a^j|)))$ for a suitable B. In particular, we may apply Theorem 12.8 to the sequence $\{a^j\} = \{-\log j\}$. We change our notation for this theorem in order to conform to the classical notation.

THEOREM 12.9. *Suppose $f(s) = \sum c_n n^{-s}$ converges absolutely in some half-plane and that f can be continued to be an entire function satisfying $f(s) = 0(\exp(\beta|s|\log|s|))$ for a sufficiently small $\beta < 1$. Then*

$$c_n = 0(\exp(-Bn))$$

for an appropriate B.

Remark 12.9. The classical Riemann ζ function shows that we cannot take $\beta = 1$. For $\zeta_1(s) = (1 - 2^{1-s})\zeta(s) = \sum(-1)^n n^{-s}$ satisfies (by the functional equation for $\zeta(s)$)

$$|\zeta_1(s)| = 0(\exp(|s|\log|s|)).$$

It is likely that a more careful analysis of the proof would show that any $\beta < 1$ would be suitable in Theorem 12.9.

Actually, it is possible to show that any $\beta < 1$ is suitable (and even more) by a somewhat different method. In order to accomplish this we shall need to use, in addition to the Fourier transform, the Mellin transform. Let $\mathscr{V}$ be a space of functions or distributions of the single variable s. Suppose that for each $z \neq 0$ we have $z^{is} \in \mathscr{V}$. Then we define the Mellin transform $\tilde{T}$ of $T \in \mathscr{V}'$ by

$$\tilde{T}(z) = T \cdot z^{is}$$

(compare (1.5)). There are some difficulties the Mellin transform has which the Fourier transform does not have because for fixed s, z^s is *not* an entire function of z but is multivalued. These difficulties are not hard to overcome; we shall work fairly formally and leave the details to the reader.

As above, let $f(s) = \sum c_n n^{-is}$. As in Example 3 of Chapter V we identify f with a distribution. Consider the Mellin transform S of f. Formally

$$S = \sum_{n>0} c_n \delta_n,$$

where δ_n is the unit mass at the point n. Now take the Fourier transform F of S. We have, clearly,

$$F(x) = \sum_{n>0} c_n \exp(inx).$$

Now $F(x)$ is clearly periodic of period 2π and is analytic in $\mathrm{Im}\, x > 0$. We want to deduce an exponential decrease in the c_n from regularity conditions on F (that is, by saying that F belongs to a suitable AU space $\mathscr{W}_3$), that is, we want to apply the theory of lacunary series to F. The simplest example of a space $\mathscr{W}_3$ which gives the conclusion is the space of functions analytic in $\mathrm{Im}\, x > -B$ where $B > 0$. (This is seen by replacing $\exp(inx)$ by z^n and using the fact that a power series is absolutely convergent inside a circle of analyticity. Another proof can be obtained from the method of proof of Theorem 12.6). We claim that if f can be continued to an entire function satisfying $f(s) = 0(\exp \beta |s| \log |s|)$ for some $\beta < 1$, then F belongs to $\mathscr{W}_3$ for any $B > 0$. To see this, let $\mathscr{W}_2$ be as in the proof of Theorem 12.8 except that now $\beta_1 < 1$ is chosen so that $f \in \mathscr{W}_2$. We now let $Ш'_2$ be the Mellin transform of $\mathscr{W}'_2$. Using the description of $\mathbf{W}'_2$ and the fact that the Mellin transform is obtained from the Fourier transform by an exponential change of variables, we see that $Ш'_2$ consists of all entire functions of order $< 1/\beta_1$. Let $\hat{Ш}_2$ denote the Fourier transform of the dual of $Ш'_2$. It is easily seen that $\hat{Ш}_2$ consists of entire functions if $\beta_1 < 1$. Since f is in $\mathscr{W}_2$, the Mellin transform of f is in $Ш_2$ so $F \in \hat{Ш}_2$. Since $\beta_1 < 1$ this implies that $f \in \mathscr{W}_3$ for any $B > 0$, which is our contention.

We could change the definition of $\mathscr{W}_3$ to the space of functions analytic in $\mathrm{Im}\, x > 0$ with indefinitely differentiable boundary conditions. This leads to a refinement of the bound

$$f(s) = 0(\exp(\beta |s| \log |s|))$$

for some $\beta < 1$ but the conclusion

$$c_n = 0(e^{-Bn})$$

is to be replaced by

$$c_n = 0(n^{-A})$$

for all $A > 0$. We leave the details to the reader.

We can go even further. We want to study ways in which ζ_1 is extremal. If we denote by G the Fourier transform of the Mellin transform of $\zeta_1(is)$, then

$$G(x) = \sum (-1)^n \exp(inx)$$

$$= \frac{1}{1 + \exp(ix)}.$$

Note that G is meromorphic with simple poles at πj for j an odd integer. Since G has simple poles, we might try to use the ideas of limitation of singularities (see Section VIII.1) to "pick out" G modulo functions with small coefficients. This can be done as follows: Let $\mathscr{W}_4$ be the AU space of functions analytic in Im $x > 0$. Call $\mathscr{W}_5$ the space of functions analytic in the interior of the angle formed by the half-lines emanating from i to π and $-\pi$ (see Fig. 38) and having distribution boundary values. Let

$$F(x) = \sum c_n \exp(inx),$$

where the series converges in the topology of $\mathscr{W}_4$. Suppose $F \in \mathscr{W}_5$. Then, since F is periodic, F is regular in the whole plane with the possible exception of the points $j\pi$ for j an odd integer. If, in addition,

$$F(x) = 0((x - \pi)^{-1})$$

as $x \to \pi$, then F can have a pole of order at most one at π. Since F is periodic we must have

$$F(x) = \alpha G(x) + \sum c_n' \exp(inx),$$

where $c_n' = 0(e^{-Bn})$ for all B.

Thus G can be characterized modulo "trivial" functions, that is, functions of the form $\sum c_n' \exp(inx)$, by

(a) Having a Fourier series convergent in $\mathscr{W}_4$.
(b) Belonging to $\mathscr{W}_5$.
(c) Satisfying $O(1/(x - \pi))$ (or even $o([1/(x - \pi)]^2)$) as $x \to \pi$.

Note that $\mathscr{W}_4$ and $\mathscr{W}_5$ are AU and so are emenable to our general Fourier analysis, but the space of functions satisfying (c) is not AU.

Although G cannot be characterized by AU conditions, nevertheless it is readily verified that any F satisfying (a) and (b) is meromorphic in the whole finite complex plane. Since F is periodic we can write

$$F(x) = \sum_{\text{finite}} \alpha_j G^{(j)}(x) + \sum c_n' \exp(inx),$$

where $c_n' = O(e^{-Bn})$ for all B.

Now, let Ш$_5$ be the inverse Mellin transform of $\mathbf{W}_5$, the dual of $\mathbf{W}_5'$. The above shows that if $f(s) = \sum c_n n^{-is}$, where the series converges for $Rs > 1$,

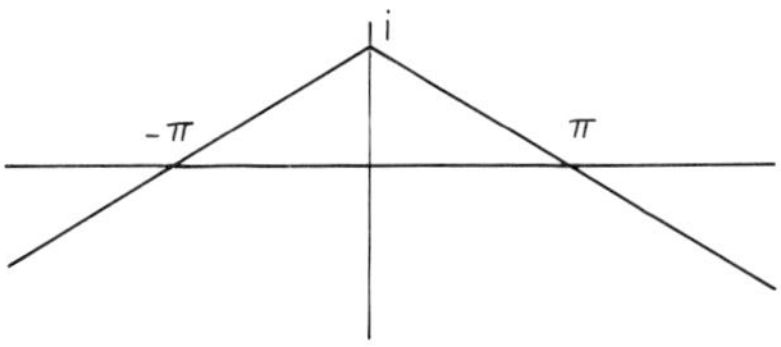

FIGURE 38

and if $f \in Ш_5$, then f is of the form

$$f(s) = \sum_{\text{finite}} \alpha_j \zeta_1(is - j) + \sum c'_n n^{-is},$$

where $c'_n = O(e^{-Bn})$ for all B. For,

$$G^{(j)}(x) = \sum (-1)^n (in)^j \exp (inx)$$

so $G^{(j)}(x)$ is the Fourier transform of the Mellin transform of

$$i^j \sum (-1)^n n^{-is+j} = i^j \zeta_1(is - j).$$

Now, clearly it follows easily from the functional equation for $\zeta(s)$ that if, as $s \to i\infty$,

$$\sum_{\text{finite}} \alpha_j \zeta_1(is - j) = o(\zeta_1(is - 1))$$

then $\alpha_j = 0$ for $j > 0$. Thus we have shown

THEOREM 12.10. *Let $f(s) = \sum c_n n^{-is}$ where the series converges in* Im $s < -1$. *Suppose $f \in Ш_5$ and*

$$f(s) = o(\zeta_1(is - 1))$$

as $s \to i\infty$ through imaginary values. Then

$$f(s) = \alpha\zeta_1(is) + \sum c'_n n^{-is},$$

where $c'_n = O(e^{-Bn})$ for all B.

We leave to the reader the problem of giving an intrinsic description of $Ш_5$.

It is possible to put the above in a somewhat more general framework: We start with the classical idea

$$\Gamma(s) = \int_0^\infty t^s e^{-t} \frac{dt}{t}$$

$$= n^s \int_0^\infty t^s e^{-nt} \frac{dt}{t}.$$

Thus,

$$f(-is)\Gamma(s) = \int_0^\infty t^s F(-it) \frac{dt}{t}.$$

This means that $\zeta(s)\Gamma(s)$ is the (inverse) Mellin transform of $G(-it)$ so $G(t)$ is the Mellin transform of $\zeta(s)\Gamma(s)$ (thought of as a function on Re $s =$ const.). The relation of this to the previous passage from f to F is that, since Γ is the Mellin transform of e^{-t}, the composition of Mellin transform with Fourier transform can be accomplished by first multiplying by $\Gamma(s)$ then taking Mellin transform. (It seems to us that by considering the two-stage passage of f to F as above we simplify the method because we eliminate the Γ function.)

Hecke [2], pp. 591–626, 644–707, has made a careful study of the operation of multiplication by Γ followed by Mellin transform. He studied functions $f(s)$ having Dirichlet series convergent Re s sufficiently large and such that $(s-k)f(s)$ can be continued to an entire function so that a functional equation of the form

$$\left(\frac{2\pi}{\lambda}\right)^{-s}\Gamma(s)f(s)=\gamma\left(\frac{2\pi}{\lambda}\right)^{s-k}\Gamma(k-s)f(k-s)$$

for some $\gamma=\pm 1$ is satisfied. He showed that for such f the Mellin transform of $\Gamma(s)f(s)$ is a certain type of automorphic function. The above functional equation allows f to grow essentially like $\exp(2|s|\log|s|)$ which is larger than the growth of $\zeta(s)$. Nevertheless, by looking at spaces defined by more subtle growth conditions than those depending only on $|s|$ we might hope to "pick out" the functions satisfying the above type of functional equations modulo "trivial" functions.

The difficulty in applying a limitation of singularities argument (above we used: $F(x)=o((x-\pi)^{-2})$ implies $F(x)=\alpha G(x)+\textit{regular}$) is illustrated by the function $\zeta(2s)$ which satisfies the Hecke conditions. The Mellin transform of $\Gamma(s)\zeta_1(2s)$ is

$$\psi(x)=\sum(-1)^n e^{-in^2x}.$$

This is now a lacunary series in the Fabry sense. Thus the real axis is a natural boundary so our above method involving singularities cannot work.

PROBLEM 12.4.

Find an analog of Theorem 12.9 where the functions satisfying functional equations of the Hecke type replace $\zeta(s)$. In this way, show how to distinguish functions satisfying Hecke type relations by growth conditions rather than by use of the explicit form of the functional equation.

A possible approach to this problem is to replace the Mellin transform of $f(s)$ by the Mellin transform of $f(s/2)$. If we denote by $\tilde{F}(x)$ the Fourier transform of this Mellin transform, then

$$\tilde{F}(x)=\sum c_n \exp(-i\sqrt{n}\,x).$$

Thus $\tilde{F}$ is no longer periodic, but it is almost periodic. Now, however, if $f(s)=O(\exp(\beta|s|\log|s|))$ with $\beta<1$ then, as above, $\tilde{F}(x)$ is entire. Thus there may be some hope of using a limitation of singularities argument.

A possible type of answer to Problem 12.4 is in the spirit of relative problems discussed in Section XII.1. Namely, let $f(s)$ be the product of one of the Dirichlet series satisfying a functional equation of the Hecke type (or perhaps some other special function defined by a Dirichlet series) by a simple factor which removes the pole. Let $g(s)$ be a Dirichlet series whose coefficients are not allowed to be much larger than those of $f(s)$. Suppose that $g(s)$ is an entire function which, in the whole s plane, is "somewhat" smaller than $f(s)$. Then we can write

$$g(s) = g_1(s) + g_2(s),$$

where $g_1(s)$ and $g_2(s)$ are Dirichlet series, where the coefficients of $g_1(s)$ are very small, and where $g_2(s)$ is an entire function which is "considerably" smaller than $g(s)$.

Such a result, if true, would mean that the Dirichelt series considered by Hecke have the only "exceptional" types of cancellation effects. One might even hope to prove a stronger result, namely, that with $g(s)$ as above, we can write

$$g(s) = \sum d_i f_i(s) + h(s),$$

where $h(s)$ has small coefficients, and the $f_i(s)$ are Dirichlet series of the Hecke type, and the d_i are complex numbers. If this is true, then we could think of this expression for $g(s)$ as analogous to the partial fraction decomposition of a meromorphic function, with the term $h(s)$ playing the role of the regular part.

Although this idea might seem appealing at first, an analysis of the method of proof of Theorem 12.10 shows that no result of this type is possible. The reason for this is that once we consider entire functions $f(s)$ which grow like $\exp((1+\varepsilon)|s| \log |s|)$ for some $\varepsilon > 0$, then the analog of the space $\mathscr{W}_3$ considered above would consist of functions which are analytic in $\operatorname{Im} z > \delta > 0$ which do not have boundary values on $\operatorname{Im} z = 0$. From this it follows without difficulty that no decomposition

$$g(s) = \sum d_i f_i(s) + h(s)$$

of the desired type can exist.

The reason for the failure of the above approach to make progress in the solution of Problem 12.4 is that we have not described all the exceptional or "cancellation" properties of the Hecke functions. Notice that the amount whereby the sum of a series $\sum a_n$ is less than the sum of the absolute values of the terms is only one type of cancellation property. A more general property is to say that, for suitable linear functions L, $L\{a_n\}$ is much less than would be expected. Thus Problem 12.4 must be attacked by finding cancellation properties of Hecke functions which differ from growth conditions.

Results of this type seem to be important for making progress in analytic number theory. For example, suppose $P(x)$ is a homogeneous polynomial of even degree in $x = (x_1, \ldots, x_n)$ which is positive definite. We form the generating function

$$g(s) = \sum_{x \text{ integer}} [P(x)]^{-s} = \sum c^j j^{-s}.$$

Here c^j is the number of integers x satisfying $P(x) = j$. According to the circle method of Hardy and Littlewood (see Landau [1] Vol. I_2) we write

$$g(s) = f(s) + h(s),$$

where $f(s)$ is a special Dirichlet series whose coefficients are defined in terms of the number of congruence solutions

$$P(x) \equiv j \pmod{m}$$

for arbitrary m, and, hopefully, the coefficients of h are small.

If the degree of P is greater than 2, it would not be expected that g or f has a meromorphic continuation to the whole plane. Nevertheless, one might expect that g and f have certain "exceptional" properties. We could then try to find some other special functions f_i with these properties and try to write

$$g(s) = \sum d_i f_i(s) + h(s),$$

where $h(s)$ has "very small" coefficients. Such a result would constitute a remarkable improvement to the circle method.

See Remarks 12.10 and 12.11.

XII.8. Example 4. An Analog of the Fabry Gap Theorem for Several Variables and Complex a^j

Let Ω be an open convex set in complex x space. By $\mathscr{W}$ we denote the space of functions analytic on Ω. Suppose for simplicity that the origin is a point in the boundary of Ω, that $\operatorname{Im} x_1 = 0$ is a plane of support of Ω at the origin, and that Ω lies below this plane. Call $\mathscr{W}_1$ the space of functions analytic on $\|x_1\| < \delta$, $\|x_2\| < \delta, \ldots, \|x_n\| < \delta$, where $\|x_j\| = \max(\operatorname{Re} |x_j|, |\operatorname{Im} x_j|)$.

Let $\vec{D}^j$ be the system $\{\partial/\partial x_i - i a_i^j\}$ for arbitrary $a^j \in C^n$. By A^j we denote the rectangular parallelepiped $\|z - a^j\| \leq \frac{1}{2} \|a^j\|$, where $\|z\|$ is the sum of $|\operatorname{Re} z_i| + |\operatorname{Im} z_i|$. Let $\{u^j\}$ be a sequence with $u^j < \varepsilon |a^j|$ for $|a^j|$ sufficiently large. We now make the important uniform discreteness condition: $v_1^j = 0$, $v_2^j = \exp(-\varepsilon |a^j|)$. It is to be noted that, unlike the case $n = 1$, all a^j real, this is *not* implied by $|a^j - a^{j'}| \geq 1$.

Let us note that the methods of Example 5 of Chapter V would not prove that $\mathscr{W}$ is AU unless Ω is a convex polyhedron. However, $\mathscr{W}$ appears in our theorems only insofar as the series $\sum c^j \exp(ia^j \cdot x)$ is required to converge in the topology of $\mathscr{W}$. We could, therefore, replace $\mathscr{W}$ by the space $\mathscr{E}(\Omega)$ of functions of $y \in R^{2n}$ which are indefinitely differentiable on Ω. $\mathscr{E}(\Omega)$ is AU for any convex Ω. In the series $\sum c^j \exp(ia^j \cdot x)$ we are thinking of x and a^j as vectors in C^n (not C^{2n} with coordinates w which is the space on which the functions of $\mathbf{E}'(\Omega)$ are defined). Thus we are only interested in the topology of $\mathbf{E}'(\Omega)$ on those points $(w_1, w_2, \ldots, w_{2n})$ with $w_{2j} = iw_{2j-1}$, since $a^j \cdot x = w \cdot y$ with $x_j = y_{2j-1} + iy_{2j}$ and $w_{2j-1} = a^j$, $w_{2j} = ia^j$. That is, as far as we are concerned, $\mathscr{W}$ behaves as though it is AU with AU structure consisting of those functions $k(z)$ of n complex variables z which dominate all $\exp[B|z|\Psi(\theta)]$ for $B < 1$, where Ψ is a suitable function on the unit sphere in C^n which is determined by Ω. (This is to be modified slightly if Ω does not have compact closure.) Actually, we shall use below (following (12.81)) a slightly simpler property of the topology of $\mathscr{W}$ (cf. Remark 12.3).

A simple modification of Chapter V, Example 1 shows that $\mathscr{W}_1$ is AU. An AU structure consists of those k_1 which dominate all

$$\exp[\alpha_1 \|z_1\| + \cdots + \alpha_n \|z_n\|]$$

where

$$\sum \alpha_i < \delta.$$

For simplicity, we shall carry out the details for the case $n = 1$, as the general case is similar. From (12.34) we have for $\|z - a^j\| \leqslant \frac{1}{2}\|a^j\|$,

$$|'G^j(z)| \leqslant \exp(\varepsilon|a^j|)(\tilde{u}^j)^{-\tilde{u}^j} c_2^{\tilde{u}^j} |a^j|^{\tilde{u}^j} \leqslant \exp(c_3 \varepsilon |a^j|) \tag{12.71}$$

as in the derivation of (12.36). (We sometimes write $|a^j|$ for $\|a^j\|$.)

From (12.34) we have for $\|z - a^j\| \geqslant \frac{1}{2}\|a^j\|$, as in the argument following (12.35)

$$\begin{aligned} |'G^j(z)| &\leqslant \exp(c_4 \varepsilon |a^j|)(\tilde{u}^j)^{-\tilde{u}^j}(|a^j| + \|z\|)^{\tilde{u}^j} \\ &\leqslant \exp(c_4 \varepsilon |a^j|)\varepsilon^{-\varepsilon|a^j|}(1 + \|z\|/|a^j|)^{\varepsilon|a^j|} \\ &\leqslant \exp[\varepsilon(c_4 + \log \frac{1}{\varepsilon})|a^j| + \varepsilon\|z\|]. \end{aligned} \tag{12.72}$$

Next we define $''G^j$ by

$$''G^j(z) = \{\exp[\varepsilon_1(z - a^j) - i\varepsilon_2(z - a^j) \operatorname{sgn} \operatorname{Im} a^j]\} \times \{\sin[(\varepsilon_3(z - a^j)/l^j)/\varepsilon_3(z - a^j)/l^j]\}^{l^j}, \tag{12.73}$$

where l^j is the greatest integer $\leqslant \varepsilon_4 |a^j|$. Note that, because of the structure of Ω, $G^j = {'G^j}\, {''G^j} \in \mathbf{W}' \cap \mathbf{W}_1'$ if ε, ε_1 and ε_2 are small enough subject to some simple restrictions which will be met below.

From (12.40) we have for all z

$$(12.74) \quad |{''G^j}(z)| \leqslant |\exp [\varepsilon_1 |a^j| + \varepsilon_1 z + \varepsilon_2 \operatorname{Im}(z - a^j) \operatorname{sgn} \operatorname{Im} a^j + \varepsilon_3 |\operatorname{Im}(z - a^j)|]|.$$

From (12.39) and (12.73) we have

$$(12.75) \quad |{''G^j}(z)| \leqslant |\exp [\varepsilon_1 |a^j| + \varepsilon_1 z + \varepsilon_2 \operatorname{Im}(z - a^j) \operatorname{sgn} \operatorname{Im} a^j + \varepsilon_3 |\operatorname{Im}(z - a^j)| - \varepsilon_4 |a^j| + 1]|$$

if $|z - a^j| \geqslant e\varepsilon_4 |a^j| / \varepsilon_3$.

Now let $\mathscr{W}_2$ be the space of functions analytic in $\|x\| < \delta - \varepsilon_5$. We claim that for ε small enough the analytic $(\mathscr{W}, \mathscr{W}_1)$ density of $\{D^j\}$ is $\leqslant \mathscr{W}_2$. As in the above examples we must verify (12.11) and (12.12).

We begin with (12.11). For this it suffices to show

$$(12.76) \qquad |G^j(z)| \leqslant \text{const} \exp (\varepsilon_5 \|z\|).$$

From (12.71) and (12.74) we have for $\|z - a^j\| \leqslant \frac{1}{2} \|a^j\|$

$$(12.77) \qquad |G^j(z)| \leqslant \exp [c_8(\varepsilon + \varepsilon_1 + \varepsilon_2 + \varepsilon_3)|a^j|]$$

so (12.76) holds if ε, ε_1, ε_2, and ε_3 are sufficiently small. Combining (12.72) and (12.75) we find for $\|z - a^j\| \geqslant \frac{1}{2} \|a^j\|$

$$(12.78) \quad |G^j(z)| \leqslant \left| \exp \left[\left(c_9 \,\varepsilon \log \frac{1}{\varepsilon} + \varepsilon_1 - \varepsilon_4 \right) |a^j| + \varepsilon_1 z + 1 + \varepsilon \|z\| + \varepsilon_2 \operatorname{Im}(z - a^j) \operatorname{sgn} \operatorname{Im} a^j + \varepsilon_3 |\operatorname{Im}(z - a^j)| \right] \right|,$$

provided that $\frac{1}{2} \geqslant e\varepsilon_4/\varepsilon_3$. We set $\varepsilon_4 = \varepsilon_3/2e$, so (12.78) holds for $\|z - a^j\| \geqslant \frac{1}{2} \|a^j\|$.

We now let $\varepsilon_4 > c_9 \varepsilon \log 1/\varepsilon + \varepsilon_1$. We also set $\varepsilon_2 = \varepsilon_3$. If $|\operatorname{Im} z| \geqslant \frac{1}{2} |\operatorname{Im} a^j|$ then

$$|\operatorname{Im}(z - a^j)| \leqslant |\operatorname{Im} z| + |\operatorname{Im} a^j| \leqslant 3 |\operatorname{Im} z| \leqslant 3 \|z\|$$

so (12.76) holds. If $|\operatorname{Im} z| \leqslant \frac{1}{2} |\operatorname{Im} a^j|$ then

$$\begin{aligned} \operatorname{Im}(z - a^j) \operatorname{sgn} \operatorname{Im} a^j &= -(\operatorname{Im} a^j) \operatorname{sgn} \operatorname{Im} a^j + (\operatorname{Im} z) \operatorname{sgn} \operatorname{Im} a^j \\ &= -|\operatorname{Im} a^j| + (\operatorname{Im} z) \operatorname{sgn} \operatorname{Im} a^j, \end{aligned}$$

while

$$|\operatorname{Im}(z - a^j)| = |\operatorname{Im} a^j| \pm |\operatorname{Im} z|.$$

Thus,

$$|\text{Im}\,(z - a^j)\,\text{sgn Im}\,a^j + |\text{Im}\,(z - a^j)|\,| \leqslant 2\,|\text{Im}\,z|$$

so (12.76) holds in this case also.

It remains to verify (12.12). In the present case it suffices to show, for ε_6, t, t' to be prescribed below:

$$|G^j(z)| \leqslant \text{const}\,\exp\,[(2\varepsilon_5 - \delta - \varepsilon_6)\,|\text{Re}\,z| + t\,|\text{Im}\,z|] \tag{12.79}$$

for $\text{Re}\,z \leqslant 0$, $\|z - a^j\| \geqslant \frac{1}{2}\,\|a^j\|$ and

$$|G^j(z)| \leqslant \text{const}\,\exp\,[t'\,|\text{Re}\,z| + t\,|\text{Im}\,z|] \tag{12.80}$$

for $\text{Re}\,z \geqslant 0$, $\|z - a^j\| \geqslant \frac{1}{2}\,\|a^j\|$. We require that $\varepsilon_6 > \varepsilon_5$ and that the set

$$|\text{Re}\,x| \leqslant t + \delta - \varepsilon_5\,,\; -t' - \delta + \varepsilon_5 \leqslant \text{Im}\,x \leqslant \varepsilon_5 - \varepsilon_6 \tag{12.81}$$

be contained in Ω. (If δ is small enough then such ε_5, ε_6, t, t' will exist satisfying the conditions to be imposed below.) The meaning of (12.79) and (12.80) is that $\{G^j\}$ be a bounded set in $\mathbf{W}_3'$, where $\mathscr{W}_3$ is the space of holomorphic functions on the rectangle

$$|\text{Re}\,x| \leqslant t,\; 2\varepsilon_5 - \delta - \varepsilon_6 \geqslant \text{Im}\,x \geqslant -t'. \tag{12.82}$$

See Fig. 39.

To derive (12.79) and (12.80) we note that since Ω contains the rectangle (12.81), the functions k for $\mathscr{W}$ dominate the corresponding functions k' belonging to an AU structure for the space of analytic functions on this rectangle. Using Example 5 of Chapter V to describe the k', we deduce (12.79) and (12.80) if we replace, in (12.12), k by k'.

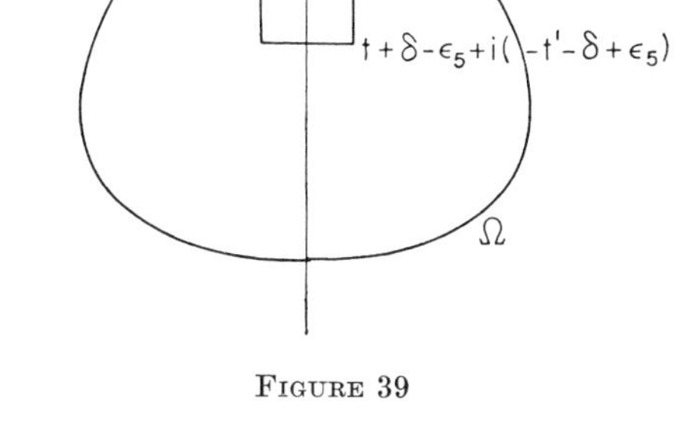

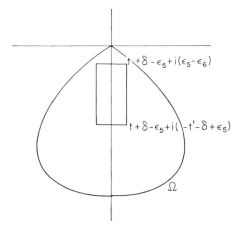

FIGURE 39

Now (12.79) follows from (12.78) if $\varepsilon_1 \geqslant \varepsilon + \delta + \varepsilon_6 - 2\varepsilon_5$ and $\varepsilon_2 = \varepsilon_3$. This is seen as follows:

Case a. $|\text{Im } z| \geqslant \frac{1}{2}|\text{Im } a^j|$. Then

$$|\text{Im } (z - a^j)| \leqslant |\text{Im } z| + |\text{Im } a^j| \leqslant 3\,|\text{Im } z|.$$

Thus, (12.79) holds if $t \geqslant 6\varepsilon_2 + \varepsilon$. (Recall that $\varepsilon_4 > \varepsilon \log 1/\varepsilon + \varepsilon_1$.)

Case b. $|\text{Im } z| \leqslant \frac{1}{2}|\text{Im } a^j|$. Then

$$\begin{aligned}\text{Im } (z - a^j) \text{ sgn Im } a^j &= -\text{Im } a^j \text{ sgn Im } a^j + (\text{Im } z) \text{ sgn Im } a^j \\ &= -|\text{Im } a^j| + (\text{Im } z) \text{ sgn Im } a^j,\end{aligned}$$

while

$$|\text{Im } (z - a^j)| = |\text{Im } a^j| \pm |\text{Im } z|.$$

Thus,

$$\big|\text{Im } (z - a^j) \text{ sgn Im } a^j + |\text{Im } (z - a^j)|\,\big| \leqslant 2\,|\text{Im } z|.$$

This implies that (12.79) holds if $t \geqslant 4\varepsilon_2 + \varepsilon$.

Finally the verification of (12.80) is similar for suitable constants.

Thus, we have proved that the analytic $(\mathscr{W}, \mathscr{W}_1)$ density is $\leqslant \mathscr{W}_2$. The same method works for $n > 1$. Applying Theorem 12.4 gives the following sharp extension of the Fabry gap theorem to $n > 1$ and to complex exponents:

THEOREM 12.11. *Let $\{a^j\}$ be a discrete sequence of points in C and let $\{a^{j_t}\}$ be a subsequence with the following properties: There is a $\delta > 0$ so that for every j_t the number $\tilde{u}^{j_t}$ of a^j in $|z - a^{j_t}| \leqslant \delta\,|a^{j_t}|$ is $\leqslant \varepsilon\,|a^{j_t}|$, where ε can be taken arbitrarily small for $|a^{j_t}|$ large enough, and these a^j satisfy $\prod |a^j - a^{j_t}| \geqslant$* $\exp\,(-\varepsilon|a^j|)(\frac{1}{2}\tilde{u}^j)^{\tilde{u}^j}$. *Suppose the series*

$$f(x) = \sum c^j \exp\,(ia^j \cdot x)$$

is absolutely convergent, uniformly on compact subsets of some open convex set Ω. Suppose f can be continued to be analytic in a neighborhood of a boundary point x^0 of Ω. Then the series

$$g(x) = \sum c^{j_t} \exp\,(ia^{j_t} \cdot x)$$

is absolutely convergent in an open convex set containing Ω and x^0.

See Remark 12.12.

XII.9. Grouping of Terms

In all the examples discussed so far, each $\mathfrak{V}^j$ (the multiplicity variety associated to $\vec{\mathrm{D}}^j$ by Theorem 4.2) has been a single point. The definition of geometric density allows us to treat also the case in which $\mathfrak{B}^j$ consists of many points, so we are really dealing with series in which convergence is by grouping of terms (that is, we think of a series $\sum f_j$ as being

$$\sum_i \left(\sum_{B_i} f_j\right)$$

where B_i are the groups) (see Schwartz [4], Ehrenpreis [4] for other examples). As shown in Schwartz [4] and Ehrenpreis [4], for $n = 1$ there is a way to define G^j other than by (12.28) and (12.23), which is actually more efficient: Call

$$J^j(z) = \prod_{a^{j'} \in A^j} (1 - z/a^{j'})^{e^{j'}} \tag{12.83}$$

and set

$$\tilde{G}^j(z) = \begin{cases} -J^j(z) \int_{\Gamma^j} d\zeta/(z-\zeta)J^j(\zeta) & \text{if } z \text{ does not lie inside } \Gamma^j \\ -\left[J^j(z) \int_{\Gamma^j} d\zeta/(z-\zeta)J^j(\zeta)\right] + 1 & \text{if } z \text{ lies inside } \Gamma^j, \end{cases} \tag{12.84}$$

where Γ^j is a contour which contains V^j in its interior, but does not contain any other point of any $V^{j'}$ for $j' \neq j$ in its interior or on it (see Ehrenpreis [4], p. 545). (The case of z on Γ^j is dealt with by shifting the contour.) Here V^j is the set of points in $\mathfrak{B}^j$. It is a simple consequence of the residue calculus that $\tilde{G}^j$ satisfies Conditions 1 and 2 of Section XII.3.

In order to estimate $\tilde{G}^j$ we have to know that the groups V^j are "nicely separated" so that we can choose Γ^j in such a way that J^j is large on Γ^j. That is, Condition (a) for geometric density of

$$\{\mathfrak{B}^j\} \leqslant \{(A^j, u^j, v_1^j, v_2^j)\}$$

is replaced by:

(a′) For each j there exists a contour Γ^j as above, with length $\Gamma^j \leqslant c^{\tilde{u}^j}$ and for all $z \in \Gamma^j$

$$|J^j(z)| \geqslant v_1^j + v_2^j(\tilde{u}^j)^{\tilde{u}^j}. \tag{12.31′}$$

We then verify easily that (12.34) holds for $\tilde{G}^j$. We can use $\tilde{G}^j$ in place of $'G^j$ and in so doing we can eliminate the uniform discreteness conditions from Theorems 12.6, 12.7, 12.8, and 12.10. For example, we can derive the classical extension of the Fabry gap theorem:

Let $n = 1$ *and let* $\sum c^j \exp(ia^j x)$ *converge uniformly on compact sets of the lower half-plane to a function* $f(x)$ *which is holomorphic in the neighborhood of* $x = 0$. *Suppose the* a^j *are real and for any* $\varepsilon > 0$ *the number of* a^j *in* $-A < x < A$ *is* $\leqslant \varepsilon A$ *for* A *sufficiently large. Then* f *is holomorphic in some half-plane* Im $z < \delta$, *where* $\delta > 0$.

We wish next to discuss an important result of Ostrowski [2]. This asserts: Let

$$w(x) = \sum c^j x^j$$

have radius of convergence 1. Suppose there exists an infinite sequence of Hadamard gaps, that is, for an infinite number of $\theta_l (\theta_l \to \infty)$ we have $c^j = 0$ for $\theta_l < j < (1 + \delta)\theta_l$, where $\delta > 0$. Then the "grouped series"

$$\sum_l \sum_{j=\theta_l}^{\theta_{l+1}} c^j x^j$$

converges in the neighborhood of every point of the unit circle at which w is regular.

We can restate Ostrowski's result as follows: If w is regular at $x = 1$, then the grouped series converges in a neighborhood of $x = 1$.

The abstract generalization is now clear: Let $\sum c^j \exp (ix \cdot a^j)$ converge in the topology of $\mathscr{W}$ to $f(x)$ which lies in $\mathscr{W}_1$. Then, under suitable lacunary conditions, a suitable grouped series converges in the topology of $\mathscr{W}_2$. Here we no longer make the assumption that $\mathscr{W}_2 \subset \mathscr{W}$ as we did before but rather that $\mathscr{W}_2 \supset \mathscr{W}_1$. We shall not go into details but presumably the results are the same as before.

See Problem 12.5.

XII.10. Natural Boundaries

Let $g(x) = \sum c^j x^j$ be regular in the unit circle of the complex x plane. Suppose the radius of convergence of the power series is 1. $|x| = 1$ is a natural boundary for g means that f is not regular at any point of $|x| = 1$. In our general form, one might say that $f(x) = \sum c^j \exp (ia^j \cdot x)$ (where the series converges in the topology of $\mathscr{W}$) has $\mathscr{W}$ as a natural boundary if f does not belong to any AU space $\mathscr{W}_1$ which does not contain $\mathscr{W}$. A little consideration shows that for most $\mathscr{W}$ no f could have this property because there are too many possible spaces $\mathscr{W}_1$.

An important result related to natural boundaries is that for an appropriate choice of $\eta^j = \pm 1$ the function $\sum \eta^j c^j x^j$ has $|x| = 1$ as a natural boundary. The general result is:

THEOREM 12.12. *Let $f(x) = \sum c^j \exp (ia^j \cdot x)$, where the series converges in the topology of $\mathscr{W}$. Let $\mathscr{W}_1$ be any AU space. Then for an appropriate choice of $\eta^j = \pm 1$ the function*

$$\tilde{f}(x) = \sum \eta^j c^j \exp (ia^j \cdot x)$$

does not belong to $\mathscr{W}_1$ *if the sequence* $\{c^j \exp(ia^j \cdot x)\}$ *is not bounded in the topology of* $\mathscr{W} \overset{a}{\cap} \mathscr{W}_1$.

PROOF. It is not hard to show that if the sequence $\{c^j \exp(ia^j \cdot x)\}$ is not bounded in $\mathscr{W} \overset{a}{\cap} \mathscr{W}_1$, then we can write $f = f_1 + f_2$, where

$$f_2(x) = \sum c^{j_t} \exp(ia^{j_t} \cdot x),$$

where $\{a^{j_t}\}$ is an infinite sequence which is $(\mathscr{W}, \mathscr{W}_1)$ lacunary and such that the series $\sum c^{j_t} \exp(ia^{j_t} \cdot x)$ does not converge in the topology of $\mathscr{W} \overset{a}{\cap} \mathscr{W}_1$. If f and $f - 2f_2$ were both in $\mathscr{W}_1$, then also $f_2 \in \mathscr{W}_1$ which would imply by lacunarity that the series $\sum c^{j_t} \exp(ia^{j_t} \cdot x)$ converges in the topology of $\mathscr{W}_1$, which contradicts our construction. Thus, either f or $f - 2f_2 \notin \mathscr{W}_1$. Clearly both f and $f - 2f_2$ are of the form $\tilde{f}$, so our assertion is proved. □

See Problem 12.6.

XII.11. The Converse Problem

Up to now we have dealt with the problem of showing that the $(\mathscr{W}, \mathscr{W}_1)$ density of $\{\vec{\mathrm{D}}^j\}$ is $\leqslant \mathscr{W}_2$. We have not shown that our method leads to anything like best possible results. We shall now outline a procedure for proving, under suitable conditions, that the $(\mathscr{W}, \mathscr{W}_1)$ density of $\{\vec{\mathrm{D}}^j\}$ is not $\leqslant \mathscr{W}_2$. We shall assume for simplicity that $n = 1$ and that $\mathfrak{B}^j = (a^j, \text{identity})$ for each j.

Our construction will use the idea of grouping of terms (see Section XII.9 above). Let $\{a^j\}$ be a very lacunary sequence of positive numbers, e.g., $a_j = \exp(j)$. Let $a_1^j = a^j + \varepsilon^j$, where $\varepsilon^j \to 0$ very rapidly, say $\varepsilon^j = \exp(-\exp j)$. Consider the series

$$f(x) = \sum (\varepsilon^j)^{-1}[\exp(ia^j x) - \exp(ia_1^j x)]. \tag{12.85}$$

The series without any grouping clearly converges absolutely for $\operatorname{Im} x > 1$, but does not converge absolutely for $\operatorname{Im} x > 1 - \delta$ for any $\delta > 0$. On the other hand,

$$\exp(ia^j x) - \exp(ia_1^j x) = \exp(ia^j x)[1 - \exp(i\varepsilon^j x)]. \tag{12.86}$$

It is clear that on every compact set of the complex plane we have $|1 - \exp(i\varepsilon^j x)| = 0(\varepsilon^j)$, as $j \to \infty$. Thus, the series (12.85) (in the grouping indicated) converges uniformly on compact sets of $\operatorname{Im} x > 0$. Hence, f is holomorphic in $\operatorname{Im} x > 0$. The Fabry gap theorem (see Section XII.9) shows that the real axis is a natural boundary for f.

The "meaning" of this construction is seen by taking the formal Fourier transform of f, say

$$\hat{f} = \sum (\varepsilon^j)^{-1}(\delta_{a^j} - \delta_{a^j+\varepsilon^j}). \tag{12.87}$$

Each term in the series (12.87) looks like a difference quotient. The reason f is holomorphic in $\operatorname{Im} x > 0$ is that the series (12.87) is required to converge when applied to suitable entire functions; the convergence results because the difference quotient of such an entire function is close to the derivative which is estimated by Cauchy's formula.

To generalize this example, we could use higher difference quotients which approximate higher derivatives. For some examples (e.g., the Fabry gap theorem) this method shows that our results are best possible. It would be of interest to show that certain general geometric density conditions are best possible; that is, they lead to the actual $(\mathscr{W}, \mathscr{W}_1)$ density of a sequence $\{\vec{\mathrm{D}}^j\}$.

There is a similar procedure which is somewhat more efficient. This involves replacing δ_{a^j} by use of Cauchy's formula. Thus, if

$$f(x) = \sum c^j \exp(ia^j x).$$

then (formally)

$$\hat{f} = \sum c^j \delta_{a^j}.$$

Since $\hat{f}$ is to be applied to entire functions g, we may write

$$\hat{f} \cdot g = \sum c^j \int \frac{g(z)}{z - a_j}\, dz. \tag{12.88}$$

where the contour surrounds a^j. Under suitable conditions we could interchange the order of summation and integration. If we call

$$F(z) = \sum \frac{c^j}{z - a^j},$$

then (12.88) becomes (formally)

$$\hat{f} \cdot g = \int_{\gamma_0} g(z) F(z)\, dz, \tag{12.89}$$

where the contour of γ_0 surrounds all the a^j. If F is large on a certain contour γ which surrounds all the a^j, we could hope to change the contour to γ. In this way we should be able to extend the class of g on which $\hat{f}$ is defined, namely, to functions g which are small on γ, thus showing that f belongs to a subspace of $\mathscr{W}$. In addition to shifting the contour to γ, we can also modify F by adding expressions which have no poles inside the contours considered.

To illustrate the method we give two examples:

EXAMPLE 1. Let us show that

$$f(x) = \sum_{n \geqslant 0} (-1)^n \exp(inx)$$

is analytic in the strip $|\mathrm{Re}\, x| < \pi$. This is of course obvious by direct calculation, since

$$f(x) = \frac{1}{1 + \exp(ix)}$$

but we want to understand why this is the case without using the explicit formula for the sum of a geometric series. By the above we want to find the natural class of entire functions g for which the right side of (12.89) has a meaning on a suitable γ, instead of γ_0. Here

$$F(z) = \sum_{n \geqslant 0} \frac{(-1)^n}{z - n},$$

and γ_0 is, for example, the contour consisting of the half lines $\mathrm{Im}\, z = \pm 1$ for $\mathrm{Re}\, z \geqslant -\frac{1}{2}$ and the vertical segment $\mathrm{Re}\, z = -\frac{1}{2}$, $-1 \leqslant \mathrm{Im}\, z \leqslant 1$. ($\gamma$ is described below.) The original space of g for which this formula is defined is $\mathbf{W}'$, where $\mathscr{W}$ is the space of functions holomorphic in $\mathrm{Im}\, x > 0$ (see Section XII.7 for the AU structure). For such g it is clear that (12.88) implies (12.89). If we replace F by

$$\sum_{n=-\infty}^{\infty} \frac{(-1)^n}{z - n} = F_1(z),$$

which is permissible by our remarks, then, since $F_1(\zeta) = \sec \pi\zeta$, we can shift the contour γ_0 to the line γ: $\mathrm{Re}\, z = -\frac{1}{2}$. It is clear that we can extend f to all of $\mathbf{W}_1'$, where $\mathscr{W}_1$ is the space of functions holomorphic in $|\mathrm{Re}\, x| < \pi$. (See Section XII.5 for the AU structure.)

EXAMPLE 2. We can apply the same construction to show that

$$\zeta_1(z) = \sum_{n \geqslant 0} (-1)^n n^{-s}$$

is an entire function which is $\leqslant$const $\exp(A\,|s| \log(1 + |s|))$ for any $A > 1$ (actually, even better estimates). We need only observe that, if $F(z)$ is the function corresponding to $\zeta_1 = \sum (-1)^n \delta_{i \log n}$ according to (12.89), then we can replace $F(z)$ in (12.89) by $F_1(z) = \sec(e^{-iz})$ which has the points $z = i \log n$ amongst its poles. (See Section XII.7 for the AU structure.)

We can now see how to interpret the nonlacunary behaviors in Examples 1 and 2 in terms of the high order difference quotient approach mentioned at the beginning of Section XII.11. For simplicity we consider Example 1. We have shown that we can represent $\hat{f}$ as $1/\sin z$ on the imaginary axis. By the definition of the Riemann integral, we can approximate $\hat{f}$ by a discrete linear combination of point masses on the imaginary axis. Since $\hat{f}$ is to be applied to entire functions, Taylor's formula shows how to write each point mass on the imaginary axis in terms of differential operators of infinite order on the real axis. That is, for g an entire function and t real, we have

$$\delta_{it} \cdot g = \sum (it)^n g^{(n)}(0)/n!.$$

These operators can be approximated by finite difference operators which is what we want

See Problem 12.7.

A slightly different way of looking at the representation of $\hat{f}$ as an infinite differential operator is the following: We note that the Cauchy integral representation for $\delta_\alpha - \delta_{\alpha+\varepsilon}$ (applied to entire functions) involves

$$\frac{1}{z-a} - \frac{1}{z-a+\varepsilon} = \frac{\varepsilon}{(z-a)(z-a+\varepsilon)}$$

which is small on suitable curves surrounding a and $a+\varepsilon$. A similar remark holds for higher difference operators, the point being that we consider contours which surround all the points in question. From this point of view it is natural to regard (12.89) as representing $\hat{f}$ as an infinite difference operator.

Remarks

Remark 12.1. See page 406.

We could generalize the main problem by not requiring that $\mathscr{W}_1$ be a topological vector space, but rather use the generalized extension described in Section VIII.1. The methods of this chapter could be modified so as to treat such problems.

Remark 12.2. See page 407.

In case $f^j(x)$ is of the form $c^j \exp(ia^j x)$, we can see that, in general, convergence of $\sum f^j$ in the topologies of $\mathscr{W}$ and $\mathscr{W}_1$ is equivalent to convergence in the topology of $\mathscr{W} \overset{a}{\cap} \mathscr{W}_1$. Thus, our notion of lacunary agrees with the heuristic notion formulated on page 397.

Remark 12.3. See page 411.

For most of our applications f^j will be of the form $c^j \exp(ia^j \cdot \quad)$. If $\{f^j\}$ is bounded, then so are all of its derivatives, since differentiation is continuous on $\mathscr{W}_2$. Thus, if $\{c^j \exp(ia^j \cdot \quad)\}$ is bounded, so is $\{|a^j|^l c^j \exp(ia^j \cdot \quad)\}$. In most of our applications there is an l such that

$$\sum' |a^j|^{-l} < \infty,$$

where Σ' means that $a^j = 0$ is to be omitted. (This is not the case in Example 3, Section 7.) Now, for any bounded set B in $\mathscr{W}'_2$, the set

$$\left\{|a^j|^l c^j S \cdot \exp(ia^j \cdot \quad)\right\}_{\substack{\text{all } j \\ S \in B}}$$

is bounded. This means that the series

$$\Sigma\, c^j \hat{S}(a^j) = \Sigma\, c^j S \cdot \exp(ia^j \cdot \quad)$$

converges uniformly for $S \in B$. By the definition of the topology of $\mathscr{W}_2$ it follows that $\Sigma\, c^j \exp(ia^j \cdot\ )$ converges in the topology of $\mathscr{W}_2$.

This means that in most applications the conclusion of Theorem 12.3, namely, "$\{f^j\}$ is bounded in $\mathscr{W}_2$," can be replaced by "Σf^j converges in $\mathscr{W}_2$." Thus, Theorem 12.3 says, roughly, that analytic $(\mathscr{W}, \mathscr{W}_1)$ density $\leqslant \mathscr{W}_2$ implies $(\mathscr{W}, \mathscr{W}_1)$ density $\leqslant \mathscr{W}_2$.

A similar approach could be applied to Example 3 in Section 7 below, except that we should use infinite differential operators (see Ehrenpreis [7]). We shall omit the details, as this result is not needed for our purposes.

Another important point to observe is the following: the space $\mathscr{W}$ is used essentially only to go from the convergence of $\sum f^j$ in $\mathscr{W}$ to an inequality on the "Fourier transform" of the f^j. For example, if $f^j = c^j \exp(ia^j \cdot\quad)$. then we obtain bounds on $\{c^j\}$. For this reason we can generally replace the space $\mathscr{W}$ by any other AU space $\mathscr{W}_3$ such that the growth conditions of $\mathbf{W}'$ and $\mathbf{W}'_3$ are the same on the union of the $\mathfrak{V}^j$. This is also apparent from (12.12) above.

Remark 12.4. See page 413.

Theorem 12.4 can be strengthened, namely, we can allow "many" $\vec{\mathrm{D}}^j$ for which the $\mathfrak{V}^j$ are close to $\mathfrak{V}^{j_t}$, as long as we make some assumptions about the additional f^j. An example in which a positivity condition is imposed is given in Remark 12.6 (which applies to the general theory).

Remarks 12.5 and 12.6. See page 419.

Remark 12.5. By sharpening the above inequalities slightly we could strengthen a result of Pólya [1]: If a suitable geometric density of $\{a^j\}$ is $\leqslant d$, then if $f = \Sigma\, c^j \exp(ia^j z)$ (as above) is analytic on an interval of the boundary of S of length $> d$, then $\Sigma\, c^{j_t} \exp(ia^{j_t} z)$ converges in a strip which is larger than S.

Remark 12.6. Theorem 12.6 can be strengthened still further. We do not need to know that the number of a^j in $|z - a^{j_t}| < \eta |a^{j_t}|$ should be $\leqslant \varepsilon |a^{j_t}|$. It suffices that the number of a^j in this interval $|z - a^{j_t}| < \eta |a^{j_t}|$ be $\leqslant \varepsilon |a^{j_t}|$ *except* for a set of a^j for which the c^j lie in a fixed angle of the complex plane $\theta_{j_t} \leqslant \arg c^j \leqslant \theta_{j_t} + \theta'$, where $\theta' < \pi$. The proof proceeds along essentially the same lines as above, except that we derive two "one-sided" bounds for the c^{j_t}. For example, if $\theta_{j_t} = \theta' = 0$, then we derive estimates of the form

$$c^{j_t} = c_1^{j_t} + c_2^{j_t} = c_3^{j_t} + c_4^{j_t},$$

where $|c_1^{j_t}|$ and $|c_3^{j_t}|$ are small and $c_2^{j_t}$ $(c_4^{j_t})$ are bounded from above (below) by a small quantity. These inequalities are obtained by making the $G^{j_t}(a^j)$ positive (negative) for those a^j in question. The first example of this type of result is due to Mandelbrojt [3]. (He treats a case in which the subsequence $\{j_t\}$ is all of $\{j\}$.) Our methods apply to the general theory, and, in particular, this type of result holds for all the examples treated below.

Remark 12.7. See page 427.

A similar result holds if the space $\mathscr{W}$ in Theorem 12.7 is replaced by the space of distributions on the real line. The same proof applies in that case.

In case $\{a^{j_i}\}$ is the whole sequence $\{a^j\}$, then the result of Theorem 12.7 can be deduced from results in Malliavin [1]. Malliavin's proof is much different from ours.

Remark 12.8. See page 424.

The reader who is not interested in going into the details of the argument can *define* $\mathscr{W}_2$ as the AU space with this AU structure. This will not affect any of the results.

Remark 12.9. See page 426.

The classical Riemann ζ function shows that we cannot take $\beta = 1$. For $\zeta_1(s) = (1 - 2^{1-s})\zeta(s) = \sum (-1)^n n^{-s}$ satisfies (by the functional equation for $\zeta(s)$)

$$|\zeta_1(s)| = 0(\exp (|s| \log |s|)).$$

It is likely that a more careful analysis of the proof would show that any $\beta < 1$ would be suitable in Theorem 12.9

Remarks 12.10–12.11. See page 432.

Remark 12.10. We may consider the property of a series that the sum of the series is small while the sum of the absolute values is large as being a cancellation effect. Thus, Theorem 12.8 shows that the series for $\zeta_1(s)$ has essentially the most possible cancellation that any Dirichlet series $\sum c_n n^{-s}$ can have.

Remark 12.11. Suppose we want to use the above to derive a characterization of the ζ function. Then we want to "eliminate" functions $f(s)$, which are $0(\exp (\beta|s| \log |s|))$ for $\beta < 1$. One such method is by requiring that $f(s) = 0$ for s a negative even integer. This can be seen by applying the Mellin transform to $\Gamma(s)f(s)$. For, no function of the form $f(s)$ in $\mathscr{W}_2$ can vanish at the negative even integers if and only if $\{\delta_{-2j}\}$ are dense in $\mathscr{W}_2'$. This is equivalent to $\{z^{2j}\}$ dense in the Mellin transform $\text{Ш}_2'$ of $\mathscr{W}_2'$ on the set of Mellin transforms of $\{f(s)\}$ which is the same as saying that all the even derivatives of $F(x)$ cannot vanish at the origin. But if $F^{(2j)}(0) = 0$ for all $j > 0$ then F would be a constant plus an odd function. This is readily seen to be incompatible with the expression

$$F(x) = \sum_{n \geqslant 0} c_n \exp (-inx)$$

by the uniqueness of the Fourier series.

Alan Taylor has pointed out to me that it is possible to improve the above, namely, $F^{(j')}(0)$ cannot vanish for any sequence of even j' of positive lower density. This is a consequence of the following theorem of Rényi [1]: If $g(x) = \sum a_k x^k$ is periodic and entire and the lower density of k for which $a_k = 0$ is $> \frac{1}{2}$, then $g \equiv 0$. We apply this result to $g(x) = F(x) - F(-x)$. (The argument following the statement of Theorem 12.9 shows that F is entire.) Since g is odd, if $F^{(j')}(0) = 0$ for a set of even j' of positive lower density, we would have $g \equiv 0$ hence, since

$$g(x) = 2c_0 + \sum_{n>0} c_n \cos nx$$

$c_n \equiv 0$. We are thus led to a refinement of the classical result of Hamburger (see Titchmarsh [1], p. 31), namely, that $\zeta(s)$ is determined by its functional equation (actually, a slightly weaker property), a mild growth condition in the strip $|\text{Re } s| \leqslant 2$, and the fact that it can be written as $\sum a_n n^{-s}$, where the series converges in the half-plane $\text{Re } s > 1$. That the ζ function can be characterized by a growth condition in the whole plane and its vanishing at the negative even integers was proved by Beurling in [5]. However, Theorem 12.8 does not make the requirement that $f(s) = 0$ at the negative even integers or that $a^j = \log j$.

Remark 12.12. See page 436.
A slight refinement of our argument would show that the series representing g converges absolutely on the convex hull of Ω and any open set into which f can be continued analytically.

Problems

PROBLEM 12.1. See page 397.
Extend our methods to convolution equations.

PROBLEM 12.2. See page 401.
Carry out this program. In particular, generalize Cantor's theorem from the following point of view: Let $\vec{D} = (D_1, \ldots, D_r)$, where the D_j are linear partial differential operators with constant coefficients. Using the notation of Section X.1, suppose there is uniqueness for the parametrization problem. Under what conditions can one find a stronger uniqueness property of the following form: Suppose $f \in \mathscr{E}(\Omega - B_1 - \cdots - B_h)$ and $\vec{D}f = 0$ on $\Omega - B_1 - \cdots - B_h$. Suppose f is "not too large" near $B_1 \ldots B_h$ and $\delta_j f(x) \to 0$ in some suitable sense as $x \to B_j$. Then $f \equiv 0$.

Riemann's theorem corresponds to the case $n = 2$, $r = 1$, $\vec{D}$ = Laplacian, Ω = unit disk, $h = 1$, B_1 = unit circle, δ_1 = identity. It is also possible to think of the results of Section IX.9 as being in the same spirit as Riemann's theorem.

PROBLEM 12.3. See page 413.
Extend the results to the higher dimensional case.

PROBLEM 12.4. See page 430.
Find an analog of Theorem 12.9 where the functions satisfying functional equations of the Hecke type replace $\delta(s)$. In this way, show how to distinguish functions satisfying Hecke type relations by growth conditions rather than by use of the explicit form of the functional equation.

PROBLEM 12.5. See page 438.
Carry out the details.

PROBLEM 12.6. See page 439.
The above construction can be modified to show that "almost all" choices of $\pm$ give functions $\hat{f}$ satisfying the desired conclusion. There is a relation between lacunary series and almost all results: In general a property of a lacunary series or of "almost all" series holds only if the "obvious" sufficient conditions are satisfied.

Give an "intrinsic" explanation of this phenomenon.

Can the results of this chapter be modified by replacing density conditions by probabilistic conditions?

PROBLEM 12.7. See page 442.
Can this approximation procedure be carried out in such a way that one can see intrinsically, that is, from the structure of the difference operators, that the limit is

$$\sum_{n \geq 0} (-1)^n \delta_n$$

and that the limit exists in the topology of $\mathbf{W}_1$?

CHAPTER XIII

General Theory of Quasianalytic Functions

Summary

We study certain questions related to the theory of quasianalytic classes. Let $n = 1$; let $B = \{b_j\}$ be a sequence of positive numbers. Denote by $\mathscr{E}_B$ the space of functions $f \in \mathscr{E}$ which, together with each of their derivatives, satisfy, on every compact set,

$$(*) \qquad |f^{(j)}(x)| \leqslant a\varepsilon^j b_j$$

for any $\varepsilon > 0$. $\mathscr{E}_B$ is given a natural topology. In Chapter V it was shown that $\mathscr{E}_B$ is LAU.

The classical Denjoy-Carleman theorem (see Paley and Wiener [1]) gives a necessary and sufficient condition on B in order that there exist a nontrivial $f \in \mathscr{E}_B$ which vanishes together with all its derivatives at the origin. $\mathscr{E}_B$ is called *quasianalytic* (QA) if no such f exists.

We discuss a more general quasianalytic problem: Let D be a linear constant coefficient operator; consider the space of f which, together with each of its derivatives, satisfies, for any $\varepsilon > 0$

$$|D^j f(x)| \leqslant a\varepsilon^j b_j$$

uniformly on compact sets. Call $\mathscr{E}_B(D)$ the class of f. Then we want to know if an $f \in \mathscr{E}_B(D)$ can vanish together with all its derivatives on a noncharacteristic of D. This problem extends in the obvious way to $r > 1$, and refines the uniqueness question of Chapter IX. It is discussed in Section XIII.1.

In Section XIII.2 we discuss a refinement of the Denjoy-Carleman theorem in which (*) is replaced by a condition of degree of approximation of f by suitable analytic functions. This type of quasianalytic problem was introduced by Bernstein.

Let $n = 1$ and denote by ${}_0\mathscr{E}_B$ the space of sequences $\{a_j\}$ which satisfy

$$|a_j| = 0(\varepsilon^j b_j)$$

for every $\varepsilon > 0$. There is a natural map φ: $\mathscr{E}_B \to {}_0\mathscr{E}_B$ defined by

$$\varphi(f) = \{f^{(j)}(0)\}.$$

For I a closed interval in R we denote by ${}_I\mathscr{E}_B$ the space of those f which are indefinitely differentiable on I and satisfy (*) there. We denote by ψ the restriction map of $\mathscr{E}_B$ into ${}_I\mathscr{E}_B$. In Section XIII.3 we prove: *If either φ or ψ is onto, so is the other. A necessary and sufficient condition that this be the case is that there should exist an $\varepsilon > 0$ so that*

$$\lambda(z) \geqslant \text{const } \gamma(\varepsilon |z|).$$

Here

$$\lambda(z) = \sum |z|^j / b_j$$

and

$$\gamma(r) = \sup_{\operatorname{Im} z > 0,\ |z| = r} \exp \alpha(\operatorname{Re} z, \operatorname{Im} z),$$

where α is the Poisson integral of $\log \lambda$ in the upper half-plane.

We treat the following type of problems in Section XIII.4: Let Ω be an open set in C which contains the origin and which is the union of connected line segments (real dimension 1) through the origin (i.e., Ω is star shaped with respect to the origin). Suppose for a suitable set of such line segments l^j we are given functions f^j on l^j which belong to a fixed QA space. Then, under suitable compatibility conditions, we can conclude the existence of a function F which is complex analytic on Ω such that $F = f^j$ on l^j for each j.

XIII.1. General Quasianalyticity and Approximation

The previous chapter dealt with lacunary sequences of operators. As we shall see, the present chapter is concerned with what may be thought of as an opposite situation, namely, dense sequences of operators.

The theory of quasianalyticity is concerned with generalizations of the uniqueness properties of analytic functions. The first type of problem considered is to find interesting classes $\mathscr{C}$ of indefinitely differentiable functions of one real variable such that if $f, g \in \mathscr{C}$ and $f^{(j)}(p) = g^{(j)}(p)$ for all j at some point p then $f \equiv g$.

The classical class $\mathscr{C}$ considered is linear and is defined as follows: Let $B = \{b_j\}$ be a sequence of positive numbers. As in Chapter V, Example 6, denote by $\mathscr{E}_B$ the space of indefinitely differential functions f on R which, together with all their derivatives, satisfy, on every compact set

$$|f^{(j)}(x)| \leqslant a\varepsilon^j b_j \tag{13.1}$$

for any $\varepsilon > 0$. As in Chapter V we shall assume that the sequence $\{b_j\}$ is convex in the sense that $b_j^2 \leqslant b_{j-1} b_{j+1}$ for all $j \geqslant 1$, though this is probably not needed for most of our results.

The classical theory of quasianalytic functions is concerned with the question of whether an $f \in \mathscr{E}_B$ can vanish together with all its derivatives at the origin, without being identically zero. This theory culminates with the theorem of Denjoy-Carleman which gives a necessary and sufficient condition on B in order that such a function exist. When no such f exists the class $\mathscr{E}_B$ is called *quasianalytic*.

It is interesting to note that, though the classes $\mathscr{E}_B$ are linear, there a nonlinear aspect to the quasianalytic problem. This is illustrated by a result of Mandelbrojt [3] and Beurling [2] which asserts that any $f \in \mathscr{E}$ can be written in the form $f = f_1 + f_2$, where $f_i \in \mathscr{E}_{B_i}$ and each $\mathscr{E}_{B_i}$ is quasianalytic. Another surprising feature of quasianalyticity is the result (see Ehrenpreis [4], Bang [1]) that the intersection of all nonquasianalytic $\mathscr{E}_B$ is the space of real analytic functions.

One might ask for the reason that classes satisfying (13.1) for suitably small b_j are quasianalytic. Note that if $\mathscr{E}_B$ is non-quasianalytic then there clearly exists an $f \in \mathscr{E}_B$ which vanishes on a half-line and is not identically zero. (Actually there even exists as $f \in \mathscr{E}_B$, $f \not\equiv 0$, f of compact support in this case.) Now if $f \in \mathscr{E}_B$, $f \not\equiv 0$, and $f(x) = 0$ for $x < 0$, then $x^p f(x)$ also vanishes for $x < 0$ for any $p > 0$. If the b_j are small then (13.1) says that the derivatives of f are small. Thus both the derivatives of f and $x^p f$ have "smallness" properties. This contradicts the spirit of the Heisenberg uncertainty relation.

PROBLEM 13.1.
Use the Heisenberg uncertainty relation to prove the sufficient condition on B given by the Denjoy-Carleman theorem for the quasianalyticity of $\mathscr{E}_B$.

From the analytic point of view, the usual proofs of the Denjoy-Carleman theorem show that it is closely related to minimum-modulus problems: Given an entire function F with certain growth restrictions, how fast can it fall off on the real axis, or, how fast can its minimum modulus $\to 0$?

As usual, let $\mathscr{E}'_B$ be the dual of $\mathscr{E}_B$. Then the question of the quasianalyticity of $\mathscr{E}_B$ is the same as the question of whether the linear combinations of the $\delta^{(j)}$, where δ is the unit mass at the origin, are dense in $\mathscr{E}'_B$. By Fourier transform this is the question of whether the polynomials are dense in $\mathbf{E}'_B$.

The road to generalization is now clear. Let n be arbitrary and let $\mathscr{W}$ be any AU space. Then we want to find linear subsets of $\mathscr{W}'$ which are dense.

The most interesting example in the spirit of the classical theory of quasianalyticity is the following: Let $\vec{\mathrm{D}} = \{D_1, \ldots, D_r\}$ and let $B = \{b_j\}$ be a multisequence of positive numbers $[j = (j_1, \ldots, j_r)]$. We define the space $\mathscr{E}_B(\vec{\mathrm{D}})$ as consisting of all indefinitely differentiable f on R such that, on every compact set, f and all its derivatives satisfy

$$(13.1^*) \qquad |(D_1^{j_1} \cdots D_r^{j_r} f)(x)| \leqslant a\varepsilon^{|j|} b_j$$

for all $\varepsilon > 0$. In the notation of Chapter IX, let $(T^d, \ldots, T^0)$ be a noncharacteristic for $\vec{\mathrm{D}}$. Then we ask the

General quasianalytic problem. What are the conditions on B in order that no $f \in \mathscr{E}_B(\vec{\mathrm{D}})$ except $f = 0$ can vanish together with all its derivatives on $T^d \cup \cdots \cup T^0$?

In the classical case $r = n$ and $\vec{\mathrm{D}} = (\partial/\partial x_1, \ldots, \partial/\partial x_n)$, $d = 0$ and $T^0 = \{0\}$.

We could extend the above problem as follows: We do not require that $(T^d, \ldots, T^0)$ be noncharacteristic but, as in Chapter IX, we add growth conditions to the definition of $\mathscr{E}_B(\vec{\mathrm{D}})$. We shall say only a few words about this problem.

See Problem 13.2.

From another point of view quasianalyticity can be thought of as the property of a space $\mathscr{W}$ that "sufficient" vanishing of $w \in \mathscr{W}$ should imply $w = 0$. Thus, for the spaces $\mathscr{E}_B$, the "vanishing" condition is that f and all its derivatives should vanish at a point. This can be restated for $n = 1$ as $x^{-j} f(x) \to 0$ as $x \to 0$ for all j, that is, $f(x) = 0(x^j)$ for all j as $x \to 0$. We could ask for a stronger decrease of f as $x \to 0$. Such a question has been treated by Turán [1] who considers the case $f(x) = 0(e^{-|x|^{-p}})$ for a suitable space $\mathscr{W}$ of "regular" functions. Mandelbrojt [1] had previously considered an analogous problem in which the regularity condition on $\mathscr{W}$ is replaced by the condition that each $f \in \mathscr{W}$ be of the form

$$f(x) = \sum c^j \exp(ia^j x),$$

where the a^j are suitably lacunary, that is,

$$\sum (a^j)^{-\delta} < \infty \quad \text{for some } \delta < 1.$$

We could naturally consider a problem in which we are given both lacunary and regularity conditions on $\mathscr{W}$. The methods of this chapter combined with those of Chapter XII could be used to treat this type of problem (see Section VIII.1 and XII.1).

A similar type of problem in which behavior at $x = 0$ is replaced by behavior at infinity was considered in Chapter IX in relation to uniqueness questions.

Let us mention briefly a connection with a generalization of quasi-analyticity for $n = 1$ due to Bernstein [1] and developed to a great extent by Beurling in [2] and [4]. These authors show how to translate the conditions (13.1) into "approximation conditions," namely, we consider those f which satisfy for each compact set $\Omega \subset R$

$$\max_{x \in \Omega} |f(x) - g_l(x)| \leqslant \varepsilon_l .$$

Here $g_l = g_l^+ + g_l^-$, where $g_l^\pm$ is analytic in the upper (lower) half-plane and bounded by constants $c_l^\pm$ on a neighborhood of Ω in the half-plane. In the case of the spaces $\mathscr{E}_B$ we may actually take $c_l^+ = c_l^-$. Beurling has treated the case where $c_l^- = 0$ for all l in [2], [4], and where the condition that f and all its derivatives vanish at a point is replaced by the vanishing of f on an interval or on a set of positive measure. From a slightly different point of view this was considered previously by Levinson [1]. In unpublished work Beurling has considered other examples.

We shall not go into the precise details of showing how this fits into our theory. We remark only that if $\mathscr{W}$ is an AU space, then every $w \in \mathscr{W}$ has a Fourier representation (see Theorem 1.5)

$$w(x) = \int \exp (ix \cdot z) \, d\mu(z)/k(z), \tag{13.2}$$

where k is in an AU structure K for $\mathscr{W}$, and where μ is a bounded measure on C. Then, by approximating C by suitable compact subsets, we get from (13.2) methods of approximating w by entire functions; the behavior of these entire functions and the degree of approximation of w depend on k, hence, on $\mathscr{W}$.

There is another point of view which relates the theory of quasi-analyticity even more closely to that of lacunary series. Let $n = 1$. Suppose for simplicity that an f satisfying (13.1) is represented by a series

$$f(x) = \sum c^j \exp (ia^j x) \tag{13.3}$$

which converges in the topology of $\mathscr{E}_B$. Then the vanishing of f on $[-1, 1]$ means that the restriction of f to $[-1, 1]$ belongs to the space $\mathscr{W}_1$ which consists of 0 only. Thus, we might expect the methods of Chapter XII to apply to show that under appropriate conditions each $c^j = 0$.

We should like to mention that there is another aspect of quasianalyticity which seems restricted to $n = 1$. Namely, let f, as in (13.1), be represented as $f = f^+ + f^-$, where $f^\pm$ is analytic in the upper (lower) half-plane and has boundary values in a suitable sense. Then the vanishing of f on $[-1, 1]$ means that f^+ can be continued to be analytic in the lower half-plane and on $[-1, 1]$. Thus, again, if f^+ has a representation as an exponential series, we could try to apply the methods of Chapter XII. (It should be noted that $\mathscr{W}_1$ is the space of functions analytic in the lower half-plane and on $[-1, 1]$ and so is *not* AU. Nevertheless, a slight variation of our technique could be used.) Since this approach seems peculiar to $n = 1$, we shall not pursue it. (See, however, Levinson [1].)

See Problem 13.3.

We wish to explain the relation between the lacunary series approach described above with another (more usual) approach to the quasianalytic problem. Let us consider the space $\mathscr{E}_B$. Suppose we want to know if there exists a $g \in \mathscr{E}_B$ of compact support. Now, G, the Fourier transform of g, is an entire function of exponential type, whereas F, the Fourier transform of f described by (13.3), has its support on the discrete set $\{a^j\}$. Thus, we seem to have two very different types of problems. However, they can be treated by a single method: Suppose G is real on the real axis and the support of g does not meet $[-1, 1]$. To show g does not exist, we could try to find a function $h \in \mathscr{E}$ whose support is contained in $[-1, 1]$ such that $HG \geqslant 0$ everywhere on R. For then $(h * g)(0) = \int h\check{g} > 0$ which contradicts the fact that the supports of h and $\check{g}$ do not meet. (Here $\check{g}(x) = g(-x)$.) Thus, essentially, H should have the same zeros as G. Of course, since G is small at infinity, we do not need $HG \geqslant 0$ exactly, but only approximately in order to deduce $\int h\check{g} > 0$. On the other hand the construction of such an H is the essential tool in showing by the methods of Chapter XII that such an f could not exist if the sequence $\{a^j\}$ is lacunary in a suitable sense. Roughly speaking, this sense is about the same as the existence of an H as above which vanishes on $\{a^j\}$. It makes $\{a^j\}$ slightly too small to be sufficient, say for $\mathscr{E}_B[-2, 2]$.

The problem of polynomial approximation in $\mathbf{W}'$, where $\mathscr{W}$ is AU, is closely related to the *Bernstein Approximation Problem* in which we ask for polynomial approximation in the same type of norms on the real part of C only.

We shall now study the General Quasianalytic Problem posed above. We shall consider only the case when each $T^j \subset T^d$. Define $\lambda(\zeta)$ for $\zeta \in C^r$ by

$$\lambda(\zeta) = \sum |\zeta_1|^{j_1} \cdots |\zeta_r|^{j_r}/b_j.$$

If the series does not converge, we set $\lambda(\zeta) = \infty$. We shall assume that $\{b_j\}$ is a product sequence, that is, there exist sequences $B^{(p)} = \{b_l^{(p)}\}$ for $p = 1, \ldots, r$ so that $b_j = b_{j_1}^{(1)} \cdots b_{j_r}^{(r)}$.

THEOREM 13.1. *Suppose that $B^{(p)}$ satisfies the hypothesis of Theorem 5.26 except that the assumption that the entire functions lie in $\mathscr{E}_B$ is replaced by $\lambda(z) \leqslant \exp(c\,|z|)$ for some c. Suppose that moreover there exists a constant* *so that for all p, j, j'*

$$b_{j+j'}^{(p)} \leqslant (\text{const})^{j+j'} b_j^{(p)} b_{j'}^{(p)}.$$

Then $\mathscr{E}_B(\vec{\mathrm{D}})$ is AU. $\mathbf{E}'_B(\vec{\mathrm{D}})$ consists of all entire functions $F(z)$ of exponential type which satisfy, for some a, c, c'

$$|F(z)| \leqslant c(1+|z|)^{c'} \exp(c'|\operatorname{Im} z|)\lambda[aP_1(z), \ldots, aP_r(z)].$$

An AU structure for $\mathbf{E}'_B(\vec{\mathrm{D}})$ consists of all k of the form

$$\tilde{\lambda}[P_1(z), \ldots, P_r(z)]k_1(|\operatorname{Im} z|)k_2(|\operatorname{Re} z|),$$

where k_1 is a continuous positive function which dominates all exponentials, and k_2 is a continuous positive function which dominates all polynomials, and $\tilde{\lambda}$ dominates $\lambda(az)$ for all $a > 0$.

Here $\vec{\mathrm{P}} = (P_1, \ldots, P_r)$ is the Fourier transform of $\vec{\mathrm{D}}'$.

See Remark 13.1.

PROOF. The fact that each $F \in \mathbf{E}'_B(\vec{\mathrm{D}})$ is of the above form is an immediate consequence of the fact that each $S \in \mathscr{E}'_B(\vec{\mathrm{D}})$ can be represented in the form

$$S = \partial \sum D_1^{'j_1} \cdots D_r^{'j_r} \mu_{j_1 \cdots j_r}/b_j l^{|j|},$$

where the $\mu_{j_1 \cdots j_r}$ are uniformly bounded measures having their supports in a fixed compact set, ∂ is a constant coefficient operator, and $l > 0$. Here $|j| = j_1 + \cdots + j_r$. The above representation is an easy consequence of the Hahn-Banach theorem and the definition of the topology of $\mathscr{E}_B(\vec{\mathrm{D}})$.

The rest of the proof of Theorem 13.1 proceeds along the same lines as the proof of Theorem 5.26. We must make the following modifications: We might first try to replace the "heat equation" construction of the proof of Theorem 5.26 by setting $y = (y_1, \ldots, y_r)$ with dual variables ζ and

$$g(y, x) = \sum_j (D_1^{j_1} \cdots D_r^{j_r} f)(x) y_1^{(m+1)j_1} \cdots y_r^{(m+1)j_r}/[(m+1)j]! \tag{13.4}$$

where $j! = j_1! \cdots j_r!$. It is clear that g satisfies the heat equations

$$D_q g = \partial^{m+1} g/\partial y_q^{m+1} \tag{13.5}$$

for $q = 1, 2, \ldots, r$.

Call V the variety in (ζ, z) space defined as the common zeros of

$$P_q(z) - (i\zeta_q)^{m+1} = 0$$

for $q = 1, \ldots, r$. Then, as in the proof of Theorem 5.26 (compare (5.118)) we find a Fourier representation of f in the form

$$f(x) = \int \exp(ix \cdot z)\, d\mu(z)/k(z), \tag{13.6}$$

where μ is a bounded measure and $k(z)$ dominates

$$(13.7) \qquad (1+|z|)^A \exp \{A\,|\mathrm{Im}\, z| + \sum_q \psi_q[a\,|\mathrm{Re}\,[P_q(z)^{1/m+1}]|]\}$$

for all A, a. Here $\Psi(\zeta) = \sum \psi_q(\zeta_q)$ is the conjugate of the function describing the growth of g. Now, this may not be good enough for our purposes because the real parts of $P_q(z)^{1/m+1}$ may not be large near the real axis; thus k may not be large so the Fourier representation (13.6) of f would not show that $f \in \mathscr{E}_B(\vec{\mathrm{D}})$.

In order to remedy this difficulty, we might attempt to proceed as follows: Let $\gamma = (\gamma_1, \ldots, \gamma_r)$ denote any of the 2^r possible vectors each of whose components is either $-i$ or one. Then for each γ we define, instead of (13.4)

(13.4*)

$$g_\gamma(y, x) = \sum_j (D_1^{j_1} \cdots D_r^{j_r} f)(x)(\gamma_1 y_1)^{(m+1)j_1} \cdots (\gamma_r y_r)^{(m+1)j_r}/[(m+1)j]!\,.$$

The heat equations (13.5) become

$$(13.5^*) \qquad D_q g_\gamma = \gamma_q^{-m-1}\, \partial^{m+1} g_\gamma / \partial y_q^{m+1}.$$

For each γ we obtain an integral representation for f of the form (13.6) where (we write k_γ for k) k_γ dominates, instead of (13.7)

$$(13.7^*) \qquad (1+|z|)^A \exp \{A\,|\mathrm{Im}\, z| + \sum_q \psi_q[a\,|\mathrm{Re}\,[\gamma_q P_q(z)^{1/m+1}]|]\}.$$

Now, no single (that is, q fixed) expression (13.7*) may be large enough to deduce from (13.6) that $f \in \mathscr{E}_B(\vec{\mathrm{D}})$. But we could modify the proof of Theorem 5.26 to deduce that $f \in \mathscr{E}_B(\vec{\mathrm{D}})$ if we could use $\max k_\gamma(z) = \tilde{k}(z)$ in place of the individual $k_\gamma(z)$. To replace the various representations (13.6) for $k =$ any k_γ by a single representation with k replaced by $\tilde{k}(z)$ is just the type of problem dealt with in Chapter XII, since $\tilde{k}$ is in the AU structure for the AU intersection of the spaces each of whose AU structures is given by k_γ. As mentioned in Section XII.1 the passage from the k_γ to $\tilde{k}$ is a "cutting up" problem. We gave one technique for solving this cutting up problem which is applicable in the present situation:

For each γ we choose a variable $x^\gamma = (x_1^\gamma, \ldots, x_n^\gamma)$; call $X = (x^1, \ldots, x^{2^r})$. We consider the tensor product $\mathfrak{E}_B(\vec{\mathrm{D}})$ of 2^r copies of $\mathscr{E}_B(\vec{\mathrm{D}})$ defined as in Section XII.1. Because of our assumption that

$$b_{j+j'}^{(p)} \leqslant (\mathrm{const})^{j+j'}\, b_j^{(p)}\, b_{j'}^{(p)}$$

for all p, j, j', it follows that

$$F(X) = f(x^1 + x^2 + \cdots + x^{2^r}) \in \mathfrak{E}_B(\vec{\mathrm{D}}).$$

We now use $F(X)$ as "initial value" for a heat equation, namely we define

$$(13.4^{**}) \qquad G(Y, X) =$$

$$\sum_j [D_1^{j^1{}_1}(1) \cdots D_r^{j^1{}_r}(1) D_1^{j^2{}_1}(2) \cdots D_1^{j^2{}_r}(2) \cdots D_1^{j^{2^r}{}_1}(2^r) \cdots D_r^{j^{2^r}{}_r}(2^r) F](X)$$

$$(\gamma_1^1 y_1^1)^{(m+1)j^1{}_1} \cdots (\gamma_r^1 y_r^1)^{(m+1)j^1{}_r} \cdots (\gamma_1^{2^r} y_1^{2r})^{(m+1)j^{2^r}{}_1} \cdots (\gamma_r^{2^r} y_r^{2^r})^{(m+1)j^{2^r}{}_r} / [(m+1)J]!.$$

Here $J = (j^1, \ldots, j^{2^r})$ and $Y = (y^1, \ldots, y^{2^r})$, and we have written $D_i(\alpha)$ to indicate that D_i acts on the variables x^α.

Now G satisfies the heat equations

$$(13.5^{**}) \qquad D_q(\alpha) G = (\gamma_q^\alpha)^{-m-1} \partial^{m+1} G / (\partial y_q^\alpha)^{m+1}$$

for all q, α. In addition, G satisfies the equations

$$(13.8) \qquad \partial G / \partial x_t^\alpha = \partial G / \partial x_t^\beta$$

for all t, α because of the special form of F. We now apply the method of the proof of Theorem 5.26 to obtain a Fourier integral representation for G using the equations (13.5**) and (13.8). Thus G has a Fourier representation where the frequencies lie in the variety V defined by all the equations

$$(13.9) \qquad P_q(z^\alpha) - (\gamma_q^\alpha)^{-m-1} (i \zeta_q^\alpha)^{m+1} = 0$$

and also

$$(13.10) \qquad z_t^\alpha - z_t^\beta = 0.$$

Equations (13.9) are used as before and present no novel points. However, equations (13.10) say that the Fourier representation of G is on the "diagonal," namely, $z^\alpha = z^\beta$ for all α, β. If we take into account the definitions this leads to an expression for G of the form (compare (5.117))

$$(13.11) \qquad G(Y, X) = \int_V \exp(iX \cdot Z + iY \cdot \overline{Z}) \, d\mu(\overline{Z}, Z) / K(\overline{Z}, Z),$$

where $K(\overline{Z}, Z)$ dominates

$$(13.12) \qquad (1 + |Z|)^A \exp\{A |\text{Im } \overline{Z}| + A |\text{Im Z}| + \sum_\alpha \Psi(a |\text{Im } \zeta^\alpha|)\}.$$

Now, $f(2^r x) = G(0, x, \ldots, x)$. Using (13.10) and (13.11) this yields

$$f(2^r x) = \int_{V'} \exp(i 2^r x \cdot z) \, d\mu(\overline{Z}, z) / K(\overline{Z}, z),$$

where we have written $z = z^1 = \cdots = z^{2^r}$ and where V' is the variety in $(\overline{Z}, z)$ space defined by (13.9) with z replacing z^a. If we now use (13.9) to solve for ζ^α in terms of z we obtain

$$(13.6^{**}) \qquad f(rx) = \int \exp(irx \cdot z)\, d\mu'(z)/\tilde{k}(z),$$

where by (13.12) $\tilde{k}(z)$ dominates

$$(13.7^{**}) \quad (1+|z|)^A \exp\{A\,|\mathrm{Im}\, z| + \sum_\alpha \sum_q \psi_q[a\,|\mathrm{Re}\,[\gamma_q^\alpha P_q(z)^{1/m+1}]|]\}.$$

This is exactly the type of representation desired, namely $\tilde{k}(z)$ is of the form $\prod k_\gamma(z)$ (or equivalently $\max_\gamma k_\gamma(z)$) with k_γ dominating the right sides of (13.7*). The proof can now be completed in the same manner as the proof of Theorem 5.26. □

We shall now change our definitions and define $\mathscr{E}_B(\vec{\mathrm{D}})$ for any B to be the AU space whose AU structure is given by the conclusion of Theorem 13.1, even if B does not satisfy the hypothesis of Theorem 13.1.

Before studying the General Quasianalytic Problem we shall consider the special case when $D_j = \partial/\partial x_j$. (We write $\mathscr{E}_B$ for $\mathscr{E}_B(\vec{\mathrm{D}})$, in this case.)

THEOREM 13.2. *Let $D_j = \partial/\partial x_j$, for all j. A necessary and sufficient condition that no $f \in \mathscr{E}_B$ can vanish together with all its derivatives at the origin is: For each line $\sum$ in R, denote by $\mathscr{E}_B(\sum)$ the subspace of $\mathscr{E}_B$ consisting of those functions which are constant in directions orthogonal to $\sum$. Then each $\mathscr{E}_B(\sum)$ is quasianalytic in the usual sense. Equivalently, the restriction λ_Ξ of λ to any real line Ξ satisfies*

$$\int \frac{\log \lambda_\Xi(\zeta)}{1+\zeta^2}\, d\zeta = \infty.$$

PROOF. If some $\mathscr{E}_B(\sum)$ is non-quasianalytic, then $\mathscr{E}_B$ clearly contains a nontrivial function which vanishes with all its derivatives at the origin.

Conversely, if $f \in \mathscr{E}_B$ vanishes with all its derivatives at the origin and if $f(x^0) \neq 0$, then we consider the function f^0 which is the restriction of f to the line $\sum$ determined by x^0 and the origin. Let g be the function which is f^0 on $\sum$ and is constant in directions orthogonal to $\sum$. Applying the Fourier representation of Theorem 1.5 to the function f, we can show that $g \in \mathscr{E}_B(\sum)$. Thus $\mathscr{E}_B(\sum)$ is non-quasianalytic.

We can use the AU nature of $\mathscr{E}_B$ to show that $\mathscr{E}_B(\sum)$ is a space $\mathscr{E}_{B'}$ of functions of a single real variable such that the associated function λ' is just λ_Σ. The equivalence of the quasianalyticity of $\mathscr{E}_B(\sum)$ with the divergence of the above integral is just the Denjoy-Carleman Theorem (see Carleman [1]). □

We pass now to the general case:

THEOREM 13.3. *Suppose B satisfies the hypothesis of Theorem* 13.1. *Let T be a principal noncharacteristic for* $\vec{\mathrm{D}}$. *Use variables* $x = (t, \xi)$ *with dual variables* (s, w) *where* ξ *is the variable on* T. *A sufficient condition that no* $f \in \mathscr{E}_B(\vec{\mathrm{D}})$ *can vanish together with all its derivates on* T *is: There exists an* $a > 0$ *so that in the region* $|s| \geqslant a(|w| + 1)$ *the function* $\lambda(\vec{\mathrm{P}}(z))$ *satisfies the following condition: For each s call*

$$\lambda^1_{\vec{\mathrm{D}}}(s) = \min_{a(|w|+1) \leq |s|} \lambda(\vec{\mathrm{P}}(s, w)).$$

For each complex line Ξ *in the complex s space which contains a real line there exists a convex function* $\lambda^2_{\vec{\mathrm{D}}}(|\zeta|; \Xi) \geqslant 1$ *such that for some* $c > 0$

$$\lambda^2_{\vec{\mathrm{D}}}(|\zeta|; \Xi) \leqslant c\lambda^1_{\vec{\mathrm{D}}}(\zeta) \exp(c|\mathrm{Im}\, \zeta|)$$

for all $\zeta \in \Xi$, *and, moreover,*

$$\int_{\zeta \text{ real}} \frac{\log \lambda^2_{\vec{\mathrm{D}}}(|\zeta|; \Xi) d\zeta}{1 + \zeta^2} = \infty$$

PROOF. It is possible, using the method of proof of Theorem 13.1, to reduce this result to Theorem 9.30 for the system of differential equations which are the "heat equations" constructed in the proof of Theorem 13.1. We can also outline a more direct proof based on the result of Theorem 13.1:

The idea of the proof is essentially the same as the corresponding result in Theorem 9.30, namely, using Theorems 13.1 and 1.5 we can write any $f \in \mathscr{E}_B(\vec{\mathrm{D}})$ as $f = f_1 + f_2$, where

$$f_j(x) = \int \exp(ix \cdot z)\, d\mu_j(z)/k(z).$$

Here μ_1 has its support in $|s| \leqslant a|w|$ and μ_2 in $|s| \geqslant a|w|$. Then the method of proof of Theorem 9.30 (for $\alpha = 1$) shows that a dense set of "averages" of f_1 are analytic in a suitable variable y. A direct calculation shows that the corresponding averages of f_2 will be quasianalytic in y. Since the sum of an analytic function, and a quasianalytic function is again quasianalytic, this gives the result. In the language Section IX.9, if we call A the set $|s| \leqslant a|w|$, then (A, C) is quasianalytically closed for $\mathscr{E}_B(\vec{\mathrm{D}})$, and C is well behaved in s outside A. ▯

PROBLEM 13.4 (Exercise).
Carry out the details of the proofs of Theorems 13.3, 13.5, 13.6, and 13.7.

THEOREM 13.4. *Let D be a single operator of degree m. Suppose* $B = \{b_j\}$ *is monotonic increasing and satisfies the hypothesis of Theorem* 13.1. *Call* B_1 *the sequence* $\{b_{[j/m]}\}$, *where* $[j/m]$ *denotes the greatest integer in* j/m.

Suppose $t = 0$ is noncharacteristic for D. A necessary and sufficient condition that no function in $\mathscr{E}_B(D)$ can vanish together with all its derivatives on $t = 0$ is that the class $\mathscr{E}_{B_1}(d/dt)$ be quasianalytic in the usual sense.

PROOF. We prove the sufficiency first. Since $t = 0$ is noncharacteristic, there exist constants a, $c > 0$ so that

$$|P(s, w)| \geqslant c\,|s|^m$$

for

$$|s| \geqslant a + a|w|.$$

Hence, in this region

$$\begin{aligned}\lambda(P(z)) &= \sum |P(s, w)|^j/b_j \\ &\geqslant \sum c^j\,|s|^{mj}/b_j \\ &\geqslant |s|^{-m} \sum c^{[j/m]}\,|s|^j/mb_{[j/m]}\end{aligned}$$

for $|s| \geqslant 1$. The right side is essentially $\lambda_1(c\,|s|)$. The last inequality comes from the fact that

$$\sum_{j=0}^{\infty} c^{[j/m]}\,|s|^j/mb_{[j/m]} = \sum_{j=0}^{\infty} (c^j\,|s|^{mj}/b_j) \sum_{l=0}^{m-1} |s|^l/m \leqslant |s|^m \sum_{j=0}^{\infty} c^j\,|s|^{mj}/b_j\,.$$

The sufficiency can now be derived from Theorem 13.3. Actually, the method of proof of Theorem 13.3 which we outlined is much easier to carry out in the present situation.

To prove the necessity we observe that a function f of t only will belong to $\mathscr{E}_B$(D) if f belongs to $\mathscr{E}_{B_1}(d/dt)$ because D^jf is a sum of $\leqslant c^j$ terms each of which, by the monotonicity of B, is majorized by $c^{mj}b_{[mj/m]} = c^{mj}b_j$. The necessity is now clear. ▯

It seems rather remarkable to us that the quasianalyticity of $\mathscr{E}_B(D)$ depends only on the order of D.

We wish to now discuss the case when $t = 0$ is a regular characteristic for the system $\vec{\mathrm{D}} = (D_1, \ldots, D_r)$. As in Section IX.9, let $\mathscr{E}(\Phi)$ be a space of rapidly increasing functions, where Φ does not involve any growth restrictions in the t variables. We shall assume that

$$\Phi(\xi) = \varphi(\xi_1) + \varphi(\xi_2) + \cdots + \varphi(\xi^d),$$

though the general case can probably be handled by use of the Oka embedding (see Section IV.5). For B as usual we define the space $\mathscr{E}_B(\Phi; \vec{\mathrm{D}})$ as consisting of all functions $f \in \mathscr{E}(\Phi)$ such that for every $\varepsilon > 0$, f and all its derivatives satisfy

$$|(D_1^{j_1} \cdots D_r^{j_r}f)(x)| \leqslant a\varepsilon^{|j|}b_j \exp\left(\Phi(\varepsilon x)\right)$$

for some $a > 0$. $\mathscr{E}_B(\Phi; \vec{D})$ is topologized in the usual fashion. For $\vec{D} = (\partial/\partial x_1, \ldots, \partial/\partial x_n)$ we shall denote $\mathscr{E}_B(\Phi; \vec{D})$ by $\mathscr{E}_B(\Phi)$.

By use of the method of proof of Theorem 13.1 we can show

THEOREM 13.5. *Suppose that Φ, B satisfy suitable regularity conditions. Then $\mathscr{E}_B(\Phi; \vec{D})$ is AU. $\mathbf{E}'_B(\Phi; \vec{D})$ consists of all entire functions $F(z)$ for which there exist $a, b > 0$ such that*

$$|F(z)| \leqslant a(1 + |z|)^b \exp [\Psi(b \operatorname{Im} w) + b |\operatorname{Im} s|]\lambda(b\vec{P}(z)).$$

An AU structure for $\mathbf{E}'_B(\Phi; \vec{D})$ consists of all k of the form

$$k(z) = \tilde{\lambda}(\vec{P}(z))k_1(\operatorname{Im} s)k_2(\operatorname{Im} w)k_3(\operatorname{Re} z),$$

where k_1 dominates all $\exp (b |\operatorname{Im} s|)$, *$k_2$ dominates all* $\exp [\Psi(b \operatorname{Im} w)]$, *$k_3$ dominates all polynomials, and $\tilde{\lambda}$ dominates all $\lambda(cz)$.*

Remark 13.2. In Theorems 13.5, 13.6, and 13.7 we shall not describe the regularity conditions explicitly as they are somewhat complicated and these results are not used elsewhere in this book.

As in Section IX.9, we could study the question of when the conditions $f \in \mathscr{E}_B(\Phi; \vec{D})$, f and all ξ derivatives $= 0$ on $\tilde{T}$ imply $f = 0$ on T. As in Theorem 13.3, we could determine sufficient conditions. Since the situation is rather complicated we shall restrict our considerations to the case $r = 1$, $\tilde{T} = T$. Our methods give

THEOREM 13.6. *Let D be of degree m and of degree m_1 in t. Suppose $\mathscr{E}(\Phi)$ is a uniqueness space for the Cauchy problem for D on $t = 0$. Let B^1 denote the sequence $\{b_{[j/m_1]}\}$. Let B^2 be defined by*

$$\lambda^2(s) = \sum |s|^j/b_j^2 = \exp \psi(|s|^{1/\alpha}), \tag{13.13}$$

where α is defined in Theorem 9.30 and ψ is the conjugate of φ for s large enough. Suppose B satisfies suitable regularity conditions. A sufficient condition that no $f \in \mathscr{E}_B(\Phi; D)$ can vanish together with all its derivatives on $t = 0$ is that no function in the space

$$\mathscr{E}_{B^1} + \mathscr{E}_{B^2}$$

can vanish together with all its derivatives at the origin.

See Remarks 13.3 and 13.4.

If, for example, D is the heat operator, $D = \partial/\partial t - \partial^2/\partial \xi^2$, then we can also deduce a converse result. We have

THEOREM 13.7. *Let D be the heat operator. A necessary and sufficient condition that no $f \in \mathscr{E}_B(\Phi, D)$ can vanish together with all its t derivatives at the origin is that no function in $\mathscr{E}_{B^1} + \mathscr{E}_{B^2}$ can vanish with all its derivatives at the origin.*

XIII.2. A Nonsymmetric Generalization of the Denjoy-Carleman Classes

We shall now discuss a (nonsymmetric) generalization of the Denjoy-Carleman classes. For $i = 1, 2, \ldots, 2^n$, let C^i be the ith orthant in C, that is the ith closed part of C in which the imaginary parts of the points have a fixed sign. C^1 is the orthant where $\operatorname{Im} z_j \geqslant 0$ for all j. Let $B_i = \{b_i^j\}$ be positive multisequences with $b_i^j \geqslant j!$. Denote by ${}^i\mathscr{E}_{B_i}$ the space of functions which are analytic on all of C^i, except possibly for R, which are indefinitely differentiable in all of C^i, and whose restrictions to R belong to $\mathscr{E}_{Bi}$. We give these spaces the "natural" topology, that is, the topology defined by the seminorms

$$\max_{x \in L, j} |\partial f^{(j)}(x)|/c^j b_j$$

for L any compact set in C^i, $c > 0$, and ∂ any linear partial differential operator with constant coefficients. Denote by $^*\mathscr{E}_B$ the space of functions f on R which can be represented by $f = \sum f^i$, where $f^i \in {}^i\mathscr{E}_{B_i}$. We define $\lambda_i(z)$ as related to B_i in the usual way. For z real let

$$\lambda(z) = \max \lambda_i(z), \tag{13.14}$$

where the maximum is taken over those i for which there exists an $x \in C^i$ with $(\operatorname{sgn} z_1, \ldots, \operatorname{sgn} z_n) = (\operatorname{sgn} \operatorname{Im} x_1, \ldots, \operatorname{sgn} \operatorname{Im} x_n)$. (Unless some $z_j = 0$ there will be only one i on the right side of (13.14).) For any z we shall write $\lambda(z)$ for $\lambda(\operatorname{Re} z)$. This is somewhat in contradiction with our previous notation in which λ was a function of $|z|$. However, this will not cause any trouble since $\lambda(z) | \leqslant \exp(|z|)$.

For example, for $n = 1$,

$$\lambda(z) = \begin{cases} \lambda_1(z) & \text{for } z > 0 \\ \lambda_2(z) & \text{for } z < 0 \\ \max[\lambda_1(0), \lambda_2(0)] & \text{for } z = 0. \end{cases}$$

Since when $n = 1$ the function $\lambda(z)$ can have a different behavior as $z \to +\infty$ from that as $z \to -\infty$, we refer to the space $^*\mathscr{E}_B$ as a nonsymmetric analog of the Denjoy-Carleman spaces.

THEOREM 13.8. *Suppose each $\mathscr{E}_{B_i}$ satisfies the hypothesis of Theorem 5.26 and $\lambda(z) = 0(\exp A|z|)$ for some $A > 0$. Then $^*\mathscr{E}_B$ is AU. $^*\mathbf{E}'_B$ consists* *of all entire functions F of exponential type which satisfy*

$$|F(z)| \leqslant c(1+|z|)^c \exp(c'|\mathrm{Im}\, z|)\, \lambda(c'z)$$

for some $c, c' > 0$. An AU structure for $^*\mathscr{E}_B$ consists of all k of the form $\tilde{\lambda}(z)k_1(|\mathrm{Im}\, z|)k_2(|\mathrm{Re}\, z|)$, where k_1 dominates all exponentials, k_2 dominates all polynomials, and $\tilde{\lambda}$ dominates all $\lambda(az)$. A similar result holds for $^1\mathscr{E}_{B_1}$ except that for the λ_i for $i > 1$ we take $\exp(|z|)$.

PROOF. We shall consider only the case $n = 1$ as the passage to $n > 1$ presents no difficulties. Let us begin with the space $^1\mathscr{E}_{B_1}$. If $g \in {}^1\mathscr{E}_{B_1}$ then $g \in \mathscr{E}_{B_1}$ so we can use Theorem 5.26 in conjunction with Theorem 1.5 to write

$$g(x) = \int \exp(ix \cdot z)\, d\mu(z)/k(z),$$

where μ is a bounded measure and k belongs to an AU structure for $\mathscr{E}_{B_1}$. Let $\mu^\pm$ be the measure μ restricted to $\mathrm{Re} \geqslant 0$ ($\mathrm{Re} < 0$). Thus

$$g(x) = g^+(x) + g^-(x),$$

where

$$g^\pm(x) = \int \exp(ix \cdot z)\, d\mu^\pm(z)/k(z).$$

It follows from the explicit expression for the AU structure for $\mathscr{E}_{B_1}$ that g^+ and g^- belong to $\mathscr{E}_{B_1}$ and $g^\pm$ is analytic in the upper (lower) half plane. Thus $g^+ \in {}^1\mathscr{E}_{B_1}$ and since $g = g^+ + g^-$ is in $^1\mathscr{E}_{B_1}$, also $g^- \in {}^1\mathscr{E}_{B_1}$, that is, g^- is regular in the upper half-plane. Since g^- is also regular in the lower half-plane and has smooth boundary values, it follows that g^- must be entire. Thus, by Theorem 1.5 combimed with Theorem 5.3 we can write

$$g^-(x) = \int \exp(ix \cdot z)\, d\mu_1(z)/k_1(z),$$

where k_1 is in an AU structure for $\mathscr{H}$, that is, k_1 dominates all linear exponentials.

Putting these results together we can write

$$g(x) = \int \exp(ix \cdot z)\, d\mu_2(z)/k_2(z),$$

where

$$\mu_2(z)/k_2(z) = \mu^+(z)/k(z) + \mu_1(z)/k_1(z).$$

It is clear that we may assume that μ_2 is a bounded measure and k_2 is of the form

$$k_2(z) = \begin{cases} \tilde{\lambda}_1(z)k_3(|\mathrm{Im}\, z|)k_4(|\mathrm{Re}\, z|) & \text{for } \mathrm{Re}\, z \geqslant 0 \\ k_3(|z|) & \text{for } \mathrm{Re}\, z \leqslant 0. \end{cases}$$

Here $\tilde{\lambda}_1(z)$ dominates all $\lambda_1(az)$, k_3 dominates all linear exponentials, and k_4 dominates all polynomials.

Thus we have verified the integral representation for elements of ${}^1\mathscr{E}_{B_1}$ which would follow, if we knew Theorem 13.8, by combining Theorems 13.8 and 1.5. It is not difficult to use this result to prove Theorem 13.8. In the first place it is easy to show from the definitions (compare the arguments in Chapter V) that any $F \in {}^1\mathbf{E}'_{B_1}$ satisfies the inequality stated in Theorem 13.8.

Conversely, if F satisfies this inequality and if $g \in {}^1\mathscr{E}_{B_1}$ is, as above, of the form

$$g(x) = \int \exp (ix \cdot z) \, d\mu_2(z)/k_2(z),$$

then we can define the inverse Fourier transform S of F by

$$S \cdot g = \int F(z) \, d\mu_2(z)/k_2(z).$$

Of course, it must be shown that $S \cdot g$ depends on g and not the particular μ_2/k_2. For this purpose we must show that we can approximate F suitably by linear combinations of exponentials. This being done, it is not difficult to show, by having μ_2/k_2 depend in a "nice" way on g, that S is continuous on ${}^1\mathscr{E}_{B_1}$, that is $S \in {}^1\mathscr{E}'_{B_1}$. (${}^1\mathscr{E}_{B_1}$ is metrizeable and reflexive.) It is clear that F is the Fourier transform of S. Thus, the set of functions in ${}^1\mathbf{E}_{B_1}$ is as stated.

The proof that the topology of ${}^1\mathbf{E}'_{B_1}$ is as stated goes as follows: In the first place, we show that, given any bounded set L in ${}^1\mathbf{E}'_{B_1}$, there are constants c, c' so that all $F \in L$ satisfy

$$|F(z)| \leqslant c(1 + |z|)^c \exp (c'|\text{Im } z|)\lambda(c'z)$$

for all $z \in C$. [Recall that in the present case $\lambda_2(z) = \exp (|z|)$.] This is proven by use of the Hahn-Banach representation for the elements of ${}^1\mathscr{E}'_{B_1}$. It follows that the set N_k of $F \in {}^1\mathbf{E}'_{B_1}$ which satisfy

$$\max_{z \in C} |F(z)|/k(z) \leqslant 1$$

for k as above swallows every bounded set. Since ${}^1\mathbf{E}'_{B_1}$ is bornologic, N_k is a neighborhood of zero.

We want to show that the N_k form a fundamental set of neighborhoods of zero in ${}^1\mathbf{E}'_{B_1}$. Given any neighborhood N of zero, there is a bounded set M in ${}^1\mathbf{E}_{B_1}$ so that N consists of all $F \in {}^1\mathbf{E}'_{B_1}$ with $|T \cdot F| \leqslant 1$ for all $T \in M$. We use the above argument to get representations for all T in the form μ_2/k_2. Since M is bounded, we can modify our above construction to show that we can choose k_2 independent of $T \in M$ and the μ_2 to vary over a set of measures with uniformly bounded total variation. It follows that

$$T \cdot F = \int F(z)\, d\mu_2(z)/k_2(z)$$

is uniformly bounded for $T \in M$ and F in N_{k_2}. Thus $aN_{k_2} \subset N$ for a suitable $a > 0$ which proves that the $\{N_k\}$ form a fundamental system of neighborhoods of zero in ${}^1\mathbf{E}'_{B_1}$.

We have thus completed the proof of the part of Theorem 13.8 which concerns the space ${}^1\mathscr{E}_{B_1}$ except for the proof that F as above can be approximated by linear combinations of exponentials. The proof of the possibility of such approximation is a rather delicate extension of our above "heat equation" technique and we shall omit it; we leave it to the interested reader as an exercise. In any case, the above shows how to derive the slightly weaker result in which the topology of ${}^1\mathbf{E}'_{B_1}$ is as stated but the functions of ${}^1\mathbf{E}'_{B_1}$ are the limits of the linear combinations of exponentials in this topology. This weaker result is sufficient for all of our applications.

The passage to the space ${}^*\mathscr{E}_B$ is fairly simple. We note that the dual of ${}^1\mathscr{E}_{B_1} + {}^2\mathscr{E}_{B_2}$ is the intersection ${}^1\mathscr{E}'_{B_1} \cap {}^2\mathscr{E}'_{B_2}$. Thus our description of the elements of ${}^*\mathbf{E}'_B$ is verified. It is not difficult to check the assertion about the topology. We leave the details to the reader. ▯

It should be remarked that, even if $\mathscr{E}_{B_1}$ is non-quasianalytic, it is clear that there are no nontrivial functions in ${}^1\mathscr{E}_{B_1}$ which are of compact support. Thus the analog of the proof of Theorem 5.19 could not apply. (It might apply to the proof of the Theorem for ${}^*\mathscr{E}_B$ for certain B, but we have not checked this.)

See Problem 13.5.

Remark 13.5. As in Section XIII.1 we shall change our notation to define ${}^*\mathscr{E}_B$, in general, as the AU space whose AU structure is as in Theorem 13.8.

The proof of Theorem 13.2 gives

THEOREM 13.9. *A necessary and sufficient condition that no function except zero in* ${}^*\mathscr{E}_B$ *can vanish together with all its derivatives at the origin is: For each line* $\sum$ *through the origin the same is true of the subspace* ${}^*\mathscr{E}_B(\sum)$ *consisting of those functions which are constant in directions orthogonal to* $\sum$. *Equivalently, for any complex line* Λ *through the origin which contains a real line and any* k *as in Theorem* 13.8, *and any real point* $p \in \Lambda$ *we can find polynomials* $P_j(z)$ *so that*

$$\max_{z \in \Lambda} |P_j(z) - \exp(ip \cdot z)|/k(z) \to 0.$$

The statement following the word "equivalently" comes from the fact that the nonexistence of a function vanishing with all its derivatives at the origin in a space is equivalent to the density of the linear combinations of the derivatives of δ in its dual space, which is equivalent, by Fourier transform, to the density of polynomials in the Fourier transform of the dual space. Since the linear combinations of the δ_p are clearly dense in the space $^*\mathscr{E}'_B(\sum)$ it suffices to approximate $\exp(ip \cdot z)$.

Theorem 13.9 shows that the problem of whether or not a function in $^*\mathscr{E}_B$ can vanish with all its derivatives at the origin is reduced to the case $n = 1$ which we shall now assume. For the space $\mathscr{E}_B$ the Denjoy-Carleman theorem shows how to translate the condition into a condition on $\lambda(z)$. If

$$\lambda(z) \leqslant \text{const}\ \lambda(c \operatorname{Re} z) \exp(c\,|\operatorname{Im} z|)$$

(which is true in most "interesting" cases), then this means that the quasianalyticity of $\mathscr{E}_B$ depends on the topology of $\mathbf{E}'_B$ of the real axis. This is remarkable, because the real axis is not a sufficient set for $\mathscr{E}_B$, for if it were then every function f in $\mathscr{E}_B$ would have a Fourier representation

$$f(x) = \int e^{ixz}\, d\mu(z)/k(z), \tag{13.15}$$

where μ is a bounded measure with support on the real axis. But such an f is bounded, whereas there certainly exist unbounded functions in $\mathscr{E}_B$, e.g., polynomials.

To understand this point better we note that for $\mathscr{E}_B$ the question of whether or not there exists a function all of whose derivatives vanish at the origin is the same as the question of whether or not there exists a function in $\mathscr{E}_B$ which vanishes on an open set or whether or not there exists a function of compact support in $\mathscr{E}_B$. Thus, if $\mathscr{E}_B$ is non-quasi-analytic in any of these senses, then there will exist an $f \in \mathscr{E}_B$ which can be represented as in (13.15). This means that the question of density of the polynomials in $\mathbf{E}'_B$ is equivalent to their density in the induced norms on the real axis. Thus, if we can approximate $\exp(iz)$ by polynomials $P_j(z)$ in the induced norms on the real axis, then we can find polynomials $P''_j(z)$ such that

$$\sup_{\text{all } z} |P''_j(z) - \exp(iz)|/k(z) \to 0$$

for all k. On the other hand, we can also find polynomials P'_j such that

$$\sup_{z \text{ real}} |P'_j(z) - \exp(iz)|/k(z) \to 0$$

for all k, but

$$\sup_{\text{all } z} |P'_j(z) - \exp(iz)|/k(z) \text{ does not} \to 0$$

for some k. The last fact comes from the fact that the real axis is not sufficient for $\mathscr{E}_B$.

We do not know of any way of proving this assertion intrinsically, that is, within the space $\mathbf{E}'_B$ without using the space $\mathscr{E}_B$.

See Problems 13.6 and 13.7.

(For $n > 1$ the proof of Theorem 13.2 shows that the existence of a function in $\mathscr{E}_B$ which vanishes with all its derivatives at the origin is equivalent to the existence of a function which vanishes on an open set. However, these conditions are not equivalent to the existence of a function of compact support, as the example $\mathscr{E}_B =$ space of functions indefinitely differentiable in x_1, analytic in $x_2, \ldots, x_n$, shows.)

For the space $*\mathscr{E}_B$ the questions of the existence of a function all of whose derivatives vanish at a point and the existence of a function which vanishes on an open set may be different. For example, if ${}^1\mathscr{E}_{B_1}$ is the space of entire functions, then for no B_2 does there exist a function in $*\mathscr{E}_B$ which vanishes on an open set of R. We make the following

DEFINITION. A set σ of real numbers is of *zero B-density* if there exists a function in $*\mathscr{E}_B$ which vanishes on σ.

Except in some very special cases we do not know how to determine whether or not a given set σ is of zero B-density.

We wish now to give some examples.

EXAMPLE 1. $B_2 = B_3 = \cdots = B_n$; B_1 is *trivial*, that is $b_1^0 = 1$ and $b_1^j = \infty$ for all $j \neq 0$. For example, if $n = 1$, then the space $*\mathscr{E}_B$ is the space of sums $f_1 + f_2$, where f_1 is an arbitrary indefinitely differentiable function on $\operatorname{Im} x = 0$ which can be extended to be analytic in $\operatorname{Im} x > 0$, and f_2 is analytic in $\operatorname{Im} x < 0$ and has boundary values belonging to $\mathscr{E}_{B_2}$.

From another point of view, $*\mathscr{E}_B$ can be thought of as the space of functions which have a given degree of approximation by functions analytic in the upper half-plane in the sense mentioned at the beginning of this chapter. The degree of approximation is measured by B_2. It should be noted that (also for $n > 1$) we are interested in the question of whether or not a function in $*\mathscr{E}_B$ can vanish on an open set. For we can always find a function which is analytic on $C^1 - R$ and indefinitely differentiable on C^1 (hence, for most "interesting" B, is in $*\mathscr{E}_B$) which vanishes with all its derivatives at some point in R. Beurling [2], [4] has shown that for $n = 1$, vanishing on a set of positive measure is equivalent to vanishing on an open set. The methods of Beurling [2], [4] or Levinson [1] for $n = 1$ combined with Theorem 13.9 give

THEOREM 13.10. *A necessary and sufficient condition that no function in* $^*\mathscr{E}_B$ *can vanish on an open set is: For every real line* Ξ *through the origin*

$$\int_{-\infty}^{0} \frac{\log \lambda_{2\Xi}(z)}{1+z^2}\, dz = \infty. \tag{13.16}$$

Here $\lambda_{2\Xi}$ is the restriction of λ_2 to Ξ.

EXAMPLE 2. *The boundary values of an analytic function.* Here we want the functions of $^i\mathscr{E}_{B_i}$ to be entire for $i > 1$. Thus $^i\mathscr{E}_{B_i} \subset {}^1\mathscr{E}_{B_i}$ for all i which means that $^*\mathscr{E}_B$ is just $^1\mathscr{E}_{B_1}$. For $n = 1$ it is clear that no function in $^1\mathscr{E}_{B_1}$ can vanish on an open set. It follows by considering complex analytic lines in C which contain real lines that the same is true for $n > 1$. The interesting quasianalytic question is therefore whether a function in $^1\mathscr{E}_{B_1}$ can vanish together with all its derivatives at the origin.

Let us first consider the case $n = 1$. Then $\lambda(z) = \lambda_1(z)$ for $z > 0$ and $\lambda(z) = \exp(|z|)$ for $z < 0$. In accordance with Theorem 13.9, we want to approximate $\exp(iz)$ by polynomials $P_j(z)$.

LEMMA 13.11. *Suppose* $\lambda(z) = 0(\exp |z|)$ *for* $z \to +\infty$. *A necessary and sufficient condition that* $\exp(iz)$ *belong to the closure of the polynomials in* $^1\mathbf{E}'_{B_1}$ *is that the polynomials be dense in the space* $\mathscr{C}_{\lambda(z^2)}$ *of continuous bounded functions on the real line in the topology defined by the seminorms*

$$\|f\|_{\varphi,\tilde{\lambda}} = \sup_{z \in R} |f(z)|/\tilde{\lambda}(z^2)\varphi(|z|^2)$$

for any continuous positive φ *which dominates all polynomials, and any* $\tilde{\lambda}$ *which dominates all* $\lambda(az)$. *This is equivalent to*

$$\int_{-\infty}^{\infty} \frac{\log \lambda(z^2)}{1+z^2}\, dz = \infty. \tag{13.17}$$

PROOF. We note first that the linear combinations of the functions $\exp(iaz^2)$ for a real are dense in the even functions in $\mathscr{C}_{\lambda(z^2)}$. For, suppose this is not so. Then for some φ which dominates polynomials and some $\tilde{\lambda}$ there is an even bounded measure μ on $[0, \infty)$ such that

$$\int_0^{\infty} \exp(iaz^2)\, d\mu(z)/\tilde{\lambda}(z)\varphi(|z|^2) = 0$$

for all real a. But this implies clearly that

$$\int_0^{\infty} \exp(iaz)\, d\mu(z^{1/2})/\tilde{\lambda}(z)\varphi(|z|) = 0$$

which shows that $\mu = 0$ because $\mu(z^{1/2})/\tilde{\lambda}(z)\varphi(|z|)$ is a bounded measure, hence it cannot be orthogonal to all $\exp(iaz)$. (Writing $\mu(z)$ as a sum of a monotonic increasing and a monotonic decreasing function shows that $\mu(z^{1/2})$ is of bounded variation.) A similar method shows that the linear combinations of $\{z \exp(iaz^2)\}$ are dense in the odd functions in $\mathscr{C}_{\lambda(z^2)}$.

Now, suppose that $P_j(z) \to \exp(iz)$. Then, by Theorem 13.8,

$$\max_{z>0} |P_j(z) - \exp(iz)|/\tilde{\lambda}(z)\varphi(|z|) \to 0$$

so that

$$\max_{z \text{ real}} |P_j(z^2) - \exp(iz^2)|/\tilde{\lambda}(z^2)\varphi(|z|^2) \to 0.$$

Similarly for $\exp(iaz^2)$ and $z \exp(iaz^2)$. Thus, by the above, the polynomials are dense in $\mathscr{C}_{\lambda(z^2)}$.

We note that, by definition, $\lambda(z^2) = \sum z^{2j}/b_j$. If we set ${}_1\lambda(z) = \lambda(z^2) + |z|\,\lambda(z^2)$, then ${}_1\lambda(z) = \sum |z|^j {}_1b_j$, where

$${}_1b_j = \begin{cases} b_{j/2} & \text{for } j \text{ even} \\ b_{(j-1)/2} & \text{for } j \text{ odd.} \end{cases}$$

We set ${}_1B = \{{}_1b_j\}$. Since $|{}_1\lambda(z)| = (1 + |z|)\lambda(z^2)$ it is not hard to show that the density of polynomials in $\mathscr{C}_{\lambda(z^2)}$ is equivalent to density in $\mathscr{C}_{{}_1\lambda(z)}$ [the factor $(1 + |z|)$ being unimportant for this question].

The convergence or divergence of the integrals

$$\int \frac{\log {}_1\lambda(z)}{1+z^2}\, dz, \qquad \int \frac{\log \lambda(z^2)}{1+z^2}\, dz$$

is clearly the same. If the first of these integrals is finite then, by the Denjoy-Carleman theorem, there exists a nontrivial function h in $\mathscr{E}_{{}_1B}$ which is of compact support, whose support does not meet $\{0\}$. Let H be the Fourier transform of h. A simple estimate shows that $f \to \int H(z)f(z)\, dz$ for $f \in \mathscr{C}_{{}_1\lambda(z)}$ defines a nontrivial element of $\mathscr{C}'_{{}_1\lambda(z)}$ which we denote again by H. Since h and all its derivatives are zero at the origin, H is orthogonal to all polynomials. Thus the polynomials are not dense in $\mathscr{C}_{\lambda(z^2)}$ if (13.17) fails.

It remains to show that (13.17) implies that the polynomials are dense in ${}^1\mathbf{E}'_{B_1}$. If (13.17) holds then the class $\mathscr{E}_{{}_1B}$ is quasianalytic so the polynomials are dense in $\mathbf{E}'_{{}_1B}$. Now $\mathbf{E}'$ is contained in $\mathbf{E}'_{{}_1B}$ so, in particular, we can approximate any $L \in \mathbf{E}'$ by polynomials in the topology of $\mathbf{E}'_{{}_1B}$. (See Example 5 of Chapter V for a description of $\mathbf{E}'$.) By Theorem 13.8 this means that we can find polynomials P_j so that

$$\max_{z \text{ complex}} |P_j(z) - L(z)|/\tilde{\lambda}((\operatorname{Re} z)^2)\varphi(|z|^2)k_1(\operatorname{Im} z) \to 0 \tag{13.18}$$

for any φ, $\tilde{\lambda}$ as above, and for any k_1 which dominates all linear exponentials.

We use the fact that for any $\varepsilon > 0$,

$$\exp(iz^2 - \varepsilon z^2) = c_\varepsilon \int_{-\infty}^{\infty} \exp[ixz - x^2/4(\varepsilon - i)]\, dx,$$

where $c_\varepsilon = 1/2\sqrt{\pi}(\varepsilon - i)^{1/2}$. To approximate $\exp(iz^2 - \varepsilon z^2)$ by functions L in $\mathbf{E}'$ we let

$$L(z) = c_\varepsilon \int_{-A}^{A} \exp[ixz - x^2/4(\varepsilon - i)]\, dx$$

for A large. It is clear that $L \in \mathbf{E}'$.

Suppose that ε is small, A large. The error in the approximation is measured by

$$\left|\int_A^\infty \exp[ixz - x^2/4(\varepsilon - i)]\, dx\right| \leqslant \int_A^\infty \exp(x|\mathrm{Im}\, z| - \varepsilon x^2/8)\, dx$$

$$= \exp(2|\mathrm{Im}\, z|^2/\varepsilon) \int_A^\infty \exp[-(\varepsilon^{1/2}x/2\sqrt{2} - \sqrt{2}\,|\mathrm{Im}\, z|\, \varepsilon^{-1/2})^2]\, dx$$

$$= \exp(2|\mathrm{Im}\, z|^2/\varepsilon) \int_{A - 4|\mathrm{Im}\, z|/\varepsilon}^{\infty} \exp(-\varepsilon x^2/8)\, dx$$

Let us observe that for $4|Iz|/\varepsilon \leqslant A/2$

$$\int_{A-4|\mathrm{Im}\, z|/\varepsilon}^{\infty} \exp(-\varepsilon x^2/8)\, dx \leqslant \int_{A/2}^{\infty} \exp(-\varepsilon x^2/8)\, dx$$

$$\leqslant c\varepsilon^{-\frac{1}{2}} \exp(-c\varepsilon A^2)$$

for a constant c. For $8|\mathrm{Im}\, z|/\varepsilon \geqslant A$ we have

$$\int_{A-4|\mathrm{Im}\, z|/\varepsilon}^{\infty} \exp(-\varepsilon x^2/8)\, dx \leqslant \int_{-\infty}^{\infty} \exp(-\varepsilon x^2/8)\, dx$$

$$= 2\sqrt{2\pi/\varepsilon}$$

$$\leqslant c'(1 + |z|)/\varepsilon^2 A.$$

Thus, for all z we have

$$|L(z) - \exp(ix^2 - \varepsilon z^2)| \leqslant c'(1 + |z|) \exp(2|\mathrm{Im}\, z|^2/\varepsilon)/A\varepsilon^2. \tag{13.19}$$

On the other hand, $\exp(iz^2 - \varepsilon z^2)$ approximates $\exp(iz^2)$ in the following sense: We have

$$|\exp(iz^2 - \varepsilon z^2) - \exp(iz^2)| = |\exp(iz^2)|\, |\exp(-\varepsilon z^2) - 1|$$

$$\leqslant \exp(-\mathrm{Im}\, z^2)\, |\exp(-\varepsilon z^2) - 1|.$$

For $|z| \leqslant \varepsilon^{-1/4}$ we have

$$|\exp(-\varepsilon z^2) - 1| \leqslant c'\varepsilon^{1/2}.$$

For all z,

$$|\exp(-\varepsilon z^2) - 1| \leqslant 1 + \exp(-\varepsilon \operatorname{Re} z^2).$$

Thus, for $|z| \geqslant \varepsilon^{-1/4}$,

$$|\exp(iz^2 - \varepsilon z^2) - \exp(iz^2)| \leqslant c'\varepsilon^{1/2}(1 + |z|^2)\exp(3|z|^2|\arg z|). \tag{13.20}$$

Here $\arg z$ is measured from $-\pi$ to π. Hence, by the above, (13.20) holds for all z.

We can now approximate $\exp(iz^2)$ by functions L in $\mathbf{E}'$. We claim that approximation is possible in the space $\mathbf{U}'$ defined by seminorms of the form

$$\|F\|_k = \sup_{z \in C} |F(z)/k(z)|$$

for any k which dominates $(1 + |z|^b)\exp(b|z|^2|\arg z|)$ for any b. Let ε_0 and k be given; then by (13.20), there is an ε so that

$$\|\exp(iz^2) - \exp(iz^2 - \varepsilon z^2)\|_k \leqslant \varepsilon_0/2.$$

Since k dominates all $(1 + |z|)^b \exp(b|\operatorname{Im} z|^2)$, we can choose N so large that for $|\operatorname{Im} z| \geqslant N$ we have

$$(1 + |z|)\exp(2|\operatorname{Im} z|^2/\varepsilon)/k(z) < \varepsilon_0 .$$

Thus, if A is large enough, the right side of (13.19) divided by $k(z)$ will be $\leqslant \varepsilon_0/2$ for all z. Hence

$$\|\exp(iz^2) - L(z)\|_k \leqslant \varepsilon_0 ,$$

which is our assertion.

Combining this approximation with (13.18) shows that we can find polynomials $P_j(z^2)$ so that

$$\max |P_j(z^2) - \exp(iz^2)|/\tilde{\lambda}((\operatorname{Re} z)^2)\varphi(|z|^2)k(z) \to 0$$

for any $\tilde{\lambda}$, φ, k as above, since k dominates k_1. (The polynomials are even because $\exp(iz^2)$ is.) Thus,

$$\max |P_j(z) - \exp(iz)|/\tilde{\lambda}((\operatorname{Re} z^{1/2})^2)\varphi(|z|)k(z^{1/2}) \to 0. \tag{13.21}$$

Here $z^{1/2}$ is defined to be $|z|^{1/2}\exp i(\arg z)/2$, with $\arg z$ measured as before; $k(z)$ is an arbitrary positive continuous function which dominates all $(1 + |z|^2)^b \exp(b|z|^2|\arg z|)$ so the multi-valuedness of $z^{1/2}$ on $\arg z = \pm\pi$ presents no difficulty, that is, we can take either value. $k(z^{1/2})$ is clearly an arbitrary positive function (which can be assumed to be continuous) which dominates $(1+|z|)^b \exp(b|z|\,|\arg z|)$.

Now

$$\exp(|z|\,|\arg z|) \leqslant \begin{cases} \exp(\pi\,|\operatorname{Im} z|/2) & \text{for } \operatorname{Re} z \geqslant 0 \\ \exp(\pi\,|\operatorname{Im} z| + \pi\,|\operatorname{Re} z|) & \text{for } \operatorname{Re} z \leqslant 0. \end{cases}$$

Moreover $(\operatorname{Re} z^{1/2})^2 \leqslant |\operatorname{Re} z|$. In the present case, $\lambda(z) = \exp(|z|)$ for $z < 0$. Thus we can replace the factor $k(z^{1/2})$ in the right hand side of (13.21) by $k_1(\operatorname{Im} z)$ where k_1 dominates all linear exponentials and replace $\tilde{\lambda}((\operatorname{Re} z^{1/2})^2)$ by $\tilde{\lambda}(\operatorname{Re} z)$. Theorem 13.8 then asserts that $P_j \to \exp(iz)$ in the topology of ${}^1\mathbf{E}'_{B_1}$ which is the desired result. □

Combining Lemma 13.11 with Theorem 13.9 yields

THEOREM 13.12. *A necessary and sufficient condition that no function in* ${}^1\mathscr{E}_{B_1}$ *can vanish with all its derivatives at a point is that for any real line* Ξ *through the origin*

$$\int_0^\infty \frac{\log \lambda_\Xi(z)}{1+z^{3/2}}\,dz = \infty.$$

See Remark 13.6.

We should like to contrast the results in Examples 1 and 2 for $n = 1$. In Example 1, λ is given to be of polynomial growth on the positive real axis, and we seek the behavior on the negative axis in order that the linear combinations of $\exp(itz)$ for $0 \leqslant t \leqslant 1$ be dense. (Polynomials can never be dense, because the positive real axis is sufficient (see Chapter I) for the topology of the space of polynomials (see Chapter V, Example 8)). In Example 2, λ is given to be exponentially increasing on the negative real axis and we seek the behavior on the positive axis in order that the linear combinations of the polynomials be dense. (It is clear by Fourier transform that the linear combinations of $\exp(itz)$ for $0 \leqslant t \leqslant 1$ are dense.)

This leads to the interesting

PROBLEM 13.8.

Let λ be fixed on the negative real axis. Determine the conditions on λ on the positive real axis in order that the polynomials (or linear combinations of $\exp(itz)$ for $0 \leqslant t \leqslant 1$) be dense.

This problem seems very difficult. There is, however, one simple case: If

$$\int_{-\infty}^0 \frac{\log \lambda(z)}{1+|z|^{3/2}}\,dz < \infty,$$

then no matter how λ is chosen on the positive axis, polynomial approximation is impossible. This follows easily from Theorem 13.12. Beurling (unpublished) has solved the problem for polynomial approximation for $\lambda(x) = \exp(|x|^{\alpha})$ for $x < 0$, if $\frac{1}{2} \leqslant \alpha < 1$. The above remark handles the case $\alpha < \frac{1}{2}$, and $\alpha = 1$ is treated in Example 2.

Relation to ideal theory in certain function algebras. Theorem 13.12 seems to have some interesting relation to ideal theory. Let Ω be an open set in C with a real analytic boundary. For the algebra $\mathscr{H}(\overline{\Omega})$ of functions holomorphic on the closure $\overline{\Omega}$ of Ω it is true that every closed ideal is the intersection of its punctual ideals, that is, the ideals it generates in the ring of formal power series at each point of $\overline{\Omega}$. The analogous result for the algebra $\mathscr{H}(\Omega) \cap \mathscr{E}(\overline{\Omega})$ of functions holomorphic on Ω and indefinitely differentiable on $\overline{\Omega}$ is false as is shown by Beurling [1] for the case $n = 1$, $\Omega =$ unit disk. In fact, in this case there exists an ideal I with two generators having no common zeros such that the closure of I is not the whole ring.

We say that *spectral synthesis* holds for a subalgebra of $\mathscr{E}(\overline{\Omega})$ if every closed ideal is the intersection of its punctual ideals. We shall be interested in the subalgebras $\mathscr{H}_B(\overline{\Omega}) = \mathscr{H}(\Omega) \cap \mathscr{E}_B(\overline{\Omega})$, where $\mathscr{E}_B(\overline{\Omega})$ is the space of functions in $\mathscr{E}(\overline{\Omega})$ whose derivatives satisfy on $\overline{\Omega}$ the bounds determined by B. (For example, if $n = 1$ and Ω is a half-plane, then $\mathscr{H}_B(\overline{\Omega})$ is just ${}^1\mathscr{E}_{B_1}$ of Example 2.)

Now, for $n = 1$, spectral synthesis fails for those algebras $\mathscr{H}_B(\overline{\Omega})$ for which there exist functions which have zeros of infinite order at points of the boundary of $\overline{\Omega}$. The reason for this is that by the very nature of punctual ideals, they capture information only about how many derivatives of a function vanish at a point. (When all the derivatives of a function vanish at a point, we might want to know more detailed information concerning how fast the function tends to zero.) This leads us to make the general

PROBLEM 13.9. CONJECTURE.

If $\mathscr{H}_B(\overline{\Omega})$ is quasianalytic in the sense that no function can vanish together with all its derivatives at a point then spectral synthesis holds.

The method of Beurling [1] shows that the conjecture is true for $n = 1$.

It is easy to extend the construction of $\mathscr{H}_B(\overline{\Omega})$ to the case when Ω has a real analytic distinguished boundary, in particular to the polydisk or to a product of half-planes. We obtain a generalization of the spaces ${}^1\mathscr{E}_{B_1}$.

We want to study the question of when $\mathscr{H}_B(\overline{\Omega})$ is quasianalytic. Suppose for simplicity that Ω is strictly convex and relatively compact. We use coordinates $x_1, \ldots, x_{2n}$ in $R^{2n} = C^n$ with complex coordinates $x_{2j-1} + ix_{2j}$. Let p be a boundary point of $\overline{\Omega}$; we want to know if the linear combinations of the derivatives of δ_p are dense in $\mathscr{H}'_B(\overline{\Omega})$. We may assume that p is the origin, $x_1 = 0$ is the tangent plane to Ω at p, and that $\Omega \subset \{x_1 > 0\}$.

By Fourier transform we want to know if the polynomials are dense in $\mathbf{H}'_B(\overline{\Omega})$. Suppose that $\mathscr{E}_B(\overline{\Omega})$ is LAU (which will presumably always be true). Then $\mathscr{H}_B(\overline{\Omega})$ is the kernel of the Cauchy-Riemann operators so, as in Chapter V, Example 5, we can describe $\mathbf{H}'_B(\overline{\Omega})$ as a space of entire functions on the variety V: $z_{2j-1} + iz_{2j} = 0$ for $j = 1, 2, \ldots, n$. The norms are of the form

$$\lambda(z) \exp(\psi(z)),$$

where $\lambda(z)$ is determined by B in the usual way and ψ is a (real) convex function determined by Ω. Since $x_1 = 0$ is the tangent plane to Ω at the origin and Ω lies in $x_1 > 0$, we see easily that ψ is bounded on the positive imaginary z_1 axis. Because Ω is strictly convex ψ is unbounded along any other real half-line in V. Thus, approximation by polynomials is "most difficult" on the positive imaginary z_1 axis on V, where the norm is of the form $\lambda(\text{Re } z_1, 0, \ldots, 0)$ in terms of coordinates $z_1, z_3, \ldots, z_{2n-1}$ on V. By taking Theorem 11.12 into account, we

PROBLEM 13.10. CONJECTURE.

A necessary and sufficient condition that $\mathscr{H}_B(\overline{\Omega})$ be quasianalytic is

$$\int_0^\infty \frac{\lambda(t, 0, 0, \ldots, 0)}{1 + t^{3/2}} = \infty.$$

Note that we cannot reduce the conjecture to Theorem 13.12 because of the factor $\exp(\psi(z))$ in the norm on V for the space $\mathbf{H}'_B(\overline{\Omega})$, which accounts for the essential difference between $\mathbf{H}'(\overline{\Omega})$ and ${}^1\mathbf{E}'_{B_1}$.

Relation to the Schrödinger equation. Suppose $n = 1$. As in Example 2 we perform the transformation $z \to z^2$, that is, we consider the space of functions of the form $F(z^2)$ for $F \in {}^*\mathscr{E}_B$. Assume that $\lambda(z) = 0(\exp |z|)$. Then from ${}^*\mathscr{E}_B$ we obtain norms on the complex plane which are essentially, that is, modulo factors of small growth, and modulo the fact that the constants are "arbitrarily large" (e.g., when we write $\lambda(cz^2)$ we mean a function which dominates $\lambda(cz)$ for all c) of the form

$$\begin{aligned} &\lambda_1(cz^2) \quad \text{for } z \text{ real} \\ &\lambda_2(cz^2) \quad \text{for } z \text{ pure imaginary} \\ &\exp(c\,|z^2 \sin 2\theta|) \quad \text{for other } z(z = |z|\, e^{i\theta}). \end{aligned}$$

The problem is to show that the polynomials, or the linear combinations of $\exp(itz^2)$ for $0 \leqslant t \leqslant 1$, are dense.

We can obtain these norms in a different way: Let $B' = \{b'_j\}_{j=0}^\infty$ and let Φ be a convex function of x_2. Let us consider the space $\mathscr{E}_{B'}(\Phi)$ (see Examples 6 and 7 of Chapter V) of indefinitely differentiable functions $f(x_1, x_2)$

such that for every $\varepsilon > 0$, and every j', and every compact set in x_1 space, there is an $A > 0$ so that for all j and for all x_1 in the compact set

$$|\partial^{j+2j'} f(x_1, x_2)/\partial x_1^{j'} \partial x_2^{j+j'}| \leqslant A \varepsilon^j b'_j \exp(\Phi(\varepsilon |x_2|)).$$

Then $\mathscr{E}_{B'}(\Phi)$ is LAU and an AU structure K is given by all continuous positive functions

$$k(z) = k_1(|z|)\tilde{\lambda}'(\text{Re } z_2) k_2(\text{Im } z_2) k_3(\text{Im } z_1),$$

where k_1 dominates all polynomials, k_2 dominates $\exp(\Psi(b \text{ Im } z_2))$ for all $b > 0$, k_3 dominates all $\exp(b \text{ Im } z_1)$, and $\tilde{\lambda}'$ dominates all $\lambda'(bz)$. Here Ψ is the conjugate of Φ.

Now let us examine the solutions f in $\mathscr{E}_{B'}(\Phi)$ of the Schrödinger equation

$$i\frac{\partial f}{\partial x_1} - \frac{\partial^2 f}{\partial x_2^2} = 0. \tag{13.22}$$

By the results of Chapters IV and VII these depend on the norms for $\mathbf{E}_{B'}(\Phi)$ on the variety $V : z_1 = z_2^2$. We write z for z_2 and identify a point $(z^2, z) \in V$ with z. By the above description of the AU structure for $\mathscr{E}_{B'}(\Phi)$, the norms of V are essentially of the form

$\lambda'(cz)$ for z real

$\exp(\Psi(cz))$ for z pure imaginary

$\exp(c|z^2| \sin 2\theta)$ for other z.

(We are assuming that $\Psi(t) = 0(t^2)$.)

Comparing this with the description of the norms in $^*\mathbf{E}'_B$ we see we have essentially the same type of norms. A more careful analysis shows that we obtain exactly the same type of norms, of course, for $\lambda'(z) = \lambda_1(z^2)$, $\exp(\Psi(z)) = \lambda_2(z^2)$. Using this correspondence we see:

The problem of polynomial approximation in $^*\mathbf{E}'_B$ is equivalent to the question of whether a solution of the Schrödinger equation in $\mathscr{E}_{B'}(\Phi)$ can vanish together with all its derivatives at a point. The problem of the density in $^*\mathbf{E}'_B$ of the linear combinations of $\exp(itz)$ *for t in an interval is equivalent to the uniqueness in $\mathscr{E}_{B'}(\Phi)$ of the solutions of the Schrödinger equation when the Cauchy data is prescribed on an interval of the initial surface* $x_2 = 0$.

The methods of Chapter IX give a way of attacking the uniqueness question for the Schrödinger equation, hence a way of attacking the problem for $^*\mathscr{E}_B$, that is, to obtain a "constructive" solution to the problem for $^*\mathscr{E}_B$. Conversely, Pollard's [1], [2] results can be translated (using the proof of Lemma 13.11) into uniqueness questions for the Schrödinger equation in a space which is a slight modification of $\mathscr{E}_{B'}(\Phi)$, namely, we require growth conditions in the x_1 direction also.

XIII.3. The Punctual and Local Images of Quasianalytic and Non-Quasianalytic Classes

By ${}_0\mathscr{E}_B$ we denote the space of sequences $\{a_j\}$ of complex numbers such that, for every $\varepsilon > 0$ there is an $A > 0$ so that

$$|a_j| \leqslant A\varepsilon^{|j|}b_j. \tag{13.23}$$

${}_0\mathscr{E}_B$ is given the natural topology; it is Fréchet.

Moreover, let I be a closed bounded convex set in R with a piecewise C^∞ boundary. By ${}_I\mathscr{E}_B$ we denote the space of functions f which are indefinitely differentiable on I and together with each of their derivatives satisfy

$$\max_{x \in I} |f^{(j)}(x)| \leqslant A\varepsilon^{|j|}b_j.$$

Again, ${}_I\mathscr{E}_B$ is a Fréchet space.

There exist natural continuous maps $\varphi\colon \mathscr{E}_B \to {}_0\mathscr{E}_B$ and $\psi\colon \mathscr{E}_B \to {}_I\mathscr{E}_B$. We call $\varphi(\mathscr{E}_B)$ the *punctual image* of $\mathscr{E}_B$ and $\psi(\mathscr{E}_B)$ the *local image* of $\mathscr{E}_B$.

We shall consider the following problems:

PROBLEM *I*. Determine the images of φ and ψ. In particular, when is φ or ψ *onto*?

PROBLEM *II*. In case φ or ψ is not onto, determine those B' for which $\varphi(\mathscr{E}_B) \supset {}_0\mathscr{E}_{B'}$ or $\psi(\mathscr{E}_B) \supset {}_I\mathscr{E}_{B'}$.

Problems I and II are closely related to the hyperbolic problems discussed in Chapter VIII. We shall give a solution of Problems I and II. In particular, φ and ψ are onto for $B = \{(lj_1)! \cdots (lj_n)!\}$ for any $l \geqslant 1$.

In case $l = 2$ this is equivalent to saying that every function $g(x)$ which is entire is the Cauchy value of a solution $h(t, x)$ of the heat equation $h_t = h_{xx}$ for all x, t. (For simplicity we consider the case $n = 2$.)

For, let $a_j = g^{(2j)}(0)$, $d_j = g^{(2j+1)}(0)$, so $\{a_j\}, \{d_j\} \in {}_0\mathscr{E}_B$, where $b_j = (2j)!$. Suppose $k(t), m(t) \in \mathscr{E}_B$ satisfies $\varphi k = \{a_j\}$, $\varphi m = \{d_j\}$. By multiplying by a suitable factor we may assume that k, m are of compact support. It is well known that there is a (unique) solution h of the heat equation defined for all x, t satisfying $h(t, 0) = k(t)$, $h_x(t, 0) = m(t)$. (This can be proved by a simple modification of Section IX.3.) Then,

$$\frac{\partial^{2j}h}{\partial x^{2j}}(0, 0) = \frac{\partial^j h}{\partial t^j}(0, 0) = a_j = g^{(2j)}(0)$$

and

$$\frac{\partial^{2j+1}h}{\partial x^{2j+1}}(0, 0) = \frac{\partial^j}{\partial t^j}\left(\frac{\partial h}{\partial x}\right)(0, 0) = d_j = g^{(2j+1)}(0).$$

This, together with the known fact (see Section VIII.2) that h is analytic in x implies that $h(0, x) = g(x)$ which is the desired result.

The converse, that is, if every entire function g is the Cauchy value of a solution of the heat equation then φ is *onto* is proved in a similar manner (see also Section IX.6).

See Remark 13.7.

Bang has shown in his thesis [1] that φ is never onto if $\mathscr{E}_B$ is quasi-analytic, but not analytic.

Using slightly different definitions, Beurling (unpublished) has determined the punctual image of a space which is similar to $\mathscr{E}_B$ for

$$B = \{(j \log j)^j\}.$$

Actually, Beurling's method and results go deeper than ours in the case he studied.

See Remark 13.8.

If $\varphi\,(\mathscr{E}_B) \supset {}_0\mathscr{E}_{B'}$, we show that there is a discrete sequence $\{t_j\}$ so that every $a \in {}_0\mathscr{E}_{B'}$ is of the form $\varphi(f)$, where $f(t) = \sum a_j \exp(-it_j \cdot x)$, where the a_j are sufficiently small. A consequence of this is that, no matter how lacunary a sequence $\{t_j\}$ is given, there exists an indefinitely differentiable function $f \neq 0, f(x) = \sum \mu_j \exp(-it_j \cdot x)$, where $\mu_j = 0(1 + |t_j|)^{-m}$ for all m, such that $f^{(j)}(0) = 0$ for all j. This should be contrasted with a generalized Fabry gap theorem (see Chapter XII) which implies that for $\{t_j\}$ sufficiently lacunary f could not vanish on an interval. Compare also the remark on Mandelbrojt's work mentioned at the beginning of this chapter which implies that f could not vanish too rapidly at 0.

In analogy to the space $\mathscr{E}_B(\vec{\mathrm{D}})$ defined at the beginning of this chapter we could define a space $\mathscr{E}_B(\vec{\mathrm{D}};\, T)$ for T a principal noncharacteristic for $\vec{\mathrm{D}}$, where $\mathscr{E}_B(\vec{\mathrm{D}};\, T)$ consists of sequences of functions a_j on T which satisfy conditions like (13.23), namely,

$$|a_j(\xi)| \leqslant A\varepsilon^{|j|}b_j \tag{13.23*}$$

uniformly for ξ in compact sets of T. Then we could define analogs of the maps φ and ψ. It would be of great interest to study these.

See Problem 13.11.

We shall now return to the problem of finding the images of φ and ψ. We shall assume first that $n = 1$. We let I be the unit interval. For $i = 1, 2$ we denote as usual by R^i the closed part of R where x has a fixed signum. In addition to the spaces $\mathscr{E}_B$ and ${}_0\mathscr{E}_B$ we shall need to use some spaces which are not AU. Let us denote by $\mathscr{S}'$ the space of tempered distributions of L. Schwartz [1], that is, $\mathscr{S}$ is the space of C^∞ functions on R all of whose derivatives fall off at infinity faster than $(1 + |x|)^{-l}$ for any l. By $\mathscr{S}^{i'}$ we denote the space of tempered distributions on R^i, that is, $\mathscr{S}^1$ is the space of indefinitely differentiable functions f on $[0, \infty)$ such that all derivatives of f fall off at infinity faster than $(1 + |x|)^{-l}$ for any l. ($\mathscr{S}^2$ is defined similarly.)

The spaces $\mathscr{S}^{i'}$ are not AU. However, it is not difficult to show from the definitions:

$\mathbf{S}^{i'}$ *consists of all functions* $H(z)$ *which are holomorphic in the interior of* C^i *and can be represented in the form* $\partial\tilde{H}(z)$ *for some constant coefficient differential operator* ∂ *and some* $\tilde{H}$ *which is holomorphic in the interior of* C^i, *and continuous on* C^i, *and bounded by* $c(1 + |z|)^c$.

It is not difficult to describe the topology of $\mathbf{S}^{i'}$ intrinsically, but we do not need it.

Let $f \in \mathscr{E}_B$. Then, by definition, for any $\varepsilon > 0$ the series

$$\sum \varepsilon^{-j} f^{(j)}(x)/b_j \tag{13.24}$$

converges in the topology of $\mathscr{E}$. We denote by ${}_i\mathscr{E}_B$ the space of those $f \in \mathscr{E}_B$ for which the series (13.24) also converge in the topology of $\mathscr{S}^i$. The topology of ${}_i\mathscr{E}_B$ is defined by the seminorms in $\mathscr{E}_B$ together with the seminorms of the form

$$\|f\|^{\varepsilon,j',l} = \max_{x \in R^i} \left|\sum \varepsilon^{-j}\, b_j^{-1}(1 + x^2)^l f^{(j+j')}(x)\right|$$

for any ε, j', l.

Finally we define $\mathscr{E}^i_B$ as the space of C^∞ functions f on $[0, \infty)$ for which the series (13.24) converge in the topology of $\mathscr{S}^i$. The topology of $\mathscr{E}^i_B$ is defined in a manner similar to that of ${}_i\mathscr{E}_B$.

It is clear that $\mathscr{E}^i_B$ and ${}_i\mathscr{E}_B$ are Fréchet spaces. Note that ${}_i\mathscr{E}_B$ differs from $\mathscr{E}^i_B$ in that the functions of $\mathscr{E}^i_B$ are required to be defined on R^i only.

We need some information about the Fourier transforms of the duals of these spaces. In case $\mathscr{E}_B$ is quasianalytic we shall assume that B satisfies the regularity conditions imposed in Theorem 5.26.

THEOREM 13.13. (a) $\mathscr{E}_B$ *is AU.* $\mathbf{E}'_B$ *consists of all entire functions* $H(z)$ *for which there exists an* $A > 0$ *so that*

$$H(z) = 0(\exp(A\,|\text{Im } z|)\lambda(Az)(1 + |z|)^A). \tag{13.25}$$

An AU structure consists of all continuous positive functions which dominate $\exp(A\,|\text{Im } z|)\lambda(Az)(1 + |z|)^A$ *for all* A.

(b) ${}_0\mathscr{E}_B$ *is AU.* ${}_0\mathbf{E}'_B$ *consists of all entire functions* $H(z)$ *for which there exists an* $A > 0$ *so that*

$$H(z) = 0(\lambda(Az)(1 + |z|)^A). \tag{13.26}$$

An AU structure consists of all continuous positive functions k which dominate $\lambda(Az)(1+|z|)^A$ *for all* A.

(c) ${}_I\mathscr{E}_B$ *is AU.* ${}_I\mathbf{E}'_B$ *consists of all entire functions* $H(z)$ *for which there exists an* $A > 0$ *so that*

$$H(z) = 0(\exp|\operatorname{Im} z|\,\lambda(Az)(1+|z|)^A) \tag{13.27}$$

for all A. *An AU structure for* ${}_I\mathscr{E}_B$ *consists of all continuous positive k which dominate the functions on the right side of* (13.27).

(d) ${}_i\mathbf{E}'_B$ *consists of all functions* $H(z)$ *which are of the form* $\partial\tilde{H}(z)$, *where* $\tilde{H}(z)$ *is holomorphic on the interior of* C^i, *continuous on the closure of* C^i *and satisfies there*

$$\tilde{H}(z) = 0[\exp(A|\operatorname{Im} z|)\lambda(Az)(1+|z|)^A] \tag{13.28a}$$

for some $A > 0$. *Here* ∂ *is a linear constant coefficient differential operator.*

For any $A > 0$, *the topology of* ${}_i\mathbf{E}'_B$ *is weaker than the topology induced by*

$$L_2[\operatorname{Im} z = 0,\ -\infty < \operatorname{Re} z < \infty;\ d(\operatorname{Re} z)/\lambda(Az)(1+z)^A \exp(A|\operatorname{Im} z|)]$$

uniformly in $\operatorname{Im} z \geqslant 0$ $(\operatorname{Im} z \leqslant 0)$ *on those* $H \in {}_i\mathbf{E}'_B$ *which have boundary values on* $\operatorname{Im} z = 0$ *and belong to this* L_2 *space. The* L_2 *space is defined by the norm*

$$\|G\| = \sup \int |G(x+iy)|^2\, dx/\lambda(Az)(1+|z|)^A \exp(A|\operatorname{Im} z|),$$

where the sup *is over all* y *with* $y \geqslant 0$ $(y \leqslant 0)$.

(e) $\mathbf{E}^{i'}_B$ *consists of all functions* $H(z)$ *of the form* $\partial\tilde{H}$ *with* $\partial, \tilde{H}$ *as above except that* (13.28a) *is to be replaced by*

$$\tilde{H}(z) = 0[\lambda(Az)(1+|z|)^A]. \tag{13.28b}$$

Assertion (a) is just Theorem 5.26. Assertions (b) and (c) can be proved in essentially the same manner. There are some complications in the proofs of (d) and (e) due to the fact that the spaces involved are not AU.

PROBLEM 13.12. (Exercise).
Carry out the details.

The precise definitions of the spaces $\mathscr{E}^i_B$ and ${}_i\mathscr{E}_B$ are not very important and there are many modifications of them which would suffice for our purposes. The use of the series of the form (13.24) to define various spaces of infinitely differentiable functions is discussed in detail in Ehrenpreis [7]. Actually, because of the result of Bang [1] mentioned above, we shall be mostly interested in the non-quasianalytic case for which the proofs are simpler (they do not rely on the heat equation) and less regularity on B must be imposed.

The following result and proof are due to I. Glicksberg:

LEMMA 13.14. *Let $\mathscr{U}$, $\mathscr{V}$, $\mathscr{W}$ be Fréchet spaces, and $T : \mathscr{U} \to \mathscr{V}$ and $S : \mathscr{W} \to \mathscr{V}$ continuous linear maps. A necessary and sufficient condition for $T(\mathscr{U}) \supset S(\mathscr{W})$ is: Whenever $v' \in \mathscr{V}'$ are such that $T'v'$ converge in the weak topology of $\mathscr{U}'$, then $S'v'$ converge in the weak topology of $\mathscr{W}'$.*

PROOF. Suppose first that $T(\mathscr{U}) \supset S(\mathscr{W})$. By passing to quotient spaces (which is permissible, since weak topologies and the Fréchet property behave well under this passage) we may suppose that T and S are one-to-one. Then we can define a map $L : \mathscr{W} \to \mathscr{U}$ by

$$Lw = T^{-1}Sw \quad \text{for } w \in \mathscr{W}.$$

It is readily verified that the graph of L is closed, so L is continuous, hence the adjoint L' of L is weakly continuous. It follows thus that if $T'v'$ converge weakly in the topology of $\mathscr{U}'$ then

$$L'T'v' = (S'(T')^{-1})T'v' = S'v' \text{ converge (weakly).}$$

Conversely, suppose the condition is satisfied and let $w \in \mathscr{W}$. Then define the linear function $\tilde{u}$ on $T'\mathscr{V}'$ by

$$T'v' \cdot \tilde{u} = S'v' \cdot w$$

for any $v' \in \mathscr{V}'$. Our assumptions imply that $\tilde{u}$ is continuous on $T'\mathscr{V}'$ in the weak topology induced by $\mathscr{U}'$, so $\tilde{u}$ has an extension to $u \in \mathscr{U}$. Clearly, $Tu = Sw$, so that result is established. ▯

We wish now to decide if the map φ is onto. Denote by φ' the adjoint of φ so $\varphi' : {}_0\mathscr{E}'_B \to \mathscr{E}'_B$. By Fourier transform we get a map $\hat{\varphi}' : {}_0\mathbf{E}'_B \to \mathbf{E}'_B$. We verify easily from the definitions that $\hat{\varphi}'$ is the natural injection of ${}_0\mathbf{E}'_B$ into $\mathbf{E}'_B$, that is, $\hat{\varphi}'(H(z)) = H(z)$.

The Hahn-Banach theorem and the reflexivity of ${}_0\mathscr{E}_B$, $\mathscr{E}_B$ tell us that φ will be onto if, whenever $F \in {}_0\mathbf{E}'_B$ and $F \to 0$ in the topology of $\mathbf{E}'_B$, then also $F(z) \to 0$ in the topology of ${}_0\mathbf{E}'_B$. By Theorem 13.13 this means that we must determine conditions on B (hence on λ) so that if $F \in {}_0\mathbf{E}'_B$ and

$$\sup |F(z)|/k(z) \to 0 \tag{13.29}$$

for all k in an AU structure for $\mathscr{E}_B$, then also

$$\sup |F(z)|/k'(z) \to 0 \tag{13.30}$$

for all k' in an AU structure for ${}_0\mathscr{E}_B$.

There are several cases to be distinguished.

Case 1. *λ dominates some exponential* $\exp(A\,|z|)$.

In this case we must have $b_j \leqslant c^j j!$ for all j. We assume the following regularity condition on B:

$$cb_{j+1} \leqslant (j+1)b_j$$

for all j. This implies that

$$\begin{aligned}\lambda'(z) &= \sum j\,|z|^{j-1}/b_j \\ &= \sum (j+1)\,|z|^j/b_{j+1} \\ &\geqslant \sum c\,|z|^j/b_j .\end{aligned}$$

Thus,

$$\lambda'(z)/\lambda(z) \geqslant c.$$

For any $z > z_0 > 1$ we integrate the above inequality to obtain

$$\log \lambda(z) \geqslant \log \lambda(z_0) + c(z - z_0).$$

Thus, setting $z = c'z_0$,

$$\log \lambda(c'z_0) \geqslant \log \lambda(z_0) + c(c' - 1)z_0$$

or

$$\lambda(c'z_0) \geqslant \exp\,[c(c' - 1)z_0]\lambda(z_0)$$

which means that exponential factors can be "absorbed" in λ, that is, the product of λ by an exponential is majorized by $\lambda(c''z)$ for some c''.

It is clear from the above and Theorem 13.13 that k and k' are the same, so φ is *onto* in this case. Similarly, ψ is *onto*.

Case 2. *The class* $\mathscr{E}_B$ *is non-quasianalytic.*

We shall consider first the case $n = 1$. We note by Theorem 13.13 that an AU structure for $\mathscr{E}_B$ can be chosen to consist of functions of the form $k_1(\operatorname{Im} z)\tilde{\lambda}(z)$, where k_1 dominates all exponentials and $\tilde{\lambda}$ dominates all $(1 + |z|)^4\lambda(Az)$. As $\int \log \lambda(\xi)\, d\xi/(1 + \xi^2) < \infty$, we see easily that given any $\tilde{\lambda}'$ we can find a $\tilde{\lambda} \leqslant \tilde{\lambda}'$ such that also $\int \log \tilde{\lambda}(\xi)\, d\xi/(1 + \xi^2) < \infty$. Thus, we may assume the finiteness of this integral for all $\tilde{\lambda}$ which we shall consider.

We introduce the Poisson integral of $\log \tilde{\lambda}$

(13.31) $$\alpha(x, y) = \frac{1}{\pi}\int_{-\infty}^{\infty} \frac{y \log \tilde{\lambda}(|x'|)\, dx'}{(x - x')^2 + y^2}$$

for $y > 0$. From (13.29) we see that

(13.32) $$\sup_{x \text{ real}}\ |F(x)|/\tilde{\lambda}(|x|) \to 0.$$

Let $\beta(x, y)$ be a conjugate of $\alpha(x, y)$ in the upper half-plane. Call $\delta(z) = \exp\,(\alpha(z) + i\beta(z))$, where $z = x + iy$. Then (13.32) implies

(13.33) $$\sup_{x \text{ real}}\ |F(x)/\delta(x)| \to 0,$$

because $|\delta(x)| = \exp \alpha(x, 0) = \tilde{\lambda}(x)$.

Since F belongs to ${}_0\mathbf{E}'_B$, the principle of harmonic majorant (see the proof of Theorem 5.5) may be applied to show that

$$\sup_{\operatorname{Im} z \geqslant 0} |F(z)/\delta(z)| \to 0. \tag{13.34}$$

Call

$$\gamma(r) = \sup_{\operatorname{Im} z \geqslant 0,\, |z|=r} |\exp \alpha(x, y)|.$$

Then the above (together with a similar reasoning for Im $z < 0$) shows that

$$\sup_{z \in C} |F(z)|/\gamma(|z|) \to 0. \tag{13.35}$$

In particular, we have

LEMMA 13.15. *Suppose for any $\tilde{\lambda}_1$ which dominates all $(1 + |z|)^A \lambda(Az)$ and satisfies $\int \log \tilde{\lambda}_1(\xi)\, d\xi/(1 + \xi^2) < \infty$ there is a $\tilde{\lambda}$ with $\gamma(|z|) = 0(\tilde{\lambda}_1(z))$. Then φ is onto.*

For, (13.35) implies that $F \to 0$ in the topology of ${}_0\mathbf{E}'_B$.

By contour integration we verify easily that the hypotheses of Lemma 13.15 are satisfied for $b_j = (lj)!$ for some $l > 1$. However, the hypotheses are not satisfied for $b_j = (j \log^2 j)^j$ (which is non-quasianalytic).

We want to go further and determine necessary and sufficient conditions in order that φ be onto. Let μ denote the natural map of ${}_i\mathscr{E}_B$ into $\mathscr{E}^i_B$.

LEMMA 13.16. *For $i = 1, 2$, a necessary condition for*

$$\mu({}_i\mathscr{E}_B) \supset \mathscr{E}^i_{B''} \tag{13.36}$$

is: For any $\tilde{\lambda}''$ there is a $\tilde{\lambda}$ so that

$$\tilde{\lambda}''(z) \geqslant \text{const}\; \gamma(|z|). \tag{13.37}$$

PROOF. Suppose (13.37) is violated. Call $\alpha_B(x, y)$ the Poisson integral of $\log \lambda$, that is, α_B is obtained by replacing $\tilde{\lambda}$ in the right side of (13.31) by λ. Let $\beta_B(x, y)$ be a conjugate of α_B in the upper half-plane. Call $\delta_B = \exp(\alpha_B + i\beta_B)$. A simple diagonal argument based on (13.37) shows that for some $\tilde{\lambda}''$ we have

$$\delta_B(z) \neq 0(\tilde{\lambda}''(z)). \tag{13.38}$$

(Actually $\delta_B(z)$ should be replaced by $\gamma_B(|z|)$ on the left side of (13.38). However, since $\tilde{\lambda}''$ depends only on $|z|$ this does not matter.)

Using a simple estimate of the Poisson integral (compare Chapter V, Example 1), we see that

$$\delta_B(z) = 0(\exp A\, |\operatorname{Im} z|)\lambda(Az)(1 + |z|)^A$$

for some A, so by Theorem 13.13 we have $\delta_B \in {}_i\mathbf{E}'_B$ (the particular i depending on whether $y > 0$ or $y < 0$). However by (13.38) and Theorem 13.13, δ_B is not in $\mathbf{E}^{i'}_{B''}$.

We claim there is a sequence $\{G_j\}$ of functions in $\mathbf{E}^{i'}_B$ which converges to δ_B in the strong topology of ${}_i\mathbf{E}'_B$. Assuming this, since δ_B is not in $\mathbf{E}^{i'}_{B''}$, $\{G_j\}$ cannot converge strongly in $\mathbf{E}^{i'}_{B''}$. For sequences, weak and strong convergence coincide, so G cannot converge weakly in $\mathbf{E}^{i'}_{B''}$. Thus, by Lemma 13.14 the result will follow.

Suppose for simplicity that δ_B is analytic in $\operatorname{Im} z > 0$. Since δ_B is expressed in terms of a Poisson integral, δ_B^{-1} is an outer function in the sense of Beurling [1]. Thus, by Beurling's theorem, we can approximate 1 in the topology of $L_2[-\infty < \operatorname{Re} z < \infty, \operatorname{Im} z = y_0; d(\operatorname{Re} z)/(1 + (\operatorname{Re} z)^2)]$ uniformly in $y_0 \geqslant 0$ by $G_j\,\delta_B^{-1}$, where G_j are polynomials in $(z - i)/(z + i)$. This means that $(G_j - \delta_B) \to 0$ in the topology of $L_2[-\infty < \operatorname{Re} z < \infty, \operatorname{Im} z = y_0; d(\operatorname{Re} z)/\delta_B^2(z)(1 + (\operatorname{Re} z)^2)]$ uniformly in $y_0 \geqslant 0$.

Now, $\delta_B(x) = \lambda(x)$ for x real.

Assumption. There exists an A so that $\lambda^2(x) \leqslant A\lambda(Ax)$.

We should like to replace δ_B^2 in the above by $\lambda(A \operatorname{Re} z) \exp (A \operatorname{Im} z)$ for some A. If this could be done then we could apply Theorem 13.13d to conclude that $\{G_j\}$ converges in the topology of ${}_i\mathbf{E}_B$. Thus we need to know that $|\delta_B^2(z)| \leqslant c\lambda(A \operatorname{Re} z) \exp (A \operatorname{Im} z)$. Since $\alpha_B = \log |\delta_B|$, it suffices to show that

$$\alpha_B(x, y) \leqslant c' + \log \lambda(Ax) + Ay.$$

Using (13.31) we can write (we suppose $x \geqslant 0$ as the case $x < 0$ is similar; we also may suppose, without loss of generality, that $\lambda(z) > 1$)

$$\alpha_B(x, y) \leqslant \frac{1}{\pi} \int_{-\infty}^{\infty} \frac{y \log \lambda(x' + x)\, dx'}{x'^2 + y^2} = \frac{1}{\pi} \int_{-\infty}^{-x} + \frac{1}{\pi} \int_{-x}^{x} + \frac{1}{\pi} \int_{x}^{\infty}.$$

We have, by the monotonicity of $\log \lambda$,

$$\left| \frac{1}{\pi} \int_{-x}^{x} \right| \leqslant \frac{\log \lambda(2x)}{\pi} \int_{-x}^{x} \frac{y dx'}{x'^2 + y^2}$$

$$\leqslant \frac{\log \lambda(2x)}{\pi} \int_{-\infty}^{\infty} \frac{y\, dx'}{x'^2 + y^2}$$

$$= \log \lambda(2x).$$

Moreover, in $-\infty < x' \leqslant -x$ and $x \leqslant x' < \infty$ we have $|x + x'| \leqslant 2|x'|$.

Thus

$$\left|\frac{1}{\pi}\int_{-\infty}^{-x}\right| \leqslant \frac{1}{\pi}\int_{-\infty}^{-x}\frac{y\log\lambda(2x')\,dx'}{x'^2+y^2}$$
$$\leqslant \frac{1}{\pi}\int_{-\infty}^{\infty}\frac{y\log\lambda(2x')\,dx'}{x'^2+y^2}.$$

Now,

$$\int_{-\infty}^{\infty}\frac{\log\lambda(2x')\,dx'}{x'^2+1} < \infty.$$

It follows that the last integral $\int_{-\infty}^{\infty}\frac{y\log\lambda(2x')\,dx'}{x'^2+y^2}$ is uniformly bounded for $0 \leqslant y \leqslant 1$ since it is the value at the point $\operatorname{Re} z = 0$, $\operatorname{Im} z = y$, of an harmonic function in $\operatorname{Im} z > 0$ having boundary value $\log\lambda(2x)$. On the other hand, for $y \geqslant 1$ the integral is bounded by

$$\frac{y}{\pi}\int_{-\infty}^{\infty}\frac{\log\lambda(2x')\,dx'}{x'^2+1} \leqslant \text{const } y.$$

The integral $\int_x^\infty$ is handled similarly.

We have thus shown that

$$\alpha_B(x, y) \leqslant c' + \log\lambda(2x) + \text{const } y$$

which is what was needed to complete the proof of Lemma 13.16.

THEOREM 13.17. (13.37) *is a necessary and sufficient condition for*

$$\varphi(\mathscr{E}_B) \supset {}_0\mathscr{E}_{B''} \tag{13.39}$$

or

$$\psi(\mathscr{E}_B) \supset {}_I\mathscr{E}_{B''}. \tag{13.40}$$

PROOF. The sufficiency of (13.37) for (13.39) is proved as in Lemma 13.15. Now, any $f \in {}_I\mathscr{E}_{B''}$ defines two punctual images $f^\pm$ at the two end-points of I. If (13.39) holds then both $f^\pm$ have extensions $g^\pm$ to all of R which belong to $\mathscr{E}_B$.

Now define h by

$$h(x) = \begin{cases} g^-(x) & \text{for } x \text{ to the left of } I \\ f(x) & \text{for } x \in I \\ g^+(x) & \text{for } x \text{ to the right of } I. \end{cases}$$

(13.39) clearly implies that the inequalities defining $\mathscr{E}_{B''}$ are more stringent than those defining $\mathscr{E}_B$; it follows that $h \in \mathscr{E}_B$. We have shown that (13.39) implies (13.40).

Thus, we have the chain of implications

$$(13.36) \Rightarrow (13.37) \Rightarrow (13.39) \Rightarrow (13.40).$$

It remains to show that (13.40) implies (13.36). Let $f \in \mathscr{E}^1_{B''}$. If (13.40) holds, then there exists $g_0 \in \mathscr{E}_B$ with $g_0 = f$ on $[0, 1]$.

Now, (13.40) clearly implies that $\mathscr{E}_B \supset \mathscr{E}_{B''}$ so that λ'' dominates λ. Thus, if we define $g(x)$ to be $g_0(x)$ for $x < 0$ and $f(x)$ for $x \geqslant 0$, we have $g \in {}_1\mathscr{E}_B$ and $\mu(g) = f$. Hence, (13.36) holds. This completes the proof of Theorem 13.17. □

See Remark 13.9.

Case 3. *$\mathscr{E}_B$ is quasianalytic but λ does not dominate all exponentials. In this case $\varphi(\mathscr{E}_B) \supset {}_0\mathscr{E}_{B''}$ if and only if λ'' dominates λ and some exponential. The same condition is necessary and sufficient for $\psi(\mathscr{E}_B) \supset {}_I\mathscr{E}_{B''}$.*

PROOF. We consider $\varphi' : {}_0\mathscr{E}'_B \to \mathscr{E}'_B$. By the definition of quasianalyticity φ is one-to-one, so the image of φ' is dense. Since $\mathscr{E}_B$ is quasianalytic, the linear combinations of the derivatives of δ are dense in $\mathscr{E}'_B$. By Fourier transform, for any t there are polynomials $P(z)$ such that $P(z) \to \exp(itz)$ in the topology of $\mathbf{E}'_B$. If $\varphi(\mathscr{E}_B) \supset {}_0\mathscr{E}_{B''}$, then Lemma 13.14 shows that $P(z)$ must converge in the (weak) topology of ${}_0\mathbf{E}'_{B''}$. Since the weak topologies of ${}_0\mathbf{E}'_{B''}$ and $\mathbf{E}'_B$ are stronger than pointwise convergence, because the exponential functions are in $\mathscr{E}_B$ and ${}_0\mathscr{E}_B$, the only limit to which it could converge is $\exp(itz)$, so ${}_0\mathbf{E}'_{B''}$ must contain all exponentials. This can be the case only if λ'' dominates some exponential.

It is a simple consequence of the assumption $\varphi(\mathscr{E}_B) \supset {}_0\mathscr{E}_{B''}$ that λ'' dominates λ.

Conversely, if λ'' dominates some exponential and λ'' dominates λ, the result follows from the result in Case 1, since

$$\varphi(\mathscr{E}_B) \supset \varphi(\mathscr{E}_{B''}) = {}_0\mathscr{E}_{B''}.$$

The proof for ψ is similar.

The above completes the study for $n = 1$. The passage to $n > 1$ presents no difficulties, namely, $\varphi(\mathscr{E}_B) \supset {}_0\mathscr{E}_{B''}$ or $\psi(\mathscr{E}_B) \supset {}_I\mathscr{E}_{B''}$ if and only if the restrictions of λ and λ'' to each real line satisfy the above conditions for $n = 1$. For if for some real line $\sum$, $\varphi[\mathscr{E}_B(\sum)] \not\supset {}_0\mathscr{E}_{B''}(\sum)$, then there is an $a \in {}_0\mathscr{E}_{B''}(\sum)$ which cannot be extended. This a clearly defines an element of ${}_0\mathscr{E}_{B''}$ which cannot be extended. A similar result holds for ${}_I\mathscr{E}_B$. The converse is proved by using inequalities obtained from the one-dimensional case to show that the inverses of $\hat{\varphi}'$ and $\hat{\psi}'$ are continuous. We leave the details to the reader. In particular, if $B = B''$ is trivial (that is, $b_0 = 1$, $b_j = \infty$ for $j \neq 0$), then we derive the classical result that every sequence $\{a_j\}$ is of the form $\{f^{(j)}(0)\}$ for some $f \in \mathscr{E}$. □

Let us note that the above proof shows that, in the nonquasianalytic case, $\varphi(\mathscr{E}_B) = {}_0\mathscr{E}_B$ if and only if R is a sufficient set (see Section I.3) for $\mathscr{E}_B$. Actually, much more is true. Using an argument of Cartwright (see Levinson [1]) we can show

THEOREM 13.18. *$\varphi(\mathscr{E}_B) = {}_0\mathscr{E}_B$ if and only if there exists a discrete set in R which is sufficient for ${}_0\mathscr{E}_B$. Moreover, given any infinite discrete set $\theta \subset R$ there is a B so that θ is sufficient for ${}_0\mathscr{E}_B$ and $\varphi(\mathscr{E}_B) = {}_0\mathscr{E}_B$. Moreover the removal of any point from θ leaves us with a ${}_0\mathscr{E}_B$ sufficient set.*

COROLLARY. *Let $\{t_l\}$ be any unbounded sequence of real numbers. Then there exists an $a \in \mathscr{E}$, $a \neq 0$, of the form*

$$a(x) = \sum v_l \exp(ixt_l), \tag{13.41}$$

where the series converges in the topology of $\mathscr{E}$, so that $a^{(j)}(0) = 0$ for all j.

PROOF. By passing to a subsequence we may assume $t_{l+1} > t_l + 1$ (or $t_{l+1} < t_l - 1$) for all l. Let B be as in Theorem 13.18 and let $a' \in {}_0\mathscr{E}_B$ so we can write

$$a'(x) = \sum v_l' \exp(ixt_l), \tag{13.42}$$

where $v_l' = 0(1/k(t_l))$ for some k in an AU structure for ${}_0\mathscr{E}_B$. Thus, the series (13.42) defines a function $a' \in \mathscr{E}$ (in fact, $a' \in \mathscr{E}_B$).

Suppose that $v_{l_0}' \neq 0$ and let $\theta' = \theta - \{t_{l_0}\}$. Then θ' is still sufficient for ${}_0\mathscr{E}_B$ by Theorem 13.18, so we can express $\{a'^{(j)}(0)\}$ in terms of exponentials whose frequencies lie in θ'. That is, we can write

$$a''(x) = \sum_{l \neq l_0} v_l'' \exp(it_l x), \tag{13.43}$$

where $a'^{(j)}(0) = a''^{(j)}(0)$ for all j and again $a'' \in {}_0\mathscr{E}_B$. We choose $a = a' - a''$. The fact that $a \not\equiv 0$ follows from the fact that $v_{l_0}' \neq 0$ and the results of Chapter XII. In fact, by the results of Chapter XII a cannot vanish on an interval if $\{t_l\}$ is, as we may suppose, sufficiently lacunary. □

Suppose we want to determine the image $\varphi(\mathscr{E}_B)$. (Similar remarks apply to $\psi(\mathscr{E}_B)$.) Let $a \in {}_0\mathscr{E}_B$. Then (see Theorem 1.5) we have a Fourier expansion

$$a(x) = \int \exp(ix \cdot z)\, dv(z)/k(z), \tag{13.44}$$

where v is a bounded measure on C and k is in an AU structure for ${}_0\mathscr{E}_B$. More precisely, (13.44) means

$$a_j = \int (iz)^j\, dv(z)/k(z). \tag{13.45}$$

Using the reflexivity of $\mathscr{E}_B$ and ${}_0\mathscr{E}_B$, we have

THEOREM 13.19. *A necessary and sufficient condition that a belong to*

$\varphi(\mathscr{E}_B)$ *is: Whenever* $H \in {}_0\mathbf{E}'_B$ *and* $H \to 0$ *in the topology of* $\mathbf{E}'_B$

$$\int H(z)\, dv(z)/k(z) \to 0.$$

XIII.4. Quasianalytic Functions on Lines

We now come to a different type of theorem about quasianalyticity. For a particular quasianalytic B and for Ω a domain in the space of one complex variable the theorem we are about to describe was proved by Beurling by an entirely different method, and his result inspired the general one.

Before stating the result let us introduce some notation. Let ${}^*\mathscr{E}_B$ be as in Section XIII.2 above for $n = 1$. For any interval I we define ${}^*_I\mathscr{E}_B$ as the analogous space of functions on I. Let l be a compact segment on $R(n \geqslant 1)$. Using the directional derivatives along l, we may identify ${}^*_l\mathscr{E}_B$ with a space of functions on l. Similarly, we define the concept of a collection of functions defined on compact line segments (perhaps all different) being bounded in ${}^*\mathscr{E}_B$. (We should perhaps more properly write "in $\{{}^*_l\mathscr{E}_B\}$.")

Let Ω be a domain in C with compact closure. By the *Šilov boundary* of Ω, denoted by $\delta\Omega$, we mean the smallest closed subset of the closure of Ω on which every function which is analytic on Ω and continuous in its closure takes its maximum. It is known (see Gelfand, Raikov, and Šilov [1]) that the Šilov boundary exists and is unique.

Let $L = \{l^j\}$ be a set of closed line segments in C (with coordinates $(x_1, \ldots, x_n)$) containing the origin. Suppose for each j we are given a function f^j on l^j which is indefinitely differentiable. We say the collection $\{f^j\}$ is *analytically compatible at the origin* if there exists a formal power series $h(x)$ such that the following condition holds: Let l^j have the direction cosines $(\operatorname{Re} v_1^j, \operatorname{Im} v_1^j, \ldots, \operatorname{Re} v_n^j, \operatorname{Im} v_n^j)$. Then for all p the pth derivative of f^j along l^j at the origin (denoted by $(d^p_{l^j} f^j)(0)$) is

$$(d^p_{l^j} f^j)(0) = \left[\left(v_1^j \frac{\partial}{\partial x_1} + \cdots + v_n^j \frac{\partial}{\partial x_n}\right)^p h\right](0), \tag{13.46}$$

where the right side is to be taken in the formal sense. Here

$$v_k^j = \operatorname{Re} v_k^j + i \operatorname{Im} v_k^j.$$

Clearly analytic compatibility at the origin is a necessary condition for the existence of an analytic function F which agrees on each l^j with f^j. The result below gives a sufficient condition for the converse.

THEOREM 13.20. *Let Ω be a domain in C whose closure is compact and which is a union of connected line segments (of one real dimension) through* *the origin. Let $\{l^j\} = L$ be a set of closed line segments containing the origin contained in the closure of Ω whose intersection with the Šilov boundary $\delta\Omega$ is dense in $\delta\Omega$. Let ${}^*\mathscr{E}_B$ be a fixed quasianalytic (no function can vanish together with all its derivatives at a point) space of functions of one real variable. Suppose for each j we are given a function f^j on l^j which agrees on l^j with a function in ${}^*\mathscr{E}_B$. Suppose the functions f^j form a bounded set in ${}^*\mathscr{E}_B$ and the f^j are analytically compatible at the origin. Then there is a function F which is analytic on Ω and continous on its closure, whose restriction to each l^j is f^j.*

PROOF. We may assume that Ω is contained in the unit sphere. Let α be a real variable and β its dual. Let $P_t(i\beta)$ be polynomials with converge to $\exp(i\beta)$ in the topology of ${}^*\mathbf{E}'_B$. The P_t exist by Theorem 13.9. By the monotonicity of the function $\lambda(\beta)$ it follows that $P_t(ia\beta)$ converge uniformly to $\exp(ia\beta)$ for $-1 \leqslant a \leqslant 1$. By Fourier transform this means that $\delta_a = \lim P'_t(a\partial/\partial\alpha)\delta_0$. By definition of ${}^*_l\mathscr{E}_B$ this means that for any $x \in l$ we have $\delta_x = \lim P'_t(\pm |x|\, d_l)$.

Let x be any point on l^j. Then certainly $x = \pm(|x|v_1^j, \ldots, |x|v_n^j)$. By the above we have

$$\begin{aligned} f^j(x) &= \lim_t\, [P_t(|x|\, d_{l^j})f^j](0) \\ &= \lim_t \left[P_t\left(x_1 \frac{\partial}{\partial s_1} + \cdots + x_n \frac{\partial}{\partial s_n}\right)\right] h(s)\bigg|_{s=0} \end{aligned}$$

because the f^j are analytically compatible at the origin. Note however, that

$$Q_t(x) = \left[P_t\left(x_1 \frac{\partial}{\partial s_1} + \cdots + x \frac{\partial}{\partial s_n}\right)\right] h(s)\bigg|_{s=0}$$

is, for each t, a polynomial in x. Now the $P_t(ia\beta)$ are uniformly bounded in ${}^*\mathbf{E}'_B$ for $-1 \leqslant a \leqslant 1$ and the f^j are uniformly bounded in ${}^*\mathscr{E}_B$. As above, the $[P_t(|x|\, d_{l^j})f^j](0)$ are uniformly bounded for $|x| \leqslant$ length of l^j. This implies that the $Q_t(x)$ are uniformly bounded on the union of the l^j, hence on the Šilov boundary of Ω. By the same reasoning, $Q_t(x)$ converge to $F(x)$ which is analytic on Ω, and continuous on the closure of Ω. Clearly $F(x) = f^j(x)$ for all $x \in l^j$, so our result is proved. ∎

See Remarks 13.10 and 13.11.

THEOREM 13.21. *Let Ω be the unit disk in the complex $x = \xi + i\eta$ plane. Let $\{l^j\}$ be a dense set of lines through the origin, say $l^0 = \xi$ axis and $l^1 = \eta$ axis. Let $f \in \mathscr{E}(\Omega)$ have the property that for each j, the restrictions of f, $\partial f/\partial\xi$, and $\partial f/\partial\eta$ to $l^j \cap \Omega$ belong to some ${}^*\mathscr{E}_B$ which is quasianalytic; suppose* *moreover, that the restrictions f^j of f to the l^j are analytically compatible at the origin. Then f is holomorphic in Ω.*

PROOF. We show that f satisfies the Cauchy-Riemann equations. Let a be a point on some $l^j \cap \Omega$ and denote by δ_a the unit mass at the point a. We claim that

$$(\partial/\partial\xi + i\partial/\partial\eta)'\delta_a \cdot f = 0. \tag{13.47}$$

This is sufficient, because the linear combinations of the δ_a are dense in $\mathscr{E}'(\Omega)$. By quasianalyticity we can write (as in the proof of Theorem 13.19), using the analytic compatibility of the f^j at 0,

$$\begin{aligned} \partial f(a)/\partial\xi &= \lim_t \,[P_t(|a|d_{l^j}(\partial f/\partial\xi)^j)](0) \\ &= \lim_t \, P_t(a\partial/\partial\xi(\partial f/\partial\xi))](0) \\ &= \lim_t \,[P_t(-ia\partial/\partial\eta)(-i\partial f/\partial\eta)](0) \\ &= \lim_t \,[P_t(|a|d_{l^j})(-i\partial f/\partial\eta)^j](0) \\ &= -i\partial f(a)/\partial\eta. \end{aligned} \tag{13.48}$$

This proves Theorem 13.21. □

It should be noted that the analytic compatibility at the origin of the f^j is implied by the analyticity of f in a neighborhood of the origin.

See Problem 13.13.

Remarks

Remark 13.1. See page 452.

Formally when all $b_j \to 0$ the space $\mathscr{E}_B(\mathrm{D})$ becomes $\mathscr{E}_{\vec{\mathrm{D}}}$, the kernel of $\vec{\mathrm{D}}$ in $\mathscr{E}$, while $\lambda(\vec{\mathrm{P}}(z)) \to \infty$ except on the variety V of common zeros of the P_j, so we can regard our theorem as being closely related to Theorems 4.1 and 4.2.

Remark 13.2. See page 458.

Remark 13.3. See page 458.

Although $\mathscr{E}_{B^1}$ and $\mathscr{E}_{B^2}$ are both quasianalytic, $\mathscr{E}_{B^1} + \mathscr{E}_{B^2}$ need not be. In fact, given any $f \in \mathscr{E}$ we can find B^1 and B^2 such that $\mathscr{E}_{B^1}$ and $\mathscr{E}_{B^2}$ are quasianalytic, but

$$f \in \mathscr{E}_{B^1} + \mathscr{E}_{B^2}.$$

This is a fairly simple consequence of our Fourier representation for $\mathscr{E}$ and $\mathscr{E}_{B^i}$. Similar results are due to Mandelbrojt [1] and to Beurling [2].

Remark 13.4. See page 458.

We suspect that the condition in Theorem 13.6 is also necessary but we have not been able to prove it. The method of proof of Theorem 13.4 shows that every function

$f_1(t)$ in $\mathscr{E}_{B^1}$, when extended to R by being constant in ξ, belongs to $\mathscr{E}_{B^1}(\Phi; D)$. Suppose that $n = 2$, that is, $d = 1$. Then, using the method of proof of Theorem 9.17 together with a study of sufficient sets, we can show that every function $f_2(t)$ in $\mathscr{E}_{B^2}$ is the restriction to the t axis of a solution of the equation $Dh = 0$, $h \in \mathscr{E}(\Phi)$. Such an h is clearly in $\mathscr{E}_B(\Phi; D)$. We deduce that if $\mathscr{E}_{B^2}$ contains a function f which vanishes together with all its derivatives at the origin, there exists an h $\in \mathscr{E}_B(\Phi; D)$ which vanishes together with all its t derivatives at the origin.

Remark 13.5. See page 462.

Remark 13.6. See page 469.

It seems that various analogs of Theorem 13.12 are "known" to experts, but we know of no published result.

Remark 13.7. See page 474.

A slightly different theorem for the heat equation is found in Ehrenpreis [7] and [8] and the method of proof in those papers is, with certain formal modifications, essentially the method of proof for the sufficiency conditions for $\varphi(\mathscr{E}_B) \supset {}_0\mathscr{E}_{B'}$.

This relation between the heat equation and the *onto* question for φ was pointed out to the author by L. Hörmander who also suggested the general question as to when φ is onto.

Remark 13.8. See page 474.

By a method which is different from the author's, Carleson [3] has derived the sufficient condition on B for $\varphi(\mathscr{E}_B) \supset {}_0\mathscr{E}_B$ for $n = 1$. Carleson's condition is the same as ours (see Theorem 13.3) and so is necessary and sufficient. It should be remarked that our results and methods were obtained independently and at about the same time as Carleson's. Originally, the manuscript was submitted to the Transactions of the American Mathematical Society at the time the results were derived, but it was decided to publish them here.

For $B = \{jl!\}$, $l > 1$ the fact that φ is onto has been announced by Mityagin [1].

Remark 13.9. See page 482.

It is easily seen that the conditions given in Theorem 13.17 allow for a small variation in the sequences B and B''. For this reason we can deduce a similar result in which the spaces $\mathscr{E}_B$ are replaced by slightly modified spaces (see Remark 5.6). The sufficiency of (13.37) for the analogs of (13.39) and (13.40) for these spaces can be proved just as above, but the necessity cannot be proved directly because we cannot apply Lemma 13.14 since the other spaces are not Frechet.

Remarks 13.10 and 13.11. See page 485.

Remark 13.10. For $n = 1$ the example $1/(z - a)$ shows that the assumption that the f^j be uniformly bounded in $^*\mathscr{E}_B$ cannot be dropped. L. Bers posed the problem as to whether it can be dropped if the closure of Ω is equal to the union of all l^j. In case $^*\mathscr{E}_B$ consists of real analytic functions this is proved simply by a Heine-Borel argument. We do not know the answer, in general.

Remark 13.11. We could replace the family of line segments by a suitable family of parts of linear varieties of higher dimension.

Problems

PROBLEM 13.1. See page 448.

PROBLEM 13.2. See page 449.

Carry out the program.

PROBLEM 13.3. See page 451.

Can the method be extended to $n > 1$?

PROBLEM 13.4. (Exercise) See page 456.

PROBLEM 13.5. See page 462.

Discuss the nonsymmetric analog of the spaces $\mathscr{E}_B(\vec{\mathrm{D}})$.

Give a proof of Theorem 13.8 using the heat equation. The difficulty with trying to apply the heat equation is that we have to replace the space $\mathscr{E}(\Phi)$ used in the proof of Theorem 5.26 by an LAU space $\mathscr{W}$ such that the growth conditions for functions $F(z) \in \mathbf{W}'$ depend in a delicate way on Re z and Im z.

PROBLEM 13.6 See page 464.

Find an intrinsic proof for this.

PROBLEM 13.7. See page 464.

It would be of interest to know whether the problem of polynomial density in $*\mathbf{E}_B'$ is the same as that in the induced norms on the real axis. This is the same as the problem of whether the existence of a function in $*\mathscr{E}_B$ which vanishes together with all its derivatives at the origin is the same as the existence of such a function which has a Fourier representation (13.15) with the support of μ on the real axis. Unfortunately, the technique of harmonic majorant does not seem to shed any light on the situation.

PROBLEM 13.8. See page 469.

Let λ be fixed on the negative real axis. Determine the conditions on λ on the positive real axis in order that the polynomials (or linear combinations of exp (itz) for $0 \leqslant t \leqslant 1$) be dense.

This problem seems very difficult. There is, however, one simple case: If

$$\int_{-\infty}^{0} \frac{\log \lambda(z)}{1 + |z|^{3/2}}\, dz < \infty$$

then no matter how λ is chosen on the positive axis, polynomial approximation is impossible. This follows easily from Theorem 13.12. Beurling (unpublished) has solved the problem for polynomial approximation for $\lambda(x) = \exp(|x|^\alpha)$ for $x < 0$, if $\frac{1}{2} \leqslant \alpha < 1$. The above remark covers the case $\alpha < \frac{1}{2}$, and $\alpha = 1$ is treated in Example 2.

PROBLEM 13.9. (Conjecture) See page 470.

If $\mathscr{H}_B(\overline{\Omega})$ is quasianalytic in the sense that no function can vanish together with all its derivatives at a point then spectral synthesis holds.

PROBLEM 13.10. (Conjecture) See page 471.

A necessary and sufficient condition that $\mathscr{H}_B(\overline{\Omega})$ be quasianalytic is:

$$\int \frac{\lambda(t, 0, \ldots, 0)}{1 + t^{3/2}}\, dt = \infty.$$

Note that we cannot reduce the conjecture to Theorem 13.12 because of the factor exp $(\psi(z))$ in the norm on V for the space $\mathbf{H}_B'(\overline{\Omega})$, which accounts for the essential difference between $\mathbf{H}_B'(\overline{\Omega})$ and ${}^1\mathbf{E}_{B1}'$.

Problem 13.11. See page 474.

Prove analogs of the results of this section for these spaces. In particular, for $r = 1$, T a noncharacteristic for D, does the question of whether φ or ψ is onto depend only on the order of D (as in Theorem 13.4)? The methods of the present section provide a solution to these questions if the order of D is one.

Study also the analogous question when $\mathscr{E}_B(\vec{\mathrm{D}})$ is mapped into $\mathscr{E}_{B'}(\vec{\mathrm{D}}^*, T)$, where $\vec{\mathrm{D}}$ and $\vec{\mathrm{D}}^*$ are two different systems, and B and B' are two sequences. This seems of particular interest when $\vec{\mathrm{D}}$ is hyperbolic with spacelike T, and $\vec{\mathrm{D}}^* = (\partial/\partial t_1, \ldots, \partial/\partial t_{n-d})$, where t is the orthogonal coordinate to T.

Problem 13.12. (Exercise) See page 476.

Problem 13.13 See page 486.

Can we remove the assumption that $\partial f^j/\partial \xi$ and $\partial f^j/\partial \eta$ belong to $^*\mathscr{E}_B$ from the hypothesis of Theorem 13.20?

Bibliography

BANG, T.

[1] Om Quasi-analytiske Funktioner (thesis), Copenhagen, 1946.

BATEMAN MANUSCRIPT PROJECT

[1] Higher Transcendental Numbers, Vol. 1, 2, 3, New York, 1953.

BERENSTEIN, C., and DOSTAL, M.

Structures analytiques uniformes dans certains espaces de distributions, *C.R. Acad. Sci.*, Vol. 268 (1969), pp. 146–49.

BERNSTEIN, S.

[1] Leçons sur les propriétés extrémales et la meilleure approximation des fonctions analytiques d'une variable réele, Paris, 1926.

BERS, L.

[1] *Lectures on Several Complex Variables*, Courant Institute Lecture Notes, 1964.

BEURLING, A.

[1] On two problems concerning linear transformations in Hilbert space, *Acta Mathematica*, Vol. 81 (1948), pp. 239–255.

[2] *Lectures on Quasi-analytic Functions*, The Institute for Advanced Study, Princeton, 1956–57 (not published).

[3] *Lectures on Balayage Problems*, The Institute for Advanced Study, Princeton, 1959–60 (not published).

[4] Lectures at Stanford, Summer 1961 (mimeographed).

[5] Notes on Dirichlet series, *J. Indian Math. Soc.*, Vol. 27 (1963), pp. 19–26.

BIRKHOFF, G., and S. MACLANE

[1] *A Survey of Modern Algebra*, New York, 1941.

BOAS, R. P.

[1] *Entire Functions*, New York, 1954.

BOCHNER, S.

[1] Partial Differential Equations and Analytic Continuation, *Proc. Natl. Acad. Sci.*, Vol. 38 (1952), pp. 227–30.

BOCHNER, S., and W. T. MARTIN

[1] *Several Complex Variables*, Princeton, 1948.

BRELOT, M.

[1] *Éléments de la théorie classique du Potentiel*, Centre de Documentation Universitaire, Paris, 1965.

CARLEMAN, T.

[1] *L'intgérale de Fourier et questions qui s'y rattachent*, 1944, Uppsala.

CARLESON, L.

[1] On Universal Moment Problems, *Mathematica Scandinavica*, Vol. 9 (1961), pp. 197–207.

[2] Interpolations by Bounded Analytic Functions and the Corona Problem, Ann. Math., (2) 76 (1962), pp. 547–559.

CARTAN, H., and J. P. SERRE

[1], *Séminaires E.N.S.* (Cartan), 1951–52.

CASSELS, J.

[1] *An Introduction to the Geometry of Numbers*, Berlin, 1959.

COURANT, R., and D. HILBERT

[1] *Methods of Mathematical Physics*, New York, 1953.

DIENES, P.

[1] *The Taylor Series*, Dover, New York, 1957, 552 pp.

DIEUDONNÉ, J., and L. SCHWARTZ

[1] La dualité les espaces $(\mathscr{F})$ et $(\mathscr{L}\mathscr{F})$, Annales de l'institut Fourier, Grenoble, Vol. I (1950), pp. 61–101.

DINGHAS, A.

[1] *Vorlesungen über Funktionentheorie*, Berlin, 1961.

EHRENPREIS, L.

[1] Solutions of Some Problems of Division I, *Am. J. Math.*, Vol. 76 (1954), pp. 883–903.

[2] Solution of Some Problems of Division II, *ibid.*, Vol. 77 (1955), pp. 286–292.

[3] Solutions of Some Problems of Division III, *ibid.*, Vol. 78 (1956), pp. 685–715.

[4] Solutions of Some Problems of Division IV, *ibid.*, Vol. 32 (1960), pp. 522–588.

[5] Solutions of Some Problems of Division V, *ibid.*, Vol. 84 (1962), pp. 324–348.

[6] Mean Periodic Functions I, *ibid.*, Vol. 77 (1955), pp. 293–328; also appendix in same journal, Vol. 77 (1955), pp. 731–733.

[7] Theory of Infinite Derivatives, *ibid.*, Vol. 81 (1959), pp. 799–845.

[8] Analytic Functions and the Fourier Transform of Distributions I, *Ann. Math.*, Vol. 63 (1956), pp. 945–946.

[9] Analytic Functions and the Fourier Transform of Distributions II, *Trans. Am. Math. Soc.*, Vol. 89 (1958), pp. 450–483.

[10] Analytically Uniform Spaces and Some applications, *ibid.*, Vol. 101 (1961), pp. 52–74.

[11] Cauchy's Problem for Linear Partial Differential Equations with Constant Coefficients, *Proc. Natl. Acad. Sci. U.S.A.*, Vol. 42 (1956), pp. 642–646.

[12] General Theory of Elliptic Equations, *ibid.*, Vol. 42 (1956), pp. 39–41.

[13] On the Theory of Kernels of Schwartz, *Proc. Am. Math. Soc.*, Vol. 7 (1956), pp. 713–718.

[14] Sheaves and Differential Equations, *ibid.*, Vol. 7 (1956), pp. 1131–1138.

[15] Some Applications of the Theory of Distributions to Several Complex Variables, in *Seminar on Analytic Functions*, Princeton, 1957, Vol. 1, pp. 65–79.

[16] The Fundamental Principle for Linear Constant Coefficient Partial Differential Equations, in *Proc. Intern. Symp. Linear Spaces*, Jerusalem, 1960, pp. 161–174. Also presented to Conference on Linear Spaces, Warsaw, 1960, and summarized in *Studia Mathematica*, special series, 1963, pp. 35–36.

[17] Lectures at Stanford, 1960 (mimeographed).

[18] Conditionally Convergent Functional Integrals and Partial Differential Equations, in *Proc. Intern. Congr. Mathematicians*, Stockholm, 1962, pp. 337–338.

[19] *Function Space Integration and Partial Differential Equations*, lectures at Harvard, 1964–1965 (notes to appear).

[20] A New Proof and an Extension of Hartog's Theorem, *Bull. Am. Math. Soc.*, Vol. 67 (1961), pp. 507–509.

[21] Complex Fourier Transform Technique in Variable Coefficient Partial Differential Equations, *J. d'Analyse* (Jerusalem), Vol. XIX (1967), pp. 75–95.

EHRENPREIS, L., V. GUILLEMIN, and S. STERNBERG

[1] On Spencer's Estimate for δ-Poincaré, *Ann. Math.*, Vol. 82 (1965), pp. 128–138.

EHRENPREIS, L., and P. MALLIAVIN

[1] Invertible Operators and Interpolation in AU Spaces, (to appear).

GÅRDING, L.

[1] Linear Hyperbolic Partial Differential Equations with Constant Coefficients, *Acta Mathematica*, Vol. 85 (1950), pp. 1–62.

[2] Some Trends and Problems in Linear Partial Differential Equations, in *Proc. Intern. Congr. Mathematicians* in Edinburgh 1958, Cambridge Univ. Press 1960.

GÅRDING, L., and B. MALGRANGE

[1] Opérateurs différentiels partiellement hypoelliptiques et partiellement elliptiques, *Mathematica Scandinavica*, Vol. 9 (1961), pp. 5–21.

GELFAND, I. M., D. RAIKOV, and G. ŠILOV

[1] Commutative Normed Rings, *Uspekhi Mat. Nauk*, Vol. 2 (1946), pp. 48–146.

GELFAND, I. M., and G. E. ŠILOV

[1] Fourier Transforms of Rapidly Increasing Functions and Questions of Uniqueness for the Solution of Cauchy's Problem, *Transl. Am. Math. Soc.*, (2), Vol. 5 (1957), pp. 221–274.

GROTHENDIECK, A.

[1] Sur les espaces (F) et (DF), *Summa Brasiliensis Mathematicae*, Vol. 3 (1954), pp. 57–123.

[2] *Produits tensoriels topologiques et espaces nucléaires*, Memoirs Am. Math. Soc., No. 16, 1955.

GUNNING, R. C., and H. ROSSI

[1] *Analytic Functions of Several Complex Variables*, Prentice-Hall, 1965.

HECKE, E.

[1] *Vorlesungen über die Theorie der Algebraischen Zahlen*, New York, 1948.

[2] Mathematische Werke, Göttingen, 1959.

HOLMGREN, E.

[1] Über Systeme von linearen partiellen Differentialgleichungen, *Öfversigt af Vetenskaps-Akad. Förh.*, Vol. 58 (1901), pp. 91–105.

HÖRMANDER, L.

[1] On the Theory of General Partial Differential Operators, *Acta Mathematica*, Vol. 94 (1955), pp. 161–248.

[2] La transformation de Legendre et la théorème de Paley-Wiener, *Comptes Rendus des Seances de l'Academie des Sciences*, Vol. 240 (1955), pp. 392–395.

[3] *Linear Partial Differential Operators*, Springer, 1963.

[4] Differentiability Properties of Solutions of Systems of Linear Equations, *Arkiv for Matematik*, Vol. 3 (1958) pp. 527–535.

[5] *An Introduction to Complex Analysis in Several Variables*, Princeton, 1966.

[6] On the Range of Convolution Operators, *Ann. Math.*, Vol. 76 (1962), pp. 148–170.

[7] Generators for some rings of analytic functions, *Bulletin of the Amer. Math. Soc.*, Vol. 73 (1967) pp. 943–949.

JOHN, F.

[1] On Linear Partial Differential Equations with Analytic Coefficients, *Commun. Pure Appl. Math.*, Vol. 2 (1949), pp. 209–253.

[2] Non-Admissible Data for Differential Equations with Constant Coefficients, *ibid.*, Vol. 10 (1957), pp. 391–398.

[3] Continuous Dependence of Data for Solutions of Partial Differential Equations, *ibid.*, Vol. 13 (1960), pp. 551–585.

[4] *Plane Waves and Spherical Means*, New York, 1955.

KELLEHER, J. J.

[1] An Application of the Corona Theorem to Rings of Entire Functions, *Am. Math Soc. Summer Institute*, San Diego, 1966.

KELLEHER, J. J., and TAYLOR, A.

[1] An application of the corona theorem to some rings of entire functions, *Bull. Amer. Math. Society*, Vol. 73 (1967), pp. 246–249.

KELLOGG, O. D.

[1] Foundations of the Potential Theory, Springer, Berlin, 1929.

KOHN, J. J.

[1] Solution of the $\bar{\partial}$-Neumann Problem on Strongly Pseudoconvex Manifolds, *Proc. Natl. Acad. Sci., U.S.A.*, Vol. 47 (1961), pp. 1198–1202.

[2] Harmonic Integrals on Strongly Pseudoconvex Manifolds I, II, *Ann. Math.*, Vol. 78 (1963), pp. 206–213 and Vol. 79 (1964), pp. 450–472.

LANDAU, E.

[1] *Vorlesungen über Zahlentheorie*, Liepzig, 1927.

LECH, C.

[1] A Metric Property of the Zeros of a Complex Polynomial Ideal, *Arkiv for Mathematik*, Vol. 3 (1958), pp. 543–554.

LERAY, J.

[1] Le calcul différential et intégral sur une variété analytique complex (Problem de Cauchy, III) Bulletin de la Société Mathématique de France, Vol. 87 (1959), pp. 81–180.

[2] Séminaire Sur Les Equations Aux Dérivées Partielles, College de France, 1961–1962.

LEVINSON, N.

[1] *Gap and Density Theorems*, New York, 1940.

LEWY, H.

[1] An Example of a Smooth Linear Partial Differential Operator Without Solutions, *Ann. Math.*, Vol. 66 (1957), pp. 155–158.

LIONS, J. L.

[1] Supports dans la transformation de Laplace, *J. d'Analyse Math.*, Vol. 2 (1952–53), pp. 369–380.

ŁOJASIEWICZ, S.

[1] Sur le probleme de division, Studia Math., Vol. 18 (1959), pp. 87–136.

MALGRANGE, B.

[1] Existence et approximation des solutions des équations aux dérivées partielles et des équations de convolution, *Ann. l'Institut Fourier* (Grenoble), Vol. 6 (1956), pp. 271–355.

[2] Sur la propagation de la régularité des solutions des équations à coefficients constants, *Bull. Math. Soc. Roumaine*, Vol. 3 (1959), pp. 433–440.

[3] Sur les ouverts convexes par rapport à un opérateur différentiel, *Comptes Rendus des Séances de l'Académie des Sciences*, Vol. 2542 (1962), pp. 614–15,

[4] Sur les systèmes differentiels à coefficients constants, in *Séminaire sur les Equations aux Dérivées Partielles*, Collège de France, 1961–62. Also in *Séminaire Bourbaki*, 1962–63, No. 246. Also in *Les Equations aux Dérivées Partielles*, Paris, 1963, pp. 113–122.

[5] Some Remarks on the Notion of Convexity for Differential Operators, in *Differential Analysis*, Oxford, 1964.

MALLIAVIN, P.

[1] Croissance radiale d'une fonction méromorphe, Illinois J. Math., Vol. 1 (1957), pp. 259–296.

MANDELBROJT, S.

[1] *Séries de Fourier et classes quasi-analytiques de fonctions*, Paris 1935.

[2] *Séries Adhérentes, Régularisation des Suites, Applications*, Paris, 1952.

[3] Sur les fonctions indéfiniment dérivables, Acta Mathematica, Vol. 72 (1940), pp. 15–29.

MARDEN, M.

[1] *The Geometry of the Zeroes of a Polynomial in a Complex Variable*, New York, 1949.

MARTINEAU, A.

[1] Sur les fonctionelles analytiques et la transformation de Fourier-Borel, J. d'Analyse Math., Vol. 9 (1963), pp. 1–164.

MILNOR, J.

[1] *Morse Theory*, Princeton, 1963.

MITYAGIN, B. S.

[1] Doklady Akademii Nauk SSSR, Vol. 138 (1961), pp. 289–292.

NAGATA, M.

[1] *Local Rings*, New York, 1962.

NEWMAN, D. J., and SHAPIRO, H. S.

[1] Fischer Spaces of Entire Functions, *Am. Math. Soc. Summer Institute*, San Diego, 1966.

OKA, K.

[1] Sur les fonctions analytiques de plusieurs variables I, *J. Sci. Hiroshima Univ.*, Vol. 6 (1936), pp. 245–255.

[2] Sur les fonctions analytiques de plusieurs variables II, *ibid.*, Vol. 7 (1937), pp. 115–130.

[3] Sur les fonctions analytiques de plusieurs variables IX, *Japanese J. Math.*, Vol. 23 (1953), pp. 97–155.

OSTROWSKI, A.

[1] Recherches sur la méthode de Graeffe et les zéros de polynômes et des séries de Laurent, *Acta Mathematica*, Vol. 72 (1940–41), pp. 99–257.

[2] On Representation of Analytic Functions by Power Series, *J. London Math. Soc.*, Vol. 1 (1926), pp. 251–263.

PALEY, R. E. A. C., and N. WIENER

[1] *Fourier Transforms in the Complex Domain*, New York, 1934.

PETROWSKY, I. G.

[1] Sur l'analyticité des solutions de systèms d'équations différentielles, *Mat. Sbornik*, Vol. 47 (1939), pp. 3–68.

[2] On the Diffusion of Waves and the Lacunas for Hyperbolic Equations, *ibid.*, Vol. 59 (1945), pp. 289–370.

POLLARD, H.

[1] Solution of Bernstein's Approximation Problem, *Proc. Am. Math. Soc.*, Vol. 4 (1953), pp. 869–874.

[2] The Bernstein Approximation Problem, *ibid.*, Vol 6 (1955), pp. 402–411.

PÓLYA, G.

[1] Über die Existenz unendlich vieler singularer Punkte auf der Konvergenz Geraden gewisser Dirichletscher Reihen, *Sitzber. Preussischen Akad. Wiss.*, 1923, pp. 45–50.

PÓLYA, G., and G. SZEGÖ

[1] *Aufgaben und Lehrsätze aus der Analyse*, Vol. II, Berlin, 1925.

RIESZ, M.

[1] L'intégrale de Riemann-Liouville et le problème de Cauchy, *Acta Mathematica*, Vol. 81 (1949), pp. 1–223.

SAMUEL, P.

[1] *Méthodes d'algebre abstraite en géometrie algébrique*, Berlin, 1955.

SCHECHTER, M.

[1] Various Types of Boundary Conditions for Elliptic Equations, *Commun. Pure Appl. Math.*, Vol. 13 (1960), pp. 407–425.

SCHWARTZ, L.

[1] *Théorie des Distributions*, Vol. I, II, Paris, 1950–51.

[2] *Etude des Sommes d'Exponentielles Réelles*, Paris, 1943.

[3] Approximation d'une fonction par des sommes d'exponentielles imaginaires, *Université de Toulouse, Annales de la Faculté des Sciences*, Vol. 6 (1942), pp. 111–174.

[4] Théorie générale des fonctions moyenne—périodiques, *Ann. Math.*, Vol. 48 (1947), pp. 857–929.

[5] Division par une fonction holomorphe sur une variété analytique complexe, *Summa Brasiliensis Mathematicae*, Vol. 3 (1955), pp. 181–209.

SERRE, J. P.

[1] Geómetrie algébrique et géometrie analytique, *Ann. l'Institut Fourier*, Vol. 6 (1955–56), pp. 1–41.

[2] Un theorem de dualité, *Comment. Math. Helv.*, Vol. 29 (1955), pp. 9–26.

STEIN, E. M., and G. WEISS

[1] On the Theory of Harmonic Functions of Several Variables I, *Acta Mathematica*, Vol. 103 (1960), pp. 25–62.

[2] On the Theory of Harmonic Functions of Several Variables II, *ibid.*, Vol. 106 (1961), pp. 137–174.

TÄCKLIND, S.

[1] Sur les classes quasianalytiques des solutions des équations aux dérivées partielles du type parabolique, *Nova Acta Regiae Societatis Scientiarum Upsaliansis*, Vol. 10 (1936), pp. 1–56.

TAYLOR, B. A.

[1] Some Locally Convex Spaces of Entire Functions, Lecture Notes at the Summer Institute on Entire Functions and Related Parts of Analysis, La Jolla, California, 1966.

TITCHMARSH, E. C.

[1] *The Theory of the Riemann Zeta-Function*, Oxford, 1951.

[2] *The Theory of Functions*, Oxford, 1932.

TURÁN, P.

[1] *Eine Neue Methode in der Analyse und deren Anwendungen*, Budapest, 1953.

VAN DER WAERDEN, B. L.

[1] *Einführung in die algebraische Geometrie*, Berlin, 1939.

[2] *Algebre*, Berlin, 1955.

Walfisz, A.
[1] *Gitterpunkte in Mehrdimensionalen Kugeln*, Warsaw, 1957.
Weil, A.
[1] *Foundations of Algebraic Geometry*, New York, 1946.
Wermer, J.
[1] Addendum to An Example Concerning Polynomial Convexity, *Ann. Math.*, Vol. 140 (1960), pp. 322–323.
Weyl, H.
[1] *Die Idee der Riemannschen Flächen*, 3rd ed., Stuttgart, 1955.
Zerner, M.
[1] Solutions de l'équation des ondes présentant des singularités sur une droite, *C. R. Acad. Sci. Paris*, Vol. 250 (1960), pp. 2980–2982.
Zygmund, A.
[1] *Trigonometric Series*, Warsaw, 1935; 2nd ed., Cambridge, 1959.

Special Notations

Symbol	*Description*	*Page*
$A = \{a_j\}$ (Chap. V)	a sequence of positive numbers	163
AU	analytically uniform	9
$B = \{b_j\}$	a sequence of positive numbers	446
B^+S	space of C^∞ maps into S	273
$B^+{}_0S(V)$	space of suitable C^∞ functions on V	273
C (or C^n)	complex Euclidian space of dimension n	8
$c(A)$	cylinder of base A	251
D	differential operator with constant coefficients	140, 175
$\vec{D}$	vector of differential operators with constant coefficients (usually with r components, sometimes l components)	141, 175
$\boxed{D}$	r by l matrix of differential operators with constant coefficients	141, 175
D' $\vec{D}'$, $\boxed{D}'$	adjoints of D, $\vec{D}$, $\boxed{D}$	140, 141, 143
∂	differential operator associated to multiplicity variety	25
d^t_{jk} (Chaps. II, III, IV)	algebroid (algebraic) differential operators	31
d (Chaps. IX, X)	dimension of a variety	220
dx	Lebesgue measure on R^n	140
dz	Lebesgue measure divided by $(2\pi)^n$	140
$\mathscr{D}$	space of C^∞ functions of compact support with Schwartz topology	148
$\mathscr{D}'$	space of distributions	148
$\mathbf{D}$, $\mathbf{D}_F$	space of entire functions of exponential type, rapidly decreasing on R	
$\mathscr{D}_F$	space of C^∞ functions of compact support with weaker topology than $\mathscr{D}$	139
$\mathscr{D}'_F$	space of distributions of finite order	139, 143, 148
$\mathscr{D}_A$	C^∞ functions of compact support whose derivatives satisfy growth conditions	163

Symbol	*Description*	*Page*
$\vec{e}$, e	fundamental solution	179, 182
$\mathscr{E}$	Space of C^∞ functions	152
$\mathscr{E}'$	Space distributions of compact support	153
$\mathbf{E}'$	space of entire functions of exponential type polynomially increasing on R	155
$\mathscr{E}(\Omega)$	C^∞ functions on the convex Ω	116, 159
$\mathscr{E}_A$, $\mathscr{E}_B$	C^∞ functions whose derivatives satisfy growth conditions	44, 163
$\mathscr{E}(\Phi)$	C^∞ functions with growth conditions	169, 247
$\mathbf{E}'(\Phi)$	entire functions with suitable growth conditions	170
$\mathscr{E}_B(\vec{D})$	Denjoy-Carleman class related to $\vec{D}$	449
${}^i\mathscr{E}_{Bi}$, ${}^*\mathscr{E}_B$	nonsymmetric Denjoy-Carleman classes	459
${}_0\mathscr{E}_B$	punctual Denjoy-Carleman class	473
${}_I\mathscr{E}_B$	local Denjoy-Carleman class	473
${}_i\mathscr{E}_B$, $\mathscr{E}_B^i$	other Denjoy-Carleman classes	475
$\mathscr{F}$	Fourier transform	
G^{ij}, $'G^j$, $''G^j$, G^j	functions used in computing density (Chap. XII)	414
$\mathscr{H}$	space of entire functions	122
$\mathbf{H}'$	space of entire functions of exponential type	122
$\mathscr{H}(\Omega)$	holomorphic functions on Ω	160
$\mathbf{H}'(\Omega)$	entire functions of exponential type satisfying suitable growth conditions	160
Im	imaginary part of	
K	AU structure	9
$\{k\}$	the functions in an AU structure	9
$\Vert\ \Vert_K$	uniform norm defined by the function k	2
l	number of columns (rows) of a matrix of differential operators (resp. analytic functions)	141
L_{ij} (Chaps. II, III, IV)	coefficients in the L'Hospital parametrization	47
$M = \{m\}$	BAU structure	96
$\mathscr{M}(V)$	meromorphic functions on V	15
$\mathscr{O}$ or $\mathscr{O}(n; z^0)$	germs of functions analytic at z^0	14
$\mathscr{O}(\lambda)$ or $\mathscr{O}(n; z^0; \lambda)$ (Chaps. I, II, III)	functions on V analytic in $\lvert z - z^0\rvert < \lambda$	14
$\mathscr{O}(\mathfrak{V})$	functions holomorphic on $\mathfrak{V}$	15
$\mathscr{O}(\mathfrak{V}; n; z^0)$	germs of functions on $\mathfrak{V}$ analytic at z^0	25
$\mathscr{O}(\mathfrak{V}, n; z^0; \lambda)$	functions on $\mathfrak{V}$ analytic in $\lvert z - z^0\rvert < \lambda$	25

Symbol	*Description*	*Page*
$\mathcal{O}(t;\alpha;\beta)$	functions holomorphic on $\gamma(t;\alpha;\beta)$	73
$\mathcal{O}(\mathfrak{V};t;\alpha;\beta)$	functions holomorphic on $\mathfrak{V}\cap\gamma(t;\alpha;\beta)$	73
$\mathcal{O}(t;\alpha;\beta;A)$	functions in $\mathcal{O}(t;\alpha;\beta)$ bounded by A	73
$\mathcal{O}(\beta)$ (Chap. IV)	cochains	99
$\mathcal{O}(\beta;\vec{F})$ (Chap. V)	cochains with special property	99
$\mathcal{O}_a(\beta)$	cochains holomorphic in a strip	108
$\mathcal{O}_M$	C^∞ functions of polynomial growth	256
P	polynomial, the Fourier transform of D'	140, 175
$\vec{P}$	vector of polynomials, $\vec{D}'$	141, 175
$\boxed{P}$	matrix of polynomials, the Fourier transform of $\boxed{D}'$	141, 175
$\mathscr{P}$	ring of polynomials	83
q_{ij}	differential operators associated to Cauchy problem	218
Q_{ij} (Chaps. II, III, IV)	basis in the L'Hospital parametrization	47
Q_{ij}	Fourier transform of q_{ij}	218
R (or R^n)	real euclidian space of dimension n	8
Re	real part of	
r	number of rows (columns) of a matrix of differential operators (resp. analytic functions)	141
$\mathscr{R}(V)$	rational functions on V	223
$s_j(z)$	roots of an analytic function or polynomial	20
$\mathscr{S}$	rapidly decreasing C^∞ functions	272
T_{ij}^q, T_{ij} (Chaps. II, III, IV)	varieties in L'Hospital parametrization plans in Cauchy problem	47 220
$\{U_j\}$ (Chaps. II, III, IV)	covering of a neighborhood of the origin	30
V	a variety or z variety	15
$\mathfrak{V}=(V_1,\partial_1;\ldots;V_a,\partial_a)$	a multiplicity variety or Z variety	24
$\|\mathfrak{V}\|\mathfrak{V}$	restriction to $\mathfrak{V}$	24
$\mathfrak{V}(\alpha;\beta)$	semilocal multiplicity variety	74
$\mathscr{W}$	generic letter for topological vector space	8
$\mathscr{W}'$	the space whose dual is $\mathscr{W}$; sometimes the dual of $\mathscr{W}$	9
$\mathbf{W}'$	the Fourier transform of $\mathscr{W}'$	9
$\mathbf{W}$	the dual of $\mathbf{W}'$	10
$\mathfrak{W}'(\beta)$	nice cochains	99
$\mathfrak{W}'(\beta;\vec{F})$	nice cochains with special property	99
$\mathfrak{W}_a'(\beta)$, $\mathfrak{W}'(\beta,\boxed{F})$	nice chochains holomorphic in a strip	108
$_W\mathfrak{W}_a'^{ij}(\beta,\boxed{F}^{j+i})$	suitable space of nice cochains	110

Symbol	*Description*	*Page*
$\boxed{D}\mathscr{W}^r$	image of $\vec{D}$ acting on $\mathscr{W}^l$	176
$\mathscr{W}_{\vec{D}}$	kernel of $\vec{D}$ acting on $\mathscr{W}$	234
$\mathscr{W} \overset{a}{\cap} \mathscr{W}_1$	AU intersection	402
$\mathscr{W} \overset{a}{\cup} \mathscr{W}_1$	AU union	402
$x = (x_1, x_2, \ldots, x_n)$	coordinate on R	
$\|x\|$ or $\|x\|$	max $\|x_j\|$	122
$z = (z_1, \ldots, z_n) = (s, w)$	coordinate on C	
$\|z\|$ or $\|z\|$	$\sum \|z_j\|$	122
$\alpha = (\alpha_1, \ldots, \alpha_{2n})$	lattice point in C	98
$\alpha, \alpha', \hat{\alpha}$	maps associated to boundary value problem	285
$\gamma(t; \alpha; \beta)$ (Chaps. III, IV)	open cube in C^t, center α, side 2β	73
γ, γ', γ'	maps associated with Cauchy problem	218
δ	Dirac mass, i.e., unit mass at the origin	179
δ_a	unit mass at the point a	179
ζ	the ζ function of Riemann	422
ζ_1	$(1 - 2^{1-s})$ times the ζ function of Riemann	423
$\lambda_{\vec{F}}, \lambda\boxed{F}$	map from functions to cochains	100
λ_a^j	map from functions to cochains	108
λ	function associated to nonquasianalytic classes	163, 447
μ	a measure on C or on a suitable subset with bounded total variation	10
ρ	restriction map	48
$\rho\mathfrak{B}$	restriction to $\mathfrak{B}$	55
ρ_B^L	local restriction map	99
τy	translation by y	141
σ	usually a sufficient subset of C	12
$\varphi(z, z^0; \eta)$	function related to PLAU	97
Φ	convex function	169
φ (Chapt. XIII)	punctual image mapping	447, 473
χ (Chaps. II, III, IV)	inverse of ψ	61
ψ (Chaps. II, III)	map associated with L'Hospital parametrization	50
$\psi(z; z^0; \eta)$ (Chaps. IV, V)	function related to PLAU	97
ψ (Chap. XIII)	local image mapping	447, 473
Ω	convex set	116
Щ′ Щ′	space of C^∞ maps	268
Щ′	Mellin transform of $\mathfrak{W}'$	427

General Notations

1. If a Roman letter represents an object the same letter with an arrow on top ($\rightarrow$) represents a vector of such objects, and the same letter in a box ($\square$) represents a matrix of such objects. For example, see page 30, where F is a function, $\vec{F}$ is a vector of functions, and $\boxed{F}$ is a matrix of functions.

2. Primes ($'$) are used for dual spaces. $\mathscr{W}$ is the dual of $\mathscr{W}'$ (see p. 9). For $S \in \mathscr{W}'$, $f \in \mathscr{W}$, $S \cdot f$, or $f \cdot S$ represent the value $f(S)$.

3. Primes ($'$) are also used for the adjoints of maps. On page 140, D' is the adjoint of D.

4. Generally, script and boldface letters are used for topological vector spaces. The spaces represented by corresponding script and boldface letters are Fourier transforms of each other. For example $\mathscr{W}'$ and $\mathbf{W}'$ are Fourier transforms (see p. 9). For an element $S \in \mathscr{W}'$ we denote by $\hat{S}$ the transform of S.

5. If $\mathscr{W}$ is a topological vector space, $\mathscr{W}^r$ is the r fold direct sum of $\mathscr{W}$ with itself, $\hat{\mathscr{W}}_D$, $\mathscr{W}_{\vec{D}}$, $\mathscr{W}^1_{\boxed{D}}$ are the kernels of D, $\vec{D}$, $\boxed{D}$ on $\mathscr{W}$, $\mathscr{W}$, and $\mathscr{W}^1$ respectively. ${}_{\boxed{D}}W^r$ is the subspace of $\mathscr{W}^r$ satisfying certain compatibility conditions (see page 176).

6. If f(x) is function, we denote by F(z) its Fourier transform, defined formally by

$$F(z) = \int f(x) \exp (i \quad x \cdot z) dx.$$

For further notation regarding Fourier transforms see pages 139–140.

7. The convolution of two functions f,g is (formally)

$$(f * g)(x) = \int f(x-y) g(y) dy.$$

For further notations regarding convolution see pages 141–142.

Index

A adjacent lattice points, 109
Above, 16, 334
Admissible Cauchy data, 235
Admissible (for V), 74
($\mathscr{W}$; A, B) Admissible, 251, 252
Algebraic differential operator, 24
Algebraic function, 17
Algebraic map, 17
Algebraic multiplicity variety, 24
Algebraic variety, 14
Algebroid differential operator, 24
Algebroid function, 17
  abstract, 16
Algebroid map, 17
Algebroid multiplicity variety, 24
Analytic, 15
Analytic connectivity, 389
Analytic density, 411
  relative, 412
Analytic function, 17
  on multiplicity variety, 24
Analytic in β outside c (A), 251
Analytic uniform intersection, 402
Analytic uniform union, 402
Analytic variety, 14
($\mathscr{W}$; A, B) Analytically admissible relative to b, 252
Analytically closed, 251
Analytically compatible, 484
Analytically uniform (AU) space, 8
  localizable (LAU), 101
  product localizable (PLAU), 96
Analytically uniform structure, 9
  bounded (BAU), 96
Approximate solutions, 328
Approximation, 450, 463
A semiadjacent lattice points, 109

Balayage, 283
Bernstein approximation problem, 451
Bessel functions, 383
Bitten, 331
Bornologic, 124
Boundary value problem, 285
Bounded analytically uniform (BAU) structure, 96

Cauchy data, admissible, 235
Cauchy hyperbolic (CH), 231
  principally (PCH), 231
Cauchy-Kowalewski (CK) space, 224
Cauchy-Kowalewski theorem, 224, 385
Cauchy parabolic (CP), 232
Cauchy problem, 217
  uniqueness, 244, 323
  well posed, 218
Cauchy zero, 272, 274
Cochain, 99
  nice, 99
Comparison theorem, 199
  general, 403
Component, essential, 18
Components of a variety, 15
Connectable, 78
Convolution, 13, 141, 317
Corona problem, 321
Covering variety, 16
Cutting up conditions, 404

Degree of approximation, 450, 463
Denjoy-Carleman class, 446
Density, 406
  analytic, 410
  geometric, 413
B-Density, 464

Derived sequence, 230
Determined system of differential equations, 176
Different, 230
Differentiation, 24
Differential operator, 24
  algebraic, 24
  algebroid, 24
Diophantine condition, 296
Dirac δ function, 179
Dirichlet problem, 295
Distinguished polynominal, 49
Distinguished variable, 49
Distributions, 148
  of finite order, 139
Domain, of dependence, 243
  of existence, 338
  of holomorphy, 338

Elliptic, 204
Essential component, 18
Exponential type, 122
Extendible, 199
Extension problem, 200, 325, 403
  local, 335, 339

Fabry gap theorem, 394, 416
Fourier transform, 9, 140
Fredholm alternative, 305
Function, algebraic, 17
  algebroid, 17
  analytic, 17
  analytic on multiplicity variety, 24
Function, integral, 16
  multivalued, 16
  regular, 17
Function elements, 16
Fundamental principle, 98
Fundamental solution, 179, 182, 240, 331

General comparison problem, 403
Geometric density, 413
Germ, 14
Goursat problem, 311
Grouping, 322, 436
Group representations, 379

Hartog's theorem, 325
Heat equation, 165, 209, 452
Hecke functional equation, 430
Holomorphic, 15
Homology theory for partial differential equations, 376
Hyperbolic, 206, 231, 339
Hypoelliptic, 204

Index, 305
Initial value problem, 231
Integral algebroid function, 17
Integral algebroid map, 17
Integral function, 16
Interpolation, 286
Intersect properly, 303
$\vec{D}$ Intersection, 346
Irreducible variety, 15

Lacuna, 243
Lacunary, 406
Lagrange interpolation formula, 54
Lattice, points, 98
  a adjacent, 109
  a semiadjacent, 109
Legendre functions, 383
L'Hospital expressible (describable), 30
L'Hospital parametrization, 47
L'Hospital representation, 29
Limitation of singularities, 201
Local extension, 335, 339
Local image, 473
Local quotient structure theorem, 40
Localizable analytically uniform space (LAU), 101
Localizable analytically uniform space, weakly, 115

Map, algebraic, 17
  algebroid, 17
  integral, 17
Mellin transform, 313, 426
Meromorphic, 15
Meromorphic solutions, 348
Metahyperbolic, 211
Module of relations, 32
Morse theory, 376
Mouth, 331
Multiplicity (Z) variety, 24
  algebraic, 24
  algebroid, 24
  vector, 24

Multiplier, 13
Multivalued, 15

Natural boundary, 438
Nevinlinna theory, 265
Newman-Shapiro problem, 311
Nibble, 331
Nice cochain, 99
Noncharacteristic, 78, 225
  principal, 226
Non-quasianalytic, 163
Nullstellensatz, 29

Obstruction, 375
Oka embedding, 116
Outer function, 40
Overdetermined system of differential equations, 176

Parabolic, 231
  Cauchy (CP), 232
Parametrix, 201, 407
Parametrization, L'Hospital, 47
Parametrization problem, 285
Principal noncharacteristic, 226
Principal regular characteristic, 248
Principally Cauchy hyperbolic (PCH), 231
Product localizable analytically uniform space, (PLAU), 96
  weakly, 115
Product of function and distribution, 143
Pseudoconvex, 344
Puiseux expansion, 7
Punctual image, 473

Quasianalytic (QA), 163, 446
($\mathscr{W}$; A, B) Quasianalytically admissible, 253
Quasianalytically closed, 253
Quasihyperbolic, 210
Quotient structure theorem, 98
  local, 40
  semilocal, 75

Regular function, 15
Regular set of planes, 220, 248
Relative analytic density, 412
Removable singularity, 330, 345
Residue, 349
Restriction (to a multiplicity variety), 24
Riemann ζ function, 422

Schrödinger equation, 471
Schwarz alternating procedure, 303
Semifundamental solution, 186, 350
Semihyperbolic, 210
Semilocal quotient structure theorem, 75
Semi-simple Lie groups, 385
Šilov boundary, 484
Singularities, 325, 351
  limitation of, 201
  removable, 330, 345
  specification of, 201
Slowly decreasing, 13, 317
Spacelike surface, 243
Space variables, 206
Special functions, 379
Spectral synthesis, 470
Sphere of influence, 243
Strong L uniqueness, 334
Sufficient, 12
Swallows, 124

Tauberian theorems, 399
Time variables, 206

Underdetermined system of differential equations, 176
Uniform discreteness, 415
Unique continuation, 326, 352
L Uniqueness, 334
L Uniqueness, strong, 334

Variable coefficient equations, 385
Variety, 14
  algebraic, 14
  algebraic multiplicity, 24
  algebroid multiplicity, 24
  analytic, 14
  components of, 15
  covering, 16
  irreducible, 15
  local (germ of), 14
  multiplicity (Z), 24
  vector multiplicity, 24
  Z, 15
Vector multiplicity (Z) variety, 24

Weakly product localizable, 115
Well behaved in β, 253
Well-posed boundary value problem, 285
Well-posed Cauchy problem, 218
Wiener-Hopf problem, 309
Wiener's tauberian theorem, 400

Z closure, 15
Z variety, 15